AF251541

CELLULAR AUTOMATA

SPECIAL ISSUES OF *PHYSICA D*

The titles in this series are paperback, readily accessible special issues of *Physica D*, produced by special agreement with Elsevier Science Publishers B.V.

Emergent Computation: Self-Organizing, Collective, and Cooperative Phenomena in Natural and Artificial Computing Networks, edited by Stephanie Forrest, 1991.

Cellular Automata: Theory and Experiment, edited by Howard Gutowitz, 1991.

CELLULAR AUTOMATA

THEORY AND EXPERIMENT

edited by
Howard GUTOWITZ

A Bradford Book
The MIT Press
Cambridge, Massachusetts
London, England

First MIT Press edition, 1991

Reprinted from *Physica D,* Volume 45, Numbers 1–3, 1990. The MIT Press has exclusive license to sell this English-language book edition throughout the world.

Printed and bound in the Netherlands.

Library of Congress Cataloging-in-Publication Data

Cellular automata : theory and experiment / edited by Howard Gutowitz. — 1st MIT Press ed.
    p.   cm. — (Special issues of physica D)
  "A Bradford book."
  Includes bibliographical references and index.
  ISBN 0-262-57086-6
  1. Cellular automata. I. Gutowitz, Howard. II. Series.
QA267.5.C45C46 1991
511.3—dc20   90-28927
                                                              CIP

Physica D 45 (1990) vii–xiv
North-Holland

INTRODUCTION

Howard A. GUTOWITZ

*Center for Nonlinear Studies and Complex Systems Group, Los Alamos National Laboratory,
MS-B258, Los Alamos, NM 87545, USA*

1. Introduction

Cellular automata (CA) are dynamical systems in which space and time are discrete. The states of cells in a regular lattice are updated synchronously according to a deterministic local interaction rule. Each cell obeys the same rule, and has a finite (usually small) number of states. Some of the articles in these Proceedings are in part reviews of cellular automaton studies, obviating an extensive review here. In particular, the articles by Jen, Toffoli and Margolus, and Culik et al. (first article) provide background and motivation. This introduction is focused on a few general questions that may help guide one's reading in this volume.

Broadly speaking, there are two types of problems concerning cellular automata: the forward problem and the inverse problem. The forward problem is: Given a cellular automaton rule, determine (predict) its properties. The inverse problem is: Given a description of some properties, find a rule, or set of rules, which have these properties. These problems are obviously strongly interrelated. In order to design a rule with specified properties, one must have a way of determining the properties of rules one finds in the course of the design process. Conversely, the forward problem is usually studied only for rules that have been found already (by some method) to have interesting properties. The contributions to these Proceedings address forward and inverse problems in various admixtures, the components of which can be only fractionally teased apart.

2. The forward problem

The forward problem presents itself when a cellular automaton has been found of interest, and one wishes to decisively grasp the nature of that rule. On the mathematical side, the forward problem often consists of finding quantities that are calculable on the basis of a rule table, and characterize in some way the behavior of the rule upon repeated iteration starting from some set of initial conditions. For instance, Jen examines the generation of aperiodic sequences by cellular automata. She exploits a property of the rule tables of some nonlinear automata in order to effectively linearize the behavior of these rules. This permits the application of linear analysis to prove that under generic conditions certain of these rules generate only aperiodic temporal sequences. The work of Voorhees is similar, in that working directly from the structure of cellular automaton rule tables, he is able to make statements concerning the large-time behavior of these rules. He decomposes nearest-neighbor rules in terms of certain basis operators that are themselves cellular automata in this set. This decomposition allows a number of behavioral characteristics of these automata to be determined algorithmically.

Vichniac investigates a function which can be applied to a rule table to reveal some aspects of its behavior. This function is called a derivative. While clearly not the same as the usual notion of a derivative for continuous systems, this derivative has many natural and familiar properties. The derivative is applied

to the study of the growth of complexity as certain rules are iterated, and to lattice gases.

Another mathematical aspect of the forward problem is the study of quantities that are calculable on the basis of the set of configurations generated by a cellular automaton and that serve to make tangible and vivid the complexities of that set. There are a number of examples here. Fisch considers a class of cyclic cellular automata: automata in which the possible states of the cells can be ordered in a circle. A salient feature of these automata is their tendency to generate clusters of sites all in a given state. Fisch describes theorems concerning the distribution and size of clusters of cell states in the limit of large time. One way to characterize the "lumpyness" of a distribution is in terms of a spectrum of generalized (fractal) dimensions. Takahashi takes on the calculation of the dimension spectra of rules which obey an additive superposition principle. He shows that the complete spectrum for some rules can be determined exactly.

Another sort of pattern generated by cellular automata are the binary difference fields of Barbé. These binary difference fields can be rather complex, but for some rules, at least, they are subject to a mathematical analysis that reveals a sort of conservation law.

Consider also in this framework the work of Aizawa et al. These authors study CA which support solitons, localized structures which can interact with each other while maintaining their identity. Their survey of the phenomenology of such automata is augmented by the calculation of various quantities, such as mutual information flow, which serve to characterize the invariant states of these automata.

Quantities of particular importance in the application of CA to physics are invariants of evolution. An example of a possible invariant is the ratio between the number of cells in state 0 to the number of cells in state 1 in a two-state per cell system. In physical applications this quantity is often called the magnetization. There are some rules that do not change the magnetization of a configuration no matter how many times they are applied to that configuration. One such rule is studied by Canning and Droz. They compare a well-know magnetization-preserving dynamics, Kawasaki spin exchange, with a cellular automaton model in the study of diffusion-controlled reactions. The CA model offers some potential computational advantages as it is completely parallel in nature.

There is at present no general method for finding invariants of CA, and one suspects that this problem is indeed undecidable (see below). Nonetheless, a good number of rules with interesting invariants have been found. Hartman and Tamayo have studied a class of reversible CA designed to model chemical turbulence. They have discovered a number of invariants of these models.

Once the invariants of a CA have been found, there still remains the problem of determining the effect these invariants have on the evolution of the CA from non-equilibrium initial conditions. This is discussed in some detail by Hartman and Tamayo. Takesue has studied a class of reversible CA for which he has found a variety of invariants. In his contribution to this volume he investigates the role these invariants play in the relaxation to equilibrium of non-equilibrium initial states. He finds that relaxation relative to some of these invariants is as to be expected from standard statistical mechanics, while for other invariants it is not.

A fresh approach to the question of invariants in CA is presented by Hasslacher and Meyer. These authors show how to use the theory of knots to establish relationships between CA and exactly solvable statistical-mechanical models at criticality. One promise of this work is bring forth a deep understanding of the origin of thermodynamic behavior in cellular automata.

An invariant of paramount importance in the application of CA to physics is time-reversal invariance. Fundamental physical laws are (microscopically) time-reversal invariant, hence a cellular automaton which is to model such physics should be time-reversal invariant as well. A rule is reversible if every configuration has a unique predecessor under the rule. This permits the automaton to be run deterministically both forward and backward in time. The full range of issues surrounding reversibility of CA is covered in the review by Toffoli and Margolus. They focus on fundamental issues concerning the use of reversible CA in physics, and connections to computation theory. This is essential background for much of the work in these Proceedings, which pursues specific investigations into properties of reversible CA.

The themes of reversibility, invariants, criticality, and generalizations of CA (see below) rejoin in the paper by Martin. In this paper a review of reversibility in the context of probabilistic CA clarifies the relationship between some of the studies discussed above with standard statistical mechanics. Martin presents a renormalization method for probabilistic CA in the universality class of the kinetic Ising model. This leads to relations between spatial and temporal correlations at criticality.

On the experimental side, the forward problem amounts to the design of good simulations which adequately reveal the behaviors under investigation. The usual problem here is that one wishes to describe the behavior of an automaton in the infinite-size and/or -time limit, though any experiment is perforce finite. Similarly, one may wish to discuss the behavior of a large or infinite set of automata on the basis of simulations of only a finite number of them (Li et al., Wootters and Langton, Gutowitz). There is a great need for flexible and reliable criteria to assess finite simulations meant to calculate behaviors in the infinite-size and -time limit.

The forward problem also arises in computation theory. Since the readers of this volume are presumed to be for the most part physicists, some general comments are in order. In computation theory, the goal is to understand cellular automata viewed as parallel computers. Parallel computation, in which many (typically identical) processors share a computational work load, is of tremendous practical interest. It also poses profound theoretical questions. Parallel computational architectures are often referred to as non-von Neumann architectures, to distinguish them from serial architectures as reduced to practice by von Neumann. It is well known, however, that von Neumann (with the aid of Stan Ulam) invented the cellular automaton architecture as well. Just as practical parallel computation is undeveloped relative to serial computation, much remains to be explored in the theory of parallel computation.

In computation theory, the forward problem of characterizing the set of configurations generated by a cellular automaton is often expressed in terms of recognition of the language represented by that set of configurations. A language is simply a subset of all possible strings of symbols from a finite alphabet. To recognize a language means to determine which strings (configurations, words) are in the language and which are not. One of the basic problems is to determine what sort of machine recognizes a given language, given that there exists such a machine. Once the type of machine has been determined, one inquires after the amount of resources (how much time, how much space) the machine requires to perform the recognition task. For example, Kim and McCloskey investigate the set of languages which can be recognized by a one-dimensional CA in constant time. Recognition of a language in constant time means that given a word of the language as input, the number of time steps which the CA requires to enter a (pre-determined) accepting state is independent of the length of the word.

For every characteristic that a cellular automaton might have, it is a priori conceivable that a machine could be constructed which decides, given a cellular automaton rule table, whether the automaton has that property. For some properties it is possible to prove that there exists no such machine. These are called undecidability results. For instance, Kari describes his proof of a long-standing conjecture that reversibility of cellular automata in two dimensions in undecidable. The proof involves mapping the reversibility problem into a problem concerning tiling of the plane, which is known to be undecidable. Another set of undecidability results are found by Sutner. He proves that there exists no machine to decide on even some rather simple properties, such as the property of mapping all spatially periodic configurations eventually to the null state. A variety of other undeciability results (indeed, all manner of results in the computation theory of CA) can be found in Culik et al.'s survey article.

Another source of forward problems is the desire to extend techniques developed for the analysis of other types of systems to encompass cellular automata as well. These might also be called lateral problems. In some cases, this means attempts to apply experience and methodology associated to the study of differentiable dynamical systems to the discrete dynamics of cellular automata. Examples here are the articles by Vichniac and Smith, among others. While Vichniac is concerned with the formal extension of the differential operator itself to CA, Smith suggests ways in which continuous change can be dealt with in the context of CA without explicit appeal to a formal notion of derivative.

In other cases lateral problems involve the use of computation theory to address problems in physics. In this vein, Fredkin issues a spirited manifesto concerning Digital Mechanics. Digital Mechanics is a hybrid of computer science and physics which uses the tools of the former to attack the problems of the latter. Central to this approach is the view that the universe itself and in its entirety should be considered to be a cellular automaton. The problem for physics, then, is to determine the rule by which this automaton operates. Picking up a theme related to Fredkin's, Svozil reviews some characterizations of chaos, a concept from physics, in terms of computation-theoretic measures of complexity. He suggests that some physical phenomena, and in particular, the emergence of time irreversibility in macroscopic systems governed by microscopically reversible laws may be understood using tools of computation theory.

A connection between computation theory and mathematical physics is taken up in the two articles by Culik et al. Their concern is to develop a relationship between the theory of dynamical systems acting on compact metric spaces and computation systems acting on strings. The connection is made via formal language theory, the study of sets of strings which are defined by the operation of various sorts of machines.

This sort of cross-fertilization may also take the form of setting up relationships between cellular automata and various generalizations of cellular automata such as neural nets. Lee et al., Qian et al., Sieburg et al., and Martin introduce structures which are not rigidly CA. Lee et al. and Qian et al. consider systems in which the local rule applied to a cell state can vary from cell to cell. Sieburg et al. allow the connectivity between cells to vary depending on the location of the cell. Martin considers rules with probabilistic rather than deterministic transitions.

Further generalizations of CA are found in chapter 6. In that chapter, Garzon compares a particular generalization of CA, discrete neural networks, with CA themselves in terms of their computational power. He presents a formalism which facilitates such comparison. Walker studies systems called Boolean nets, which, like neural nets, can have spatial variation in connectivity. However, the cells of Boolean nets have only two possible states, while the cells of traditional neural nets have continuous states. Walker focuses on the basins of attractions of these Boolean nets, in particular the change in the size and dominance of these basins with change in system size. A final generalization of CA discussed in this volume, by Livi et al., is the coupled map lattice. The only difference between a CA and a coupled map lattice is that the cells in a coupled map lattice have continuous states, while the cells in a CA have only a finite number of discrete states. The coupled map lattices studied by Livi et al. display a rich variety of chaotic and periodic behavior, depending on the values of a control parameter.

It is essential for the cellular automaton theorist to contemplate these various generalizations. On one hand, a generalized CA may exhibit behavior with character and origin similar to some CA behavior of interest. This behavior may well be more tractably analyzed in the variant CA, and the experience gained in that analysis may be brought back to illuminate the cognate CA behavior. On the other hand, elucidation of fundamental *differences* in behavior between a CA and some variant may help sharpen focus on the domain in which CA may be expected to perform well – as models, as devices, as sources of mathematical theory.

3. The inverse problem

It is only quite recently that any general methods to address the inverse problem have been developed. This area is sure to gain in prominence. This gain will be driven primarily by applications of cellular automata in the natural sciences. The long-term goal is to develop a set of techniques which allow one to find a rule, or set of rules, which quantitatively reproduce some set of observations of a physical system.

If one has complete knowledge of the underlying structure and laws of a system under study, it may be possible to directly engineer a rule with the desired properties. A relevant example in this regard is von Neumann's seminal construction of a cellular automaton which can reproduce itself. He wanted to build a universal computer in a cellular space and knew what the component parts of such a computer must

be. What was required then was the ingenuity to design these components and deploy them in the space so that they could properly interact.

Conway's invention of the Game of *Life* was undoubtedly driven by a fairly clear intuition concerning the type of behavior to be desired in a life-like rule, and how to obtain that behavior. He wanted a rule which was marginally stable, that is, neither too many live cells nor too many dead cells should be generated, and the configurations generated should have both local cohesiveness and global unpredictability. It is these criteria, coupled with a good deal of experimentation, which could have potentially lead to a rule with complex behavior like the Game of *Life*.

In most modeling situations of interest, complete knowledge of the underlying dynamics is not available. Otherwise, a model might hardly be necessary. Hence the core of the inverse problem is (1) to use whatever knowledge is available about the system to be modeled to constrain the set of automata which must be considered, and (2) to search the remaining set of automata in some efficient way.

In order to search a set in an efficient way, it is helpful to have some knowledge of the organization of that set. This topic is taken up in the chapter entitled Structure of the Space of Cellular Automata, which is concerned with the nature of the set of cellular automata as a whole. That is, these authors investigate how properties of cellular automata change as one moves from one part of the space of cellular automata to another, by making changes in cellular automaton rule tables. This is a forward problem in the sense of having characterization as a goal, but it leads to a framework for attack on the inverse problem. For instance, Li et al. investigate phase transitions in the space of cellular automata. In the case of phase transitions in physical systems, at the critical values of some control parameter (typically temperature) the system undergoes a marked change in its properties, changing, say, from liquid to gas. The control parameter, λ, employed by Li et al. is not a temperature, but changes in this parameter produces changes in cellular automaton behavior strongly reminiscent of physical phase transitions. Wootters and Langton analyze some of the phenomena discussed by Li et al. in terms of the so-called mean field theory for cellular automata. One of their concerns is to find sharp transitions, such as are observed in physical systems, in sets of CA. Sharp transitions are found in the limit where the number of possible cell states approaches infinity. McIntosh extends the study of phase transitions in CA rule space in terms of mean field theory by focusing on one well-known example: Conway's Game of *Life*. He argues that the complex behavior of this rule may in part be attributed to the fact that the mean field curve for this rule indicates marginal stability at the fixed point. It is in this sense that *Life* lies at a transition in the space of rules. A different approach to the study of rules at a phase transition in the space of CA is taken by Chaté and Manneville. They approximate continuous-state systems known to undergo phase transitions by sequences of CA. When the sequence crosses a critical surface in the space of CA, rules of particularly complex character are found. At the close of that chapter, Gutowitz presents a classification for CA that incorporates the λ-parameter, the mean field theory, and still higher-order schemes for parameterizing the space of CA in a uniform framework. These schemes compose a strict hierarchy. As one descends through levels of the hierarchy, longer-range correlations between cell states are accounted for, and hence, finer distinctions can be made between otherwise similar rules.

It is clear that understanding the structure of CA rule space, as discussed above, could aid in the design of rules with specified properties. Nonetheless, Lee et al., Qian et al., and Richards et al. show that much can be done using a proper optimization method alone. Lee et al. present the theory of, and Qian et al. an application of a scheme for learning CA to perform specific tasks. These are not CA in the pure sense, as the rule used to update a cell state may vary from cell to cell. The site-specific rules are, however, drawn from the same specified *distribution* of rules. Learning takes place in the modification of this distribution. The particular application chosen – balancing a pole – has been well studied by a variety of other methods. That it yields to a CA approach is sure to generate quite some interest in adaptive-learning community. Richards et al. discuss an alternate method for finding CA with given properties, the genetic algorithm. The genetic algorithm is a method for searching a space of strings that is inspired by biology. It is used here to investigate a physical problem: dendrite crystal growth. Using this technique, Richards et al. are

able to build, directly from experimental data, CA which generate patterns very much the same as those generated by growing crystals.

Many of the articles in the chapter entitled Cellular Automata and the Natural Sciences deal, at least implicitly, with the inverse problem. In application of cellular automata to the natural sciences, one is presented with a collection of phenomena which, due to their spatial regularity, locality of interactions etc., suggest a cellular automaton model. The problem then is to construct an automaton to reproduce the given phenomena.

The first two articles of that chapter concern biology, but at rather different levels of abstraction and generality. The first of these, by Victor, concerns the fundamental contributions the theory of cellular automata might make to a biological understanding of brain function. Despite the wide-spread resistance to formal approaches in the biological community, there is hope that if theorists work with sensitivity to the nature of empirical inquiry in biology, then some more abstract investigations may soon bring light in this area. One of the problems confronting the cellular automaton theorist in biology is that biological entities are never exactly uniformly arrayed, and seldom interact only with their neighbors according to unvarying rules. When constructing a model of a particular biological system, one is tempted to relax the strict definition of cellular automata in order to match as well as possible the design of the system under study. The trade-off is always between the realism and the tractability of the model. Sieburg et al. present a case study for this kind of modeling. Their model deviates from a strict cellular automaton, yet retains certain formal handholds. It is a vast simplification of the system under study (the immune system), yet displays certain salient features of the system's behavior.

Chemical and physical systems are in some regards simpler than biological ones, and hence may be expected to be more readily representable by cellular automata. Further, the formal implications of the fundamental laws of physics are better understood than the formal implications of the basic tenets of biology. These can be used as the basis of engineering cellular automata to model physical processes. In particular, insistence that a cellular automaton possess a certain invariant, i.e. obeys a certain conservation law, seriously constrains the design of the automaton. As pointed out by Toffoli and Margolus, it may in general be easier to construct a CA with a given invariant (inverse problem) than it is to determine the invariants of a given CA (forward problem). Another sort of physically motivated inverse problem concerns time-reversal invariance. Not all cellular automata possess this property; in some sense only a vanishingly small fraction of them do. However, the Fredkin construction allows one to take an arbitrary cellular automaton and make a reversible one corresponding to it. The Fredkin construction is described in Toffoli and Margolus and used in most contributions in this volume which deal with reversibility in CA.

The inverse problem in computation theory is to be understood as the problem of constructing machines to perform various tasks, or to have certain characteristics. Hurd, as discussed in Culik et al., has constructively solved a number of inverse problems by exhibiting CA whose limit languages have various degrees of computational complexity.

The construction of CA to perform specific computations, i.e. programming, has thus far been approached in only a piecemeal fashion. For a few examples, see McIntosh. Search through the space of CA for rules which compute specified patterns, such as discussed by Richards et al. is a sort of programming. However, one does not have the same control over the genetic operators used in this search as one does over operators in a programming language. The ideal mechanism for solution of inverse problems would be one that is as rigid as a traditional programming language concerning aspects of a correct solution of which one is certain, but that also, like the genetic algorithm, generates its own initiatives where the intuition and ingenuity of the programmer fail.

4. Cellular automata and the Interesting

I would now like to propose a tentative synthesis of forward and inverse problems in a question which underlies both. This discussion is stimulated by a footnote in A.M. Turing's classic paper, "Systems of logic based on ordinals". In this paper, consideration of the purpose of ordinal logics leads to an analysis of mathematical thought itself. Turing states: "Mathematical reasoning may be regarded rather schematically as the exercise of a combination of two faculties‡ which we may call *intuition* and *ingenuity*." (Turing's italics.) The footnote is critical:

‡ *We are leaving out of account that most important faculty which distinguishes topics of interest from others; in fact, we are regarding the function of the mathematician as simply to determine the truth or falsity of propositions.*

Here I wish to assume intuition and ingenuity as understood, and ask what challenge is posed by cellular automata to that faculty which Turing leaves out of account.

Hence, if there is one central question in the field of cellular automata it is this: "What makes a cellular automaton interesting?" Of course, that which is interesting to a mathematician may not be the same as that which is interesting to a biologist. Still, the question can be posed in some generality. If one generates cellular automata at random, that is, by filling cellular automaton rule tables by flipping an unbiased coin, chances are that the cellular automata thus produced will have no interest from any perspective. They will be boring from a statistical mechanical viewpoint since it is likely that these automata will have an unbiased, uncorrelated invariant measure. They will be boring from a computational-theoretic viewpoint since they will not perform any significant computation. They will be boring from an applications viewpoint since they will not be a good model for any physical system, except, perhaps, an ideal gas at equilibrium. The cellular automata which are interesting relative to any specific question are likely to form a Lebesque measure-zero set in the set of all cellular automata. This is not to advise dispair. There are an infinite number of interesting cellular automata. The challenge is to find them and characterize them.

Steven Wolfram (for references see Appendix II) dealt, in part, with the question of interesting cellular automata with his well-known classification of cellular automata. The most interesting cellular automata compose class IV in Wolfram's scheme. The elements of class IV are those rules which elude adequate characterization by the techniques of dynamical systems theory, computation theory, etc., techniques which serve to pin down the properties of rules in the other three classes (apart from the decision problems discussed in the Culik et al. survey article.) That is, rules in class IV are those rules which demand new theoretical and experimental tools, rules which may be expected to drive future research.

All the same, rules may be complex, subject to quite sophisticated analysis, and yet still be of interest. In this case, however, the interest typically derives from questions from other disciplines to which study of these cellular automata may pertain. Lattice gas automata are prime examples as they are the most successful application of CA in the natural sciences to date. Lattice gas automata are for the most part studied as a means to an end: understanding the nature of physical fluid flow. If lattice gas automaton behavior were to refuse to correspond to the behavior of real fluids, then they would fail to generate the same excitement in the physics community. This regardless of the interest these automata might possess qua automata. (For recent work on lattice gas automata, see the companion volume to this one, to appear in Physica D, and edited by G. Doolen.)

We have seen that cellular automata have applications in many fields, and conversely, that the analytical techniques of many disciplines can be productively trained on cellular automata. One is justified to ask, "what is it about cellular automata that gives their study a distinct identity, separate from the multiple reflections from other fields?". Part of the answer lies in the simplicity, transparency, and naturalness of the definition of cellular automata. Despite this simplicity, the space of CA is populated by rules with enormously complex and intriguing behavior.

Another part of the answer lies in the intrinsic versatility of cellular automata, their ability to support a wide range of interpretations. It may happen, as some hold, that the concept of a cellular automaton will become so interwoven into general patterns of intellectual activity that their study will never develop a unique character. I believe this to be unlikely. Rather, it will become ever more apparent where the boundaries of the effectiveness of cellular automata lie. It could be that in a sense, a sense comparable to Church's thesis, that any phenomenon can be represented in the behavior of a cellular automaton. Nonetheless, there are systems for which one *must* invoke global controls, continuous states etc. in order to build a convincing and tractable model. Hence it is crucial for the health of the field of CA research to clarify the limits of that domain of problems for which CA are the tool of choice.

Acknowledgements

The Conference on which these Proceedings are based was called Cellular Automata: Theory and Experiment, and was held at Los Alamos National Laboratory September 9–12, 1989. It was supported by the Center for Nonlinear Studies (CNLS) of the Los Alamos National Laboratory.

I would like to thank David Campbell and Gary Doolen of CNLS for generously sharing their wisdom and experience with me during all phases of the organization of this meeting. Marian Martinez and Lisa Crane of CNLS and Susan Spach of the Theoretical Division at Los Alamos expertly handled all of the administrative burden. It was a great deal of fun to work with them.

I am grateful to Margreet den Haan and Mark Eligh of the North-Holland staff for their professionalism and patience during the preparation of this volume. I am also indebted to the many reviewers who took such care in enforcing quality control on these Proceedings. I hope they are pleased with the result.

CONTENTS

Chapter 4   Cellular automata and the natural sciences

Biology

Physics and chemistry

Chapter 5   Computation theory of cellular automata

MATHEMATICAL ANALYSIS OF CELLULAR AUTOMATA

Physica D 45 (1990) 3–18
North-Holland

APERIODICITY IN ONE-DIMENSIONAL CELLULAR AUTOMATA

Erica JEN

Theoretical Division, Los Alamos National Laboratory, Los Alamos, NM 87545, USA

Received 25 March 1990
Revised manuscript received 16 April 1990

Cellular automata are a class of mathematical systems characterized by discreteness (in space, time, and state values), determinism, and local interaction. A certain class of one-dimensional, binary site-valued, nearest-neighbor automata is shown to generate infinitely many aperiodic temporal sequences from arbitrary finite initial conditions on an infinite lattice. The class of automaton rules that generate aperiodic temporal sequences are characterized by a particular form of injectivity in their interaction rules. Included are the nontrivial "linear" automaton rules (that is, rules for which the superposition principle holds); certain nonlinear automata with injectivity properties similar to those of linear automata; and a wider subset of nonlinear automata whose interaction rules satisfy a weaker form of injectivity together with certain symmetry conditions. The evolution of this last subset of automata can be viewed as consisting of multiple domains within which behavior exactly mimics that of a linear automaton. A linearization technique is used to establish that these domains coalesce in finite time to produce at most two domains, and thereby to assert the aperiodicity of their temporal sequences.

1. Introduction

Cellular automata are a class of mathematical systems characterized by discreteness (in space, time, and state values), determinism, and local interaction. A cellular automaton consists of a finite-dimensional lattice of sites whose values are restricted to a finite (typically small) set of integers $Z_k = \{0, 1, \ldots, k - 1\}$. The value of each site at any time step is then determined as a function of the values of the neighboring sites at the previous time step. The general form of a one-dimensional cellular automaton, for example, is given by

$$x_i^{t+1} = f(x_{i-r}^t, \ldots, x_i^t, \ldots, x_{i+r}^t),$$

$$f : Z_k^{2r+1} \to Z_k, \tag{1.1}$$

where x_i^t denotes the value of site i at time t, f represents the "rule" defining the automaton, and r is a non-negative integer specifying the radius of the rule. The simplest cellular automata are those with $r = 1$ and $k = 2$; designated by Wolfram [1] as "elementary," these automata are defined on a one-dimensional spatial lattice, and consist of binary-valued sites evolving in time according to a nearest-neighbor interaction rule.

Cellular automata's distinctive set of features has attracted, in recent years, substantial attention as simple models for complex physical and biological phenomena [2–6]. In particular, cellular automata can be viewed as prototypical models for systems consisting of a large number of simple, identical, and locally connected components. Examples of phenomena that have been modeled using cellular automata include turbulent flow resulting from the collisions of fluid molecules, dendritic growth of crystals resulting from aggregation of atoms, and patterns of electrical activity in simple neural networks resulting from neuronal interaction. Such problems are conventionally studied using continuous models based on partial differential equations. Models based on cellular automata differ from, but are qualitatively and to some extent quantitatively consistent with, those obtained from a continuous approach. Simulations based on cellular automata may provide an increase in computational efficiency, as well as insight into the relation between continuous and dis-

crete modeling of the phenomena being studied.

In addition to providing modeling tools, cellular automata represent intriguing and little-understood mathematical systems. At present, few tools exist for the analysis of their behavior. Problems in cellular automata research pose special difficulties since they often fall outside the purview of traditional continuous mathematics; cellular automata problems typically reflect, in both their formulation and their solution, the features of discreteness and local interaction that make these systems distinctive. Although some cellular automata are clearly equivalent to other standard mathematical constructs, including shift-commuting maps and finite-difference schemes for solving partial differential equations [3,7], others are not. Many automata, such as that defined by

$$x_i^{t+1} = x_{i+1}^t \dotplus \max(x_i^t, x_{i-1}^t), \qquad (1.2)$$

where "$\dotplus$" denotes addition modulo 2, cannot easily be identified as discretizations of a continuous system.

In fact, the evolution of a cellular automaton is governed typically not by a function expressed in closed-form, but by a "rule table" consisting of a list of the discrete states that occur in an automaton together with the values to which these states are to be mapped in one iteration of the rule. (The rule table can be converted to a function involving Boolean expressions [1,8], but such reformulation is useful in only limited contexts.) Not much of calculus applies to such systems.

Thus, while it is natural to view cellular automata as dynamical systems, problems arise in that many concepts standard to that field (for example, "stability", "attractors", "sensitive dependence", "chaotic behavior") do not have unambiguous analogues in this new context. In particular, the lack of a natural metric in either cellular automaton state space or cellular automaton rule space makes it difficult to translate these concepts for cellular automata and other spatially extended systems defined with a finite set of state values. In recent years, a number of theoretical and numerical studies [1,2,9,10] have been undertaken to identify appropriate statistical quantities for cellular automata – analogous to, say, dimensions, entropies, and Lyapunov exponents – and to use these quantities to characterize the dynamical behavior of these systems.

This paper focuses on an aspect of cellular automata for which the analogy with continuous systems is relatively clear; namely, the periodicity of the behavior they generate. The question to be considered is, more precisely, the *aperiodicity* of the sequence of values $\{x_i^t, t = 0, 1, \ldots\}$ assumed by a particular site x_i under successive iterations of the automaton rule. The purpose of the paper is not to provide a systematic or exhaustive study of aperiodicity, but to present a collection of analytical results representative of the phenomenon for these systems. A major theme throughout is the distinction between *linearity* and *nonlinearity*, and the development of techniques that permit results on aperiodic behavior in linear automata to be extended to certain nonlinear cases.

The organization of the paper is as follows. Section 2 provides definitions and other introductory material on elementary cellular automata. Section 3 considers the behavior of linear rules and a certain subset of nonlinear but "injective" rules applied to finite initial conditions (that is, initial conditions with a compact support) on an infinite lattice, and in particular establishes that these rules generate infinitely many aperiodic temporal sequences. Section 4 outlines an exact linearization technique that leads to the extension of this result to a wider class of nonlinear automata. Concluding remarks are made in section 5.

2. Preliminaries

This section contains a review of definitions and concepts preparatory to the discussion of aperiodicity in one-dimensional automata.

An "elementary" [1] (that is, nearest-neighbor, binary site-valued) cellular automaton is defined by

$$x_i^{t+1} = f(x_{i-1}^t, x_i^t, x_{i+1}^t), \quad f : \{0,1\}^3 \to \{0,1\},$$

where x_i^t denotes the value of site i at time t, and f represents the "rule" defining the automaton.

Since the domain of f is the set of 2^3 possible 3-tuples, the rule function f is completely defined by specifying the "rule table" of values $a_i \in \{0,1\}$ with $i = 0, 1, \ldots, 7$ such that

$$000 \to a_0, \ 001 \to a_1, \ \ldots, 111 \to a_7,$$

where $xyz \to a_i$ indicates that $f(xyz) = a_i$. There is a total of $2^8 = 256$ distinct elementary rules.

The conventional labelling scheme [1] assigns the integer

$$R = \sum_{i=0}^{i=7} a_i 2^i$$

to the elementary rule defined by f. The rule number thus assumes an integer value between 0 and 255.

The major focus of the sections that follow will be on the aperiodic behavior of temporal sequences generated by automata on infinite lattices. Note that in the case of an automaton rule operating on finite spatial lattices, the discreteness of state values and determinism of the interaction rule imply that all initial conditions will be attracted into "limit cycle" behavior; that is, aperiodic behavior cannot occur. Here, a limit cycle of period p on a cylinder size n is defined to be a set of spatial sequences $\{x_i^t; i = 0, 1, \ldots n - 1; t = T, T + 1, \ldots, T + p - 1\}$ such that $x_i^t = x_i^{t+p}$ for all $t \geq T$; that is, a set of spatial sequences on the cylinder that repeat themselves periodically in time.

For clarity, notational distinctions will be made between an automaton's *spatial sequences* S^t with components $\{S_i^t \equiv x_i^t; -\infty < i < \infty\}$ at time t, and the *temporal sequences* W_i with components $\{W_i^t \equiv x_i^t, t = 0, 1, \ldots\}$ representing the values assumed by the site x_i with successive iterations of the automaton rule.

Definition 1: The temporal sequence W_i is periodic of period $0 < p < \infty$ with transience $0 < T < \infty$ if $x_i^{t+p} = x_i^t$ for $t > T$.

It will be assumed that the automaton rules are operating on infinite lattices with arbitrary finite initial conditions; that is, arbitrary spatial sequences of finite length with compact support.

Definition 2: A finite initial condition on an infinite lattice is an initial condition $\{x_i^0, -\infty < i < \infty\}$ such that for some finite M, N, $x_i^0 = 0$ for $i < M$ and $i > N$ with $x_M^0 = x_N^0 = 1$. The length of such an initial condition is defined to be $N - M + 1$.

Finally, a distinction will be made throughout between "linear" and "nonlinear" automaton rules.

Definition 3: A rule R is defined to be linear if it satisfies the additivity condition; that is, for any 3-tuples y and z, the function f defining the rule R satisfies

$$f(y) \dot{+} f(z) = f(y \dot{+} z),$$

where "$\dot{+}$" denotes binary addition.

Linear elementary automata include rules 0 (zero rule), 15 (right-shift with toggle rule), 51 (toggle rule), 60, 90, 105 (sum rule with toggle), 150 (sum rule), 170 (left-shift rule), and 140 (identity rule); together with their equivalents under symmetry transformations.

From the point of view of dynamical systems, most linear rules generate rather simple behavior. Extensive work [11,12] has been done, however, on the behavior of the "nontrivial" linear rules (that is, rules 60, 90, and 150) operating on finite lattices with periodic or fixed boundary conditions. Techniques have been developed for enumeration of their limit cycles, and computation of maximal limit cycle periods and maximal transience length. It has further been shown [12] that a given spatial sequence appears in limit cycles for linear rules iff its values satisfy a linear recurrence relation defined by the automaton rule, and thus the considerable mathematical machinery of recurring sequences over finite fields can be used for the analysis of the detailed structure of limit cycle sequences.

Nonlinear cellular automata are not susceptible to the approaches of refs. [11,12] since many of those results derive from algebraic properties of the linear operator representing the automaton rule, and further, the inability to invoke the superposition principle cripples the analysis of nonlinear rules for arbitrary initial conditions. In the following sections, a simple result establishing aperiodic behavior for a linear rule on infinite lattices will be shown to hold for certain nonlinear rules as well.

In the extension of results from linear to nonlinear rules, a central concept will be that of a special type of "injectivity" as a feature of certain nonlinear as well as linear rules. An automaton

rule will be said to be *injective in a particular component* if, for every 3-tuple, the component uniquely determines the value assigned by the rule tables to that tuple (assuming fixed values for the other components). The injectivity of a rule table is shown elsewhere [13,14] to be a critical factor in many features of the automaton's dynamical behavior, including the generation of domain walls (that is, temporal sequences that serve to divide the automaton into "independent" spatial domains); the scaling of the number of preimages for arbitrary spatial sequences with sequence length; and the distribution of limit cycles on finite lattices with periodic boundary conditions.

The precise definition is as follows:

Definition 4: A rule R is injective in the $(i + k)$th component $(k = -1, 0, 1)$ if for every tuple $(x_{i-1} x_i x_{i+1})$, the rule table for R represents a one-to-one mapping between x_{i+k} and $f(x_{i-1} x_i x_{i+1})$ when the two other components $x_{i+j}, j \neq k$ are fixed.

Thus, for example, rule 15, defined by

$$\{100, 101, 110, 111\} \rightarrow 0,$$

$$\{000, 001, 010, 011\} \rightarrow 1,$$

representing the action of a right-shift with toggle, is injective in the $(i - 1)$th component; and rule 150, defined by

$$\{000, 011, 101, 110\} \rightarrow 0,$$

$$\{001, 010, 100, 111\} \rightarrow 1,$$

is injective in all three components.

The following proposition asserts that all linear automaton rules (with the exception of rule 0) are injective in at least one component. The result is stated for elementary automata, but applies to rules of arbitrary neighborhood size with sites assuming values in $F_q = \{0, 1, \ldots, q - 1\}$, where q is a prime power.

Proposition 1. Let R be a linear automaton rule. Then either R is injective in at least one component, or R is the rule that maps all tuples to 0.

Proof: Suppose R is not injective in any component. Then for instance there must exist $x, y \in \{0, 1\}$ such that

$$f(xy0) = f(xy1),$$

where f is the interaction rule defining R. The additivity principle of linear rules implies that

$$\begin{aligned} f(001) &= f(xy0 \dot{+} xy1) \\ &= f(xy0) \dot{+} f(xy1) \\ &= 0. \end{aligned}$$

Similarly, $f(010) = f(100) = 0$, from which it follows using superposition that all tuples are mapped to 0. $\qquad\square$

The converse is not true. Rule 30, for example, defined by

$$\{000, 101, 110, 111\} \rightarrow 0,$$

$$\{001, 010, 011, 100\} \rightarrow 1,$$

is injective in the $(i - 1)$th component, but does not satisfy the definition of linearity.

3. Aperiodic behavior of injective rules on infinite lattices

In this section, it will be shown that, starting from arbitrary finite initial conditions on infinite lattices, a certain class of elementary automaton rules generates infinitely many aperiodic temporal sequences; that is, sequences that are not periodic of any period. This class of rules is characterizable as exhibiting injectivity together with a feature that ensures infinite-time "propagation" of the automata.

The aperiodicity by temporal sequences generated of a particular nontrivial linear rule will be singled out for discussion here. The rule is defined by

$$\{000, 010, 101, 111\} \rightarrow 0,$$

$$\{001, 011, 100, 110\} \rightarrow 1,$$

alternatively specified by the functional form

$$x_i^{t+1} = x_{i-1}^t \dot{+} x_{i+1}^t, \tag{3.1}$$

to be referred to hereafter as rule 90. Fig. 1 provides examples of the rule's evolution on finite

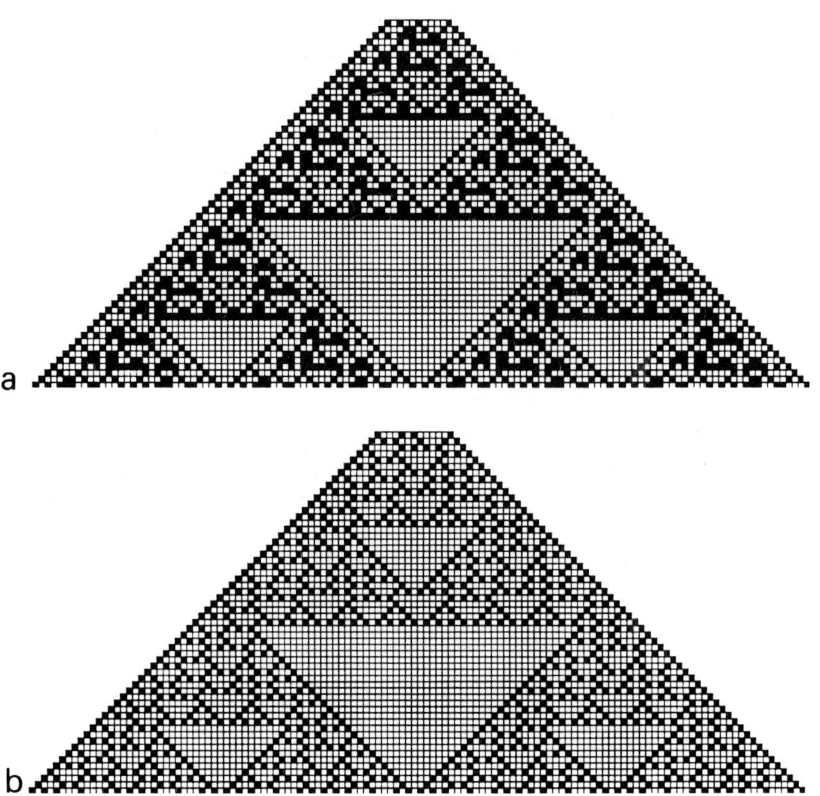

Fig. 1. Evolution of rule 90 defined by $\{000, 010, 101, 111\} \to 0$, $\{001, 011, 100, 110\} \to 1$, from two different finite initial conditions on an infinite lattice. In the figures, each row of dots represents the spatial sequence generated at time t, with the top row corresponding to $t = 0$. A black square represents a site value of 1; a white square represents a site value of 0. Proposition 2 establishes the aperiodicity of temporal sequences ("columns" in this figure) generated by rule 90 from arbitrary finite initial conditions.

initial conditions. Note that rule 90 is injective in both its $(i - 1)$th and $(i + 1)$th components. A simple argument (easily extended to the case of the other nontrivial linear automata, including rules 60 and 150) establishes the aperiodicity of all temporal sequences generated by rule 90 (with the exception of a trivial case to be described below). The proof rests on showing that there exists a K such that every temporal sequence contains a subsequence of 2^k 0's for all $k > K$.

Proposition 2: With the exception of the trivial case, every temporal sequence generated by rule 90 with arbitrary finite initial conditions on an infinite lattice is aperiodic. The trivial case is the temporal sequence of all 0's generated by rule 90 from an initial condition that is spatially symmetric and of odd length.

Proof: Let the initial condition be specified by $\{x_i^0, -\infty < i < \infty\}$ with $x_M^0 = x_N^0 = 1$ being the "leftmost" and "rightmost" 1's. For convenience, assume $M < 0 < N$. From (3.1), it can be seen that the left and right "borders" of 1's propagate with unit speed. Also from (3.1), it follows that

$$x_i^t = x_{i-t}^0 \dot{+} x_{i+t}^0 \tag{3.2}$$

whenever t is a power of 2. Consider an arbitrary temporal sequence W_j and assume, without loss of generality, $j > 0$. Define K to be a positive integer such that

$$j - 2^K < M \quad \text{and} \quad j + 2^K > N.$$

Then (3.2) implies that $x_j^t = 0$ for

$$2^k \le t \le 2^k + \min[2^k - j - M, 2^k + j - N],$$

with k such that $k \ge K$, since the values of the temporal sequence W_j in this range are given

as sums of the values of the temporal sequences W_{j-2^k} and W_{j+2^k}, each of which consists of 0's until terminated by the propagation of the left and right borders. (Termination on the left occurs at $t = 2^k - j - M$ and on the right at $t = 2^k + j - N$.) Moreover the temporal sequence W_j cannot have all values equal to 0 since the spatial sequence generated at time step 2^k consists of two "copies" of the initial condition with a block of 0's in the center (see fig. 1), and the two 1's bordering the block propagate toward the center with unit speed, thereby generating a 1 at each site (except possibly the center). Since the result holds for k arbitrary large, it follows that the temporal sequence W_j is aperiodic except in the trivial case where it is the "center" sequence of all 0's generated by an odd-length spatially symmetric initial condition.   □

The above result depends on the use of the (essentially) linear representation of rule 90. In what follows, it will be shown that by discarding linearity but retaining injectivity, certain nonlinear rules exhibit aperiodic behavior as well.

Consider now the class of automata injective in either their $(i-1)$th or their $(i+1)$th component. Each such class contains a total of 2^4 elementary rules, with linear rules 90 and 150 the only automata belonging to both classes.

In order to establish aperiodicity of temporal sequences generated by these injective rules, a lemma [13] is needed. The lemma asserts that the existence of two periodic temporal sequences anywhere in an automaton implies the periodicity of every temporal sequence in between. Note that the lemma makes no assumption on the automaton rule (in other words, it is not restricted to injective rules).

Lemma 1: Let W_i and W_j, $i < j$, be two temporal sequences periodic of periods p_i, p_j, respectively. Then for $i < k < j$, the temporal sequence W_k is periodic of period $p|(p_i, p_j)$ (where $x|y$ indicates that x divides y evenly), and of transience bounded by $T < \infty$, where T is a constant independent of k.

On the basis of the above lemma, the following proposition [13] asserts that (propagating) automaton rules belonging to class A and B generate

at most one aperiodic temporal sequence.

Proposition 3: Let R be a rule injective in its $(i+1)$th component with $\{100\} \to 1$ (or injective in its $(i-1)$th component with $\{001\} \to 1$). Then with arbitrary finite initial conditions, there can exist at most one periodic temporal condition.

Proof: The argument will be given for rules injective in their $(i+1)$th components. The fact that $\{100\} \to 1$ implies that the "rightmost" 1 in the initial condition propagates with unit speed to the right. Suppose that within the "borders" of the automaton there exists a periodic temporal sequence W_i with transience T_i, and a periodic sequence W_j with transience T_j, with $i < j$. Then the lemma establishes that every sequence W_k is also periodic with some transience $T < \infty$ for $i < k < j$. In particular, W_{j-1}, the temporal sequence immediately to the "left" of W_j, is periodic with transience T. The fact that R is injective in its $(i+1)$th component (the "mirror" argument holds for injectivity on the left) implies that the values of the two adjacent temporal sequences W_{j-1} and W_j determine the values (and hence the periodicity properties) of the temporal sequence W_{j+1}. Hence that temporal sequence is periodic with transience T, as is in fact *every* temporal sequence to the right of W_j. A contradiction results however since there must exist beyond the right-hand "border" a temporal sequence that is not periodic with transience T.   □

The conditions of the above theorem hold for elementary automata rules 30, 86, 90, 150, 154, 210. The nature of the aperiodicity of temporal sequences generated by these automaton rules differs significantly from case to case. In the case of rule 90 (the purely linear rule) evolving on, say, an initial condition consisting of a single non-zero site, the aperiodicity of the temporal sequences takes the form of an orderly sequence of 1's separated by sequences of 0's with lengths that are powers of 2. By contrast, the temporal sequences generated by rule 30 (depicted in fig. 2) are aperiodic in a much stronger sense, and have been proposed by Wolfram [15] as highly efficient pseudo-random number generators.

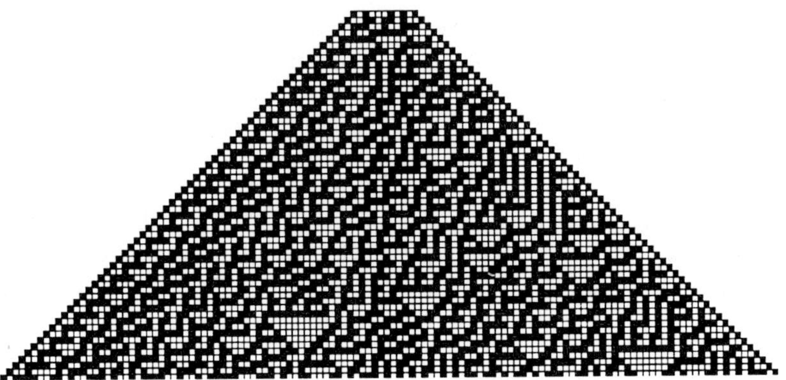

Fig. 2. Evolution of rule 30 defined by $\{000, 101, 110, 111\} \to 0$, $\{001, 010, 011, 100\} \to 1$. Proposition 3 establishes that, with arbitrary finite initial conditions, at most one temporal sequence can be periodic.

4. Aperiodicity in nonlinear rules that mimic rule 90

As has been noted elsewhere [16,17], several nonlinear rules – in particular, rules 18, 22, 122, 146, 182 – "simulate" rule 90 in that their behaviors coincide when restricted to certain spatial subsequences. In addition, as observed in ref. [16], the evolution of these rules on an infinite lattice can be viewed as consisting of multiple "domains" within which the system emulates rule 90. In the infinite case, the positions of the domain walls vary in an ostensibly random fashion, with collision of walls leading to annihilation and merging of domains. This annihilation process leads gradually to the emergence of a more and more "ordered" configuration, much as in the case of spontaneous symmetry breaking [16].

A technique is outlined here for the exact linearization of nonlinear one-dimensional automaton rules that "mimic" rule 90 in a sense to be defined below. The technique may be applied to the analysis of nonlinear rules evolving on either finite or infinite lattices. In the finite case, it permits the analysis of the limit cycle structure (including features such as transience length and maximal limit cycle period) [18]; in the infinite case, as will be discussed here, it leads to an extension of the aperiodicity results of sections 2 and 3 to this wider class of elementary cellular automata.

For nonlinear automata evolving on infinite lattices, the linearization procedure provides a rigorous characterization of the phenomenon of symmetry breaking for these rules. In particular, it

will be shown that the propagation and annihilation of domain walls for the nonlinear systems are governed by a mapping defined on an associated linear system. Analysis of this mapping then establishes that for arbitrary finite initial conditions on an infinite lattice, evolution under the nonlinear rule gives rise in finite time to at most one domain wall to the right and left of which the automaton exactly reproduces the linear rule 90.

The linearization procedure essentially consists of four steps:

(i) identification of the subsequences on which the action of the nonlinear rule deviates from linearity;

(ii) transformation of the spatial lattice (specifically, insertion of new site values) so as to preclude the occurrence of these subsequences while leaving the evolution of the rest of the system undisturbed;

(iii) analysis of the resultant linear system;

(iv) reconstruction of the original nonlinear system.

The procedure will be discussed here in detail for rule 18, defined by

$$\{000, 010, 011, 101, 110, 111\} \to 0,$$

$$\{001, 100\} \to 1. \tag{4.1}$$

The first step in linearization is to identify the subsequences on which rule 18 deviates from, say, the linear rule 90. Like rule 90 and the automaton rules of the previous section, rule 18 exhibits injective behavior on the subsets

$\{00*\}, \{10*\}$

and

$\{*00\}, \{*01\},$

where "$*$" denotes the two values $\{0, 1\}$, and as before, "injective" implies that the tuples in each subset are mapped to pairwise differing values. Unlike those rules, however, rule 18 maps

$\{01*\}, \{11*\}$

and

$\{*10\}, \{*11\}$

to pairwise equal values; it is in the values to which it maps these subsets – in particular, the tuples $\{011, 110\}$ – that rule 18 differs from rule 90. (The mapping of the tuple 111 can be ignored here since it is easy to show that the rule never generates more than two adjacent 1's.)

Hence for any sequence consisting only of isolated 1's (separated by 0's), rule 18 mimics exactly the behavior of rule 90. Fig. 3 depicts the evolutions of rules 18 and 90 on various finite initial conditions. It is easy to see that a pair of adjacent 1's occurs in the evolution of rule 18 if and only if a block of 0's of even length has appeared at some previous time step. In the discussion that follows, such a block will be termed a "irregular" block.

The next step in the linearization procedure is to devise a means of transforming the spatial lattice for rule 18 so as to preclude the occurrence of irregular blocks without perturbing the evolution of the entire system. Suppose that one iteration of the rule is applied to an arbitrary spatial sequence, and assume for the moment that the irregular blocks in the sequence are sufficiently separated (in space) so as not to affect one another. The actions of rule 18 and rule 90 are indistinguishable on all but the irregular blocks. Moreover, the definition of rule 18 specifies that

$$\{000, 010, 011, 101, 110\} \rightarrow 0,$$

and hence the rule possesses the property that a subset of its rule table is injective (and in fact identical to the linear rule 90), and the remainder – the part that deviates from linearity – collapses

onto 0. The implication is that the desired transformation can be achieved by merely inserting an extra 0 in each irregular block. The action of rule 18 on the transformed spatial sequence is identical to that of rule 90, and yet faithfully reproduces – except with the retention of the extra 0 – the effect of applying rule 18 to the original sequence. For example, suppose that the sequence contains the block 0001101, which under iteration by rule 18 generates 01000. The transformed sequence is given by 00010101, which under iteration by either rule 18 or rule 90, generates $010\bar{0}00$, where $\bar{0}$ denotes the inserted 0.

To summarize the discussion so far, the linearization procedure transforms the nonlinear system's spatial lattice so as to induce the rule to mimic the evolution of a linear rule on the transformed lattice while preserving its own evolution on the unperturbed lattice. The transformation has the effect of suppressing the occurrence of spatial subsequences upon which the action of the rule deviates from linearity ("irregular blocks"), and can be defined whenever these subsequences possess certain symmetry properties related to the injectivity properties of the nonlinear rule table.

Before continuing with the analysis of the linear system constructed via the transformation introduced above, a more precise statement of the results is needed. As suggested by the preceding comments, the linearization of rule 18 depends critically on the definition of a quantity, to be denoted c_t, representing the number of irregular blocks in a spatial sequence S^t. Precisely, define

$$c_t = \text{number of even-length blocks of 0's in } S^t,$$

noting that 11 is encompassed in this definition.

Further, define a transformation $G(S)$ that acts upon a spatial sequence S^t so as to insert an extra zero into each of the irregular blocks, thereby converting it into an odd-length block. The transformation G applied to the sequence, for example,

$$S \equiv \ldots 0100011010010000010 \ldots$$

produces the sequence

$$G(S) = \ldots 010001010100010000010 \ldots .$$

Note that the transformation is many-to-one, implying that given a sequence such as $G(S)$ above

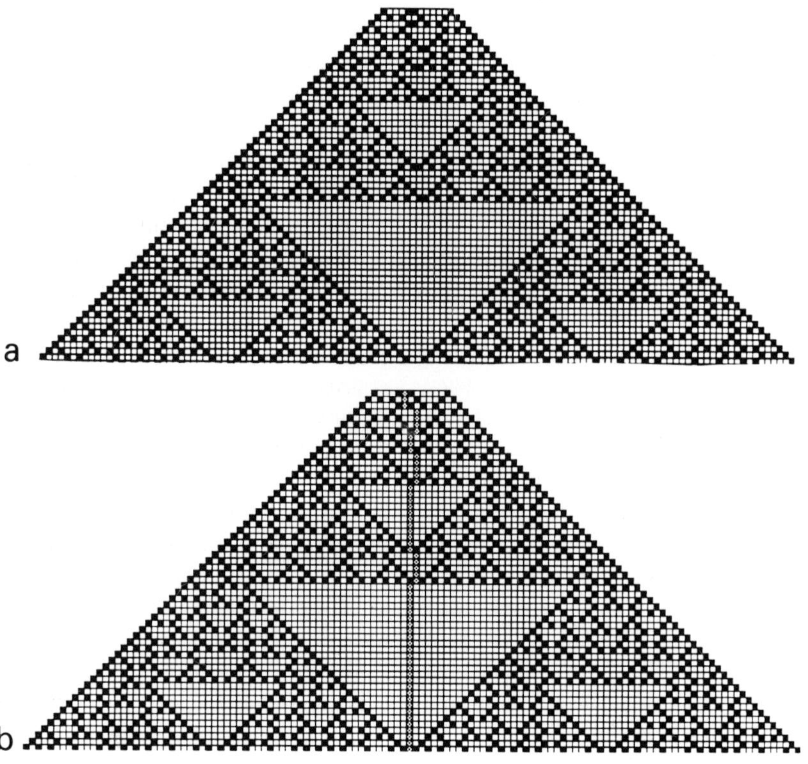

Fig. 3. Evolutions of rule 18 defined by $\{000, 010, 011, 101, 110, 111\} \rightarrow 0$, $\{001, 100\} \rightarrow 1$. The corollary to proposition 5 establishes the aperiodicity of temporal sequences generated by rule 18 from arbitrary finite initial conditions. (a) The initial condition for this automaton contains one "irregular block"; that is, a 0-block of even length. Note that each subsequent spatial sequence also contains a single irregular block. (b) By inserting an extra 0 (shown as a black-bordered white square) into the irregular block in the initial condition, the transformation G induces rules 18 to "mimic" rule 90. In this case, the transformed rule 18 automaton mimics the automaton of fig. 1b.

and the information, say, that the original sequence S contains exactly one irregular block, it is in general not possible to reconstruct S unambiguously.

Next, for any spatial sequence S, let $R_\alpha^\beta[S]$ denote the spatial sequence generated by iteration of the automaton with rule number α for β time steps on the sequence S. For example, $R_{18}^T[S]$ denotes the result of T iterations of rule 18 applied to the sequence S.

In ref. [18], the following proposition is established.

Proposition 4. For rule 18 evolving on (either finite or infinite) lattices with arbitrary initial conditions:

(i) the quantity c_t is monotonically non-increasing as a function of t;

(ii) for any range of iterations $[T_0, T_0 + T]$ for which the quantity c_t is conserved, the sequences S^t evolving under rule 18 "mimic" the sequences $G(S^t)$ evolving under rule 90; that is,

$$G(R_{18}^t[S^{T_0}]) = R_{90}^t[G(S^{T_0})],$$

for $0 \le t \le T$.

The above proposition asserts that in any range of iterations for which the number of irregular blocks is conserved, the evolution of rule 18 may be mapped onto the evolution of rule 90 by inserting an extra 0 into each of the irregular blocks. The analysis of rule 18 is hence achieved by characterizing the behavior of the appropriate linear automaton, and then "inverting" the transformation to re-capture the features of the original non-linear system; the circuitry thus is as shown below:

$$S^{T_0} \xrightarrow{\quad G \quad} G(S^{T_0})$$

$$\Big\downarrow \text{rule 18} \qquad\qquad \Big\downarrow \text{rule 90}$$

$$R_{18}^T[S^{T_0}] \underset{(G^{-1})^\bullet}{\xleftarrow{\quad G \quad}} R_{90}^T[G(S^{T_0})]$$

The many-to-one nature of the transformation G, and its implications for the limit cycle structure for rule 18, are the subject of much of the discussion in ref. [18]. In the context of this paper, however, a full discussion of the inversion procedure is not necessary to characterize the periodicity properties of temporal sequences generated by rule 18. As will be shown below, analysis of the linearized system itself suffices to establish that arbitrary initial conditions evolve under a finite number of iterations of rule 18 to spatial sequences for which the quantity c_t has converged to either 0 or 1; that is, the sequences contain at most one irregular block. Moreover, it will be shown that the spatial position of the irregular block in these spatial sequences is fixed; and thus there is no ambiguity in the reconstruction of the nonlinear system, and hence no obstacle to the extension of the aperiodicity results.

Some additional notation and definitions are needed for the discussion that follows. Let S^0 be an arbitrary finite spatial sequence (beginning and ending with 1), and let Q be the number of blocks of 0's separated by isolated 1's contained in S^0. Label these blocks $b_1^0, \ldots, b_Q^0$ (adopting henceforth the convention of moving left to right), and let $\mathrm{len}(b_i^0)$ denote the number of 0's in the ith block. (Note that the requirement of isolated 1's implies that a sequence such as 1011 has 2 blocks of 0's, with $\mathrm{len}(b_2) = 0$.)

Now apply rule 18 to the sequence S^0 and consider the evolution of its blocks of 0's. At each time step, the number of 0-blocks in the spatial sequences generated will fluctuate; nevertheless the objective is to "tag" the blocks b_i^0 of 0's in the initial condition S^0 and to track their "paths of influence" in the automaton. The "path of influence" of a block b_i^0 is defined so as to link b_i^0 to the spatial blocks $b^t, t = 1, 2, \ldots$ characterized by the requirement that if b_i^0 is irregular, then b^t is irregular as well under evolution by rule 18. Proposition 4 asserts that the number of irregular blocks is non-increasing with time, implying that for each time step t, at most one block b^t lies in b_i^0's path of influence. Fig. 4a depicts an example of a path of influence for $t = 64$ iterations of rule 18 applied to a finite initial condition.

In what follows, a mapping ϕ will be defined that formalizes this notion of a path of influence. Although defined for the linear rule 90, the mapping can be used, as a consequence of proposition 4, to study the evolution of the nonlinear automata – such as rule 18 – for which rule 90 automata serve as templates.

As outlined above, the effect of ϕ is to map each block b_i^t in a sequence S^t into the block b_i^{t+1} in S^{t+1} that belongs to the path of influence of b_i^t. In the later part of this section, the mapping will be interpreted in exactly this fashion; however, for the purposes of the lemma that follows, ϕ is defined as acting not on the blocks themselves, but on values representing the spatial positions of the 1's serving as the left borders of the blocks. More precisely, for an arbitrary spatial sequence S^t, construct the associated sequence $G(S^t)$. Denote the value of site j in the sequence $G(S^t)$ by x_j; the ith block of 0's in $G(S^t)$ by b_i^t; and the site position of the block's leftmost 1 by k_i. Then define

$$\phi(k_i) = k_i + 1, \quad \text{if } x_{k_i+2} = 0,$$
$$= k_i - K, \quad \text{otherwise,}$$

where $K \geq 1$ is the odd integer such that

$$x_{k_i-j} = 0 \quad \text{for } j = 1, 3, \ldots, K,$$
$$= 1 \quad \text{for } j = 2, 4, \ldots, K-1,$$
$$= 0 \quad \text{for } j = K+1.$$

The effect of ϕ then is to map the site position k_i onto $k_i + 1$ if there are more than three 0's in the block b_i^t, and otherwise to map it onto the site position of the leftmost 0 in the contiguous sequence of alternating 0's and 1's to the left of b_i^t. Note that, by construction, all 0-blocks in $G(S^t)$ are of odd length.

Lemma 2: Let S^t be an arbitrary finite spatial sequence containing a irregular block b_i^t whose left

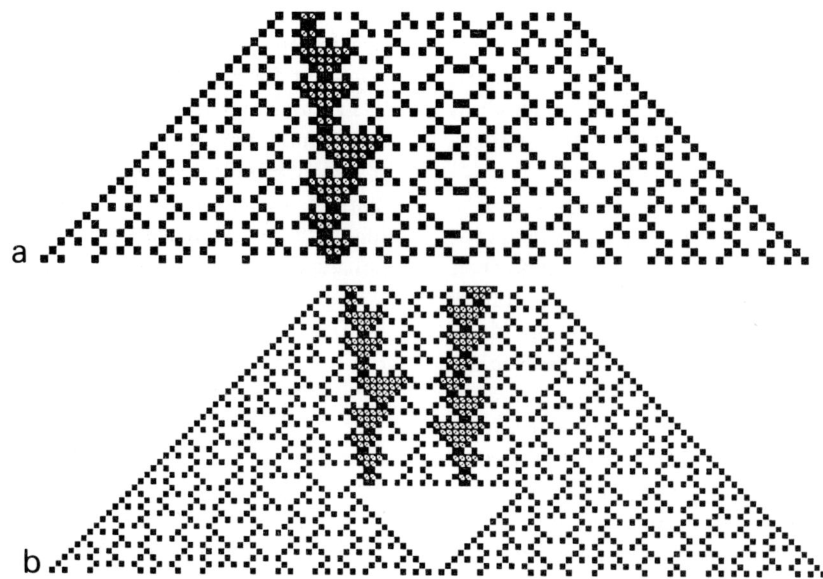

Fig. 4. Evolution of rule 18 from an initial condition with two irregular blocks. (a) The path of influence of the left irregular block is shown with a textured shading. The path links the block in the initial condition to the blocks that are irregular at subsequent time steps. (b) The paths of influence of both irregular blocks are shown with a textured shading. Upon collision, the blocks coalesce to form a regular block.

bordering 1 is at site k_i, and let S^{t+1} be the spatial sequence obtained from S^t with one iteration of rule 18. If there is no other irregular block b_j^t (with left bordering 1 at k_j) such that $\phi(k_i) = \phi(k_j)$, then the block b_i^{t+1} in S^{t+1} is irregular, and has its left bordering 1 at site $\phi(k_i)$.

Proof: The transformation ϕ has been defined to exhibit precisely the feature described in the lemma. In particular, suppose the irregular block in S^t is of length $\operatorname{len}(b_i^t) \geq 2$. Since rule 18 specifies

$$\{000\} \to 0, \qquad \{001, 100\} \to 1,$$

the block is mapped onto a irregular block of length $\operatorname{len}(b_i^t) - 2$. The corresponding block in $G(S^t)$ is of length ≥ 3, and $\phi(k_i) = k_i + 1$. On the other hand, if $\operatorname{len}(b_i^t) = 0$ in the original sequence S^t, then the rule table

$$\{010, 011, 101, 110\} \to 0, \qquad \{001, 100\} \to 1,$$

effects a coalescence of the block with all contiguous blocks to the left (and right) of length 1 or less. If this string of blocks contains no block of length 0 other than b_i^t, then the resultant block at time $t + 1$ is irregular and its left bordering 1 is

situated at the site corresponding to the leftmost 0 of the string of alternating 1's and 0's. □

Note that if the condition in the lemma is violated – in other words, if there exists at least one block b_j^t with $\phi(k_i) = \phi(k_j)$ – then the irregular blocks annihilate each other in pairs, and the coalescence produces a block that is either irregular or regular depending on the parity of the number of such blocks. Fig. 4b depicts an example in which two irregular blocks coalesce to produce a regular block.

The following corollary to the above lemma asserts that the left bordering 1 of a block of 0's must have speed of propagation less than 1; that is, it cannot propagate to the right faster than does a right diagonal sequence of the automaton. A symmetric result holds for propagation to the left of the right bordering 1 of a block of 0's; namely, it cannot propagate to the left faster than does a left diagonal sequence of the automaton.

Corollary 1: Let k_i be the position of the left bordering 1 in a block b_i of 0's. Then, with ϕ^t denoting the tth iterate of the mapping ϕ,

$$\phi^t(k_i) \leq k_i + t, \quad \text{for all } t \geq 0.$$

The transformation ϕ provides the tool for clarifying the behavior of the quantity c_t introduced above. Suppose that rule 18 is applied to a spatial sequence in which there are exactly two irregular blocks b_i and b_j. If $\phi^t(k_i) \neq \phi^t(k_j)$ for all t, then the paths of influence of the two blocks never intersect and $c_t = 2$ is a conserved quantity. If, however, $\phi^t(k_i) = \phi^t(k_j)$ for some t, then the paths of influence of the two blocks collide and annihilate – producing a single regular block – and the quantity c_t drops to 0. In particular, note that a decrease in the value of c_t is a necessary consequence of the collision of two paths; it is impossible, in other words, for the paths to "cross" or otherwise survive intact.

The preceding discussion permits the derivation of the main result of this section; namely, that for arbitrary finite initial conditions evolving under rule 18, collision of paths of influence leads in finite time to annihilation as complete as possible for all irregular blocks. Specifically, for any initial condition of odd length, the evolution of the automaton generates in finite time a spatial sequence with no irregular blocks; initial conditions of even length evolve to a spatial sequence with exactly one irregular block anchored in the "center" of the automaton. (The position of the irregular block in fact regularly drifts slightly from the center for a fixed number of iterations at increasingly long time intervals; this drift is, however, immaterial to the argument at hand.) In the language of ref. [16], the result establishes that for odd-length initial conditions, all domain walls are annihilated in finite time. For even-length initial conditions, a finite number of iterations leads to the emergence of two domains – inside each of which the behavior exactly reproduces that of rule 90 – separated by a wall running through the center of the lattice.

Proposition 5: Suppose rule 18 is applied to an arbitrary finite initial condition. If the length of the initial condition is even, then the quantity c_t (the number of irregular blocks at time t) converges in finite time to $c^* = 1$. If the length is odd, c_t converges in finite time to $c^* = 0$.

Proof: Let S^0 be an arbitrary finite initial condition of length L with a total of Q blocks of 0's labelled $b_1^0, \ldots, b_Q^0$. Suppose that c_0 of the Q blocks are irregular, with $c_0 > 1$, and that $c_t = c_0$ is constant for all t.

From proposition 4, it follows that the evolution of S^0 under rule 18 mimics its evolution under rule 90. In particular, for any t with $T = 2^t \geq L$, the spatial sequence S^T consists of a 0-sequence of length $2^{t+1} - L$ bordered on the left and right by subsequences S_ℓ^t and S_r^t such that

$$G(S_\ell^T) = G(S_r^T) = G(S^0),$$

where G is the transformation that converts irregular to regular blocks. Note that while the mimicking ensures that both S_ℓ^T and S_r^T reproduce S^0 as described in proposition 2 (and in particular possess exactly Q blocks of 0's), the two sequences S_ℓ^T and S_r^T nevertheless may differ since ambiguity has now arisen as to exactly which blocks in these sequences are irregular. Label the blocks in S_ℓ^T as $_\ell b_i^T$; the blocks in S_r^T as $_r b_i^T$; and the center 0-block as b_0^T. Suppose that some block b_i^0 in the initial condition is irregular. The path of influence generated by b_i^0 then determines the block b_j^T (either on the left or right) – where j may differ from i – that is irregular at time step T. (The blocks $_\ell b_i^T$ and $_r b_i^T$ that "reproduce" b_i^0 will be irregular only if they lie in the path of influence of some irregular block b_k^0.)

The paths of influence of the blocks (both irregular and regular) in S^0 are determined by the transformation ϕ of the lemma. That transformation will in the subsequent discussion be considered as acting on the block labels themselves, rather than (equivalently) on the site positions of their left-bordering 1's. It follows from the definition of ϕ that there exist integers q_1, q_2 with $1 \leq q_1 \leq q_2 \leq Q$ such that blocks $b_1^0, \ldots, b_{q_1}^0$ are mapped by ϕ^T to a subset of the blocks $_\ell b_1^T, \ldots, _\ell b_Q^T$ on the left in S_ℓ; blocks $b_{q_1+1}^0, \ldots, b_{q_2}^0$ are mapped by ϕ^T to the center 0-block b_0^T; and the remaining blocks $b_{q_2+1}^0, \ldots, b_n^0$ are mapped to a subset of the blocks $_r b_1^T, \ldots, _r b_Q^T$ on the right in S_r. Fig. 5 depicts an example of multiple paths of influence decomposed into these three sets.

Define now the mapping Φ^T to be the combination of ϕ^T and the mapping of b_0^T into itself. The remainder of the proof rests on showing that Φ^T in fact governs the evolution of the paths of influence for all time; in other words, at each time step

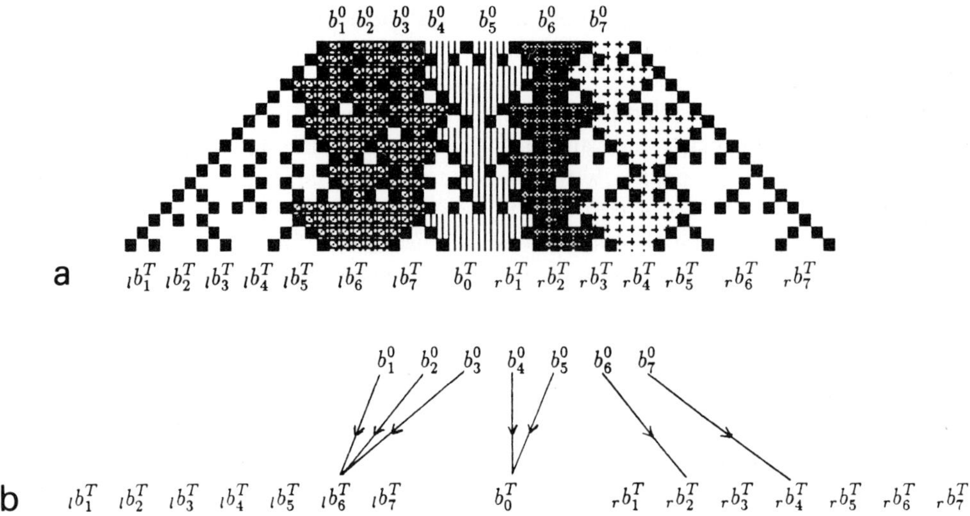

Fig. 5. Evolution of rule 90 depicting the paths of influence for $T = 16$ iterations from an initial condition with 7 blocks of 0's. (a) The paths of influence that coalesce in the time interval of the figure are colored with the same textures. Note that the spatial sequence at time T consists of two copies of the initial condition separated by a center block of 0's. (b) The mapping ϕ^T maps a subset of the blocks in the initial condition into blocks on the left, a subset into the center 0-block, and the remaining into blocks on the right. The proof of proposition 5 establishes that no "fixed point" (that is, a block b_i mapped by ϕ into the block on the left or right with the same subscript) of the mapping exists. The implication is that the path of influence from any irregular block evolving under rule 18 must annihilate itself through collision with another path, or be attracted into the center 0-block.

t that is a multiple of T, the spatial sequence generated by the automaton will consist of "copies" of the original initial condition separated by blocks of 0's, and the paths of influence for the blocks of 0's in these copies of the initial condition are determined by Φ^T. In particular, it will be shown that the assumption that $c_t = c_0$ for all t implies that there must be exactly c_0 fixed points of the mapping Φ^T, where by a "fixed point" is meant either the center 0-block b^T_0, or a block b_i such that either

$$\Phi^T(b^0_i) = {}_lb^T_i \quad \text{or} \quad \Phi^T(b^0_i) = {}_rb^T_i .$$

(The above two relations cannot both be true for any i.)

To show that Φ^T must have c_0 fixed points, consider an arbitrary irregular block b^0_i mapped by Φ^T into, say, $_rb^T_j$ on the right-hand side in the spatial sequence S^T. At each future time step t a multiple of T, the automaton again resolves itself into right- and left-hand copies of S^0. If at any such time step the path of influence from $_rb^T_j$ veers to the left rather than to the right, then the irregular block will either coalesce with an-

other irregular block to form a regular block, or be "attracted" into the center block of 0's. (Note that the corollary to the lemma prevents the block from propagating at a speed sufficient to escape this attractor.) The same argument holds if b^0_i is mapped by Φ^T into $_lb^T_j$ on the left-hand side of S^T, and hence the conclusion is that for c_t to be conserved, the path of influence for each irregular block in S^0 must maintain a consistent directional orientation (whether left or right) for all time. It therefore follows that for any time step t a multiple of T, the c_0 irregular blocks necessarily occur in the subsequences S^t_ℓ and S^t_r that are the copies of S^0 appearing on the extreme left or right of the automaton.

For any such subsequence S^t_ℓ and S^t_r on the extreme edges of the automaton, the backward light-cone of the subsequence coincides exactly for T time steps with the backward light-cone of S^T_ℓ and S^T_r. Thus the paths of influence linking S^t_ℓ and S^t_r back to S^{t-T}_ℓ and S^{t-T}_r are exactly described by the mapping Φ^T. Moreover for c_t to be conserved, these paths must be attracted to fixed points of Φ^T since the center 0-block b^T_0 is an attractor from

which there is no escape, and the monotonically non-decreasing nature of the mapping precludes the possibility of limit cycles.

It is impossible, however, for a fixed point other than the center block b_0^T to exist since any such block would be required to propagate with constant speed equal to 1; in other words, the path of influence generated by the block must follow exactly the path of the appropriate right or left diagonal. The corollary to the lemma restricts the speed of propagation to be less than or equal to 1, and the definition of the mapping Φ implies that the speed to the right, say, is less than 1 whenever a 0 appears in the right diagonal emanating from the left-bordering 1 of the block. Since a 0 is generated whenever a sequence of diagonal 1's encounters a sequence of diagonal 1's travelling in the opposite direction – this occurs for example for any block b_i^t after $\frac{1}{2}\text{len}(b_i^t)$ iterations – it follows that a irregular block's speed of propagation to the right must be less than 1. The same result holds for propagation to the left, and therefore the conclusion is that fixed points (other than the center 0-block) cannot occur.

The contradiction implies that c_t cannot be conserved if $c_0 > 1$. Moreover convergence time to either $c^* = 0$ or $c^* = 1$ is bounded above by the quantity $c_0 T$ representing the product of the reproduction time step T and the number of 0-blocks in the initial condition. Parity arguments then establish that $c^* = 0$ for initial conditions of odd length, and $c^* = 1$ for initial conditions of even length. $\qquad\square$

The consequences of the above proposition for the periodicity properties of temporal sequences generated by rule 18 are stated in the corollary below.

Corollary 2: For arbitrary finite initial conditions of even length on an infinite lattice, every temporal sequence generated by rule 18 is aperiodic. For arbitrary finite initial conditions of odd length, every temporal sequence – with the exception of the trivial case – is aperiodic. The trivial case is the center temporal sequence of all 0's generated by rule 18 from a finite spatial sequence that is spatially symmetric, with all 0-blocks of odd length.

Proof: For initial conditions of odd length, the result follows immediately from propositions 2, 4, and 5. For initial conditions of even length, further arguments are needed to ensure that the existence of a irregular block does not transform an aperiodic temporal sequence for rule 90 into a periodic temporal sequence for rule 18. Proposition 5 establishes that for initial conditions of even length, c_t converges in finite time to $c^* = 1$; in other words, that the spatial sequences generated will have exactly one irregular block. The proof of proposition 5 further establishes that in finite time, say, by time step T^*, this irregular block will necessarily gravitate to the center 0-block of the automaton. The "self-reproducing" nature of the rule 90 template automata then ensures that at time steps representing powers of 2, the spatial sequence generated will exhibit an increasingly long center sequence of 0's; moreover, at each such time step, this center 0-block is irregular. After a number of iterations equal to half the length of this center 0-block, the irregular block will "drift" slightly away from the center (unless the spatial sequence is itself symmetric, in which case the irregular block will never shift position); nevertheless it must return to the center block by the next time step representing a power of 2. (See fig. 6 for an example of this behavior.) Thus in a precise sense the irregular block remains anchored in the center of the automaton.

Assume now that some temporal sequence in rule 18 is periodic of period p. Choose a time step t^* sufficiently large that the temporal sequence is no longer undergoing transient behavior. Further require t^* to be sufficiently large that the "triangle" of 0's generated by the center 0-block of the automaton at time t^* completely encloses p terms of the temporal sequence. The preceding arguments establish that in this regime, the values of the temporal sequence in rule 18 coincide exactly with those in rule 90; it follows that the sequence must be identically equal to 0. The contradiction implies that no such periodic sequence can exist. $\qquad\square$

A similar result can be obtained for the other nonlinear rules (including rules 22, 126, 146, 182) that mimic rule 90.

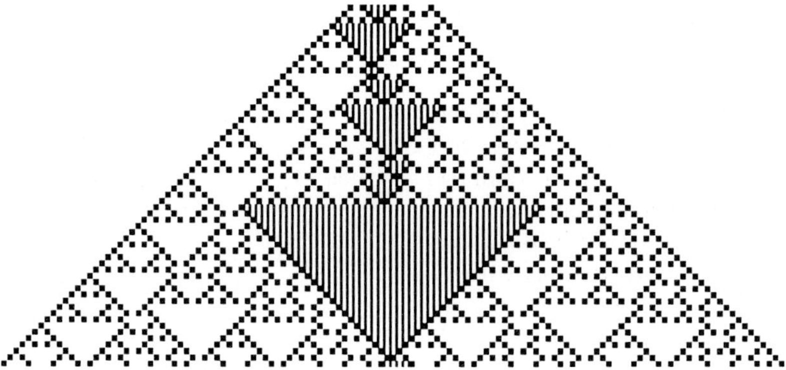

Fig. 6. Evolution of rule 18 from an initial condition with a single irregular block (textured). The corollary to proposition 5 establishes that the irregular block is anchored to the center of the automaton but drifts slightly for a fixed number of iterations at increasingly long time intervals. The resemblance between these systems and rule 90 automata is therefore strong enough to establish the aperiodicity of temporal sequences generated by rule 18 from arbitrary finite initial conditions.

5. Concluding remarks

The preceding sections have established the aperiodicity of the temporal sequences generated by a certain class of one-dimensional, binary site-valued, nearest-neighbor automata evolving from arbitrary finite initial conditions on an infinite lattice. Included in this class are those linear (here "linearity" implies that the superposition principle holds) automaton rules nontrivial in the sense of being more complicated than the identity, simple shift, or the zero rule ; certain nonlinear rules exhibiting an "injective" property together with a feature that ensures infinite-time propagation; and a subset of nonlinear automata whose evolution can be viewed as consisting of multiple domains within which the behavior is strictly injective. For this last subset of automata, a linearization procedure is used to show that collision and coalescence of domains produce in finite time a system with at most two domains, and thus to establish the aperiodicity properties of the temporal sequences generated by these automaton rules.

It should not be concluded from the above that aperiodicity is in any sense generic to cellular automaton rules. As mentioned earlier, all automaton rules operating on finite lattices (with either fixed or periodic boundary conditions) generate limit cycle (that is, periodic) behavior. Even on infinite lattices, the systematic simulation studies performed by Wolfram [1] and others indicate that many automata evolve either to a homoge-neous state or to limit cycle behavior (specifically, the generation of fixed domain walls within which the automaton undergoes periodic behavior) or to behavior that is extremely complicated but possibly not aperiodic. (Moreover, the periodicity properties of temporal sequences for these rules may well depend on the details of the initial conditions used. A simple example is that of a shift rule.) Loosely speaking, such automata may be viewed as possessing "dissipative" characteristics that force the contraction of arbitrary initial conditions onto limit cycle attractors. Again loosely speaking, the automaton rules that generate aperiodicity are distinctive in that the special "injective"-like features of their rule tables preclude such contractive behavior.

The connection between the aperiodicity of temporal sequences in cellular automata and chaotic behavior in dynamical systems is as yet unclear. (Aperiodicity of course encompasses a much broader set of behavior.) Given a continuous dynamical system that exhibits chaotic behavior, it is probably correct to assume that aperiodicity should appear as a feature of any cellular automaton representing an "appropriate discretization of that system. Since procedures do not yet exist (and may in general not be feasible) for construction of continuous systems corresponding to arbitrary cellular automata, the implications in reverse are not known.

Acknowledgements

This research was supported in part by the Applied Mathematics Program of the Department of Energy under Contract No. KC-07-01-01. Support was also provided by the Center for Nonlinear Studies at Los Alamos National Laboratory. The presentation and clarity of the proofs in this paper were much improved by discussions with G. Zweig. Figures were generated using the *Cellsim* software package written by D. Hiebeler and C. Langton.

References

[1] S. Wolfram, Theory and Applications of Cellular Automata (World Scientific Press, Singapore, 1986).

[2] D. Farmer, T. Toffoli and S. Wolfram, eds., Cellular automata: Proceedings of an interdisciplinary workshop, Physica D 10 (1, 2) (1984).

[3] E. Domany and W. Kinzel, Equivalence of cellular automata to Ising models and directed percolation, Phys. Rev. Lett. 53 (1984) 311.

[4] U. Frisch, B. Hasslacher and Y. Pomeau, Lattice-gas automata for the Navier–Stokes equation, Phys. Rev. Lett. 56 (1986) 1505.

[5] B. Boghosian and D. Levermore, A cellular automaton for Burgers' equation, Complex Systems 1 (1987) 17.

[6] Cellular Automata and the Modeling of Complex Physical Systems, Proceedings of the 1989 Les Houches Workshop (Springer, Berlin, 1989).

[7] G.A. Hedlund, Endomorphisms and automorphisms of the shift dynamical system, Math. Systems Theor. 3 (1969) 320.

[8] G.Y. Vichniac, Boolean derivatives on cellular automata, Physica D 45 (1990) 63–74, these Proceedings.

[9] H.A. Gutowitz, A hierarchical classification of cellular automata, Physica D 45 (1990) 136–156, these Proceedings.

[10] W. Li, N. Packard and C. Langton, Transition phenomena in cellular automata rule space, Physica D 45 (1990) 77–94, these Proceedings.

[11] O. Martin, A. Odlyzko and S. Wolfram, Algebraic properties of cellular automata, Commun. Math. Phys. 93 (1984) 219.

[12] E. Jen, Linear cellular automata and recurring sequences in finite fields, Commun. Math. Phys. 119 (1988) 13.

[13] E. Jen, Global properties of cellular automata, J. Stat. Phys. 43 (1986) 219–242.

[14] E. Jen, Preimages and forecasting for cellular automata, in: 1989 Lectures in Complex Systems (Addison–Wesley, New York, 1990).

[15] S. Wolfram, Random sequence generation by cellular automata, Adv. Appl. Math. 7 (1986) 123.

[16] P. Grassberger, Chaos and diffusion in deterministic cellular automata, Physica D 10 (1983) 52.

[17] S. Wolfram, Cellular automata, Los Alamos Science (1985).

[18] E. Jen, Exact linearization of nonlinear cellular automata, Los Alamos National Laboratory preprint.

Physica D 45 (1990) 19–25
North-Holland

CYCLIC CELLULAR AUTOMATA AND RELATED PROCESSES

Robert FISCH

Department of Mathematics, Colby College, Waterville, ME 04901, USA

Received 9 January 1990
Revised manuscript received 28 March 1990

A cyclic cellular automaton is a discrete-time process defined on the state space $\{0, 1, \ldots, N-1\}^{Z^d}$. Starting from a random initial condition, at each time step, each site looks at its nearest neighbors. If the state at site x is i, then site x will change state to $i+1 \pmod{N}$ if it locates state $i+1 \pmod{N}$ among its neighbors; otherwise, site x will remain in state i.

Cyclic cellular automata have properties akin to cyclic particle systems, introduced by Bramson and Griffeath, which are continuous-time Markov processes on $\{0, 1, \ldots, N-1\}^{Z^d}$ with a stochastic evolution rather than a deterministic one. In one dimension, some clustering results about cyclic cellular automata are presented; the analogous problems in the context of cyclic particle systems are still unresolved. Also, we discuss the consequences of these clustering results in the context of one-dimensional systems of particles with various interaction mechanisms. Finally, we touch upon the work currently being done on two-dimensional cyclic particle systems and cellular automata.

1. Definition of the cyclic cellular automaton

A cyclic cellular automaton is a discrete-time process defined on the d-dimensional integer lattice, Z^d. At any given time, each lattice point is assigned one of a finite number of values, so that the state space for such a process is $\{0, 1, \ldots, N-1\}^{Z^d}$. A key feature of cyclic cellular automata is that a structure is imposed on the N values which a lattice site may attain, and the rules of evolution depend on this structure. We shall refer to the N values as *colors*, because this is how these values are displayed on a computer monitor. The colors are arranged in a *cyclic hierarchy*, which may be thought of as a color wheel. Color 0 is followed by color 1, then color 2, and so on up to color $N-1$, which is followed by color 0. We say each color may *eat* the color below it in this heirarchy. In other words, color i may eat color j if $i - j = 1 \pmod{N}$.

The initial configuration of a cyclic cellular automaton is product measure which is symmetric over the colors. Thus, each lattice site is independently assigned one of the N colors, and the prob-

ability of a particular color being assigned to a site is $1/N$. To update, each site looks at its $2d$ nearest neighbors (site x looks at the set $\{y: |y-x| = 1\}$), and if it sees the color which eats it among those sites, it takes on that color at the next time step; otherwise, its color remains unchanged. For example, if the color of site x is 2 and if the color of a neighbor is 3, then the color of x will be 3 in the next configuration, but if no 3 is among the neighbors, then x will remain color 2 in the next configuration. All sites update in this manner synchronously to yield the next configuration.

2. Cyclic particle systems

The cyclic cellular automaton is similar to a process studied by Bramson and Griffeath [1] known as the *cyclic particle system*. The state space is $\{0, 1, \ldots, N-1\}^{Z^d}$, and the N colors have the same cyclic heirarchy previously described. But this process is a continuous-time Markov process which has a random evolution. Using a graphical representation to define this process, each site waits for its exponentially distributed, mean-1

alarm clock to go off, and then a random neighbor is chosen. If the neighboring color can eat the color of the site in question, then it does so; otherwise no change in the process occurs. In the case of $N = 2$, this process is the same as the voter model of Clifford and Sudbury [2] and Holley and Liggett [3].

The main theorem of ref. [1] concerns the one-dimensional N-color cyclic particle system. The process is said to *fluctuate* if each lattice site changes state infinitely often through the evolution of the process, and it is said to *fixate* if each site has a final state it attains. One may think of fixating as "getting stuck". The statement of the theorem is as follows. The abbreviation *a.s.* stands for "almost surely", meaning that the event occurs with probability 1.

Theorem 1 *(Bramson and Griffeath)*
One-dimensional N-color cyclic particle systems fluctuate a.s. if $N \leq 4$ and fixate a.s. if $N \geq 5$.

Among the questions unanswered in ref. [1] are the following:

Q1: Does clustering occur when $N \leq 4$?

Q2: If so, how does the average cluster size grow?

Q3: What can be said in higher dimensions, in particular, $d = 2$?

Due to the difficulty in answering such questions about cyclic particle systems, it is natural to look at corresponding questions about cyclic cellular automata in hopes that the lack of randomness in the rules of evolution may make the analysis of such questions tractable.

3. Theorems concerning one-dimensional cyclic cellular automata

This section contains results about one-dimensional cyclic cellular automata which are proved elsewhere. (See refs. [4] and [5].) In fig. 1, instances of this one-dimensional system are pictured for the cases of 3, 4, and 5 colors. The first theorem is the analogy to the cyclic particle system result of Bramson and Griffeath, and the reader should observe that this theorem describes the behavior exhibited in fig. 1. Let us note that it makes sense to make statements about cyclic cellular automata

in the *a.s.* sense because the initial configuration is random, even if the evolution thereafter is not.

Theorem 2 *One-dimensional N-color cyclic cellular automata fluctuate a.s. if $N \leq 4$ and fixate a.s. if $N \geq 5$.*

Now let us focus on the 3-color cyclic cellular automaton, which we shall denote as $\{\eta_t\}$. The state space is $\{0, 1, 2\}^{\mathbb{Z}}$, and the cyclic heirarchy specifies that color 1 eats color 0, color 2 eats color 1, and color 0 eats color 2. The initial configuration is product measure, and each site updates by looking at its two nearest neighbors to see if the color of either site can eat its color.

We say that $\{\eta_t\}$ *clusters* if at large times t, there is overwhelming probability that any interval we specify is all one color. Formally, $\{\eta_t\}$ clusters if for any $x < y$, $\lim_{t\to\infty} P(\eta_t(x) = \eta_t(x+1) = \ldots = \eta_t(y)) = 1$. The following theorem answers question **Q1** in the context of the 3-color cyclic cellular automaton.

Theorem 3 $\{\eta_t\}$ *clusters.*

Define the *mean cluster size* of a configuration $\eta \in \{0, 1, 2\}^{\mathbb{Z}}$ to be

$$C(\eta) = \lim_{n \to \infty} \frac{2n}{\# \text{ of clusters } \eta \text{ has in } [-n, n]}. \qquad (1)$$

Question **Q2** concerns the behavior of $C(\eta_t)$ for large t. The next theorem addresses this. The symbol $\sim$ indicates that the limit of the quotient of the left-hand side to the right-hand side goes to 1 as t approaches infinity.

Theorem 4 $C(\eta_t) \sim \sqrt{3\pi t/2}$, *a.s.*

No results along the lines of theorems 3 and 4 are known for the 4-color cyclic cellular automaton. However, computer simulation (or rather calculation, since the evolutions are computed exactly) indicates that the cluster rate of such a system is unlikely to be proportional to the square root of time.

4. Systems of moving particles with collision rules

The proofs of theorems 3 and 4 involve examining the evolution of the edges between blocks of

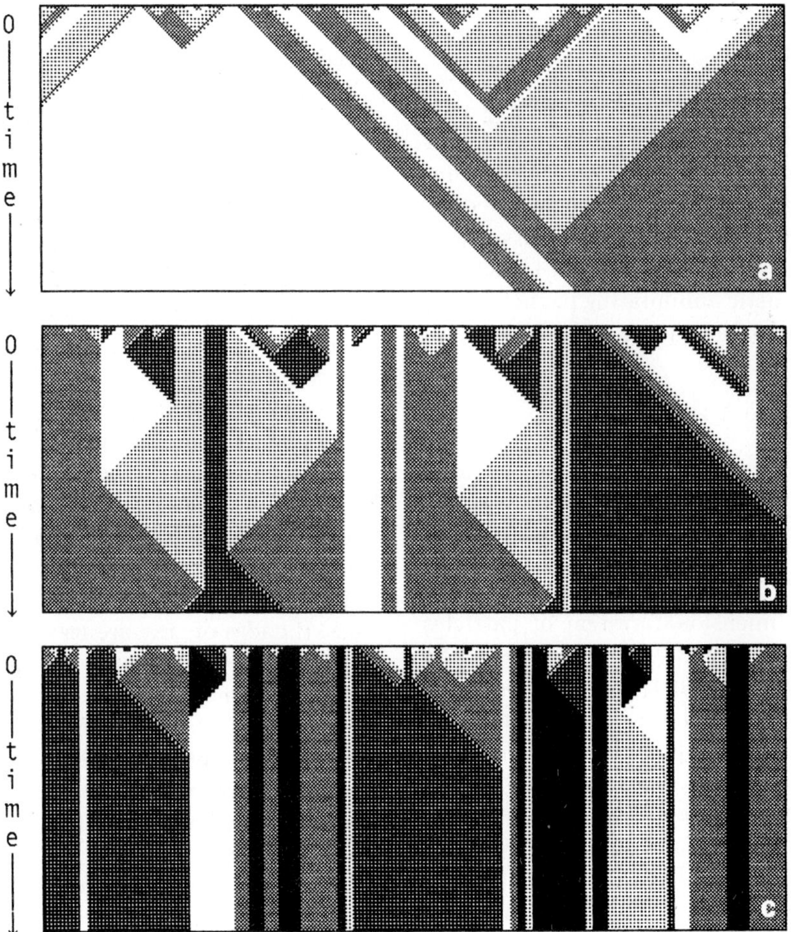

Fig. 1. The one-dimensional cyclic cellular automaton in the instances of (a) 3, (b) 4, and (c) 5 colors. The initial configuration is random, and time moves downward. Angled edges represent one color eating another, whereas vertical edges represent an interface between colors which do not interact. Eating persists in the cases of 3 and 4 colors, but fixation occurs in the process with 5 colors. (Figure reprinted with permission of the Journal of Theoretical Probability.)

colors. This process of edge motions is called the edge cellular automaton, and it can be defined in the following way. Define two types of particles: ℓ particles, which move to the left deterministically one site per time unit, and r particles, which do the same to the right. As for the initial condition, at each site of Z, independently flip a fair three-sided coin and either place an ℓ, an r, or no particle, which we shall denote by 0. At each time step, move the particles one site in their indicated direction, unless this causes two particles to occupy the same site or pass through each other. In either of these cases, we specify that the particles annihilate each other, so that neither particle appears in the update of the process. From fig. 1a, one can see that the edge cellular automaton keeps track of the edges between blocks of color in the 3-color cyclic cellular automaton.

The quantity of interest for the edge cellular automaton is the *mean interparticle distance* of a configuration $\zeta \in \{\ell, r, 0\}^{Z}$, defined as

$$D(\zeta) = \lim_{n \to \infty} \frac{2n}{\# \text{ of particles } \zeta \text{ has in } [-n, n]}. \quad (2)$$

If we couple $\{\eta_t\}$ with its edge cellular automaton, which we shall call $\{\zeta_t\}$, then we have that $D(\zeta_t) = C(\eta_t)$. This gives rise to a reformulation of theorem 4.

Theorem 5 $D(\zeta_t) \sim \sqrt{3\pi t/2}$, *a.s.*

This theorem may be generalized in the following way. The edge cellular automaton $\{\zeta_t\}$ may also be considered a system of deterministic annihilating particles. By changing the initial condition, we can define a family of such processes which no longer correspond to edges from a cyclic cellular automaton. So, for each $p \in (0, \frac{1}{2}]$, define a system of deterministic annihilating particles $\{\zeta_t^p\}$ with the same rules of evolution as $\{\zeta_t\}$, but with initial condition

$$P(\zeta_0^p = \ell) = P(\zeta_0^p = \text{r}) = p,$$

$$P(\zeta_0^p = 0) = 1 - 2p. \tag{3}$$

The following generalizes theorem 5.

Theorem 6 $D(\zeta_t^p) \sim \sqrt{\pi t/2p}$, *a.s.*

A related one-dimensional system of particles was suggested by Bennett [6]. In this system, the initial condition is specified by (3), and particles move deterministically as above. However, upon collision, two particles coalesce into one. The resulting particle randomly chooses to become an ℓ or an r with equal probability and thereafter moves deterministically in the appropriate direction one site per time unit. Let $\{\xi_t^p\}$ be this system of deterministic coalescing particles. The following theorem about the interparticle distance for this process holds. (See ref. [7].)

Theorem 7 $D(\xi_t^p) \sim \frac{1}{2}\sqrt{\pi t/p}$, *a.s.*

The proof of this theorem follows the same lines as that of theorem 6 but is somewhat more complicated. A key difference is that annihilation is symmetric with respect to the colliding particles, whereas coalescence is not.

5. Two-dimensional systems

Let us now turn attention to two-dimensional cyclic processes, the topic of question **Q3**. Only limited results have been obtained about such systems, but much empirical work has been done with the aid of a cellular automata machine (CAM). (For more information about CAMs and their applications, see ref. [8].) This section discusses work currently being performed with Janko Gravner and David Griffeath.

The two-dimensional N-color cyclic particle system starts with product measure on $\mathbb{Z}^2$. Each site waits for its exponentially distributed mean-1 alarm clock to go off, then randomly picks one of its four neighbors and allows the color of the neighbor to eat its color, if possible. Simulation of this process on the CAM gives rise to spectacular dynamics which are quite different from the dynamics of the one-dimensional system. None of what is mentioned here has been proved; all is conjecture. Extrapolating from what we have observed, the process goes through the following stages regardless of the number of colors N. The number of colors merely serves to change the appropriate scale from which to observe the process.

Fig. 2 portrays the evolution of the two-dimensional cyclic particle system in the case of 12 colors. (Shades of gray are used to represent the colors.) In the early stages of the process, eating goes on throughout the plane, but does not last long in most places, leaving *debris* scattered about. However, due to large deviation effects, certain regions remain active enough to create large formations called *critical droplets*. The boundary of a critical droplet is large enough that the likelihood of finding new food remains high. Critical droplets grow until they merge with others. When critical droplets collide, the interface is characterized by robust eating about regions of debris, which is an environment ripe for the formation of *spirals*. The conjecture is that these spirals expand into all available space, until the entire plane is engulfed. For each number of colors, spirals seem to have a preferred density range – too many spirals and some will die off, too few spirals and "unstable" regions of the plane beget new spirals.

There are two key features of this dynamic. First, unlike the one-dimensional processes, these systems never fixate regardless of the number of colors. Second, clusters do not form, so that given a large box, the likelihood of it being one solid color at large times is small. Although clustering results are not known about the one-dimensional random processes which fluctuate (other than the voter model), empirical evidence indicates that clustering does occur. For a survey article about the investigation into two-dimensional cyclic par-

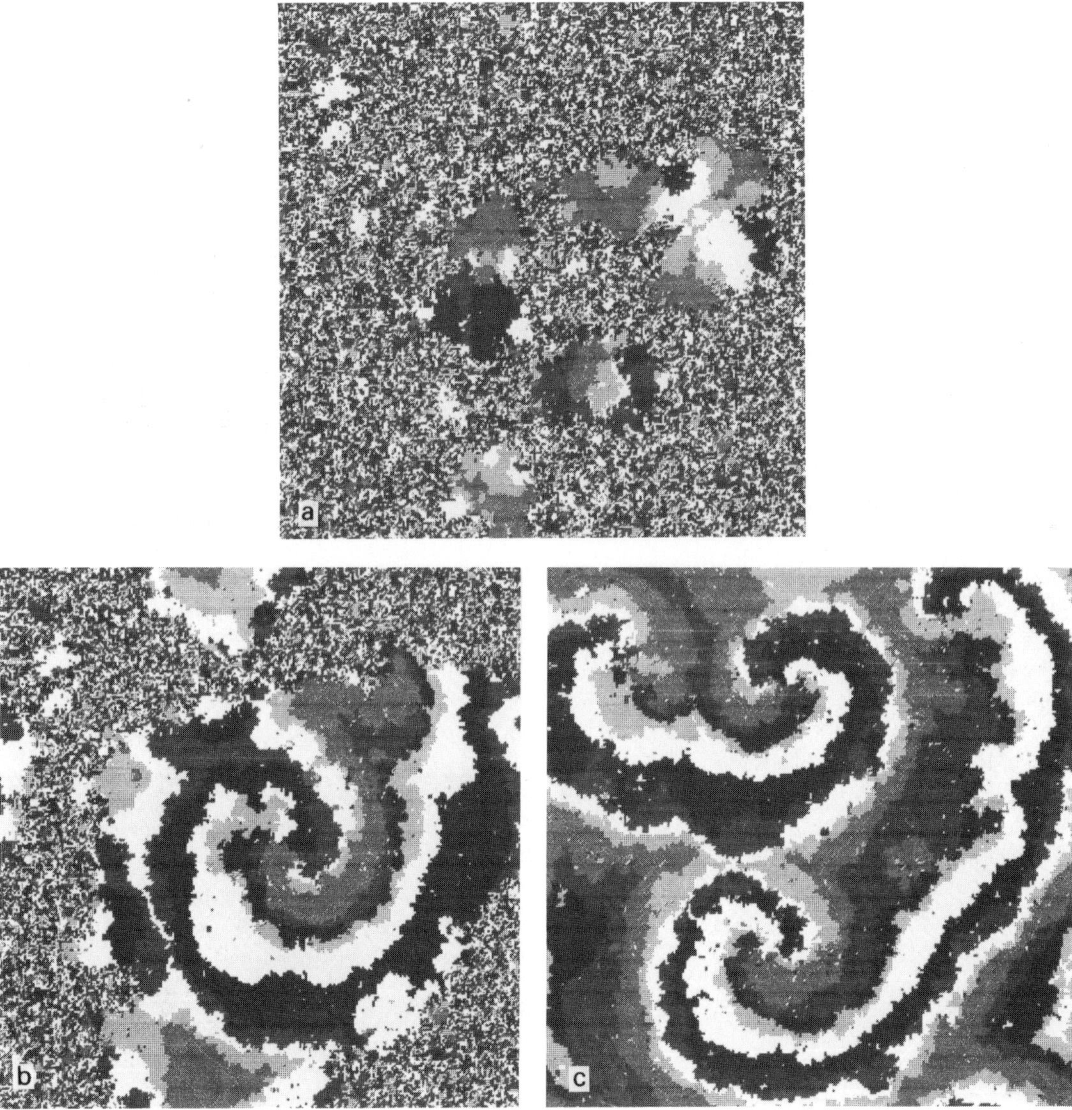

Fig. 2. The two-dimensional 12-color cyclic particle system. (a) ($t = 2000$) Most of the plane has run out of food to eat, leaving debris, but some areas have been fortunate to grow large enough that the likelihood of finding food on the boundary remains high. These critical droplets expand until they collide with each other. (b) ($t = 5000$) When these droplets collide, spiral formations often arise. (c) ($t = 10\,000$) These spiral formations eventually engulf the entire plane.

ticle systems, see ref. [9].

As for two-dimensional cyclic cellular automata, we have been able to prove theorems about how the evolution proceeds. The process starts from product measure on $\mathbf{Z}^2$, and at each time step, each site looks at its four neighbors for a color that can eat its color. Fig. 3 is an illustration of the two-dimensional 18-color cyclic cellular au-

tomaton. Just as for cyclic particle systems, debris covers most of the plane at the outset, but critical droplets form and grow. As droplets collide, the interface provides the opportunity for the formation of loops of lattice points about which the colors cycle. These stable formations are called *defects* and lead to the creation of spiral structures. Finally, defects of minimal size give rise to *demons*, which

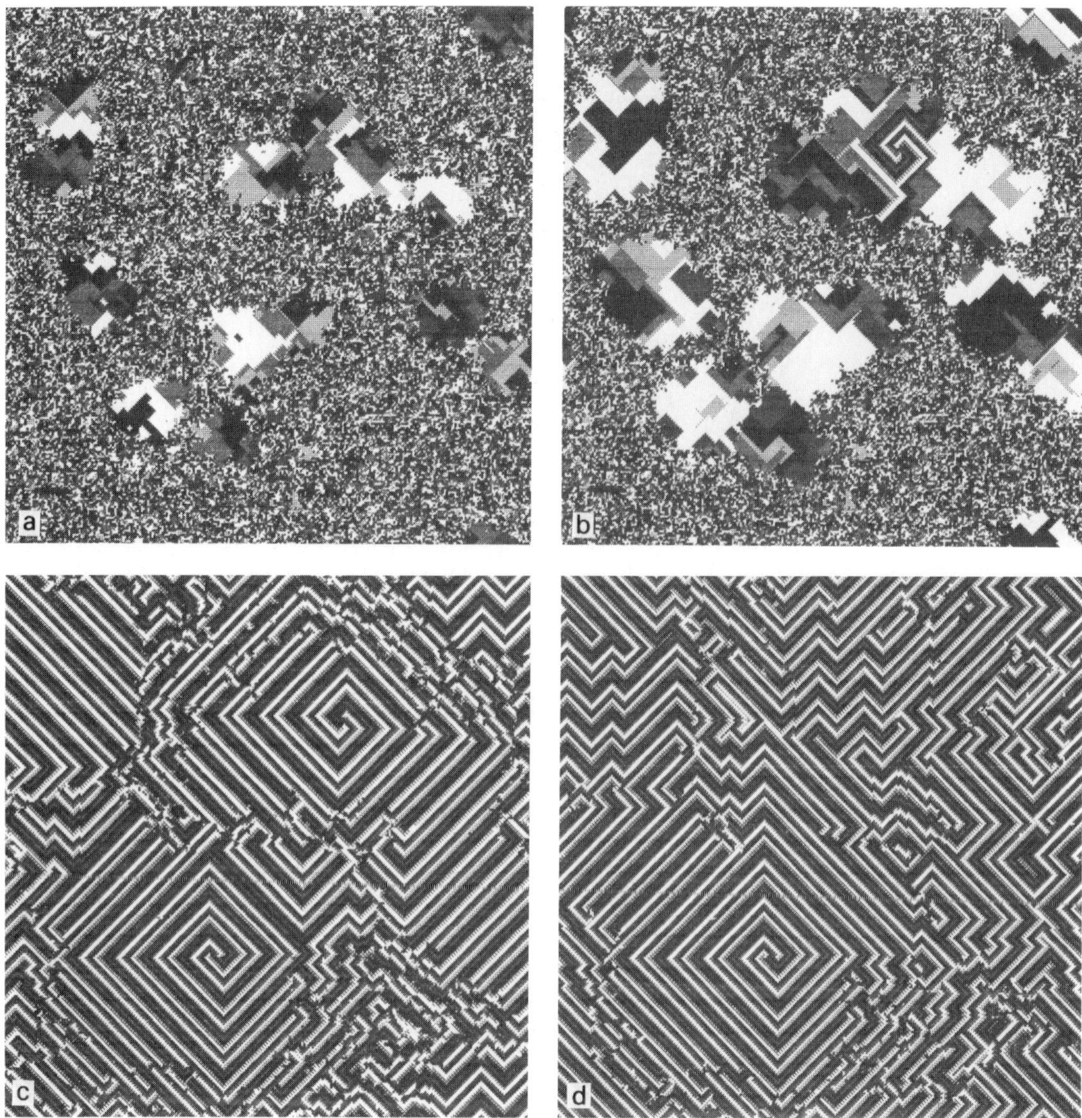

Fig. 3. The two-dimensional 18-color cyclic cellular automaton. (a) ($t = 300$) Debris dominates the plane, but several critical droplets have arisen. (b) ($t = 450$) A defect has formed (towards the upper right) which gives rise to a spiral formation. (c) ($t = 1000$) Spirals take over the plane; the defect from (b) is still apparent, but another defect, which is actually a demon, has also formed. A demon is a more "efficient" spiral, that is, each site is changing color every time step. (d) ($t = 3000$) Demons eventually dominate the plane. The less efficient defect in the upper right has been overrun by the demon.

are regions of the plane where each site changes color every time step. Eventually, each point of the plane becomes part of some demon. In particular, this implies that the two-dimensional cyclic cellular automaton fluctuates *a.s.* In refs. [10] and [11], one can find more discussion about this process.

For both the two-dimensional N-color cyclic

particle system and cellular automaton, we have investigated the structure of the initial configuration in the locations where droplets arise in order to understand how fluctuation occurs. (The proof of fluctuation for the two-dimensional cyclic cellular automaton is nonconstructive.) This has led us to look at processes defined on finite boxes whose

sides are of length L. One result shows the existence of a constant c such that if $L < e^{cN}$, then the probability of any site ever becoming part of a critical droplet goes to zero as N goes to infinity. The reader is referred to ref. [12] for more information about this work.

Acknowledgements

This work was supported in part by NSF Grant DMS-89-05418 and by funds from the Foundation of The University of North Carolina at Charlotte and from the State of North Carolina.

References

[1] M. Bramson and D. Griffeath, Flux and fixation in cyclic particle system, Ann. Probability 17 (1989) 26–45.

[2] P. Clifford and A. Sudbury, A model for spatial conflict, Biometrica 60 (1973) 581–588.

[3] R. Holley and T. Liggett, Ergodic theorems for weakly interacting systems and the voter model, Ann. Probability 3 (1975) 643–663.

[4] R. Fisch, The one-dimensional cyclic cellular automaton: a system with deterministic dynamics which emulates an interacting particle system with stochastic dynamics, J. Theor. Probability 3 (1990) 311–338.

[5] R. Fisch, Clustering in the one-dimensional 3-color cyclic cellular automata, to appear.

[6] C.H. Bennett, On the nature and origin of complexity in discrete, homogeneous, locally-interacting systems, Found. Phys. 16 (1986) 585–592.

[7] R. Fisch, in preparation.

[8] T. Toffoli and N. Margolis, Cellular Automata Machines: A New Environment for Modeling (MIT Press, Cambridge, MA, 1987).

[9] D. Griffeath, Cyclic random competition: a case history in experimental mathematics, in: Computers and Mathematics, ed. J. Barwise, AMS Notices 35 (1988) 1472–1480.

[10] A.K. Dewdney, Computer recreations, Sci. Am. 261 (August 1989) 102–105.

[11] R. Fisch, J. Gravner and D. Griffeath, Cyclic cellular automata in two dimensions, in: T.E. Harris Festschrift (Birkhauser, Basel, 1990).

[12] R. Fisch, J. Gravner and D. Griffeath, in preparation.

Physica D 45 (1990) 26–35
North-Holland

NEAREST NEIGHBOR CELLULAR AUTOMATA OVER Z_2 WITH PERIODIC BOUNDARY CONDITIONS

Burton VOORHEES

Faculty of Arts and Sciences, Athabasca University, Box 10000, Athabasca, Alberta, Canada T0G 2R0

Received 23 February 1990
Revised manuscript received 29 March 1990

One-dimensional nearest-neighbor cellular automata defined over Z_2 are characterized in terms of a set of eight non-additive basis operators which act on the automaton state space. Every evolution rule for such automata can be expressed as an operator which is a direct sum of the basis operators. This approach allows decomposition of automata rules into additive and non-additive parts. As a result it is simple to determine fixed points (those states for which the rule reduces to the identity), and shift cycles (sets of states on which the rule reduces to a shift). Sets of states on which any given nearest-neighbor automaton reduces to an identity or a shift are characterized, and maximum tree heights are computed for certain cases. The formalism is used to explore and generalize a relationship recently studied by Ito. We conclude by indicating a connection to the invariance matrix approach to cellular automata which has been developed by Jen.

1. Introduction

Much current interest in cellular automata (CA) is evidenced by the large number of recent publications. Review of this literature shows a wide variety of approaches: empirical, heuristic, and statistical studies [1 5]; analytic treatment of additive CA [6–10]; specific studies of certain tractable non-additive CA [11,12]; and abstract analytical results [13–15]. Much remains to be accomplished, however, both in general, and in the analysis of specific cases.

This paper deals with nearest-neighbor cellular automata defined over Z_2. In addition, although the formalism to be presented is not so restricted, considerations will be limited to finite cellular automata with periodic boundary conditions. Taxonomies of these CA have been proposed, but they are either overly general (e.g. possessing only two catagories [16]), or merely phenomenological. The most generally known classification is that of Wolfram [1], which defines four phenomenological types on the basis of time evolution behavior. A major failing of this classification, however, is that it is undecidable to which class a given cel-

lular automata belongs [17]. Recently, however, a generalized mean-field approach has been introduced by Gutowitz [18], which shows a number of advantages.

This paper presents an analysis of nearest-neighbor cellular automata based on the separation, for each automaton rule, of additive and non-additive parts. Expressions for each rule are formalized in terms of a set of seven additive, and eight non-additive operators defined on the space of possible automata states. On the basis of this formalization a number of results concerning properties of these CA can be determined. This includes numbers of fixed points, conditions for states to be on a shift cycle (i.e. a cycle on which the rule is equivalent to a shift) and certain relationships between different CA rules.

2. Finite nearest-neighbor cellular automata with periodic boundary conditions

A finite cellular automata with periodic boundary conditions (cylindrical automata in the termi-

nology of Jen [13]), defined over Z_2, is pictorially represented as a set of n cells located on a circle. Each cell contains a 0 or a 1. The set of all possible states for such an automaton is denoted E_n.

If the automaton rule is such that the value in cell i at time $t + 1$ is determined only on the basis of the values in cells $i - 1$, i, and $i + 1$ at time t then the automaton follows a nearest-neighbor rule. Denote this automaton (Q, E_n). If, for all $\mu, \mu' \in E_n$, $Q(\mu + \mu') = Q(\mu) + Q(\mu')$, the automaton rule is said to be additive.

If μ_i denotes the ith component (i.e. the value in cell i) of a state μ then the general component form for the action of the operator representing an additive nearest-neighbor rule is

$$\mu_i(t + 1) = [Q(\mu(t))]_i$$
$$= x\mu_{i-1}(t) + y\mu_i(t) + z\mu_{i+1}(t), \qquad (1)$$

where $x, y, z \in Z_2$ and all sums and products are reduced mod(2), while all component indices are reduced mod(n).

Notationally, the state consisting of all 0's will be indicated by $\mathbf{0}$ and the state consisting of all 1's by $\mathbf{1}$.

Eq. (1) defines seven distinct additive operators, excluding the trivial zero operator which maps all states to $\mathbf{0}$. If I, σ, and σ^{-1} denote respectively the identity, left shift, and right shift on E_n then (1) is equivalent to the operator equation

$$Q = x\sigma^{-1} + yI + z\sigma. \qquad (2)$$

The explicit forms for the non-trivial additive nearest-neighbor operators are given in table 1, using the labeling scheme initiated by Wolfram [1].

In order to include non-additive operators we define a set of eight non-additive operators on E_n as indicated in table 2.

Since whenever a site maps to 1 under one of these operators it maps to 0 under the remaining seven there is no interference, and every operator representing a nearest-neighbor rule over Z_2 can be uniquely expressed as a direct sum of these eight operators. Thus the operators of table 2 provide a basis for the non-linear algebra of operators defined by the set of nearest-neighbor rules over Z_2. Expression of an operator Q in terms of these

basis operators will be called the *canonical representation of Q*.

To determine the canonical representation of an operator its numeric label is written in powers of 2 and the appropriate substitutions are made from table 2. For example, the canonical representation of rule 176 is obtained via $176 = 128 + 32 + 16 = \eta^- + \chi + \theta$.

Lemma 1 *For all $\mu \in E_n$*

(i) $\qquad \beta^\pm(\mu) = \eta^\pm(\mathbf{1} + \mu)$.
(ii) $\qquad \theta(\mu) = \iota(\mathbf{1} + \mu)$.
(iii) $\qquad \chi(\mu) = \kappa(\mathbf{1} + \mu)$.
(iv) $\qquad (\beta^\pm)^2(\mu) = 0$.
(v) $\qquad \iota^2(\mu) = \iota(\mu)$.

The results of this paper relating to shift cycles and fixed points are based on the canonical representations of the seven non-trivial additive operators. These are given in table 3.

Canonical representations can be added with coefficients reduced mod(2). For example, $D = I + \sigma = (\beta^+ + \beta^- + \chi + \iota) + (\beta^- + \eta^+ + \theta + \chi) = \beta^+ + \eta^+ + \theta + \iota$. This means that every nearest-neighbor rule can be decomposed into an additive and a non-additive part, although it turns out that this decomposition is not generally unique. Rule 172, for example, has canonical representation $\beta^- + \theta + \chi + \iota$, which can also be written as $\sigma + \eta^+ + \iota = I + \beta^+ + \theta = D + \beta^+ + \beta^- + \eta^+ + \chi = D^- + \eta^- + \chi$. We will be particularly interested in those cases in which the additive part of a rule is the identity, or a shift.

If Q^* is a non-additive operator with decomposition $Q^* = Q + F$, where Q is an additive operator, and if Q^* reduces to Q on some non-empty subset of E_n-$\{\mathbf{0},\mathbf{1}\}$, the decomposition will be called *legal*. If the canonical representation of Q^* contains κ then Q^* will be said to be *generative*. For simplicity most of what follows will be restricted to non-generative rules.

Two CA with periodic boundary conditions are called *conjugate* if each is transformed into the other by a reversal of orientation. Under such reversal the basis operators transform according to $\eta^+ \leftrightarrow \eta^-, \beta^+ \leftrightarrow \beta^-$ while $\chi, \theta, \iota, \kappa$ remain invariant. This shows up as a part of the "folding" of CA space, as described by Li and Packard [19].

Table 1
Nearest-neighbor additive cellular automata over Z_2.

(x, y, z)	Operator	Component form	Rule
$(0, 0, 0)$	$0*$	$[0 * (\mu)]_i = 0$	0
$(1, 0, 0)$	σ^{-1}	$[\sigma^{-1}(\mu)]_i = \mu_{i-1}$	240
$(0, 1, 0)$	I	$[I(\mu)]_i = \mu_i$	204
$(0, 0, 1)$	σ	$[\sigma(\mu)]_i = \mu_{i+1}$	170
$(1, 1, 0)$	D^-	$[D^-(\mu)]_i = \mu_{i-1} + \mu_i$	60
$(0, 1, 1)$	D	$[D(\mu)]_i = \mu_i + \mu_{i+1}$	102
$(1, 0, 1)$	δ	$[\delta(\mu)]_i = \mu_{i-1} + \mu_{i+1}$	90
$(1, 1, 1)$	Δ	$[\Delta(\mu)]_i = \mu_{i-1} + \mu_i + \mu_{i+1}$	150

Table 2
Basis operators for canonical representation.

Automaton	Operator	Rule
1	κ	$\{000\} \to 1$, all others $\to 0$
2	η^+	$\{001\} \to 1$, all others $\to 0$
16	η^-	$\{100\} \to 1$, all others $\to 0$
32	θ	$\{101\} \to 1$, all others $\to 0$
4	ι	$\{010\} \to 1$, all others $\to 0$
8	β^-	$\{011\} \to 1$, all others $\to 0$
64	β^+	$\{110\} \to 1$, all others $\to 0$
128	χ	$\{111\} \to 1$, all others $\to 0$

Table 3
Canonical representation of additive operators.

Automaton	Operator	Rule
0	$0*$	all map to 0
170	$\sigma = \beta^- + \eta^+ + \theta + \chi$	$\{011, 001, 101, 111\} \to 1$ $\{110, 100, 010, 000\} \to 0$
240	$\sigma^{-1} = \beta^+ + \eta^- + \theta + \chi$	$\{110, 100, 101, 111\} \to 1$ $\{011, 001, 010, 000\} \to 0$
204	$I = \beta^+ + \beta^- + \chi + \iota$	$\{011, 110, 111, 010\} \to 1$ $\{100, 001, 101, 000\} \to 0$
102	$D = \beta^+ + \eta^+ + \theta + \iota$	$\{110, 001, 101, 010\} \to 1$ $\{011, 100, 111, 000\} \to 0$
60	$D^- = \beta^- + \eta^- + \theta + \iota$	$\{011, 100, 101, 010\} \to 1$ $\{110, 001, 111, 000\} \to 0$
90	$\delta = \beta^+ + \beta^- + \eta^+ + \eta^-$	$\{110, 011, 001, 100\} \to 1$ $\{101, 111, 010, 000\} \to 0$
150	$\Delta = \eta^+ + \eta^- + \chi + \iota$	$\{001, 100, 111, 010\} \to 1$ $\{110, 011, 101, 000\} \to 0$

Canonical representations, together with their legal decompositions for all non-generative nearest-neighbor rules over Z_2 can be obtained from the author. These decompositions contain a great deal of information; in particular, concerning states which map to **0** or **1**, and the existence of fixed points, and shift cycles. The information represented in the appendix can be represented taxonomically as indicated in fig. 1.

In this figure Q stands for any one of the seven non-trivial additive operators. Thus seven hierarchical orderings can be constructed which, taken together, contain all of the non-generative operators.

3. Fixed points and shift cycles

All non-generative CA have **0** as a trivial fixed point, and all generative CA either have **1** as a fixed point or have $0 \leftrightarrow 1$ as a binary cycle. In this section we consider fixed points, and cycles on which automata rules act as right or left shifts. Analysis is based on representation of automata rules in terms of the additive operators I (for fixed points), σ (for left shifts), and σ^{-1} (for right shifts) as indicated in fig. 1. Letting Q stand for any one of these three operators, the state **1** maps to **0** if the Q-representation of a rule contains χ, while **1** is a fixed point if this representation does not contain χ. If the rule is generative (i.e. the Q-representation contains κ) the state **0** maps to **1**. Hereafter consideration will be restricted to states in E_n-$\{\mathbf{0},\mathbf{1}\}$. A detailed analysis is given only for non-generative automata. Extension to generative cases is relatively simple.

Suppose that Q^* is a non-generative non-additive automaton with legal decomposition $Q^* = Q + F(\beta^\pm, \eta^\pm, \theta, \chi, \iota)$ with Q being I, σ, or σ^{-1}. We ask what constraints must be satisfied by a state μ if the equation $Q^*(\mu) = Q(\mu)$ is to be true. That is, we look for the subset of E_n which maps to **0** under F.

For given Q^* the set of all legal decompositions partitions into four subsets. The first consists of those automata for which Q^* reduces to Q only on a highly restricted subset of E_n. These decompositions will be called weakly legal. The non-additive part F in a weakly legal decomposition will con-

tain combinations of basis operators which include $\beta^\pm + \eta^\pm$, $\eta^\pm + \chi$, or $\eta^\pm + \chi + \iota$, and if the automaton is generative, combinations involving $\beta^\pm + \kappa$ and $\beta^\pm + \theta + \kappa$.

The weakly legal non-generative forms reduce to Q either on states $\underline{01}$ and $\underline{10}$ (underlining indicates repetition) (subset a, requiring n even); on states $\underline{110}$, $\underline{101}$, and $\underline{011}$ (subset b, requiring $n = 3m$); or on states consisting of single or double 1's separated by isolated 0's (subset c).

The remaining three sub-classes of Q-decompositions, labeled A, B, and C, are obtained similarly.

Operators in sub-class A reduce to Q on the set of states having only isolated ones (those involving $\beta^\pm$), or only isolated zeroes (those involving $\eta^\pm$). Operator forms in sub-class B reduce to Q on states having only isolated ones separated by two or more zeroes (those involving $\beta^\pm$), or only isolated zeroes separated by two or more ones (those involving $\eta^\pm$). The constraints are more complicated for sub-class C: no isolated zeroes ($Q + \theta$); no isolated ones ($Q + \iota$); no isolated zeroes or ones ($Q + \theta + \iota$); only single or double ones ($Q + \chi$); only single or double ones, and no isolated zeroes ($Q + \theta + \chi$); only double ones ($Q + \chi + \iota$); only double ones, and no isolated zeroes ($Q + \theta + \chi + \iota$). A full listing of these automata can be obtained from the author.

For an operator $Q^* = Q + F(\beta^\pm, \eta^\pm, \theta, \chi, \iota)$ we denote the set of states on which Q^* reduces to Q by $E_n(Q, F)$. Elementary inclusion arguments yield:

Theorem 1

(i) $E_n(Q, a\eta^+ + \beta^-) = \mathbf{1} + E_n(Q, c\beta^+ + d\beta^- + e\chi)$ (a, b and c, d not both 0).

(ii) $E_n(Q, \theta) = \mathbf{1} + E_n(Q, \iota)$.

(iii) $E_n(Q, \chi + \iota) \cap E_n(Q, \theta + \chi) = E_n(Q, \chi + \iota) \cap E_n(Q, \theta + \iota) = E_n(Q, \theta + \chi) \cap E_n(Q, \theta + \iota) = E_n(Q, \theta + \chi + \iota)$.

(iv) $E_n(Q, \theta) \cap E_n(Q, \chi) = E_n(Q, \theta + \chi)$; $E_n(Q, \theta) \cap E_n(Q, \iota) = E_n(Q, \theta + \iota)$.

(v) $E_n(Q, \chi) \subset E_n(Q, a\beta^+ + b\beta^- + c\theta + d\chi)$ (a, b, c, d not all 0).

It is now possible to determine the fixed point, and single shift behavior of all nearest-neighbor cellular automata defined over Z_2, although we confine ourselves here to only the non-generative automata.

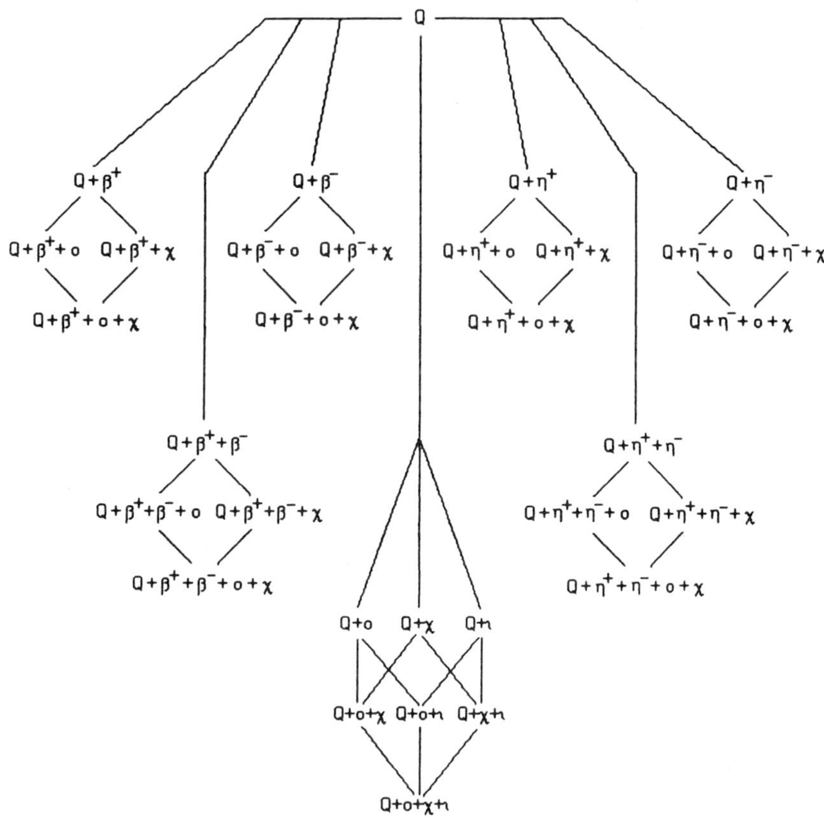

Fig. 1. Hierarchy of representations in terms of a linear operator Q.

A complete list of all non-generative automata having fixed points contained in E_n-$\{0,1\}$ has been computed, as well as a similar breakdown for those automata which reduce to right or left shifts.

There are several special cases:

(i) Automata (184, 226) have decompositions $\sigma^{-1} + \beta^+ + \beta^- = \sigma + \eta^+ + \eta^-$, and $\sigma + \beta^+ + \beta^- = \sigma^{-1} + \eta^+ + \eta^-$ indicating that for automaton 184 all states having only isolated 1's lie on σ^{-1}-cycles, while all states having only isolated zeroes lie on σ-cycles; with conjugate results for automaton 226.

(ii) Automata (172, 228) have as fixed points all states having only isolated ones separated by two or more zeroes, while all states having only isolated zeroes separated by two or more ones lie on σ^{-1}-cycles (228), or σ-cycles (172).

(iii) Automata (202, 216) have as fixed points all states consisting only of isolated zeroes separated by two or more ones, while all states con-sisting of isolated ones separated by two or more zeroes lie on either σ^{-1}-cycles (216), or σ-cycles (202). Thus automata (202, 216) may be viewed as complimentary to automata (172, 228) in the sense that fixed points of (202, 216) lie on cycles of (172, 228), and vice versa.

The following theorem indicates that most of the automata which have strongly legal shift representations have only single shift cycles:

Theorem 2 *All automata with strongly legal shift representations except (100, 44), (102, 60), (120, 106), (88, 74), (212, 142), and (84, 14) have only single shift cycles.*

Proof: We present a proof for the automaton 248. Proof for all other cases follows a similar pattern.

The canonical representation for 248 is $\beta^+ + \beta^- + \eta^- + o + \chi$. If a state μ has only isolated 1's then the representation $\sigma^{-1} + \beta^-$ indicates that rule 248 acts as a right shift on μ, hence μ must

lie on a shift cycle with period $d|n$ where d is the spatial period of μ.

On the other hand, suppose that μ contains a string of ones of the form ...011...110... After a single iteration of the automaton rule this becomes either ...111...111... or ...011...111.... That is, the length of this string has irreversibly increased by at least one. Thus any μ containing two or more adjacent ones will eventually iterate to the fixed point 1. $\qquad\square$

For automata in classes A and B there is a simple recursion relation which given the number of states on which the rule reduces to the additive operator Q. Let $S_n(r;s)$ stand for the subset of E_n consisting of states composed of blocks of exactly r ones separated by blocks of s or more zeroes.

Lemma 2 *Let $N_n(r;s)$ be the number of elements contained in $S_n(r;s)$. Then, for $1 \leq m < r$, $N_n(r;s) = N_n(r-m;s+m)$.*

Lemma 3 $N_{n+1}(1;s) = N_n(1;s)+N_{n-s}(1;s)+1$.

We note that if the role of 1's and 0's are interchanged in the above lemmas the results still obtain.

Corollary 1 *For $n > 1$ the number of states in E_n having only isolated ones, or only isolated zeroes, is given by $L_n - 1$ where L_n is the nth Lucas number.*

These results allow enumeration of fixed points or states on cycles for all those automata in classes A and B and those in class C having the form $Q+\chi+\iota$ or $Q+\theta+\chi+\iota$; excepting those automata listed in theorem 2, which have cycles other than single shift cycles.

4. Relations between certain automata

Ito [11] proves that the CA labeled 18 and 126 have many similarities in their state transition diagrams. In particular, he demonstrates that there is a transformation $T: E_n \to E_n$ such that if $\mu \to Q^{(18)}(\mu)$ then $T(\mu) \to Q^{(126)}(T(\mu)) = T(Q^{(18)}(\mu))$. In our language these automata are:

18: $\qquad \eta^+ + \eta^-$

126: $\qquad \beta^+ + \beta^- + \eta^+ + \eta^- + \theta + \iota$

Ito defines the transformation T by the rule that "...Every local sequence of successive 1 sites sandwiched in between two 0 sites (such as 010 or 011...110) in the configuration... should be transformed into a new sequence (such as 011 or 011...111, respectively) by changing the site value 0 on the right side into 1."

This rule corresponds to the operator $I + \eta^- + \theta$ which, in our notation, is rule 252: $\beta^+ + \beta^- + \eta^- + \chi + \theta + \iota$. Thus Ito's results are equivalent to:

Theorem 3 *(Ito). For all μ in E_n,*

$$(I + \eta^- + \theta)(\eta^+ + \eta^-)(\mu)$$

$$= (\beta^+ + \beta^- + \eta^+ + \eta^- + \theta + \iota)(I + \eta^- + \theta)(\mu).$$

Proof: First, note that $\iota(I + \eta^- + \theta)(\mu) = \mathbf{0}$ since $(I + \eta^- + \theta)(\mu)$ contains no isolated 1's. Also, it is clear that all sequences of a state μ which consist of blocks of any number of 1's separated by only isolated 0's (e.g. 101110110111101101) become sequences consisting of all 1's in $(I + \eta^- + \theta)(\mu)$, and sequences of all 0's in $(\eta^+ + \eta^-)(\mu)$. It is only necessary, then, to consider concatanations of such sequences, separated by two or more 0's. There are three cases: when two such sequences are separated by two 0's; by three 0's; and by more than three 0's. These cases, and their behavior, are shown in fig. 2. Proof of the theorem follows by inspection of this figure. $\qquad\square$

In general, a relation in which two automata representations $Q^{(r)}$ and $Q^{(s)}$ are related by a third representation $Q^{(k)}$ according to the formula $Q^{(k)}Q^{(r)} = Q^{(s)}Q^{(k)}$ will be termed an Ito relation. Several Ito relations can be given in terms of the generative automata I^* defined by $I^* = \eta^+ + \eta^- + \theta + \kappa$. It is not difficult to demonstrate that this operator satisfies $I^*(\mu) = \mathbf{1} + \mu$ for all μ. Making generous use of lemma 1; and, in the case of rules (180, 166) and (210, 154) applying the equality $\delta = D + D^-$ to demonstrate that $(\beta^- + \chi)(\mathbf{1} + \mu) = \mathbf{1} + (\beta^- + \chi)(\mu) + D(\mu)$, various identities can be obtained, such as $I^*(\Delta + \theta) = (\Delta + \iota)I^*$, and $I^*(D + \beta^+ + \chi) = (\delta + \beta^+ + \chi)I^*$.

Denote a cycle plus its basin of attraction by $\mathbf{B}$, and by $1 + \mathbf{B}$ the set obtained by adding 1 to every state of $\mathbf{B}$. Denote a tree rooted on a cycle or fixed

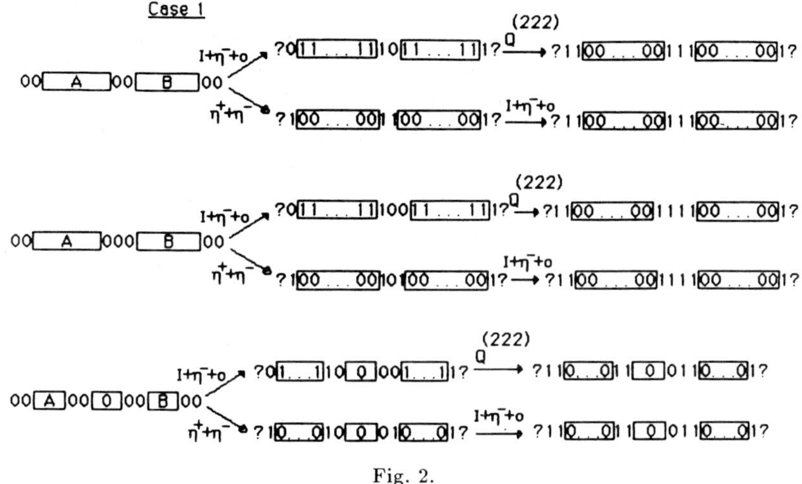

Fig. 2.

point by **T**. Then these identities can be stated in a way which makes clear the relation which they imply between state transition diagrams:

Theorem 4

(i) **B** *is a cycle plus basin of attraction for, respectively, rules 182, (214, 158), (180, 166), or (224, 168) if and only if* $1 + $ **B** *is a cycle plus basin of attraction for, respectively, rules 146, (148, 134), (210, 154), or (248, 234).*

(ii) *If* **B** *is a cycle plus basin of attraction for, respectively, rules (242, 186), (194, 152), or (216, 202) then* $1 + $ **B** *is a cycle plus basin of attraction for, respectively, rules (176, 162), (188, 230), or (228, 172).*

(iii) *If* **T** *is a tree of rules (220, 206) then* $1 + $ **T** *is a tree of rules (196, 140).*

5. Generative automata

The generative nearest-neighbor CA are obtained from the set of non-generative operators listed in the appendix by adding κ to each representation, and adding 1 to the decimal designation of each non-generative rule. The effect of this, in terms of fixed-point or shift-cycle behavior, is to reduce the number of automata which exhibit such behavior for relatively unrestricted subsets of E_n.

A number of Ito-type relations can be demonstrated for the combined set of generative and non-generative CA. Most generally, consider any

CA having a representation of the form $Q^* = Q + F(\beta^\pm, \eta^\pm, \chi, \theta, \iota, \kappa)$ where Q is a linear operator satisfying $Q(\mathbf{1}) = \mathbf{1}$. Define a second CA by $R^* = Q + F'(\beta^\pm, \eta^\pm, \chi, \theta, \iota, \kappa)$, where $F'(\beta^\pm, \eta^\pm, \chi, \theta, \iota, \kappa) = F(\eta^\pm, \beta^\pm, \chi, \theta, \iota, \kappa)$.

Theorem 5 *Let* Q^* *and* R^* *be as above. Then* $I^*Q^* = R^*I^*$.

Proof: By definition $I^*Q^*(\mu) = \mathbf{1} + (Q + F)(\mu) = \mathbf{1} + Q(\mu) + F(\mu)$. But, making use of the additivity of Q, $R^*I^*(\mu) = (Q + F')I^*(\mu) = (Q + F')(\mathbf{1} + \mu) = \mathbf{1} + Q(\mu) + F'(\mathbf{1} + \mu)$ and, making use of lemma 1, $F'(\mathbf{1} + \mu) = F(\mu)$. $\qquad\square$

6. Some results on trees

A question of interest in analysis of any CA is the maximum number of iterations required to reach a cycle or fixed point. Martin et al. [10] provide an answer to this question for additive CA but results of similar generality are difficult to come by for non-additive rules. On the other hand, the formalism introduced in this paper allows particular results for many of the non-additive automata. The technique is to identify the states which have no predecessors under a given rule, determine which of those is furthest from a cycle or fixed point, and compute how many iterations are required in order to reach the cycle or fixed point.

For example, rule 76 has representation $I + \chi$. Fixed points other than $\mathbf{0}$ have only single or double 1's, and $(I + \chi)(\mathbf{1}) = \mathbf{0}$. Consider any state μ possessing a string of $k \geq 3$ ones: ...0111...1110.... Then $\chi(\mu)$ will contain a corresponding string of $k - 2$ ones with the form ...0011...1100.... Hence when $I(\mu) = \mu$ is added to this the result will be a string with the form ...0100...0010... with at least one 0 separating the two 1's. Thus after a single iteration of the rule there are no remaining strings of three or more 1's and we have a fixed point. That is, the maximum tree height is 1.

7. Invariance matrices

Jen [13] defines the invariance matrix of an automaton, and shows that the trace of the kth power of this matrix equals the number of fixed points which that automaton possesses when $n = k$. Table 4 gives the invariance matrices for the basic non-linear automata of table 2, and table 5 gives the invariance matrices for the linear automata of table 3.

Other invariance matrices of interest are those for I^* defined as $I^* = \eta^+ + \eta^- + \theta + \kappa$ which has the property that $I^*(\mu) = (1+\mu)$, 0^* (which maps all states to 0), and 1^* (which maps all states to 1). These are

$$A(I^*) = \begin{pmatrix} 0 & 0 & 0 & 0 \\ 0 & 0 & 0 & 0 \\ 0 & 0 & 0 & 0 \\ 0 & 0 & 0 & 0 \end{pmatrix},$$

$$A(0^*) = \begin{pmatrix} 1 & 1 & 0 & 0 \\ 0 & 0 & 0 & 0 \\ 1 & 1 & 0 & 0 \\ 0 & 0 & 0 & 0 \end{pmatrix},$$

$$A(1^*) = \begin{pmatrix} 0 & 0 & 0 & 0 \\ 0 & 0 & 1 & 1 \\ 0 & 0 & 0 & 0 \\ 0 & 0 & 1 & 1 \end{pmatrix}.$$

Comparison with table 5 indicates that $A(I) = A(0^*) + A(1^*)$, suggesting a curious relationship between the non-linear operator algebra in which 0^* is the zero element, and the invariance matrix algebra in which $A(I^*)$ is the zero element. One

consequence of this relation is found in the following theorem, which allows computation of the invariance matrix of an operator from its canonical representation.

Theorem 6 *If the canonical form for the operator representing an automaton rule has an odd number of terms then the invariance matrix for this rule is the sum, modulo 2, of the invariance matrices of each term. If there are an even number of terms the invariance matrix is the mod(2) sum of the invariance matrices for each term, plus $A(0^*)$.*

Proof: Let $Q = \sum_{i=1}^{8} a_i q_i$ where the q_i are the basis operators of table 2 and $a_i = 0$ or 1. Writing the expression $\sum a_i A(q_i)$ yields the matrix

$$\begin{pmatrix} \sum_{i \neq 8} a_i & \sum_{i \neq 3} a_i & 0 & 0 \\ 0 & 0 & a_7 & a_2 \\ \sum_{i \neq 4} a_i & \sum_{i \neq 5} a_i & 0 & 0 \\ 0 & 0 & a_1 & a_6 \end{pmatrix}. \tag{3}$$

On the other hand, $A(Q)$ is easily computed to be

$$A(Q) = \begin{pmatrix} 1 + a_8 & 1 + a_3 & 0 & 0 \\ 0 & 0 & a_7 & a_2 \\ 1 + a_4 & 1 + a_5 & 0 & 0 \\ 0 & 0 & a_1 & a_6 \end{pmatrix}. \tag{4}$$

If an odd number of the a_i are non-zero then $\sum a_i = 1$ and $\sum_{i \neq j} a_i = 1 + a_j$. Thus in this case $\sum a_i A(q_i) = A(\sum a_i q_i)$. If an even number of the a_i are non-zero the $\sum a_i = 0$. We have $\sum_{i \neq j} a_i = a_j$. Noting the form of $A(0)$ we see that $\sum a_i A(q_i) + A(0) = A(\sum a_i q_i)$. $\square$

8. Discussion

The purpose of this paper has been to introduce a natural formalism for one-dimensional nearest-neighbor cellular automata in terms of a basis set of eight non-linear operators. Making use of this formalism it has been possible to classify nearest-neighbor CA and, for a large subset of CA, to determine significant properties such as fixed points, single-shift cycles, maximum tree heights, and certain Ito-like relations between different CA. This approach can be extended to more general cases

Table 4
Invariance matrices for basic non-linear automata.

Automaton	Operator	Invariance matrix
$2, 16$	η^+, η^-	$A(\eta^+) = \begin{pmatrix} 1 & 0 & 0 & 0 \\ 0 & 0 & 0 & 0 \\ 1 & 1 & 0 & 0 \\ 0 & 0 & 0 & 0 \end{pmatrix}, \quad A(\eta^-) = \begin{pmatrix} 1 & 1 & 0 & 0 \\ 0 & 0 & 0 & 0 \\ 0 & 1 & 0 & 0 \\ 0 & 0 & 0 & 0 \end{pmatrix}$
32	θ	$A(\theta) = \begin{pmatrix} 1 & 1 & 0 & 0 \\ 0 & 0 & 0 & 0 \\ 1 & 0 & 0 & 0 \\ 0 & 0 & 0 & 0 \end{pmatrix}$
128	χ	$A(\chi) = \begin{pmatrix} 1 & 1 & 0 & 0 \\ 0 & 0 & 0 & 0 \\ 1 & 1 & 0 & 0 \\ 0 & 0 & 0 & 1 \end{pmatrix}$
$64, 8$	β^+, β^-	$A(\beta^+) = \begin{pmatrix} 1 & 1 & 0 & 0 \\ 0 & 0 & 0 & 0 \\ 1 & 1 & 0 & 0 \\ 0 & 0 & 1 & 0 \end{pmatrix}, \quad A(\beta^-) = \begin{pmatrix} 1 & 1 & 0 & 0 \\ 0 & 0 & 0 & 1 \\ 1 & 1 & 0 & 0 \\ 0 & 0 & 0 & 0 \end{pmatrix}$
4	ι	$A(\iota) = \begin{pmatrix} 1 & 1 & 0 & 0 \\ 0 & 0 & 1 & 0 \\ 1 & 1 & 0 & 0 \\ 0 & 0 & 0 & 0 \end{pmatrix}$
1	κ	$A(\kappa) = \begin{pmatrix} 0 & 1 & 0 & 0 \\ 0 & 0 & 0 & 0 \\ 1 & 1 & 0 & 0 \\ 0 & 0 & 0 & 0 \end{pmatrix}$

as well. It is only necessary to write out the appropriate set of basis operators, and to express additive operators of interest (e.g. shifts and the identity) in terms of these basis operators. The only difficulty which arises is that the number of basis operators grows exponentially. For example, for a one-dimensional neighborhood of radius 2, or for a two-dimensional von Neumann neighborhood, there are $2^5 = 32$ basis operators and a total of 2^{32} evolution rules.

A general problem with this approach is that it is difficult to compute cycle behavior other than shift cycles because of the non-linear nature of the formalism. That is, $Q^*(\mu + \mu') \neq Q^*(\mu) + Q^*(\mu')$, hence iteration formulas cannot be easily computed.

Jen [13] has shown that all cycles will be multiple-shift cycles, so one approach may be to study equations of the form $Q^{*k}(\mu) = \sigma^r(\mu)$ in order to determine, for any given Q^*, the values of k and r for which non-trivial solutions exist. Some work along these lines for the case of linear automata is underway.

Another possible approach which shows some promise is to consider neighborhoods of size $2r + 1$ ($r = 1, 2, \ldots$), defining non-linear operator formalisms for each, and constructing an embedding of $2r + 1$ neighborhood rules into $2(r+1) + 1$ rules. Whether the payoff from such a program is worth the computational effort involved remains to be seen.

Table 5
Invariance matrices for non-trivial additive automata.

$$A(I) \;=\; \begin{pmatrix} 1 & 1 & 0 & 0 \\ 0 & 0 & 1 & 1 \\ 1 & 1 & 0 & 0 \\ 0 & 0 & 1 & 1 \end{pmatrix}$$

$$A(\sigma) \;=\; \begin{pmatrix} 1 & 0 & 0 & 0 \\ 0 & 0 & 0 & 1 \\ 1 & 0 & 0 & 0 \\ 0 & 0 & 0 & 1 \end{pmatrix} \qquad\qquad A(\sigma^{-1}) \;=\; \begin{pmatrix} 1 & 1 & 0 & 0 \\ 0 & 0 & 0 & 0 \\ 0 & 0 & 0 & 0 \\ 0 & 0 & 1 & 1 \end{pmatrix}$$

$$A(D) \;=\; \begin{pmatrix} 1 & 0 & 0 & 0 \\ 0 & 0 & 1 & 0 \\ 1 & 0 & 0 & 0 \\ 0 & 0 & 1 & 0 \end{pmatrix} \qquad\qquad A(D-) \;=\; \begin{pmatrix} 1 & 1 & 0 & 0 \\ 0 & 0 & 1 & 1 \\ 0 & 0 & 0 & 0 \\ 0 & 0 & 0 & 0 \end{pmatrix}$$

$$A(\delta) \;=\; \begin{pmatrix} 1 & 0 & 0 & 0 \\ 0 & 0 & 0 & 1 \\ 0 & 1 & 0 & 0 \\ 0 & 0 & 1 & 0 \end{pmatrix}$$

$$A(\Delta) \;=\; \begin{pmatrix} 1 & 0 & 0 & 0 \\ 0 & 0 & 1 & 0 \\ 0 & 1 & 0 & 0 \\ 0 & 0 & 0 & 1 \end{pmatrix}$$

Acknowledgements

This work has benefited from conversations with Al Weiss. I also thank a referee who provided a detailed and very useful critique of an earlier manuscript.

References

[1] S. Wolfram, Rev. Mod. Phys. 55 (1983) 601–644.

[2] S. Wolfram, Physica D 10 (1984) 1.

[3] W.J. Wilbur, D.J. Lipman and S.A. Shamma, Physica D 19 (1986) 397–410.

[4] S. Wolfram, Adv. Appl. Math. 7 (1986) 123.

[5] S. Wolfram, Physica D 22 (1986) 385.

[6] B. Voorhees, Physica D 31 (1988) 135–140.

[7] B. Voorhees, Commun. Math. Phys. 117 (1988) 431–439.

[8] S.J. Wilson, Physica D 24 (1987) 179–189, 190–206.

[9] P. Grassberger, Physica 10 D (1984) 52–58.

[10] O. Martin, A.M. Odlyzko and S. Wolfram, Commun. Math. Phys. 93 (1984) 219–258.

[11] H. Ito, Physica D 31 (1988) 318–338.

[12] P. Grassberger, J. Phys. A 20 (1987) 4039.

[13] E. Jen, Commun. Math. Phys. 118 (1988) 569–590.

[14] D.A. Lind, Physica D 10 (1984) 36–44.

[15] R. Cordovil, R. Dilao and A. Noronha da Costa, Disc. Comput. Geo. 1 (1988) 277–288.

[16] M. Dubois-Violette and A. Rouet, Commun. Math. Phys. 112 (1987) 627–631.

[17] K. Culik and S. Yu, Complex Systems 2 (1988) 177–190.

[18] H.A. Gutowitz, A hierarchical classification of cellular automata, Physica D 45 (1990) 136–156, these Proceedings.

[19] W. Li and N. Packard, The structure of the elementary cellular automata rule space, CCSR, University of Illinois, reprint (1989).

Physica D 45 (1990) 36–48
North-Holland

CELLULAR AUTOMATA AND MULTIFRACTALS:
DIMENSION SPECTRA OF LINEAR CELLULAR AUTOMATA

Satoshi TAKAHASHI [1]

Department of Biophysics, Kyoto University, Kyoto 606, Japan

Received 9 January 1990
Revised manuscript received 26 April 1990

The dimension distribution of space patterns of linear cellular automata are studied. A dimension spectrum derived from the dimension distribution is shown to correspond to a singularity spectrum of a multifractal. Using this relation, the dimension spectrum is efficiently calculated.

1. Introduction

Cellular automata are systems consisting of lattice points. Each lattice point (cell) has a state with discrete time steps by a local rule. A number of relationships between linear cellular automata and fractals have been given by Willson [11–15]. He showed that a series of space–time patterns of a prime-state linear cellular automaton converges to a fractal, which is called the limit set. He also devised an efficient computation method for the Hausdorff dimension of the limit set. Previously, I extended Willson's result to more general linear cellular automata and showed the relationship between the transition matrix (defined by Willson [14]) and the structure of the limit set [19].

In this paper, we concentrate on the dimensions of space patterns of a linear cellular automaton. In section 3, we define a dimension spectrum, which represents the distribution of the dimensions of space patterns. The notion of multifractal is useful to compute the dimension spectrum. The relation between the dimension spectrum of a linear cellular automaton and the singularity spectrum of the correspond-

ing multifractal is shown in section 6. Using this relation, we present an approximation method for calculating the dimension spectrum of linear cellular automata (Markov weight transformation approximation) in section 8, and the efficacy of this method is shown in section 9.

2. Preliminaries

In this section, we introduce some results of my previous work on linear cellular automata (LCA) [9].

Let $\mathbb{Z}$, $\mathbb{N}$, and d be the set of integers, the set of natural numbers and a natural number, respectively. Consider a d-dimensional lattice, each site of which is referred to as a *cell*. $i \in \mathbb{Z}^d$ indicates the location of a cell and $t \in \mathbb{N}$ denotes the time. Each cell is specified as an integer in the range 0 through $M-1$, where M is a natural number, representing the number of the states. We denote by $a_i^t \in \{0, ..., M-1\}$ the state of the cell located at the point i and time t. Consider a neighbourhood of cell i, $\{i+r_1, ..., i+r_m\}$, where $(r_1, ..., r_m) \in (\mathbb{Z}^d)^m$ is called a *neighbourhood index*. If the states of the cells at time $t-1$ determine the states of the cells at time t by the following *transition rule*,

---

[1] Present address: Department of Mathematics, Osaka City University, Sugimoto, Sumiyoshi, Osaka, 558, Japan.

$$a_i^t = c_1 a_{i+r_1}^{t-1} + \ldots + c_m a_{i+r_m}^{t-1} \quad \mathrm{mod}\, M , \qquad (2.1)$$

where $c_1, \ldots, c_m$ are natural numbers, then we call this system an *M-state linear cellular automaton* (abbreviated *M-state LCA*).

We consider the space–time pattern of an LCA in a $(d+1)$-dimensional Euclidean space and define the limit set of this space–time pattern as follows.

Let t_n be a monotone series of natural numbers, and $S(t_n)$ be a subset of the $(d+1)$-dimensional Euclidean space defined by

$$S(t_n) = \{(i, t) \mid t < t_n, a_i^t \neq 0\} .$$

$S(t_n)$ represents the space–time pattern of the LCA by successive steps up to time $t_n - 1$. Multiplying each element of $S(t_n)$ by $1/t_n$, we obtain a contracted set $S(t_n)/t_n$, i.e. $(i/t_n, t/t_n) \in S(t_n)/t_n$, when $(i, t) \in S(t_n)$.

If there exists a time sequence t_n, for which $S(t_n)/t_n$ converges to a set, we call that set the *limit set* of the LCA.

The following assumption is made for the initial states of cells:

$$a_i^0 = 1 \quad i = 0 ,$$

$$ = 0 \quad \text{otherwise} .$$

Main results of my previous work [9] are summarized as follows:

Theorem 1. Consider a p^k-state LCA, where p is a prime and k is a natural number. If t is divisible by p^{k-1}, then $a_{pi}^{pt} = a_i^t$. If t is divisible by p^k and at least one of the elements of i is indivisible by p, then $a_i^t = 0$.

Theorem 2. For a p^k-state LCA, if we take $t_n = p^n$, then the limit set $\lim_{n \to \infty} S(t_n)/t_n$ exists.

Theorem 3. The limit set of a p^k-state LCA can be represented as a union of some members of a family of sets X_j which have the following property: Each X_j is composed of n_{qj} $1/p$-scaled X_q's. If a *transition matrix* is defined by $A = (n_{qj})$, then the Hausdorff dimension of the limit set is given by $\log_p \lambda$, where λ is the maximum eigenvalue of the matrix A.

The following example clarifies the ideas of this theorem.

Example 1. A one-dimensional 2-state LCA with a transition rule,

$$a_i^t = a_{i-1}^{t-1} + a_i^{t-1} + a_{i+1}^{t-1} \quad \mathrm{mod}\, 2 .$$

Figs. 1a–1e show the sets $S(2^n)/2^n$ for $n = 1$–5, respectively, which converge to a limit set (see fig. 1f).

As shown in fig. 2, the limit set is composed of two triangle patterns designated X, if the symmetric figures on the right and left halves are regarded as equivalent. X is composed of two $1/2$-scaled X's and one $1/2$-scaled square pattern designated Y. Y is composed of four $1/2$-scaled X's. Therefore, we have $n_{11} = 2$, $n_{12} = 4$, $n_{21} = 1$ and $n_{22} = 0$, and the transition matrix is given by

$$\begin{pmatrix} 2 & 4 \\ 1 & 0 \end{pmatrix} .$$

Since the maximum eigenvalue of the transition matrix is $1 + \sqrt{5}$, the Hausdorff dimension of the limit set is given by $\log(1 + \sqrt{5})/\log 2$.

3. Definition of dimension spectrum

Consider a limit set of an LCA. From the definition of the limit set, the time coordinate of the limit set lies between 0 and 1. We denote the time of the limit set by $\tau \in [0, 1]$. Let us cut the limit set perpendicular to the time axis at τ and consider the dimension of the cutting surface (space pattern of the limit set).

A *dimension spectrum* $H(\delta)$ of the limit set is defined by the dimension of the set of time τ's at which the space patterns have a dimension δ. The dimension spectrum is calculated by the following procedure.

We first count the number of non-zero cells of the LCA at time t:

$$N(t) = \#\{i \mid a_i^t \neq 0\} , \qquad (3.1)$$

where $\#$ denotes the number of the elements of a set.

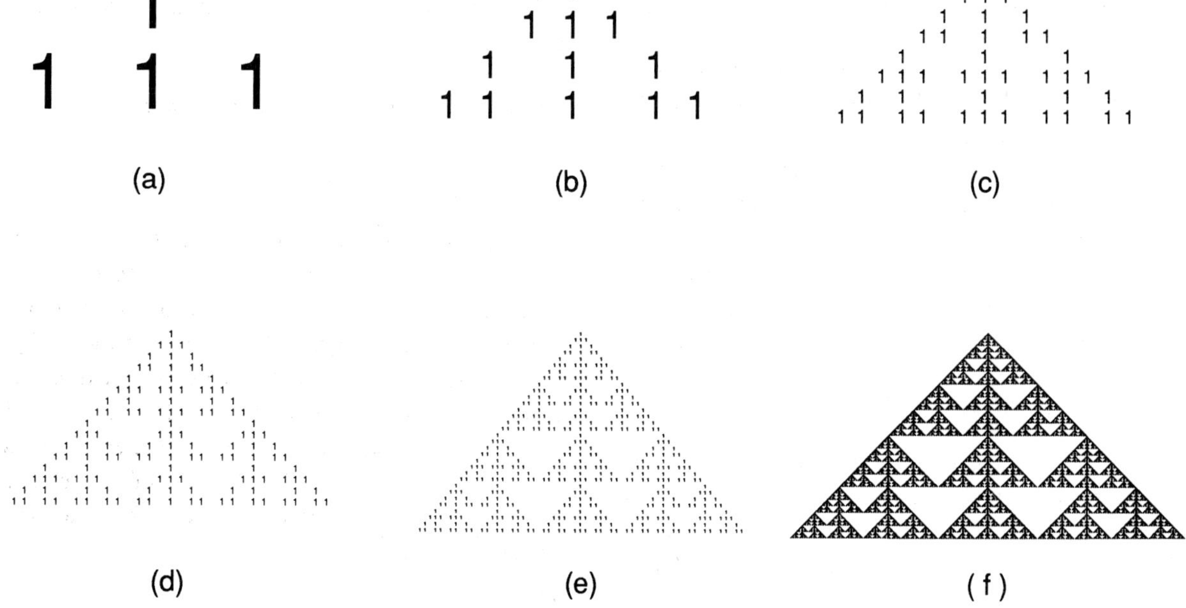

Fig. 1. $S(2^n)/2^n$ for the LCA with the rule $a_i^t = a_{i-1}^{t-1} + a_i^{t-1} + a_{i+1}^{t-1}$ mod 2. 1 represents the points in set $S(2^n)/2^n$. They converge to the limit set. (a) $n=1$, (b) $n=2$, (c) $n=3$, (d) $n=4$, (e) $n=5$, (f) limit set.

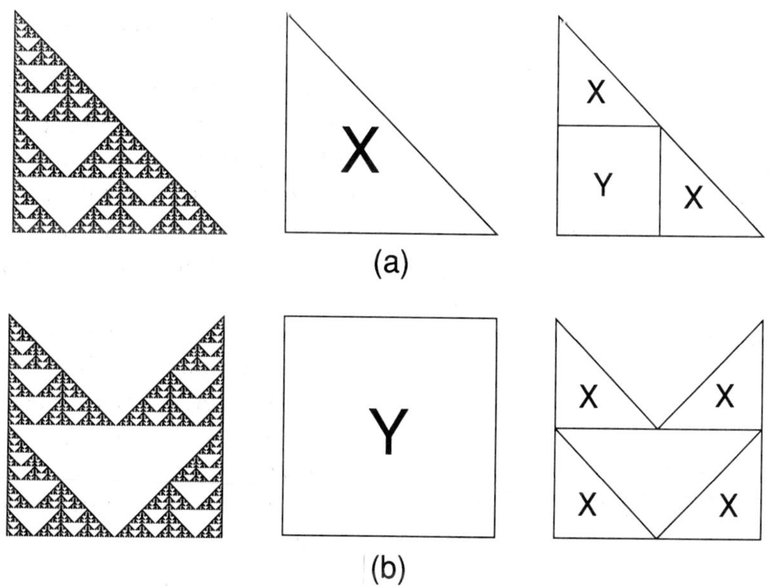

Fig. 2. The limit set of the LCA with the rule $a_i^t = a_{i-1}^{t-1} + a_i^{t-1} + a_{i+1}^{t-1}$ mod 2 is composed of two patterns X and Y as shown in (a) and (b), respectively. X is composed of two 1/2-scaled X's and one 1/2-scaled Y, and Y is composed of four 1/2-scaled X's.

We then take the time series t_n for which a limit set of the LCA exists, i.e. the series of the sets $S(t_n)/t_n$ converges. Let us define the following quantity:

$$D(t, t_n) = \frac{\log N(t)}{\log t_n} \quad \text{for } t < t_n. \tag{3.2}$$

Thus the dimension spectrum $H(\delta)$ is given by

$$H(\delta) = \lim_{\epsilon \to 0} \lim_{n \to \infty} \Delta(\delta, t_n, \epsilon), \tag{3.3}$$

where,

$$\Delta(\delta, t_n, \epsilon)$$

$$= \frac{\log \#\{t \mid \delta - \epsilon < D(t, t_n) < \delta + \epsilon\}}{\log t_n}. \tag{3.4}$$

Above procedure gives $H(\delta)$ from the following reason. Consider a space pattern of the limit set at an arbitrary time τ ($\in [0,1]$). For a given n, take a time $t(n)$ such that

$$\frac{t(n)}{t_n} < \tau \le \frac{t(n)+1}{t_n}.$$

Then we can cover the space pattern at time τ by $N(t(n))$ boxes

$$\{(\tau', i') \mid t(n)/t_n \le \tau' \le [t(n)+1]/t_n,$$

$$(i_s - r)/t_n \le i'_s \le (i_s + r)/t_n, i = (i_1, ..., i_d),$$

$$i' = (i'_1, ..., i'_d), a_{i}^{t(n)} \neq 0\},$$

where r is the range of neighbourhood and is given by $\max\{|r_1|, ..., |r_m|\}$. As n tends to infinity, the dimension of the space pattern at time τ is evaluated by

$$\lim_{n \to \infty} \frac{\log N(t(n))}{\log t_n} = \lim_{n \to \infty} D(t(n), t_n) = \delta.$$

We next consider an interval $[t/t_n, (t+1)/t_n]$ ($t = 0, ..., t_n - 1$) and assign $D(t, t_n)$ to that interval. The argument of the logarithm in the numerator of the r.h.s. of (3.4) indicates the number of the intervals in which $D(t, t_n)$ lie between $\delta - \epsilon$ and $\delta + \epsilon$. As n tends to infinity, $D(t, t_n)$ approaches the dimension of the space pattern and $\Delta(\delta, t_n, \epsilon)$ converges to the dimension of the set of τ's at which the dimension of the space pattern lies between $\delta - \epsilon$ and $\delta + \epsilon$. At the limit of ϵ to zero, $H(\delta)$ gives the dimension of the set of τ's at which the dimension of the space pattern is δ.

Note. To prove eq. (3.3) rigorously, we need the law of large numbers and theorem 14.1 in ref. [2]. We omit the proof due to limitations of space and time.

4. Computation of $H(\delta)$

Example 2. Consider a 2-state LCA satisfying the following rule:

$$a_i^t = a_{i-1}^{t-1} + a_{i+1}^{t-1} \quad \text{mod } 2.$$

Taking the time series $t_n = 2^n$, we have the limit set of this LCA as shown in fig. 3. This limit set is called Sierpiński gasket, a typical fractal figure.

The dimension spectrum of this limit set is given by

$$H(\delta) = \frac{-\delta \log \delta - (1-\delta) \log(1-\delta)}{\log 2}. \tag{4.1}$$

The derivation of (4.1) is given as follows.

We can show from theorem 1 that

$$N(2t) = N(t) \tag{4.2}$$

and

$$N(2t+1) = 2N(t). \tag{4.3}$$

Let $t = \sum_{j=0}^{n-1} t_j 2^j$, where $t_j = 0$ or 1 (i.e. $t = t_{n-1} t_{n-2} ... t_0$

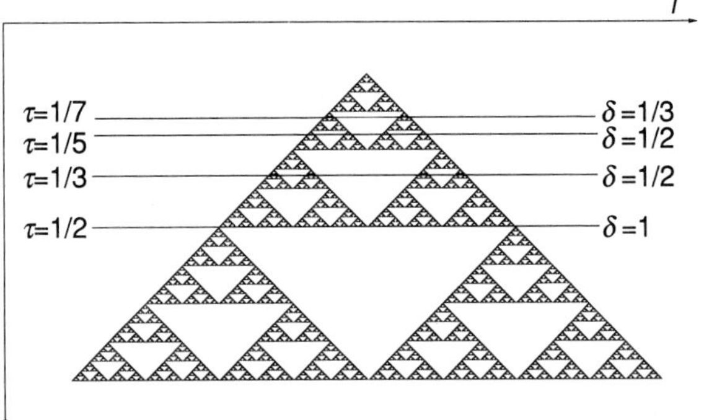

Fig. 3. The limit set of the LCA with the rule $a_i^t = a_{i-1}^{t-1} + a_{i+1}^{t-1}$ mod 2. δ indicates the dimension of the space pattern at time τ.

in diadic expansion). Then by iterating (4.2) and (4.3) we have $N(t)=2^{\Sigma t_j}$. Let $m=\Sigma t_j$ and $\delta=m/n$, then from (3.2), we obtain $D(t, 2^n)=\delta$. Substituting this relation into (3.4), we have

$$\Delta(\delta, 2^n, \epsilon) = \frac{\log_n C_m}{n \log 2}$$

for sufficiently small ϵ. Applying Stirling's formula leads to

$$\Delta(\delta, 2^n, \epsilon) = \frac{-\delta \log \delta - (1-\delta) \log(1-\delta)}{\log 2}$$

for sufficiently large n and for sufficiently small ϵ. Thus we obtain (4.1).

5. Multifractals and singularity spectra

Singularity spectrum $f(\alpha)$ represents the distribution of measure on a multifractal [5–7]. To correlate the dimension spectrum of an LCA with the singularity spectrum of a multifractal, here we consider a set with weight in stead of a measure. In a set with weight, total weight varies depending on the way of division of the set.

Consider a set which is divided into N pieces. Let l_i and m_i be the size and the weight of the ith piece, respectively. If the set has a uniform structure to which a uniform measure is assigned, then m_i is proportional to l_i^D, where D is the dimension of the set. In analogy, we define the singularity of the ith piece, α_i, by

$$m_i \sim l_i^{\alpha_i} \qquad (l_i \to 0). \tag{5.1}$$

α_i indicates the dimension of the ith piece in the limit of l_i to zero.

Assume that all the pieces have the same size l, and the distribution of α_i's has the following form:

$$\#\{i \mid \alpha < \alpha_i < \alpha + d\alpha\} \sim d\alpha\, l^{-f(\alpha)}. \tag{5.2}$$

In the limit of l tending to zero, $f(\alpha)$ represents the dimension of the set of points with a singularity α. Thus we call $f(\alpha)$ *singularity spectrum*.

Direct numerical calculation of $f(\alpha)$ by using (5.2) is a formidable task. However, the calculation can be considerably simplified by introducing a *partition function* [5]:

$$\Gamma(q, \tau, l) = \sum m_i^q / l^\tau. \tag{5.3}$$

Assume that in the limit of l to zero, the following relation holds:

$$\sum m_i^q \sim l^{\tau(q)}, \tag{5.4}$$

where $\tau(q)$ is given by

$$\tau(q) = \sup\{\tau \mid \lim_{l \to 0} \Gamma(q, \tau, l) = 0\}$$

$$= \inf\{\tau \mid \lim_{l \to 0} \Gamma(q, \tau, l) = \infty\}. \tag{5.5}$$

On the other hand, from (5.1) and (5.2), we have

$$\sum m_i^q \sim \int d\alpha\, l^{-f(\alpha)} l^{q\alpha} \qquad (l \to 0). \tag{5.6}$$

Because l is very small, only $\alpha(q)$ that minimizes $q\alpha - f(\alpha)$ contributes to this integral. Therefore we have

$$\left. \frac{df(\alpha)}{d\alpha} \right|_{\alpha = \alpha(q)} = q. \tag{5.7}$$

Then, comparing (5.4) and (5.6) yields

$$\tau(q) = q\alpha(q) - f(\alpha(q)). \tag{5.8}$$

Combining (5.7) and (5.8), we have

$$\frac{d\tau(q)}{dq} = \alpha + q\frac{d\alpha}{dq} - \frac{df(\alpha)}{d\alpha}\frac{d\alpha}{dq} = \alpha. \tag{5.9}$$

Therefore, if we obtain $\tau(q)$ from (5.5), $f(\alpha)$ is calculated by using (5.8) and (5.9).

Notice that an approximate value of $\tau(q)$ is usually calculated from the equation $\Gamma(q, \tau(q), l) = 1$ for a finite l (<1) with $\sum m_i = 1$. If the measure is distributed in a self-similar manner, this procedure gives an exact $\tau(q)$, because $\Gamma(q, \tau, l^n)$ is given by $\Gamma(q, \tau, l)^n$ (see ref. [5]).

Example 3. The multifractal corresponding to the LCA of example 2 is generated by the following procedure (see fig. 4). Take a unit interval and assign weight 1 to it. Divide it into two same-sized intervals and assign weight 1 to the left interval and 1/2 to the right interval. Continue infinitely the same process

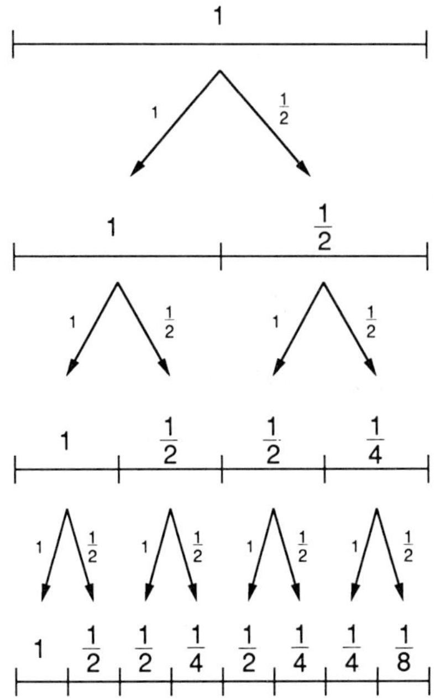

Fig. 4. The multifractal of example 3. Number assigned to each interval indicates the weight of that interval.

of dividing intervals and assigning the weight of the previous interval to the left half and 1/2 of that value to the right half. Since this multifractal has a self-similar structure, the singularity spectrum is calculated by letting (5.3) be equal to unity at first division, thus we have

$$(1/2)^q + 1 = (1/2)^\tau .$$

Taking logarithms of both sides, we have

$$\tau(q) = -\frac{\log[(1/2)^q + 1]}{\log 2} .\tag{5.10}$$

From (5.9), α is given by

$$\alpha = -\frac{(1/2)^q}{(1/2)^q + 1} ,\tag{5.11}$$

or

$$q = -\frac{\log \alpha - \log(1 - \alpha)}{\log 2} .\tag{5.12}$$

Using (5.11), (5.10) is rewritten as

$$\tau(q) = \frac{\log(1 - \alpha)}{\log 2} .\tag{5.13}$$

Substituting (5.12) and (5.13) into (5.8), we have the singularity spectrum as

$$f(\alpha) = \frac{-\alpha \log \alpha - (1 - \alpha) \log(1 - \alpha)}{\log 2} .$$

The same result as above can also be obtained by the method used in section 4.

6. Dimension spectra and singularity spectra

We calculated the dimension spectrum $H(\delta)$ for the case of example 2. However, it is not possible to calculate analytically dimension spectra for general LCA. In this section, we show how to construct a multifractal whose singularity spectrum $f(\alpha)$ is the same as the dimension spectrum $H(\delta)$ of a given LCA.

Consider a LCA in which a set $S(t_n)/t_n$ (defined by (2.2)) converges to a limit set for a time series t_n. To generate a multifractal corresponding to this limit set, we take a unit interval [0, 1], and divide this interval into t_n equivalent intervals. Assign weight $m_t = 1/N(t)$ to the interval $[t/t_n, (t+1)/t_n]$, where $N(t)$ is the number of non-zero cells at time t in the LCA as defined in (3.1). In the limit of n tending to infinity, we have a multifractal.

Relation (5.1) for this multifractal is written as

$$1/N(t) \sim (1/t_n)^\alpha .\tag{6.1}$$

Comparing (3.2) and (6.1), we can see that $D(t, t_n)$ in (3.2) corresponds to α in (6.1). It thus follows from definition (3.4) and (5.2) that the singularity spectrum of this multifractal is given by the same function as the dimension spectrum of the LCA.

The relationship between the dimension spectrum of an LCA and the singularity spectrum of the corresponding multifractal is summarized in table 1. Fig. 5 shows the relation between the limit set of the LCA given in example 2 and the corresponding multifractal shown in example 3. Using this relationship, we

Table 1
The relationship between the parameters of the dimension spectrum and the singularity spectrum.

Dimension spectrum	$1/t_n$	$1/N(t)$	$D(t, t_n)$	δ	$H(\delta)$
Singularity spectrum	l	m	$\log m/\log l$	α	$f(\sigma)$

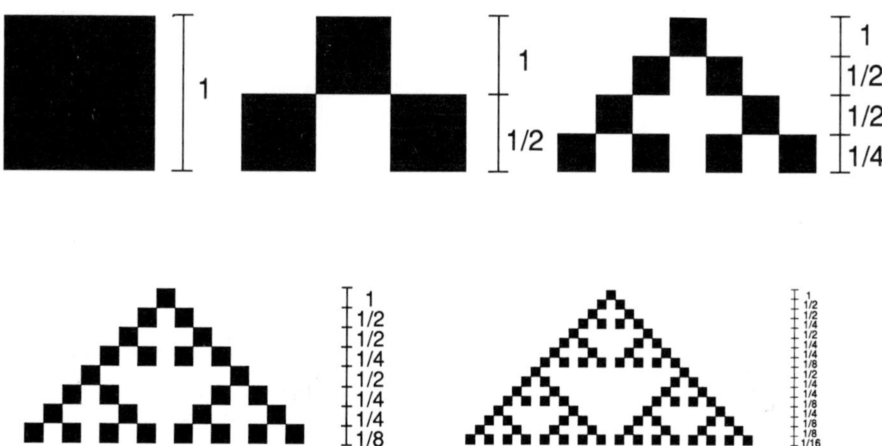

Fig. 5. Correspondence between example 2 and example 3.

will find a good approximation of $H(\delta)$ for general LCA as will be shown in section 8.

7. Partially self-similar limit sets

Example 4. In general, the limit set of an LCA consists of some elementary patterns, which in turn, comprise contracted patterns of those elementary patterns. This type of fractal structures is called to be *partially self-similar* [4], while fractals composed of single elementary patterns are called to be self-similar (see fig. 3 for an example of a self-similar set).

As an example of a partially self-similar set, we consider the LCA in example 1, whose limit set is composed of two elementary patterns X and Y (see fig. 2). To obtain the dimension spectrum of this LCA, we divide the elementary patterns, X and Y, into upper and lower halves by a horizontal center line. The upper part of X contains one 1/2-scaled X, and the upper part of Y contains two 1/2-scaled X's (fig. 6a). The lower part of X contains one 1/2-scaled

X and one 1/2-scaled Y, and the lower part of Y contains two 1/2-scaled X's. Let A_0 (A_1) denote a matrix whose ijth ($i, j = 1, 2$) element $(A_0)_{ij}$ ($(A_1)_{ij}$) is given by the number of the 1/2-scaled ith pattern contained in the upper (lower) part of the jth pattern, where the ith pattern represents X for $i = 1$ and Y for $i = 2$. Thus we have

$$A_0 = \begin{pmatrix} 1 & 2 \\ 0 & 0 \end{pmatrix}, \qquad A_1 = \begin{pmatrix} 1 & 2 \\ 1 & 0 \end{pmatrix}. \tag{7.1}$$

Using the matrices A_0 and A_1, the number of nonzero cells at time t, $N(t)$, as defined by (3.1) is given by

$$N(t) = \tfrac{1}{2} c A_{t_1} A_{t_2} \dots A_{t_s} \boldsymbol{u}, \tag{7.2}$$

where $\boldsymbol{u} = \binom{2}{0}$, $\boldsymbol{c} = (1\ 2)$ and $t = t_s t_{s-1} \dots t_1$ in diadic expansion. A brief sketch of the derivation of (7.2) is given below. This derivation is similar to that given by Willson [14].

Suppose that a neighbouring pair of cells (say i and $i+1$ cells) at time t are at state $(a_i^t\, a_{i+1}^t) = (1\, 0)$. We can see from theorem 1 that $(a_{2i}^{2t}\, a_{2i+1}^{2t}\, a_{2i+2}^{2t})$

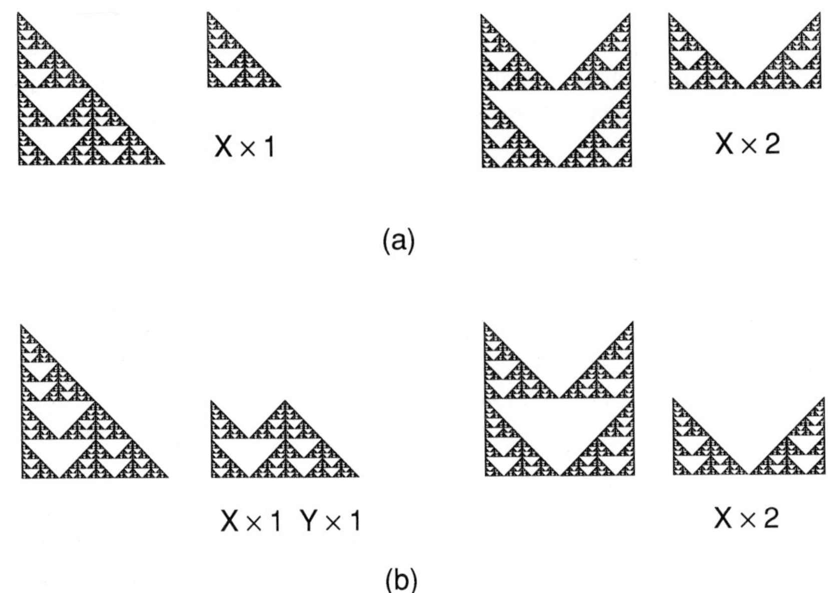

Fig. 6. Numbers of the 1/2-scaled X's and 1/2-scaled Y's contained (a) in the upper part of X and Y, (b) in the lower part of X and Y.

$=(1\,0\,0)$. At the next time $2t+1$, $(a_{2i}^{2t+1}\,a_{2i+1}^{2t+1}\,a_{2i+2}^{2t+1})$ is calculated as $(1\,1\,0)$ (fig. 7a). If we call a neighbouring pair of cells a block, the above result can be restated that a block $(1\,0)$ at time t yields one $(1\,0)$ at time $2t$ and one $(1\,0)$ and one $(1\,1)$ at time $2t+1$, when we count blocks allowing overlaps between them. Similarly, a block $(1\,1)$ at time t yields two $(1\,0)$ at time $2t$ and two $(1\,0)$ at time $2t+1$, if we regard block $(0\,1)$ as equivalent to block $(1\,0)$, because of the symmetry of the transition rule (fig. 7b). Therefore we see that the ijth element of the matrix A_0 (A_1) in (7.1) equals the number of block i at time $2t$ $(2t+1)$ generated from block j at time t, where block 1 represents block $(1\,0)$ and block 2 block $(1\,1)$. Let us now focus on the configuration at time t and count the numbers of blocks

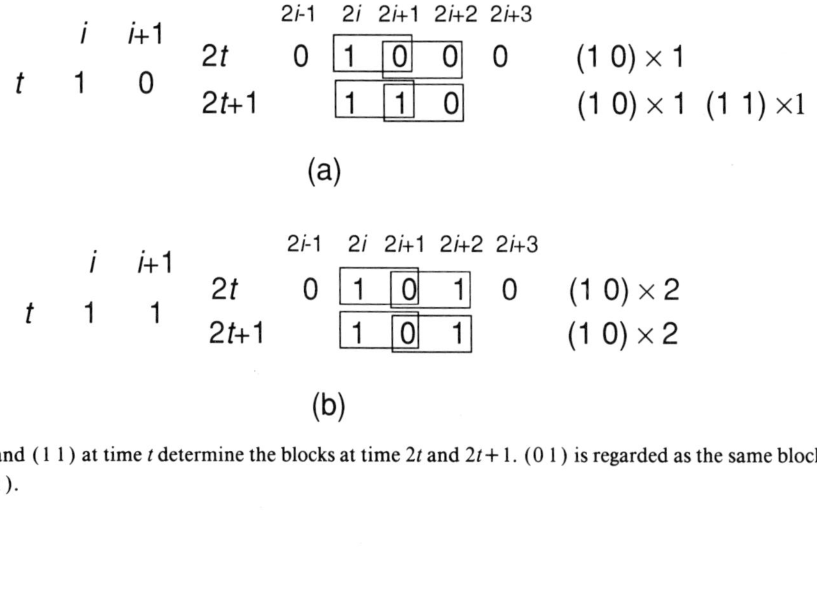

Fig. 7. Blocks $(1\,0)$ and $(1\,1)$ at time t determine the blocks at time $2t$ and $2t+1$. $(0\,1)$ is regarded as the same block as $(1\,0)$. (a) block $(1\,0)$, (b) block $(1\,1)$.

(1 0), (0 1) and (1 1), allowing overlaps between different blocks. Take a vector $\boldsymbol{v}$ whose first element represents the sum of the numbers of the blocks (1 0) and (0 1), and the second element represents the numbers of the block (1 1) at time t. Then $A_0\boldsymbol{v}$ and $A_1\boldsymbol{v}$ give the numbers of blocks at time $2t$ and at time $2t+1$, respectively. Let a counting vector $\boldsymbol{c}=(1\,2)$, which represents the numbers of 1's contained in each block, and let a unit vector $\boldsymbol{u}=\binom{2}{0}$, which indicates the number of blocks at time 0. Then for a time $t=t_s t_{s-1}...t_1$ in diadic expansion, $N(t)$ is given by $\frac{1}{2}\boldsymbol{c}A_{t_1}A_{t_2}...A_{t_s}\boldsymbol{u}$, where the coefficient $\frac{1}{2}$ is introduced to cancel the effect due to overlaps of blocks.

Finally we show why the matrices A_0 and A_1 originally defined for the transition of the patterns X and Y (fig. 6) also specify the transition of blocks. Recall the procedure of constructing the limit set of the LCA: we take the space–time pattern up to time 2^n and contract it by a rate $1/2^n$. Choose an arbitrary block (1 0) in a contracted space–time pattern of time 2^n. We focus on the square region under this block (see fig. 8a). This square region is occupied by a pattern

$$\begin{array}{ccc} 1 & 0 & 0 \\ 1 & 1 & 0 \end{array}$$

at time 2^{n+1}. Iterating the same procedure for each block in the square region, the pattern of 1 in the square region converges to the $1/2^n$-scaled X. Similarly, a square region under a block (1 1) becomes the pattern Y (fig. 8b). Therefore the blocks (1 0) and (1 1) in the space–time pattern up to 2^n time steps precisely correspond to $1/2^n$-scaled X's and Y's in the limit set. Thus the number of $1/2^s$-scaled X's and Y's contained in the region between two lines $\tau=(t+1)/2^s$ and $\tau=(t+1)/2^s$ is given by $A_{t_1}A_{t_2}...A_{t_s}\boldsymbol{u}$.

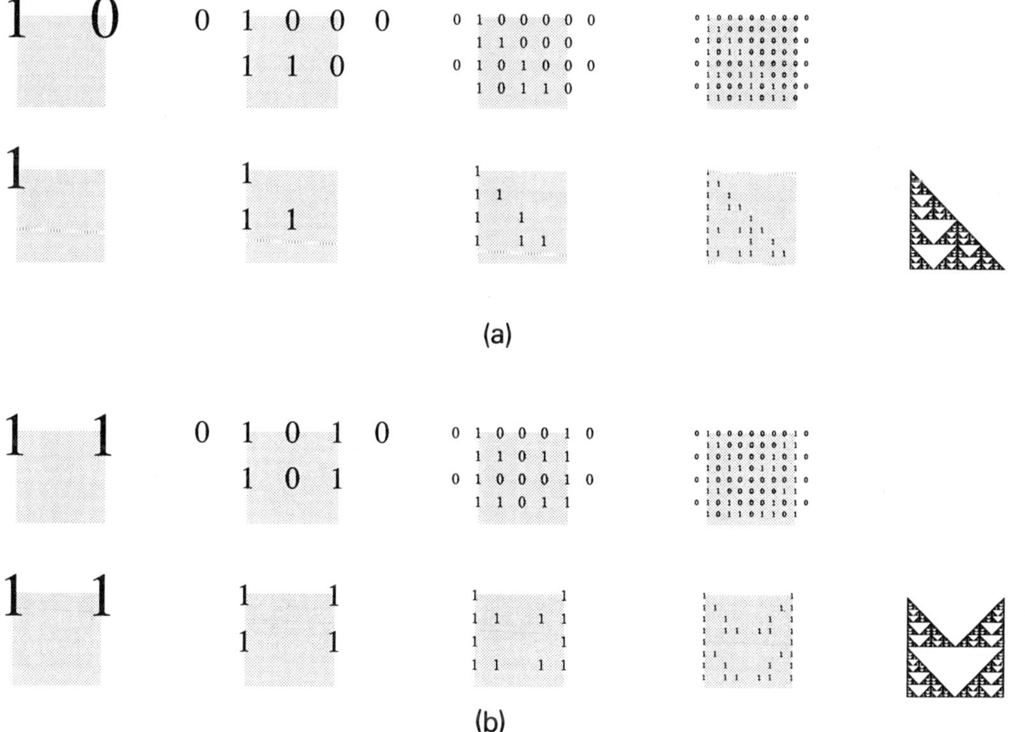

(a)

(b)

Fig. 8. Block (1 0) generates the pattern X, and block (1 1) generates the pattern Y. Figures correspond to $S(t)/t$, $S(2t)/2t$, $S(4t)/4t$, $S(8t)/8t$ from left to right.

It has been shown in the theory of random matrices that $\log N(t(n))/\log t_n$ defined by (3.1) converges to a certain value δ_0 for almost all time sequence [3]. Therefore almost all space patterns take a dimension δ_0. For the case of example 2, space patterns almost certainly take a dimension $1/2$, since $H(\delta) < 1$ ($\delta \neq 1/2$). In general, however, it is difficult to obtain δ_0 analytically.

8. Markov weight transformation approximation

We have shown in secion 5 that if the limit set of an LCA is self-similar, then the weight of the corresponding multifractal is distributed in a self-similar manner, and hence the dimension spectrum of the LCA is calculated exactly (see example 3). This does not apply to partially self-similar limit sets like example 4, because the weight is not distributed self-similarly in the corresponding multifractals.

When we divide an interval into two intervals for constructing the corresponding multifractal, the transformation of weight is achieved by multiplying the matrices A_0 or A_1 given by (7.2).

In this section, we devise an approximation method to calculate the dimension spectra of partially self-similar limit sets of LCA. We explain the approximation method by applying it to example 4.

Consider $N(t)$ defined by (7.1). Let $j = j_s j_{s-1} \dots j_1$ in diadic expansion. The $N(j)$ is given by

$$N(j) = \tfrac{1}{2} c A_{j_1} A_{j_2} \dots A_{j_s} \boldsymbol{u} . \tag{8.1}$$

(8.1) is approximated by

$$N(j) \simeq b \tfrac{1}{2} c A_{j_1} A_{j_2} \dots A_{j_s} \begin{pmatrix} 1 \\ 1 \end{pmatrix} , \tag{8.2}$$

where b is given by

$$b = \frac{(1\ 1) A_{j_1} \dots A_{j_s} \boldsymbol{u}}{(1\ 1) A_{j_1} \dots A_{j_s} \binom{1}{1}} .$$

This approximation is valid for large s, since directions of $A_{j_1} \dots A_{j_s} \boldsymbol{u}$ and $A_{j_1} \dots A_{j_s} \binom{1}{1}$ approach each other as s increases (see fig. 9):

$$A_{j_1} \dots A_{j_s} \boldsymbol{u} \cong A_{j_1} \dots A_{j_s} \begin{pmatrix} 1 \\ 1 \end{pmatrix} \quad \text{for large } s . \tag{8.3}$$

Let $i = i_s i_{s-1} \dots i_1$ in diadic expansion. Then $N(t)$ at $t = i + 2^s j$ is given by

$$N(i + 2^s j) = \tfrac{1}{2} c A_{i_1} A_{i_2} \dots A_{i_s} A_{j_1} A_{j_2} \dots A_{j_s} \boldsymbol{u} . \tag{8.4}$$

By using relation (8.3) again, we approximate (8.4) as

$$N(i + 2^s j) \cong B_{ij} b \tfrac{1}{2} c A_{i_1} A_{i_2} \dots A_{i_s} \begin{pmatrix} 1 \\ 1 \end{pmatrix} , \tag{8.5}$$

where B_{ij} is given by

$$B_{ij} = \frac{(1\ 1) A_{i_1} \dots A_{i_s} A_{j_1} \dots A_{j_s} \binom{1}{1}}{(1\ 1) A_{i_1} \dots A_{i_s} \binom{1}{1}} .$$

Similarly $N(k + 2^s i + 2^{2s} j)$ is approximated by

$$N(k + 2^s i + 2^{2s} j) \cong B_{ki} B_{ij} b \tfrac{1}{2} c A_{k_1} A_{k_2} \dots A_{k_s} \begin{pmatrix} 1 \\ 1 \end{pmatrix} , \tag{8.6}$$

where $k = k_s k_{s-1} \dots k_1$ in diadic expansion and B_{ki} is given by

$$B_{ki} = \frac{(1\ 1) A_{k_1} \dots A_{k_s} A_{i_1} \dots A_{i_s} \binom{1}{1}}{(1\ 1) A_{k_1} \dots A_{k_s} \binom{1}{1}} . \tag{8.7}$$

Since $b \tfrac{1}{2} c A_{i_1} A_{i_2} \dots A_{i_s} \binom{1}{1}$ is bounded for any i, only the product by B_{ij} contributes to the dimension of a space pattern, δ, and hence $H(\delta)$. Therefore, it is reasonable to assume that the ratio of $N(j + 2^s t)$ and $N(i + 2^s j + 2^{2s} t)$ for any t is given by B_{ij}.

Recalling the relationship between the number of the elementary patterns of an LCA and the weight of the corresponding multifractal as shown in table 1, we define a $2^s \times 2^s$ *weight transformation matrix M* whose element m_{ij} is given by the inverse of B_{ij}. We construct a multifractal using this weight transformation matrix by the following procedure.

Consider a unit interval, which we label as 0 (singularity spectrum $f(\alpha)$ of the interval does not depend on the first choice of the label). Divide the interval into 2^s intervals with length $l = (1/2)^s$ and assign to the jth interval $[j/2^s, j/2^s + l] \equiv [l_j, l_j + l]$ ($j = 0, \dots, 2^s - 1$) weight m_{j0} and label that interval as j. The jth interval is further divided into 2^s equiva-

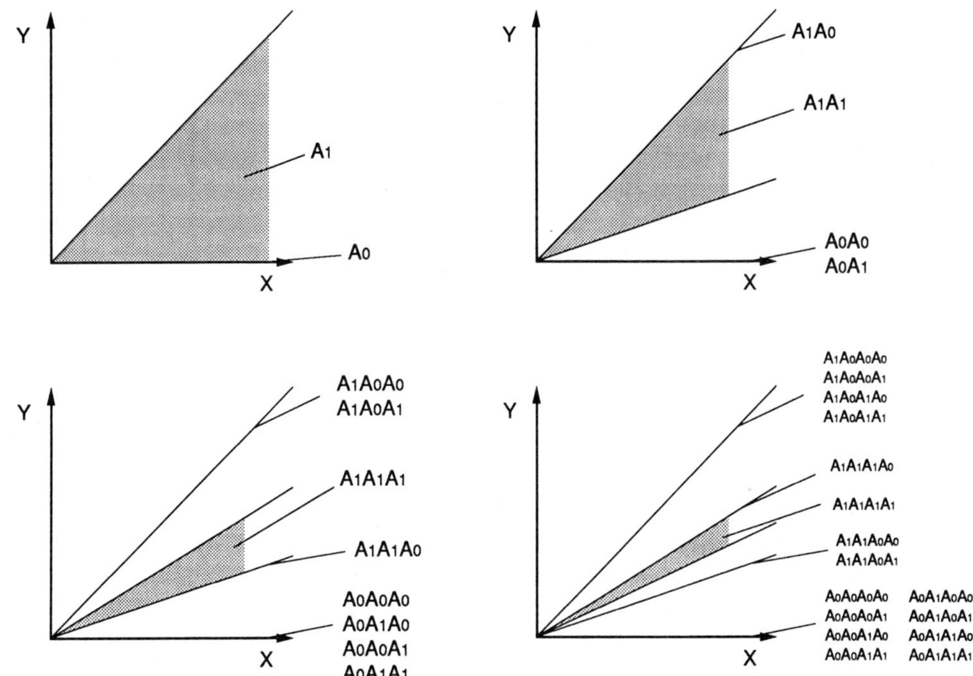

Fig. 9. Operation of multiple products of the matrices A_0 and A_1 confines the first quadrant to lines, if the sequence of A_i contains at least one A_0, and the dotted region converges to a line as multiplicity increases.

lent intervals. We assign to the ith interval $[l_j + (i/2^s)l, \; l_j + [(i+1)/2^s]l]$ $(i=0, \ldots, 2^s-1)$ weight $m_{ij}m_{j0}$ and label that interval as i. Repeating this procedure, we obtain the desired multifractal in the limit (see fig. 10). The weight transformation in this multifractal is not self-similar, and depends on both label i and label j like a Markov process. Therefore we call this weight transformation a *Markov weight transformation*.

The partition function of this multifractal is obtained by using a *partition matrix*, $\Gamma(q, \tau)$, whose ijth element is given by $\Gamma_{ij} = m_{ij}^q/l^\tau$, where $l = 1/2^s$. The partition function with 2^{ns} equivalent intervals is given by

$$(1\,1\,...1)\,(\Gamma(q,\tau))^n \begin{pmatrix} 1 \\ 0 \\ \vdots \\ 0 \end{pmatrix}.$$

From the definition of $\tau(q)$ given by (5.5), we have $\tau(q)$ such that the maximum eigenvalue of $\Gamma(q, \tau(q))$ is unity. Substituting $\tau(q)$ for (5.8) and (5.9), we obtain $f(\alpha)$, which approximates the dimension spectrum $H(\delta)$.

9. Simulation results

Here, we demonstrate efficacy of the Markov weight transformation approximation. We first calculate the dimension spectrum $H(\delta)$ of the LCA given in example 4. Fig. 11 shows the result of the calculation of $\Delta(\delta, 2^n, \epsilon)$. These graphs approach a real $H(\delta)$ as we take more time points t_n. However, they converge very slowly and still remain far from $H(\delta)$, even if we take millions of time points.

If we use the Markov weight transformation approximation, even the second step (i.e. $s=2$ in sec-

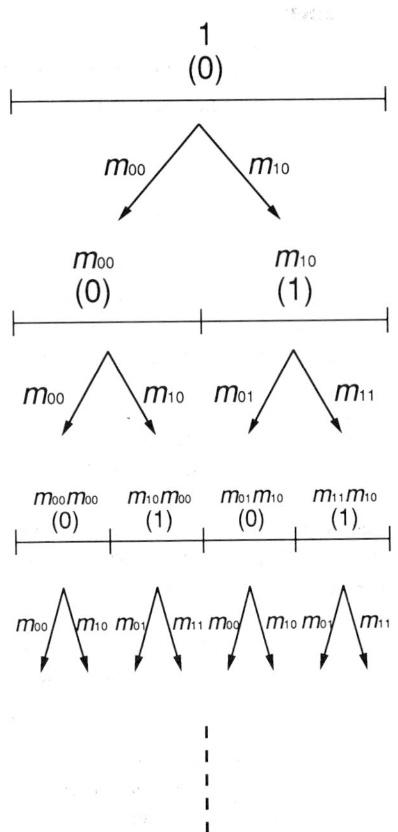

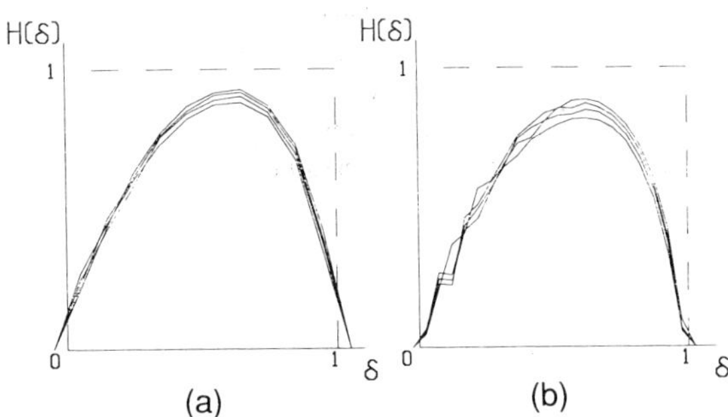

Fig. 10. A multifractal generated by the Markov weight transformation.

tion 8) gives a good approximation of $H(\delta)$ and it seems to converge almost completely at the third or fourth step (fig. 12a). The Markov weight transformation approximation gives quite good results for other prime-state LCA (figs. 12b, 12c).

We also calculate the dimension spectrum of a p^k-state LCA, where p is a prime and k is an integer greater than 1. In this case the Markov weight transformation approximation does not converge rapidly, as shown in fig. 12d. However, we can show that the dimension spectrum of the corresponding p-state LCA that evolves according to the transition rule (2.1) with mod p instead of mod p^k gives the same dimension spectrum as that of the original p^k-state LCA.

Singularity spectra of multifractals have often been calculated on the assumption that the measure is distributed in a self-similar manner as discussed in section 5. We compare this conventional method (here

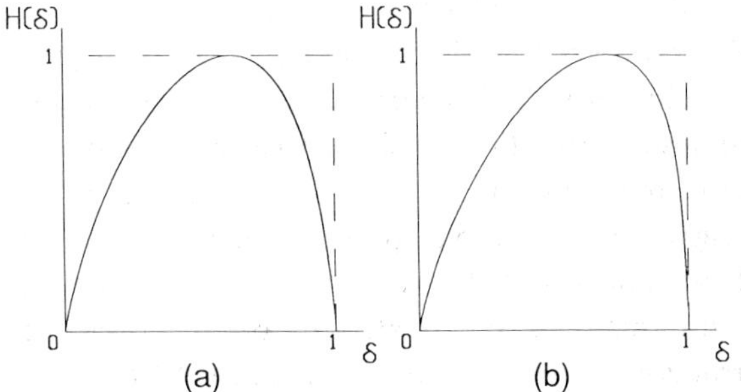

Fig. 11. Direct calculations of $\Delta(\delta, 2^n, \epsilon)$ of the LCA in example 1. $n=15, 17, 19, 21$. As n increases, the peak of the bell-shaped curves tend to shift upwards. (a) $\epsilon=0.1$, (b) $\epsilon=0.05$.

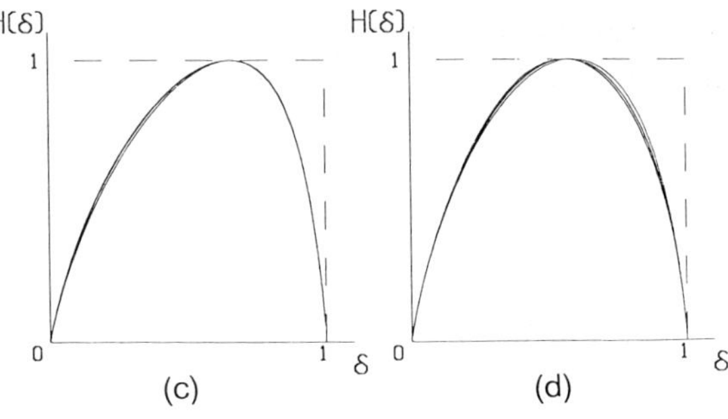

Fig. 12. Calculation of $H(\delta)$ by the Markov weight transformation approximation. $s=2$–5 in (a), (b), (c) and $s=2, 5, 8$ in (d). (a) $a_i^t = a_{i-1}^{t-1} + a_i^{t-1} + a_{i+1}^{t-1}$ mod 2; (b) $a_i^t = a_{i-1}^{t-1} + a_i^{t-1} + 2a_{i+1}^{t-1}$ mod 3; (c) $a_i^t = a_{i-1}^{t-1} + a_i^{t-1} + a_{i+2}^{t-1}$ mod 2; (d) $a_i^t = a_{i-1}^{t-1} + a_{i+1}^{t-1}$ mod 4. In (d), the graph shifts to the left as s increases.

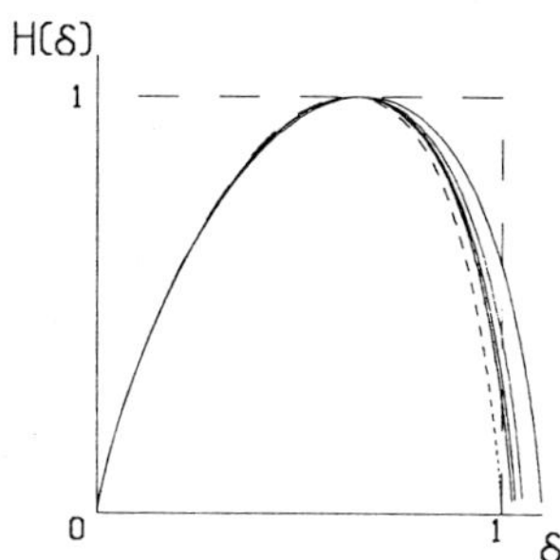

Fig. 13. Calculation of $H(\delta)$ by the self-similar weight transformation approximation for the LCA in example 1. $s=4, 8, 12, 16$. The graph shifts to the left as s increases. Calculation by the Markov weight transformation approximation at $s=5$ is also shown with a dashed line, which takes the correct value 0 at $\delta=1$.

called the *self-similar weight transformation approximation*) with the Markov weight transformation approximation. To calculate the dimension spectrum of an LCA by using the self-similar weight transformation approximation, the weight of the ith interval of the sth step is defined by the inverse of the number of non-zero cells at time i, i.e. $m_i = N(i)^{-1}$. $\tau(q)$ is given by $\sum m_i^q / l^\tau = 1$, where $l = 1/t_n = 1/2^s$. In fig. 13, the results from the self-similar weight transformation approximation at $s=7-16$ and the Markov weight transformation approximation at $s=5$ are shown. For small δ, the error of the self-similar weight transformation approximation is not large, while the error is notable for large δ. Convergence of $H(\delta)$ with respect to s is also not rapid in the self-similar weight transformation approximation. Therefore the Markov weight transformation approximation is more suitable to calculate dimension spectra of LCA.

Acknowledgement

I am very grateful to Dr. Nanako Shigesada for her careful reading of the manuscript and many helpful suggestions.

References

[1] S. Amoroso and G. Cooper, Tessellation structures for reproduction of arbitrary patterns, J. Computer System Sci. 5 (1971) 455–464.

[2] P. Billingsley, Ergodic Theory and Information (Wiley, New York, 1965).

[3] H. Furstenberg and H. Kesten, Products of random matrices, Ann. Statist. 31 (1960) 457–469.

[4] W.J. Gilbert, Complex bases and fractal similarity, Ann. Sci. Math. Québec 11 (1987) 65–77.

[5] T.C. Halsey, M.H. Jensen, L.P. Kadanoff, I. Procaccia and B.I. Schraiman, Fractal measures and their singularities: the characterization of strange sets, Phys. Rev. A 33 (1986) 1141–1151.

[6] H.G. Hentschel and I. Procaccia, The infinite number of generalized dimensions of fractals and strange attractors, Physica D 8 (1983) 435–444.

[7] B.B. Mandelbrot, An introduction to multifractal distribution of functions, preprint.

[8] T.J. Ostrand, Pattern reproduction in tessellation automata of arbitrary dimension, J. Computer System Sci. 5 (1971) 623–628.

[9] S. Takahashi, Self-similarity of linear cellular automata, J. Computer System Sci., to appear.

[10] Y. Tsujii, personal communication.

[11] S.J. Willson, Growth rates and fractional dimensions in cellular automata, Physica D 10 (1984) 69–74.

[12] S.J. Willson, Cellular automata can generate fractals, Discrete Appl. Math. 8 (1984) 91–99.

[13] S.J. Willson, The equality of fractional dimensions for certain cellular automata, Physica D 24 (1987) 179–189.

[14] S.J. Willson, Computing fractal dimensions for additive cellular automata, Physica D 24 (1987) 190–206.

[15] S.J. Willson, Calculating growth rates and moments for additive cellular automata, preprint.

Physica D 45 (1990) 49–62
North-Holland

A CELLULAR AUTOMATON
RULED BY AN ECCENTRIC CONSERVATION LAW

André M. BARBÉ

Department of Electrical Engineering, Katholieke Universiteit Leuven, K. Mercierlaan, 94, 3030 Leuven, Belgium

Received 9 January 1990
Revised manuscript received 19 March 1990

Binary difference fields are introduced. They are isomorphic to the state-time evolution of the rule-102 binary cellular automaton. A particular conservation law controls the evolution. The fields that emerge may exhibit highly structured complex patterns. An isomorphic structure involving a walk on a certain integer-valued fractal matrix is introduced.

1. Randomness and binary difference fields

The kind of CA-mechanism that will be presented here may seem a bit strange, but it originated in the information-theoretic problem of classifying binary sequences according to some degree of randomness [1]. The particular randomness-characteristic under consideration was based on the concept of "differentiating" a binary sequence. The *difference sequence* $D(S)$ of a binary sequence S is hereby defined as the sequence obtained by replacing every pair of successive values in S by a 1 if the binary values in the pair differ, and by a 0 if not. This is illustrated in the following example showing consecutive difference sequences $D_k(S)$ of a basic sequence S:

$$
\begin{aligned}
S = D_0(S): &\quad 1\ 0\ 0\ 1\ 1\ 0\ 1\ 1\ 1\ 0 \\
D_1(S): &\quad 1\ 0\ 1\ 0\ 1\ 1\ 0\ 0\ 1 \\
D_2(S): &\quad 1\ 1\ 1\ 1\ 0\ 1\ 0\ 1 \\
D_3(S): &\quad 0\ 0\ 0\ 1\ 1\ 1\ 1
\end{aligned}
$$

A complete set of difference sequences arranged as in the example above will be called a *binary difference field* (BDF). We will particularly be interested in difference fields of *unbounded size*.

With N_k^1 indicating the total number of 1-values in $D_k(S)$, the vector $(N_0^1, N_1^1, N_2^1, N_3^1, \ldots)$ can be considered as a characteristic related to a degree of randomness. A question arising quite nat-

urally in this context is whether there are different fields with the same characteristic vector $(N_0^1, N_1^1, N_2^1, \ldots)$. Or, stated otherwise, are there sequences S for which a *rearrangement* of some of their elements induces a mere rearrangement of elements on all difference levels? Or still, what fields do *conserve* $(N_0^1, N_2^1, \ldots)$ under a certain change ?

BDFs satisfying this kind of invariance are interesting for different reasons. First, they form a minority of cellular automata. Second, they may show intriguingly complex but structured patterns. Third, they are difficult to grasp: some deep mathematical problems seem to emerge. Last, they possibly provide a prototype of a broader class of problems. The production of such BDFs is dealt with in section 3. Existential and structural properties are discussed in section 4. Section 5 links these BDFs with a constrained walk on a particular fractal dust. But first, the problem under study will be stated as a simple generalization of the conservation law mentioned above.

2. Nets for taming BDFs and the problems they cause

The rowwise downward evolution of the difference sequences is isomorphic to the state-time evolution of the classical one-dimensional binary cel-

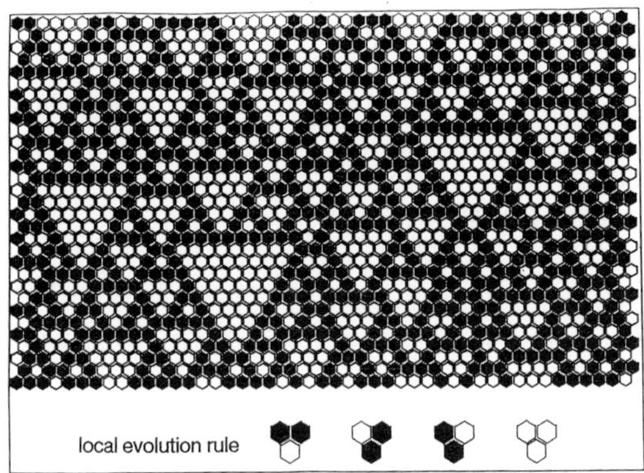

Fig. 1. A binary difference field.

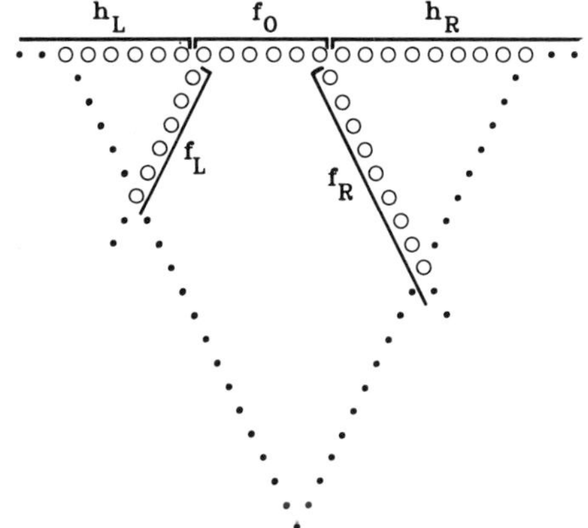

Fig. 2. The triples $(h_{\mathrm{L}}, f_0, h_{\mathrm{R}})$ and $(f_{\mathrm{L}}, f_0, f_{\mathrm{R}})$ that completely determine a BDF.

lular automation obeying rule 102 [2].

Fig. 1 shows an example, showing *occupied* (black) and *unoccupied* (white) cells in a triangular lattice. Black and white correspond to the binary values 1 and 0 respectively. Notice that the local evolution rule is of utmost simplicity.

Fig. 2 shows that a BDF is also completely determined by the state of a finite segment f_0 on the top row, together with the state of all cells on the adjacent *left-skew* and *right-skew columns* f_{L} and f_{R}. When f_{L} and/or f_{R} are forced to be periodic, the overall field has some striking *periodic* and *fractal properties* [3].

Now, select some cells on the top row of the BDF and change their value. If is required that the underlying field has to remain a BDF, it is clear that every row will have to change the value of some of its cells. The set of all these changing

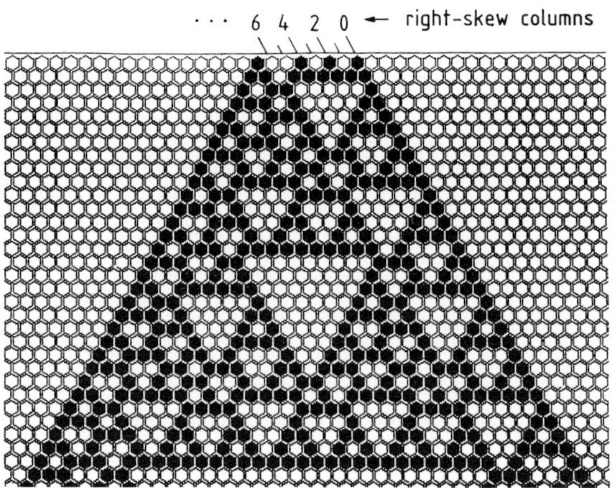

Fig. 3. A change propagation net (the set of black cells). The CPN seed $g_0 = 10010101$.

cells is called the *change propagation net* (CPN).

Fig. 3 gives an example. It shows that a CPN can be imbedded in a new difference field that is obtained by replacing all changing cells in the original field by occupied cells, and the nonchanging cells by empty ones. Notice that the complete CPN is determined by its top-row configuration g_0. This will be called the *CPN seed*.

We will now consider a CPN that *covers* a BDF. An example is shown in fig. 4. Only the part of the BDF confined to the CPN domain is shown. For convenience, the binary values $1/0$ have been replaced by $+/-$. On each row k, the numbers of occupied and nonoccupied BDF cells that are covered by the CPN are counted. These are represented by N_k^+ and N_k^- respectively. The difference $e_k \triangleq N_k^+ - N_k^-$ is the *occupational excess* of the BDF under the given CPN at row k. The CPN is meant to "tame" BDFs so that they will exhibit *rowwise constant* excess e under this CPN.

When $e = 0$, the original invariance problem presented in section 1 reappears. Making $e \neq 0$ seems to be the simplest possible generalization. The excess e is always an *even integer*, as the number of CPN cells on any row $k \neq 0$ is even.

Formalizing a bit, we state two *main problems*:

(i) Find the CPNs (specified by their seed g_0) for which there are binary difference fields of unbounded size with rowwise constant occupational excess e.

$$
\begin{array}{lll}
& & \text{free choice} \\
\text{row 0:} & \oplus\ \oplus\ \oplus\ +\ -\ -\ -\ \ominus\ \ominus\ +\ \ominus\ \ominus\ -\oplus & \downarrow \\
\text{row 1:} & \oplus\ -\ -\ \ominus\ +\ -\ -\ \ominus\ -\ \oplus\ \oplus\ -\ \ominus\ \oplus\ \ominus & * \\
& \ominus\ \oplus\ -\ \ominus\ \oplus\ +\ -\ \ominus\ \ominus\ \oplus\ -\ \oplus\ \ominus\ +\ +\oplus & * \\
& \oplus\ +\ \oplus\ \ominus\ +\ \ominus\ +\ \ominus\ -\ +\ \oplus\ \oplus\ +\ \oplus\ -\ \ominus\ \ominus & * \\
& \oplus\ \ominus\ \ominus\ +\ \oplus\ \oplus\ \oplus\ \oplus\ \ominus\ +\ \ominus\ -\ \ominus\ \ominus\ \oplus\ \ominus\ -\oplus & \\
& \ominus\ +\ -\ \oplus\ \ominus\ -\ -\ -\ +\ \oplus\ \oplus\ \ominus\ \ominus\ -\ +\ +\ \ominus\ \oplus\ \oplus & * \\
& \oplus\ \oplus\ +\ \oplus\ +\ \ominus\ -\ -\ +\ \ominus\ -\ +\ -\ \ominus\ +\ -\ \oplus\ +\ -\ominus & * \\
& \odot\ -\ \ominus\ \ominus\ \ominus\ \oplus\ \ominus\ -\ +\ \oplus\ \ominus\ +\ +\ \ominus\ \oplus\ +\ \oplus\ \ominus\ +\ \ominus\ \odot & : \text{ impossible to obtain} \\
& & e = 0
\end{array}
$$

Fig. 4. Evolutionary procedure for finding a partial solution to the (g_0, f_0, e) problem. The CPN cells are encircled. $+/-$ represent the occupational states of the BDF cells. $g_0 = 1\,1\,1\,0\,0\,0\,0\,1\,1\,0\,1\,1\,0\,1$, $f_0 = 1\,1\,1\,1\,0\,0\,0\,0\,0\,1\,0\,0\,0\,1$, $e = 0$.

(ii) Find all BDFs of unbounded size that have rowwise constant excess e under a given CPN.

Where the first problem concerns existential conditions, the second one is about producing solutions.

A couple (g_0, e) that produces a BDF satisfying the constant excess condition will be called a *productive seed*. Any particular BDF that is a solution to problem (ii) will be characterized by a triple (f_0, f_L, f_R) (recall fig. 2). f_0 is selected to be the top-row sequence of the BDF restricted to the domain of the covering CPN seed g_0. Such a *particular solution* will be represented by $P(f_0, f_L, f_R; g_0, e)$. Equivalent is the problem of finding particular solutions $P(f_L, f_R; g_0, f_0, e)$, where now (g_0, f_0, e) is considered as a given seed. Problem (i) can then be reinterpreted as a decision problem about what seeds (g_0, f_0, e) are indeed productive.

As yet, we have not been able to formulate a general answer to the existential problem. To the second problem, it is possible to obtain restricted-size solutions to a given (g_0, f_0, e) by the procedure that follows next.

3. Taming the BDFs

Consider fig. 4 of a combined (CPN, BDF) field. On any row k, the state of the CPN can be represented by the *binary string* g_k that is formed by placing a 1 at each cell that belongs to the CPN, and a 0 at the cells within the CPN domain that do not belong to the CPN. The state g_k evolves according to

$$g_{k+1} = 1 \circ Dg_k \circ 1, \tag{1}$$

where $\circ$ denotes concatenation, and Dg_k is the difference sequence of g_k.

In the same way, f_k is the binary string representing the occupational state of the BDF cells at row k that are confined to the CPN domain. With Df_k the difference sequence of f_k, the transition equation is

$$f_{k+1} = [0|1] \circ Df_k \circ [0|1]. \tag{2}$$

$[0|1]$ denotes that a selection has to be made between 0 and 1. Indeed, the occupational state of the two outer CPN cells has to be chosen so as to obtain the proper excess-value e for all (g_k, f_k) pairs. g_k and f_k will be called *costates*.

Consider the first transition from (g_0, f_0, e) to (g_1, f_1, e). The selection of the state of the outer cells at row 1 depends on the excess-contribution of the "inner" (Dg_0, Df_0) sequences. If this already equals the required e value, there is one degree of freedom for the outer cells: they have to be of opposite sign. In that case the choice is made randomly, by making the leftmost cell occupied with a certain probability. If the excess in (Dg_0, Df_0) equals $e \pm 2$, the choice is not free as the outer cells must both be assigned the same state. In case the excess of (Dg_0, Df_0) is $e \pm k$ with $k > 2$, there is no solution.

If a proper (g_1, f_1, e) is found, we can proceed in the same way to row 2, and so on.

When it is impossible to obtain the right excess-value at a certain row, one should go back to the last row where there was a free choice and swap the states of the edge cells on that row. This makes the choice there unfree. Then proceed further downwards until the next conflicting situation arises, whereby the retrograde procedure is initiated again.

Fig. 5. (a) A constant excess quaternary field (confined within CPN area); (b) the corresponding binary field; (c) the corresponding completed field. $(g_0, f_0, e) = (10111, 01011, 0)$. Size: 320×300 cells (width $\times$ depth).

Observe that the whole evolution of the CPN and the underlying BDF can also be considered as the evolution of a *quaternary cellular automaton* with possible states $-, +, \ominus, \oplus$. The excess is then just the difference between the number of $\oplus$ cells and $\ominus$ cells on each row.

Fig. 5a shows the structure of a CPN-confined field generated by a given productive seed (g_0, f_0). The CPN information is then removed, uncovering a part of the underlying BDF (fig. 5b). This field is then completed to the left and to the right (fig. 5c).

Numerous computer experiments suggest that there are two kinds of responses to an arbitrary (g_0, f_0, e) seed:

(i) *Finite-depth solutions*:

For *most* (g_0, f_0, e) seeds, the search procedure returns quickly to the top row, mostly after excursions to a row depth which is of the order of the seed size or even less. These *abortive* solutions will not be considered further.

(ii) *Potentially unbounded solutions*:

For some (g_0, f_0, e) configurations, a full-screen-size solution (about 300 rows) is found. In most cases, this involves a search time that may fairly exceed the number of rows in the image, even by several orders of magnitude. Although we still have only solutions of limited size, we are tempted to conjecture that they show a part of a potentially unbounded solution. This is corroborated by two facts:

(1) For initial seeds of comparable size, there is a clear distinction between the finite-depth solutions and the potentially unbounded solutions. The former are very limited in size as compared to the latter. There seems to exist a boundary level for the size of the finite-depth solutions. Once this level is crossed, an unbounded solution seems to emerge, albeit possibly with long times of indecisive wandering back and forth at certain zones before proceeding downward. Also, retrogressions may become quite important on several occasions, but they seem to be overcome in the long run. Fig. 6 shows a typical search history.

(2) Some of the obtained partial solutions show particular periodic patterns. Just this suggests that they can be extrapolated inductively.

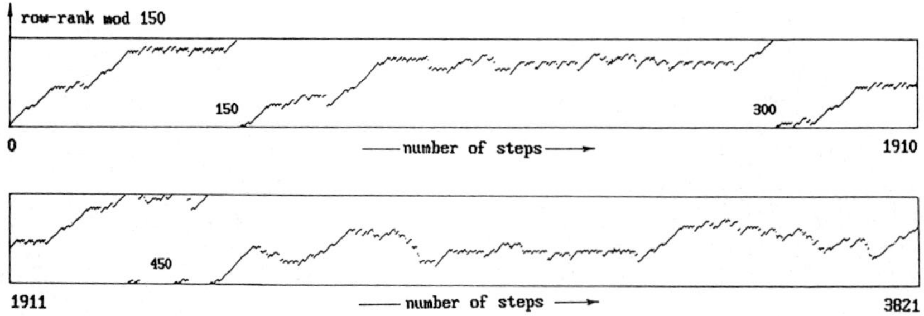

Fig. 6. A typical search history: evolution of solution size versus the number of steps.

4. The nature of the tamed BDFs

Fig. 7 shows some solutions for excess-values that are *nonzero*. The first three fields represent the complete set of solutions to one particular (g_0, e) seed. Observe that the left-skew CPN boundary f_L (recall fig. 2) in all these examples is either completely occupied (when $e > 0$) or empty (when $e < 0$). (This could have been the case for f_R instead, but looking at the field from the backside makes f_R interpretable as f_L or vice versa.) The other boundary f_R is periodic. It follows that the overall pattern is periodic along any skew column in a very particular way. More about this can be found in ref. [3]. No counterexample has been found to this feature of one-sidedness and full periodicity.

Fig. 8 shows solutions for different seeds $(g_0, f_0, e = 0)$. Fields with rowwise constant excess $e = 0$ under a given CPN will be called *rowwise balanced*. Apart from the completely arbitrary random solution in fig. 8a, the other solutions seem to be highly structured. Some of the solutions (figs. 8b, 8c, 8e, 8f) give rise to *periodic patterns* in the left- and right-hand outer field, after perhaps an initial *transient* zone. In these cases, the whole pattern seems to emerge from the collision of two periodic half-infinite sequences on the top row, creating a kind of interference pattern. Also, the eventually induced periodic patterns along f_L and f_R have the same period length. Notice that figs. 8a, 8c, 8e are solutions to the same CPN seed g_0. Moreover, figs. 8c and 8e are two solutions related to the same (g_0, f_0) seed. For the examples shown in figs. 8d and 5c, no overall periodicity seems to emerge, at least not for the limited-size solutions available. Nevertheless, these solutions too

are highly structured, showing a strange interweaving between *locally periodic* (crystalline) and *locally amorphous* structures. One really wonders whether this feature is not manifest on ever increasing scales, and whether it has anything in common with self-similar and overall nonperiodic sequences such as the *Morse–Thue* sequence which will be considered more closely in the next section. Or are they just part of a long transient ?

As for the *number* of excess-invariant BDFs that may show up, the observations can be summarized as follows:

(i) For most seeds (g_0, e), there is no solution (whatever the value of the *costate* f_0).

(ii) If there are solutions for (g_0, e), the number of costates f_0 for which (g_0, f_0, e) is productive is rather limited.

(iii) A productive (g_0, f_0, e) seed produces a unique solution if $e \neq 0$. This is not so when $e = 0$. In the latter case the solutions depend on the particular random sequences generated during the search procedure. As yet, nothing is known about their actual number.

A *general characterization* of productive seeds seems difficult to obtain, but some partial knowledge is available. A few of the properties with highest impact on the existential problem will be presented next. A more complete treatment and proofs can be obtained from the author.

Property 1 *For any solution to a productive CPN seed* (g_0, f_0), *there is a solution for the corresponding* $(g_0, -e)$ *seed. This is obtained by the actual change of all CPN-covered cells in the field produced by* (g_0, e).

Observe that the (g_k, f_k) pair at row k in a

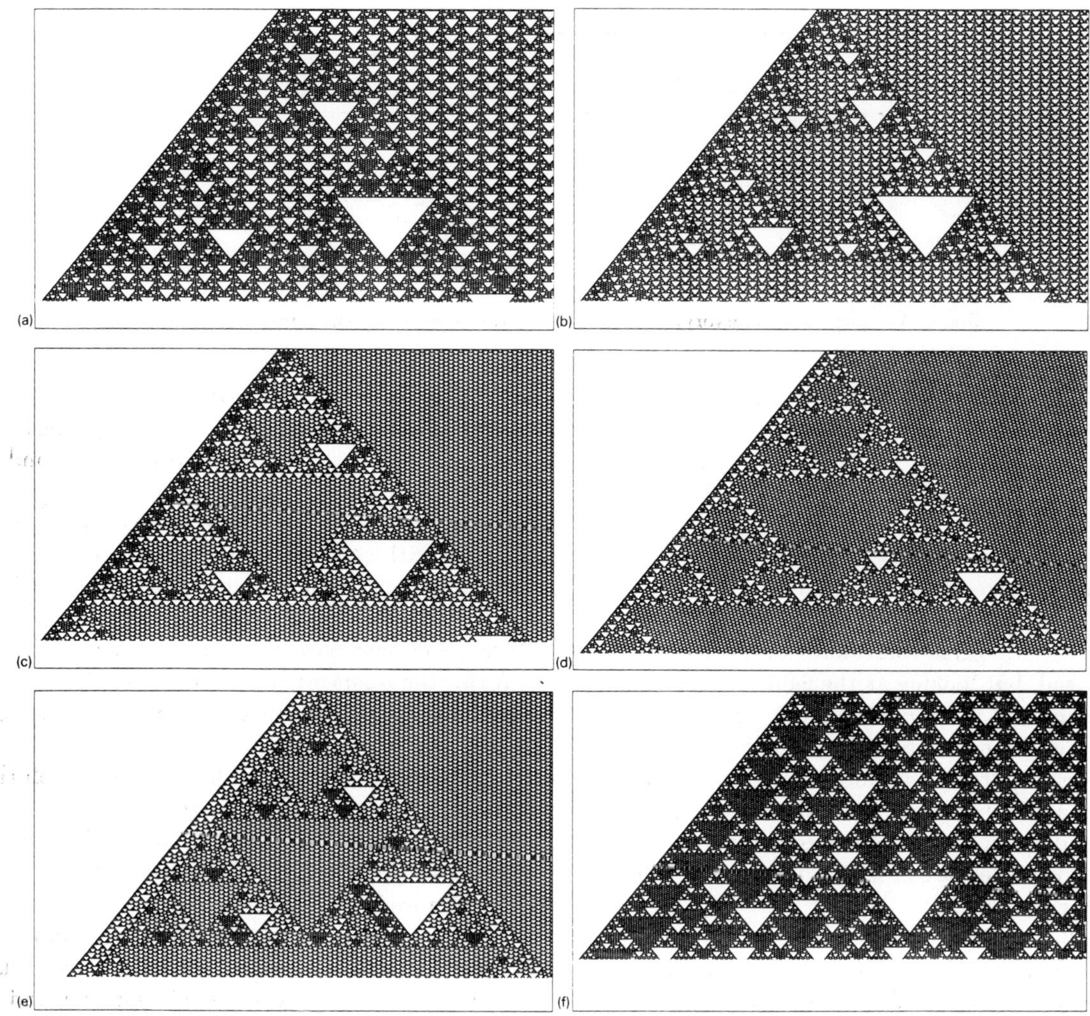

Fig. 7. Partial solutions for excess $e \neq 0$. (a)–(c) The solutions for $g_0 = 1(001)^7$, $e = 8$. (a) $f_0 = (100)^5 11101^4$; (b) $f_0 = (100)^4 101^6 01$; (c) $f_0 = (100)^4 1^4 001^4$, (d) $(g_0, f_0, e) = (10011000111, 10011001110, 4)$, (e) $(g_0, f_0, e) = (1(001)^9, (0)^{16}(110000)^2, -10)$, (f) $(g_0, f_0, e) = (1(001)^{15}, (100)^{10} 10110110(1)^8, 16)$, exponents refer to repeated concatenation.

solution field can be considered as the seed that produces the field obtained by removing all rows above row k in the original field. So, any productive seed initiates a chain of productive seeds of increasing rank and size.

But, is the converse also true? Can the solutions to any productive seed (g_k, f_k, e) be obtained from a possible *preseed*, i.e. from some (g_l, f_l, e) with $l < k$? If it were, we would only have to look for *primitive* seeds (i.e. the preseeds of smallest size that can generate a given seed (g_k, f_k, e)). Unfor-

tunately, we have to state :

Property 2 *The search for solutions produced by primitive seeds is insufficient to provide all the solutions for nonprimitive seeds.*

Property 3 *If $(g_0, e \neq 0)$ is a productive CPN seed, then $(g_0, e = 0)$ is also productive. A field produced by the first contains a solution to the last as a subset.*

The converse of property 3 seems to be true

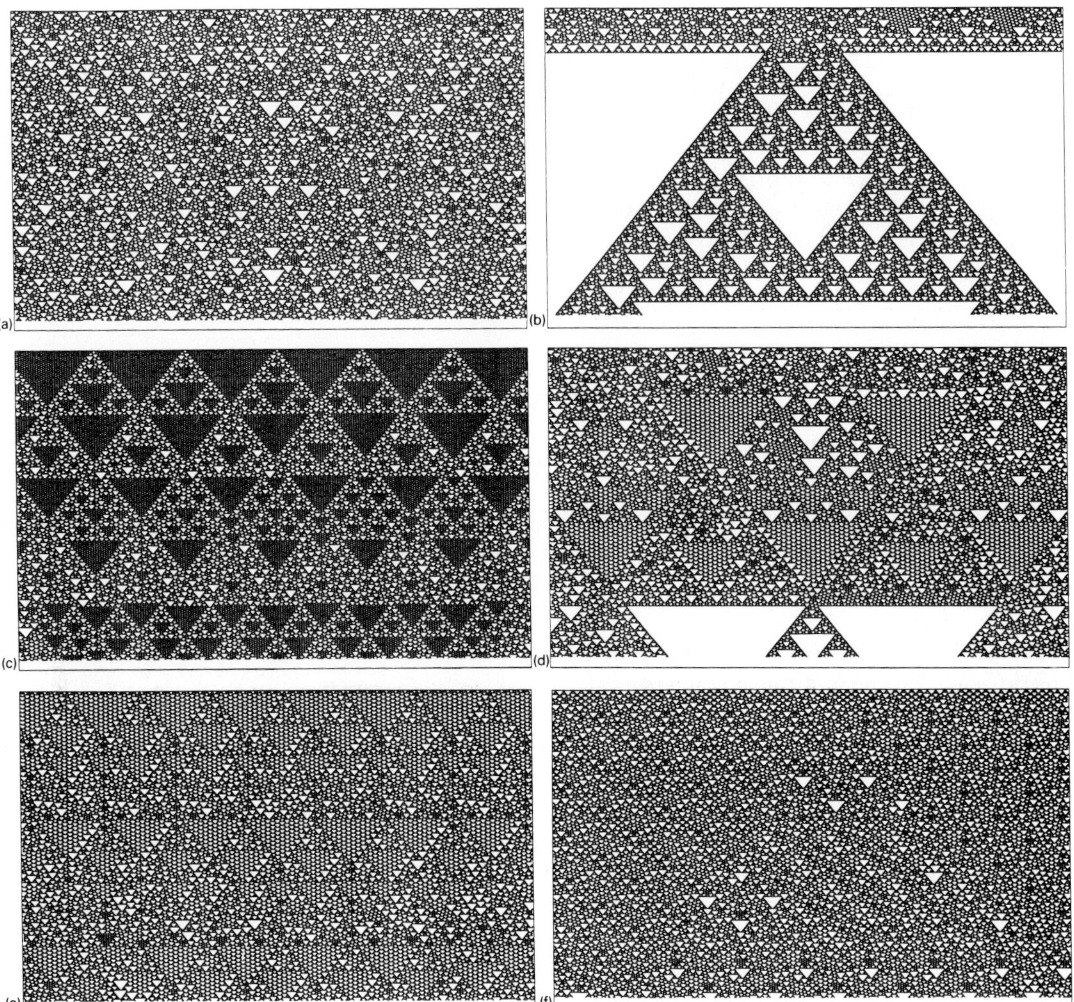

Fig. 8. Examples of rowwise balanced fields. (a) $(g_0, f_0) = (1001, 1000)$ random solution; (b) $(g_0, f_0) = (1001101, 0011110)$; (c) $(g_0, f_0) = (1001, 1010)$; (d) $(g_0, f_0) = (10111, 01011)$; (e) $(g_0, f_0) = (1001, 1010)$; (f) $(g_0, f_0) = (100001, 100010)$.

also: if $(g_0, e = 0)$ is productive, then there is *some* excess-value $e \neq 0$ for which $(g_0, e \neq 0)$ is productive.

Property 4 *If*

(a) $e = 0$,

(b) g_0 is a symmetric CPN seed,

(c) the CPN cells in g_0 show an antisymmetric occupation,

(d) the non-CPN cells in g_0 show a symmetric occupation,

then there are an infinite number of completely arbitrary solutions in the sense that there is a free choice at each row. This means that f_L in the solution $P(f_L, f_R; g_0, f_0, 0)$ is completely arbitrary, while f_R is the complement of f_L. It also follows that there is complete symmetry for the BDF cells not covered by the CPN.

The symmetry conditions mentioned together under the points (a), (b), (c), (d) above will be called the *cosymmetry* conditions on (g_0, f_0). This property is simply a consequence of the fact that cosymmetry on the top row induces cosymmetry on all subsequent rows. It seems natural then to speak about a *cosymmetric field*.

Table 1
Auxiliary sequences

n	0	1	2	3	4	5	6	7	8	9	10	11	12	13	14	15	...
$M(n)$	0	1	1	0	1	0	0	1	1	0	0	1	0	1	1	0	...
$T(n)$	2	4	6	2	6	4	2	4	6	4	2	6	2	4	6	2	...
$S(n)$	0	0	0	1	0	1	0	0	0	1	0	0	0	0	0	1	...
$I(n)$	1	0	1	0	0	1	0	1	1	0	1	0	0	1	0	1	...
$O(n)$	1	1	0	0	1	1	0	0	0	0	1	1	0	0	1	1	...
$\Gamma(n)$	3*	5	9*	15	17	23*	27	29*	33*	39	43*	45	51	53*	57	63*	...
$\Phi(n)$	0	2	6	4	12	14	10	8	24	26	30	28	20	22	18	16	...
$Q(n)$	0	1	1	2	1	2	2	3	1	2	2	3	2	3	3	4	...

The following example may clarify the situation.

$\downarrow$ symmetry axis

$$\oplus \ominus - + \oplus - \ominus + - \oplus \ominus$$
$$\ominus + \ominus + \oplus \oplus \ominus \oplus + \oplus + \oplus$$
$$\oplus \oplus \oplus \oplus \oplus + - + \ominus \ominus \ominus \ominus$$
$$\oplus - - - - \ominus + + \oplus - - - - \ominus$$

The example in fig. 8a was produced from a cosymmetric seed. Property 4 has a converse:

Property 5 *If there exists an arbitrary solution to $(g_0, f_0, 0)$, then (g_k, f_k) is cosymmetric for all k.*

The rule-102 CA is *additive*, and as a consequence the BDFs and CPNs are additive too. Let the couple (CPN, BDF) denote an arbitrary BDF covered by an arbitrary CPN. Two such couples can then be added by separately adding the CPNs and the BDFs modulo 2, and this results in a new (CPN, BDF) couple. A *k-shift self-addition* ($k \neq 0$), will be defined as the addition of a (CPN, BDF) couple with a horizontally shifted version of itself over k cells. In contrast to the joint (CPN, BDF) additivity, it is unfortunate that we have to formulate

Property 6 *(CPN, BDF) fields with rowwise constant excess are generally neither additive, neither k-shift self-additive in the sense that rowwise excess-values should remain constant after addition.*

There is one exception to this: cosymmetric (CPN, BDF) fields preserve rowwise constancy of excess under k-shift self-addition.

Concluding this section, we state:
– Cosymmetric fields completely lack "personality", and show no mystery (everything is provable).
– Fields with excess-values $e \neq 0$ are one-sided and fully periodic.
– Rowwise balanced fields that are not cosymmetric are most exotic and difficult to grasp.

5. Tamed BDFs walk on a fractal dust

In this section, the evolution of the costates (g_k, f_k) will be looked at through just a different pair of glasses: the binary numbers g_k, f_k will be seen as their equivalent decimal representations γ_k and φ_k. Or, to put it more abstractly, the evolution of the *ranking* of (g_k, f_k) will be observed. A dramatic change of the scene will be witnessed, and a complex but fascinating dance will be performed.

We will first look at the path that (γ_k, φ_k) can follow in a corresponding two-dimensional discrete phase space. Then the paths that *must* be taken will be considered. A number of evolutionary rules will be presented without proofs (these are rather cumbersome).

It will be helpful to introduce the *auxiliary sequences* as given in table 1. $M(n)$ is the sequence representing the parity of the set of digits in the binary representations of n (= modulo 2 sum of the digits). It is known in the literature as the *Morse–Thue* sequence. A multitude of references discussing properties of this sequence and its generalizations can be found in refs. [4,5]. $M(n)$ can be constructed recursively starting with 0. Each following step then appends the complement of the already existing part, thus doubling the sequence length.

$S(n)$ can be reconstructed from $M(n)$ by substituting each 0 in $M(n)$ by 0001 and each 1 in $M(n)$ by 0100. $T(n)$ is derived from $M(n)$ by replacing each 01, 11, 10, 00 in $M(n)$ respectively by 2, 4, 6, 4. $I(n)$ starts with 1, and in each size-doubling step, the already existing part is appended, alternatively with and without complementation. $O(n)$ to $Q(n)$ will be described further.

5.1. The possible steps of γ_k

Recall eq.(1), which governs the evolution of g_k as a binary number:

$$g_{k+1} = 1 \circ Dg_k \circ 1, \tag{3}$$

The number of CPN cells on any row being even, g_k has an even number of 1's in it, hence its parity is 0. As the rightmost digit in g_k is always 1, γ_k must be odd. It follows that the set Γ of *allowable* CPN states is given by those *odd* natural numbers n for which $M(n) = 0$. These are precisely the numbers n for which $S(n) = 1$. So $\Gamma = \{n : S(n) = 1\} = \{3, 5, 9, 15, \ldots\}$. The corresponding sequence is $\Gamma(n)$. Observe that $\Gamma(n+1) = 3 + \sum_{i=0}^{n} T(n)$ for $n > 0$, with $\Gamma(0) = 3$. CPN states γ_k that have no prestates in Γ (i.e., the γ_k for which a possible prestate g_{k-1} has an odd number of ones) will be called *irreducible*. The irreducible states in $\Gamma(n)$ are found to be those for which $I(n) = 1$. These have been marked by an asterisk in $\Gamma(n)$. Only the initial CPN state can be an irreducible one. It will be convenient to consider pairs of states in $\Gamma(n)$, as indicated below:

$$\Gamma(n) = (3^* \ 5)(9^* \ 15)(17 \ 23^*)(27 \ 29^*)(33^* \ 39)$$
$$(43^* \ 45)(51 \ 53^*)(57 \ 63^*) \ldots$$

Observe that each pair consists of just one reducible and one irreducible state, but in changing order. If the order (a^*, b) is denoted by 1, and (a, b^*) by 0, the orientations vary according to the sequence $O(n) \stackrel{\Delta}{=} I(2n)$. Define *block* B_j as the interval of integers from 2^{j-1} to $2^j - 1$, for $j = 1, 2, 3, \ldots$. Block $B_0 = [0]$, and j is the *block-rank*. The number of states in B_j is $\#B_j = 2^{j-1}$. Define $B\Gamma_j$ as the *set of allowable states* γ in block B_j. It has a number of states $\#B\Gamma_j = 2^{j-3}$.

We will now present an algorithm to find the

successor γ_{k+1} of γ_k. This will be as complex as the transition from g_k to g_{k+1} given by (3) is simple. First observe that if $\gamma_k \in B\Gamma_j$, then $\gamma_{k+1} \in B\Gamma_{j+1}$ (as a consequence of eq. (3)). From $B\Gamma_2$ up to $B\Gamma_6$, the following transitions are found:

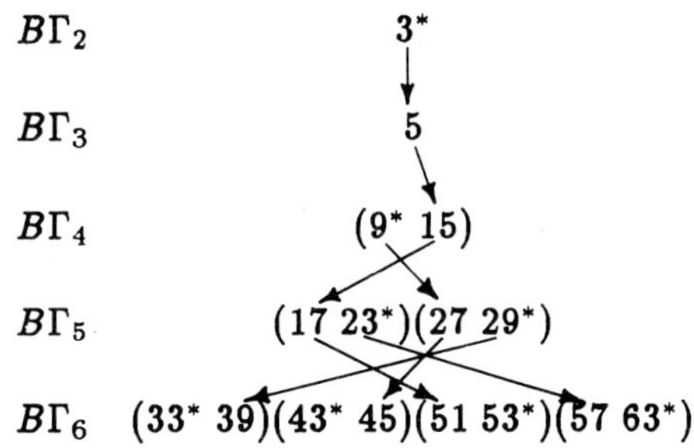

Now, each element in $B\Gamma_j$ will be mapped into the reducible state of a pair in $B\Gamma_{j+1}$. This mapping is obtained by a rather complex *shuffle* involving $\log_2(\#B\Gamma_j) = j - 3$ steps.

All operations to be introduced next will work upon sequences with a cardinality equal to an integer power of 2, so that they can be divided into 2, 4, 8, ... subsequences of equal length. Consider such a sequence S. A *splitting operator of rank r*, $D_r(S)$, will divide S into 2^r equally sized parts. Define an *ordering sequence Ω_r of rank r*. $\Omega_0 = L$, $\Omega_1 = LR$, $\Omega_2 = LRRL$, etc. In general, Ω_r corresponds to the first 2^r elements of the Morse–Thue sequence $M(n)$ where 0 has been replaced by L and 1 by R. L and R operate on sequences. L means that the elements of the sequence on which it operates should be rearranged in increasing order from right to left. R will rearrange the elements in increasing order to the right.

A *shuffle of rank r* on a sequence S is the combined operation $\Omega_r(D_r(S))$, whereby the 2^r elements of Ω_r operate in their order of appearance on the 2^r subsequences in $D_r(S)$, also in their proper order. Such a shuffle will be denoted by $\mathrm{Sh}_r(S)$.

The procedure for mapping $B\Gamma_j$ into $B\Gamma_{j+1}$ then goes as follows (it is illustrated in fig. 9 for the map $B\Gamma_7 \to B\Gamma_8$):
– Step 1: Consider the sequence S_j of *pairs* in $B\Gamma_j$. This sequence shows its natural order. Apply $\mathrm{Sh}_0(S_j)$, i.e. reverse the order of the pairs (the order of the γ-values inside the pairs should remain unchanged). Call the resulting sequence S_j^0.

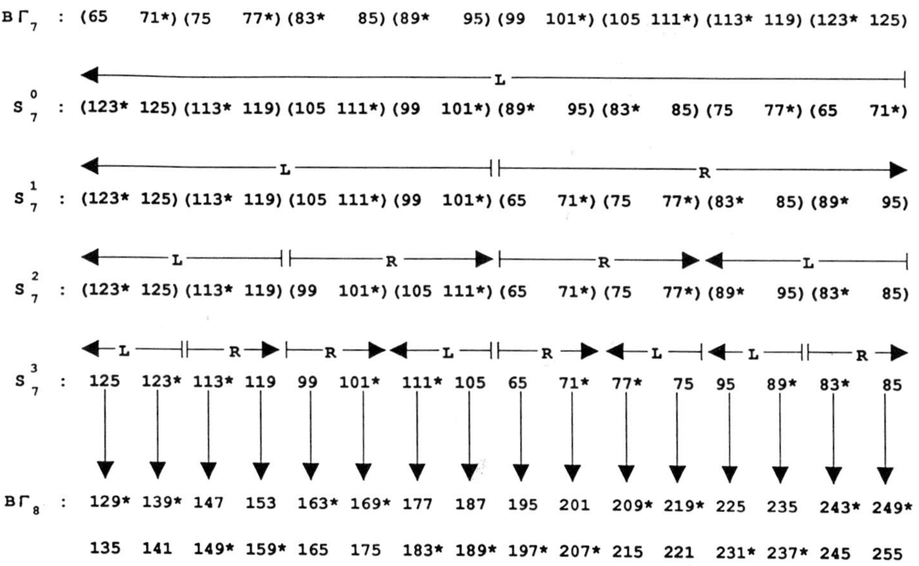

Fig. 9. The map from $B\Gamma_7$ into $B\Gamma_8$. The pairs in $B\Gamma_8$ are represented vertically. The actual mapping is into the reducible states (without an asterix).

— Step s: ($s = 2$ to $j - 4$). Obtain $S_j^{s-1} = \mathrm{Sh}_{s-1}(S_j^{s-2})$ (in-pair orientation unchanged).
— Step $j-3$: Reorient the elements inside the pairs by performing $\mathrm{Sh}_{j-4}(S_j^{j-5})$ to obtain $\hat{S}_j$.

$\hat{S}_j$ now represents a permutation of the *elements* in $B\Gamma_j$. They map in their order of appearance into *the pairs* in $B\Gamma_{j+1}$ in their natural order.

The mapping $B\Gamma_j \rightarrow B\Gamma_{j+1}$ has another property: when splitting up $B\Gamma_j$ and $B\Gamma_{j+1}$ in 4 equally sized subsequences indexed $\Gamma_j^1, \Gamma_j^2, \Gamma_j^3, \Gamma_j^4$ and $\Gamma_{j+1}^1, \Gamma_{j+1}^2, \Gamma_{j+1}^3, \Gamma_{j+1}^4$, it will always show the *constant permutation*

$$
\begin{array}{cccc}
\Gamma_j^1 & \Gamma_j^2 & \Gamma_j^3 & \Gamma_j^4 \\
\downarrow & \downarrow & \downarrow & \downarrow \\
\Gamma_{j+1}^3 & \Gamma_{j+1}^4 & \Gamma_{j+1}^2 & \Gamma_{j+1}^1
\end{array}
$$

Such constancy does not exist for the permutations of smaller subsequences in a shuffle. We learn from this that there is a large variation in position of the state γ inside the blocks $B\Gamma_j$ when it moves through the consecutive steps (i.e. when the CPN evolves along the rows). The state moves from a larger (smaller) value in $B\Gamma_j$ to one of the smaller (larger) values in $B\Gamma_{j+1}$. This will be referred to further as the *agitation* property.

5.2. The possible steps of φ_k

Recall that g_k is the CPN state at row k. Its number of digits is $l(g_k) \overset{\Delta}{=} \lfloor \log_2 \gamma_k \rfloor + 1$. f_k is the binary string representing the state of the underlying BDF cells at row k confined to the g_k range. f_k also has $l(g_k)$ digits. But, as opposed to g_k, f_k may have zeroes everywhere. Where $\gamma_k \in [2^{l(g_k)-1} + 1, 2^{l(g_k)} - 1]$ and belongs to block $B_{l(g_k)}$, φ_k may range from 0 to $2^{l(g_k)} - 1$. φ_k belongs to block B_{ρ_k}, with $\rho_k \overset{\Delta}{=} \lfloor \log_2 \varphi_k \rfloor + 1$. It follows that $\rho_k \leq l(g_k)$. When $\rho_k = l(g_k)$, the costates (γ_k, φ_k) have the *same block rank*. It will be said that φ_k is *block-aligned* with γ_k, and it occurs when f_k has a 1 as its leftmost digit.

As opposed to γ_k, which has to be a non-starred element of $\Gamma(n)$, φ_k may be *any* natural number. Recall that the evolution of its binary equivalent is given by eq. (2):

$$
f_{k+1} = [0|1] \circ D f_k \circ [0|1]. \tag{4}
$$

For φ_k, this will once again translate into a more complex scheme. First define the $\Phi(n)$ sequence (cf. table 1). It can be recursively constructed in a size-doubling way:
— step 0: start with 0;

− step j: take the already existing sequence, add 2^j to all its elements and append this in reverse order to the already existing part.

The transition from φ_k to φ_{k+1}, and from ρ_k to ρ_{k+1} is then obtained as follows:

− If φ_k is *not block-aligned* with γ_k, (i.e. $\rho_k < l(g_k)$), then

$$\varphi_{k+1} = [0|2^{l(g_k)}] + \Phi(\varphi_k) + [0|1], \qquad (5)$$

where the particular choice must correspond to the one in eq. (4). Selection of $2^{l(g_k)}$ in the first term causes alignment:

$$\rho_{k+1} = l(g_{k+1}) \quad (= l(g_k)+1). \qquad (6)$$

If 0 is selected instead, non-alignment remains:

$$\rho_{k+1} = \rho_k + 1 < l(g_{k+1}) \quad (= l(g_k)+1). \qquad (7)$$

− If (φ_k, γ_k) *are block-aligned* (i.e. $\rho_k = l(g_k)$), then

$$\varphi_{k+1} = [0|2^{l(g_k)}] + \Phi(2^{\rho_k} - \varphi_k - 1) + [0|1]. \qquad (8)$$

Selection $2^{l(g_k)}$ preserves alignment. If 0 is selected, the alignment is undone and

$$0 \leq \rho_{k+1} \leq \rho_k. \qquad (9)$$

We are now in a position that enables us to trace the path of (γ_k, φ_k), starting from an initial seed. A typical trajectory over the *block ranks* for increasing k is shown in fig. 10.

Among all possible paths that *can* evolve from a given initial seed, we now want to find those that guarantee constant excess.

5.3. The stepping-stones on which to walk

This section will mark the places of constant excess in the (γ_k, φ_k) space. These will serve as the stepping-stones on which the BDFs have to walk in order to survive.

The costates (g_k, f_k) can be considered as a pair of binary strings of equal length. N_k^+, the *number of occupied cells* on row k that are covered by the given CPN, can be found by counting the number of 1's in $g_k \otimes f_k$, where $\otimes$ denotes the bitwise AND-ing operation of g_k and f_k. In the same way, N_k^-, the number of nonoccupied CPN-covered cells is found by counting the 1's in $g_k \otimes \bar{f}_k$, where $\bar{f}_k$ is the bitwise complement of f_k.

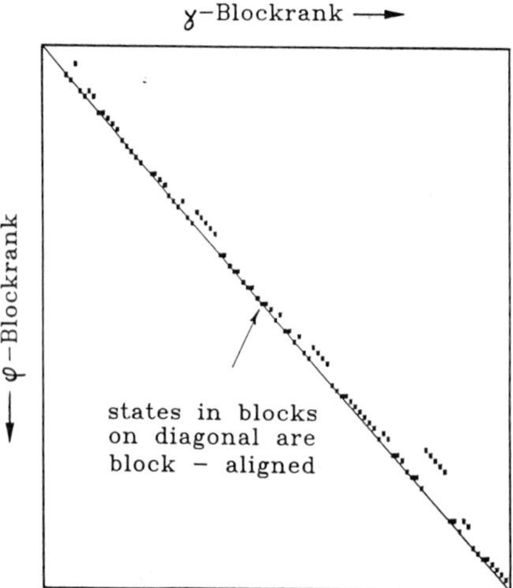

Fig. 10. Possible motion over the (γ, φ) blocks in the state space. Its direction is from left to right with a steadily increasing γ block rank. The motion is only possible in the right upper triangular part.

Table 2
AND-ing two binary sequences in decimal form.

φ	γ 0	1	2	3	4	5	6	7	...
0	0	0	0	0	0	0	0	0	...
1	0	1	0	1	0	1	0	1	...
2	0	0	2	2	0	0	2	2	...
3	0	1	2	3	0	1	2	3	...
4	0	0	0	0	4	4	4	4	...
5	0	1	0	1	4	5	4	5	...
6	0	0	2	2	4	4	6	6	...
7	0	1	2	3	4	5	6	7	...
⋮	⋮	⋮	⋮	⋮	⋮	⋮	⋮	⋮	⋱

How does this AND operation present itself on the decimal scene? The results of this operation for any two numbers (γ, φ) can be found in a matrix C as the element at the corresponding column- and row entries. The matrix C has the particular structure shown in table 2. It can be constructed in a size-doubling recursive way:

$$C_{m+1} = \begin{bmatrix} C_m & C_m \\ C_m & C_m + 2^m \end{bmatrix} \quad \text{with } C_0 = [0]. \quad (10)$$

$C_m + 2^m$ means that 2^m must be added to *all* elements in C_m. The size of C_m equals $2^m \times 2^m$.

The number of 1's in the binary representation

Table 3
The number of 1's in the sequence resulting from an AND operation.

φ	γ 0	1	2	3	4	5	6	7	...
0	0	0	0	0	0	0	0	0	...
1	0	1	0	1	0	1	0	1	...
2	0	0	1	1	0	0	1	1	...
3	0	1	1	2	0	1	1	2	...
4	0	0	0	0	1	1	1	1	...
5	0	1	0	1	1	2	1	2	...
6	0	0	1	1	1	1	2	2	...
7	0	1	1	2	1	2	2	3	...
$\vdots$	$\vdots$	$\vdots$	$\vdots$	$\vdots$	$\vdots$	$\vdots$	$\vdots$	$\vdots$	$\ddots$

of a number n can be found as the corresponding element in the sequence $Q(n)$ (cf. table 1). $Q(n)$ can also be obtained in a recursive way. Start with $Q_0 = 0$, and at each next stage, append the previous sequence augmented by 1. If all odd numbers in $Q(n)$ are replaced by 1 and the even ones by 0, the Morse–Thue sequence appears once more.

Performing the substitution $n \to Q(n)$ in the AND matrix C, results in the matrix C^+, which shows the number of 1's in the result of an $\otimes$ operation. C^+ can also be obtained using the recursion

$$C_{m+1}^+ = \left[\begin{array}{cc} C_m^+ & C_m^+ \\ C_m^+ & C_m^+ + 1 \end{array} \right] \quad \text{with } C_0^+ = [0]. \quad (11)$$

C^+ is shown in table 3.

The corresponding matrices for the $g \otimes \bar{f}$ operation are found by turning the former matrices upside down, as $\bar{f}$ takes the complementary value of f. This results in

$$C_{m+1}^- = \left[\begin{array}{cc} C_m^- & C_m^- + 1 \\ C_m^- & C_m^- \end{array} \right] \quad \text{with } C_0^- = [0]. \quad (12)$$

The excess-value of occupied cells over nonoccupied cells corresponding to a (γ, φ) pair can then be found in the excess-matrix $E \stackrel{\Delta}{=} C^+ - C^-$. Its recursive form is

$$E_{m+1} \stackrel{\Delta}{=} C_{m+1}^+ - C_{m+1}^-$$

$$= \left[\begin{array}{cc} E_m & E_m - 1 \\ E_m & E_m + 1 \end{array} \right] \quad \text{with } E_0 = [0].$$

Part of the excess-matrix is shown in table 4. E_m displays all integers from $-m$ to m, and these are distributed over the matrix in a very intricate way. Fig. 11 shows the distributions for different excess-values in E_9 (matrix dimension 512×512). The *self-similarity* is apparent, and these distributions remain quite similar for growing matrix dimensions. In the unbounded limit, these subsets form a *fractal dust* [6] with fractal dimension 2 (!). The overall matrix exhibits several interesting scaling and recursive properties. Notice the symmetry between upper half and lower half

Table 4
The excess-matrix E_4.

φ	γ 0	1	2	3	4	5	6	7	8	9	10	11	12	13	14	15
0	0	-1	-1	-2	-1	-2	-2	-3	-1	-2	-2	-3	-2	-3	-3	-4
1	0	1	-1	0	-1	0	-2	-1	-1	0	-2	-1	-2	-1	-3	-2
2	0	-1	1	0	-1	-2	0	-1	-1	-2	0	-1	-2	-3	-1	-2
3	0	1	1	2	-1	0	0	1	-1	0	0	1	-2	-1	-1	0
4	0	-1	-1	-2	1	0	0	-1	-1	-2	-2	-3	0	-1	-1	-2
5	0	1	-1	0	1	2	0	1	-1	0	-2	-1	0	1	-1	0
6	0	-1	1	0	1	0	2	1	-1	-2	0	-1	0	-1	1	0
7	0	1	1	2	1	2	2	3	-1	0	0	1	0	1	1	2
8	0	-1	-1	-2	-1	-2	-2	-3	1	0	0	-1	0	-1	-1	-2
9	0	1	-1	0	-1	0	-2	-1	1	2	0	1	0	1	-1	0
10	0	-1	1	0	-1	-2	0	-1	1	0	2	1	0	-1	1	0
11	0	1	1	2	-1	0	0	1	1	2	2	3	0	1	1	2
12	0	-1	-1	-2	1	0	0	-1	1	0	0	-1	2	1	1	0
13	0	1	-1	0	1	2	0	1	1	2	0	1	2	3	1	2
14	0	-1	1	0	1	0	2	1	1	0	2	1	2	1	3	2
15	0	1	1	2	1	2	2	3	1	2	2	3	2	3	3	4

for opposite excess-values. Values in one column are either all odd or all even. Column γ shows a periodic pattern with period $2^{\lceil \log_2(\gamma+1) \rceil}$. A more complete treatment can be found in ref. [7].

5.4. Watching the reflected scene

Observing the trajectory (γ_k, φ_k) only, gives but a partial view on the scene. Indeed, a (CPN, BDF) couple can be looked at from both the front and the backside. If the binary-shaped costates (g_k, f_k) appear at one side, the other side shows the *mirror image* (g_k^m, f_k^m), where g^m denotes the left–right reflection of g. It is clear that both decimal-form *mirror states* (γ_k, φ_k) and $(\gamma_k^m, \varphi_k^m)$ have to walk on the same excess-matrix. They both move ac-

cording to the same mechanisms studied in sections 5.1 and 5.2, but they generally start from different initial positions.

Both motions are not independent: the one completely determines the other. Nevertheless, they may behave quite differently in a way that depends on the specified excess-value.

First, consider a *positive* excess (e.g. fig. 11, $e = 4$). Because the concentration of the dust is highest along the main diagonal blocks, and the (γ, φ)-motion is restricted to blocks in the right upper-triangular part (diagonal blocks included), it seems reasonable to expect that the motion of proper solutions should occur along the diagonal blocks. This would indeed maximize the chance of staying on a dust particle. So, we expect the

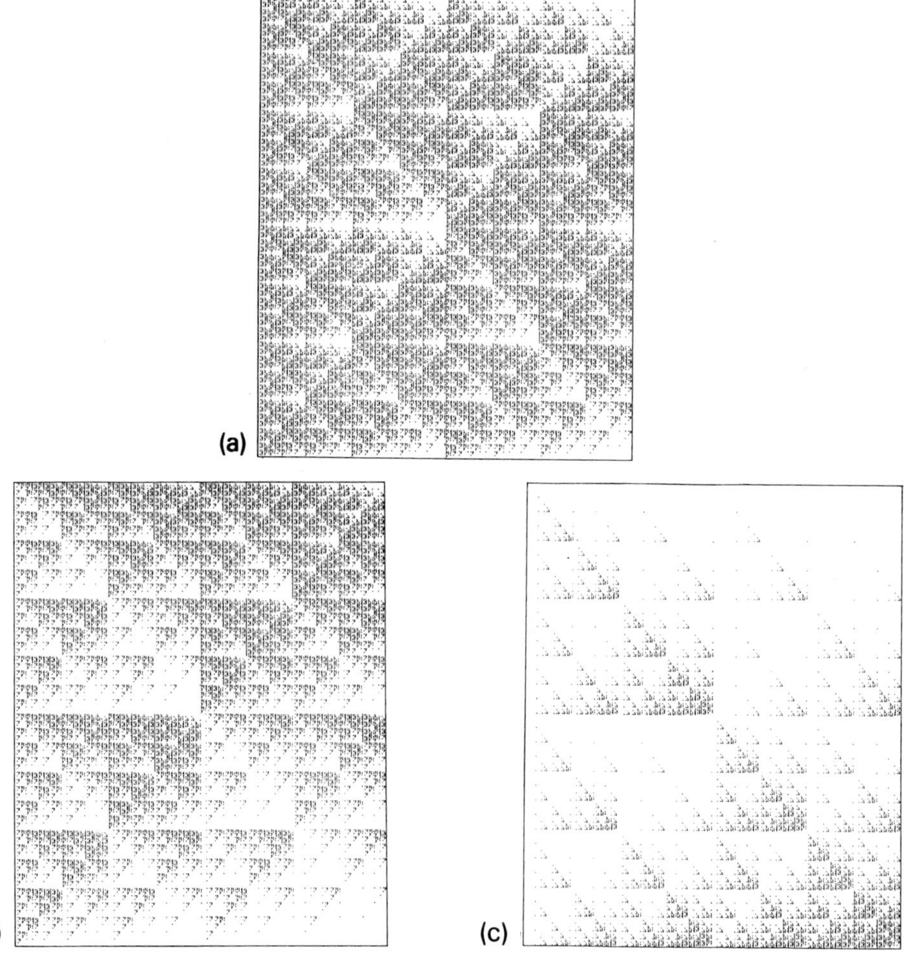

Fig. 11. Distribution for different excess-values: (a) $e = 0$, (b) $e = -2$, (c) $e = 4$.

solutions to have aligned costates at any time.

This is exactly what happens, but only for one of the mirror states. The motion of the second mirror state shows a periodic alignment. The symmetry is completely broken, just as in the BDF patterns themselves.

In the same way, it seems plausible to find no alignment of the costates when $e < 0$. Indeed, as the example of fig. 11 ($e = -2$) shows, the density of the dust is highest above the diagonal blocks. But here again, this only occurs for (γ_k, φ_k). $(\gamma_k^m, \varphi_k^m)$ is again periodically aligned.

For an *excess* $e = 0$, there is a more regular spread of the dust in all regions of the (γ, φ) space. This makes it less surprising that costates are not exclusively aligned or not. Moreover, symmetry seems to be reestablished here.

In this context, cosymmetric seeds have identical γ_k and γ_k^m. The solutions they produce have the property that costates (γ, φ) may become aligned or not in a completely arbitrary way and this in a complementary manner for the two mirror states: they will always remain on the dust associated with $e = 0$.

6. Concluding remarks

The nature of the evolutionary patterns of a very simple binary cellular automaton that also satisfies the more global condition of rowwise excess conservation has been discussed.

It has been shown that solutions exist for a minority of particular initial seeds. In the most general case (absence of cosymmetry), the emerging fields are complex, but highly structured. This is what makes them particularly interesting.

A number of properties could only be presented in a conjectural status, as they were based on observations of partial computer results. One should keep in mind that the solutions shown have negligible size as compared to the conjectured unboundedness. It is presently not known if it will ac-

tually be possible to extend the solutions beyond any limit, nor how the partial solutions already obtained will change eventually while continuing to search for an extended solution. The possibilities for thorough mathematical analysis have to be investigated further.

Finding a general answer to the formulated problems is not made easier by the fact that excess-values on a given row starting from an arbitrary initial (g, f) seed are not readily expressible, and that solutions are generally non-additive. An attempt for a new attack on the problem was given by the isomorphism with an evolution on a fractal dust, which gave a certain intuitive understanding of some results.

So, apart from possible generalizations to other automata rules, finding more complete answers to the excess-invariant problem remains a challenging task. Meantime, obtaining views of the overall structure of partial solutions much beyond the size available yet, especially for solutions which do not seem strictly periodic, remains a fascinating undertaking.

References

[1] A. Barbé, in: Proceedings of the 8th Symposium on Information Theory in the Benelux, ed. D. Kleima (Werkgemeenschap voor Informatie- en Communicatietheorie, Enschede, The Netherlands, 1987) p. 21.

[2] S. Wolfram, Statistical mechanics of cellular automata, Rev. Mod. Phys. 55 (1983) 601–644.

[3] A. Barbé, Periodic patterns in the binary difference field, Complex Systems 2 (1988) 209–233.

[4] M.R. Schroeder, Number Theory in Science and Communication, 2nd. Ed. (Springer, Berlin, 1986) pp. 315–324.

[5] M. Queffélec, Substitution Dynamical Systems – Spectral Analysis, Lecture Notes in Mathematics, Vol. 1294 (Springer, Berlin, 1980).

[6] B. Mandelbrot, The Fractal Geometry of Nature (Freeman, San Francisco, 1982).

[7] A. Barbé, Fractal matrices, Internal Report ESAT, Katholieke Universiteit Leuven (1990).

Physica D 45 (1990) 63–74
North-Holland

BOOLEAN DERIVATIVES ON CELLULAR AUTOMATA

Gérard Y. VICHNIAC

Plasma Fusion Center, Massachusetts Institute of Technology, Cambridge, MA 02139, USA

Received 13 March 1990
Revised manuscript received 28 March 1990

Boolean derivatives are applied to cellular automata. They extract the linear part of arbitrary rules, and provide an intrinsic classification. Derivatives are applied to tracing complexity growth and to lattice gases. A complete table of derivatives is given for one-dimensional elementary cellular automata.

1. Introduction

This paper investigates the applications of the notion of Boolean derivative to cellular automata. It stems from two motivations. First, one is interested in exploring all possible connections between cellular automata and concepts and tools of calculus. This interest is enhanced by the recent success in modeling with cellular automata certain solutions of the Navier–Stokes equations [5,9,1,8]. The surprising ability of binary objects of representing in a non-numerical way paradigms of continuum mathematics (e.g. differential equations and wave motion) encourages the search for new fruitful connections between discrete and continuum mathematics.

A second motivation for this study is provided by an interest in new tools for cellular automata "zoology," i.e. for the study of cellular automata for their own sake. In particular, the notion of Boolean derivative permits some characterization of the global behavior of cellular automata in terms of their local definition. In more precise way, a Boolean cellular automaton with N cells is a fully discrete dynamical system whose evolution is given by the iterations of a *global mapping F*:

$$F : \{0,1\}^N \mapsto \{0,1\}^N. \tag{1}$$

The homogeneity and the locality of cellular automata permit a compact description of F in terms of a *local transition rule f*,

$$f : \{0,1\}^n \mapsto \{0,1\}, \tag{2}$$

that maps the occupancies at time t of a *neighborhood* (of size n, with $n \ll N$) around any cell i to the next value x_i^{t+1} of that cell. For example, in peripheral one-dimensional cellular automata, the neighborhood is of size $n = 3$ and it consists of the considered cell and its two immediate neighbors. One has then

$$x_i^{t+1} = f(x_{i-1}^t, x_i^t, x_{i+1}^t). \tag{3}$$

Since in general the rule f is fully described by a look-up table of length 2^n, all its properties can be readily obtained by simple inspection of this table. Deriving properties of F from those of f is in general a difficult problem. However, the Boolean partial derivatives of f are non-trival attributes that can be used for characterizing F. We can indeed define the *Boolean derivative F'* of F as the $N \times N$ Jacobian matrix with elements

$$F'_{ij} = \partial x_i^{t+1} / \partial x_j^t \,, \tag{4}$$

where x_i^{t+1} is expressed with a local rule f as in eq. (3). The partial derivative in this expression is in turn defined as

$$\partial f / \partial x_j = f(x_1, \ldots, \bar{x}_j, \ldots x_n) \oplus f(x_1, \ldots, x_j, \ldots x_n), \tag{5}$$

where $\oplus$ is the Exclusive OR (XOR) Boolean operation, and $\bar{x}_j = x_j \oplus 1$ is the binary complement of x_j. In words, the last two equations define a matrix element F'_{ij} as 1 if varying site j at time t affects site i at time $t+1$, and as 0 otherwise.

2. Examples: rules 150 and 30

Before going into the properties of the Boolean derivative in more detail, let us give right away two representative examples.

Rule 150

One of the best known cellular automata is the "sum mod 2", or "XOR" automaton, introduced in the 1960's by Fredkin (see e.g. ref. [7]). The new value x_i^{t+1} of each cell is the parity, or sum modulo 2 of values in the pre-defined neighborhood. In peripheral one-dimensional cellular automata, the XOR rule reads

$$x_i^{t+1} = x_{i-1}^t \oplus x_i^t \oplus x_{i+1}^t. \tag{6}$$

In Wolfram's labeling convention [19], the rule has number 150. The partial derivative (5) of this rule with respect with x_{i-1} is (omitting the time superscript)

$$\partial f_{(150)} / \partial x_{i-1} = (\bar{x}_{i-1} \oplus x_i \oplus x_{i+1}) \oplus (x_{i-1} \oplus x_i \oplus x_{i+1}). \tag{7}$$

The terms x_i and x_{i+1} appear twice and cancel out, leaving $\partial f_{(150)} / \partial x_{i-1} = \bar{x}_{i-1} \oplus x_{i-1} = 1$. By symmetry, the two other partial derivatives are also 1. For a finite chain with circular boundary conditions, the Jacobian matrix (4) takes the form (where only nonzero elements are indicated):

$$F'_{(150)} = \begin{pmatrix} 1 & 1 & & & & & 1 \\ 1 & 1 & 1 & & & & \\ & 1 & 1 & 1 & & & \\ & & & \ddots & & & \\ & & & & 1 & 1 & 1 \\ 1 & & & & & 1 & 1 \end{pmatrix}. \tag{8}$$

The matrix is tridiagonal because the neighborhood is of size $n = 3$ – an expression of the *local* nature of cellular automata – except for the off-diagonal corner elements, which embody the circular boundary conditions. The 1's express the linear dependence of the rule with respect to each of its arguments (if an argument is complemented, so is $f_{(150)}$). As in the continuum case, a linear mapping is identical to its first derivative and its Hessian vanishes.

Rule 30

The rule defined as

$$f_{(30)} = x_{i-1} \oplus (x_i + x_{i+1}) \tag{9}$$

has been studied in detail by Wolfram for its remarkable and not yet fully understood ability to generate random-looking sequences [20]. As above, $\partial f_{(30)}/\partial x_{i-1} = 1$ and $\partial f_{(30)}/\partial x_i$ is obtained by expressing the $+$ (Boolean OR) in the parentheses in terms of $\oplus$'s:

$$x_i + x_{i+1} = x_i \oplus x_{i+1} \oplus x_i x_{i+1}, \tag{10}$$

and definition (5) gives

$$\partial f_{(30)}/\partial x_i = (x_{i-1} \oplus \bar{x}_i \oplus x_{i+1} \oplus \bar{x}_i x_{i+1}) \oplus (x_{i-1} \oplus x_i \oplus x_{i+1} \oplus x_i x_{i+1}) = \bar{x}_{i+1}. \tag{11}$$

(The last equality is obtained by canceling the terms appearing twice, factorizing out x_{i+1}, and using $\bar{x}_i \oplus x_i = 1$.) By symmetry, one has $\partial f_{(30)}/\partial x_{i-1} = \bar{x}_i$. For a chain of length N with circular boundary conditions, the Jacobian matrix for this rule is thus

$$F'_{(30)} = \begin{pmatrix} \bar{x}_2 & \bar{x}_1 & & & & & 1 \\ 1 & \bar{x}_3 & \bar{x}_2 & & & & \\ & 1 & \bar{x}_4 & \bar{x}_3 & & & \\ & & & \ddots & & & \\ & & & & 1 & \bar{x}_N & \bar{x}_{N-1} \\ \bar{x}_N & & & & & 1 & \bar{x}_1 \end{pmatrix}. \tag{12}$$

This expression means that x_i^{t+1} is unconditionally sensitive to a complementation of x_{i-1}^t, but it is sensitive to a complementation of x_i^t only if x_{i+1}^t is zero, and *vice versa*.

3. Interpretation of the Boolean derivative

Let us find out, by analogy with the continuum case, on what objects does the Jacobian matrix F' apply. In analysis, the derivative at $\boldsymbol{x}$ of a mapping $F : \mathrm{R}^N \mapsto \mathrm{R}^N$ is defined as the linear operator F' such that

$$F(\boldsymbol{x} + \boldsymbol{h}) - F(\boldsymbol{h}) = F'(\boldsymbol{x}) \cdot \boldsymbol{h} + o(\boldsymbol{h}), \tag{13}$$

where, as indicated by the Landau symbol o, the last term vanishes faster than $\|h\|$ when $\boldsymbol{h} \to \boldsymbol{0}$. In this formula, F' acts on an increment $\boldsymbol{h}$. By analogy, we can state that the Boolean F' acts on an "small" Boolean increment $\boldsymbol{h}$, made, say, of a single 1 in a background of 0's. For fully linear (or rather, additive) rules such as 150, evolutions of difference patterns (from two initial conditions differing only by $\boldsymbol{h}$) are the same as evolutions starting with a single seed – cf. tables 4 and 5 in the Appendix of Wolfram's book [21]. As we saw in the last section, $F = F'$ for rule 150; linear rules are identical with their derivatives. More generally, F' can be viewed as extracting the linear part of arbitrary mappings.

4. Derivatives of Boolean functions

The notion of derivative for Boolean functions is not new. It has been introduced by Akers in 1959 as a tool for the synthesis of Boolean functions and circuits. The monograph by Thayse [14] gives an extensive bibliography of works prior to 1981. Recent applications of Boolean derivatives include Fogelman's analysis of Kauffman's notion of *forcing structures* in random Boolean networks [4], Robert's study of convergence properties of iterated mappings [12], Reif's simulation of stuck-at faults in VLSI circuits [11], and Boyack's analysis of the complexity and cryptographic value of Boolean functions [2].

As we shall now review the elementary properties of Boolean derivatives, the present paper is self-contained. We shall give in section 6 several methods of computing derivatives for cellular automata. The following properties derive immediately from the definition (5) of the partial derivative:

(i) *The derivative of a constant function is zero.*

(ii) $\partial(x_k)/\partial x_k = x_k \oplus \bar{x}_k = 1$.

(iii) *Additivity:* $\partial(f_1 \oplus f_2)/\partial x_k = \partial f_1/\partial x_k \oplus \partial f_2/\partial x_k$.

(iv) *Derivative of AND:* $\partial x_k(x_k x_l)/\partial x_k = (x_k \oplus \bar{x}_k)x_l = x_l$.

(v) $\partial f/\partial x_k$ is independent of x_k. Indeed, $\partial^2 f/\partial x_k^2 = 0$ because f is obviously invariant under *two* complementations of any variable. Moreover, mixed second derivative are equal. In other words, the Hessian of f has a zero diagonal and it is symmetric.

(vi) *Derivation and complementation:*

– complementation of the function:

$$\frac{\partial}{\partial x_k}\bar{f} = \frac{\partial}{\partial x_k}(f \oplus 1) = \frac{\partial}{\partial x_k}f, \tag{14}$$

– derivation with respect to the complemented variable:

$$\frac{\partial}{\partial x_k}f = \frac{\partial}{\partial \bar{x}_k}f, \tag{15}$$

– derivation and variable complementation commute:

$$\frac{\partial}{\partial x_k}f(x_k, \bar{x}_l, x_m) = \left[\frac{\partial}{\partial x_k}f(x_k, x_l, x_m)\right]_{x_l \leftarrow \bar{x}_l}. \tag{16}$$

(vii) *Chain rule*: it applies along the "light cone", but not in general – see section 7.

(viii) *Leibniz rule*: it takes the form

$$\frac{\partial}{\partial x_k}(fg) = \left(\frac{\partial f}{\partial x_k}\right)g \oplus f\left(\frac{\partial g}{\partial x_k}\right) \oplus \left(\frac{\partial f}{\partial x_k}\right)\left(\frac{\partial g}{\partial x_k}\right). \tag{17}$$

Notice the correction term and the $\oplus$-arithmetic – see section 10 for a use of this rule.

5. Weights of derivatives

It is natural to define the *gradient* of a function as the vector with elements made of partial derivatives in the proper order [2]. For example, the gradient of a rule defined with eq. (3) is

$$\nabla f = \left(\frac{\partial f}{\partial x_{i-1}}, \frac{\partial f}{\partial x_i}, \frac{\partial f}{\partial x_{i+1}}\right). \tag{18}$$

The *weight* of a Boolean function is usually defined as the cardinality of the inverse image of 1 (i.e., the number of 1's in table 1). In studying symmetric rules (cf. section 9), it is convenient to define the *density* of a gradient as the average weight of the partial derivatives. Heavy rules generate many 1's, rules with heavy derivatives are sensitive to the initial condition; densities are thus naturally related to Lyapunov exponent [13]. This rule characterization is intrinsic in that it does not rely on averages over many initial conditions.

6. Derivatives of 1D peripheral cellular automata

Table 1 gives the gradients of the Boolean functions of three variables. These functions can in particular be identified with one-dimensional peripheral (three neighbors) two states per cell (Boolean) cellular automata. Because of right–left reflection symmetry and transformation under complementations of all three variables, the gradients of only 88 "minimal representative" rules (out of 256) are listed.

This table shows explicitly how the new state x_i^{t+1} depends on the present states x_{i-1}^t, x_i^t and x_{i+1}^t of the neighbors. It can be compared with the column "dep" in Hurd's table 1 in the Appendix of Wolfram's book [21]. A 1 indicates linear dependence with the corresponding variable (corresponding to Hurd's "○"); a 0 indicates independence (cf. Hurd's "-"); finally the Boolean expressions details Hurd's notation "●" for complex dependence.

There are numerous methods of calculating Boolean derivatives; some of them, indicated below, shed light on general properties of classes of rules.

– *Method 1: Applying the definition.*   We saw in section 2 how the direct application of eq. (2) gave the gradients of rules 30 and 150.

– *Method 2: "Boolean calculus".*   Properties (i)–(iv), together with the ring properties of XOR ($\oplus$) and AND ($\cdot$), allow the use of the usual rules of calculus in computing partial derivatives of Boolean functions expressed as the exclusive OR ($\oplus$) of products (AND's) of variables (ring sum expansions[1]). For example, the gradient of rule 30 is readily obtained first by expressing the OR ($+$) in terms of $\oplus$'s:

$$x_{i-1} \oplus (x_i + x_{i+1}) = x_{i-1} \oplus (x_i \oplus x_{i+1} \oplus x_i x_{i+1}), \tag{19}$$

and then applying the standard rules of calculus. One gets

$$\nabla f_{(30)} = (1, \bar{x}_{i+1}, \bar{x}_i). \tag{20}$$

Rule 30 illustrates a consequence of properties (i) and (v) of section 4:

If one partial derivative is 0 or 1, then the two other partial derivatives have at most one literal.

See also rules 3, 5, 10, 12, 34, 45, 51, 57, 60, 90, 105, 106, 108, 136, 150, 154, 156, 160, 170, and 204.

Ring sum expansions themselves can simply be obtained from the rule table. Each term of the polynomial will correspond to the variable combinations for which the rule returns a 1.

– *Method 3: Using symmetries.*   Complementing all the three arguments and the rules yields, by property (vi), to conjugate rules and their gradients from the listed minimal ones. Moreover, complementing one or two variables only can generate a string of minimal gradients from a known one. Start for example with rule 1, which returns a 1 when and only when its three arguments are all 0. Its table is 00000001 and its Boolean expression is:

$$f_{(1)} = \bar{x}_{i-1}\bar{x}_i\bar{x}_{i+1}. \tag{21}$$

Using any of the methods above gives immediately:

$$\nabla f_{(1)} = (\bar{x}_i\bar{x}_{i+1}, \bar{x}_{i-1}\bar{x}_{i+1}, \bar{x}_{i-1}\bar{x}_i). \tag{22}$$

Complementing x_{i+1}, x_i, or x_{i-1} in the last two expressions leads respectively to rules 2, 4, and 16 and their gradients:

---

[1] *Bona fide* ring sum expansions do not involve complemented variables. Property (vi), however, allows the economy of the expansions $\bar{x}_l = x_l \oplus 1$ of complemented variables.

Table 1
Gradients of peripheral one-dimensional cellular automata. Because of symmetries, only the 88 minimal representative rules (out of 256) are listed, see section 6.

Rule f	Table	$(\partial f/\partial x_{i-1}, \partial f/\partial x_i, \partial f/\partial x_{i+1})$
0	00000000	$(0, 0, 0)$
1	00000001	$(\bar{x}_i\bar{x}_{i+1},\ \bar{x}_{i-1}\bar{x}_{i+1},\ \bar{x}_i\bar{x}_{i-1})$
2	00000010	$(\bar{x}_i x_{i+1},\ \bar{x}_{i-1}x_{i+1},\ \bar{x}_{i-1}\bar{x}_i)$
3	00000011	$(\bar{x}_i,\ \bar{x}_{i-1},\ 0)$
4	00000100	$(x_i\bar{x}_{i+1},\ \bar{x}_{i-1}\bar{x}_{i+1},\ \bar{x}_{i-1}x_i)$
5	00000101	$(\bar{x}_{i+1},\ 0,\ \bar{x}_{i-1})$
6	00000110	$(x_i\bar{x}_{i+1}+\bar{x}_i x_{i+1},\ \bar{x}_{i-1},\ \bar{x}_{i-1})$
7	00000111	$(\bar{x}_i+\bar{x}_{i+1},\ \bar{x}_{i-1}x_{i+1},\ \bar{x}_{i-1}x_i)$
8	00001000	$(x_i x_{i+1},\ \bar{x}_{i-1}x_{i+1},\ \bar{x}_{i-1}x_i)$
9	00001001	$(x_i x_{i+1}+\bar{x}_i\bar{x}_{i+1},\ \bar{x}_{i-1},\ \bar{x}_{i-1})$
10	00001010	$(x_{i+1},\ 0,\ \bar{x}_{i-1})$
11	00001011	$(\bar{x}_i+x_{i+1},\ \bar{x}_{i-1}\bar{x}_{i+1},\ \bar{x}_{i-1}x_i)$
12	00001100	$(x_i,\ \bar{x}_{i-1},\ 0)$
13	00001101	$(x_i+\bar{x}_{i+1},\ \bar{x}_{i-1}x_{i+1},\ \bar{x}_{i-1}\bar{x}_i)$
14	00001110	$(x_i+x_{i+1},\ \bar{x}_{i-1}\bar{x}_{i+1},\ \bar{x}_{i-1}\bar{x}_i)$
15	00001111	$(1, 0, 0)$
18	00010010	$(\bar{x}_i,\ x_{i-1}\bar{x}_{i+1}+\bar{x}_{i-1}x_{i+1},\ \bar{x}_i)$
19	00010011	$(\bar{x}_i x_{i+1},\ \bar{x}_{i-1}+\bar{x}_{i+1},\ x_{i-1}\bar{x}_i)$
22	00010110	$(\bar{x}_i+\bar{x}_{i+1},\ \bar{x}_{i-1}+\bar{x}_{i+1},\ \bar{x}_{i-1}+\bar{x}_i)$
23	00010111	$(x_i\bar{x}_{i+1}+\bar{x}_i x_{i+1},\ x_{i-1}\bar{x}_{i+1}+\bar{x}_{i-1}x_{i+1},\ x_{i-1}\bar{x}_i+\bar{x}_{i-1}x_i)$
24	00011000	$(x_i x_{i+1}+\bar{x}_i\bar{x}_{i+1},\ x_{i-1}\bar{x}_{i+1}+\bar{x}_{i-1}x_{i+1},\ x_{i-1}\bar{x}_i+\bar{x}_{i-1}x_i)$
25	00011001	$(x_i x_{i+1},\ \bar{x}_{i-1}+\bar{x}_{i+1},\ \bar{x}_{i-1}+\bar{x}_i)$
26	00011010	$(\bar{x}_i+x_{i+1},\ x_{i-1}\bar{x}_{i+1},\ \bar{x}_{i-1}+\bar{x}_i)$
27	00011011	$(x_{i+1},\ \bar{x}_{i+1},\ x_{i-1}\bar{x}_i+\bar{x}_{i-1}x_i)$
28	00011100	$(x_i+\bar{x}_{i+1},\ \bar{x}_{i-1}+x_{i+1},\ x_{i-1}\bar{x}_i)$
29	00011101	$(x_i,\ x_{i-1}\bar{x}_{i+1}+\bar{x}_{i-1}x_{i+1},\ \bar{x}_i)$
30	00011110	$(1,\ \bar{x}_{i+1},\ \bar{x}_i)$
32	00100000	$(\bar{x}_i x_{i+1},\ x_{i-1}x_{i+1},\ x_{i-1}\bar{x}_i)$
33	00100001	$(\bar{x}_i,\ x_{i-1}x_{i+1}+\bar{x}_{i-1}\bar{x}_{i+1},\ \bar{x}_i)$
34	00100010	$(0,\ x_{i+1},\ \bar{x}_i)$
35	00100011	$(\bar{x}_i\bar{x}_{i+1},\ \bar{x}_{i-1}+x_{i+1},\ x_{i-1}\bar{x}_i)$
36	00100100	$(x_i\bar{x}_{i+1}+\bar{x}_i x_{i+1},\ x_{i-1}x_{i+1}+\bar{x}_{i-1}\bar{x}_{i+1},\ x_{i-1}\bar{x}_i+\bar{x}_{i-1}x_i)$
37	00100101	$(\bar{x}_i+\bar{x}_{i+1},\ x_{i-1}x_{i+1},\ \bar{x}_{i-1}+\bar{x}_i)$
38	00100110	$(x_i\bar{x}_{i+1},\ \bar{x}_{i-1}+x_{i+1},\ \bar{x}_{i-1}+\bar{x}_i)$
40	00101000	$(x_{i+1},\ x_{i+1},\ x_{i-1}\bar{x}_i+\bar{x}_{i-1}x_i)$
41	00101001	$(\bar{x}_i+x_{i+1},\ \bar{x}_{i-1}+x_{i+1},\ \bar{x}_{i-1}+\bar{x}_i)$
42	00101010	$(x_i x_{i+1},\ x_{i-1}x_{i+1},\ \bar{x}_{i-1}+\bar{x}_i)$
43	00101011	$(x_i x_{i+1}+\bar{x}_i\bar{x}_{i+1},\ x_{i-1}x_{i+1}+\bar{x}_{i-1}\bar{x}_{i+1},\ x_{i-1}\bar{x}_i+\bar{x}_{i-1}x_i)$
44	00101100	$(x_i+x_{i+1},\ \bar{x}_{i-1}+x_{i+1},\ x_{i-1}\bar{x}_i)$
45	00101101	$(1,\ x_{i+1},\ \bar{x}_i)$
46	00101110	$(x_i,\ x_{i-1}x_{i+1}+\bar{x}_{i-1}\bar{x}_{i+1},\ \bar{x}_i)$
50	00110010	$(\bar{x}_i\bar{x}_{i+1},\ x_{i-1}+x_{i+1},\ \bar{x}_{i-1}\bar{x}_i)$
51	00110011	$(0, 1, 0)$
54	00110110	$(\bar{x}_{i+1},\ 1,\ \bar{x}_{i-1})$
56	00111000	$(\bar{x}_i+x_{i+1},\ x_{i-1}+x_{i+1},\ \bar{x}_{i-1}x_i)$
57	00111001	$(x_{i+1},\ 1,\ \bar{x}_{i-1})$
58	00111010	$(x_i x_{i+1}+\bar{x}_i\bar{x}_{i+1},\ x_{i-1},\ \bar{x}_{i-1})$

Table 1. Continued.

Rule f	Table	$(\partial f/\partial x_{i-1}, \partial f/\partial x_i, \partial f/\partial x_{i+1})$
60	00111100	$(1,\,1,\,0)$
62	00111110	$(\bar{x}_i+x_{i+1},\ x_{i-1}+\bar{x}_{i+1},\ \bar{x}_{i-1}\bar{x}_i)$
72	01001000	$(x_i,\ x_{i-1}\bar{x}_{i+1}+\bar{x}_{i-1}x_{i+1},\ x_i)$
73	01001001	$(x_i+\bar{x}_{i+1},\ \bar{x}_{i-1}+\bar{x}_{i+1},\ \bar{x}_{i-1}+x_i)$
74	01001010	$(x_i+x_{i+1},\ x_{i-1}\bar{x}_{i+1},\ \bar{x}_{i-1}+x_i)$
76	01001100	$(x_ix_{i+1},\ \bar{x}_{i-1}+\bar{x}_{i+1},\ x_{i-1}x_i)$
77	01001101	$(x_ix_{i+1}+\bar{x}_i\bar{x}_{i+1},\ x_{i-1}\bar{x}_{i+1}+\bar{x}_{i-1}x_{i+1},\ x_{i-1}x_i+\bar{x}_{i-1}\bar{x}_i)$
78	01001110	$(x_{i+1},\ \bar{x}_{i+1},\ x_{i-1}x_i+\bar{x}_{i-1}\bar{x}_i)$
90	01011010	$(1,\,0,\,1)$
94	01011110	$(\bar{x}_i+x_{i+1},\ \bar{x}_{i-1}\bar{x}_{i+1},\ x_{i-1}+\bar{x}_i)$
104	01101000	$(x_i+x_{i+1},\ x_{i-1}+x_{i+1},\ x_{i-1}+x_i)$
105	01101001	$(1,\,1,\,1)$
106	01101010	$(x_i,\ x_{i-1},\ 1)$
108	01101100	$(x_{i+1},\ 1,\ x_{i-1})$
110	01101110	$(\bar{x}_ix_{i+1},\ x_{i-1}+\bar{x}_{i+1},\ x_{i-1}+x_i)$
122	01111010	$(x_i+\bar{x}_{i+1},\ x_{i-1}x_{i+1},\ \bar{x}_{i-1}+x_i)$
126	01111110	$(x_ix_{i+1}+\bar{x}_i\bar{x}_{i+1},\ x_{i-1}x_{i+1}+\bar{x}_{i-1}\bar{x}_{i+1},\ x_{i-1}x_i+\bar{x}_{i-1}\bar{x}_i)$
128	10000000	$(x_ix_{i+1},\ x_{i-1}x_{i+1},\ x_{i-1}x_i)$
130	10000010	$(x_{i+1},\ x_{i+1},\ x_{i+1}x_i+\bar{x}_{i+1}\bar{x}_i)$
132	10000100	$(x_i,\ x_{i-1}x_{i+1}+\bar{x}_{i-1}\bar{x}_{i+1},\ x_i)$
134	10000110	$(x_i+x_{i+1},\ \bar{x}_{i-1}+x_{i+1},\ \bar{x}_{i-1}+x_i)$
136	10001000	$(0,\ x_{i+1},\ x_i)$
138	10001010	$(\bar{x}_ix_{i+1},\ x_{i-1}x_{i+1},\ \bar{x}_{i-1}+x_i)$
140	10001100	$(x_i\bar{x}_{i+1},\ \bar{x}_{i-1}+x_{i+1},\ x_{i-1}x_i)$
142	10001110	$(x_i\bar{x}_{i+1}+\bar{x}_ix_{i+1},\ x_{i-1}x_{i+1}+\bar{x}_{i-1}\bar{x}_{i+1},\ x_{i-1}x_i+\bar{x}_{i-1}\bar{x}_i)$
146	10010010	$(x_i+x_{i+1},\ x_{i-1}+x_{i+1},\ x_{i-1}+x_i)$
150	10010110	$(1,\,1,\,1)$
152	10011000	$(\bar{x}_{i+1}\bar{x}_i,\ x_{i-1}+x_{i+1},\ x_{i-1}+x_i)$
154	10011010	$(\bar{x}_i,\ x_{i-1},\ 1)$
156	10011100	$(\bar{x}_{i+1},\ 1,\ x_{i-1})$
160	10100000	$(x_{i+1},\ 0,\ x_{i-1})$
162	10100010	$(x_ix_{i+1},\ \bar{x}_{i-1}x_{i+1},\ x_{i-1}+\bar{x}_i)$
164	10100100	$(x_i+x_{i+1},\ \bar{x}_{i-1}\bar{x}_{i+1},\ x_{i-1}+x_i)$
168	10101000	$(\bar{x}_ix_{i+1},\ \bar{x}_{i-1}x_{i+1},\ x_{i-1}+x_i)$
170	10101010	$(0,\,0,\,1)$
172	10101100	$(x_ix_{i+1}+\bar{x}_i\bar{x}_{i+1},\ \bar{x}_{i-1},\ x_{i-1})$
178	10110010	$(x_ix_{i+1}+\bar{x}_i\bar{x}_{i+1},\ x_{i-1}\bar{x}_{i+1}+\bar{x}_{i-1}x_{i+1},\ x_{i-1}x_i+\bar{x}_{i-1}\bar{x}_i)$
184	10111000	$(\bar{x}_i,\ x_{i-1}\bar{x}_{i+1}+\bar{x}_{i-1}x_{i+1},\ x_i)$
200	11001000	$(x_i\bar{x}_{i+1},\ x_{i-1}+x_{i+1},\ \bar{x}_{i-1}x_i)$
204	11001100	$(0,\,1,\,0)$
232	11101000	$(x_i\bar{x}_{i+1}+\bar{x}_ix_{i+1},\ x_{i-1}\bar{x}_{i+1}+\bar{x}_{i-1}x_{i+1},\ x_{i-1}\bar{x}_i+\bar{x}_{i-1}x_i)$

$$\nabla f_{(2)} = (\bar{x}_i x_{i+1}, \bar{x}_{i-1} x_{i+1}, \bar{x}_{i-1} \bar{x}_i), \tag{23}$$

$$\nabla f_{(4)} = (x_i \bar{x}_{i+1}, \bar{x}_{i-1} \bar{x}_{i+1}, \bar{x}_{i-1} x_i), \tag{24}$$

$$\nabla f_{(16)} = (\bar{x}_i \bar{x}_{i+1}, x_{i-1} \bar{x}_{i+1}, x_{i-1} \bar{x}_i). \tag{25}$$

In the same vein, complementing both variables x_i and x_{i+1} in eqs. (21)–(22) gives

$$\nabla f_{(8)} = (x_i x_{i+1}, \bar{x}_{i-1} x_{i+1}, \bar{x}_{i-1} x_i). \tag{26}$$

– *Method 4: Using additivity.* The eight unit weight rules (with numbers $2^m, m = 0, \dots, 7$) span the whole rule space. For example, the gradient of rule 30 (=16+8+4+2) is obtained by $\oplus$-adding the four last equations:

$$\nabla f_{(30)} = (1, \bar{x}_{i+1}, \bar{x}_{i-1}). \tag{27}$$

Note in particular how the four partial derivatives with respect to x_{i-1} add up to the constant $\partial f_{(30)}/\partial x_{i-1}$ = 1. This example illustrates another general property:

A rule that depends linearly on one or more variables has weight four.

This property also derives from the fact that half of the table must be the complement of the other half.

– *Method 5: Using the table.* The rule's dependence on its individual arguments can be derived directly from table 1. The leftmost digit is the value returned by the rule when its three arguments are 1, the second digit corresponds to $x_{i-1} = x_i = 1$ and $x_{i+1} = 0$, and so on (see ref. [21]). The dependence upon x_{i-1}, for example, is given by differences in the first and last halves of the table. For rule 1, these halves vary only at the fourth and eighth digits, i.e., for $x_i = x_{i+1} = 0$, yielding $\partial f_{(1)}/\partial x_{i-1} = \bar{x}_i \bar{x}_{i+1}$. Rule 30 is simpler: the two halves differ everywhere, i.e., irrespective of x_i and x_{i+1}, giving again $\partial f_{(30)}/\partial x_{i-1} = 1$.

7. Propagation of information and higher derivatives

Let us consider here the propagation of information over time. Using the notation of eq. (3), we shall look at the influence over two time steps of a value x_i^t upon its first and second neighbors. The state at step $t + 2$ of site i's right neighbor is

$$x_{i+1}^{t+2} = f(x_i^{t+1}, x_{i+1}^{t+1}, x_{i+2}^{t+1}), \tag{28}$$

or, in terms of the original values at time t:

$$x_{i+1}^{t+2} = f(f(x_{i-1}^t, x_i^t, x_{i+1}^t), f(x_i^t, x_{i+1}^t, x_{i+2}^t), f(x_{i+1}^t, x_{i+2}^t, x_{i+3}^t)). \tag{29}$$

Similarly, we have for the second neighbor:

$$x_{i+2}^{t+2} = f(x_{i+1}^{t+1}, x_{i+2}^{t+1}, x_{i+3}^{t+1}), \tag{30}$$

and

$$x_{i+2}^{t+2} = f(f(x_i^t, x_{i+1}^t, x_{i+2}^t), f(x_{i+1}^t, x_{i+2}^t, x_{i+3}^t), f(x_{i+2}^t, x_{i+3}^t, x_{i+4}^t)). \tag{31}$$

This site $i + 2$ will "know" at $t + 2$ that site i has flipped if and only if ($\Longleftrightarrow$) site $i + 1$ is affected at $t + 1$ and ($\wedge$) can in turn influence site $i + 2$ at $t + 2$. In terms of Boolean derivatives:

$$\frac{\partial x_{i+2}^{t+2}}{\partial x_i^t} = 1 \iff \left(\frac{\partial x_{i+2}^{t+2}}{\partial x_{i+1}^{t+1}} = 1 \quad \wedge \quad \frac{\partial x_{i+1}^{t+1}}{\partial x_i^t} = 1 \right). \tag{32}$$

Expressing $\wedge$ with the Boolean multiplication ($\cdot$):

$$\frac{\partial x_{i+2}^{t+2}}{\partial x_i^t} = \frac{\partial x_{i+2}^{t+2}}{\partial x_{i+1}^{t+1}} \cdot \frac{\partial x_{i+1}^{t+1}}{\partial x_i^t}. \tag{33}$$

The last equation is nothing but the chain rule for the derivative of a composite function; the rule extends readily for Boolean derivatives when information can take only one path (along the "light cone"), as is the case between x_i^t and x_{i+2}^{t+2} (the two values are separated by two units in both space and time).

The chain rule is not valid, however, when information can propagate along multiple paths, as between x_i^t and x_{i+1}^{t+2}. In that case, both x_i^{t+1} and x_{i+i}^{t+1} mediate the influence of x_i^t upon x_{i+1}^{t+2}. It can be shown [16] that

$$\frac{\partial x_{i+1}^{t+2}}{\partial x_i^t} = \frac{\partial x_{i+1}^{t+2}}{\partial x_i^{t+1}} \cdot \frac{\partial x_i^{t+1}}{\partial x_i^t} \oplus \frac{\partial x_{i+1}^{t+2}}{\partial x_{i+1}^{t+1}} \cdot \frac{\partial x_{i+1}^{t+1}}{\partial x_i^t} \oplus \frac{\partial^2 x_{i+1}^{t+2}}{\partial x_{i+1}^{t+1} \partial x_i^{t+1}} \cdot \frac{\partial x_i^{t+1}}{\partial x_i^t} \cdot \frac{\partial x_{i+1}^{t+1}}{\partial x_i^t}. \tag{34}$$

This formula express the dependence as an exclusive sum over all possible paths of inheritance, with possible destructive interferences, because of the $\oplus$'s.

In a similar way, $\partial x_i^{t+2} / \partial x_i^t$ is expressed in terms of third derivatives as well second and first derivatives.

The derivatives express consequences of changes of x_i^t. But if the initial configuration is all 0's (the "quiescent state"), a change of x_i^t from 0 to 1 ignites a unit source, and we have a kind of response function or discrete Green function. An arbitrary value x_j^{t+m} is now simply given by the derivative $\partial x_j^{t+m} / \partial x_i^t$ if only site i is "on" at time t. For example,

$$x_{i+1}^{t+2} = \frac{\partial x_{i+1}^{t+2}}{\partial x_i^{t+1}} \cdot \frac{\partial x_i^{t+1}}{\partial x_i^t} \oplus \frac{\partial x_{i+1}^{t+2}}{\partial x_{i+1}^{t+1}} \cdot \frac{\partial x_{i+1}^{t+1}}{\partial x_i^t} \oplus \frac{\partial^2 x_{i+1}^{t+2}}{\partial x_{i+1}^{t+1} \partial x_i^{t+1}} \cdot \frac{\partial x_i^{t+1}}{\partial x_i^t} \cdot \frac{\partial x_{i+1}^{t+1}}{\partial x_i^t}. \tag{35}$$

In this formula all the derivatives are *time constants*, they are to be evaluated at the unperturbed (i.e., all $x_i = 0$) configuration, once and for all. The first derivatives, for example, can be obtained by plugging $x_{i-1} = x_i = x_{i+1} = 0$ in the appropriate entry of table 1. (This holds, of course, only for even-numbered rules that leave the quiescent state invariant.) The generic value x_j^{t+m} is given by a generalization of (35), involving a sum over all "causal" paths connecting (i, t) and $(j, t+m)$. The expansion in terms of Boolean derivatives is (unlike Taylor series) always finite: derivatives of order higher than the neighborhood size n vanish identically (cf. property (v) of section 4).

8. Derivatives and complexity

One generally characterizes an evolution as truly complex when it is believed to be *computationally irreducible*, i.e. when there is *no shortcut* to the knowledge of a state x_j^{t+m} of an arbitrary site j in the future. The computation of x_{i+1}^{t+2} with (29) involves four evaluations the rule f. In general, for peripheral cellular automata, it costs m^2 evaluations of f to compute the value of an arbitrary site at time $t+m$ from the knowledge of the array at time t. Irreducibility means that this cost reflects the minimal computational effort to predicting an arbitrary value x_j^{t+m}.

It is further often claimed that for complex rules, the *only* way to arrive at an arbitrary value in the future is the repeated applications of the definition, as in (29); no closed mathematical expression can yield x_j^{t+m} if it does not invoke m^2 explicit evaluations of the rule. Eq. (35), by its very existence, shows that this is not the case, at least for evolutions starting from a single seed. Such evolutions can be very

complex, Wolfram's rule 30, for example, generates pseudo-random sequences of very high quality [20] out of a single 1.

Does the existence, for a generic value x_j^{t+m}, of a *finite* expansion involving $\oplus$'s and $\wedge$'s of *constants* defeat a chief motivation for studying cellular automata, namely, their ability to generate complexity out of simple sources? Not necessarily. For truly complex rules, the computation of x_j^{t+m} via a generalization of (35) should incur a computational effort of the order of that required for performing m^2 iterations of f, as in (29). At any rate, expression (35) is arguably more pleasant and lucid than the blind iteration (29). It makes explicit the inheritance from the source at (i, t), and for some f's it can simplify and reveal previously unsuspected structure. Since irreducibility of computational cost is a necessary condition of true complexity, the Boolean derivative offers a test for complexity.

9. Derivatives of symmetric rules

Much attention has been devoted to the special class of rules in one and two dimensions that depend on total number of the ones in the neighborhood, but not on their particular positions. Despite this simplification, these rules seem to exhibit the most general behaviors (except, of course, asymmetric evolution from a symmetric starting configuration). These rules, called "totalistic" [18] and "counting" [15] are invariant under permutation of their arguments. In fact, the converse is also true. A theorem due to Shannon states that all such invariant functions (called "symmetric" in the literature on Boolean functions), merely "count" the number of zeroes and ones among their arguments, see e.g. ref. [10].

These rules can be described by specifying the neighborhood and the values of the totals for which a one is returned. In turn, these values can conveniently be characterized in terms of blocks of ones in the counting space. Each block r is delimited by an inclusive lower threshold $T_{\min}^{(r)}$ and a maximal total $T_{\max}^{(r)}$. These two inclusive limits can coincide if the block reduces to a isolated peak. For example, rule 150 is defined by two blocks of length one with $T_{\min}^{(1)} = T_{\max}^{(1)} = 1$ and $T_{\min}^{(2)} = T_{\max}^{(2)} = 3$.

In terms of these thresholds, symmetric rules can be simply described by stating that f be one if the total of ones is within a 1-block and zero otherwise, i.e.,

$$f = \sum_r \left(\Theta\left(\sum_{j=1}^n x_j - T_{\min}^{(r)} \right) \cdot \Theta\left(T_{\max}^{(r)} - \sum_{j=1}^n x_j \right) \right), \tag{36}$$

where the step function Θ is 1 if its argument is nonnegative and 0 otherwise. The partial derivatives take the form

$$\frac{\partial}{\partial x_k} f = \sum_r \left(\delta\left(\sum_{j \neq k} x_j, T_{\min}^{(r)} - 1 \right) + \delta\left(T_{\max}^{(r)}, \sum_{j \neq k} x_j \right) \right), \tag{37}$$

where the Kronecker $\delta(\ ,\ )$ is 1 if its two arguments are equal and 0 otherwise. This formula is valid for neighborhoods of any size n and shape in any number of dimensions.

Eq. (37) treats blocks of zeroes and blocks of ones on the same footing. The δ's are peaked right below any discontinuity between zeroes and ones in the counting space; their positions are not affected if the zeroes and ones are interchanged. This agrees with the invariance of the derivative with respect to the complementation of the rule (eq. (14)). (Note that with the notations of (36) and (37), the derivative of Θ's involve δ's.) The more numerous the discontinuities are, the heavier the derivatives will be. The extreme examples are the constant rules 0 and 255, with no discontinuity and therefore with vanishing derivatives, and rules 150 and its complement 105, which have 0–1 changes at every point in the counting space and therefore have density one gradients (1,1,1).

A two-dimensional example illustrates the dramatic effect of increasing the derivative density by adding discontinuities in the counting space. The majority rule in the von Neumann neighborhood (made of the central cell and its four adjacent neighbors in the square lattice) returns a one if and only if 3, 4, or 5 cells of the neighborhood have the value one. This rule has been called [15,7] V345 and "three-out-of-five (3/5)". Its derivative with respect to the center cell is "Q2"; it returns a one if there are exactly two ones among the four adjacent neighbors. The weight of this partial derivative is $\binom{4}{2}/2^4 = 0.375$. (By symmetry, the weight of the other partial derivatives is the same.) Under V345, random configurations evolve after a very short transient into stable percolation-like patterns that are very robust under small perturbations [15,16]. But if we "twist" the definition of the rule around its unique 0–1 discontinuity, one gets V245 (where now a total of 2 instead of 3 returns a one), with three 0–1 discontinuities. According to formula (37), the derivative with respect to the center is Q123, which returns a one for counts of 1, 2, or 3 ones among the four adjacent neighbors, of weight $(\binom{4}{1} + \binom{4}{2} + \binom{4}{3})/2^4 = 0.875$. This rule, which has a heavier derivative, generates richer dynamics

10. Derivatives of lattice-gas automata

Boolean derivatives provide insight on the logic of lattice-gas automata rules. For almost all lattice-gas automata, the rule that applies at each lattice site can be expressed under the generic form (see, e.g. ref. [17]):

$$n_i^{t+1} = n_i^t \bar{A}_i \oplus C_i, \tag{38}$$

where n_i^{t+1} is the Boolean variable for a particle in direction i exiting the considered site, and A_i and C_i are particle creation and annihilation operators for the incoming i direction. These operators are functions of occupation numbers of particles entering the site (denoted here as n^t or without superscript). This equation simply says that a particle will leave the site if it entered it and was not removed through some collision (advection), or if direction i becomes populated as a result of a collision. In the original HPP model [6], the operators A_i and C_i read

$$C_i = n_{i+1} n_{i+3} \bar{n}_i \bar{n}_{i+2} \tag{39}$$

and

$$A_i = n_i n_{i+2} \bar{n}_{i+1} \bar{n}_{i+3}. \tag{40}$$

Using Leibniz rule (17) and the last three equations, one finds

$$\frac{\partial n_i^{t+1}}{\partial n_i^t} = \overline{n_{i+2} \bar{n}_{i+1} \bar{n}_{i+3}} \oplus \bar{n}_{i+2} n_{i+1} n_{i+3}. \tag{41}$$

This equation shows that the derivative $\partial n_i^{t+1}/\partial n_i^t$ can be viewed as a predicate taking the value true (1) or false (0) depending on whether or not the following holds:

$$\mathrm{NOT}[n_{i+1} = n_{i+3} = \bar{n}_{i+2}]. \tag{42}$$

This expression of the derivative as a predicate determines under what condition direction i's new occupancy n_i^{t+1} will be independent of its present value n_i^t. If the variables in the brackets above are all 0 (1), then $n_i^{t+1} = 0$ (1), independently of n_i^t. In both cases, the derivative $\partial n_i^{t+1}/\partial n_i^t$ vanishes. Conversely, a particle ($n_i = 1$) or a hole ($n_i = 0$) will be advected ($n_i^{t+1} = n_i^t$) unless a collision occurs. In this case, the derivative is 1.

In a similar way, derivatives $\partial n_i^{t+1}/\partial n_{i+1}^t$ and $\partial n_i^{t+1}/\partial n_{i+2}^t$ with respect to other incoming directions are equivalent, respectively, to the predicates

$$[n_i = n_{i+2} = \bar{n}_{i+3}] \tag{43}$$

and

$$[n_{i+1} = n_{i+3} = \bar{n}_i]. \tag{44}$$

This lattice-gas example points to the fact that Boolean derivatives are *functional* derivatives (and thus could be denoted $\delta n^{t+1}/\delta n^t$) rather that ordinary derivatives with respect to space and time. This feature makes using Boolean derivatives for constructing differential equations indirect.

Acknowledgements

This work was supported in part by the Dean of the School of Engineering, MIT, and by a grant from AFOSR, No. 88-0119.

References

[1]   Complex Systems 1, No. 4 (1987).
[2]   S.W. Boyack, The robustness of combinatorial measures of Boolean matrix complexity, Ph. D. Thesis, Department of Mathematics, Massachusetts Institute of Technology (1985).
[3]   D. Farmer, T. Toffoli and S. Wolfram, eds., Cellular Automata (North-Holland, Amsterdam, 1984), special issue Physica D 10 (1984).
[4]   F. Fogelman-Soulié, Contribution à une théorie du calcul sur réseaux, Thèse, Grenoble (1985).
[5]   U. Frisch, B. Hasslacher and Y. Pomeau, Phys. Rev. Lett. 56 (1986) 1505.
[6]   J. Hardy, O. de Pazzis and Y. Pomeau, Phys. Rev A 13 (1976) 1949.
[7]   B. Hayes, Sci. Am. 250 (March 1984) 12.
[8]   P. Manneville, N. Boccara, G. Y. Vichniac and R. Bideaux, eds., Cellular Automata and Modeling of Complex Physical Systems (Springer, Berlin, 1989).
[9]   N. Margolus, T. Toffoli and G. Vichniac, Phys. Rev. Lett. 56 (1986) 1694.
[10] R.E. Miller, Switching Theory (Wiley, New York, 1965).
[11] J.H. Reif, Efficient VLSI fault simulation, Harvard Univerity Center for Research in Computing Technology TR-03-85 (1985).
[12] F. Robert, Discrete Iterations: A Metric Study (Springer, Berlin, 1986).
[13] S. Ruffo and G.Y. Vichniac, in preparation.
[14] A. Thayse, Boolean Calculus of Differences (Springer, Berlin, 1981).
[15] G.Y. Vichniac, Physica D 10 (1984) 96.
[16] G.Y. Vichniac, in: Chaos and Complexity, eds. R. Livi, S. Ruffo, S. Ciliberto and M. Buatti (World Scientific, Singapore, 1988).
[17] G.Y. Vichniac, in: Instabilities and Nonequilibrium Structures II, eds. E. Tirapegui and D. Villarroel (Kluwer, Dordrecht, 1989).
[18] S. Wolfram, Rev. Mod. Phys. 55 (1983) 601.
[19] S. Wolfram, Physica D 10 (1984) 1.
[20] S. Wolfram, Adv. Appl. Math. 7 (1986) 123.
[21] S. Wolfram, ed., Theory and Applications of Cellular Automata (World Scientific, Singapore, 1986).

CHAPTER 2

STRUCTURE OF THE SPACE OF CELLULAR AUTOMATA

Physica D 45 (1990) 77–94
North-Holland

TRANSITION PHENOMENA IN CELLULAR AUTOMATA RULE SPACE

Wentian LI[a], Norman H. PACKARD[a, b, 1] and Chris G. LANGTON[b]

[a] *Santa Fe Institute, 1120 Canyon Road, Santa Fe, NM 87501, USA*
[b] *Complex Systems Group, Theoretical Division, Los Alamos National Laboratory, Los Alamos, NM 87545, USA*

Received 1 March 1990
Revised manuscript received 21 March 1990

We define several qualitative classes of cellular automata (CA) behavior, based on various statistical measures, and describe how the space of all cellular automata is organized. As a cellular automaton is changed by varying entries in its rule table, abrupt changes in qualitative behavior may occur. These abrupt changes have the character of bifurcations in smooth dynamical systems, or of phase transitions in statistical mechanical systems. The most complex CA rules exhibit many of the characteristics of second-order transitions, suggesting an association between computation, complexity, and critical phenomena.

1. Introduction

Cellular automata were originally invented by von Neumann [18] for a particular task, to prove the existence of a self-reproducing universal computer. They have since caught the imagination of many, partly because of their rich and diverse phenomenology. One cellular automaton rule that has particularly rich phenomenology is Conway's Game of Life. Many other rules which display interesting phenomena are presented in ref. [11].

In the space of all cellular automata, not all rules are interesting, of course. We will consider the question of locating and characterizing those cellular automata rules that give rise to "interesting" behavior. In this case, there are two reasons for rules to be "uninteresting", and both reasons are based on the simplicity of a typical space–time configuration. One kind of simplicity is that the space–time configuration exhibits a repetitive – or periodic – structure (possibly shifting in time). The other kind of simplicity is that the space–time configuration is completely random, lacking

any discernible pattern.

The term *bifurcation* has come to be used in a general sense within the context of continuous dynamical systems to denote any change in a qualitative behavior as a parameter in the system is varied, such as a change from periodic to chaotic behavior.

Bifurcations have been studied in lattice systems with continuous variables at each lattice site [4], and in cellular automata obtained by discretizing continuous variables [2]. Cellular automata are discrete dynamical systems; there are no continuous parameters to vary to move from one cellular automaton rule to an arbitrarily nearby rule. The space of rules is totally discrete; any change in a rule entails a "quantum jump" away from the rule.

Nevertheless, there will generally be changes in the qualitative behavior of a CA as a rule is changed. In spite of the discrete nature of cellular automata, there is one unambiguous notion of proximity in the space of rules given by Hamming distance. This work is partly an exploration into the question of whether observables behave continuously in the limit that the size of the rule table becomes large and small changes (unit Hamming distance) are made to the rules.

Following Langton [5], we will use a control pa-

---

[1] Permanent address: Physics Department and the Center for Complex Systems Research, Beckman Institute, University of Illinois, 405 N. Mathews St., Urbana, IL 61801, USA. Email n@complex.ccsr.uiuc.edu

rameter λ, which is defined as the percentage of all the entries in a rule table which map to non-zero states. We will use the term bifurcation in the present work to include the qualitative change in the asymptotic behavior of a cellular automaton rule when one entry in its rule table is changed.

Suppose we are looking at one-dimensional k-state, r-radius (or $(2r + 1)$-neighbor) cellular automata. Each rule is specified by a rule table with k^{2r+1} entries, determining which state each possible neighborhood configuration is mapped to at the next time step. If m out of the total k^{2r+1} neighborhood configurations map to a non-zero state, then λ is defined as

$$\lambda = m/k^{2r+1}. \tag{1}$$

The λ parameter can be compared with – though it is not equivalent to – the temperature in statistical physics, or the degree of non-linearity in dynamical systems.

The rule spaces for cellular automata with 2 states per cell have a symmetry with respect to the point $\lambda = 0.5$. This is because rules with $\lambda = x$ are equivalent to rules with $\lambda = 1 - x$ for $0 \leq x \leq 0.5$; the roles of states 0 and 1 are simply reversed.

Most of the calculations included in this paper are for 2-state cellular automata, and we will only examine rules for which $\lambda \leq 0.5$. For cellular automata with $k > 2$ states, one can define a $(k - 1)$-dimensional parameter $\vec{\rho}$, which exhibits symmetry around the point at which each of the k states occurs with frequency $1/k$ in the rule table [7].

The simplest 1D cellular automaton rule space is that for 2-state, 3-neighbor cellular automata (called "elementary rules" in ref. [19]). As an illustration of the way in which the dynamics of these rules changes as λ changes, we plot the percentage of rules belonging to four different classes as a function of λ (fig. 1). (The overall classification will be discussed in the next section.) As λ is increased from 0 to 0.5, more and more rules exhibit chaotic, rather than periodic, dynamics. Since the elementary rule space is small (only $2^8 = 256$ possible rules in all), many features exhibited by this rule space are non-generic, and may not be applicable to larger rule spaces. For example, the existence of a rather large percentage of periodic rules at $\lambda = 0.5$ is not observed in larger CA rule spaces.

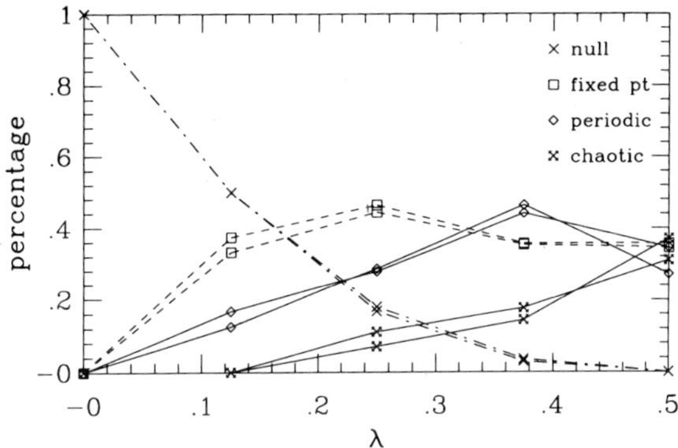

Fig. 1. Percentage of elementary rules which are null, time-invariant, periodic, and chaotic as a function of λ. There are two curves for each class of behavior for the original rule space and the folded rule space respectively (see ref. [10] for the definition of folded rule space).

This paper is devoted to the study of "large" cellular automata rule spaces. For more details about the elementary rule space, see ref. [10].

The main point of this paper is to summarize experiments suggesting that, for sufficiently large CA rule spaces, one observes a *phase transition* between *ordered* and *disordered* dynamics as λ is varied from 0 to $1 - 1/k$. Furthermore, the various qualitative classes are ordered with respect to this transition regime, with the simplest behaviors (periodic and random) being located *away* from the transition regime while the most complex behaviors are found *within* the transition regime.

The paper is organized as follows: Section 2 discusses the classification of cellular automata rules. Section 3 discusses transition-like phenomena observed using difference patterns. Sections 4, 5, and 6 discuss the transition events in terms of entropy, mutual information, and fluctuations, respectively. Section 7 discusses the limit of large neighborhoods and section 8 discusses the inferred structure of CA rule space. Appendix A gives the definitions of the quantities that are used in characterizing the transitions; and appendix B presents a mean-field theory estimation of the spreading rates of difference patterns.

2. Classification

Like most classification schemes, classification

of cellular automata dynamics depends on the particular interests of the researcher. Wolfram developed a classification scheme consisting of four qualitative classes – homogeneous fixed point (class I), periodic (class II), chaotic (class III), and complex (class IV) [20]. As an example of how these might be further refined, we list below six classes of behavior typically observed in the dynamics of 1D cellular automata. One can provide quantitative descriptions that distinguish these classes in terms of various statistical measures, such as difference pattern growth rates, entropy, and mutual information (see appendix A for definitions). Other approaches to the classification of CA rules are detailed in the papers by Chaté and Manneville [1], Gutowitz [3], and McIntosh [12] in this volume.

However, *classification alone is not enough!* What is needed is a deeper understanding of the structure of cellular automata rule space, one which provides an explanation for the existence of the observed classes and for their relationship to one another. Choosing an appropriate parameterization of the space of cellular automata rules, such as λ, allows direct observation of the way in which different statistical measures are related as a function of the parameter(s), and these relationships in turn provide an explanation for the existence and ordering of the various qualitatively distinguishable classes of cellular automata dynamics. Once one understands the "deep-structure" of cellular automata rule space, it is easy to see that many different classification schemes are possible and equally "correct," depending on which aspects of cellular automata dynamics one is interested in.

In the numerical experiments reported here, all measures are computed on cellular automata starting from random initial conditions, letting transients die away, and gathering statistics on a large enough space–time volume to achieve convergence. The question of statistical convergence is nontrivial, especially for the class 6 rules to be defined below.

Boundaries of classes are not necessarily sharp, and there is no known single quantitative measure which can be used to distinguish perfectly between all classes. To decide empirically which class a rule belongs to on the basis of the statistical measures used here, one must inevitably choose threshold values for the measures, thus the classes are not sharp.

One can readily distinguish six classes on the basis of differences in the various measures as applied to the asymptotic behavior of one-dimensional cellular automata:

(1) *Spatially homogeneous fixed points.* The difference pattern spreading rate and all entropy and mutual information measures are identically zero.

(2) *Spatially inhomogeneous fixed points*, or a uniform global shift of a fixed pattern. For this class, the difference pattern spreading rate is zero. Spatial entropy is finite, since the fixed pattern is random to a certain degree (as a result of the randomness of the initial condition). Entropy in other directions in space–time will be less than the spatial entropy, and goes to zero in the direction of periodicity, which is the direction of the net shift of the fixed pattern (the time direction, if the fixed pattern is unshifted). Mutual information will go to zero for all space–time directions as separation between space–time patches becomes large.

(3) *Periodic behavior*, or shifted periodic behavior. Typically this means regions with periodic behavior with unmoving walls between them. The behavior of statistical quantities will typically be the same as for the previous class, except that entropy and mutual information will go to nonzero values in the space–time direction of periodicity.

(4) *Locally chaotic behavior.* These rules produce chaos between walls, where the walls can either be fixed or move with an overall shift. In two dimensions, this class is characterized by fixed or oscillating boundaries separating chaotic domains. In the one-dimensional case, since there is only a finite number of states at each site, the space–time pattern between the walls must be periodic, but the hallmark of these rules is that the period increases exponentially with the distance between the walls. For the one-dimensional case, entropy is positive in all directions, mutual information is zero in all directions, and the difference pattern spreading rate is zero (because the difference pattern is stopped by the walls).

(5) *Chaotic behavior.* The difference pattern spreading rate is high, entropy is high in all directions, and mutual information is zero in all directions.

(6) *Complex behavior.* This class is character-

ized by long transients and complex space–time patterns, including both oscillating and propagating structures. This class is also characterized by a lack of statistical convergence, for it is not clear that the assumptions needed for computation of statistics hold for this class of rules. When computation of statistics is attempted, entropy is moderate, the spreading rate is roughly zero, and the mutual information is large.

This classification scheme refines Wolfram's four classes. Roughly, class 1 above corresponds to Wolfram's class I, classes 2, 3, and 4 constitute Wolfram's class II, class 5 is equivalent to Wolfram's class III, and class 6 is Wolfram's class IV.

3. The transition using difference patterns

One statistical quantity that distinguishes chaotic behavior from ordered behavior is obtained from the average asymptotic motion of the difference pattern (see appendix A.1 for a definition). This motion generally describes how two configurations that are different on part of the lattice and the same on another part, become either increasingly different or increasingly the same under the action of the cellular automaton rule.

We compute the difference pattern spreading rate, γ, along a path in the space of rules, where each successive rule on the path has a higher value of λ (more 1's in the rule table) than the previous one. We may choose an arbitrary threshold for γ, above which we will say the rule is chaotic. The first rule on the path having γ above the threshold is the "transition point" for that path.

As discussed below, not all paths undergo a transition at the same value of λ. Fig. 2 shows the transition to chaos averaged over many paths, where these paths are aligned by their transition points. Most paths exhibit a sharp jump in γ at the transition point. However, a few paths exhibit intermediate values of γ; these are class 6 rules (complex rules), which exhibit the most irregular statistics.

For small λ, perturbations hardly ever spread, for large λ perturbations always spread at roughly the same rate. Thus, away from the transition region, any particular value of λ is associated with a very narrow range of spreading rates. However,

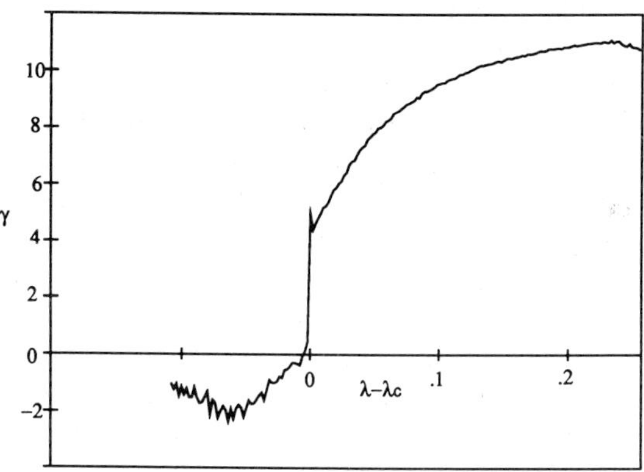

Fig. 2. Difference pattern spreading rate averaged over 200 $r = 7$ rules, with each path into chaos aligned to the point on the path where γ is larger than a threshold value of 2.

at the transition, one sees a *wide* range of possible spreading rates.

4. Transition to chaos using entropy

The most commonly used quantity for measuring randomness is entropy (see appendix A.2 for the definition). For spatially homogeneous fixed point rules, the entropy calculated from the spatial temporal pattern is zero. For periodic rules, the entropy value is non-zero but low. For random rules, all possible configurations can occur and the entropy reaches its maximum.

Fig. 3a shows the single site entropy as a function of λ for 50 different paths through the space of possible 2D, 8-state, 5-neighbor cellular automata. The square lattice has 64 cells on a side, the first 500 transients are discarded, and the next 500 patterns are used to accumulate single site probabilities. There are some restrictions on the rules being chosen: the all-zero neighborhood maps to zero, and the rules are symmetric with respect to planar rotations [6]. As expected, the entropy generally increases with increasing λ, and one usually observes a sharp jump in the entropy at the transition from regular to chaotic dynamics.

The entropy reaches its maximum value $S_{\mathrm{max}} = \log_2(8) = 3.0$ at $\lambda = 7/8$, when the rule tables are filled randomly and uniformly with respect to all 8 states. (For 2-state cellular automata, $S_{\mathrm{max}} =$

$\log_2(2) = 1.0$ at $\lambda = 1/2$, since only two symbols are used to fill the rule table randomly.)

Fig. 3a is redrawn in fig. 3b by aligning paths according to the particular λ_c at which a transi-

tion from regular dynamics to chaotic dynamics occurs on each path. In this case, the transition point was chosen to be the λ value at which no cycle of length less than 500 was detected.

For most paths, the entropy jumps discontinuously, an indication of the sudden transition from regular dynamics (either fixed point or periodic) to chaotic dynamics. In statistical physics, such a discontinuous change in entropy accompanies a first-order phase transition. By contrast, a smooth change in entropy – such as is observed for a small percentage of the paths – is associated with a second-order phase transition.

An interpretation of this is that some paths through CA rule space pass through a region of "critical" rules, which are interposed between periodic rules and chaotic rules. When such a region is encountered, the entropy appears to change smoothly, i.e., the transition seems second-order. These "critical" rules are obviously not dense, if they were, *every* path would exhibit a smooth change in entropy. Fig. 3b suggests that a first-order transition is more likely than a second-order transition, due to the low probability of passing through a "critical" region by chance.

The functional form of the entropy versus λ curve in fig. 3a can be explained by a simple "mean-field theory" [21]. For cellular automata without any restrictions, a crude estimate of the probability of non-zero sites is simply equal to λ, with each of the $k - 1$ non-zero states appearing with equal probability. The estimate of entropy is obtained from

$$S_{\mathrm{mf}} = -(1 - \lambda)\log(1 - \lambda) - \lambda \log\left(\frac{\lambda}{k - 1}\right). \quad (2)$$

As mentioned above, for the cellular automata whose entropy is plotted in fig. 3a, there is the restriction that the all-zero neighborhood maps to zero. A modified mean-field theory can be used to estimate S_{mf} in this case (for more details see ref. [21]). Notice that in the mean-field theory approximation entropy changes continuously, suggesting a second-order transition rather than the first-order transition which is observed more often in numerical simulations.

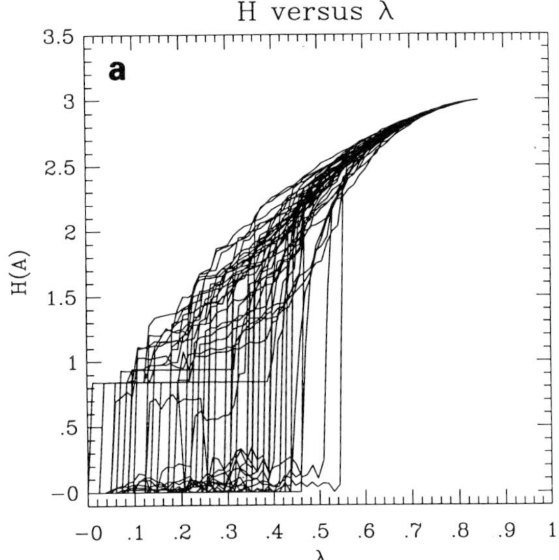

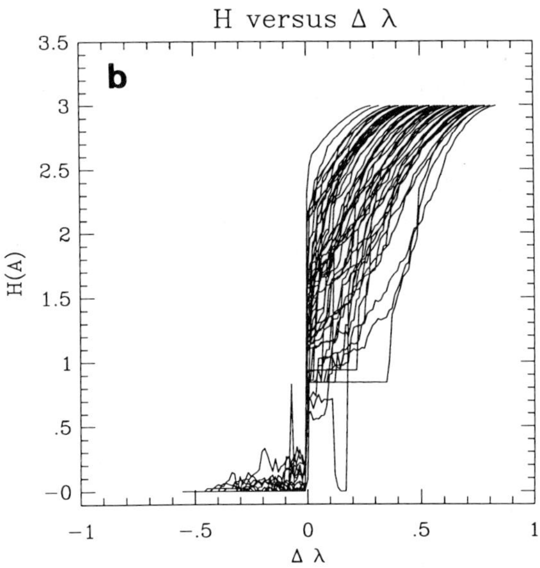

Fig. 3. Single site entropy H for two-dimensional 8-state 5-neighbor cellular automata. The single site probabilities $\{p_i\}$ ($i = 1, 2, \ldots, 8$) are accumulated from patterns at 500 consecutive times. The first 500 transient patterns are discarded. The square lattice has size 64×64. (a) H as a function of λ for 20 different paths in the rule space. (b) H as a function of $\lambda - \lambda_c$.

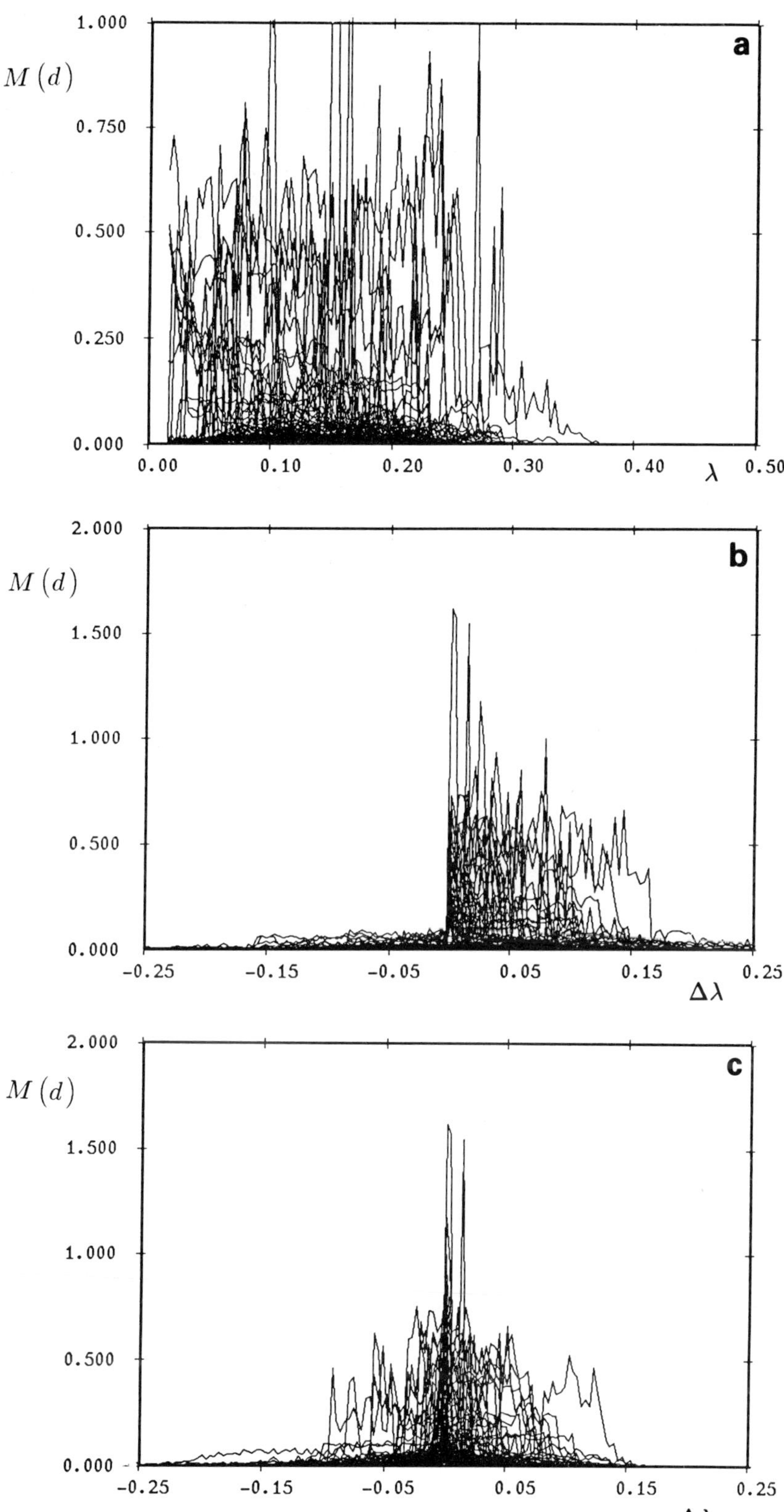

5. Transition to chaos using mutual information

Mutual information is a quantity for measuring correlations. If two random variables are statistically independent, the mutual information between them is zero. On the other hand, if the two are strongly correlated (e.g. one is a copy of another), the mutual information between them is large (see appendix A for the definition of mutual information).

The spatial mutual information for homogeneous fixed point rules is zero, and low for both periodic (including inhomogeneous fixed point rules) and random rules, because these rules do not create spatial structures. On the other hand, most complex rules give rise to highly correlated structures and consequently large spatial mutual information.

As an example, we compute the mutual information for 1D, 2-state, 11-neighbor cellular automata. In order to suppress fluctuations, we calculate the spatial mutual information between two blocks of sites (rather than between two individual sites). Also, many copies of the spatial configurations at different time steps (instead of just one configuration at a fixed time) are used in accumulating the probability distribution.

Fig. 4a shows the spatial mutual information for rules along 120 paths moving toward larger values of λ. Each step corresponds to adding 7 non-zero transitions to the rule table, an increase in λ of $7/2^{11} = 0.0034$ per step. The mutual information is computed between two blocks of 3-sites each, separated by a distance of 10 sites. Lattice size is 601, 257 transient configurations are discarded, and the next 51 configurations are used to accumulate the probabilities.

From fig. 4a, we can see that the spatial mutual information for different paths jumps at different values of λ, i.e., they have different values for λ_c. In order to characterize the behavior near the transition point, we may align the different paths by the relative λ value, $\Delta\lambda \equiv \lambda - \lambda_c$.

There are two ways to determine λ_c. The first method defines λ_c as the value of λ at the first jump of mutual information over some threshold value. The resulting plot is shown in fig. 4b.

The second method defines λ_c as the value of λ at which the maximum value of mutual information is observed. The second method is equivalent to the first if the threshold value is raised to the maximum value of mutual information for that path. Fig. 4c illustrates the second method.

As the second method results in a curve which appears to be symmetric with respect to the critical point, the situation often observed at phase transitions, we will examine the second method in more detail. The first thing we want to know is whether the mutual information at the critical point diverges as a power law or as an exponential. The data of fig. 4c is re-plotted in fig. 5 using semi-log scales. This plot gives a better fit than a log–log plot (which is not included here). This means that the divergence is exponential rather than power law: $M(|\Delta\lambda|) \sim e^{-\alpha|\Delta\lambda|}$, with $\alpha \approx 12$.

If the mutual information, which is basically a difference between entropies, can be considered as a quantity analogous to specific heat in statistical physics (the derivative of entropy with respect to temperature) [1], the behavior of the mutual information at the critical point can be used to determine the order of the phase transition. As has been seen in the previous section, for most paths the entropy jumps discontinuously at the transition point, so the transition is usually first-order. Notice that second-order phase transitions in statistical physics typically have power law divergence of specific heat, instead of exponential divergence. Unlike specific heat, the mutual infor-

---

[1] A similar comparison in the context of two-dimensional Ising model is made in ref. [16].

Fig. 4. Spatial mutual information M as a function of λ for one-dimensional 2-state 11-neighbor cellular automata rules for 120 paths in the rule space. The lattice size is 601, 257 transient configurations are discarded and the next 51 configurations are used to accumulate the probabilities. The mutual information is between two 3-blocks separated by spatial distance 10. Each data point is M for a fixed rule averaging over two different initial conditions. (a) M versus λ. (b) M versus $\Delta\lambda = \lambda - \lambda_c$ where λ_c is defined by the first λ value at which a threshold value of M is exceeded. (c) M versus $\Delta\lambda = \lambda - \lambda_c$ where λ_c is defined by the λ value with the maximum M.

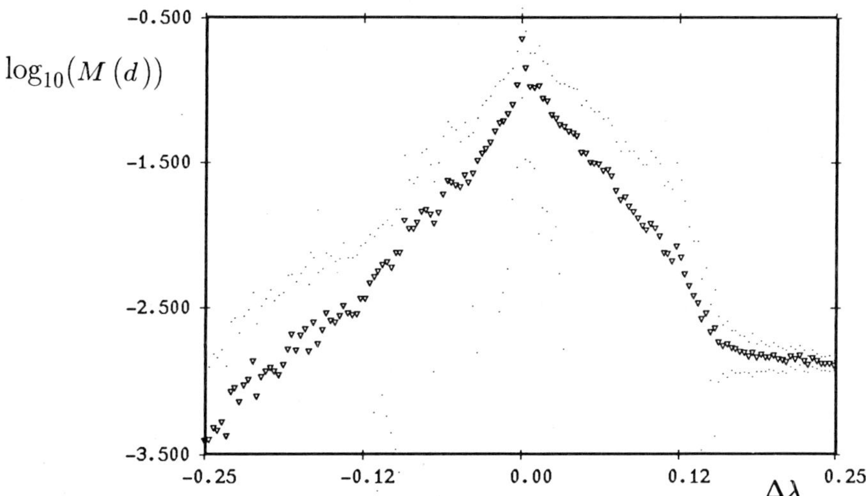

Fig. 5. Same data points as those in fig. 4c are drawn in semi-log scales. The data can roughly be fit by $M(|\Delta\lambda|) \sim e^{-\alpha|\Delta\lambda|}$, with $\alpha \approx 12$.

mation will not diverge to infinity if the number of states is finite.

Fig. 6b shows the spatial mutual information in semi-log scale similar to that in fig. 5, except that the distance between the two 3-site-blocks is only 4 sites instead of 10. (Fig. 6a shows the original M versus λ plot similar to fig. 4a.) The average maximum value of the mutual information is more or less unchanged. This is probably because the spatial structures generated by complex rules have very long correlation lengths, so a change of distance from 10 to 4 does not change the mutual information value very much. On the other hand, the mutual information is greatly reduced at longer distances for periodic and random rules, as can be seen in the plots. As a result, the exponential decay rate around the maximum mutual information is different: $M(|\Delta\lambda|) \sim e^{-\alpha|\Delta\lambda|}$, where $\alpha \approx 8$.

Fig. 7b shows the spatial mutual information in semi-log form for 15-neighbor rules. (Fig. 7a is the original M versus λ plot.) The separation between the two 3-site-blocks is again 10 sites. The exponential decay from the maximum mutual information seems to be faster than for $r = 5$ rules. It is not clear how the decay rates – which are somewhat analogous to the critical exponent – are related to the rule radius, block-length, or the separation between blocks.

6. Fluctuations at the transition to chaos

The transition to chaos appears to be marked by singular behavior in statistical measures such as entropy and difference pattern spreading rate. At the transition, however, there are problems computing these measures because of a lack of statistical convergence.

Rules at the transition have been conjectured to be capable of universal computation [5,6,14,20]. If this is the case, for generic initial conditions, then the computation of these statistical measures is problematic, because the limit set of the automaton is uncomputable, and therefore all statistical quantities are uncomputable. Rules at the transition exhibit very long transients, possibly infinite.

One can attempt to compute statistical quantities in any case, with the usual prescription of filling probability histograms over volumes of space–time. Lack of convergence is then apparent in a large spread of values for the measures when computed over an ensemble of initial conditions. The distribution of values for a measure should be close to a delta function if the empirical estimate is valid, i.e. if statistical convergence is attained. The width of the distribution of values is a measure of lack of convergence.

The poor convergence of statistical properties for complex rules is illustrated in fig. 8, which plots

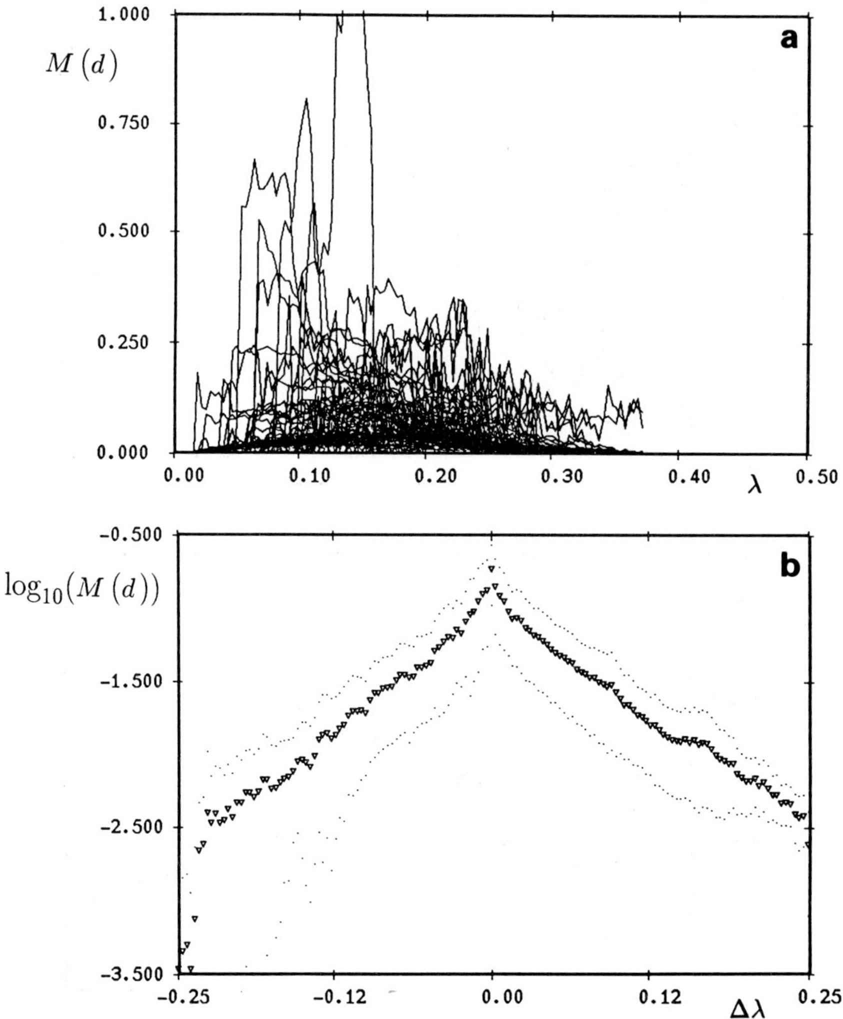

Fig. 6. Spatial mutual information M between two 3-blocks separated by distance 4 instead of 10. (a) M as a function of λ. (b) M as a function of $\Delta\lambda = \lambda - \lambda_c$, where λ_c is defined by the maximum value of mutual information, in semi-log scale. The data can roughly be fit by $M(|\Delta\lambda|) \sim e^{-\alpha|\Delta\lambda|}$, with $\alpha \approx 8$.

the distribution of the spreading rate of the difference pattern for elementary rule 110 (a complex rule) and for elementary rule 30 (a chaotic rule). The distribution is accumulated over an ensemble of different initial conditions. It is clear that rule 110 has a wider distribution of spreading rate.

7. Large rule table limit

When more neighbors are involved in updating each site, site values become increasingly sensitive to sites at larger distances. One might suppose that this increased inter-dependence among sites would make random dynamics more likely. In this section, we will show that this is indeed the case, and that the critical point λ_c which separates regular and chaotic rules approaches zero in the large neighborhood limit (the "thermodynamic limit").

Since spatial mutual information is a measure of correlation among the parts of a configuration, as discussed in section 5, searching the peak in the mutual information relation is an effective way to locate the transition region. In order to study the

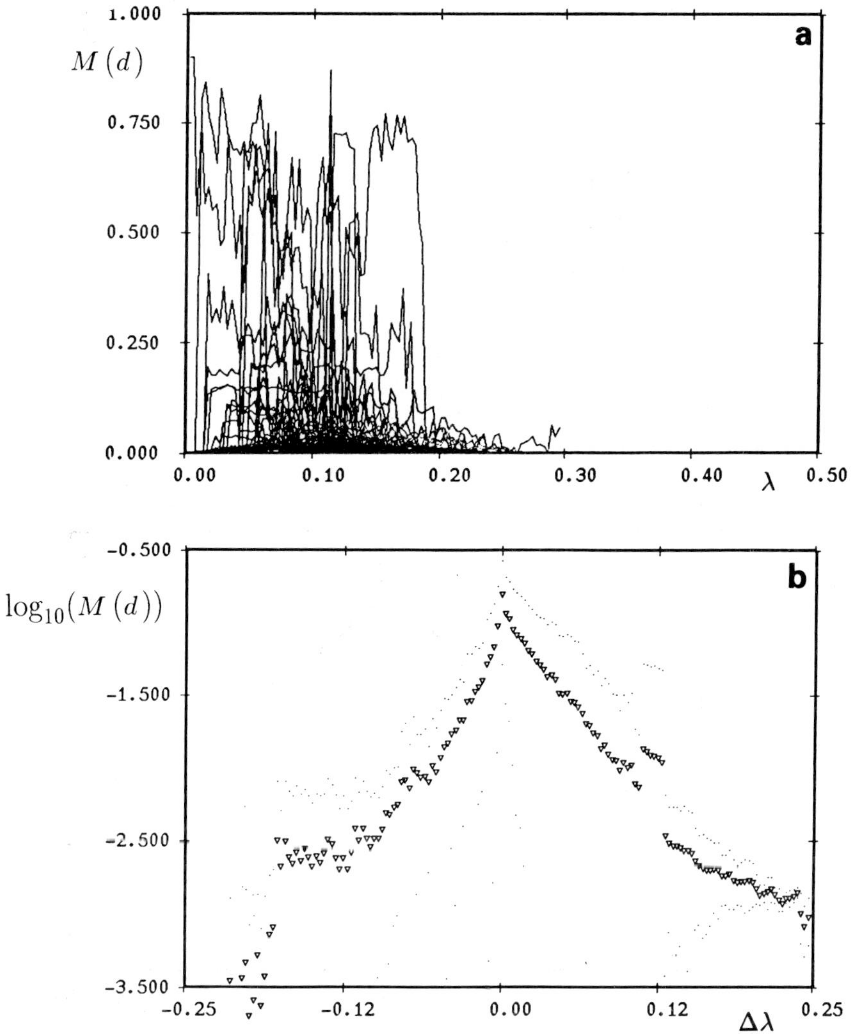

Fig. 7. Spatial mutual information M between two 3-blocks separated by distance 10 for $r = 7$, instead of $r = 5$, rule space. (a) M as a function of λ. (b) M as a function of $\Delta\lambda = \lambda - \lambda_c$, where λ_c is defined by the maximum value of mutual information, in semi-log scale.

effect of neighborhood size on the location of the transition region, we calculate the spatial mutual information versus λ for different radius rules.

The results for rules with radius $r = 3, 5, 7, 9,$ and 11 are shown in fig. 9 (the plot is taken from ref. [8]). The mutual information is computed between two blocks of 3 sites each, separated by a distance of 8 sites. The lattice size is 167, 118 transient configurations are discarded, and the next 89 configurations are used to accumulate the probability. Rules are picked randomly for each λ value,

rather than by repeatedly adding non-zero values in the rule table.

Fig. 9 shows clearly that the transition region – as indicated by the peak of mutual information – moves toward smaller values of λ as r is increased. In other words, a larger proportion of cellular automata rules generate random dynamics as the radius of coupling r is increased. If we determine λ_c as the location of the peak in the spatial mutual information, then λ_c as a function of the number of neighbors $n = 2r + 1$ is shown in fig. 10.

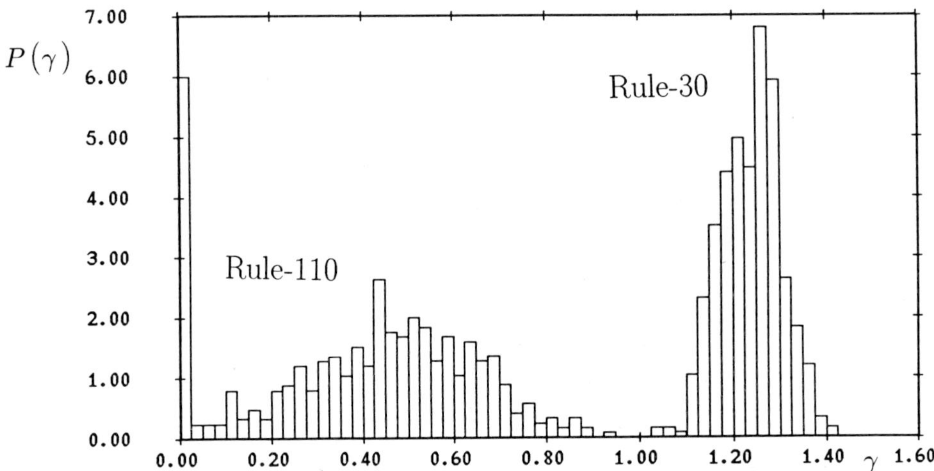

Fig. 8. The distribution of spreading rates for elementary rules 110 (on the left side of the plot) and 30 (on the right side).

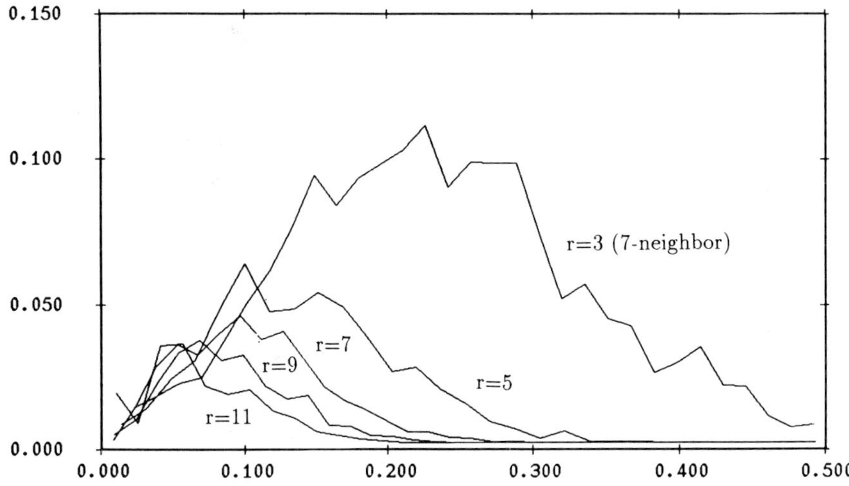

Fig. 9. Spatial mutual information for randomly sampled rules with $r = 3, 5, 7, 9$, and 11 respectively. For each r value and fixed λ value, 51 rules are sampled. The lattice size is 167, 118 transient configurations are discarded, and the next 89 configurations are used to accumulate the probability distribution. The mutual information is between two 3-blocks separated by distance 8.

How does λ_c change in the $n \to \infty$ limit (fully connected cellular automata)? A numerical simulation becomes more and more difficult since the size of the rule table expands as 2^{2r+1}, which quickly reaches the limit of storage space. Nevertheless, some simple considerations lead to the conclusion that λ_c approaches 0 in the limit of large n.

There are many other ways to define λ_c. Another definition of λ_c is by the onset of non-zero spreading rates for difference patterns. By mean-field calculations, included in appendix B, we can easily derive λ_c as a function of the radius r (eq. (B.8)):

$$\lambda_c = \frac{1}{2} - \frac{1}{2}\sqrt{1 - \frac{2}{r+1}}. \tag{3}$$

In the large r limit, this λ_c goes to zero:

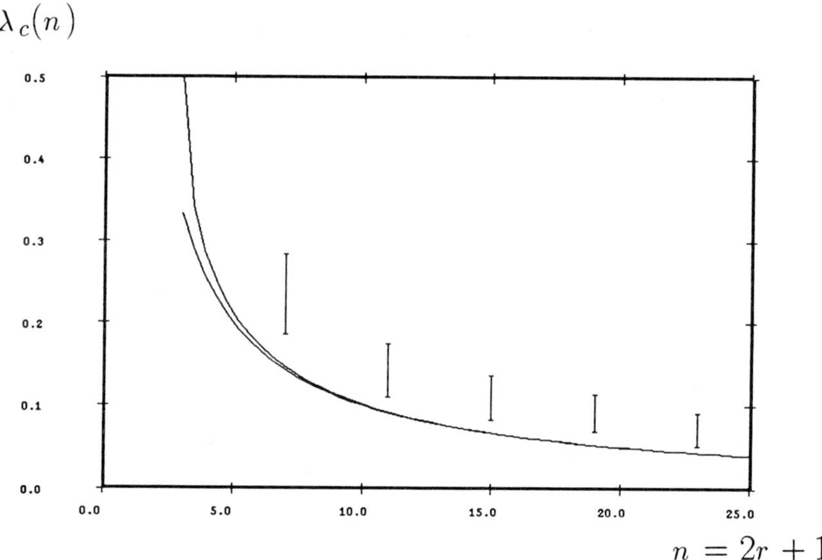

Fig. 10. Critical region λ_c as determined by the maximum value of the spatial mutual information plotted as a function of the neighborhood size $(2r + 1)$ for one-dimensional 2-state cellular automata rule spaces. Also plotted are two functions for the onset of non-zero entropy $(1/(2r + 1))$ and non-zero expansion rate for difference patterns $(\frac{1}{2} - \frac{1}{2}\sqrt{1 - 2/(r + 1)})$ according to mean field theory.

$$\lambda_c \approx \frac{1}{2(r + 1)} + \mathrm{O}\left(\frac{1}{r^2}\right) . \tag{4}$$

The case for $\lambda_c = 0$ only means that it becomes increasingly hard to find rules with non-random dynamics in the large neighborhood limit. This does *not* imply that there are no non-random rules at all. We will discuss this point in the last section.

λ_c as determined by eq. (4) is also plotted in fig. 10, as well as $\lambda_c = 1/(2r + 1)$ for the onset point of non-zero entropy estimated by mean-field theory [21]. Numerical estimates of the transition point always exceed the mean-field estimates.

8. Structure of rule space

Piecing the above results together, a clear picture of the fundamental structure of cellular automata rule spaces emerges, although there are still some details that need to be worked out.

There are two primary regimes of rules – *periodic* and *chaotic* – separated by a *transition regime*. This transition regime is *not* simply a smooth surface separating the other two domains, but itself has a complicated structure. Most of this transition regime seems to be simply a boundary between periodic and chaotic rules, containing no rules within it. Crossing the transition regime at such a boundary gives rise to a discrete jump in statistical measures of the dynamics, as is seen in first-order transitions.

However, other parts of this transition regime seem to have some "thickness," in the sense that they contain the so-called "critical" rules. Crossing the transition regime through these areas gives rise to smooth changes in statistical measures, suggesting a second-order transition.

This basic picture is illustrated schematically in fig. 11. The existence of this phase transition separating a domain of ordered dynamics from a domain of disordered dynamics, coupled with the behavior of CA within each of the domains as one approaches the phase transition, provides a fundamental explanation for the various qualitative classes of cellular automata behaviors identified by us and by Wolfram.

More importantly, however, this picture also provides an explanation for the relationship that obtains *between* these various qualitative classes, which has been lacking previously. The relative locations of these various classes are indicated in fig.

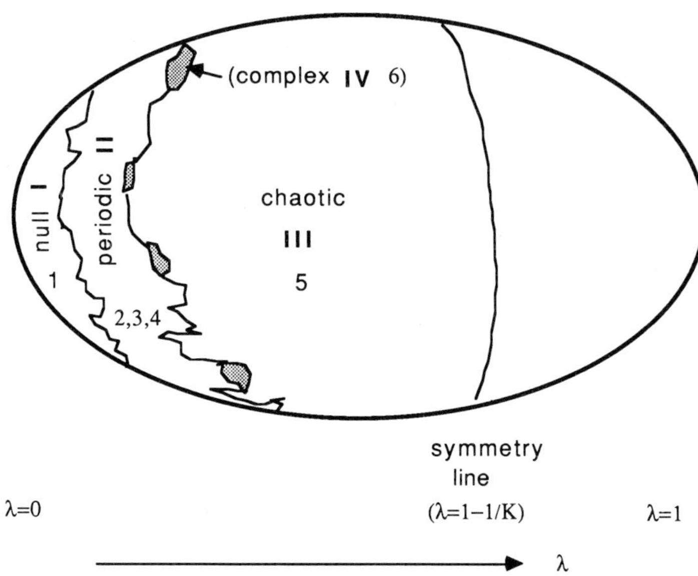

Fig. 11. A schematic picture of the structure of cellular automata rule-space, indicating the relative location of the various qualitative classes. The λ parameter is plotted from left to right, and there is no special meaning associated with the vertical axis. Complex rules are asociated with the small, shaded patches along the boundary between periodic and chaotic rules.

11. The complex rules (Wolfram's class IV and our class 6) are found in the transition regime separating the periodic rules (Wolfram's classes I and II, and our classes 1, 2, 3, and 4) from the chaotic rules (Wolfram's class III and our class 5).

Fig. 12 plots one of our statistical measures (mutual information) over the $\vec{\rho}$ parameter space (discussed earlier) for a five-neighbor, $k = 3$, 2D CA, in order to illustrate empirically the relative locations and sizes of the various regimes of CA behaviors. The chaotic rules, the largest class, occupy the vast central depression. The fixed point and periodic rules lie to the outside of the peaks in the mutual information surface, towards the vertices of the base-triangle. The "complex" rules lie in the vicinity of the peaks in the mutual information surface. In this plot, the λ parameter would constitute a verticle slice through the surface, running from one of the vertices of the base-triangle to the center of the triangle (the middle of the central depression), with $\lambda = 0$ at the vertex and $\lambda = 1 - 1/k$ at the center. The mutual information surface has been smoothed in this figure by averaging over neighboring ρ-points.

The relative location of the different qualitative

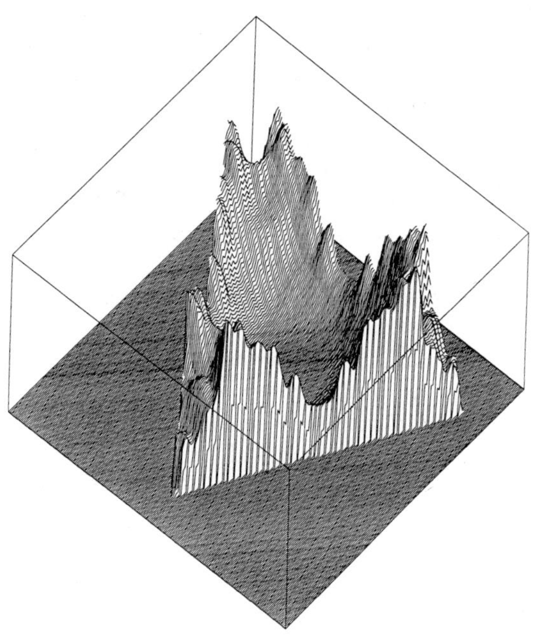

Fig. 12. The smoothed mutual information surface plotted over the ρ parameter space for $k = 3$, 2D CA. The chaotic rules constitute the vast central depression, fixed-point and periodic rules lie between the vertices of the base-triangle and the peaks in the mutual information surface, and the complex rules are found in the vicinity of these peaks.

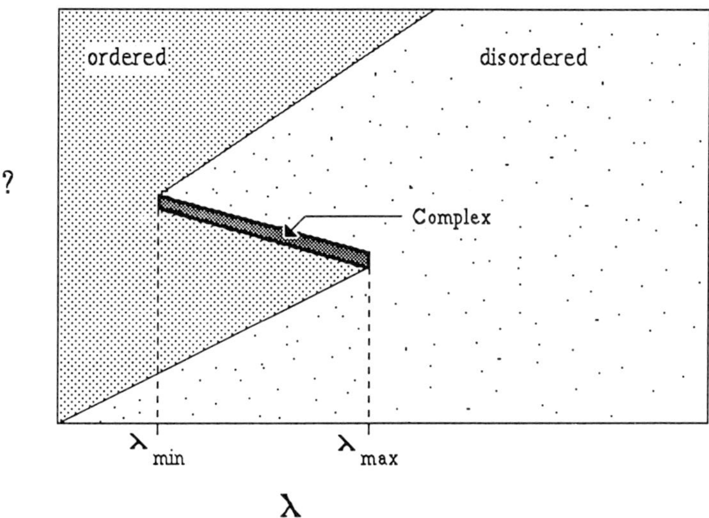

Fig. 13. A schematic diagram of a 2D phase-diagram for CA and its projection onto the λ parameter. The dark-shaded region in the middle is supposed to constitute the "critical" region. Without knowing the value of the "mystery" parameter on the ordinate, one would see both first and second order transitions over a range of λ values, as we observe experimentally. If we find other appropriate parameters governing CA dynamics, we should be able to locate the boundaries between the various dynamical regimes precisely.

classes leads immediately to a fundamental conjecture about complex behavior in general. Since we have uncovered an apparent association between the most complex rules and second order transitions, and since these complex rules seem to be associated with a capacity for computation – even universal computation – we have a fortiori uncovered a fundamental connection between complex behavior, computation, and phase transitions – especially second-order transitions. Indeed, one can identify analogs for characteristic properties of computing systems in the phenomenology of phase transitions [7]. Examples include the existence of different "complexity classes" associated with increases in transient time as one approaches a phase transition, and an analog of Turing's famous halting problem in the "critical slowing down" associated with second-order transitions [#2].

Another observation that can clearly be made is that the λ parameter alone is insufficient for locating specific dynamical regimes precisely. For many physical systems exhibiting phase transitions, more than a single parameter is required to

accurately reveal the phase-transition structure. For instance, the transition point from a solid to a fluid is not captured precisely by temperature alone, one must also control the pressure. For any specific pressure, there is a unique melting temperature, but if pressure is not being controlled carefully in an experiment, one will observe a range of temperatures at which melting will be observed to occur.

This suggests that we will have to find at least one more parameter affecting the dynamics of cellular automata before we can fill out all of the details of the transition or bifurcation structure of cellular automata rule spaces.

The situation is illustrated schematically in fig. 13. In this figure there are two primary domains of behavior, an ordered domain – the periodic rules – and a disordered domain – the chaotic rules. These two domains are separated by a transition regime with a complicated structure. Part of this transition regime – the dark-shaded region in the figure – contains the "critical" rules. The λ parameter controls the location along the abcissa, while some unknown parameter controls the location along the ordinate. We can see that, without controlling the specific value of the "mystery" pa-

---

[#2] This latter association was observed independently by Vichniac et al. in ref. [17].

rameter, we would expect to see transitions – both first and second order – over a wide range of λ values, with the critical rules restricted to lie within the boundaries delimited by λ_{min} and λ_{max}.

There are various proposals for the "mystery" parameter(s), including the possibility of defining generalized versions of the thermodynamic quantities of temperature, pressure, density, energy, and so forth [7]. Other proposals involve various quantities deriveable from mean-field theory [21], or replacing the one-dimensional λ parameter by a subset of the mean-field theory parameters [3]. So far, none of the proposals provide a completely satisfactory theoretical framework.

9. Conclusion

The space of cellular automata rules seems to be divided into regions of similar behavior. Division into regions makes sense only with a metric on the space of rules. We use Langton's λ parameter, which is simply the percentage of ones in the cellular automaton rule table (in the case of binary state automata).

Since the space of cellular automaton rules is finite and discrete, λ can take on only a finite number of values. One can, nevertheless, move in the space of rules along paths where λ increases monotonically in minimal increments. For a binary state automaton, this is accomplished by incrementing the number of ones in the cellular automaton rule table. This is a discrete analog of moving along a bifurcation arc in a space of smooth dynamical systems.

We find that as we increase the λ parameter, statistical quantities, such as entropy and difference pattern spreading rate, often undergo abrupt changes at the transition between ordered and disordered behavior. These discontinuous changes involve a jump, and have the character of a first-order phase transition. Occasionally, however, a smooth change in statistical measures is observed, indicating that something like a second-order phase transition is possible. In these latter transitions, a phenomenon analogous to "critical slowing down" is observed.

For "critical" rules – rules at or very near such a second-order transition – statistical quantities do not converge well. We propose a measure of the lack of convergence for such rules: the width of the distribution of values of a statistic. Using this measure, rules away from the transition have narrow distributions, while rules near the transition have wide distributions in statistical measures. This provides evidence for the hypothesis that rules at the transition are capable of nontrivial (possibly even universal) computation, even with random initial conditions. At the transition, a mixture of stability with respect to some local configurations and instability with respect to others may well provide a cellular automaton rule with the necessary requirements to both store and transmit information, two requirements crucial to computation [5–7,14].

Although we have presented a general picture of cellular automaton rule space, the detailed structure is complicated. Since an elementary rule (1D, 3-neighbor) can be considered as a 5-neighbor rule with no effects from the outer 2 sites, the elementary rule space is contained in the larger rule space of 5-neighbor cellular automata, which again is contained in even larger rule spaces. We know from fig. 1 that there are periodic elementary rules at $\lambda = 0.5$, and these rules will presist at $\lambda = 0.5$ in larger rule spaces, even as $\lambda_c \rightarrow 0$. The implication is that as the rule space becomes larger, it becomes more difficult, but not impossible, to find periodic rules at $\lambda > \lambda_c$.

Although the λ parameter is not sufficient for locating the different dynamical regimes within cellular automata rule spaces precisely, it is sufficient to reveal important structural details of these spaces. Furthermore, the structure revealed explains not only the existence of, but also the relationship between, previously proposed qualitative classes of cellular automata dynamics. Perhaps most importantly, the association of complex rules with second-order transitions in the space of rules suggests a fundamental connection between computation, complex dynamics, and phase transitions.

Acknowledgements

We would like to thank J. Crutchfield, D. Farmer, H. Gutowitz, H. Hartman, E. Jen, S.

Kauffman, R. Shaw, and B. Wootters for discussions. The work was supported in part by NSF Grant PHY-87-14918 and DOE Grant DE-FG05-88ER25054 at SFI, by NSF Grant PHY-86-58062 and ONR Grant N00014-88-K-0293 at CCSR of University of Illinois.

Appendix A

Definition of statistical quantities

A.1. Spreading rate of difference patterns

The left-moving difference pattern is obtained by first taking a configuration $a = \{\ldots a_{-1}, a_0, a_1, \ldots\}$ and constructing another configuration $a' = \{\ldots a'_{-1}, a'_0, a'_1, \ldots\}$ with $a_i = a'_i$ for $i < 0$, $a_i \neq a'_i$ for $i = 0$, and a'_i chosen at random (independent of the a_i) for $i \geq 0$. The difference pattern at time t is then defined as $\delta^t = 0$ if $a^t = a'^t$ and $\delta^t = 1$ otherwise, where a and a' evolve in time according to the action of the cellular automaton rule. The location of the front of the difference pattern is $i^t_f = \min\{i | \delta^t_i = 1\}$. The left-moving difference rate is then defined as

$$\gamma_{\text{left}}(a) = \lim_{t,\tau \to \infty \text{ with } t/\tau \to 0} \frac{i^\tau_f - i^{\tau+t}_f}{t - \tau} \qquad (\text{A.1})$$

providing the limit exists. The right-moving difference rate is defined similarly, with δ initially zero on the other half of the lattice, $\delta_i = 0$ for $i > 0$, and now letting $i^t_f = \max\{i | \delta^t_i = 1\}$

$$\gamma_{\text{right}}(a) = \lim_{t,\tau \to \infty \text{ with } t/\tau \to 0} \frac{i^{\tau+t}_f - i^\tau_f}{t - \tau} \qquad (\text{A.2})$$

providing the limit exists. The total difference spreading rate is then defined to be

$$\gamma(a) = \gamma_{\text{left}}(a) + \gamma_{\text{right}}(a). \qquad (\text{A.3})$$

Here, we will define a rule to be chaotic if $\gamma > 0$. $\gamma(a)$ is observed to be independent of a, if a is chosen at random, and we will often simply refer to the generic asymptotic spreading rate simply as γ. γ is measured empirically by taking a and a' to differ initially at only one site, and measuring the rate that both the left-moving and right-moving

fronts move simultaneously. Transients are allowed to die out and an average is taken over many initial values of a and a'.

A.2. Entropy

Given a probability distribution $\{p_i\}$, the definition of entropy is well known:

$$S = -\sum_i p_i \log(p_i). \qquad (\text{A.4})$$

There are many ways entropy can be applied to the spatial-temporal patterns of cellular automata. For example, the single site spatial-temporal entropy is calculated by counting the occurrence of all symbols $\{c_i\}(i = 1, \ldots, k)$ from the spatial configurations at many time steps, and the probability distribution is $\{p_i\} = \{c_i / \sum_j c_j\}$.

Sometimes, we are interested in characterizing the randomness in a particular direction in the spatial-temporal pattern, then the probability distribution is accumulated along that direction, rather than the whole spatial-temporal pattern. For example, the spatial entropy, when we only analyze the configuration at a fixed time step, and temporal entropy, when only the time sequence on a particular site is studied. These entropies are called "directional entropies" [13].

A.3. Mutual information

Mutual information is a function of two probability distributions $\{p_i\}$ and $\{p_j\}$. In order to characterize the correlation between the two distributions, we also need the joint probability $\{p_{ij}\}$ for both event i in the first variable and event j in the second variable to occur. The mutual information between the two distributions is defined as [15]

$$M = \sum_i \sum_j p_{ij} \log \frac{p_{ij}}{p_i p_j}. \qquad (\text{A.5})$$

To apply mutual information to the spatial-temporal patterns of the cellular automata, the two probability distributions can be two probabilities of having k values on two sites separated

by distance d [9]; they can also be probabilities of having k^l values on two l-blocks separated by d. If the distance is a spatial distance, we are calculating the spatial mutual information. If the distance is a time delay on the same site, we are calculating the temporal mutual information. In general, mutual information between two sites (or two blocks) separated by any spatial-temporal distances can be determined.

Appendix B

Estimation of the spreading rate of difference pattern by mean-field theory

Considering one-dimensional, 2-state, $(2r + 1)$-neighbor cellular automata, the maximum possible spreading rate of the difference pattern to the left (and right) is r, the range of the coupling, and the total maximum spreading rate is

$$\gamma_{\max} = 2r. \tag{B.1}$$

Typically, this maximum spreading rate of difference patterns will not be reached because expansions of perturbation with smaller spreading rates can also occur with non-zero probability. Suppose we are looking at rules with a particular λ value, i.e., λ percent of the $(2r + 1)$-blocks map to symbol 1, and $1 - \lambda$ percent of them map to symbol 0. Randomly choose two $(2r + 1)$-blocks, the probability that they map to the same symbol is

$$P_{\text{same}} = \lambda^2 + (1 - \lambda)^2 \tag{B.2}$$

and the probability that they map to different symbols is

$$1 - P_{\text{same}} = 2\lambda(1 - \lambda). \tag{B.3}$$

The probability that the left spreading rate is r is equal to the probability that block $(a_{-r}a_{-(r-1)} \cdots a_{r-1}1)$ and $(a_{-r}a_{-(r-1)} \cdots a_{r-1}0)$ map to different symbols, which is the same as the probability that two randomly picked $(2r + 1)$-blocks will map to different symbols: $1 - P_{\text{same}}$. The contribution from the maximum spreading to the average left spreading rate is

$$(1 - P_{\text{same}})r. \tag{B.4}$$

Similarly, the probability that the left spreading rate is $r - 1$ is equal to the probability that block

$(a_{-r}a_{-(r-1)} \cdots a_{r-1}1)$ and $(a_{-r}a_{-(r-1)} \cdots a_{r-1}0)$ map to the same symbol, which is P_{same}, multiplied by the probability that blocks $(a_{-r}a_{-(r-1)} \cdots 1a_r)$ and $(a_{-r}a_{-(r-1)} \cdots 0a_r)$ map to different symbols, $1 - P_{\text{same}}$. After counting all possibilities, including the contribution of the large negative spreading rates, the average left spreading rate of the difference pattern is

$$\gamma_{\text{left}} = \sum_{i=0}^{\infty} (r - i)P_{\text{same}}^i(1 - P_{\text{same}})$$

$$= r - \frac{P_{\text{same}}}{1 - P_{\text{same}}}. \tag{B.5}$$

The average spreading rate taking into account both left and right expansions is (expressing P_{same} in terms of λ):

$$\gamma = \frac{(2r + 2)\lambda - (2r + 2)\lambda^2 - 1}{\lambda(1 - \lambda)}. \tag{B.6}$$

There are two interesting results which can be derived from the above formula. First, at $\lambda = 0.5$,

$$\gamma_{\text{random}} = 2(r - 1) < 2r = \gamma_{\max}, \tag{B.7}$$

which says that even in the most random cases, the maximum spreading rate will not be reached. Second, the critical λ_c can be defined as the onset of the non-zero value of γ. Setting $\gamma(\lambda_c) = 0$, we can express λ_c as a function of r:

$$\lambda_c = \frac{1}{2} - \frac{1}{2}\sqrt{1 - \frac{2}{r + 1}}. \tag{B.8}$$

References

[1] H. Chaté and P. Manneville, Criticality in cellular automata, Physica D 45 (1990) 122–135, these Proceedings.

[2] J.P. Crutchfield and N.H. Packard, Bifurcations in discretized spatially-extended systems, Abstract, Cellular Automata'86 Workshop (1986).

[3] H.A. Gutowitz, A hierarchical classification of cellular automata, Physica D 45 (1990) 136–156, these Proceedings.

[4] K. Kaneko, Pattern dynamics in spatio-temporal chaos: pattern selection, diffusion of defect and pattern competition intermittency, Physica D 34 (1989) 1–41.

[5] C.G. Langton, Studying artificial life with cellular automata, Physica D 22 (1986) 120–140.

[6] C.G. Langton, Computation at the edge of chaos: phase transitions and emergent computation, Physica D 42 (1990) 12–37.

[7] C.G. Langton, Computation at the edge of chaos, Ph.D. Thesis, University of Michigan (1990).

[8] W. Li, Problems in complex systems, Ph.D. Thesis, Columbia University (1989); University Microfilm International, Ann Arbor, MI.

[9] W. Li, Mutual information functions versus correlation functions, J. Stat. Phys. 60 (1990) 823–837.

[10] W. Li and N. Packard, Structure of the elementary cellular automata rule space, Complex Systems 4 (3) 1990, in press.

[11] N. Margolus and T. Toffoli, Cellular Automata Machines (MIT Press, Cambridge, MA, 1987).

[12] H. McIntosh, Wolfram's class IV automata and a good Life, Physica D 45 (1990) 105–121, these Proceedings.

[13] J. Milnor, Directional entropies of cellular automaton-maps, in: Disordered Systems and Biological Organizations, eds. E. Bienstock et al. (Springer, Berlin, 1986)

[14] N. Packard, Adaptation toward the edge of chaos, Center for Complex Systems Research Technical Report, University of Illinois, CCSR-88-5 (1988).

[15] C.E. Shannon, The mathematical theory of communication, Bell Syst. Tech. J. 27 (1948) 379–423.

[16] R. Shaw, Information density near a phase transition, Abstract, Cellular Automata'86 Workshop (1986).

[17] G. Vichniac, P. Tamayo and H. Hartman, Annealed and Quenched Inhomogeneous Cellular Automata (INCA), J. Stat. Phys. 45 (1986) 875–883.

[18] J. von Neumann, Theory of self-reproducing automata, ed. A. Burks (Univ. Illinois Press, Champaign, IL, 1966).

[19] S. Wolfram, Statistical mechanics of cellular automata, Rev. Mod. Phys. 55 (1983) 601–644.

[20] S. Wolfram, Universality and complexity in cellular automata, Physica D 10 (1984) 1–35.

[21] W.K. Wootters and C.G. Langton, Is there a sharp phase transition for deterministic cellular automata?, Physica D 45 (1990) 95–104, these Proceedings.

Physica D 45 (1990) 95–104
North-Holland

IS THERE A SHARP PHASE TRANSITION FOR DETERMINISTIC CELLULAR AUTOMATA?

William K. WOOTTERS [a,b,c] and Chris G. LANGTON [b]

[a] *Santa Fe Institute, 1120 Canyon Road, Santa Fe, NM 87501, USA*
[b] *Theoretical Division and Center for Nonlinear Studies, Los Alamos National Laboratory, Los Alamos, NM 87545, USA*
[c] *Department of Physics, Williams College, Williamstown, MA 01267, USA*

Received 2 April 1990
Revised manuscript received 26 April 1990

Previous work has suggested that there is a kind of phase transition between deterministic automata exhibiting periodic behavior and those exhibiting chaotic behavior. However, unlike the usual phase transitions of physics, this transition takes place over a range of values of the parameter rather than at a specific value. The present paper asks whether the transition can be made sharp, either by taking the limit of an infinitely large rule table, or by changing the parameter in terms of which the space of automata is explored. We find strong evidence that, for the class of automata we consider, the transition does become sharp in the limit of an infinite number of symbols, the size of the neighborhood being held fixed. Our work also suggests an alternative parameter in terms of which it is likely that the transition will become fairly sharp even if one does not increase the number of symbols. In the course of our analysis, we find that mean field theory, which is our main tool, gives surprisingly good predictions of the statistical properties of the class of automata we consider.

1. Introduction

Of the four Wolfram classes [1] of deterministic cellular automata – homogeneous, periodic, chaotic, and complex – those exhibiting complex behavior are for many purposes the most interesting, but they also seem to be the most rare, particularly when the rule table for the automaton is large [2]. Thus, when the rule table is large, one can think crudely of the set of all rules as being divided into two large classes: those leading to periodic behavior (we count a static homogeneous automaton as trivially periodic), and those leading to chaotic behavior.

Langton [3–5], Li and Packard [6] and Li, Packard and Langton [7] have shown that if one uses an appropriate parameter, which we call λ, to characterize cellular automaton rules, then one can see a kind of phase transition between these two modes of behavior as the value of the parameter is varied: For small values of λ the behavior is typically periodic; for large values it is typically chaotic. However, there

is not a unique value of λ at which the transition occurs, but rather a range of possible transition values. In this respect the transition differs from the usual phase transitions of physics, which occur at definite values of, say, temperature or magnetic field strength. The present paper asks whether in some appropriate limit – we consider particularly the limits of large neighborhood size and large number of symbols – or with a different choice of parameter, the value at which the transition occurs becomes unique. Our results indicate that the transition does become sharp in the limit of an infinite number of symbols. We also propose a manageable experiment in which one might hope to see a reasonably sharp transition with a small set of symbols.

The question of the sharpness of the transition is interesting for a number of reasons. First, there is a good deal of evidence that automata exhibiting complex behavior are located near the transition region. If there are situations in which the transition is sharp, then in those situations one could conceivably be bet-

ter able to find complex automata, because one would know precisely where to look. Also, Packard has obtained data suggesting that cellular automaton rules, evolving under selective pressure towards performing a particular computational task, tend to gravitate to the transition region [8]. It would be interesting to see, in such an evolution, if the final evolved rules continue to be confined to the transition region as the latter becomes narrower and narrower. Finally, there are some intriguing parallels between the theory of computation and the theory of physical phase transitions [4,5], and these parallels would be given added significance if one could find in deterministic cellular automata the sort of sharp transition that one finds in physics.

The approach taken in this paper is the following. The location of the transition point is a difficult quantity to treat theoretically. But there is a more accessible quantity, the single-site entropy, which also exhibits a range of values and which is related to the location of the transition point. We study the entropy in detail, asking whether for fixed λ it approaches a unique value in the limit of large neighborhood size or large number of symbols. If it does, then it seems likely that the location of the transition point becomes unique in the same limit. We supplement our analysis of finite-state automata with numerical results bearing particularly on the case of an infinite number of symbols. Throughout this paper, we focus exclusively on a specific set of data for a particular class of automata. Let us therefore begin by describing this class.

2. Our automata

The automata we consider are two-dimensional, with a square lattice, and with a five-cell neighborhood shaped like a "+". Each site can be in one of eight possible states; in other words, each site can be occupied by one of eight possible symbols, say, the digits 0 through 7. The symbol 0 plays a special role: if a neighborhood consists entirely of zeros, then its central site is always assigned the value zero in the next step. The rules are required to be invariant under rotations. That is, if a certain neighborhood configuration sends its central site to the state x, then any rotated version of that configuration must also send its central site to x. Except for these restrictions, all rules are allowed. The lattices actually used to generate the data were 64 units on a side, with periodic boundary conditions.

The rule table can be pictured as a long column of cross-shaped neighborhoods, with each possible neighborhood configuration represented exactly once. To the right of each configuration one places the symbol to which that particular configuration is mapped. The parameter λ is defined as the fraction of neighborhood configurations that are mapped to something other than zero. Thus a rule with very small λ has a tendency to generate many zeros, and ultimately such an automaton is likely to end up in the quiescent state consisting of nothing but zeros. As λ moves away from zero, one has the possibility of more interesting behavior.

We will be looking particularly at the dependence of entropy on λ. The data were generated as follows: Fix a value of λ, and choose at random a rule table with that particular λ value. For this rule table, start with a random initial configuration, and without taking any data, let the automaton run for 500 time steps to give it a chance to reach its asymptotic behavior. Then let it continue to run (for 1000 steps), and count the frequencies of occurrence of the eight possible symbols. From these frequencies p_s, compute the entropy of the automaton, defined as

$$H = - \sum_{s=0}^{7} p_s \log p_s , \qquad (1)$$

where the logarithm is base 2. Then plot the point (λ, H). Now do this one hundred times for each value of λ, with a different randomly chosen rule table each time. Finally, repeat the whole process for many different values of λ ranging from zero to 7/8. ($\lambda = 7/8$ is the value for which all symbols are equally represented in the rule table. For larger values one does not expect particularly interesting behavior.) The result is the graph shown in fig. 1.

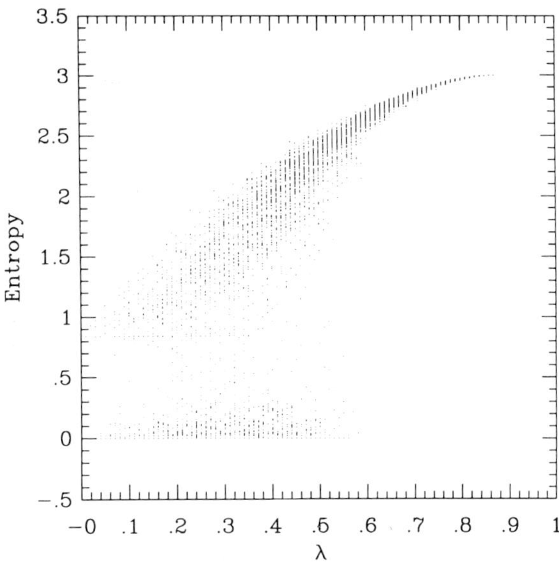

Fig. 1. Measured values of single-site entropy for automata with different values of λ. (λ is the fraction of nonzero entries in the rule table.) For each λ, one hundred rule tables were constructed at random. Each automaton was allowed to run for 500 time steps before the entropy was measured.

Note that in the graph, the automata can be grouped loosely into two categories. There is a band of low-entropy automata running along the bottom of the graph, and most of the other automata have entropy above a fairly well defined line at 0.84 bits. Roughly speaking, the low-entropy automata are typically periodic, and those in the high-entropy category are typically chaotic. If one starts with the $\lambda = 0$ automaton and slowly adds more nonzero elements to the table, one usually finds the following: The entropy remains near zero as λ increases, and then at some point it abruptly jumps to a value greater than 0.84. Occasionally the entropy jumps to an intermediate value before going above 0.84, and these intermediate automata are represented by the scatter of points lying between the two larger clumps. The value of λ at which the jump occurs varies from one trial to the next.

Thus there is a certain sense in which the transition is sudden – in a given trial it is sudden – but in another sense the transition is spread over a range of λ values. Our question is whether this range can be made to shrink to a single value.

It is not clear what will happen to the graph of fig. 1 as the size of the neighborhood (n) or the number of symbols (k) grows. In particular, it is not clear whether the spread in entropy for each value of λ will shrink or will stay essentially constant. In a certain sense, n and k are already quite large: there are 8^5 entries in the rule table, and the number of possible rule tables is roughly equal to $8^{8^5/4}$. (The requirement of rotational invariance reduces the number of choices by a factor of about 4.) Thus it is conceivable that we are already close to the infinite-rule-table limit. The next four sections are devoted to accounting for the observed spread in entropy and determining how it is likely to depend on n and k.

3. Simple mean field theory

Before trying to account for the spread, let us illustrate our method by finding a simple theoretical estimate of the average entropy as a function of λ.

Let ν_t be the density of zeros in the pattern after t time steps. We can also think of ν_t as the probability that a randomly chosen site will have a zero in it. We now compute ν_{t+1} as follows. For a given site, there are two ways in which that site can get a zero in the next step: Its whole neighborhood could be filled with zeros, in which case the site gets a zero with probability 1, or its neighborhood could contain at least one nonzero symbol, in which case the site will get a zero with probability $1-\lambda$. We assume that the probability of the neighborhood being all zeros at time t is ν_t^5. Thus the probability that the site will get a zero in the next step is

$$\nu_{t+1} = \nu_t^5 + (1 - \nu_t^5)(1-\lambda) . \tag{2}$$

Once the automaton has reached its steady state, the density ν should not change, so ν should satisfy the equation

$$\nu = \nu^5 + (1 - \nu^5)(1-\lambda) . \tag{3}$$

For $\lambda \le 0.2$ the only stable solution of this equation is $\nu = 1$ (all zeros), and for $\lambda > 0.2$ there is again exactly one stable solution, which is less than 1. The pre-

dicted density of nonzero symbols, $1 - \nu$, is plotted as a function of λ in fig. 2. To estimate the entropy, we simply assume that all seven nonzero symbols appear equally often, so that the entropy is related to ν by the equation

$$H = -\left[\nu \log \nu + 7 \left(\frac{1-\nu}{7} \right) \log \left(\frac{1-\nu}{7} \right) \right]. \qquad (4)$$

This estimate of the average entropy is plotted against λ in fig. 3. Notice that the curve does go roughly through the middle of the actual data. But to account for the spread in the data, we clearly need to be more sophisticated.

4. More sophisticated mean-field theory

In the above calculation, the rule table was characterized by the value of a single parameter, λ. Following Wolfram [9] and Gutowitz et al. [10], we now make a more refined characterization of the rule table. Let b_j be defined as follows: For a given rule table, consider all the neighborhood configurations containing exactly j zeros; b_j is the fraction of these

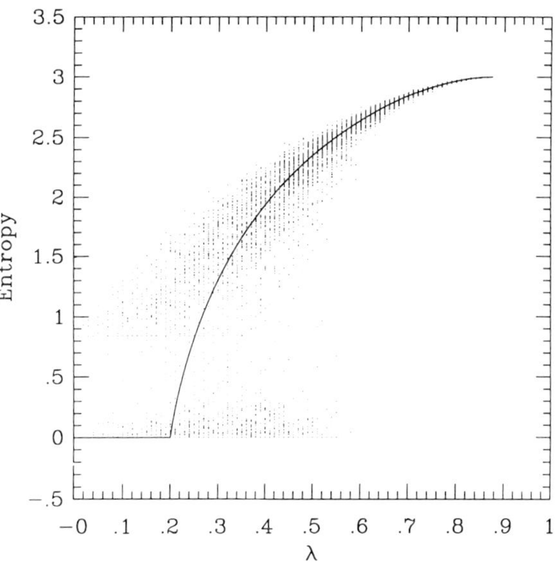

Fig. 3. The solid curve is the single-site entropy as predicted by simple mean-field theory. The dots are the observed values, just as in fig. 1.

which are mapped to zero. In effect we are breaking up the rule table into six sections ($j=0$ to 5), such that in each section all the neighborhood configurations have the same number of zeros. The quantity $1 - b_j$ is like a "local" λ for the jth section. Once one knows the set $\{b_j\}$, one can easily compute the value of λ for the whole rule table, but the converse is not true. A given value of λ is consistent with many possible sets $\{b_j\}$.

Our strategy now is simple. Knowing that for each value of λ the rules were chosen at random, we can find the distribution of the b_j's for each λ. For each set $\{b_j\}$ we can compute the mean-field-predicted value of the entropy. Because there are many sets $\{b_j\}$ for each value of λ, there will be a range of entropy values for each λ. Our question is whether this range agrees with the observed range of values.

First let us write down the mean-field equation for the density ν of zeros, using the more refined estimates of probabilities provided by the b_j's. We write the equation in terms of arbitrary values of n and k, since we are ultimately interested in knowing how things depend on these quantities. For a given site,

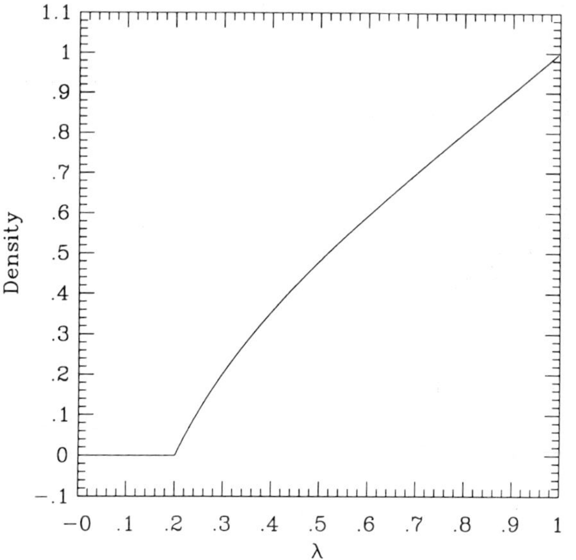

Fig. 2. Density of nonzero symbols as a function of λ, as predicted by simple mean-field theory.

the probability that its neighborhood has exactly j zeros is equal to $v^j(1-v)^{n-j}\binom{n}{j}$. (Here we make the usual mean-field approximation that the probabilities for different sites are independent.) Given that the neighborhood has exactly j zeros, the probability that the site will get a zero is b_j. Thus the equation for a stationary value of v is

$$v = v^n + \sum_{j=0}^{n-1} v^j(1-v)^{n-j}\binom{n}{j}b_j . \tag{5}$$

Here we have used the fact that b_n is equal to unity; that is, a neighborhood containing nothing but zeros is certain to yield a zero. The contribution from such a neighborhood is the first term on the right-hand side. (We use mean-field theory here only to find the stationary density. For an example of what can be done with the full dynamical mean-field theory, see ref. [11].)

It is now a matter of mathematics to find the distribution of the b_j's for each λ, and from that to find the distribution of values of v, and finally to find the distribution of values of the entropy. Let us begin by finding the distribution of the b_j's

In the actual experiment, the rule tables were constructed by proceeding through all rotationally inequivalent neighborhood configurations, and assigning to each one the symbol zero with probability $1-\lambda$, and a nonzero zymbol with probability λ. There are approximately $8^5/4$ independent choices in the construction of such a rule table. Therefore, if one constructs many rule tables in this way, the standard deviation in the actual value of λ is given approximately by $\Delta\lambda = \sqrt{\lambda(1-\lambda)/N}$, where $N = 8^5/4$. This is the usual formula for the standard deviation of a binomial distribution. For $\lambda = 1/2$, the value of $\Delta\lambda$ is 0.006, which is much too small to explain the observed spread in entropy. However, the spread in the b_j's is significantly larger, since each b_j is determined by fewer choices. Because of the rotational invariance, finding the standard deviation of each b_j requires some careful counting, but for a fixed neighborhood size and fixed number of symbols it is a straightforward matter to do the necessary arithmetic. It is con-

venient to express the results in terms of the quantities c_j defined by

$$c_j = \binom{n}{j}^2 \frac{\Delta b_j^2}{\lambda(1-\lambda)} . \tag{6}$$

For our case ($n=5$, $k=8$), one finds that

$$c_4 = 2.4286 , \quad c_3 = 0.8047 , \quad c_2 = 0.1150 ,$$

$$c_1 = 0.0083 , \quad c_0 = 0.0002 . \tag{7}$$

(The quantity b_5 trivially has no spread; it is always equal to unity.) Thus, for example, when $\lambda = 1/2$, the spread in b_4 is equal to 0.156, considerably larger than the spread in λ. Each of the b_j's is distributed according to a binomial distribution, which can be well approximated by a Gaussian distribution.

The next step is to use our knowledge of the distributions of the b_j's to find the distribution of v. The equation relating the b_j's to v is the mean-field equation (5). It is convenient to re-express this equation in terms of the difference δb_j between b_j and its average value $1-\lambda$. Thus $b_j = (1-\lambda) + \delta b_j$. In terms of δb_j, eq. (5) becomes

$$f(v) = \sum_{j=0}^{n-1} g_j(v)\,\delta b_j , \tag{8}$$

where the functions f and g are defined by

$$f(v) = (v - v^n) - (1 - v^n)(1 - \lambda) , \tag{9}$$

$$g_j(v) = v^j(1-v)^{n-j}\binom{n}{j} . \tag{10}$$

To find the distribution of v, we make one simplifying assumption, namely, that for each set $\{b_j\}$, there is a unique stable solution to eq. (5). This may not be strictly true, but it appears to be the case for all those sets $\{b_j\}$ which have any appreciable probability. If this assumption is granted, then one can express the distribution ρ of v as follows:

$$\rho(v) = \int \rho_0(b_0)...\rho_{n-1}(b_{n-1})$$

$$\times \delta(v - v_0(b_0, ... , b_{n-1}))\,\mathrm{d}b_0...\mathrm{d}b_{n-1} . \tag{11}$$

Here ρ_j is the distribution of b_j, δ is the Dirac delta function, and v_0 is the solution of eq. (5). Assuming Gaussian distributions for the b_j's, one can carry out

the integrals in eq. (11), and the result is an explicit expression for the distribution of ν:

$$\rho(\nu) = \frac{1}{\sqrt{2\pi}} \left| \frac{d}{d\nu}\left(\frac{f}{\sigma}\right) \right| \exp\left(-\frac{f^2}{2\sigma^2}\right), \quad \nu < 1 ; \qquad (12)$$

(probability that $\nu = 1$) = 1 − (probability that $\nu < 1$). Here σ is defined by

$$\sigma = \left(\sum_{j=0}^{n-1} g_j^2(\nu)\, \Delta b_j^2 \right)^{1/2} . \qquad (13)$$

Eqs. (12) and (13) are the main result of this section. They give us the mean-field prediction for the distribution of the density of zeros in terms of known quantities. From this distribution one can easily obtain the distribution of values of the entropy via eq. (4). Note that the distribution $\rho(\nu)$ depends on λ through the function f and through the Δb_j's.

It is of course very interesting to compare this theoretical distribution with the observed distribution. The graph in fig. 4 was constructed as follows: For each value of λ, one hundred values of ν were chosen at random in accordance with the distribution $\rho(\nu)$.

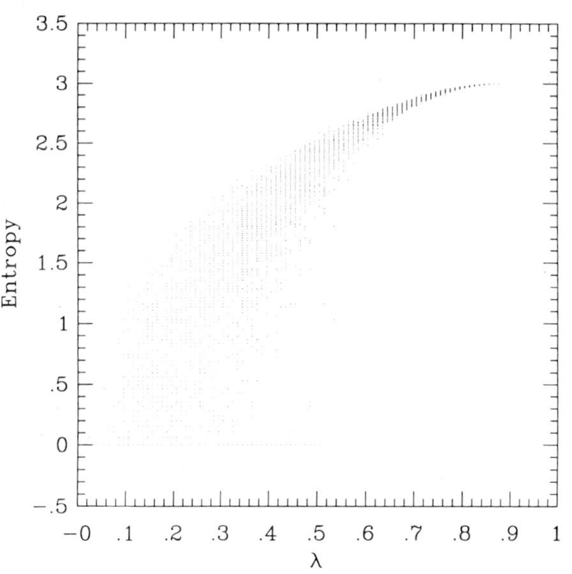

Fig. 4. Theoretical distribution of entropy values versus λ. For each λ, one hundred values were chosen at random in accordance with the predicted distribution, eq. (12). The distribution is similar to the actual distribution shown in fig. 1.

Then for each of these values, a corresponding value of the entropy was computed via eq. (4). The resulting entropy values were plotted. Note that for λ larger than about 1/2, the distribution of entropy values appears to match well the distribution seen in the data (fig. 1). Even for smaller values the match is not bad, except for one feature: the theoretical prediction does not show the gap just below $H = 0.84$ that one sees in the actual data. This disagreement is not surprising. The gap is associated with the transition from periodic to chaotic behavior, and one does not expect the predictions of mean-field theory to apply to periodic automata. Only in chaotic automata could one hope to find enough mixing of the symbols for the assumptions of mean-field theory to be approximately valid. (See, however, the paper by McIntosh in these Proceedings [12], which relates automaton behavior not to the prediction of mean-field theory but rather to the form of the mean-field equation.)

5. A quantitative comparison

It is satisfying that the theoretical graph looks something like the actual data, but one would also like to have a more quantitative comparison. An obvious way to make such a comparison would be to compute the average entropy and the mean deviation as functions of λ, both for the actual data and for the theoretical distribution. However, the low-entropy periodic automata contribute significantly both to the average and to the deviation, and we do not expect agreement when such automata are involved. Therefore, we would like to be able to eliminate these periodic automata from the average, to see whether the theoretical prediction at least works for the chaotic automata.

On the experimental side, it is straightforward to pick out all the periodic automata and to exclude them from the average. In order to get the corresponding results on the theoretical side, we need to tell the theory how to recognize periodic automata. To do this, we adopt the following *cut-off hypothesis*: Any automaton for which the mean-field-predicted entropy

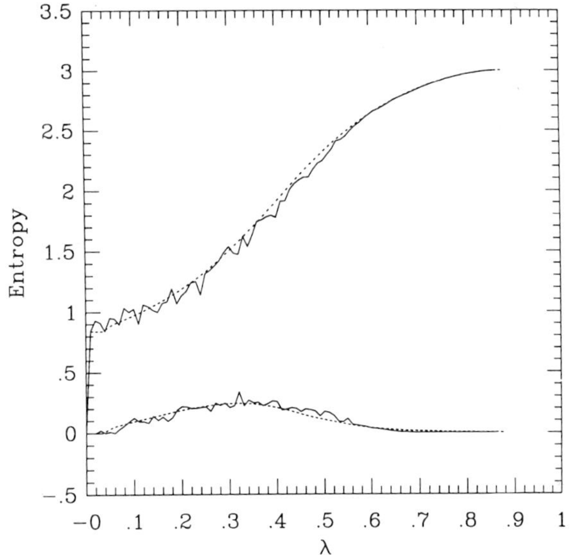

Fig. 5. The solid curve shows the average entropy (upper curve) and the mean deviation of the entropy when periodic automata are dropped from the average. The dashed curves show the theoretical predictions for the same quantities, when all automata with predicted entropies less than 0.84 are dropped from the average.

is less than 0.84 is a periodic automaton, and all others are chaotic. This is a very crude assumption, but it will be interesting to see how well it works. In fig. 5, we have plotted the theoretical predictions for $\langle H \rangle$ and ΔH, having dropped from the average all automata with a predicted entropy less than 0.84. Also plotted in fig. 5 are the experimental results with the periodic automata dropped from the average. The agreement is very good. From this agreement we draw two conclusions: First, for chaotic automata, the variation in the mean-field parameters b_j is the principal source of the variation in the entropy. Second, the cut-off hypothesis has some truth to it. (However, it is not strictly true, as we discuss in section 8.)

6. What happens as *n* or *k* gets large

Now that we have good evidence that the source of the variation is in the b_j's, we can predict how this variation will change as we change the values of n and

k. Let us estimate the behavior of the spread in two limits: large n with fixed k, and large k with fixed n. In the latter case, the values of H will typically go as $\log(k)$ – the entropy is greater when there are more symbols – so it makes sense to consider the normalized quantity $\mathscr{H} = H/\log(k)$.

The order of magnitude of $\Delta\mathscr{H}$ is the same as that of the quantity σ defined in eq. (13). In order to estimate σ, we need to estimate the Δb_j's for arbitrary n and k. A reasonable estimate is given by

$$\Delta b_j = \left[\binom{n}{j}(k-1)^{n-j} \right]^{-1/2}.$$

(The quantity in square brackets is the number of ways of filling an n-cell neighborhood, when there are k symbols available and when exactly j of the cells must contain zero.) Using this estimate, and for simplicity letting ν equal $1/2$, one finds that

$$\sigma = \left(\frac{1}{2}\right)^n \left[\sum_{j=0}^{n-1} \binom{n}{j} \frac{1}{(k-1)^{n-j}} \right]^{1/2}. \tag{14}$$

If k is held fixed, then this expression decreases exponentially with increasing n. If n is held fixed and k is allowed to get large, the expression is dominated by the term with $j = n - 1$, and the whole expression goes as $1/\sqrt{k}$. Thus in both limits, the spread in the entropy goes to zero.

We now have to imagine what happens to the graph of fig. 1 in each of these limits. Consider first the case in which k gets large while the neighborhood size stays the same. For values of λ to the right of the transition region, the spread in H will presumably get smaller and smaller, in accordance with the above prediction, probably approaching zero as k goes to infinity. Thus the scatter of points will begin to look like a curve. Assuming that this curve *remains* a curve for all λ, the transition must occur at a definite value of λ if it occurs at all. In the following section we present evidence that there is indeed a sharp phase transition in this limit, occurring around $\lambda = 0.27$.

The limit $n \to \infty$ is likely to be less interesting. As before, the scatter of points will probably become a line, but in this case one may lose the transition altogether. Mean-field theory predicts a transition from

the quiescent state to a chaotic state at $\lambda = 1/n$. (This follows from eq. (3), with the exponent 5 replaced by n.) Thus it is likely that the transition point will be pushed to $\lambda = 0$ as n approaches infinity. Numerical evidence for this conclusion can be found in ref. [7].

7. The infinite k limit: an experimental test

Let us focus now on the case of a large number of symbols. It is difficult to make a large increase in the size of k experimentally, because the rule table expands as k^5. Moreover, the spread in $\mathscr{H}$ is predicted to go as $1/\sqrt{k}$, as we have just seen, and this is a rather slow convergence. If we used 70 symbols instead of 8, the spread would be reduced only by a factor of 3.

However, it is actually possible to do the experiment when k equals infinity. If there are an infinite number of symbols, then as long as the lattice is finite and the automaton runs for a finite time, there is zero probability that any nonzero symbol will appear more than once in the entire run. (The symbol zero can appear frequently because one fixes λ at some value less than 1.) The automaton is therefore constantly exploring new parts of the rule table. Moreover, the rule table was constructed at random, so in effect one is simply running a probabilistic automaton with two symbols, zero and "other". An all-zero neighborhood is mapped to zero with probability 1, and any other configuration is mapped to "other" with probability λ.

We have run this binary probabilistic automaton many times with values of λ ranging from 0 to $1/2$. (The size and shape of the neighborhood are the same as in our finite-k case.) Fig. 6 shows the resulting values of the entropy. There does appear to be a sharp phase transition at $\lambda = 0.27$, and the points fall on a fairly well defined curve, in agreement with our expectations. Note that there does not appear to be a discontinuity in the entropy itself as a function of λ, but rather a discontinuity in the derivative. The existence of this sharp transition, combined with our prediction that ΔH decreases with increasing k, pro-

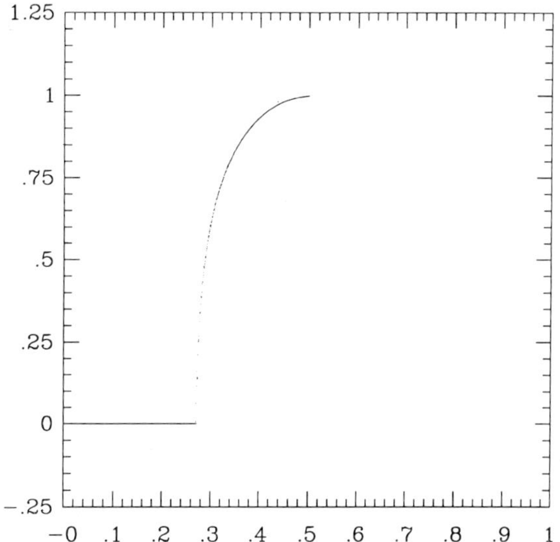

Fig. 6. Entropy versus λ for a probabilistic automaton intended to simulate our deterministic automata in the infinite k limit. The lattice was 128×128 with periodic boundary conditions. Each automaton was allowed to run for 1000 time steps before data were taken. The location of the transition point shifts slightly with larger lattices and longer waiting times.

vides strong evidence that the transition region itself shrinks as k increases, the width of the region probably going as $1/\sqrt{k}$.

In collecting the above data, we let the automaton run for 1000 time steps before collecting data. The exact location of the transition changes slightly if one uses a longer waiting time or a larger lattice. (The dependence of a related transition on lattice size has been discussed in ref. [13].) Our aim here is not so much to locate the transition point precisely as to show that the transition does become sharp in the infinite k limit. It is interesting to note that as one approaches the transition point from the low-λ side, it takes longer for the transients to die out. Thus the transition appears to be of second order.

Sharp phase transitions in probabilistic automata have been observed before. (See for example refs. [14,15].) They have also been observed in deterministic inhomogeneous automata with random assignments of rules to sites. (See for example refs. [16–18,13].) Our results show how homogeneous deter-

ministic automata can approach such behavior.

8. A sharper transition with $k=8$?

Even if the transition region becomes narrower as k gets larger, there will still be a sizable range of transition points for any realizable finite value of k. If possible, one would like to find a parameter other than λ in terms of which the transition is sharper, even when the number of symbols is held fixed.

Such a parameter is suggested by our cut-off hypothesis. Recall that the hypothesis asserts that an automaton will be periodic or chaotic, depending on whether its mean-field-predicted entropy H_{mf} is smaller or larger than 0.84 bits. Suppose this hypothesis were strictly correct for our automata. Then if one were to plot the actual entropy H against H_{mf}, one would see the following: For values of H_{mf} less than 0.84, the actual entropy would be very small; at $H_{mf}=0.84$, the entropy would suddenly jump to a value close to 0.84 and would rise gradually thereafter. Thus we would have a sharp phase transition. Or instead of using H_{mf}, one could use ν_{mf}, the mean-field-predicted density of zeros. Either of these quantities is computable from the rule table without having to run the automaton.

We therefore propose the following experiment. Let ν_{mf} take values from 0 to 1 in steps of, say, 0.01. For each value, choose at random some number of automata whose b_j's satisfy eq. (5) with the given value of ν_{mf}. Run these automata and compute their entropy. If the cut-off hypothesis were correct, the resulting plot of entropy versus ν_{mf} would show a sharp jump at the value $\nu_{mf}=0.885$, which corresponds to $H_{mf}=0.84$.

In fact the cut-off hypothesis is not strictly correct. For example, in fig. 1, one sees a number of low-entropy periodic automata around $\lambda=0.5$ whose existence is not predicted by the theoretical result shown in fig. 4. The values of H_{mf} for these automata are surely greater than 0.84, and yet the automata have low entropy. However, these automata are relatively rare, so one might still hope to see a reasonably sharp transition in the proposed experiment.

9. Conclusions

Our strongest conclusions are these: As the number of symbols increases, the transition region for our automata almost certainly shrinks, approaching a unique transition point around $\lambda=0.27$ as k goes to infinity. On the other hand, if one increases the size of the neighborhood, the transition region probably gets pushed closer and closer to $\lambda=0$, so that there is in fact no transition at all in the infinite-neighborhood limit, at least none that can be seen by varying λ.

These conclusions depend on the fact that for the automata we are considering, an all-zero neighborhood always maps to the symbol zero. If this property were removed, it is not clear whether there would be a sharp phase transition in the infinite-k limit. It may also be the case that our results depend crucially on the two-dimensionality of the lattice. To determine how these features of the automata influence the nature of the transition will require more experimental and theoretical work.

Our results also suggest that even for finite k, one can probably sharpen the transition by using a different parameter, ν_{mf} instead of λ, to explore the space of automata. This alternative parameter may be useful for finding complex automata near the transition point, which was one of our motivations for looking for a sharper transition. However, there is nothing natural about ν_{mf}, so we do not expect any deep insights to emerge through this particular parameterization of the space of automata. In order to see a closer parallel between the cellular automaton transition and physical phase transitions, it may be necessary to use a set of quantities none of which can be computed without actually running the automaton. For example, one might define the "temperature" as the exponential of the average expansion rate. One could then plot entropy against temperature. Both entropy and temperature (by this definition) are measured

properties of the automaton's behavior and, unlike λ or ν_{mf}, cannot be easily controlled. It would nevertheless be of great theoretical interest, and ultimately of practical value, if one found a sharp phase transition in such a plot.

Another more fundamental question is certainly worth investigating: If there is indeed an analogy between cellular automata and thermodynamics, how many independent thermodynamic variables are there? That is, how many different macroscopic statistical properties of an automaton does one have to specify in order that all other macroscopic statistical properties are then determined? In order to see a sharp phase transition, one would have to be able to vary separately all the independent variables. For example, in order to see a sharp phase transition between liquid water and water vapor, it is not sufficient to plot entropy against temperature while paying no attention to wild variations in the pressure. In a similar way, we may need to pay attention to two or more independent variables in order to see a sharp transition.

Acknowledgements

The ideas presented in this paper have emerged from conversations with a number of people, including David Cai, Doyne Farmer, Howard Gutowitz, Stuart Kauffman, Wentian Li, John Miller, Norman Packard, and Steen Rasmussen. We thank them for their insights and questions. We would also like to thank Stuart Kauffman and Gerard Weisbuch for directing us to some of the work on inhomogeneous automata. Finally, we would like to thank the two groups in Los Alamos' Theoretical Division that supported this work: the Complex Systems Group (WKW and CGL) and the Theoretical Astrophysics Group (WKW).

References

[1] S. Wolfram, Physica D 10 (1984) 1.

[2] S. Wolfram, Physica Scripta T9 (1985) 170.

[3] C.G. Langton, Physica D 22 (1986) 120.

[4] C.G. Langton, Computation at the edge of chaos, Physica D 42 (1990) 12.

[5] C.G. Langton, Ph. D. Thesis, University of Michigan (1990).

[6] W. Li and N. Packard, Complex Systems, submitted for publication.

[7] W. Li, N. Packard and C. Langton, Physica D 45 (1990) 77, these Proceedings.

[8] N. Packard, Adaptation toward the edge of chaos, Technical Report CCSR-88-5, Center for Complex Systems Research, University of Illinois (1988).

[9] S. Wolfram, Rev. Mod. Phys. 55 (1983) 601.

[10] H.A. Gutowitz, J.D. Victor and B.W. Knight, Physica D 28 (1987) 18.

[11] Z. Burda, J. Jurkiewicz and H. Flyvbjerg, preprint NBI-HE-90-04, The Niels Bohr Institute (1990).

[12] H.V. McIntosh, Physica D 45 (1990) 105, these Proceedings.

[13] D. Stauffer, Physica D 38 (1989) 341.

[14] P. Grassberger, F. Krause and T. von der Twer, J. Phys. A 17 (1984) L105.

[15] W. Kinzel, Z. Phys. B 58 (1985) 229.

[16] B. Derrida and D. Stauffer, Europhys. Lett. 2 (1986) 739.

[17] G. Weisbuch and D. Stauffer, J. Phys. (Paris) 48 (1987) 11.

[18] L. de Arcangelis, J. Phys. A 20 (1987) L369.

Physica D 45 (1990) 105–121
North-Holland

WOLFRAM'S CLASS IV AUTOMATA AND A GOOD LIFE

Harold V. McINTOSH

*Departamento de Aplicación de Microcomputadoras, Instituto de Ciencias, Universidad Autónoma de Puebla,
Apartado Postal 461, 72000 Puebla, Puebla, Mexico*

Received 10 January 1990
Revised manuscript received 10 April 1990

A comprehensive discussion of Wolfram's four classes of cellular automata is given, with the intention of relating them to Conway's criteria for a good game of *Life*. Although it is known that such classifications cannot be entirely rigorous, much information about the behavior of an automaton can be gleaned from the statistical properties of its transition table. Still more information can be deduced from the mean field approximation to its state densities, in particular, from the distribution of horizontal and diagonal tangents of the latter. In turn these characteristics can be related to the presence or absence of certain loops in the de Bruijn diagram of the automaton.

1. Introduction

It all began with John Horton Conway's search for an "interesting" cellular automaton; one which would perhaps have the complexity of John von Neumann's universal constructor, but with far fewer states, maybe only two. Having observed that the field of activity of an automaton tended either to increase without limit or to dwindle away to nothing, he settled for a delicately balanced combination which gained widespread publicity through its introduction in Martin Gardner's monthly column in *Scientific American* as the game of "*Life*" [1,2].

Interest in the game persisted several years, leading to a most surprising series of artifacts – oscillators, shuttles, glider guns, puffer trains, very orderly collisions and eminently predictable transformations – culminating in the demonstration that *Life* was capable of universal computation [3]. Even though *Life*'s computer shares an extraordinarily sprawling layout with such predecessors as von Neumann's [4] and Codd's [5], nevertheless they all provided concrete testimony that arbitrarily complicated mathematics could be performed within a system whose basic organization was thoroughly rudimentary.

There was no lack of awareness that those cellular automata which seemed to have interesting or useful properties had been plucked out of an environment populous beyond any normal concept of multitudes. Nor is it surprising that comparative studies of all the automata possible within some given class were not undertaken until computing facilities commensurate with the task were available, and even then not until the study of chaotic systems had gained a certain popularity [6]. But it was just this combination which enabled Stephen Wolfram to study and attempt to classify all possible automata into the four well known categories which now bear his name [7]:

class I: evolution to a constant state;
class II: evolution to isolated periodic segments;
class III: evolution which is always chaotic;
class IV: evolution to isolated chaotic segments.

A recently published reprint volume [8] contains a rather complete account of his work on automata, several closely related papers and an appendix showing sample evolutions. His scheme for numbering rules of evolution is now generally followed, as is the notation (k, r) for a k-state linear cellular automaton whose cells are surrounded by r neighbors on each side. When occasion demands a neighborhood with an even number of cells, r can be taken to be half-integral.

Even though relatively rare, the fourth class was the one which attracted attention because it was the one whose evolution could be regarded as accomplishing some purpose. It was thereby a generalization of the concept of a good *Life*, which is supposed to be one with a quiescent state, for which bounded regions of nonquiescent cells would neither die out nor grow without limit.

Other rules, variants of Conway's, have been tried without any having been reported as being worthy of further attention; it is not unnatural to inquire whether the one game is in fact unique. For example, Packard and Wolfram [9] studied two-dimensional cellular automata after the spirit of Wolfram's survey of one-dimensional automata, declaring there were no class IV automata with the exception of "trivial variants on *Life*".

Preston and Duff [10] mention experiments with hexagonal lattices, some using a rule proposed by Marcel Golay; others with a rule of their own devising. William Poundstone [11] reports a "3–4 *Life*", which had been occasionally mentioned in Robert T. Wainwright's newsletter [12]. The newsletter also carried intermittent reports of a variant of *Life* with three states; the live cells were colored red and blue, with supplementary rules for determining the color of an offspring.

The most recent variation on this theme has been a series of articles by Carter Bays [13–15] seeking a worthy three-dimensional variant of *Life*. His latest article [16] discusses the interrelation of Conway's criteria and Wolfram's classes.

2. Statement of the problem

The novelty of Wolfram's classification sufficed for a time, but further scrutiny raised some interesting questions. Increasing the number of states, the size of the neighborhoods, or even just the length of the row of cells can lead to configurations which do not reach quiescence until larger and larger numbers of generations have elapsed.

Not only do randomly selected automata exhibit this tendency; careful representation of the states and selection of the evolutionary transitions allows the fabrication of automata which can perform simple computations. Configurations of an automaton which behaves as a counter can be de-

signed to become quiescent after an exponential number of generations relative to the number of its cells, thereby approaching the upper limit for the lengths of transients in a finite automaton. Thus it could require a very long time for an experimental decision as to whether an automaton belonged to class I or not.

Even worse, as Karel Culik and Sheng Yu [17] have shown, given that it is possible to simulate a universal Turing machine within a linear cellular automaton having a sufficient number of states, and given that it is undecidable whether such a machine will ever halt with a blank tape, it only requires a certain amount of care in arranging the details of the proof to see that membership in class I is undecidable for arbitrary automata.

Continuing with class II, there can be uncertainty whether an assignment should not be made to class IV instead, if the period is very long. That is, the activity in a fairly large patch of the automaton's field may look chaotic, but if such patches are bounded, the activity must ultimately become periodic. Therefore a true class IV automaton would have to contemplate the possibility of unlimited growth, accompanied by either coalescence of neighboring chaotic patches or some kind of coordinated dilation.

Even class III is not immune to further scrutiny. Not all states need occur with comparable frequencies, nor are complex but regular textures excluded. Indeed, if the hallmark of class IV is taken to be the occurrence of quiescent intervals of arbitrary length, there is no reason to exclude similar stretches composed of other states or even of intricate designs.

Finally, there are many interesting fringe cases, even with the simplest automata. If a binary rule consists primarily of complementation, every second generation may remain relatively constant, so that the apparent class of interleaved generations may be different from the one that was originally evident. A variant of this theme has the uniform neighborhoods alternating generations (or following a longer cycle when there are sufficient states) rather than one of them being quiescent. Interesting activity may develop at the interfaces between constant patches in a configuration, which one might have classified as class IV were it not for the alternation of backgrounds.

The problem, then, is that Wolfram's classification is really not susceptible to a precise definition; even if it were, it would be better derived directly from descriptive parameters than from behavioral observations. Conversely, it would be nice to deduce the parameters from observation. The experimental data which are easiest to observe and accumulate are statistical properties, both regarding frequencies of cell states, and of the patterns into which they can organize themselves.

Consequently there has been a growing feeling that Wolfram's classification is really a scale measuring the distribution of periodic cycles and their possible interconnections, class IV forming a transition region between class II and class III. Thus class I automata admit only cycles of length and period 1, those of class II very short periods and lengths, class IV requires the quiescent state plus long periods, while class III rejects a quiescent state or analogous cycles.

The observable characteristics of the probability distributions, which would supposedly follow the presence or absence of this substructure, would be the occurrence stable fixed points, limiting configurations lying on the border between stability and instability, or some similar arrangement.

In order to discuss any of these concepts adequately, it is necessary to review their bases, first probability and then de Bruijn diagrams.

3. Probabilities and measures

At first, interest in automata consisted in observing their evolution and collecting examples of interesting configurations, but eventually the precepts of information theory were applied to *Life*, first in a 1975 paper by Dresden and Wong [18], later in 1978 by Schulman and Seiden [19].

The simplest procedure would be to judge the probability of a state by the number of neighborhoods generating it. Assuming them all to be equally likely often yields good results; an estimate which can be sharpened by weighting each neighborhood according to the frequency of its constituents, especially if the evolution is used to solve for self consistency. Since the neighborhoods from which adjoining cells evolve overlap, there are doubts as to whether their probabilities are independent; mean field theory results from suppressing such doubts, while other theories arise from the way in which they are taken into account.

An alternative to working out the cumulants required by Schulman and Seiden is to use the probabilities of sequences of cells rather than those of individual cells; just recently Wilbur, Lipman and Shamma [20] calculated some densities in linear cellular automata using this approach. Gutowitz, Victor and Knight [21] made a more detailed analysis along the same lines, basing their approach on Kolmogoroff's theory of probability measures, justifying their estimates in terms of Bayes theorem, and even describing an extension applicable in two dimensions and beyond [22].

In order to fix our ideas, consider the one-dimensional (2,1) cellular automaton evolving by Wolfram's rule 22, which also happens to be the rule of evolution of a cross section of Conway's *Life*. Moreover, it is totalistic rule 2 by the same scheme of reckoning, which means that a cell lives in the next generation if there is a single live cell anywhere in the neighborhood during the present generation. Since three of eight transitions produce a live cell, the *a priori* probability of live cells would be 0.375.

According to mean field theory, if p is the probability that a cell lives, $q = 1 - p$ is the probability that it is dead or inactive, and altogether $3pq^2$ that the neighborhood meets the requirement for a live cell in the following generation.

The requirement for self-consistency is then

$$p = 3pq^2$$

or equivalently, $p(1 - 3q^2) = 0$, whose roots are $p = 0.0$ and $q = 1/\sqrt{3}$ or $p = 0.42$. The derivative of the right-hand side at the origin is 3.0; an unstable fixed point there is clearly consistent with the fact that at low densities isolated cells persist and give birth to two neighbors, multiplying the total population by three. The derivative at the other fixed point is approximately -0.46, which implies a moderate alternating approach to stability.

Fig. 1 shows a graph of this probability (the upper curve) together with the diagonal line of unchanged probability; the two fixed points and their properties are quite clearly visible. Displaying the self-consistency equations graphically is always illuminating; even more so is plotting the empiri-

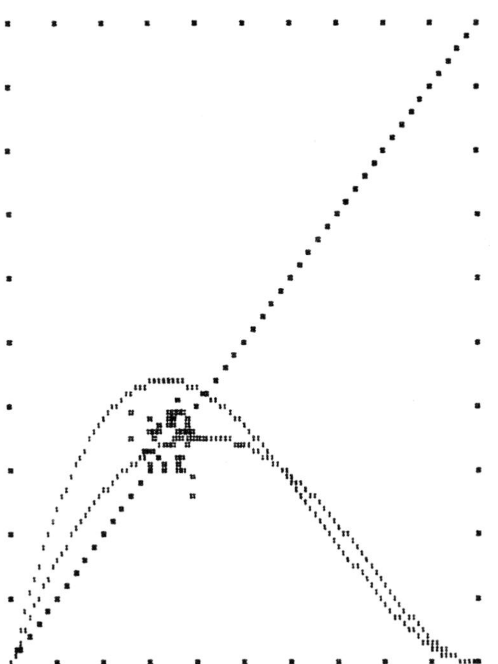

Fig. 1. Self-consistent mean field probability for rule 22. The top curve is for one generation of evolution, the second for two generations. Empirical evolutionary pairs for a 320-cell ring form the cloud of dots.

cal pairs representing successive generations, such would be encountered in an iterative approach to the fixed point.

The most striking observation is that the empirical points do not strictly follow the mean field curves, tending to cluster about a sort of average value, frequently near a fixed point; fig. 1 exhibits a cloud of 50 pairs, showing this effect clearly. As might be expected, in similar environments with unstable fixed points two separate clusters straddle the diagonal.

It can also be noted that the variance per unit length of automaton is quite large, strips of several thousand cells being required before the variance in the mean of the large sample allows predictions of three-figure accuracy to be obtained. Perhaps the intrinsic variance for a given rule could somehow be computed, but serviceable estimates can be obtained from the general knowledge that the variance of a sample of size n scales according to $1/\sqrt{n}$. Thus rows with hundreds of cells, typical of video displays, have enough variance to be conspicuous, yet not overwhelming; automata of fifty

cells or less, common when results are presented on a printed line, leave considerable doubt that any probabilistic influences are at work at all.

Careful simulations of rule 22 agree with the stability of its fixed points, but not with the precise location of 0.42. Gutowitz et al. examine this situation at considerable length, concluding that block probabilities for blocks of length six or more are required before the self-consistent numerical probability agrees with empirical observations, which favor a value of 0.35.

Fig. 2 shows a contour plot for the self-consistency of pair, or 2-block, probabilities according to their formulae. Often the equations have multiple solutions; many can be identified with specific periodic configurations, but one will be related to a limiting measure. Longer blocks require so many parameters that the overview afforded by a graphical presentation is no longer feasible; nevertheless numerical solutions by iteration can always be sought.

Another manifestation of correlations lies in the difference between iterated one-generation probabilities (having the same fixed points) and genuine two-generation probabilities taken from the compounded rule of evolution (with their own fixed points); the lower curve in fig. 1 shows the mean field probability corresponding to two generations of rule 22.

4. Further interpretation

Since there are several levels of refinement available for working with probabilities, one wonders which to use. The *a priori* estimate is surely the easiest to obtain; in any event it is the first step in iterating the mean field approximation, starting out with initially uniform probabilities.

Mean field theory works with polynomials; indeed the Bernstein polynomials of probability and spectral theory constitute an appropriate basis. Strictly speaking, a "Bernstein polynomial" is one which has been derived from an arbitrary polynomial by a process of dissection intended to facilitate the calculation of moments and their convergence. When applied to ordinary monomials typical probabilistic expressions result, such as $p^i q^{n-i}$ with $p + q = 1$; these special cases yield what

Fig. 2. Contours showing the degree of self-consistency for the evolution of pairs of cells according to rule 22, using local structure theory. Horizontal: probability of a live cell; vertical: probability of a live pair. Only the lower right triangular region is physically significant.

is here called a Bernstein polynomial. The general theory is described in a book by G.G. Lorentz [23]; applications to the moment problem can be found in the memoir of Shohat and Tamarkin [24].

As regards accuracy and convenience, block probability theories, committed as they are to rational fractions, are much more complicated and time consuming; negative factors to be balanced against their superior accuracy. Mean field theory is not only a workable compromise, it is recommended by numerous antecedents from statistical mechanics by which difficult probabilistic situations have been satisfactorily approximated by the same technique.

This does not mean that what one understands by mean field theory is the same for automata as for disordered phenomena (it is not), only that the kinds of approximation involved are similar; not only are they regarded as successful, they are well calibrated through comparison with more exact theories. It would lead too far afield to discuss probabilistic automata, where the analogy is

closer; but they can be borne in mind as another source of support for mean field theory.

Whatever the reasons for choosing to work with mean field theory, it turns out that the results are generally quite reasonable; even when evidently lacking in precision, they still lie within the range of 10% accuracy. No less impressive are the qualitative features, such as the observation that some automata have a variety of fixed points while others do not.

For example, it is hardly surprising that automata which would be judged to lie in class III tend to possess stable fixed points away from the origin (the origin, if fixed, would be unstable), in contrast to the others. Consequently the details of any analysis tend to focus on the degree to which the empirical densities actually correspond to values given by the fixed points, and the interpretation of observed behavior when they do not.

It is likewise not unexpected that a quiescent state occupies a stable fixed point for class I automata; it is often observed that class IV automata are associated with tangencies to the diagonal of the fixed point graphs, and that class II sometimes goes with near misses. However these are mere visual impressions, although formed by prolonged observation, whose validity one would like to either explain or disprove.

The easiest aspect to describe is the ocurrence of the quiescent state. As far as binary linear cellular automata go, quiescence evidently requires the absence of the term q^n from the evolution equation for live states (zero is traditionally a quiescent state, n is the size of the neighborhood). The term pq^{n-1} arises two ways – if isolated live cells survive, and if cells are born from contagion: having a single live cell nearby. The first alternative alone produces a curve with a diagonal tangent at the origin; denying both possibilities means that regions of live cells will not expand and that the curve has a horizontal tangent at the origin.

The distance of contagion is an important parameter. It will always extend to the radius of the neighborhood, the number of cells thereby affected varying with the dimension of the automaton. Gutowitz and Victor [25] have noticed this effect while comparing two seven-cell neighborhoods, one taken from a hexagonal lattice with nearest neighbor interactions, the other a linear lattice

with third neighbor interactions. In the hexagonal case, it is sufficient to exclude the terms q^7 and pq^6 from the self-consistent probability, but in the linear automaton p^2q^5 and p^3q^4 also have to be excluded to observe class II or class IV behavior. They found similar contrasts between the nine-cell neighborhoods representing two-dimensional *Life* and (2,4) linear automata, respectively.

Bays [13] has proven theorems regarding the behavior of three-dimensional automata with various constraints on the neighborhoods (in his semi-totalistic notation). Thus there seems to be a general agreement that the evolution of certain neighborhoods containing large numbers of quiescent states must preserve quiescent states, which is half of Conway's original argument that isolated or relatively isolated states must die out in order to provide an interesting automaton.

Although the other half of his argument – how cells die from overpopulation – still has to be taken into account, the implications of quiescent domains and similar structures can be explored even further.

5. De Bruijn diagrams

Whether working with probabilities, measures, or evolution, overlap between adjoining neighborhoods is the greatest technical obstacle to computations; unless this problem is resolved adequately, further progress is almost impossible. Fortunately, a diagrammatic technique lying at the heart of shift register theory [26] saves the situation; the diagrams are called de Bruijn diagrams, but are merely simple graphs showing the possible ways for neighborhoods to overlap [27]. Erica Jen has shown how many properties of automata can be extracted from such diagrams, especially the static and periodic configurations on a cylinder of fixed circumference [28]. Wolfram used these subdiagrams earlier in his theoretical analysis [29] of cellular automata; although he called it a "regular expression diagram," Wentian Li [30] used exactly the same concept to locate the periodic strings in an automaton; there is no doubt that the ideas have also been used elsewhere with greater or lesser degrees of formality.

In principle, such a diagram could be extended to automata of higher dimensions, but a problem arises from selecting partial neighborhoods that will join to form full neighborhoods in all directions. The straightforward approach of building up strips of successively higher dimension runs afoul of Post's correspondence principle when arbitrary intermediate strips have to be matched to form the layers of the next higher dimension. If only periodic solutions are required, the problem is still soluble. Hopcroft and Ullman [31, ch. 8] give a good explanation of the difficulties involved, which are related to the fact that the halting problem for a Turing machine can be worked into a similar context. This, of course, is the prototypical example of a problem without an algorithmic solution; Minsky [32, pp.273ff] also contains a short description.

At least in one dimension, there is nothing difficult about a de Bruijn diagram; as applied to cellular automata it is simply a graph in which partial neighborhoods are the nodes, with links connecting those which may overlap to form a full neighborhood. Given this correspondence between links and full neighborhoods, each link is also associated with the evolved cell belonging to the neighborhood. Consequently characteristics of the evolution can be used to select subgraphs of the de Bruijn diagram; for example, there is a subgraph composed of the neighborhoods whose central cell never changes. Global properties of the automaton can be read off from the chains inherent in the subdiagram; in this example, the chains yield all the static configurations.

De Bruijn diagrams transform automata problems into known path tracing problems. For instance, no loop can be longer than the total number of nodes in the graph without repeating some segment; but the way is open for still other loops in which the repeated segment is traversed any arbitrary number of times. As an example, since a binary automaton depending upon nearest neighbors has eight distinct neighborhoods, representable as eight links connecting four nodes, it follows that no static configuration can be more than five cells long without repeating some two-cell partial neighborhood.

Many more characteristics than the static, or "still life", configurations of an automaton can be deduced from its de Bruijn diagram. Translational

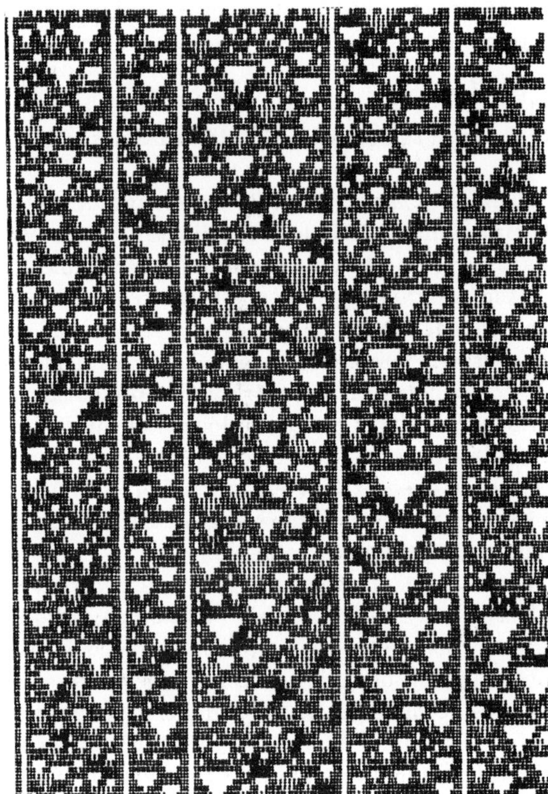

Fig. 3. Sample evolution of a typical (3,1) macrocell rule. For this rule, the state sequence 102 creates a barrier, visible as a vertical streak. Macrocells form in the intervals between barriers; small ones quickly become periodic.

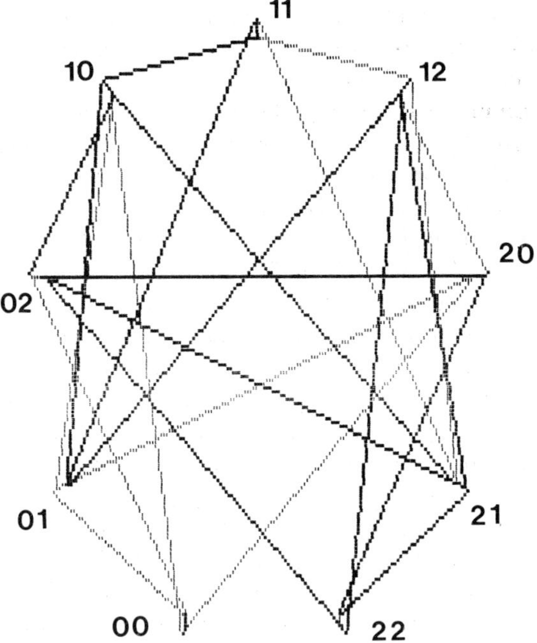

Fig. 4. A de Bruijn diagram which results in macrocells: (3,1) rule 02222121122121101222211120. The sequence 102 defines a cell membrane. Darker lines represent static links; all incoming links at node 10 and all outgoing links at node 02 define still lifes.

uniformity having been built into the very definition of cellular automata, shifting configurations which cover a distance d each generation are as easily found as the static ones. By working instead with the compound rule governing several generations of evolution, and enlarging the "neighborhood" under consideration, the characteristic can be extended to a shift of d cells every g generations; among these are the very interesting class of *superluminal* configurations – those which move further than an average of r cells per generation.

Although rarely repeating beyond a single generation, any Boolean (for $k = 2$) combination of the states of a neighborhood and its evolutes can be detected via the de Bruijn diagram; for example configurations which evolve into constants, or into their complement.

Sometimes the de Bruijn diagram reveals in-

formation about localized aspects of a configuration. If an acceptable path terminates at a node in which *all* the outgoing links are acceptable, it need continue no further. Likewise if all the incoming links are acceptable, the path may begin just as though it had been part of a loop. Thus semi-infinite structures may be located, or even finite ones if both ends have such universal terminations. This leads to the phenomenon of membranes and macrocells which Wolfram noticed during his investigations. David Brown [33] has also shown a nice example of such barriers and sheltered regions. That is, an automaton may have patches which are isolated from one another by static regions; the patches evolve quite independently.

That the (3,1) automaton whose evolution is depicted in fig. 3 meets the requirements may be discerned through examination of the de Bruijn diagram presented in fig. 4.

Conversely, a rule of evolution of an automaton is definable by postulating that the de Bruijn diagram have prescribed properties; for example that the sequence $(01)^*$ must be static. To the extent

the requirements do not contradict one another, and all the alternatives are covered, automata may be created to fulfill ones wishes.

Enumerating the paths through a graph is a classical task, to which many papers have been devoted, but which has a particularly elegant solution in terms of regular expressions. Conway's book on regular algebra [34] expounds the technique; a later article of Backhouse and Carré [35] gives a very thorough presentation.

The only requirement for partial neighborhoods is that all be similar, yet overlap sufficiently to create complete neighborhoods. The usual choice of a linear segment just one cell short of a (k, r) automaton's quota of $2r + 1$ cells per neighborhood – creating the maximum overlap possible – yields k^{2r} different partial neighborhoods. Each defines a node in the de Bruijn diagram, all of them interconnected by k^{2r+1} links, associated with the neighborhoods themselves. Omitting the right cell from a link reveals its source, excluding its left cell shows the destination node, making the relationship to a $2r$-stage shift register apparent.

Many attempts at artistic representations of de Bruijn diagrams have been made; extensive line crossings cannot be avoided except for the very simplest of them. The most systematic representation may be through selected chords of a regular k^{2r}-gon; arranged around the circumference of a circle, the vertices can be numbered sequentially as though the states of the cells were digits in a k-adic number. The connectivity matrix of the diagram would have the matrix elements

$$M_{ij} = 1, \quad j = \begin{cases} ki \\ ki + 1 \\ \dots \\ ki + k - 1 \end{cases} \pmod{k^{2r}}$$

$$= 0, \quad \text{otherwise.}$$

Traces of powers of the de Bruijn matrix readily show that there are k^n loops of length n, counting each loop once for each distinct node. However, a loop for which no node is repeated can have at most k^{2r} links. Long sequences necessarily repeat some of their subsequences, rendering really long loops redundant. Such counting arguments yield rapid derivations of various interesting results, such as the bounds obtained by Guan and

He [36] for the number of cycles in border decisive automata.

6. Loops, tangencies, Wolfram classes

There are many kinds of subdiagrams in the de Bruijn diagram; indeed the diagram's most important role is that of a common upper bound for all the diagrams, whether constructed from formal language theory or for other reasons, that are then used in automata theory. Fairly evident for subdiagrams which are nearly full subsets, the relationship can be readily overlooked or missed entirely in the case of sparse diagrams, especially if they have been extensively simplified. Yet this is the situation in which knowledge of their existence may well be the most valuable.

For example, it is generally understood that finite cellular automata must evolve into periodic cycles, whose temporal length is bounded by the total number of configurations, albeit exponentially large in terms of the numbers of cells and states. The finiteness of the de Bruijn diagram similarly bounds the spatial periodicity of an infinite or periodic automaton, in the sense that the period must be constructed from a closed loop within the diagram appropriate to the period under consideration. Again the bound may be exponentially large in terms of the number of states and length of the partial neighborhoods; in practice both bounds are extremely liberal.

So it is that many kinds diagrams are pertinent to automata theory, not a few arising from the description of configurations or their evolution via a formal language. Thus diagrams are often constructed to realize some regular expression; really the two concepts are coextensive. The particular role of *de Bruijn* diagrams lies in their ability to describe overlapping sequences; for example, all the segments of a regular expression of a given length. In turn, the algebraic properties of the de Bruijn matrix – the connectivity matrix of a de Bruijn diagram – lead to precise numerical estimates of such things as the number of possible sequences of given type, or to the bounds which have been mentioned.

One application consists of marking all the links evolving into one specific state, say the quiescent

state. Varying the evolved state partitions the links into equivalence classes. Interestingly, such a subdiagram consists exclusively of closed loops when the evolutionary rule is totalistic. A totalistic rule depends only on the sum of the values of the cells in a neighborhood, causing all its permutations to evolve the same way. Because passing from one link to the next implies dropping a cell from the left and adding another to the right, a cyclic shift will always yield a continuation preserving the given sum.

The precursors of a static configuration would be described by a path through the appropriate subdiagram, but the individual links can be used for other purposes. For example the probabilities represented by the links evolving to live cells could simply be summed, to obtain the Bernstein polynomial to be used in self-consistent comparisons. Many other arrangements of links into subdiagrams are also possible, such as those which correspond to a static central cell (still lifes), or to uniformly shifting configurations (some of which are called gliders).

If there is to be either a class II or a class IV region, the still life diagram should contain a self loop to the quiescent neighborhood, together with some other loop. If the self loop is not present, quiescent regions of arbitrary length cannot exist, and if there are no loops at all, there are no extended still lifes. If there are additional loops connected to the quiescent self loop, there is a correspondingly greater variety of still life possible; otherwise every island of still life would be identical and no one would call it chaotic. Since it is a property of diagrams that a loop which may be traversed at all may be traversed an arbitrary number of times, the variable separation between live regions in these classes of automata becomes understandable, as well as the minimum separation which is sometimes required.

The self loop to the quiescent partial neighborhood corresponds to the Bernstein monomial q^n; incoming and outgoing links to the rest of the diagram belong to the monomials pq^{n-1}. The self loop cannot be associated with the live state, by definition. Connecting links establish whether contagion is to be avoided, and in some cases how isolated cells behave. If the decision is for quiescence, all these monomials must be excluded from the self-consistent probability of the live state. Thus requiring a horizontal tangent at the origin of the probability curve relates directly to some of the finer details of the de Bruijn diagram.

The quiescent state places zero as a fixed point of the probability curve, but expansivity or virulent contagion render it unstable. Stability, maybe even superstability, results whenever quiescent regions are not so readily invaded, which is surely one of the hallmarks of Wolfram's class IV; but the rest of the probability curve has to be accounted for, as well as the degree of internal chaos necessary to distinguish class II from class IV. Whether there are additional fixed points, and the nature of their stability should be investigated.

If the mean field probabilities were reliably followed during the evolution of an automaton, well developed theories of functional iteration could be applied. One of the best known of these utilizes the Mandelbrot set and related concepts to parameterize the behavior of iteration in the complex plane as well as the real line. As it is, the empirical cell frequencies observed in each generation do not track the functional iteration of the probability curve very well. Nevertheless gross properties of the fixed points appear to be valid; where the slope of the curve is negative, higher and lower densities tend to alternate, for example. Points on the borderline of stability, which is to say, those for which there is a diagonal tangency or near tangency at the fixed point, seem to be especially significant. In such instances, there is more of a fixed interval than a fixed point, permitting a greater variety of permanent structures than otherwise.

If there are not many ways of producing or conserving a live cell, the probability curve will never reach very high values, especially if it is already constrained to begin with a horizontal tangent. Thus even its maximum value may not even be sufficient to sustain its own density for the next generation, a reasonable conclusion if the curve does not even cross the diagonal. Alternatively, the production of live cells in the new generation may just barely meet the break even point, represented by a tangency. Finally it may be that modest densities proliferate freely, as would be expected from a curve whose values lay well on the other side of the diagonal.

This reasoning could well describe the transi-

tion from class II, where maintaining a live population is difficult and the regions which survive are nonexpansive, through class IV, where it is just barely possible over a limited range but with quiescent regions freely interspersed, until the opposite extreme of a class III, in which quiescent regions do not last long, and never dominate the field of evolution.

7. A survey of small binary automata

In summary, it is proposed to explain Wolfram's classes by a mixture of probability theory and de Bruijn diagrams, resulting in a classification based on readily visible properties of the mean field theory curve:

class I: monotonic, entirely on one side of
           diagonal;
class II: horizontal tangency, never reaches
           diagonal;
class IV: horizontal plus diagonal tangency,
           no crossing;
class III: no tangencies, curve crosses diagonal.

Classes being undecidable in principle, and mean field theory being approximate, the utility of this classification must be judged on empirical evidence. Although reasons have been given for its justification, the part concerning horizontal tangencies at the origin has the best foundation, being an accurate description of both automata and desirable mean field curves, whatever the actual relationship between the two. To illustrate the procedure, consider some simple examples:

7.1. (2,1/2) automata

The simplest automata are probably the sixteen of radius $1/2$ corresponding to the sixteen primitive Boolean functions of two variables. Their Bernstein polynomials are formed from a basis of p^2, pq, and q^2, which does not provide much variety for tangencies; nevertheless much of the mischief possible in rendering classifications is already visible in this elementary example. In terms of the mean field curves, there are essentially four groups to consider:

eight noncrossing rules, 0, 2, 4, 8, 11, 13, 14, 15;

four diagonal and antidiagonal rules, 3, 5, 10, 12;

two superstable rules, 6, 9;

two with unstable crossings rules, 1, 7.

Visual inspection places the noncrossing rules in class I or class II according to whether some nonquiescent cells remain after a few generations or not. Diagonals result from copying one of the ancestors, antidiagonals to its complementation; the former produce shifts, the latter dramatic differences from generation to generation which could be assigned to class III.

The four remaining rules qualify for class III through fixed points interior to the interval (0,1); in two cases the fixed point is stable, in the other two it is not. Assignments to class III on the basis of the widespread complementation due to an unstable fixed point are always dubious; Wolfram [7] would probably sidestep the issue by classifying the composite rule for which enough generations have elapsed to possess a quiescent state.

By this criterion, only rule 6 (EXCLUSIVE OR) and rule 9 (EXCLUSIVE NOR) belong strictly to class III, none to class IV, the rest to class I or class II.

7.2. (2,1) automata

The next simplest comparisons are worthwhile because the (2,1) automata have been extensively studied. Arranging the eight possible neighborhoods in sequence corresponding to Wolfram's codes, the transitions required by quiescence and guaranteed contagion can be inserted.

111	110	101	100	011	010	001	000
.	.	.	1	.	.	1	0

Thirty two rules arise from the five free choices remaining:

18	22	26	30	50	54	58	62
82	86	90	94	114	118	122	126
146	150	154	158	178	182	*186	*190
210	214	*218	*222	*242	*246	*250	*254

Computer simulation or examination of Wolfram's tables [6,8], shows that these are Wolfram's class III automata, although not quite. The rules whose numbers are prefixed with stars resemble

class II more closely than class III, quite in agreement with the failure of their curves to cross the diagonal. Rule 58 might not be considered part of class III because of evolving towards a very strongly shifting pattern, albeit space filling; its curve has a superstable fixed point.

Among the remaining rules, there is a group complementary to the ones just considered, for which live cells would constitute the quiescent background. Beyond this, there are many nonsymmetrical with respect to contagion which can mostly be classified according to the combination of their one-sided behaviors, leaving the nonexpansive rules to fall into Wolfram's class I or class II, with very little remarkable behavior.

The de Bruijn diagram shows how to form macrocells, for example with rule 73. All one-cell environments of the segment 0110 yield still lifes, yet not all form loops; instead they form barriers enclosing macrocells. Similarly all single cell extensions to the segment 11 in rule 154 shift left, bounding a moving macrocell. Boundaries which were not originally present may jell during the course of evolution, remaining thereafter. For automata with more than two states, partial barriers as well as complete barriers may exist, be formed, or dissolve, according to how many terminal links share the properties of the interior of a segment.

As an example of automaton design, one could postulate that the quiescent loop 0^* be linked to the loop $(01)^*$, fixing five of the eight links in the de Bruijn diagram. Of the three links and eight choices remaining, rule 4, of class II, results from rejecting any additional still lifes. Its Bernstein polynomial, pq^2, always lies below the diagonal, with a diagonal tangency at the origin. The rule is nonexpansive, but low concentrations of live cells are conserved, so that it is noncontractive as well. The discrepancy with the proposed classification arises from the ocurrence of pq^2 through the survival of isolated cells rather than from contagion. This distinction must often be made to reconcile the correct classification.

Some of the other extensions of rule 4 incorporate additional loops, up to the extreme of the identity rule, rule 204, which is best left unclassified. Is it class II for extensive still life, or class IV for its blatant tangency? Its Bernstein polynomial is $p = p(p + q)^2 = pq^2 + 2p^2q + p^3$.

Rule 104 is totalistic rule 4 for (2,1) automata, which places it in an interesting category. For (k, r) automata with integer r, totalistic rule $pk^{p(r+1)}$ has an isolated still life consisting of the string $0^r p^{r+1} 0^r$, as well as for all rules whose Wolfram number is a nonzero multiple (mod k) of this basic number. For example, (2,1) totalistic rules 4 and 12, which are rules 104 and 232. A similar effect is even seen in Conway's *Life*, for which any large region in which every 3×3 square contains exactly four live cells is a still life, and will persist until its border is eaten away.

Let us call this particular block an α-block. A rule whose still lifes include α-blocks should be expected to behave anomalously with respect to block probabilities calculated for blocks shorter than $2r + 1$-blocks, a conclusion which is borne out by experience; in this instance, significant probabilities for live cells do not appear until 4-blocks are considered.

The Bernstein polynomial for an α-block rule would be

$$\frac{(2r + 1)!}{(r + 1)!r!} p^{r+1}q^r \, ,$$

which is slightly asymmetrical and agrees with class II by approaching but not crossing the diagonal. Its presence in the de Bruijn diagram is manifested by a closed loop with incoming and outgoing links to the quiescent loop.

7.3. (2,3/2) automata

Four cells per neighborhood signify an automaton which can be one-dimensional with radius $3/2$ or two-dimensional with radius $1/2$, for which the effects of a difference in dimensionality can be observed. The neighborhoods of automata with half-integer radii lack a central cell, so the survival of isolated cells is not a direct issue, although "still lifes" can span two generations.

Interestingly, totalistic rule 4 yields the Bernstein polynomial p^2q^2, shown in fig. 5, which almost displays a diagonal tangency. In two dimensions, this rule possesses several still lifes of period 2 which coincide with actual *Life* still lifes on alternate generations. Gliders have not yet been observed; the five cell gliders which work with *Life* do not glide in this environment.

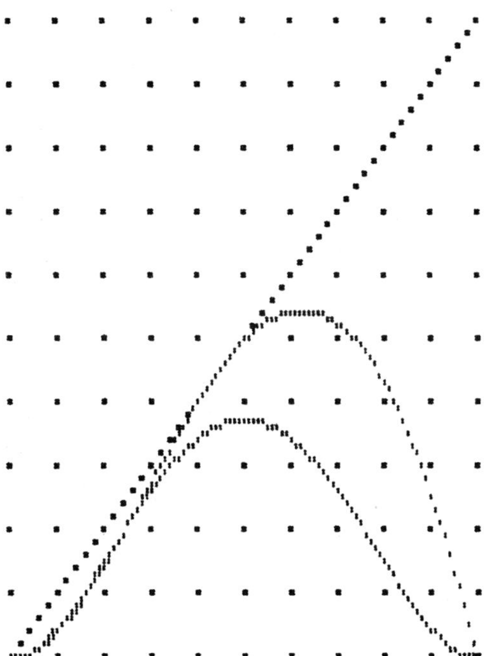

Fig. 5. Mean field probability curves. Upper: generic rule 30376, lower: totalistic rule 4 (rule 5736).

A systematic way to find diagonal tangencies is to expand the Lagrange interpolation polynomials of numerical analysis in a basis of Bernstein polynomials, or better yet, to use Lagrange–Hermite interpolation. Lagrange polynomials are defined as polynomials of nth degree which vanish at n points and take unit value at one additional point. With Lagrange–Hermite interpolation, derivatives may be specified in addition. Unfortunately, all such polynomials are extremely non-orthogonal, exposing the determination of coefficients to a very sensitive dependence on parameters. Furthermore, positive integral coefficients are required, creating a problem in integer programming.

All these difficulties beset the selection of a polynomial whose interpretation is not all that precise; much better results are obtained empirically by graphing the polynomial belonging to a given rule, and then tinkering with individual neighborhoods to alter the tangencies which are visible. Since the monomial $p^i q^{n-i}$ has a single, not too broad, maximum at i/n, it is not hard to decide which values to adjust, while simultaneously evaluating the position of the neighborhood in the de Bruijn diagram. Indeed, the classical rea-

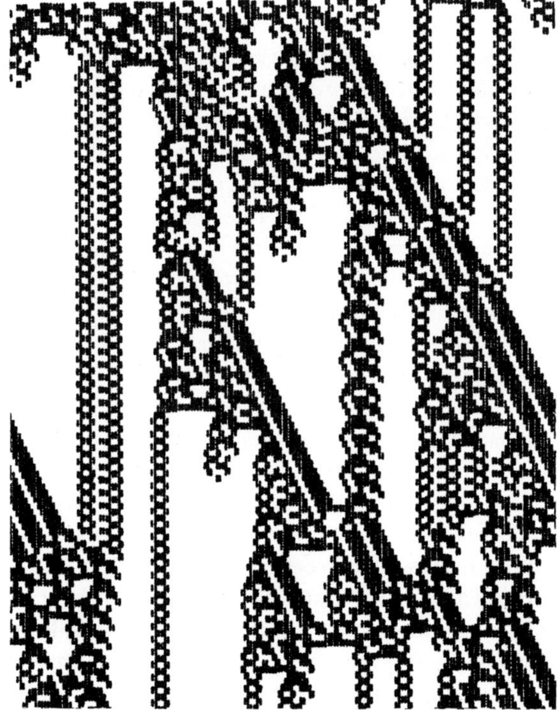

Fig. 6. The one-dimensional $(2,3/2)$ totalistic rule 4 belongs to class II, its mean field curve falling short of the diagonal. The temporal evolution of the nearby class IV rule 30376 is shown.

Fig. 7. Sample field of evolution for the two-dimensional $(2,1/2)$ (just barely) class IV totalistic rule 4, while rule 30376 is class III.

son for using Bernstein polynomials, in their general sense, was the ability to work in a given interval with functions which were entirely positive and well localized, becoming even more so in the limit of high degree.

Fig. 6, several generations of one-dimensional evolution, and fig. 7, a typical two-dimensional field, contrast the differences in dimensionalities. Low dimensions require high degrees of tangency at the origin to produce class IV. The classes are

also sensitive to the closeness of the diagonal tangency.

7.4. (2,2) automata

Wolfram has classified the 32 quiescent totalistic (2,2) linear automata [7] as follows:

class I: 0, 4, 16, 32, 36, 48, 54, 60, 62;
class II: 8, 24, 40, 56, 58;
class III: 2, 6, 10, 12, 14, 18, 22, 26, 28, 30, 34, 38, 42, 44, 46, 50;
class IV: 20, 52.

By and large, class I follows Wolfram's description, although many of the rules listed have simple structures of low period. For example, the regular expression $(0011)^*$ describes a still life for totalistic rule 4 and some of the others, while $(01)^*$ and $(000111)^*$ have period 2. While not evolution to a quiescent state, neither is it evolution into disconnected periodic regions, the province of class II. Wolfram dismisses such anomalies by invoking exceptions "of measure zero", but one suspects that they are inherent in the classification system and part of the reason for its approximate nature.

Class II consists principally of the rules with α-blocks, except for rule 58, and including rule 24, which has an exemplary class IV tangency. Its excellence notwithstanding, graphing the ninth-degree polynomial for *two* generations of evolution confirms the assignment to class II.

The class III automata behave as expected; their rules are all expansive. Whenever long gaps arise they begin to fill up immediately.

The two class IV rules are similar in their behavior, but rule 52 is especially interesting for being self-complementary; its evolution can resemble certain Escher prints at times. On account of nonorthogonality its mean field polynomial, $10p^2q^3 + 5p^4q + p^5$, gives a fairly good representation of the function p; its graph is symmetric by $180°$ rotation. All automata have similar rules, especially those of larger radii.

7.5. (2,7/2) automata

Eight cells per neighborhood allows another dimensional comparison: one-dimensional automata of radius 7/2, or three-dimensional of radius 1/2.

The three-dimensional diagonally tangent totalistic rule yields better candidates for class II automata than class IV, but it is possible to locate such artifacts as a cube–octahedron pair which oscillates with period 2, following totalistic rule 24. Larger composite structures of similar form also exist. Variants on the theme of semitotalistic rules yield other candidates for class IV.

In any event, the experience of Gutowitz and Victor [25], that similar patterns do not transfer from one dimension to another, and that a higher degree of tangency is required to obtain class IV in the lower dimension, is confirmed. Likewise Bays' experience of finding small artifacts is repeated, without encountering the larger structures characteristic of *Life*.

8. Other automata

8.1. Totalistic rules

Totalistic rules give some fairly strange results, until it is understood that they do not form a representative sampling of all the possible rules. For binary automata, a totalistic sum can only remain constant, increase, or decrease by 1 from one neighborhood to the next, so that adjoining cells assuredly do not enjoy independent probabilities. Were the probabilities truly independent, the sums would approximate a Gaussian distribution, but with such a restriction they would follow a much flatter distribution characteristic of the dominant eigenvector of a Markov chain. By favoring quiescent domains, this distortion of statistics enhances class IV automata among totalistic rules relative to the general population.

Another characteristic of totalistic rules is the fact that they lead to closed loops in the de Bruijn diagram. Summing the cells of a neighborhood defines an equivalence relation for neighborhoods, thus for links also. Cyclic shift assures every link of a continuation with the same sum in either direction, so that an equivalence class contains only closed loops. Totalistic rules map equivalence classes, and the result follows. In turn there is a closer affinity between totalistic rules and mean field probabilities, because there is no need to choose between permutations in assigning a Bern-

stein polynomial to an entire equivalence class.

8.2. Near totalistic rules

When totalistic rules do not allow for sufficient variation in parameterizing an automaton, the so-called semitotalistic rules are useful, sharing many of the statistical characteristics of totalistic rules. One reason for their use is that they arise naturally when small neighborhoods are submitted to symmetry requirements. Then the central cell may distinguish two kinds of evolution according to its value, as in *Life*. Automata with half-integer radii lack such a central cell, but there are substitutes, such as summing separately the even-parity cells and the odd-parity cells, or choosing a two-cell central region.

8.3. Automata which are not binary

When there are more than two states per cell, the Bernstein polynomials become multinomials, and the probabilities for all the states have to be fit at once, complicating the discussion of tangencies. Two correctives seem to be efficacious; one of them is to calculate a sum of squares of differences between the individual probabilities from one generation to the next – a distance between vectors of probabilities. Even so, it is not a function which is easily graphed except perhaps for trinary automata and with considerably greater difficulty of visual presentation for quaternary automata.

The second technique is to use the parameter λ introduced by Christopher Langton [37], which is the average probability of all the nonquiescent states. If the probability space is supposed to be a simplex, λ is the distance from the quiescent vertex to the opposite face, measured along an altitude. As such it may miss some of the twistings and turnings of the contours of constant deviation; it has the advantage of being a single representative parameter, but interpreting crossings of the diagonal must take into account its following a direct line rather than the gradients. Experience shows that it is a reasonable guide, but that it must be used rather cautiously. Fig. 8 shows an instance of a curved arc in a class IV automaton.

8.4. Probabilistic automata

Variants on the theme of cellular automata are interesting topics of study, either for modeling some particular physical process or for more intrinsic reasons. One is the coupled map lattice, another the probabilistic automaton. In the former an assemblage of iterative functions of a real variable (not necessarily probabilities) is coupled together in some way, typically as lattice neighbors. In the latter the state of the cells of the lattice is not known, only the probability that each is in one of its possible states.

In both cases, the law of evolution is exactly known, not approximated by mean field theory or one of the block structure theories. This was how Schulman and Seiden [19] brought estimates of the density of live cells in *Life* into closer agreement with their calculations. Recently Bidaux, Boccara, and Chaté [38] used the same approach to study the effect of dimension on phase transitions in a class of probabilistic automata. If rigorously probabilistic automata behave reasonably, any discrepancies observed in cellular automata can be pinpointed as having arisen from correlations among

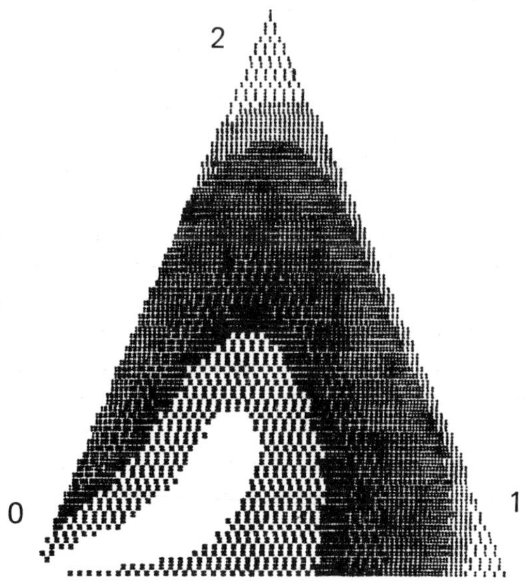

Fig. 8. Contours for mean field probability differences after one generation for (3,1) totalistic rule 792. There is an arc of tangency, which will be poorly sampled by Langton's parameter λ.

discrete states; useful limiting behavior might also be expected.

8.5. Automata designed to order

It is instructive to invert the process of seeking out interesting automata by examining the statistical properties of automata with known characteristics. This reverse process is heavily biased toward quiescent automata with large inactive regions because those are the ones which our experience tells us how to handle; highly parallel processes interacting closely are not at all commonplace. It is easy enough to devise simple automata which behave as counters, parenthesis balancers, or the like. Comparison of their statistics with the description of class IV allows a judgement of whether that is an appropriate niche for them; many examples do meet the requirements.

9. Discussion

There does not seem to be a single, infallible criterion as to what makes a good "*Life*", in part because the concept itself is somewhat subjective. Nevertheless there is a certain amount of experience in several contexts which indicates that the existence of independent isolated structures is fundamental, and that the structures which do exist should have a certain stability.

Isolation is easily tied to the existence of certain loops in the de Bruijn diagram for the automaton. Knowing the de Bruijn diagram for the still lifes does not determine the diagrams for higher periods, but their behavior is obviously part of the criterion that one would like to establish. In practice, the still lifes seem to give good guidance; moreover if the still life diagram does not behave properly it is assured that those for longer periods will not do so either.

Tying the stability of extant structures to tangencies in the mean field probability curve likewise seems to be fairly speculative. Nevertheless this curve seems to give a fairly good approximation to probabilities calculated via more complex approaches. Additionally, it is possible to alter the rule of evolution slightly and to observe a certain continuity of behavior as the probability curve changes from tangency to greater stability for one of the fixed points. For example, the (2,3) totalistic rule 88 seems to stretch the limits of what one ought to call a class IV automaton, which is partly why it supports a glider gun and not just gliders. Also note that tangency of the probability curve only implies constant density – not constancy of evolution – but the contrary assures a varied evolution.

The question of the stability of classification against minor changes is an important one. Kunihiko Kaneko [39] has discussed variations in the basin of attraction of a configuration whose individual cells have been altered; Li, Packard and Langton [40] consider the effect of alterations at the level of the de Bruijn diagram. An alteration in the evolution of one neighborhood results in reassigning its link from one subdiagram to another. This will break some loops and close others, but will not affect badly fractured loops. Some rules will be more sensitive to such shifting than others, depending on the fracture patterns arising when the subdiagrams are created; relocating one or a few links could affect many loops or none at all.

10. Conclusion

Conway's *Life* is the prototypical example of a class IV automaton, and it remains an interesting question as to why it shows such remarkable behavior relative to variants which have been studied. In a sense it is almost unique: if a rotationally and reflectively invariant rule is required, there are 102 symmetry classes of *Life* neighborhoods, and it is possible to tabulate the number of these which are used in the evolution of any of *Life*'s artifacts. If gliders, glider guns, and the two shuttles used in forming glider guns are counted, the evolution of all but a few neighborhoods is determined. What is surprising is that all the transitions required have such a succint description, via an easily described semitotalistic rule.

Of course there is no principle so far known excluding the existence of alternative collections of artifacts, with still another style of operation, so the uniqueness of *Life* is only relative to the present state of our experience. Maybe the search for alternatives has been insufficiently vigorous, or

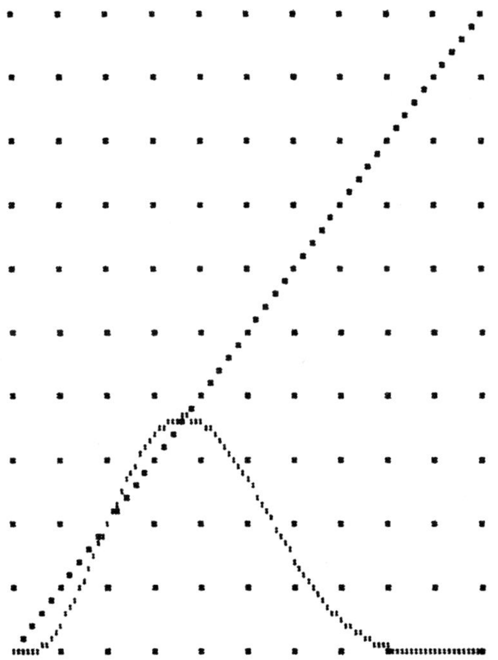

Fig. 9. Mean field curve for Conway's *Life*; crossing the diagonal slightly more than a tangency, quadratic at the origin.

possibly binary automata with close neighbors operate in just one way. Knowledge and systematic usage of de Bruijn diagrams has spread slowly; heretofore the discovery of artifacts has largely been one of ingenuity and inspiration coupled with detailed analysis. Even so, the exponential growth of the diagrams still makes it impractical to search for very large structures.

With present equipment, only first-generation patterns in strips of width less than ten cells are feasible for the nine-cell two-dimensional neighborhoods that *Life* or its variants utilize, yet reproducing results already known would need four generations. That is one reason to work with half-integer neighborhoods of four cells, whereby two generations can be searched. One pays for the privilege of a more comprehensive search by working in an arena with fewer possibilities.

The foregoing analysis may well have to be judged by the degree to which it applies to this one case, *Life*, whose curve is shown in fig. 9. It would have been nice to say that a clear cut class of automata had been identified, in which some new

specimens had been found exhibiting outstanding behavior. They may yet be found; meanwhile progress is more likely to come from inventing ingenious mechanisms which can be implemented as cellular automata. And above all, from carrying out some of the searches which are now possible.

Acknowledgements

The author is grateful to the Organizers for their kind invitation to attend Workshop CA89.

References

[1] M. Gardner, Mathematical games. The fantastic combinations of John Conway's new solitaire game "Life", Sci. Am. (October 1970) pp. 120–123.

[2] M. Gardner, Wheels, Life, and Other Mathematical Amusements (Freeman, San Francisco, 1983).

[3] E.R. Berlekamp, J.H. Conway and R.K. Guy, Winning Ways for your Mathematical Plays, Vol. 2 (Academic Press, New York, 1982) ch. 25.

[4] J. von Neumann, in: Theory of Self-reproducing Automata, ed. A.W. Burks (University of Illinois Press, Champaign, IL, 1966).

[5] E.F. Codd, Cellular Automata (Academic Press, New York, 1968).

[6] S. Wolfram, Statistical mechanics of cellular automata, Rev. Mod. Phys. 55 (1983) 601–644.

[7] S. Wolfram, Universality and complexity in cellular automata, Physica D 10 (1984) 1–35.

[8] S. Wolfram, ed., Theory and Applications of Cellular Automata (World Scientific, Singapore, 1986).

[9] N.H. Packard and S. Wolfram, Two-dimensional cellular automata, J. Stat. Phys. 38 (1985) 901–946.

[10] K. Preston Jr. and M.J.B. Duff, Modern Cellular Automata (Plenum, New York, 1984).

[11] W. Poundstone, The Recursive Universe (Morrow, New York, 1985).

[12] R.T. Wainwright, ed., Lifeline, a quarterly newsletter with 11 issues published between March 1971 and September 1973.

[13] C. Bays, The game of three-dimensional life, 20 November 1986, unpublished (available as a supplement to A.K. Dewdney's February 1987 column).

[14] C. Bays, Candidates for the game of life in three dimensions, Complex Systems 1 (1987) 373–400.

[15] C. Bays, Patterns for simple cellular automata in a universe of dense-packed spheres, Complex Systems 1 (1987) 853–875.

[16] C. Bays, Classification of semitotalistic cellular automata in three dimensions, Complex Systems 1 (1987) 373–400.

[17] K. Culik II and S. Yu, Undecidability of CA classification schemes, Complex Systems 2 (1988) 177–190.

[18] M. Dresden and D. Wong, Life games and statistical models, Proc. Natl. Acad. Sci. US 72 (1975) 956–960.

[19] L.S. Schulman and P.E. Seiden, Statistical mechanics of a dynamical system based on Conway's game of Life, J. Stat. Phys. 19 (1978) 293–314.

[20] W.J. Wilbur, D.J. Lipman and S.A. Shamma, On the prediction of local patterns in cellular automata, Physica D 19 (1986) 397–410.

[21] H.A. Gutowitz, J.D. Victor and B.W. Knight, Local structure theory for cellular automata, Physica D 28 (1987) 18–48.

[22] H.A. Gutowitz and J.D. Victor, Local structure theory in more than one dimension, Complex Systems 1 (1987) 57–68.

[23] G.G. Lorentz, Bernstein Polynomials (University of Toronto Press, Toronto, 1953).

[24] J.A. Shohat and J.D. Tamarkin, The Problem of Moments (Am. Math. Soc., Providence, RI, 1943).

[25] H.A. Gutowitz and J.D. Victor, Local structure theory: calculation on hexagonal arrays, and interaction of rule and lattice, J. Stat. Phys. 54 (1989) 495–514.

[26] S.W. Golomb, Shift Register Sequences (Holden-Day, San Francisco, 1967).

[27] A. Ralston, De Bruijn sequences – a model example of the interaction of discrete mathematics and computer science, Math. Mag. 55 (1982) 131–143.

[28] E. Jen, Cylindrical cellular automata, Commun. Math. Phys. 118 (1988) 569–590.

[29] S. Wolfram, Computation theory of cellular automata, Commun. Math. Phys. 96 (1984) 15–57.

[30] W. Li, Power spectra of regular languages and cellular automata, Complex Systems 1 (1987) 107–130.

[31] J.E. Hopcroft and J.D. Ullman, Introduction to Automata Theory, Languages, and Computation (Addison–Wesley, Reading, MA, 1979).

[32] M.L. Minsky, Computation: Finite and Infinite Machines (Prentice-Hall, Englewood Cliffs, NJ, 1967).

[33] D.B. Brown, Competition of cellular automata rules, Complex Systems 1 (1987) 169–180.

[34] J.H. Conway, Regular Algebra and Finite Machines (Chapman and Hall, London, 1971).

[35] R.C. Backhouse and B.A. Carré, Regular algebra applied to path finding problems, J. Inst. Math. Appl. 15 (1975) 161–186.

[36] P. Guan and Y. He, Upper bound on the number of cycles in border-decisive cellular automata, Complex Systems 1 (1987) 181–186.

[37] C.G. Langton, Studying artificial life with cellular automata, Physica D 22 (1986) 120–149.

[38] R. Bidaux, N. Boccara and H. Chaté, Order of the transition versus space dimension in a family of cellular automata, Phys. Rev. A 39 (1989) 3094–3105.

[39] K. Kaneko, Attractors, basin structures and information processing in cellular automata, in: Theory and Applications of Cellular Automata, ed. S. Wolfram (World Scientific, Singapore, 1986) pp. 367–399.

[40] W. Li, N. Packard and C.G. Langton, Transition phenomena in cellular automata rule space, Physica D 45 (1990) 77–94, these Proceedings.

Physica D 45 (1990) 122–135
North-Holland

CRITICALITY IN CELLULAR AUTOMATA

Hugues CHATÉ

*AT&T Bell Laboratories, Murray Hill, NJ 07974-2070, USA
and Institut de Recherche Fondamentale, DPh-G/PSRM, CEN-Saclay, 91191 Gif-sur-Yvette Cedex, France*

and

Paul MANNEVILLE

Institut de Recherche Fondamentale, DPh-G/PSRM, CEN-Saclay, 91191 Gif-sur-Yvette Cedex, France

Received 15 January 1990
Revised manuscript received 27 February 1990

Using recent results obtained for the transition to turbulence via spatiotemporal intermittency in extended dynamical systems, *critical* cellular automata rules are built. Thanks to a systematic procedure, the continuous phase transition observed in a coupled map lattice is translated into a sequence of cellular automata rules for which the critical properties of the original system are reproduced as precisely as desired. It is shown that criticality, as understood from the point of view of statistical mechanics, is intimately related to the various characteristics of Wolfram's class IV rules. This suggests in turn the picture of a "critical surface" in the space of rules and provides the basis for a discussion of the problem of classification schemes. We review recent cellular automata studies where questions linked to criticality arise and argue that they are unified in light of our results, leaving the relationship between computational and statistical characterizations of critical rules as the central problem for future studies on this subject.

1. Introduction

Whereas complexity is a notion often invoked in the field of cellular automata (CA) [1], "criticality" is much less commonly encountered, except perhaps in the study of the phase transitions of probabilistic cellular automata (PCA) such as directed percolation [2]. For deterministic cellular automata (DCA), partly due to their lack of continuous "control parameters", phase transitions and critical phenomena in the sense of statistical mechanics have only recently become topics of interest, along with the problem of the structure of the space of rules. This latter problem is obviously connected to the various attempts to define classifications of CA, either after their observed behavior (phenotypic classification) or after their rule table (genotypic classification). Indeed, once defined, classes of CA can be seen as phases, bringing up the question of the transitions between these phases, i.e. non-arbitrary paths in the space of rules that connect CA belonging to different classes.

It is well known that the traditional classification of Wolfram [3] suffers drawbacks, the most serious of which probably being its non-decidability [4]. It seems to us that the main problem is the exact status of class IV, and it may well be that its best definition remains founded on a negative statement: class IV rules are the rules which do *not* belong to any of the three other classes. Let us briefly recapitulate the (positive) definitions of these classes for infinite-size CA [3].

Class I automata evolve to a unique, homogeneous, stationary state after a finite transient from almost any initial condition.

Class II rules generate spatially inhomogeneous, stationary or periodic (with a short period), asymptotic states after a finite transient.

Class III automata evolve to a unique, chaotic,

statistically well-defined state from almost any initial configuration in a finite number of timesteps. These qualitative definitions can be made quantitative using various statistical quantities.

Class IV was initially defined by various criteria (existence of propagative structures, arbitrary long transients, no smooth infinite volume limit, influence of low probability configurations on average quantities, etc.) but was mainly linked to the property of a rule to be a universal computer, i.e. of being able to perform any digital computation given a suitable initial configuration. But until now, the equivalence between this computational property and various statistical characterizations of class IV rules has remained an open question, mainly because proving that a rule is a universal computer is usually very difficult in the absence of any systematic approach.

Without a positive definition of class IV CA based on statistical properties, the completeness of Wolfram's classification cannot be ensured. Also, and particularly as defined above, the first three classes are not "closed": for example, it is practically impossible to decide membership in the case of very long transients.

Moreover, this phenotypic classification, together with other similar attempts to distinguish CA from their behavior [5,6], also suffers from its intrinsic inability to provide clues on the structure of the space of CA that would relate to the rule tables themselves. Ideally, one would like to at least be able to make an educated guess on the general behavior of a CA given some features of its rule table, and in particular see whether it is close to a transition point, possibly showing some critical dynamical properties.

In this paper, we take advantage of our recent work on the transition to spatiotemporal disorder in dynamical systems with many degrees of freedom to attack the above-mentioned problems, focusing on the possibility of critical behavior in CA and the relative importance of the corresponding models in the space of rules. Our point of view will often be that of statistical mechanics and we will mostly use the tools common in this domain of physics.

In a series of articles [7], we have shown that spatially extended dynamical systems such as partial differential equations, coupled ordinary differential equations, and coupled map lattices (CML), can exhibit transitions to disorder akin to phase transitions as defined in statistical mechanics. For all these models, these *transitions to turbulence via spatiotemporal intermittency* are observed when varying a continuous parameter, and the eventual critical points are well-characterized by means of the physical quantities similar to those used in the field of critical phenomena.

Here, after a brief presentation of this set of results, we focus on one particular such transition observed for a CML (section 2). Then, we connect those results to CA themselves, thanks to a systematic approximation scheme that enables us to construct CA rules with well-controlled critical properties (section 3). Section 4 is devoted to a general discussion of the relationships between various approaches to criticality in CA, with special attention to the aspects unveiled by our own results and their implications for the problem of the structure of the space of rules.

2. Critical behavior of a coupled map lattice

Before presenting the continuous phase transition observed in a CML that we will translate to a sequence of rules in the space of CA, let us briefly review the physical background that led us to this problem.

2.1. Transition to turbulence via spatiotemporal intermittency

Hydrodynamical turbulence is a phenomenon that may seem far away from the concerns usually at play in CA studies. Lattice gas models represent one way of connecting the two fields at a "microscopic" level. Recently, Pomeau suggested a higher level, macroscopic connection between some hydrodynamical situations and simple PCA such as directed percolation [8].

Among the flows most defiant of both analysis and experiments are those which exhibit the coexistence of a regular (laminar) state together with a disordered (turbulent) one. The corresponding dynamical regimes are characterized by patches of both states bordered by well-defined fronts evolv-

ing in space and time. Such *spatiotemporal intermittency* regimes are ubiquitous (e.g. turbulent spots in the Blasius boundary layer [9]), and Pomeau argued that their essential feature lies in the asymmetry between the two states. As a matter of fact, the intrinsic fluctuations of the disordered state allow the nucleation of the laminar state in a turbulent region, whereas the absence of such fluctuations in the regular state prevents the spontaneous emergence of disorder within a laminar patch. In the language of PCA, this is qualitatively equivalent to saying that the laminar state is *absorbing* and that the propagation of disorder can only happen through the contamination/recession process governing the dynamics of the local fronts separating two patches. It is then natural to propose a link between the relevant hydrodynamical situations and directed percolation which can be seen as one of the simplest two-state PCA with one absorbing phase.

Directed percolation is best known for exhibiting a continuous (second order) phase transition with rather well-understood critical properties in the infinite-size limit [2]. Such critical behavior is observed when varying a parameter governing the local transition probabilities defining the rule. Pursuing the analogy between PCA of the directed percolation type and systems showing spatiotemporal intermittency, one is led to imagine the occurence of similar scaling regimes in dynamical systems with many degrees of freedom when varying a control parameter through some criticial point.

This general picture was first confirmed by a study of a simple partial differential model believed to have some relevance to hydrodynamics [10]. We then introduced a simple CML designed to retain only the dynamical features necessary for exhibiting spatiotemporal intermittency [11]. This "minimal" model, to be presented below, was studied for space dimensions $1 \leq d \leq 4$ [11,12]. It was shown that the transition to turbulence via spatiotemporal intermittency can be of different types depending on the details of the system and the lattice dimension d. For $d \leq 2$, transitions are either discontinuous (first-order-like) or continuous (second-order-like) with possible complications due to the superposition of deterministic features on the general picture of a phase transi-

tion [13]. The equivalence with directed percolation suggested by Pomeau is thus not complete, but the conjecture is valid at a qualitative level, and remarkably anticipated the novel relevance of statistical mechanics in the field of spatially extended dynamical and physical systems brought to light by our findings.

In the next section, we exemplify these results by detailing a continuous phase transition observed in our minimal CML.

2.2. A typical continuous phase transition

A coupled map lattice consists of a collection of N identical local maps f of one or more real variables sitting on a lattice of space dimension d. The sites are updated synchronously by the iteration of the local map f and a coupling procedure involving the sites of a regular neighborhood. Most of the time, the evolution can be decomposed into two steps, iteration and coupling. This can be written as

$$\cdots \xrightarrow{g} X^n \xrightarrow{f} Y^n \xrightarrow{g} X^{n+1} \xrightarrow{f} Y^{n+1} \xrightarrow{g} \cdots$$

where g is the coupling function involving a site and its neighbors, and the superscripts symbolize the (discrete) time. Although we will briefly mention other situations, we restrict ourselves here to the most studied case of a chain ($d = 1$) of maps of one real variable ($X, Y \in \mathbf{R}$), coupled to their nearest neighbors by the linear "diffusive" coupling function g:

$$g(Y_{i-1}, Y_i, Y_{i+1}) = (1 - \varepsilon)Y_i + \tfrac{1}{2}\varepsilon\,(Y_{i-1} + Y_{i+1}),$$

where ε is the coupling strength, and the subscripts denote the spatial indices of the sites.

The local map f designed to fulfill the minimal requirements for exhibiting spatiotemporal intermittency reads [11]:

$$\begin{aligned}
f(X) &= sX & &\text{if } X \in [0, 1/2], \\
f(X) &= s(1 - X) & &\text{if } X \in [1/2, 1], \\
f(X) &= X & &\text{if } X > 1,
\end{aligned}$$

with $s > 2$.

When uncoupled to its neighbors ($\varepsilon = 0$), each map eventually experiences a chaotic transient in the "turbulent" part of its phase space, $X \leq 1$, before reaching a fixed point in the laminar region,

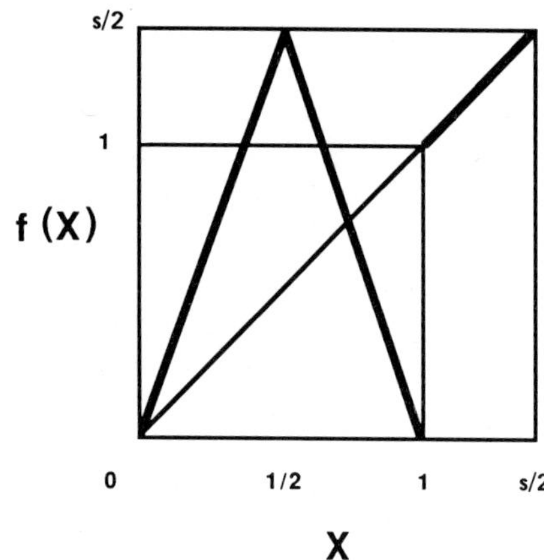

Fig. 1. Local map f of the minimal CML showing spatiotemporal intermittency. The turbulent state is the chaotic tent map ($X \leq 1$), and the laminar state is the continuum of marginally stable fixed points ($X > 1$).

$X > 1$ (see fig. 1). It is easy to show that the laminar state is absorbing in the coupled system. For strong enough ε and when starting from inhomogeneous initial conditions, the finite-amplitude perturbations introduced by the coupling may be large enough to take one quiescent site adjacent to an active one from the laminar state back to the turbulent state, and the CML then exhibits sustained regimes of spatiotemporal intermittency. There exists a threshold value ε_c (well-defined in the infinite size limit) above which this occurs, and the transition region shows scaling properties characteristic of continuous phase transitions (fig. 2).

The spatiotemporal intermittency regimes observed above threshold ($\varepsilon > \varepsilon_c$) possess well-defined statistical properties. For example, the average density of "turbulent" sites ($X > 1$), often considered as an order parameter, is the same whether the averaging is done in space, in time, or both. The following scaling properties are observed in the critical region:
— continuous decrease of the average density of turbulent sites when approaching the threshold from above (exponent β);
— divergence of the fluctuations of the instantaneous density of turbulent sites at threshold;
— divergence at threshold of the average transient time before a steady regime is reached;
— continuous decrease of the largest Lyapunov ex-

ponent when approaching threshold from above (preliminary result);
— algebraic distribution of the sizes and durations of clusters of laminar sites at threshold (exponents $\zeta_\perp$ and $\zeta_\parallel$);
— divergence of the coherence length $\xi_\perp$ and the coherence time $\xi_\parallel$ as defined from the characteristic scales extracted from the exponential distributions of sizes and durations of *laminar* clusters above threshold (fig. 3).

Most of these criteria will be used in the next section to characterize the critical CA rules emerging from our approximation of this CML. The scales mentioned in the last point are not the traditional quantities used to measure the coherence of a many-body system. Although this is still a point of ongoing controversy [14], let us mention here that autocorrelation functions do not provide clear-cut estimates of correlation scales for this system or for most systems of this type. The coherence scales defined above, on the other hand, are based on a reduction of the states of the CML deeply rooted in its dynamical properties and are relatively easy to interpret.

The equivalence with directed percolation initially proposed by Pomeau is not verified even in the case of a well-defined continuous transition. For example, the exponent $\zeta_\perp$ characterizing the distribution of sizes of laminar clusters at threshold takes two different values for $s = 2.1$ and $s = 3$ ($\zeta_\perp = 1.78$ and $\zeta_\perp = 1.99$) which are different from the corresponding value for directed percolation ($\zeta_\perp = 1.75$). This nonuniversality was part of our motivation to translate the problem into the field of DCA.

3. Sequences of CA rules approximating a continuous transition

In a previous work [15], we introduced an approximation of CML by CA in order to obtain insights on the various problems posed by the transition to turbulence via spatiotemporal intermittency and in particular the question of its nonuniversality. Here, we present a slightly different approximation scheme designed to produce CA as close as desired to a given critical point and study in some detail the sequences of rules equivalent to

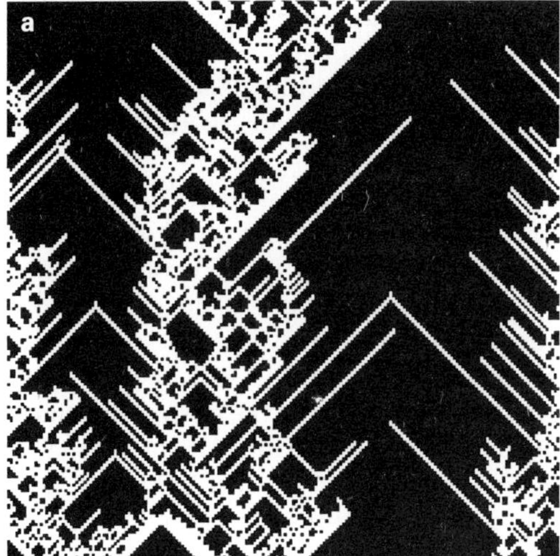
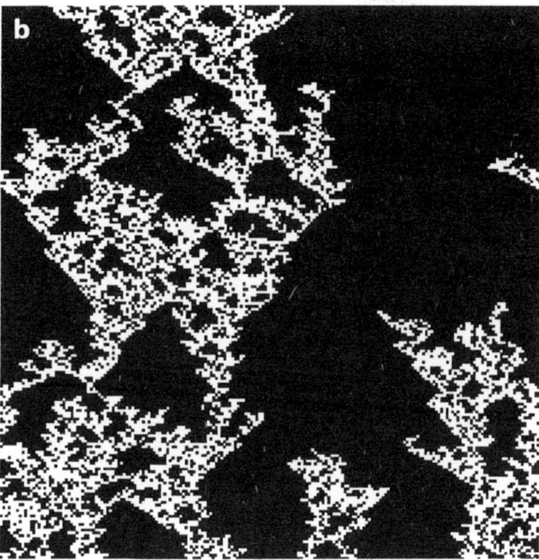

Fig. 2. Spatiotemporal intermittency in the critical regime for the minimal CML. Sites in the laminar (turbulent) state are black (white), following the natural binary reduction built in the local map f. The chain of $N = 200$ maps with periodic boundary conditions is shown just above the spatiotemporal intermittency threshold, and time is running upward. (a) $s = 3$, $\varepsilon = 0.360$, 200 iterations; (b) $s = 2.1$, $\varepsilon = 0.0047$, 200×16 iterations.

the continuous transition described in the previous section.

3.1. Approximating CML by CA

Only the discreteness of their local phase space distinguish DCA from CML. Indeed, a CML whose local map takes only discrete values (i.e. a step function) is equivalent to a DCA defined on the same lattice, the same neighborhood and a number of possible states equal to the number of steps of different heights in the step function (if the automaton is "observed" after the iteration step, cf. section 2.2). The approximation simply consists of choosing a step function $\tilde{f}$ preserving the essential features of the original local map f.

This very simple idea provides a systematic way of constructing sequences of rules approximating the variation of (continuous) parameters in the CML. Having an underlying "physical" signification, these sequences form non-arbitrary paths in the space of CA.

Virtually all parameters can be varied, including the space dimension and the size of the coupling neighborhood, so that the sequences of rules are not limited to a subset of the space of CA. For example, constructing CML based on the same lo-

cal map f but with coupling functions g approximating the diffusion operator at higher orders (on larger and larger neighborhoods) creates a sequence of rules of increasing radius whose dynamics are very similar [16]. Thus, although obviously not a systematic way of exploring the structure of the space of CA, this type of approach produces well-controlled paths along all the dimensions of this space.

3.2. Previous results

In ref. [15], we constructed the step function $\tilde{f}$ approximating the local map f of our minimal CML by considering the preimages of the laminar state $(X > 1)$ backward in time. This solution provided insights into the processes at play in the spatiotemporal intermittency regimes of the original CML. In particular, the cases $s = 2.1$ and $s = 3$ were found to produce very different sequences of rules when varying the coupling strength ε.

For $s = 2.1$, the rules of the sequence were all of class I and II, failing thus to reproduce the spatiotemporal intermittency regimes, whereas for $s = 3$, complex rules (class III and IV) were found above a threshold value ε_c depending on the order of the approximation. This indicated that

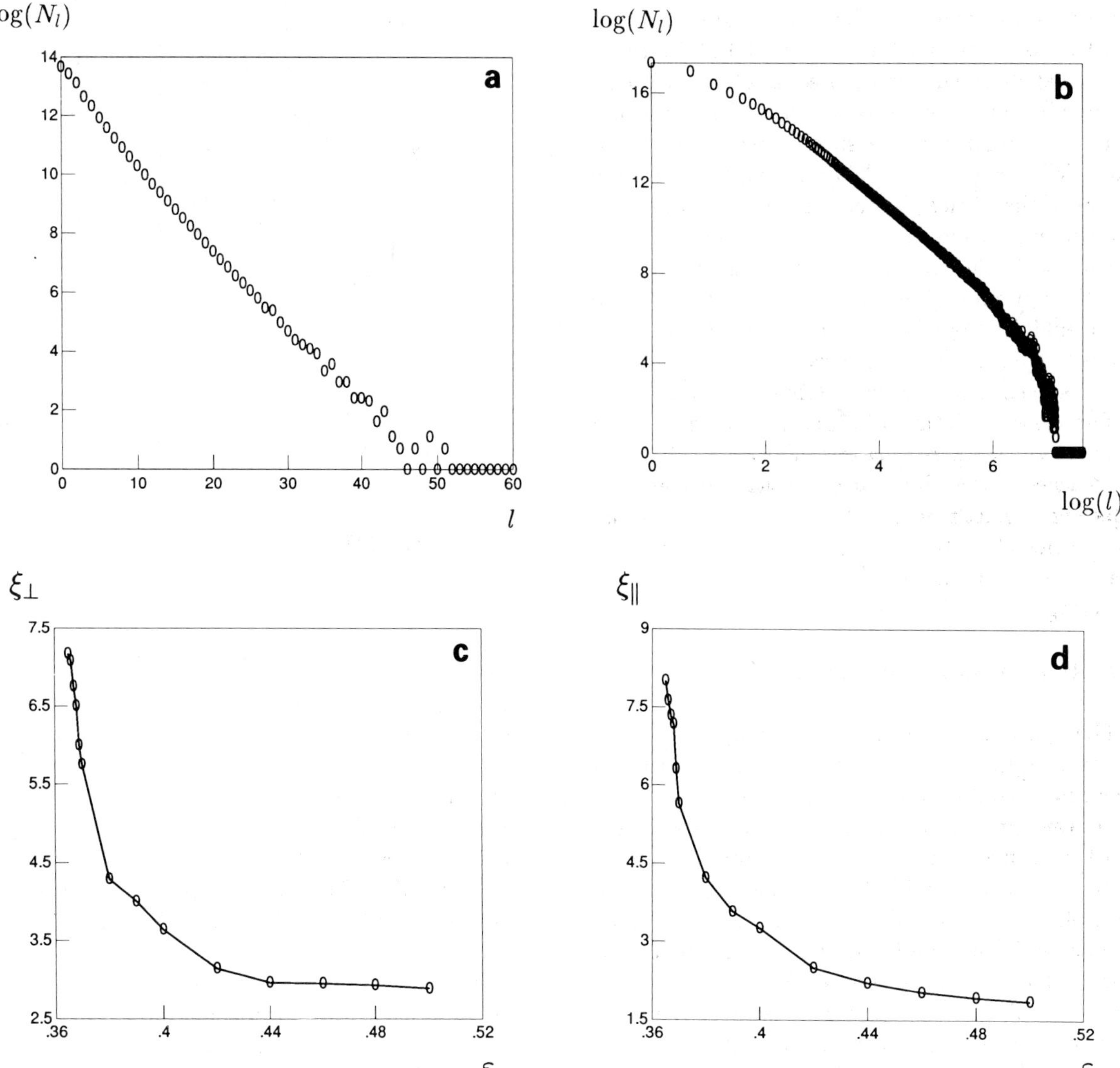

Fig. 3. (a) Histogram of the sizes of the clusters of laminar sites for the minimal CML above the spatiotemporal intermittency threshold ($s = 3$, $\varepsilon = 0.400 > \varepsilon_c = 0.360$). Statistics cumulated over space and time for a $N = 5000$ site chain with periodic boundary conditions during 5000 iterations. Exponential distribution with characteristic scale $\xi_\perp \sim 3.7$ sites. (b) Same as (a) but at the spatiotemporal intermittency threshold ($\varepsilon = 0.360$) for a lattice of 10000 sites during 10000 iterations. Algebraic distribution with critical exponent $\zeta_\perp$. (c) Variation with ε of the coherence length $\xi_\perp$ defined in (a). (d) Variation with ε of the coherence time $\xi_\parallel$ defined in (a).

for $s = 2.1$ the local processes at the origin of the spatiotemporal disorder are due to the quasi-probabilistic mixing occuring during the long excursions of each site in the turbulent state. For $s = 3$, on the other hand, the spatiotemporal intermittency regimes are similar to those produced by class III and class IV rules, i.e. quasi-deterministic processes characterized by propagating structures and triangular clearings (fig. 2). The transitions, although both continuous for the CML, are very different, as already indicated by the nonuniversality of the critical exponents.

For $s = 2.1$, the critical regimes of the CML are close to those exhibited by a PCA rule interpolated between trivial (class I and II) DCA rules. Note that directed percolation falls into this category and indeed exhibits no particular structure in its spatiotemporal evolution, just as the minimal CML for $s = 2.1$. This may explain why the values of the critical indices of the transitions are close to each other for these two systems.

For $s = 3$, the transition is qualitatively different: the critical regimes of the CML are similar to those exhibited by class IV and class III rules, and one is led to imagine that there exist DCA rules as close as desired to this well-defined critical point.

The general picture of a *critical surface* in the space of rules emerges, a critical surface that can be reached either by interpolation between two rules lying apart from it (PCA) or by constructing DCA rules close to it. It is this latter possibility that interests us here, with the final aim of constructing critical CA models.

3.3. An equal-step approximation

The approximation of the local map f based on the preimages of the laminar state [15] creates step functions $\tilde{f}$ that do *not* converge to f when the order of the approximation increases. In order to ensure this convergence, we use here step functions $\tilde{f}_k$ ($k \geq 2$) which divide the turbulent part of the local phase space of f ($0 \leq X \leq 1$) in $2k - 1$ steps of equal width having k different heights. The laminar part of the phase space ($1 < X \leq r/2$) is then also divided in steps whose heights match those previously defined. Fig. 4 shows a step function $\tilde{f}_3$ defined this way. The CML built on the step functions $\tilde{f}_k$ *is* a k-state DCA. For the one-dimensional case presented in section 2, the DCA have radius $r = 1$ (nearest-neighbor coupling in the CML) and their particular rule table depends on the parameters of the CML, s and ε.

The transition observed when varying continuously ε is translated into a discrete sequence of rules. The absorbing property of the laminar state is conserved in a global way: a configuration with all sites in the states corresponding to $X > 1$ will produce a site in one of those states. This justifies the continuing use of a binary representation (see fig. 2) even for the DCA with more than two states

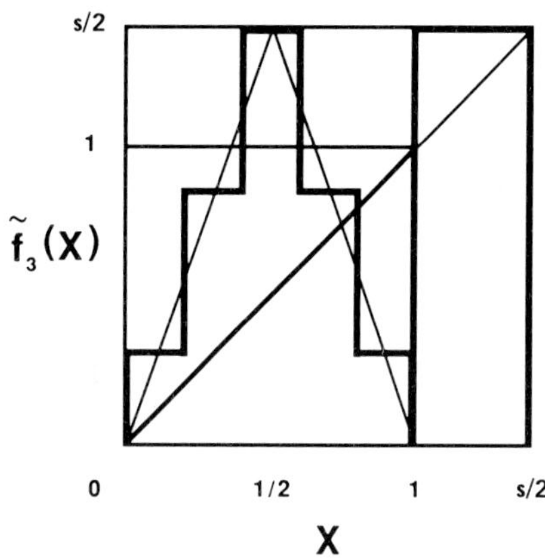

Fig. 4. Equal-width step function $\tilde{f}_3$ approximating the local map f of the minimal CML at order $k = 3$.

($k > 2$). This also allows a simple comparison with the original CML.

3.4. Results for the minimal CML with $s = 3$

Here we describe the sequences of rules produced by the approximation for the minimal CML defined in section 2.2 with $s = 3$ and ε varying from 0 to 1.

At first order ($k = 2$), the approximation is exactly equivalent to the one introduced in ref. [15]: independently of s ($2 < s \leq 3$), the sequence is composed of 7 of the 32 two-state three-neighbor legal rules studied in detail by Wolfram [17] (table 1).

Already at such a crude level, the transition to spatiotemporal intermittency observed for the CML is recovered, with a threshold ε_c separating class I and II rules from class III ones.

Increasing the order k of the approximation, more and more features of the original transition are recovered. Given the cardinal of the set of possible rules at fixed k, it is actually of no interest to define exactly the rules of the sequences for $k > 2$, so that we study them mostly through numerical simulations.

The number of rules in the sequence for $0 \leq \varepsilon \leq 1$ varies slightly with s but increases strongly with the order k of the approximation. Similarly, the mean normalized Hamming distance between consecutive rules in a given sequence decreases when k increases (fig. 5).

Table 1
Equivalent rules for the deterministic cellular automata approximating the minimal coupled map lattice at order $k = 2$ when the coupling ε is varied between 0 and 1 (results valid for $2 \leq s \leq 3$ only).

ε		$2/s - 4/s^2$	$1 - 2/s$	$4/s - 8/s^2$	$2 - 4/s$	$2/s$	$1 - 2/s + 4/s^2$
Rule	32	36	4	76	94	90	122
Dynamics			trivial (class I and II)			complex (class III)	

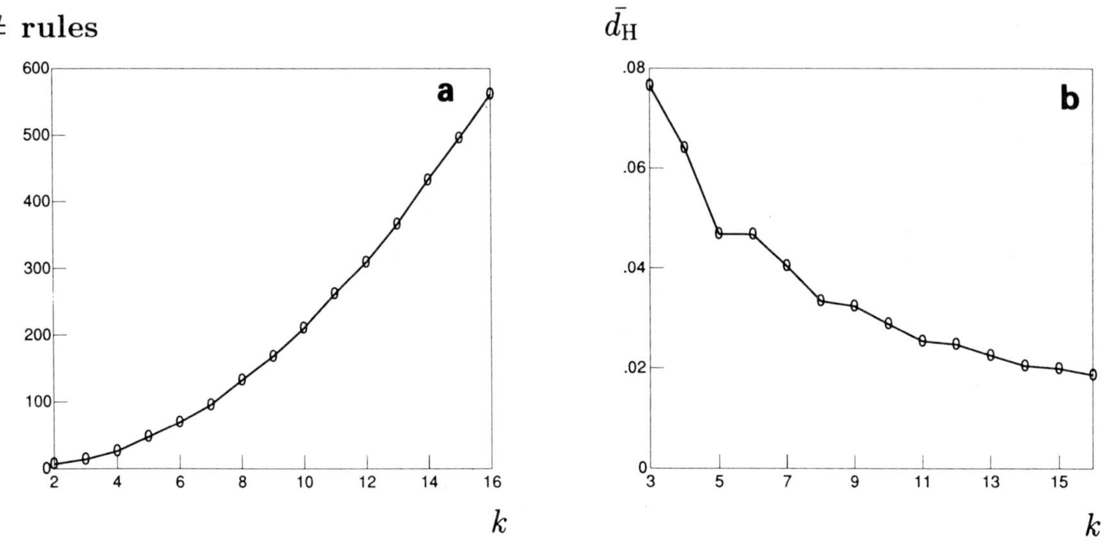

Fig. 5. (a) Variation with k of the number of rules in the sequences approximating the CML. (b) Variation with the order of the approximation k of the mean normalized Hamming distance $\overline{d_{\mathrm{H}}}$ between rules in a sequence. Rules yielding different states for every parental neighborhood lie at distance 1 from each other after normalization.

This is not surprising since the phase space is then less and less coarse-grained, and the scheme converges to the original CML as $k \rightarrow \infty$. At a given order, there is a threshold value ε_{c} separating trivial from complex rules. Table 2 shows that this threshold rapidly converges to the corresponding value for the original CML. At high orders, say $k > 5$, the contiguity in the space of rules does imply a contiguity in the dynamics. This confirms the general remark that rules with similar rule tables are likely to exhibit similar behaviors, in relation with the observation that nevertheless certain configuration outputs can qualitatively modify the dynamics ("hot bits" [5]).

One of the most remarkable facts is that even at low orders of the approximation the rules in the transition region appear to have most of the properties of class IV rules. Fig. 6 shows such rules with characteristic propagating structures, very long transients, and seemingly unpredictable asymptotic states.

In the next section, we study the sequence generated by the approximation scheme at order $k = 10$ and compare it directly with the original transition to spatiotemporal intermittency in the minimal CML.

3.5. The sequence of rules for s = 3 and k = 10

At order $k = 10$, the transition to spatiotemporal intermittency of the minimal CML with $s = 3$ and $0 < \varepsilon < 1$ is reproduced by a sequence of 210 DCA rules. At this moderately high order, there are enough rules to be able to compare with the oiginal transition of the minimal CML presented in section 2.2.

The dynamics of these 210 rules show the existence of a threshold value $\varepsilon_{\mathrm{c}} \simeq 0.39$ below which only class I and II rules are observed. The rules above the transition region possess the triangular clearings and the well-defined statistical prop-

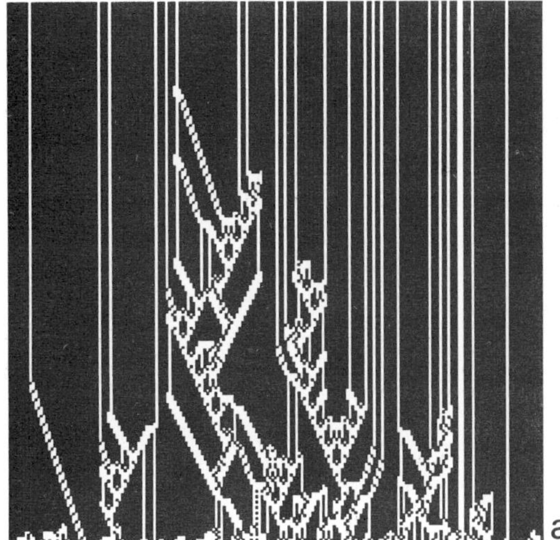
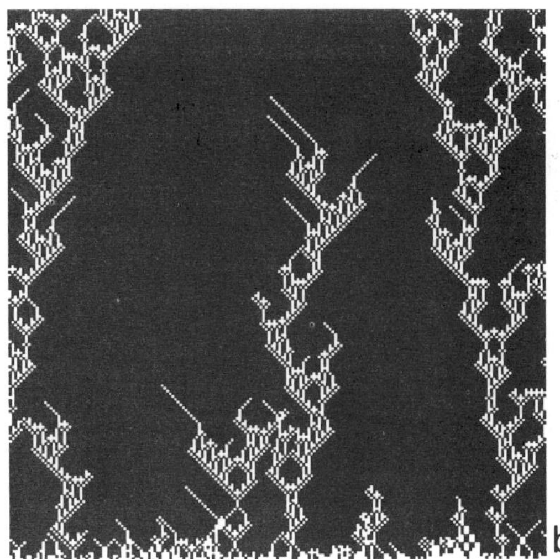

Fig. 6. Rules in the critical region of the sequences approximating the transition to turbulence via spatiotemporal intermittency in the minimal CML. Evolution of a lattice of 200 sites shown during 200 iterations under the binary reduction separating "turbulent" (white) from "laminar" (black) local states. Periodic boundary conditions, random initial conditions; time is running upward. (a) $k = 6$, $\varepsilon = 0.33$; (b) $k = 8$, $\varepsilon = 0.28$.

Table 2
Variation with order k of the threshold value ε_c separating trivial from complex rules in the sequences generated by the approximation. The parameter region comprising possibly critical rules is indicated within brackets. For higher orders of the approximation, there is no sharp limit defining the transition region. The spatiotemporal intermittency threshold for the minimal CML is $\varepsilon_c = 0.360$.

k	ε_c	Transition region
2	0.67	
3	0.50	
4	0.33	
5	0.50	[0.33–0.50]
6	0.42	[0.25–0.42]
7	0.472	[0.32–0.47]
8	0.434	[0.28–0.43]
9	0.42	[0.28–0.42]
10	0.39	[0.33–0.39]
11	0.376	
12	0.37	
13	0.364	

erties characteristic of class III CA. The transition region consists of highly complex, seemingly unpredictable rules. These rules exhibit irregular and sometimes very long transients when evolving from random initial conditions to asymptotic states, which usually depend on these initial conditions. Most of the features generally attributed to class IV rules are in fact observed. Fig. 7 shows

the spatiotemporal evolution of four typical rules of the sequence.

A somewhat more quantitative comparison can be made with the continuous phase transition of the original CML. The characteristic properties discussed in section 2.2 are easily verified for the rules along the sequence, at least at a crude level, thanks to the natural binary reduction rooted in the dynamics of both the CML and the DCA produced by the approximation scheme. The triangular clearings of the class III rules appearing above threshold correspond to the laminar clusters of the spatiotemporal intermittency regimes of the CML. Indeed the corresponding states of the DCA are absorbing. The coherence scales $\xi_\perp$ and $\xi_\parallel$ defined in section 2.2 for the CML are equivalent to the characteristic length and time extracted from the exponential distribution of the sizes of triangular clearings for class III DCA evolving from random initial conditions [3].

This property of class III rules is conserved here when collapsing all absorbing states into a unique "laminar" state under the binary reduction. Fig. 8b shows the increase of $\xi_\perp$ for the rules of the sequence when approaching ε_c from above.

As for the critical exponents $\zeta_\perp$ and $\zeta_\parallel$ given by the algebraic distribution of the sizes of the lam-

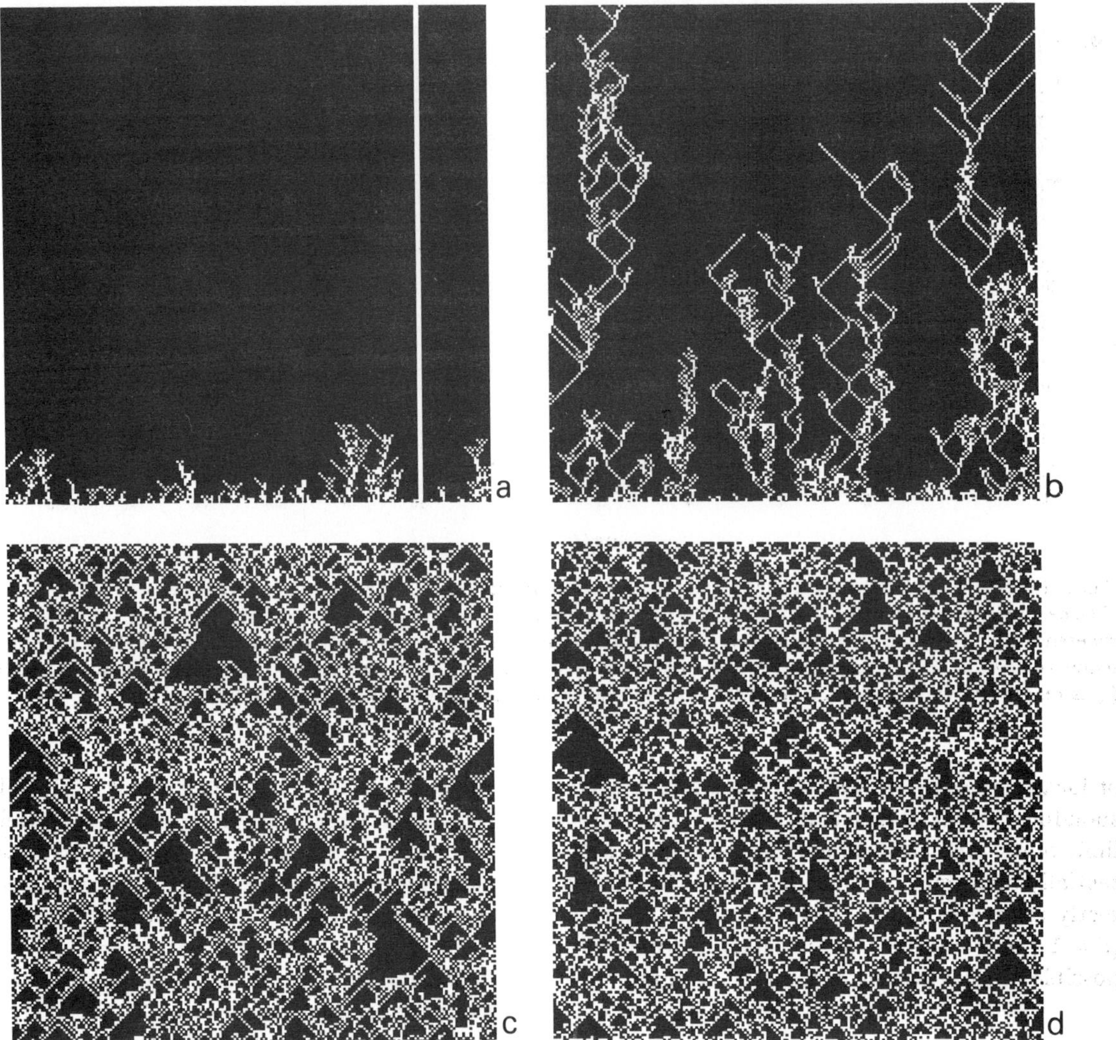

Fig. 7. Spatiotemporal evolution of four typical rules of the sequence generated by the approximation at order $k = 10$ for the transition of the minimal CML at $s = 3$. The lattice of 200 sites with periodic boundary conditions is shown under the natural binary reduction discussed in the text during the 200 iterations following random initial conditions. As in fig. 6, black sites are in one of the laminar (absorbing) states, white sites are "turbulent". (a) Class I rule well below threshold ($\varepsilon = 0.23$). (b) Rule slightly below threshold ($\varepsilon = 0.34$) (class IV ?) showing extremely long transients. (c) At threshold ($\varepsilon = 0.39$), first rule with a unique chaotic asymptotic state showing very long transients and quasi-algebraic distribution of laminar cluster sizes. (d) Class III rule above threshold ($\varepsilon = 0.50$). Note that the characteristic size of the triangular clearings is smaller far from threshold (d) than just above it (c).

inar clusters at threshold, it is not clear whether, increasing the order of the approximation, they "continuously" reach the corresponding value measured for the CML. Actually, as seen from fig. 8a, the rules in the transition region of the $k = 10$ sequence do not show a clear algebraic distribution of laminar cluster sizes, but rather lie in a crossover regime since exponential fits are not valid either. At higher orders of the approximation, the rules located around ε_c have a low mean density of "turbulent" sites, and the spatiotemporal patterns they develop locally resemble the fractal ones shown by class III rules evolving from simple seeds. For a DCA with a small number

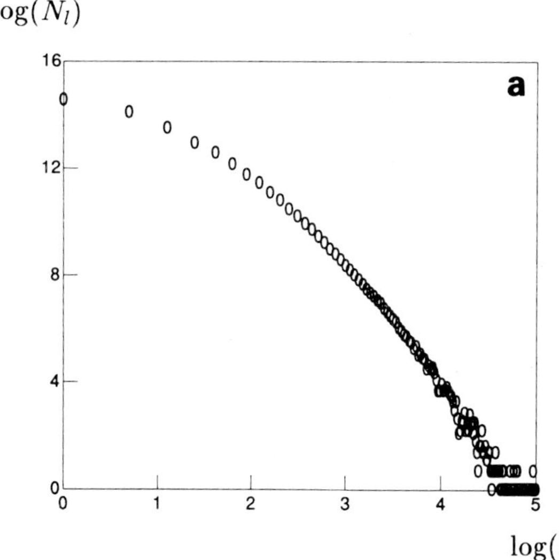
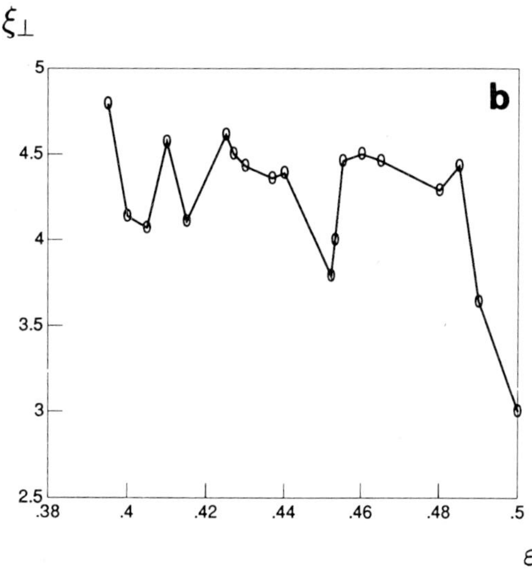

Fig. 8. (a) Histogram of the sizes of the "laminar" clusters for the rule of the sequence at order $k = 10$ and $\varepsilon = \varepsilon_c \simeq 0.39$ obtained on a lattice of 5000 sites during 5000 iterations following a long transient. (b) Variation along the sequence of the coherence length $\xi_\perp$ extracted from the exponential distribution of sizes of laminar clusters for the class III rules of the sequence at order $k = 10$ above threshold. Note that at this level of discretization, only the rough qualitative varation of $\xi_\perp$ is recovered, which indicates the still relatively important role of "hot bits" in the rule table.

of local states, there are but few possible corresponding fractal dimensions. But for the large-k limit of our approximation scheme, there are potentially many such dimensions. Since they are directly related to the critical exponents $\zeta_\perp$ and $\zeta_{\parallel}$ ($\zeta = 1 + d_f$), one is led to envision a continuum of possible critical indices in the $k \to \infty$ limit.

4. Discussion

4.1. Critical surface in the space of CA

The results given by our approximation of CML by CA provide insights into the general problem of the structure of the space of rules. In particular, the translation of the transition to turbulence via spatiotemporal intermittency (cf. section 2) into paths of DCA rules leads us to picture the existence of a critical surface on which lie "critical rules" defined by the usual criteria for critical points in statistical mechanics.

We have shown that the transition of the one-dimensional minimal CML defined in section 2.2 exhibits very different paths of rules under the ap-

proximation for $s = 2.1$ and $s = 3$. This indicates that there exist essentially two qualitatively distinct ways of approaching the critical surface and suggests that any rule close to it combines these two aspects in some proportion.

For $s = 2.1$, none of the rules produced by the approximation at low orders resides near the critical surface, and the critical regimes of the CML correspond to the probabilistic interpolation between the class I or class II rules appearing in the sequences, i.e. to a PCA of the type of directed percolation. The critical regimes of the CML can presumably be recovered only at very high orders, when the strong chaotic mixing occurring in the turbulent part of the phase space of the maps is reproduced by the complex interplay between the numerous local states of the DCA generated by the approximation. In this limit, these deterministic rules become equivalent to probabilistic ones, very much as the diadic map is related to a coin toss process [18].

In contrast, results from the approximation scheme in the $s = 3$ case pointed out that there are simple (small k) rules lying close enough to the critical surface to exhibit critical properties

(over a range of scales depending on the distance to the critical point). In this case, the complexity observed in the spatiotemporal evolution arises from very elementary features of these rules rather than from the mixing of a quasi-continuum of local states.

A priori, we can conceive intermediate cases for which these two aspects could be present in any proportion. These cases could be generated for example by the approximation of the minimal CML for $2.1 < s < 3$, although the variation of s would not produce a continuous and monotonous variation in the properties of the critical rules (for example, ε_c does not vary continuously with s for the minimal CML [19]). Nevertheless, given the potential richness of large-k DCA, there may exist a continuous set of possible critical properties for the rules on the surface.

4.2. Classifications of CA

The results gathered also shed light on Wolfram's classification. Critical rules, if they are to be defined by the statistical criteria used here, appear as limit cases of class I, II and III CA. This is easily seen, for example, from the point of view of the mean duration of transients separating random initial conditions from an eventual asymptotic state. If this quantity or the corresponding r.m.s. diverge, as is the case when approaching the transition region under our approximation scheme (from below for class I and II, from above for class III), the rules can no longer be classified. The critical surface can thus be seen as the limit of class I, II and III rules having infinite transients, fluctuations, coherence scales, etc.

This conclusion also stems from the iterative parametric genotypic classification provided by local structure theory [20]. This mean-field type of analysis has to be pushed to higher and higher orders if one wants to account satisfactorily for the statistical properties of rules with increasing correlation scales. Eventually, the approach cannot be made precise enough for rules lying very close to the critical surface. The breakdown of local structure theory can thus be a clear signal of criticality. Incidentally, the coherence scales $\xi_\perp$ and $\xi_\parallel$ introduced here directly provide estimates of the order below which the analysis will fail.

For truly critical rules, other methods should be considered, such as a renormalization-group treatment, which precisely tries to take advantage of the scaling properties [21].

4.3. Criticality and computational properties

If critical rules are defined as limit cases of class I, II or III CA, a central question is still the relationship between the existence of a critical surface and the universal computer property often used to characterize class IV rules. There is no systematic way of attacking this problem, but some comments already stated elsewhere [22,23] can be made. Critical rules may be characterized by their ability to undergo arbitrarily long transients. This is equivalent to stating the unpredictability of these rules. The undecidability of the finiteness of transients is also equivalent to the problem of the halting of a Turing machine.

Under our approximation scheme, critical rules often lie at the border between class II and class III CA. Following Wootters and Langton [23], this can be expressed in terms of the minimal requirements for a system to be able to perform calculations. Class III rules have good "communication" properties between sites (as seen from their positive Lyapunov exponents [17]) but almost no ability to store information (they reach a unique chaotic asymptotic state for almost all initial conditions). Class II rules, on the other hand, possess very good storage properties, but no communication is possible after they reach their spatially complex temporally trivial asymptotic state. In this context, critical rules seem to fulfill the necessary compromise between storage and communication at the root of any complex digital calculation.

4.4. Criticality and marginal stability

Recently, a number of rather simple CA models have been introduced and shown to exhibit scaling properties "spontaneously" [24]. We shall not discuss here their relevance to the physical phenomena they are taken as models of, but merely make the observation that these models of *self-organized criticality* may be seen as critical rules from our point of view. Given the picture of the critical sur-

face of rules developed here, it is not surprising that such models exist, so that the main question is why they sit close to the critical surface.

A key concept for self-organized criticality is *marginal stability*. Typically, a system is marginally stable when in the state for which the propagation of a local perturbation is just able to go to infinity in time and in space. How does the notion of marginal stability relate to the critical CA rules constructed here?

In the context of the continuous transition to turbulence via spatiotemporal intermittency, the transition point can be seen as the threshold in parameter space beyond which disorder will propagate to infinity in space and time (cf. the analogy with directed percolation [7]). The critical regimes of spatiotemporal intermittency can thus be seen as marginally stable states. Moreover, for the minimal CML presented in section 2, the continuous character of the transition (i.e. its criticality) seems to be intimately related to the fact that the laminar region is made of a continuum of marginally stable fixed points. Indeed, in the case of a unique stable fixed point, the transition is discontinuous [13]. Criticality is linked to the possibility for an infinitesimal local perturbation to bring a site from the laminar to the turbulent state and thus allow propagation of disorder at a negligible cost. Marginal stability appears to be an essential ingredient for an extended system to undergo a continuous transition to spatiotemporal intermittency. Coming back to the approximation of CML by CA presented here, it is clear that marginal stability is also inscribed in the critical rules generated in the transition region.

This point of view has also been recently discussed by Gutowitz [25] and McIntosh [26] at the level of mean-field theory. In this context, class I and class II rules can easily, although not strictly, be related to the existence of a unique stable fixed point at the origin of their mean-field map, whereas class III rules are characterized by the existence of a non-trivial stable fixed point. Critical rules are then naturally conjectured to represent, here again, the intermediate cases, among which the presence of a marginally stable fixed point for the mean-field map is generic.

Although mean-field theory in known to be unable to account for the details of the statistical properties of CA, it is usually believed to provide good estimates of certain important quantities [27]. This justifies the above conjecture, but it is not an absolute criterion to decide whether a rule will show critical properties. The indications given by the mean-field analysis are never quantitative (for example, the mean-field map overestimates the mean densities). Moreover, their use for critical rules seems particularly daring since these rules are characterized by their long range correlations whereas the mean-field analysis consists primarily of neglecting them. Nevertheless, the work of McIntosh does bring evidence that a marginally stable fixed point in the mean-field map is a good indication of critical behaviors for CA.

5. Conclusion

The approximation scheme first introduced in [15] and presented here in a slightly different version has permitted the construction of CA rules whose critical properties are well controlled. Although not a tool for a systematic and comprehensive exploration of the space of CA, it allows the design of sequences of rules mimicking phase transitions occurring in extended dynamical systems and ensures the existence of critical rules, pictured to lie on a critical surface.

From the results, a certain unity in the various statistical characterizations of Wolfram's class IV, now seen as a limit case of the other classes, is recovered.

A central problem remains, then, which is the nature of the relationship between the statistical characterizations of critical rules and the computational properties discussed in section 1. When does a rule with scaling properties have the property of being a universal computer? Conversely, how critical is such a rule? These are the key questions for a further progress in understanding criticality in cellular automata.

Acknowledgements

We have benefited from fruitful exchanges with R. Bidaux, P.C. Hohenberg, Li Wentian, O. Martin, Y. Pomeau and B.S. Shraiman. We are par-

ticularly indebted to Howard Gutowitz and Chris Langton for friendly discussions on most of the aspects of this work. H.C. thanks the Aspen Center for Physics and the Center for Nonlinear Studies (Los Alamos) where most of these discussions have taken place.

References

[1] P. Grassberger, Problems in quantifying self-generated complexity, preprint (1989);
C.H. Bennett, On the nature and origin of complexity in discrete, homogeneous, locally interacting systems, Found. Phys. 16 (1986) 585–592;
K. Lindgren and M.G. Nordahl, Complexity measures and cellular automata, Complex Systems 2 (1988) 409–440.

[2] W. Kinzel, Directed percolation, in: Percolation Structures and Processes, eds. G. Deutscher et al., Ann. Israel Phys. Soc. 5 (1983) 425–445; Phase transitions of cellular automata, Z. Phys. B 58 (1985) 229–244.

[3] S. Wolfram, Universality and complexity in cellular automata, Physica D 10 (1984) 1–35.

[4] K. Culik and Sheng Yu, Undecidability in classification schemes, Complex Systems 2 (1988) 177–190.

[5] W. Li and N. Packard, The structure of the elementary cellular automata rule space, preprint, 1989;
W. Li, N.H. Packard and C.G. Langton, Transition phenomena in cellular automata rule space, Physica D 45 (1990) 77–94, these Proceedings.

[6] R.W. Gerling, Classification of three-dimensional cellular automata, Physica A 162 (1989) 187–195;
S.S. Manna and D. Stauffer, Systematics of transitions of square-lattice cellular automata, Physica A 162 (1990) 176–186.

[7] H. Chaté, Transition vers la turbulence via intermittence spatiotemporelle, Thèse de l'Université Pierre et Marie Curie (1989), and articles therein (in particular refs. [10–13; 15–16; 27] hereafter).

[8] Y. Pomeau, Front motion, metastability and subcritical bifurcations in hydrodynamics, Physica D 23 (1986) 3–11.

[9] D.J. Tritton, Physical Fluid Dynamics (Van Nostrand Reinhold, New York, 1977).

[10] H. Chaté and P. Manneville, Transition to turbulence via spatiotemporal intermittency, Phys. Rev. Lett. 58 (1987) 112–115.

[11] H. Chaté and P. Manneville, Spatiotemporal intermittency in coupled map lattices, Physica D 32 (1988) 409–422.

[12] H. Chaté and P. Manneville, Continuous and discontinuous transition to spatiotemporal intermittency in two-dimensional coupled map lattices, Europhys. Lett. 6 (1988) 591–595.

[13] H. Chaté and P. Manneville, Role of defects in the transition to turbulence via spatiotemporal intermittency, Physica D 37 (1989) 33–41.

[14] T. Bohr and O.B. Christensen, Size dependence, coherence and scaling in turbulent coupled map lattices, Phys. Rev. Lett. 63 (1989) 2161–2164.

[15] H. Chaté and P. Manneville, Coupled map lattices as cellular automata, J. Stat. Phys. 56 (1989) 357–370.

[16] H. Chaté and P. Manneville, Using coupled map lattices to unveil structures in the space of cellular automata, in: Cellular Automata and Modeling of Complex Physical Systems, eds. P. Manneville et al., Springer Proceedings in Physics, Vol. 46 (Springer, Berlin, 1989).

[17] S. Wolfram, Statistical mechanics of cellular automata, Rev. Mod. Phys. 55 (1983) 601–644; Theory and Applications of Cellular Automata (World Scientific, Singapore, 1986).

[18] J. Ford, How random is a coin toss?, in: Long-time Prediction in Dynamics, eds. C.W. Horton et al. (Wiley, New York, 1983).

[19] H. Chaté and P. Manneville, unpublished.

[20] H.A. Gutowitz B.W. Knight and J.D. Victor, Local structure theory for cellular automata, Physica D 28 (1987) 18; Local structure theory in more than one dimension, Complex Systems 1 (1987) 57; Local structure theory: calculations on hexagonal arrays and the interaction of rule and lattice, J. Stat. Phys. 54 (1989) 495;
H.A. Gutowitz, A hierarchical classification of cellular automata, Physica D 45 (1990) 136–156, these Proceedings.

[21] O. Martin, Critical dynamics of one-dimensional irreversible systems, Physica D 45 (1990) 345–354, these Proceedings.

[22] S. Wolfram, Twenty problems in the theory of cellular automata, Physica Scripta T9 (1985) 170–183.

[23] W.K. Wootters and C.G. Langton, Is there a sharp phase transition for deterministic cellular automata?, Physica D 45 (1990) 95–104, these Proceedings.

[24] P. Bak, Chao Tang and K. Wiesenfeld, Self-organized criticality: an explanation of $1/f$ noise, Phys. Rev. Lett. 59 (1987) 381–384.

[25] H.A. Gutowitz, in preparation.

[26] H.V. McIntosh, Wolfram's class IV automata and a good life, Physica D 45 (1990) 105–121 , these Proceedings.

[27] R. Bidaux, N. Boccara and H. Chaté, Order of transition vs space dimension for a family of cellular automata, Phys. Rev. A 39 (1989) 3094.

Physica D 45 (1990) 136–156
North-Holland

A HIERARCHICAL CLASSIFICATION OF CELLULAR AUTOMATA

Howard A. GUTOWITZ

*Center for Nonlinear Studies and Complex Systems Group, Los Alamos National Laboratory,
MS-B258, Los Alamos, NM 87545, USA* [1]

Received 13 September 1989
Revised manuscript received 1 November 1989

A central issue in the theory of cellular automata is *classification*. A classification is a structure imposed on the space of automata which groups together cellular automata with related properties. In this paper a classification based on the action of cellular automata on n-step Markov measures is presented. This classification is hierarchical and parametric. An algorithm is described which efficiently constructs all rules in a given class at a given level of the hierarchy. The utility of the classification is explored in the study of the invariant statistical properties of next-nearest neighbor cellular automata on two states per cell.

1. Introduction

One of the most important problems in the theory of cellular automata is to understand how cellular automata can be meaningfully grouped according to their structure and behavior. In general, cellular automata with very similar transition rules may behave quite differently, while cellular automata with very different transition rules may behave identically.

The approach taken here is to classify cellular automata according to their action on probability measures. It was shown previously [1–3] that the action of cellular automata on probability measures can be approximated by their action on n-step Markov measures. An n-step Markov measure is a measure generated by a Markov process with n steps of "memory". The n-step Markov measures we will consider describe the *spatial* correlations between cells at a given time step in the evolution of the automaton. In the approximation a cellular automaton is viewed as mapping the set of n-step Markov measures, for fixed n, into itself. As n increases this approximation typically becomes an increasingly faithful representation of the action of the cellular automaton on general measures.

This sequence of approximations for one-dimensional cellular automata was introduced in ref. [1] under the name of local structure theory. It was extended to cellular automata in higher dimensions in refs. [2,3]. To emphasize the Markovian nature of the approximation in one dimension, the local structure theory will be referred to here as the Markov approximation. While the present discussion parallels those previous, the direction is quite opposite. In previous work, the Markov approximation was used to address a *forward* problem: given a cellular automaton, (approximately) describe its behavior. Here the method is used to address an *inverse* problem: given an approximation of cellular automaton behavior, find the set of automata which correspond to that approximation. The set of automata corresponding to each valid approximation will be referred to as an (nth-order) class of cellular automata. Since at each order n there is a unique approximation of a given cellular automaton according to its action on n-step Markov measures, this approximation breaks the set of automata into a disjoint collection of classes.

The key fact which makes this approximation scheme a non-trivial classification is that *if the spatial memory is small, many cellular automata*

<hr>

[1] Address from August 15, 1990: Institut de Recherche Fondamentale, DPh-G/PSRM, CEN-Saclay, F 91191 Gif-sur-Yvette Cedex, France.

will be indistinguishable in Markov approximation. In essence, if the differences between two cellular automata can only be resolved on the basis of spatial correlations beyond a distance of n cells, then these automata are considered to be in the same nth-order class.

A physical motivation for this approach is as follows. Consider making measurements of a physical system which has properties: spatial and temporal discreteness, spatial homogeneity, local interactions etc. which suggest that a cellular automaton model for the system would be appropriate. One measures, for instance, how often each cell is in a given state, and how often small groups of cells are in given configurations of states. It is not possible to measure frequencies of occurrence for all possible configurations of all possible sizes. On the basis of these limited measurements, one would like to construct an explicit cellular automaton rule which may have generated these data. Because one has only been able to collect a finite amount of information, a model cannot be uniquely specified. There will be a large set of automata which are equally valid models of this process. The classification scheme presented here is a method for analyzing the entire set of such rules on the basis of their shared characteristics.

This paper is composed of two parts. In the first part the classification scheme is presented. In the second part, the classification is used to guide a numerical exploration of the invariant statistical properties of next-nearest neighbor automata. In the first part of the paper, the classification at orders $n = 0 - 2$ is presented in detail. Each of these orders has some special features which allow a simplied treatment. Components of the general theory are introduced as needed. Then a description of the theory for general n is given. The numerical work of the second part of the paper is intended mainly to illustrate how the classification can aid investigations of the properties of large sets of cellular automata. Nonetheless, this work sheds some light on the classification scheme itself. It suggests that as the order of approximation increases, the accuracy with which the classification predicts the behavior of rules in a class increases as well.

2. The Markov approximation

2.1. The action of a cellular automaton on a probability measure

The probability measure $\tau\mu$ which results from the application of a cellular automaton τ to a probability measure μ is given by

$$\tau\mu(\mathrm{E}) = \mu(\tau^{-1}(\mathrm{E})), \tag{1}$$

where E is a μ-measurable open set of configurations [4]. The usual topology on the set of configurations is derived from the metric $d(x, y) = \sum_{-\infty}^{\infty} 2^{-|i|}|x_i - y_i|$, where x and y are configurations, and i indexes position with respect to a fixed origin. Any measurable open set of configurations can be represented as a union of sets of configurations which have a specified contiguous sequence of cell states in a specified location. These sets are know as *cylinder sets* or *blocks*. Since cellular automata are shift invariant, one need only consider probability measures which are shift-invariant: the probability of a block depends only on the sequence of cell states which define the block, not on its location. Let r be the radius of the rule, τ, $P^t(\mathrm{B})$ be the probability of a block B at time t, and $P^{t+1}(\mathrm{b})$ be the probability of a block b at time $t + 1$. Let $\delta(\tau(\mathrm{B}), \mathrm{b})$ be 1 or 0 if the application of the rule τ to B does respectively does not result in a block b. Further, let $|.|$ denote block length and $\mathbf{B}_{|\mathrm{b}|+2r}$ the set of blocks of length $|\mathrm{b}| + 2r$. Now eq. (1) can be rewritten as an infinite system of equations of the form

$$P^{t+1}(\mathrm{b}) = \sum_{\mathbf{B}_{|\mathrm{b}|+2r}} \delta(\tau(\mathrm{B}), \mathrm{b}) \, P^t(\mathrm{B}) \,. \tag{2}$$

The system (2) states that the probability of a block b at time $t+1$ is the sum of the probabilities at time t of the blocks B which lead to b under the rule. If the probabilities of all blocks of length $|\mathrm{b}|+2r$ are known, then the system (2) determines the probability of all blocks of length $|\mathrm{b}|$ at time $t+1$. To use the system (2) to find the probability of n-blocks t time steps in the future, one must have at time 0 the probability of all blocks of length $n + 2rt$. Since the number of blocks is exponential in the length of the blocks, it is impractical to use the system (2) directly to determine the probability

measure which arises when a cellular automaton is applied many times to some initial measure.

2.2. n-step Markov measures

The Markov approximation handles this problem by using knowledge of probabilities of blocks of a fixed length n to estimate the probabilities of longer blocks. The combinatorial structure of the rule, as represented by δ in eq. (2) is preserved for the map onto blocks up to a fixed size n. However the exact information concerning block probabilities, represented in eq. (2) by the functions P^t and P^{t+1}, is replaced by an approximating *n-step Markov measure*. An n-step Markov measure is generated by a Markov process with n steps of "memory". Here the memory is not temporal, but spatial. The "states" of an n-step Markov process are blocks of n cell states. It is well known (see e.g. ref. [5]), that the assignment of probability in this way to long blocks is the assignment of *maximum entropy* consistent with the given n-block probabilities. By *Markov measure* we mean the assignment of probability by a Markov process to blocks of all sizes. In the following we will typically refer to Markov measures by the assignment of probability to the (n cell) Markov states which define the measure. Context will resolve any confusion this might engender.

2.3. 0th-order Markov approximation

In the 0th-order Markov approximation the system equation (2) is radically simplified. There is only one 0-step Markov measure: the so-called *standard measure*, which assigns equal probability to all blocks of the same length. As a nontrivial dynamics cannot be constructed on a singleton set, we will formulate the 0th-order approximation as a restricted map from the set of 1-step Markov measures into itself. One begins only with the standard measure, and allows the map to be applied only once. In the 1st-order approximation described below, these restrictions are removed.

In 0th-order approximation P^t on the right-hand side of eq. (2) is replaced by the constant $1/2^{|B|}$. The only combinatorial information retained is the number of neighborhoods (blocks of length $d = 1 + 2r$) which lead to a 1 under the

rule. Given this information, P^{t+1} is determined as follows. The equation from system (2) which yields the probability of a cell in state 1, P_1, is written as

$$P_1^{t+1} = \sum_{\mathbf{B}_{1+2r}} \frac{\delta(\tau(\mathbf{B}),1)}{2^{1+2r}} \, . \tag{3}$$

Note that (i) the probability of a 1 at time $t + 1$, P_1^{t+1}, is independent of $t + 1$, since P^t is assumed to be a constant independent of t, and (ii) $\sum_{\mathbf{B}_{1+2r}} \delta(\tau(\mathbf{B}),1)$ serves to count the number of neighborhoods which yield 1 under the rule. Let λ be the number of neighborhoods which yield 1 under the cellular automaton, and let $d = 2r + 1$ be the diameter of a cellular automaton rule. Eq. (3) can then be rewritten simply as

$$P_1 = \lambda/2^d \, . \tag{4}$$

Every rule yields a particular value of λ. Conversely, to each value of λ is associated many rules. The lowest order of Markov approximation neglects all structure of a cellular automaton rule but for one property: the number of neighborhoods which yield 1 upon application of the rule. The density (fraction of cells in state 1) of a configuration at any time is just the density of the rule table itself. In particular, the prediction of the 0th-order approximation for the invariant density of a rule is the density of the rule table.

Though crude, the 0th-order approximation can yield good qualitative predictions. Langton [6] used the 0th-order approximation to explore information processing in two-dimensional cellular automata. λ of eq. (4) follows his notation. Li et al. [7] and Wootters and Langton [8] have employed similar methods to locate boundaries in the space of cellular automata between rules with different kinds of behavior.

Already at 0th-order we can introduce two concepts which will be crucial in the sequel, *types* and *classes*. Blocks will be said to be of the same (*nth-order*) type if they are by necessity assigned the same probability in nth-order approximation. All blocks of the same length will be said to be of the same 0th-order type since they all must be assigned the same probability in the 0th-order approximation. Cellular automaton rules will be said to be in the same (*nth-order*) class if they have the

same nth-order approximation. Rules of a given radius will be said to be in the same 0th-order class if they yield the same value for λ. For d-diameter rules, λ can take any integer value in the range $0\ldots 2^d$.

2.4. 1st-order approximation (mean field theory)

The next order of Markov approximation is also known as the mean field theory [9,10]. The mean field theory, like the 0th-order approximation, encodes only the combinatorial information contained in the cellular automaton map from neighborhood blocks to the states of single cells. The mean field theory, like the 0th-order approximation, represents the action of a cellular automaton on general measures by its action on 1-step Markov measures. The mean field theory is superior to the 0th-order approximation in two respects: (i) The initial measure is allowed to be any 1-step Markov measure, while in 0th-order theory the initial measure is always the standard measure, and (ii) any number of applications of the map from the set of 1-step Markov measures into itself can be considered, while in 0th-order theory only one application is allowed.

Let $\#0(\mathrm{B})$ and $\#1(\mathrm{B})$ be the number of 0's and 1's respectively in a block B. In the mean field theory, the probability of a block B is given by

$$P(\mathrm{B}) = P_1^{\#1(\mathrm{B})} \, P_0^{\#0(\mathrm{B})}. \tag{5}$$

Eq. (5) is exact in the case in which the states of different cells are completely uncorrelated. Two blocks B and B' which have the same number of cells in states 0 and 1 will be said to be of the same *1st-order type*. Blocks of the same 1st-order type are assigned the same probability by eq. (5).

Substituting eq. (5) into the equation of the form (2) for the evolution of the probability of a 1, we have the mean field equation

$$P_1^{t+1} = \sum_{\mathbf{B}_d} \delta(\tau(\mathrm{B}), 1)(P_1^t)^{\#1(\mathrm{B})}(P_0^t)^{\#0(\mathrm{B})}. \tag{6}$$

Observe that any two blocks B and B' contribute the same probability to the sum if (i) they both lead to a 1 under the rule, and (ii) they are of the same 1st-order type. Let a_i be the number of neighborhood blocks which lead to a 1 under a rule

and also contain i 1's (are of the same 1st-order type). Eq. (6) can now be rewritten as

$$P_1^{t+1} = \sum_{i=0}^{d} a_i (P_1^t)^i (1 - P_1^t)^{d-i}. \tag{7}$$

The coefficients a can have any integer value in the range $0-\binom{d}{i}$ inclusive. This polynomial equation is a model of the evolution of any cellular automaton which yields the coefficient values a. A fixed point P_1^* in the range $[0,1]$ of eq. (7) is an estimate of the invariant density of any cellular automaton which yields the coefficient values a.

Note that many different rules of a given radius can have the same values for the a coefficients. Such rules are indistinguishable at the level of mean field theory. A collection of rules with the same mean field coefficient values will be referred to as a *1st-order (or mean field) class* determined by the coefficient values a.

All rules in a given mean field class also lie in the same 0th-order class. That is, the mean field theory supplies a classification of cellular automata which is a strict refinement of the 0th-order classification. The value λ for the 0th-order class which contains a mean field class determined by coefficient values a is given by $\lambda = \sum_{i=0}^{d} a_i$.

2.5. 2nd-order Markov approximation

In mean field theory one assumes that correlations between the states of different cells are not generated by application of a cellular automaton to a configuration. Under this assumption, the probability of a large block is estimated as the product of the probabilities of the states of cells it contains. The mean field theory will fail to accurately model a cellular automaton if correlations are generated as the rule is iterated. The Markov approximation for $n > 1$ takes spatial correlations explicitly into account. In the Markov approximation the probabilities of large blocks are estimated in terms of the probability of smaller blocks which they contain. This leads to a systematic generalization of the mean field theory. In this section, the first step of this generalization is examined.

In 2nd-order Markov approximation, correlations are introduced in terms of the probabilities of contiguous pairs of cells. Pair probabilities are

used to define a 2-step Markov process. Longer block probabilities are estimated as follows. Let $s_i \in \{0,1\}$ be the possible states of a cell in position i in a block. Let $(s_1 s_2 \ldots s_n)$ be an n-block, and $P_{s_1 s_2 \ldots s_n}$ be the probability of an n-block. If the probabilities of all 2-blocks are known, the probability of an n-block, $n > 2$, is estimated by

$$P_{s_1 s_2 \ldots s_n} = \prod_{i=1}^{n-1} P_{s_i s_{i+1}} \bigg/ \prod_{i=2}^{n-1} P_{s_i}, \qquad (8)$$

where the 1-block probabilities are found by appropriate summation of the 2-block probabilities. That the extension of small block probabilities defined by eq. (8) is a consistent assignment of probabilities to larger blocks is discussed in ref. [1]. That this extension is the extension of maximum entropy consistent with the given assignment of 2-block probabilities is discussed in ref. [5]. Blocks which always have the same probability according to eq. (8) are said to be of the same *2nd-order type*.

It is useful to refer to 2nd-order types by a code. 1st-order types are coded by a single index, i, which is the power to which P_1 is raised in the 1st-order estimate for the probability of a block (eq. (5)). Given the value of this index and the length of the block, the other relevant exponent, that attached to $P_0 = 1 - P_1$, can be found. The 2nd-order estimate (eq. (8)) involves a number of different block probabilities raised to different, interrelated, powers. A good code should specify the minimum number of exponent values such that the rest can be found by appeal to the Kolmogorov consistency conditions on block probabilities [11]. These state that

$$\sum_{s_1 s_2 \ldots s_n} P_{s_1 s_2 \ldots s_n} = 1, \qquad (9)$$

$$\sum_{s_1} P_{s_1 s_2 \ldots s_n} = P_{s_2 s_3 \ldots s_n}, \qquad (10)$$

$$\sum_{s_n} P_{s_1 s_2 \ldots s_n} = P_{s_1 s_2 \ldots s_{n-1}}. \qquad (11)$$

2nd-order types are coded here by a triple $(x, y, z)_n$ where x is the total number of 10 and 01 sub-blocks counting overlaps, y is the number of 11 sub-blocks again counting overlaps, and z is the number of cells in state 1 in the central $n - 2$ region of the n-block. The number of other 1- and 2-blocks in the n-block can be found using the Kolmogorov consistency conditions. As an example, 10010 and 10100 are of the same 2nd-order type, coded by $(3, 0, 1)_5$. The probability of this type is $P_{(3,0,1)_5} = P_{00}^1 P_{01}^3 / P_1^1 P_0^2$.

The Kolmogorov consistency conditions were just used to produce a compact code for 2nd-order types. In much the same way, we again apply the consistency conditions to find a compact parameterization of block probabilities themselves. Here the probabilities of 1- and 2-blocks will be parameterized by P_1 and P_{11}. Any other pair of linearly independent 1- and/or 2-block probabilities could also serve as parameters. As will become clear, however, it is desirable to chose parameters in a hierarchical fashion. That is, so that parameters for small-block probabilities are a subset of the parameters for long-block probabilities. The other 2-block probabilities can be found from the parameters chosen here using the Kolmogorov consistency conditions, e.g. $P_{01} = P_{10} = P_1 - P_{11}$.

The 2nd-order Markov approximation preserves the combinatorial information contained in both the cellular automaton map from neighborhood blocks to single cells and the map from $(d + 1)$-length blocks to 2-blocks. The 2nd-order approximation is constructed by substitution of the probability estimate given by eq. (8) into equations of the form (2) for the evolution of P_1 and P_{11}. Then, as was done in the derivation of the mean field equation (7), the sum is rearranged so that blocks of the same type are collected together. A coefficient $b_{(x,y,z)_d}$ is associated to each 2nd-order type of d-block, and a coefficient $c_{(x,y,z)_{d+1}}$ is associated to each 2nd-order type of $(d+1)$-block. The b coefficients count the number of d-blocks of the given 2nd-order type which lead to a 1 under the cellular automaton, and the c coefficients count the number of $(d+1)$-blocks of a given 2nd-order type which lead to 11. Let $P^t_{(x,y,z)_n}$ be the probability at time t of a block of 2nd-order type $(x, y, z)_n$ according to eq. (8). The 2nd-order equations are then

$$P_1^{t+1} = \sum_{(x,y,z)_d} b_{(x,y,z)_d} P^t_{(x,y,z)_d}, \qquad (12)$$

$$P_{11}^{t+1} = \sum_{(x,y,z)_{d+1}} c_{(x,y,z)_{d+1}} P_{(x,y,z)_{d+1}}^t, \qquad (13)$$

where the sums run over the 2nd-order types of d- and $(d+1)$-blocks respectively.

As was the case for 0th- and 1st-order theories, many rules may lead to the same 2nd-order coefficient values. Thus each allowed set of 2nd-order coefficients defines a 2nd-order class of cellular automata. 2nd-order classes determined by b and c coefficient values are strict refinements of mean field classes. The a coefficient values of the mean field class containing a 2nd-order class are the sum of b coefficient values which count blocks of the same 1st-order type.

2.6. Construction of a Markov class

Above the association of a system of Markov equations with a given cellular automaton was discussed. It was noted that many rules may give rise to the same system of equations. Now the inverse problem will be considered: given a system of Markov equations, find all members of the class of rules defined.

2.6.1. 0th-order theory

It is particularly simple to find all rules in a 0th-order class. Each class is defined by the number λ of 1's in the rule table. For d-diameter rules, λ can take any integer value from 0 through 2^d. A class defined by value λ consists of all rules which map exactly λ neighborhoods to 1, and hence correspond to a size λ subset of 2^d elements.

2.6.2. 1st-order (mean field) theory

It is not difficult to find all rules in a mean field theory class. Recall that the mean field theory for d-diameter rules has $d+1$ coefficients. The ith-coefficient counts the number of neighborhood blocks which contain i 1's and lead to a 1 under the rule. There are $\binom{d}{i} \equiv A_i$ d-blocks which contain i 1's, so the maximum value of a_i is A_i. Consider a class defined by a set of coefficient values where each coefficient value is in the range 0–A_i inclusive. The number of ways of achieving the value a_i is $\binom{A_i}{a_i}$. Each way of achieving each coefficient value defines the rule table of a rule in the given mean field class. To find all rules in a mean

field class (i) list all possible ways of achieving the ith-coefficient value with d-blocks which contain i 1's, and (ii) select from the list in all possible combinations one way of achieving each coefficient value. The d-blocks selected are assigned the value 1 under the rule and the d-blocks not selected are assigned the value 0 under the rule.

2.6.3. 2nd-order Markov approximation

At 2nd-order the construction of a Markov class becomes slightly involved. It may be difficult to directly infer a rule table from a specification of theoretical coefficients values because each neighborhood block of length d may be part of several $d+1$ blocks each controlled by a different c coefficient. This means that the values of the c coefficients may interact in a complicated way to determined which transitions in the rule table are consistent with a specification of coefficient values. Below a two step process which handles these complications is outlined. In appendix B, direct construction of all reflection-symmetric rules in a given class of $r=2$ rules is described.

The first step of the construction of a 2nd-order class relies on the observation that both the a coefficients of the mean field theory and the b coefficients of the 2nd-order Markov approximation for d-diameter rules control blocks of the neighborhood size d. By employing exactly the method described above for the construction of a mean field class, a set of rules with potential membership in a 2nd-order class may be found. Such rules have the desired b coefficient values, but their c coefficients values have yet to be determined.

The second step of the construction determines the c coefficient values. The forward map from a rule table to a set of Markov coefficients is easily computed. In the second step of construction, the forward map is used to determine the c coefficient values of all cellular automata isolated in the first step. These values are then checked against the c coefficient values which define the class in question.

The number of 2nd-order types of n-blocks is greater than the number of 1st-order types of n-blocks, $n > 2$. Hence, for fixed radius, the number of b coefficients in the 2nd-order approximation is greater than the number of a coefficients

in the 1st-order approximation. However, as noted above, each of these sets of coefficients collectively control the same set of d-blocks. This implies that for fixed radius the typical values of the b coefficients will be smaller than the typical values of the a coefficients, and, a fortiori, that the number of rules isolated in the first step of the construction of a 2nd-order class will typically be much smaller than the number of rules in a mean field class.

2.7. Markov equations at order n

Each order of approximation 0–2 has some special features not present at higher orders. These low orders of approximation were presented concretely and in sequence so that details of the construction of classes could be introduced as needed. In this section the construction of classes will be discussed for general n. To construct nth-order classes for d-diameter rules it is necessary to format the Markov equations properly. In outline, one must (i) identify nth-order types of d- through $(n+2r)$-blocks, (ii) choose a hierarchical basis for n-block probabilities, and (iii) assign coefficients to types the values of which determine how many blocks of each type map to blocks in the chosen basis. Finding all rules which have a given set of coefficient values is a matter of finding all ways of filling d-diameter rule tables so that the logical implications of the given set of coefficient values are satisfied. While this is in general a hard problem, the hierarchical format of the equations permits an efficient algorithm to be given.

2.7.1. n-block types

Informally, blocks are of the same mth-order type if they contribute in the same way to an mth-order Markov equation. To describe this formally, we will need the following notation. Let b_n be an n-block in the set of all n-blocks, $\mathbf{B}_n$. Let P_{b_n} be the probability of the n-block b_n and $\#[b_m^I](b_n)$ be the number of blocks b_m in the central $n-2$ cells of the n-block (the interior of the n-block) counting overlaps and $\#[b_m^E](b_n)$ be the number of m-blocks in the full n-block (the exterior of the n-block), again counting overlaps. If the probability of all m-blocks is known, the probability of n-blocks, $n \geq m$ may be estimated by

$$P_{b_n} = \prod_{\mathbf{B}_m} P_{b_m}^{\#[b_m^E](b_n)} \bigg/ \prod_{\mathbf{B}_{m-1}} P_{b_{m-1}}^{\#[b_{m-1}^I](b_n)} . \quad (14)$$

Two n-blocks b_n and b_n' are said to be of the same *mth-order type $m \leq n$* if (i) the Kolmogorov consistency conditions demand that they have the same probability or if (ii) they are assigned the same probability by extension of m-block probabilities according to eq. (14), for any consistent assignment of probability to m-blocks. Apart from details contained in appendix A, types are specified combinatorially, that is, by a list of values for the exponents which appear in the interior and exterior products in eq. (14).

2.7.2. Parameterization of block probabilities

Due to the linear constraints imposed by the Kolmogorov consistency conditions, only 2^{n-1} basis elements are necessary to parameterize the space of consistent probability assignments to blocks of n-states. For present purposes, it is desirable to choose basis elements in a hierarchical fashion. That is, the 2^{n-1} basis elements for n-block probabilities should form the first half of the 2^n basis elements for $(n+1)$-block probabilities. The basis employed in this paper is the set of blocks whose left-most and right-most cell is in state 1, e.g. 1, 11, 101, 111, 1001, ... Using the consistency conditions the probability of any other block may be found from the probabilities of blocks in this basis set. An alternative basis may be useful in some situations. For instance, in order to minimize the values of the coefficients in the Markov equations for particular rules, one should choose basis blocks which are in a certain sense minimal [12].

2.7.3. Assignment of coefficient values

The nth-order Markov consists of 2^{n-1} equations each of which describes the evolution of the probability of a block in the basis set as chosen above. 2^{n-2} of these equations describe the evolution of the probability of an n-block, 2^{n-3} equations describe the evolution of the probability of an $(n-1)$-block and so on, down to one equation for the evolution of the probability of a single cell state. Let m be the size of the target block of a particular equation. Coefficients in the equation count the number of $(m+2r)$-blocks which (i) are

of the same nth-order type and (ii) lead to the target block of the equation under any rule described by that system of equations. Given a rule of radius r, it is a simple matter to determine the coefficient values which define the nth-order class to which the rule belongs. Algorithmically, to find the coefficient values for an equation for an m-block, (i) each $(m + 2r)$-block is mapped to an m-block under the rule, (ii) if the resultant m-block is the target block for the equation then the value of the coefficient which is associated with the type of the $(m + 2r)$-block is incremented.

2.7.4. Construction of the rules in an nth-order class

There are a finite number of r-radius rules and a finite number of nth-order equations for r-radius rules. Hence, in principle, it is a trivial matter to find all rules which have a given nth-order Markov approximation. One simply evaluates the nth-order coefficient values for each r-radius rule and then checks these against the coefficient values in question. This naive approach requires effort which is doubly exponential in the radius of the rules. However, the hierarchical format of the equations as given above allows the search for rules in a given nth-order classes to be limited first to rules in a subset of a given 0th-order class and then to a subset of a given 1st-order class etc. For the next-nearest neighbor rules considered here, a fast, practical algorithm can be given. Bounds on the computational complexity of this hierarchical search have yet to be elucidated.

The exact implementation of this strategy will depend on the application. One approach is (i) find all rules which have the correct coefficients values for the nth-order equation which describes the evolution of the probability of a 1, and then (ii) evaluate the higher-order coefficient values for each of these rules and finally (iii) check these coefficient values against those which define the class in question. The method employed here, as just outlined, takes typically less than a minute of Sun-3 CPU time to locate all rules in an nth-order class of radius-2 rules, $n > 1$.

3. Empirical results

3.1. Statistical properties of Markov classes of $r = 2$ rules

In the following the relationship between the statistical properties of the equations which define an Markov class and the statistical properties of the rules contained in an Markov class will be explored. These investigations will concern cellular automata with two states per cell and next-nearest neighbor ($r = 2$) interaction rules. The number of cellular automaton rules with radius 2 on two states per cell is 2^{32}. Obviously, a complete survey of $r = 2$ rules is not feasible. The large number of $r = 2$ rules, however, presents the opportunity to use a statistical analysis on the space of all such automata. The main questions to be addressed below are: (i) to what degree of accuracy do various orders of approximation predict the statistical properties of rules in a class? and (ii) how homogeneous are the classes defined at various orders?

These questions are distinct, but related. Consider using the classification to find a cellular automaton with some specified statistical property. Ideally, the equations which specify classes should accurately predict the expected properties of rules in the classes defined. Further, the rules in a class should have closely related properties. It could happen, for instance, that classes have wide variability over their members, yet the Markov approximation accurately predicts the average over a class of some statistical property. In this case, the variability renders the theoretical prediction useless for any practical question concerning individual cellular automaton rules. In the opposite extreme, it could happen that rules within a class have very similar statistical properties, but that these shared properties are different from those predicted by the equation which defines the class. In this case it would be difficult to find rules with some desired property by solving Markov equations. Once *one* such rule was found, however, many other rules with the same property could be constructed by inverting the equations corresponding to the given rule.

It is demonstrated below that both the accuracy with which the Markov approximation pre-

dicts the properties of rules in a class, and the homogeneity of classes increases with order of approximation. This is shown by a statistical analysis of samples of Markov classes of orders 0–2 of $r = 2$ rules. A sample of rules from each class selected is examined empirically. Empirical estimates are made of invariant 1- and 2-block probabilities of these rules. The same estimates are made in Markov approximation. The primary tool used to assess accuracy and homogeneity is the computation of the distance in measure space between various Markov measures estimated in this way.

All of the data to be discussed are presented in the form of *distributions*. It must be clearly understood that the distribution of measured properties depends on the way samples are chosen. An attempt was made to sample uniformly in the parameter space of the Markov approximation. It is anticipated that this should correspond to a sampling which is uniform as well over the set of automata themselves. Consider an ideal case in which all classes at a given order are sampled, all rules in each class have exactly the same properties (perfect homogeneity), and these properties are exactly those predicted by that order of approximation (perfect accuracy). In this case, the distribution of some property over all rules could be found by determining that property using the approximation for each class, and weighting the result by the number of rules in the class. We will see below that this ideal does not obtain for $r = 2$ rules evaluated at low orders of approximation. Beyond 0th-order, not all classes can be sampled. At any order 0–2, not all rules in a class necessarily have the same properties. These properties are not necessarily exactly predicted by the Markov approximation. Below, we will obtain *approximate* distributions over the entire set of $r = 2$ rules by theoretically and empirically evaluating the properties of a sample of classes, and weighting the results by the number of rules in the classes.

3.2. Empirical methods

3.2.1. Monte Carlo estimation of invariant measures

The invariant statistical properties of $r = 2$ rules were empirically estimated as follows. A configuration of length 4×10^4 was pseudorandomly generated so that its density was 0.5. Each rule (as selected below) was iteratively applied to the configuration. Periodic boundary conditions were imposed. After every 60 generations, P_1 and P_{11} were determined by sampling for an additional 20 generations. The 20 determinations of P_1 and P_{11} were then averaged (time averaging). Iteration was continued until the time-averaged P_1 differed by less than 0.002 from the previous time-averaged P_1. This typically occurred in well less than 1000 generations. The final time-averaged values of P_1 and P_{11} were taken as estimates of the probabilities of the blocks 1 and 11 under the invariant measure of the rule. Where empirical results are discussed below, "the invariant measure" or "the fixed-point measure" refers to these block probability estimates, or, equivalently, the 1- or 2-step Markov measures determined by them.

3.2.2. Selection of Markov classes

An attempt was made to obtain a sample of Markov classes which was uniform in the parameter space of the Markov approximation. Markov classes were selected for study as follows. *0th-order*: all 0th-order classes were considered. There are 33 of these. *1st-order*: From the set of 17424 mean field classes, 8778 classes unrelated to each other by interchange of state labels were arranged lexicographically by coefficient value. Every eighth class in this list which had at least 50 members was selected. In this way 1031 1st-order classes were selected. *2nd-order*: From each 1st-order class selected as just described, three 2nd-order classes were selected at random. All 3rd- and 4th-order classes in each of the 2nd-order classes selected as just described are included in the study on class sizes below (section 3.3).

3.2.3. Selection of rules from Markov classes

A sample of rules was taken from each of the classes selected as described above, orders 0–2.

0th-order: All rules in the 0th-order classes defined by $\lambda = 0, 1, 31, 32$ were considered. 200 rules were selected at random from each of the other 0th-order classes. *1st-order*: 50 rules were taken at random from each 1st-order class selected as above. *2nd-order*: All rules in each class selected as above were considered. In total, more than 1×10^5 rules were studied.

3.2.4. Solution of Markov equations

Markov equations were solved either by Newton's method or by simple iteration, or a combination of both. The initial probability measure was unbiased and correlated as in the Monte Carlo simulations. In this paper "the invariant measure" refers to the fixed-point measure which results from an unbiased, uncorrelated initial condition.

3.3. Size and number of Markov classes

In general, for fixed radius rules increase in order of Markov approximation results in smaller and more numerous classes of rules. For fixed order of approximation, the larger the radius the larger and more numerous the Markov classes. The number of rules in a 0th-order class defined by coefficient value λ is

$$\binom{2^d}{\lambda}.$$

The number of such classes is $2^d + 1$. The number of rules in a mean field theory class of d-diameter rules defined by coefficient values a is

$$\prod_{i=0}^{d} \binom{\binom{d}{i}}{a_i}.$$

The number of mean field classes is

$$\prod_{i=0}^{d} \binom{d}{i} + 1.$$

There is no known general formula for the size of nth-order classes, $n > 1$, or the number of such classes. However, if $B_{(x,y,z)_d}$ is the maximum value of a coefficient $b_{(x,y,z)_d}$ for d-diameter rules then there are at least $\prod_{(x,y,z)_d}(B_{(x,y,z)_d}+1)$ 2nd-order classes. A generous upper bound for the number of rules in a class defined by coefficient values $b_{(x,y,z)_n}$ and $c_{(x,y,z)_{n+1}}$ is

$$\prod_{(x,y,z)_d} \binom{B_{(x,y,z)_d}}{b_{(x,y,z)_d}}.$$

Markov classes vary widely in size. At each order of approximation there are classes, such as the class which contains the rule which maps all neighborhoods to 0, which have but a single member. The largest class at 0th-order is the class which contains rules which map exactly half of the neighborhoods to 1. For $r = 2$ rules, this class contains approximately 6×10^8 members. The largest 1st-order classes for $r = 2$ rules have approximately 6×10^6 members. There are 8 of these. It is not known how large the largest 2nd-order classes of $r = 2$ rules are. The largest 2nd-order classes found in the sample studied here have 120 members.

The distribution of $\log_2$ of the sizes of 0th- through 3rd-order classes sampled as above is shown in fig. 1. All 4th-order classes sampled contained only 1 rule. This distribution is not shown. Clearly, as the order of approximation is increased the typical size of the classes defined decreases sharply. The average class size is approximately 1.3×10^8, 2.5×10^5, 13.0, 2.3, and 1.0 for 0th- through 4th-order respectively.

Each order of approximation provides a strict refinement of classes defined at previous orders, until the order of approximation exceeds the diameter of the rules in question. This is a consequence of the fact that blocks of the same nth-order type are also of the same mth-order type, when $m \leq n$. When the order of approximation is equal to the diameter of the rule, each d-block is of a distinct type. Hence there can be only one rule in a dth-order class of d-diameter rules. This implies that the 5th-order approximation must completely classify $r = 2$ rules. As can be seen in fig. 1, 3rd-order classes of $r = 2$ rules are too small to permit meaningful statistical analysis of the distribution of properties of rules in a class. Hence, only 0th- through 2nd-order classes are included in the studies below.

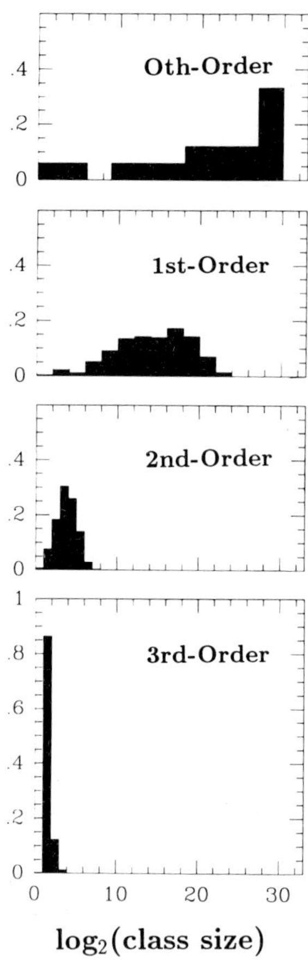

Fig. 1. Distribution of $\log_2$(class size) for 0th-order through 3rd-order classes for $r = 2$ rules. The total number of such rules is 2^{32}. Total bar height in each panel is 1.

3.4. Distribution over the set of $r = 2$ rules of P_1^* and P_{11}^*

The *class-invariant measure* is the mean of the distribution in measure space of the invariant measures of rules in a class. An empirical estimate of the class-invariant measure was obtained for each class by averaging over the rules in the class selected the empirical estimates of P_1^* and P_{11}^*. A theoretical estimate for the class-invariant measure was obtained for each class by solving for a fixed point of the Markov equations which define the class. As discussed above, by weighting these estimates by the number of rules in each class, we estimate the distribution of P_1^* and P_{11}^* over the entire set of $r = 2$ rules. These estimates are shown in fig. 2. The theoretical distribu-

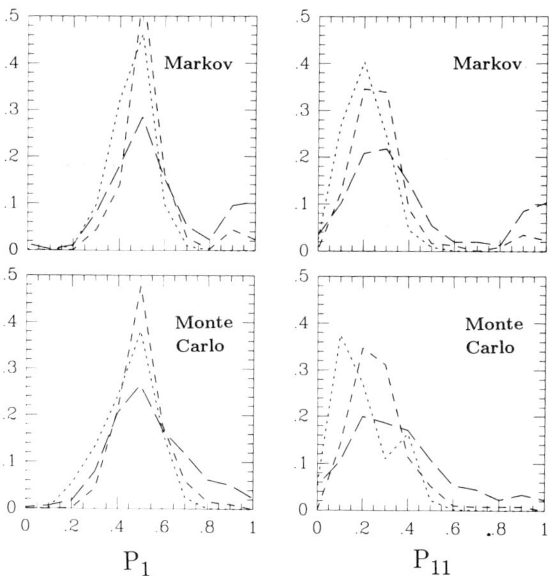

Fig. 2. Distribution of P_1^* (first column) and P_{11}^* (second column). The theoretical curves (first row, Markov) were obtained by solving for the fixed-point of the Markov equations for each class sampled. The empirical curves (second row, Monte Carlo) were obtained by averaging over the rules in a class large-time block probabilities. These are the block probabilities under the empirical class-invariant measure. The contribution of each class to a distribution is weighted by the number of rules in the class. The curves are then normalized so that the total bin height is 1. Orders of approximation 0–2 are shown. Increasing dash length corresponds to increasing order of approximation.

tions closely resemble the empirical distributions. More detailed comparisons will be made below. It should be observed here that each distribution at orders 0–2 peak near 0.5 for the estimate of P_1^* and near 0.25 for the estimate of P_{11}. It thus appears that the sampling which was roughly uniform over the parameter space of the Markov approximation corresponds to a peaked distribution over the set of automata themselves. Recall that the *standard measure* is the unbiased, uncorrelated measure. Under the standard measure all blocks of the same length have the same probability. For this particular method of sampling, most rules sampled have invariant measures which are near the standard measure. As order increases the fraction of rules with invariant measures near the standard measure represented in these samples decreases. As we will see below, distance from the standard measure is an important determinate of accuracy of

Markov approximation. The distributions shown in fig. 2 establish a baseline against which the succeeding distributions should be judged.

3.5. Theoretical versus empirical estimates of P_1^* and P_{11}^*

The ability of the Markov approximation to predict the invariant statistical properties of a cellular automaton depends in part on the nature of those invariant statistical properties. Roughly, the stronger the correlations in the invariant measure of a cellular automaton, the less accuracy can be expected from the Markov approximation. If the invariant measure of a cellular automaton is close to the standard measure then the Markov approximation tends to estimate this invariant measure well. This is suggested by the results of fig. 3. Here the empirical class P_1^* and P_{11}^* are divided into bins according to the corresponding theoretical values. The average and standard deviation over each bin are computed. The light solid curve connects the bin averages, the error bars give $\pm$ one standard deviation over the bin. The heavy solid line indicates where the theoretical estimate equals the empirical estimate.

Under the standard measure P_1 has the value 0.5 and P_{11} has the value 0.25. It is near these values that the best estimates are obtained. That is, at each order 0–2, it is at these values that the bin averages are nearest the theoretical prediction, and the bin standard deviations are smallest. In general, the standard deviation over a bin decreases as order increases. At each order, bin standard deviations for P_1^* are less than those for P_{11}^*.

There are some classes which contain rules with invariant measures far from the standard measure, and yet good estimates of these invariant measures can be made with the Markov approximation. Consider, for instance, the 0 rule under which all neighborhoods map to 0. The invariant measure of this rule is far from the standard measure. Still, the Markov approximation at all orders is exact for this rule. The converse situation may also occur: there could be rules with invariant measures close to the standard measure whose Markov approximation at any order is poor. Nevertheless, distance from the invariant measure of

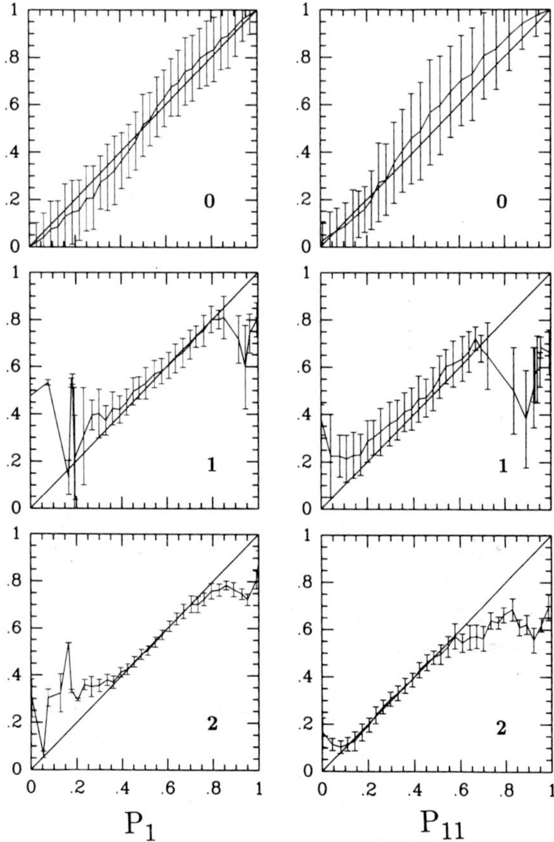

Fig. 3. Theoretical versus empirical P_1^* and P_{11}^*. The empirical class invariant P_1 and P_{11} were divided into bins according to the corresponding theoretical values. The average and standard deviation over each bin was computed. The light solid curve connects the bin averages, the error bars give $\pm$ one standard deviation over the bin. The heavy solid line indicates where the theoretical estimate equals the empirical estimate.

a rule or class of rules to the standard measure does condition the expected ability of the Markov approximation to characterize that invariant measure.

3.6. Distance in probability measure space

In the section above, the notion of distance in measure space was used informally. Here we will make this notion more precise, and use distance in measure space to further discuss the accuracy of Markov approximation. There are many ways to define a distance between probability measures. Ideally, one would compare two measures on the

basis of their assignment of probability to blocks of all sizes. This, however, is impractical. Since the measures we will discuss are 1- and 2-step Markov measures, we opt to compare these measures on the basis of the 1- and 2-block probabilities which define them. Among the most straightforward distances is the following. Let μ and ν be measures, and let $|.|$ denote absolute value. Now define the metrics

$$d_n(\mu,\nu) = \tfrac{1}{2} \sum_{\mathbf{B}_n} |\mu(\mathbf{B}) - \nu(\mathbf{B})|,$$

where the sum is taken over all n-blocks B, $n = 1$ or 2. The maximum distance between any two measures under these metrics is 1, independent of n.

In fig. 4, the distribution of the d_1 and d_2 distance between the empirical class-invariant measure and the fixed-point measure of the Markov equations which define the class is shown. In this and succeeding figures, the contribution of each class to a distribution is weighted by the number of rules in the class. The curves are then normalized so that the total bin height is 1. Orders of approximation 0–2 are shown. Increasing dash length corresponds to increasing order of approximation.

Under both the d_1 and d_2 metrics, the typical distance from empirical to the theoretical invariant measures is < 0.05. These distances are small compared to the maximum distance of 1. They are not insignificant, however, compared to the empirical convergence tolerance of 0.002 (see section 3.2.1). As order increases, the distributions peak at smaller values of the distance. Also, as order increases the tails of the distributions become longer. This reflects the fact, as shown also in fig. 3, that though accuracy typically improves with order, for some classes of rules, prediction actually deteriorates with increase in order. This phenomenon has been observed previously [1,3] in studies of individual cellular automaton rules.

Rules in a given class have a distribution of invariant statistical properties. From the results of fig. 4, this distribution is typically centered near the fixed-point measure of the system of equations which define the class. Rules within a class vary according to structure not specified by the system of equations. These differences tend to cancel each other when averages over a class are taken.

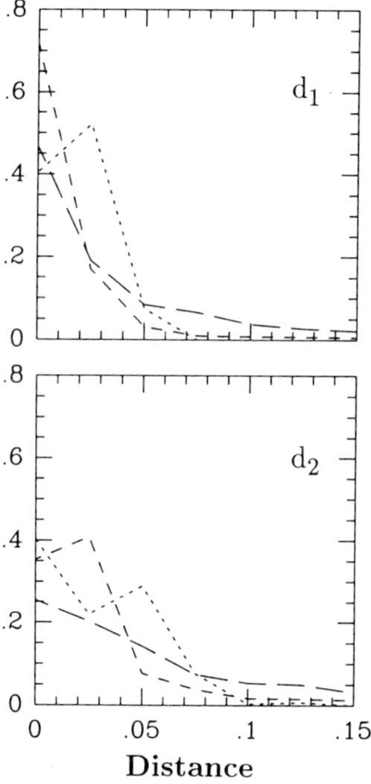

Fig. 4. Distribution of the d_1 and d_2 distance between the the empirical class-invariant measure and the fixed-point measure of the Markov equations which define the class.

Consider a process by which a measure is stepped forward by a rule selected at random from a class at each time step. One may anticipate that (in the limit of large class size) the invariant measure of such a process should be close to the fixed-point measure of the system of equations which define the class.

Since d_2 appears to be somewhat better than d_1 at resolving differences between behavior of the various orders of Markov approximation, d_2 will be used exclusively in the following.

3.7. Average-case analysis

3.7.1. Class-average distance

How far *on average* is the empirically determined invariant measure of a cellular automaton in a given class from the value predicted by the Markov equations which define the class? Above, we first averaged the empirical invariant measures of rules in a class, and then determined the distance of this average to the fixed-point measure of the equations which define the class. We will

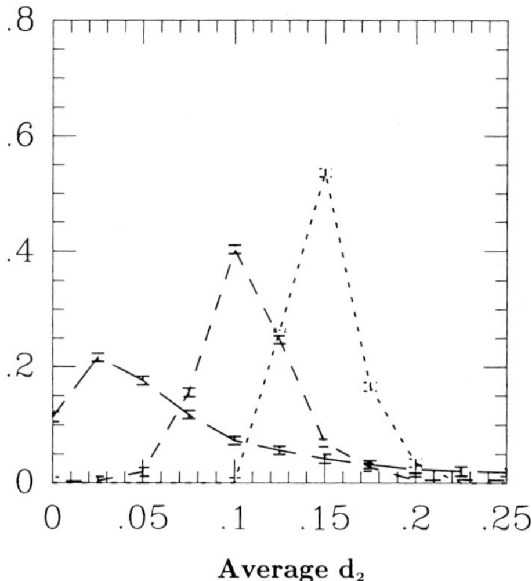

Average d_2

Fig. 5. Distribution of the average d_2 distance of the empirical invariant measure of rules in a class from the empirical class-invariant measure. Increasing dash length corresponds to increasing order of approximation.

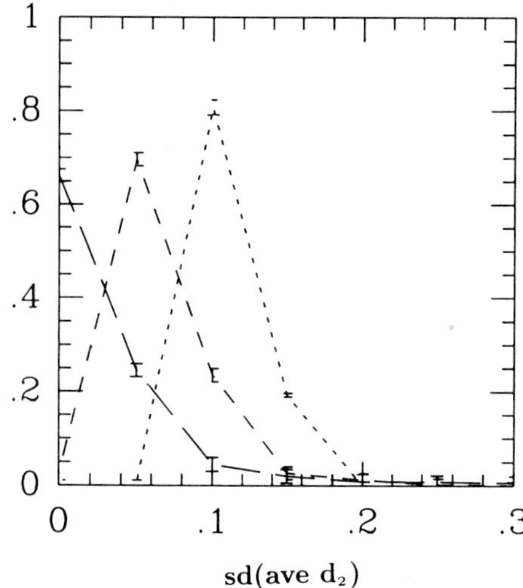

sd(ave d_2)

Fig. 6. Distribution of the class standard deviation of the d_2 distance of the invariant measures of rules in a class to the empirical class-invariant measure. Increasing dash length corresponds to increasing order of approximation.

now first compute the distance from the empirical invariant measure of each rule in the class to the fixed-point measure of the equations which define the class, and then average over the class. This gives a class average distance. For each class, the class average distance must be less than or equal to the distance between the empirical and theoretical class-invariant measure.

Fig. 5 shows the average d_2 distance of the empirical invariant measure of rules in a class from the empirical class-invariant measure. As the order of approximation is increased, the typical average distance decreases significantly. The distributions for 1st- and 2nd-order theories have long tails. It is clear, however, that average predictions tend to improve with increase in order of approximation. This may be somewhat surprising in view of the results of previous subsections. Recall that in subsection 3.4 it was demonstrated that the larger the order, the smaller the fraction of rules represented in these samples which are close to the standard measure. Then it was demonstrated in subsection 3.5 that the nearer the invariant measure of a class is to the standard measure the better, on average, the Markov estimates of that invariant measure. Hence, for this sampling distribution, lower orders of theory should yield smaller values for the dis-

tribution over rules of average distance in fig. 3. In fact, increase in order results in smaller average distances. This implies that the nature of the sampling distribution does not mask the increase with order of the predictive power of the Markov approximation.

3.7.2. Class standard deviation

For each statistical property there is a *class standard deviation* of that property. Here the class standard deviation of the d_2 distance to the empirical class-invariant measure as defined above is studied. In fig. 6 the class standard deviation of the d_2 distance is computed for each class. These results suggest that variability of statistical properties among rules in a class is small, and becomes smaller as the order of approximation is increased. Some of the variability may be due to variability in the empirical measurement of rule invariant statistical properties.

The class standard deviation of the distance will be small if rules within a class have similar properties, regardless of how well or poorly the Markov approximation serves to predict these shared properties. It has been demonstrated above that as the order of approximation increases, so does the typically accuracy of predictions. The results pre-

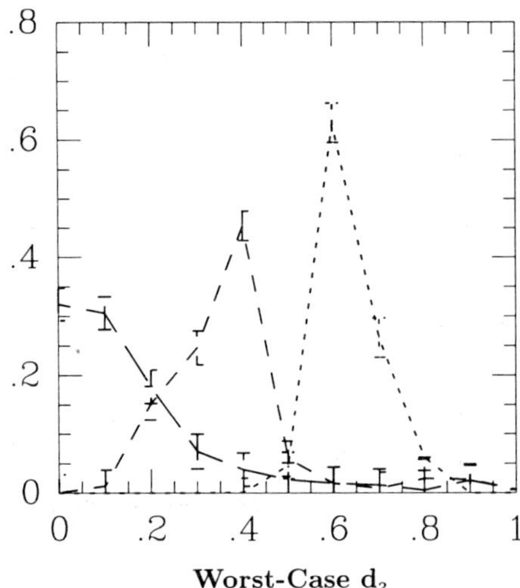

Worst-Case d₂

Fig. 7. Distribution of the worst d_2 distance of the empirical invariant measure of a rule in a class from the empirical class-invariant measure. Increasing dash length corresponds to increasing order of approximation.

sented in this subsection suggest that as the order of approximation increases, so does the tightness with which the properties of rules in a class cluster about their predicted value.

3.8. Worst-case analysis

A possible source of occasional poor average predictions is the presence of rules in a class whose behavior is rather different from that predicted by the approximation. Such rules could produce averages over a class which indicate poor performance of the approximation even though most rules in the class in fact conform well with the approximation. This possibility was explored by finding, for each class, the rule whose empirical invariant measure was furthest from that predicted by Markov approximation.

The distribution of the worst d_2 distance of the empirical invariant measure of a rule in a class from the empirical class-invariant measure is shown in fig. 7. The improvement of predictions with increase in order is more evident when worst- rather than average- cases are considered. At all orders 0–2 the worst-case distribution is shifted toward higher values relative to the corresponding average-case distribution, as should be ex-

pected. The 2nd-order distribution has a long tail, as was the case with average-case predictions (fig. 5). While the 0th- and 1st-order distributions for average-case predictions overlap considerably, the 0th- and 1st-order worst-case distributions overlap but slightly.

Recall that the 0th- and 1st-order curves are a result of *sampling* rules from classes, while the 2nd-order curve represents all rules in a class. This implies that the distributions shown in fig. 7 for 0th- and 1st-order are lower bounds for the actual worst-case performance. That is, until all rules in a class are examined, it may be possible to find a rule whose invariant measure lies at a larger distance from the theoretical prediction than that of any of those rules already sampled. Hence the improvement of performance of the 2nd-order approximation over the 0th- and 1st-order theories is actually better than indicated in fig. 7.

Still, the lower-bound distributions represented by the 0th- and 1st-order curves of fig. 7 are probably close to the actual worst-case distributions. That is, the outliers in a 0th- or 1st- order class are sufficiently numerous that the samples chosen for these experiments (200 and 50 rules/class for 0th- and 1st-order approximation respectively) typically include some of them. This is indicated by the results of fig. 8. Here, subsamples are taken at random from the samples of rules from each class, and the worst-case distance computed from rules contained in these subsamples. In the 0th-order figure, curves labeled with filled triangle, square, and pentagon represent sampling of 13, 50, and 100 rules respectively from each class. The solid curve is the same as the 0th-order curve in fig. 7 (200 rules sampled from each class). In the 1st-order figure, curves labeled with filled triangle, square, and pentagon represent sampling of 13, 20, and 33 rules respectively from each class. The solid curve is the same as the 1st-order curve in fig. 7.

These subsample sizes allow comparisons across order of approximation to be made in which sample size is kept invariant. There are typically approximately 13 rules in a 2nd-order class, hence the 2nd-order distribution of fig. 7 can be compared with the distributions in fig. 8 computed on the basis of sample size 13 (these curves are labeled by filled triangles.) These curves all rep-

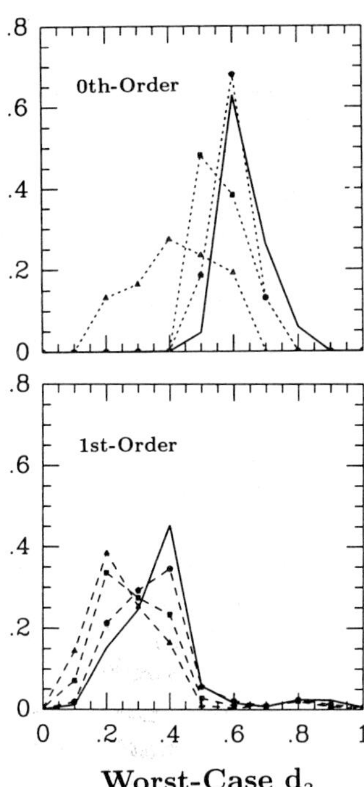

Fig. 8. Dependence of the worst distance as in fig. 7 on the number of rules sampled from a class. In the 0th-order figure, curves labeled with filled triangle, square, and pentagon represent sampling of 13, 50, and 100 rules respectively from each class. The solid curve labeled is the same as the 0th-order curve in fig. 7 (200 rules sampled from each class). In the 1st-order figure, curves labeled with filled triangle, square, and pentagon represent sampling of 13, 20, and 33 rules respectively from each class. The solid curve is the same as the 1st-order curve in fig. 7.

resent samples of roughly the same size. Still, as the order is increased, the typical worst-case performance improves considerably. In the same way, the distribution based on sub-samples of size 50 of 0th-order classes (the curve labeled with filled squares in the top panel of fig. 8) may be compared with the distribution based on samples of size 50 of the 1st-order classes (Solid curve in the bottom panel of fig. 8). From this comparison one again concludes that increase in order typically results in improved worst-case performance. For both 0th- and 1st-order approximation, as the sub-sample size is increased, the indicated worst-case performance worsens. It appears, however, that the distributions in fig. 8 are approaching a limit with increased sub-sample size. Hence, the curves based

on the largest sample size are probably a good estimate of the actual worst-case performance for these orders of approximation.

4. Discussion

A classification scheme for general one-dimensional cellular automata was presented. This classification is based on (i) Markovian (maximum-entropy) extension of block probabilities, and (ii) the combinatorial structure of one-time-step cellular automaton maps.

Markov classes at each order are strict refinements of classes at lower orders. Thus this classification is a (highly ramified) tree structure. Numerical evidence was presented which indicates that as one descends levels in this tree (i) the typical accuracy with which statistical properties of the rules in a class are predicted increases, and (ii) classes typically become more homogeneous in their statistical properties. From a combinatorial viewpoint, increase in order leads to an exponential refinement of classes. The improvement of prediction of invariant statistical properties with order is less dramatic. That there should be *any* improvement with order is in some sense remarkable. The approximations are based on the cellular automaton map on small blocks of states. As the map is iterated an infinite number of times, the range over which cell states potentially interact becomes infinite as well. These long-range interactions could affect even short-range statistics. Nonetheless, it appears that approximation by short-range interactions produces good estimates for short-range statistics for typical cellular automata.

The numerics presented here illustrated how the Markov classification can be used to explore the distribution of statistical properties in the space of cellular automata. Much work remains to be done before the kind and quality of information supplied by the Markov classification is fully understood. What are the properties of cellular automata which make them easy or difficult to characterize by Markov approximation? The intuitively correct statement is that very complex rules may escape adequate characterization by the Markov approximation to rather high order. Thus the complexity of rules should be reflected in the accuracy of theoretical predictions and homogeneity of Markov classes. The sharpness with which the Markov approximation is able to cleave the space of cellular automata varies from region to region. This in itself should indicate where "interesting" cellular automata are to be found.

The Markov classification is based on fairly general considerations. Particular applications may demand that Markov classes be either merged or split. For instance, an investigation of cellular automata which obey certain conservation laws, have a particular rule structure, or have additional relationships between them (such as blocking transformations) may be best served by the merger of some of the classes in the current scheme. It is certainly possible for rules which fall into different Markov classes to have some or all of the same statistical properties.

In most applications it should be possible to limit attention to a subset of the possible Markov classes. Consider, for example, so-called totalistic rules [10], under which the state of a cell depends only on the sum of the states in a neighborhood. Totalistic rules occupy mean field classes for which each coefficient has either the value 0 or its maximum allowed value.

Beyond nearest-neighbor cellular automata with two states per cell, it is impractical to seek interesting cellular automata by brute empirical methods. If, however, the properties of cellular automata vary in a regular way with the specification of the rules, it should be possible to conduct the exploration of the space of cellular automata systematically. It was demonstrated here that the Markov approximation of low order reflects strong regularities in the variation of statistical properties with the specification of a rule. Specification of a rule in turn specifies the values of the coefficients in the Markov approximation to the rule. Conversely, the specification of coefficient values in some order of Markov approximation defines a class of cellular automata which yield these coefficient values. We have seen that the members of such a class of rules tend to share statistical properties.

Research is under way to use the Markov classification to direct searches for cellular automata with specified properties [13]. In the present work,

the coefficients values in the Markov approximation were considered fixed, and the properties of the corresponding rules examined. In work in progress, the coefficients are treated as parameters. Properties of the equations are studied as the parameter values vary. It appears that Markov classes defined by nearby parameter values typically have similar behavior. This suggests that methods such as gradient descent could be used to find Markov equations with specified properties. Then, using the map from Markov equations to the corresponding cellular automata described here, it may be possible to find explicit cellular automaton rules with these properties.

Though the classification presented here is based only on the one-time-step cellular automaton map, its predictive power extends to properties of many iterations of that map, indeed to an infinite number of iterations. The studies undertaken here indicate that invariant statistical properties of cellular automata may be largely accounted for in terms of one-time-step approximations.

This classification scheme is trivially modified to encompass probabilistic cellular automata. Assume a set of coefficient values in a given order of approximation is specified. For deterministic cellular automata, these coefficient values specify the sum of the transitions from blocks of a given type to the target block of the equation. For probabilistic cellular automata, these coefficient values specify the sum of the transition *probabilities* from blocks of a given type to the target block of the equation. In the case of deterministic cellular automata, the coefficients take only integer values. In the case of probabilistic cellular automata, the coefficients may be real-valued.

0th- and 1st-order Markov approximation do not take into account the structure of the lattice on which a cellular automaton operates. Hence, the 0th- and 1st-order classification applies directly to cellular automata in more than one dimension. Higher-order approximations *do* take lattice structure into account, which makes these approximations more complicated in higher dimensions than they are in a single dimension [2,3]. Nonetheless, a higher-order classification of cellular automata in more than one dimension should be possible to construct.

The most well-known classification scheme for cellular automata is due to Wolfram [16]. An advantage of the Wolfram classification scheme is that it attempts to account for a variety of cellular automaton properties, including statistical, dynamical and computational properties. A further advantage is that the classification is largely independent of lattice structure and dimension, number of states of cells and the radius of the rules. The present classification scheme is designed only to account for the statistical properties of cellular automata. This classification depends in general on lattice structure and dimension, number of states of cells and the radius of the rules.

A disadvantage of the Wolfram classification scheme is that class membership is undecidable [17]. Membership in an Markov class is simply decided. Given a rule, one may decide which class at a given order it belongs to by evaluating the coefficient values for the rule in the given order of approximation. It must be noted that this does not decide the *behavior* of the rule, even in that order of approximation. For that, the resulting equations must be solved. Solution of low-order Markov equations is typically not only possible, but fast compared to any reasonable Monte-Carlo experiment. Building a collection of rules in a Wolfram class involves determination of the behavior of rules on a case-by-case basis. Here an algorithm has been give to construct all members of an Markov class.

One way to understand the Markov classification of cellular automata is through the physical notion of temperature. Schulman and Seiden [9] found that the mean field theory successfully predicted the behavior of the Game of Life in the limit of high temperature. Temperature in their studies was represented by random perturbation of cell states as the rule was iterated. One may anticipate that as the temperature is decreased, so that more correlations between cells survive over time, one should be better able to distinguish between rules which in fact differ in detail. In this sense, decreasing temperature corresponds in increasing order of approximation. While the mean field theory only encodes the gross features of cellular automata, i.e. those features not destroyed by high temperature, successive orders of approximation capture successively finer and more temperature-sensitive features.

Low orders of approximation provide a rather crude classification of cellular automata. On the other hand, a high order of approximation encodes structural details which may not be physically relevant. Since any physical process is subject to noise, this comment applies generally. Typically, for physics at a finite temperature, if one cellular automaton describes observed phenomena, then many cellular automaton rules will equivalently describe these phenomena. It should be possible to analyze such rules as a class. The present method is one way in which such analysis can be carried out.

Acknowledgements

I have benefited from discussions with M. Feigenbaum, E. Jen, B. Knight, C. Langton, Y.C. Lee, N. Packard, H. Rose, and J. Victor. This work was supported in part by an IBM postdoctoral fellowship, and a postdoctoral fellowship at the Center for Nonlinear Studies, Los Alamos National Laboratory.

A preliminary version of this paper was presented at the workshop "Learning and Recognition – A Modern Approach", Peking University, 1988.

Appendix A

Special constraints on types due to Kolmogorov consistency

In the text it is stated that two n-blocks b and b′ are said to be of the same mth-order type $m \leq n$ if (i) the Kolmogorov consistency conditions demand that they have the same probability or if (ii) they are assigned the same probability by extension of m-block probabilities according to eq. (14), for any consistent assignment of probability to m-blocks.

For the most part, types are specified by a list of values for the exponents which appear in eq. (14). However, in view of condition (i) we need only specify the *sum* of certain exponent values rather than the exponent values themselves in order to specify a type.

Condition (i) for blocks to be of the same type applies only to certain blocks. In general, blocks which have different labelings of cells may have different probabilities and still satisfy the Kolmogorov consistency conditions. Let $s \in \{0,1\}$ and $b_{n-1}s$ and sb_{n-1} be the n-block formed by adding a cell in state s on the right, respectively left of the block b_{n-1}. Let P_B be the probability of a block B. Consistency demands that

$$P_{b_{n-1}0} + P_{b_{n-1}1} = P_{0b_{n-1}} + P_{1b_{n-1}}. \qquad \text{(A.1)}$$

If $[0]^n$ denotes the block of n cells in state 0, then in particular

$$P_{[0]^n} + P_{[0]^{n-1}1} = P_{[0]^n} + P_{1[0]^{n-1}}, \qquad \text{(A.2)}$$

which implies that $P_{[0]^{n-1}1} = P_{1[0]^{n-1}}$. Similarly, $P_{[1]^{n-1}0} = P_{0[1]^{n-1}}$. Hence, $[0]^{n-1}1$ is of the same type as $1[0]^{n-1}$ for any $n > 0$, and similarly $0[1]^{n-1}$ is of the same type as $[1]^{n-1}0$ for any $n > 0$.

Condition (ii) for blocks to be of the same type does not require anything of the probability assignments to m-blocks except that they be consistent. Two blocks b and b′ will be assigned the same probability if all of the exponents which appear in eq. (14) have the same value for the two blocks. In view of condition (i) however, if the exterior exponents satisfy

$$\#[[0]^{m-1}1](b) + \#[1[0]^{m-1}](b)$$
$$= \#[[0]^{m-1}1](b') + \#[1[0]^{m-1}](b') \qquad \text{(A.3)}$$

and

$$\#[[1]^{m-1}0](b) + \#[0[1]^{m-1}](b)$$
$$= \#[[1]^{m-1}0](b') + \#[0[1]^{m-1}](b'), \qquad \text{(A.4)}$$

and the interior exponents satisfy

$$\#[[0]^{m-2}1](b) + \#[1[0]^{m-2}](b)$$
$$= \#[[0]^{m-2}1](b') + \#[1[0]^{m-2}](b') \qquad \text{(A.5)}$$

and

$$\#[[1]^{m-2}0](b) + \#[0[1]^{m-2}](b)$$
$$= \#[[1]^{m-2}0](b') + \#[0[1]^{m-2}](b'), \qquad \text{(A.6)}$$

then b and b′ will have the same probability by eq. (14) if the other exponents have the same values for the two blocks. Hence, types are specified by the sums of the exponent values as given above rather than by the exponent values themselves for these particular blocks.

Appendix B

Construction of the reflection-symmetric rules in a class

It will now be described how to construct the reflection-symmetric "core" of 0th- through 2nd-order classes of $r = 2$ rules. It will be seen that reflection symmetry significantly simplifies direct inference of rule tables from a specification of theoretical coefficients. This exercise is undertaken for two reasons. First, it illustrates a more direct method for the construction of a Markov class than that outlined in the text. Second, symmetry is often present in physical problems cellular automata may be used to address. Hence the construction of the symmetric rules in a class is an important special case.

B.1. 0th-order

Determination of the reflection-symmetric rules in a 0th-order class is entirely trivial and is described only for the sake of completeness. If an asymmetric neighborhood block leads to a given state s under a symmetric rule, then its reflection must also lead the same state s. Given a value of the 0th-order coefficient λ, one finds all ways of filling a rule table so that the total number of 1's in the table is λ, and the symmetry condition is satisfied.

B.2. Mean field theory

The symmetry condition is a more serious restriction at 1st- than at 0th-order. To satisfy the symmetry condition, each a coefficient value must be achieved by taking blocks in pairs related by reflection from the set controlled by the coefficient. Both members of the pair must map to the same state s under the rule constructed. Odd values are possible only for those coefficients which control at least one symmetric block. All mean field classes for $r = 2$ rules contain at least one symmetric rule.

B.3. 2nd-order

There is at most one symmetric rule in a 2nd-order class of $r = 2$ rules. This can be seen as follows. b coefficients for $r = 2$ rules control one to four neighborhood blocks each. For those coefficients which control one to three blocks there is only one way to achieve those coefficient values allowed by symmetry. There are two b coefficients which control four blocks. The first, coded by $[3, 0 , 1]$, controls the blocks 00101, 10100 and 01001, 10010. The second, coded by $[3, 1, 2]$, controls the blocks 01011, 11010 and 01101, 10110. For each of these coefficients there are two ways to achieve the value 2 while respecting symmetry. The other allowed values can be achieved in only one way while respecting symmetry.

Assume that $b_{[3,0,1]} = 2$. Which of these two possible ways of achieving this value actually describes the symmetric rule in the class can be determined by examining the value of the c coefficient coded by $[4, 0, 2]$. $c_{[4,0,2]}$ controls the blocks 001010, 010100 and 010010. $c_{[4,0,2]}$ cannot have the values 0 or 3 since this would imply that $b_{[3,0,1]}$ does not have the value 2, contrary to assumption. If $c_{[4,0,2]} = 1$ then 010010 leads to 11, which in turn implies that 01001, 10010 lead to 1 and 00101, 10100 lead to 0. If $c_{[4,0,2]} = 2$ then 001010, 010100 both lead to 11 which in turn implies that 00101, 10100 lead to 1 and 01001, 10010 lead to 0.

A similar argument decides between the two ways of achieving $b_{[3,1,2]} = 2$ by examination of the value of $c_{[4,1,2]}$ which controls the blocks 101011, 110101 and 101101. Hence, there is at most one reflection-symmetric rule in a 2nd-order class of $r = 2$ rules.

Appendix C

Differentiation of the Markov equations

Derivatives of the Markov equations with respect to a block probability were used above in the Newton's method solution for fixed points of the Markov equations. Once nth-order types have been identified and block probabilities properly parameterized (section 2.7), an exact formula for the derivative of the Markov equations with respect to a block probability can be derived. The mean field theory (eq. (7)) is a polynomial in P_1, the differentiation of which is trivial. The 2nd-order Markov equations (eq. (12) and (13)) are a sum of rational functions of P_1 and P_{11}. Using logarithmic differentiation, and the Kolmogorov consistency conditions, it is easy to see that the derivative of each term $R = P_{(r_1, r_2, r_3)_n}$ with respect to P_1 is

$$R \left(\frac{r_1}{P_{01}} + \frac{n - (2 + r_3)}{P_0} \right.$$

$$\left. - \frac{r_3}{P_1} - \frac{2[n - (r_1 + r_2)]}{P_{00}} \right) \tag{C.1}$$

and the derivative of each term with respect to P_{11} is

$$R \left(\frac{r_2}{P_{11}} + \frac{n - (2 + r_3)}{P_{00}} - \frac{r_1}{P_{01}} \right) . \tag{C.2}$$

The derivatives of the system of equations (12) and (13) are then the sum of these term-wise derivatives, weighted appropriately by b and c coefficient values.

Similar formulas may be derived for orders of approximation > 2.

References

[1] H.A. Gutowitz, J.D. Victor and B.W. Knight, Local structure theory for cellular automata, Physica D 28 (1987) 18–48.

[2] H.A. Gutowitz and J.D. Victor, Local structure theory in more than one dimension, Complex Systems 1 (1987) 57–68.

[3] H.A. Gutowitz and J.D. Victor, Local structure theory: Calculation on hexagonal arrays, and the interaction of rule and lattice, J. Stat. Phys. 54 (1989) 495–514.

[4] D.A. Lind, Applications of ergodic theory and systems to cellular automata, Physica D 10 (1984) 36–44.

[5] A. Schlijper, On some variational approximations in two-dimensional classical lattice systems, Thesis, University of Groningen, The Netherlands (1985).

[6] C.G. Langton, Computation at the edge of chaos, in: Emergent Computation, Proceedings of the Ninth Annual International Conference of the Center for Nonlinear Studies, Los Alamos, May 1989, Physica D 42 (1990) 12–37.

[7] W. Li, N.H. Packard and C.G. Langton, Transition phenomena in cellular automata rule space, Physica D 45 (1990) 77–94, these Proceedings.

[8] W.K. Wootters and C.G. Langton, Is there a sharp phase transition for deterministic cellular automata?, Physica D 44 (1990) 95–104, these Proceedings.

[9] L.S. Schulman and P.E. Seiden, Statistical mechanics of a dynamical system based on Conway's Game of Life, J. Stat Phys. 19 (1978) 293.

[10] S. Wolfram, Statistical mechanics of cellular automata, Rev. Mod. Phys. 55 (1983) 601.

[11] M. Denker, C. Grillenberger and K. Sigmund, Ergodic Theory on Compact Spaces, Lecture Notes in Mathematics, No. 527 (Springer, Berlin, 1976).

[12] E. Jen, Preimage scaling in cellular automata, Complex Systems 2 (1988) 1046.

[13] H.A. Gutowitz, Design of cellular automata with specified statistical properties, in preparation.

[14] S. Wolfram, Theory and Applications of Cellular Automata (World Scientific, Singapore, 1986).

[15] S. Wolfram, Approaches to complexity engineering, in: Theory and Applications of Cellular Automata, ed. S. Wolfram (World Scientific, Singapore, 1986).

[16] S. Wolfram, Universality and complexity in cellular automata, Physica D 10 (1984) 1–35.

[17] K. Culik II and S. Yu, Undecidability of CA classification schemes, Complex Systems 2 (1988) 177–190.

CHAPTER 3

LEARNING RULES WITH SPECIFIED PROPERTIES

Physica D 45 (1990) 159–180
North-Holland

ADAPTIVE STOCHASTIC CELLULAR AUTOMATA: THEORY

Y.C. LEE[1], S. QIAN
Center for Nonlinear Studies, Los Alamos National Laboratory, Los Alamos, NM 87545, USA

R.D. JONES, C.W. BARNES, G.W. FLAKE, M.K. O'ROURKE, K. LEE
Los Alamos National Laboratory, Los Alamos, NM 87545, USA

H.H. CHEN, G.Z. SUN, Y.Q. ZHANG, D. CHEN
Department of Physics and Astronomy and
Institute for Advanced Computer Studies, University of Maryland, College Park, MD 20742, USA

and

C.L. GILES
NEC Research Center, Princeton, NJ 08540, USA

Received 28 February 1990
Revised manuscript received 1 April 1990

The mathematical concept of cellular automata has been generalized to allow for the possibility that the uniform local interaction rules that govern conventional cellular automata are replaced by nonuniform local interaction rules which are drawn from the same probability distribution function, in order to guarantee the statistical homogeneity of the cellular automata system. Adaptation and learning in such a system can be accomplished by evolving the probability distribution function along the steepest descent direction of some objective function in a statistically unbiased way to ensure that the cellular automata's dynamical behavior approaches the desired behavior asymptotically. The proposed CA model has been shown mathematically to possess the requisite convergence property under general conditions.

1. Introduction

In recent years there has been an explosive growth of research on various types of artificial learning systems, ranging in scope from the more traditional machine learning which relies heavily on explicit symbolic rules and symbol manipulation, to genetic algorithms which put strong emphasis on simulating natural selection and genetic evolution, to artificial neural networks which attempt to model natural biological learning through synaptic modification. It is safe to say that the three aforementioned approaches are the current major players in the game.

Machine learning has its root in traditional artificial intelligence research which is built on the assumption that real intelligence can be simulated by abstract symbol manipulation [1]. The most distinguishing feature between the classical machine learning approach and the other two is the former's extensive use of so-called prior knowledge to combat the combinatorial explosion typical of any nontrivial learning problems. In this sense, machine learning can be considered merely as a tool to speed up learning tasks human beings already perform well. Therefore machine learning techniques are more of an extension of inherent human learning skill than an independent "emergent" phenomenon.

---

[1] Also at: Department of Physics and Astronomy and Institute for Advanced Computer Studies, University of Maryland, College Park, MD 20740, USA.

In addition, it is often difficult for researchers to decide where human knowledge aids should stop and machines should take over the task of learning; the consequence being that many machine learning problems give the feeling of being overly contrived. In fairness though, there are machine-learning inspired techniques which do not rely overly on prior knowledge, but instead are based on tried and true statistical inference techniques. Examples of the latter class are the information-theoretic decision tree approach and the conceptual clustering approach, both of which have enjoyed considerable success when used as the knowledge acquisition extension of an expert system.

Genetic algorithms, on the other hand, attempt to address the issue of the combinatorial nature of the search problem which invariably accompanies all but the most trivial learning tasks without the aid of prior knowledge [2–6]. Supporters of this line of research believe that the process of natural selection, reproduction, and genetic modification can be imitated artificially in a computer, thereby allowing searches to be done in parallel and potentially unprofitable paths pruned while potentially profitable new paths opened for exploration.

It is debatable whether genetic algorithms can indeed find good paths in general, and there are examples of landscapes which are GA-hard, meaning that for those landscapes, the genetic algorithm will not even find good solutions except by pure chance. Another major disadvantage is that a large population size is usually required in order for the genetic algorithm to function efficiently, which typically means that the given learning system will have to be replicated to generate a sufficiently large population in order for the genetic algorithm to work properly. For example, genetic algorithm researchers have attempted to train a simple neural network to learn by using more than ten thousand identical neural networks [7]. Considering the kind of tasks the neural network was asked to perform, this is clearly excessive.

The promise of artificial neural computing is that through the use of a network of simple computing elements which communicate to one another via weighted connections, one can imitate the inner function of the brain sufficiently to allow the network to behave like a massively parallel learning machine [8–19]. Owing to the extensive use of distributed representations which are less familiar and harder to think about than the more customary local symbolic representations, this approach certainly takes on the air of being distinctly non-algorithmic and more "brain-like" in the way the nets learn to perform certain tasks. This is not actually the case.

There is nothing new about distributed representations; the well known Fourier representation clearly can be considered as a kind of distributed representation, yet it has been around for more than a hundred years. The learning methods typically used for training neural nets are of either the error-correcting or reinforcement type, both of which can be put in the framework of the traditional function optimization techniques, clearly placing them in the same category as any other algorithmic technique.

What artificial neural net research *has* shown, however, is that a lot of "smart" tasks that computers can perform with the help of AI-style programming can be done by artificial neural networks simulated on a computer, with the advantage that the latter need not be programmed for the specific tasks, but rather are taught to perform them in much the same way human beings and animals are instructed to do certain things. Neural net learning algorithms, being closely related to traditional optimization techniques, are found to be quite capable of reducing the performance error even for relatively complex tasks.

Predominantly being on-line techniques, which are often suboptimal, neural learning algorithms can not match the conventional off-line optimization methods either in the speed of learning or in accuracy, especially for small tasks. However, the on-line nature of neural learning techniques renders it possible to train the neural nets for complex tasks which require the use of very large training sets. Neural nets thus hold the promise as a powerful alternative to traditional symbolic AI techniques for intelligent tasks.

Despite many recent successes, the biologically inspired approach to generating intelligent learning models still has practical limitations due to the network's more limited computational capabilities relative to general computing devices such as finite state machines, push-down automata, and Turing machines. Although recently, several groups, including our own, have started to look at the problem of getting neural networks to learn to perform like a finite state machine or push-down automaton, the problems tested have remained predominantly in the "toy"category. Hence it can be fairly stated that at present it is still too early to tell whether the present generation of recurrent neural networks has sufficient computational power to cope with quite complicated grammatical inference problems.

Cellular automata are a class of discrete non-linear dynamical systems constructed from a large number of identical finite-state automata each of which receives inputs from other automata within a predefined neighborhood. Each uses the input pattern and its current state information to update its own state as well as to determine the output it will send to other automata within its neighborhood [20–23]. The automata are typically arranged in a cellular pattern, hence the name.

What makes such distributed, nonlinear mathematical systems interesting is the fact that even with a relatively simple automaton and simple logical rules, the CA system is usually capable of generating a wide variety of complex behavior. Von Neumann, and Ulam, among others, have been able to show that certain classes of CA are capable of self-reproduction [24]. It has also been widely conjectured that even rather simple CA are able to perform universal computation [25–27]. Cellular automata, in other words, are powerful computing systems well suited for complex inference problems.

The simplest types of cellular automata have logical interaction rules which are uniform throughout the lattice. In this paper however, we consider nonuniform local interaction rules. The reason for this is that in order for the CA to be able to evolve in an adaptive manner, a sufficiently large population of logical rules will have to be maintained to reduce sampling fluctuation. The statistical rule-space homogeneity is automatically ensured by allowing local rules to randomly migrate in the lattice space. As long as there is diversity in the rule population, random migration will inject uncertainty into the CA dynamics. These kind of CA systems are said to have extrinsic stochasticity.

Training of such stochastic cellular automaton systems is done by a reinforcement technique which tries to increase or decrease the populations of sets of rules based on the estimated fitness of the rules in question as determined from environmental feedback. The learning procedure bears a strong resemblance to the genetic algorithm. The major difference is that instead of a whole population of stochastic cellular automata, it is the entire population of logical rules within a single cellular machine that is the subject of the simulated evolutionary procedure. Since a cellular automaton presumably derives much of its power from its structure, the cellular-automaton rule space that the population of rules needs to explore can be considerably smaller.

2. Stochastic cellular automata networks

2.1. Deterministic cellular automata

Deterministic cellular automata were originally introduced by von Neumann [24,28] to study the possibility that an array of simple robots (automata) can be programmed to replicate itself given that the robots could only perform certain limited tasks. A deterministic cellular automaton $\mathbf{A}$ is specified by a discrete cellular space $\mathbf{U}$, four finite sets $\mathbf{X}$, $\mathbf{Y}$, $\mathbf{Q}$, $\mathbf{N}$, and four functions Υ, δ, β, and Ω, where:

(i) The cellular space $\mathbf{U}$ is usually a discrete N-torus such that each point in $\mathbf{U}$ is indexed by an N-tuple z, where

$$z = (\mu_1, \mu_2, \ldots, \mu_N) \mod(n_1, n_2, \ldots, n_N),$$

(ii) $\mathbf{X}$ is the set of *input* symbols,
(iii) $\mathbf{Y}$ is the set of *output* symbols, and
(iv) $\mathbf{Q}$ is the set of (internal) *state* symbols.

(v) $\mathbf{N}$ is the list of *neighborhood* relations which determine the relative positions of the neighboring lattice sites from any given site in $\mathbf{U}$.

(vi) Υ, the *neighborhood state configuration function*, is such that it maps any given site $\boldsymbol{u}$ in the lattice to the neighborhood of $\boldsymbol{u}$ as specified by the neighborhood relationship defined by $\mathbf{N}$

$$\mathbf{Q} \to \otimes_{\mathbf{N}} \mathbf{Q}.$$

(vii) δ, the *next state function*, which uses the current input received at any given site together with the current neighborhood state configuration as defined by $\mathbf{N}$ to determine the state q that this particular site will be in at the next time step

$$\mathbf{X} \times \otimes_{\mathbf{N}}\mathbf{Q} \to \mathbf{Q}.$$

(viii) β, is the *output function* such that the neighborhood state configuration $\Upsilon(u)$ of any given lattice site u in $\mathbf{U}$ always yields output $\beta(\Upsilon u)$

$$\otimes_{\mathbf{N}}\mathbf{Q} \to \mathbf{Y}.$$

(ix) Ω, the *global transition and output function*, acts on the entire lattice $\mathbf{U}$, transforming any global state configuration $\mathbf{S}$ at the current time step t into a new global configuration $\mathbf{S}'$ at the next time step $t+1$ together with the global output $\mathbf{O}$

$$\otimes_{\mathbf{N}}\mathbf{Q} \to \otimes_{\mathbf{N}}(\mathbf{Q} \times \mathbf{Y}).$$

Note that the above definition of deterministic cellular automata differs from the usual description in that the dynamics of the cellular automata now depend on a possibly continuous stream of input data in addition to the current global state information. The output function is also introduced for convenience since it is required in most practical applications. However, it does not affect the overall dynamics.

2.2. Stochastic cellular automata

A *stochastic cellular automaton* $\mathbf{A}$ on a quintuplet $(\mathbf{U}, \mathbf{X}, \mathbf{Y}, \mathbf{Q}, \mathbf{N})$ of sets is defined by the neighborhood state configuration function Υ as well as by two probabilistic functions F and G, where $\mathbf{U}, \mathbf{X}, \mathbf{Y}, \mathbf{Q}, \mathbf{N}$ and Υ retain their respective meaning in the deterministic case:

(i) F: $\mathbf{X} \times \otimes_{\mathbf{N}}\mathbf{Q} \to \mathbf{Q}$ is a *stochastic next state mapping* such that it maps the current input and neighborhood configuration $x \otimes n$ into the next state q at any given site with a probability which we will denote as $f(x, n, q)$. Clearly, the PDF $f(x, n, q)$ verifies:

$$\forall x \in \mathbf{X}, \quad \forall n \in \mathbf{N}, \quad \sum_{q \in \mathbf{Q}} f(x, n, q) = 1.$$

(ii) G: $\otimes_{\mathbf{N}}\mathbf{Q} \to \mathbf{Y}$ is the *stochastic output mapping*, such that the neighborhood configuration n

at any given site will yield the output y with a probability $g(n, y)$. Conservation of probabilistic measure again stipulates that:

$$\forall n \in \mathbf{N}, \quad \sum_{y \in \mathbf{Y}} g(n, y) = 1.$$

A global stochastic transition and output function Ω can also be defined with its attendant PDFs. It follows directly from F, and G, and is therefore nonessential for the definition of the stochastic cellular automata.

Stochastic cellular automata with fixed input values are called *autonomous* stochastic cellular automata. Clearly any automaton with constant inputs is equivalent to an automaton with no input symbols. It is interesting to note that the formulation of an autonomous stochastic cellular automaton can be used to describe the dynamical behavior of a deterministic cellular automaton with inputs which are *independent random variables*, or, more generally, a finite Markov process [29]. This can be seen as follows: Let the state transition (*next state*) function for the deterministic cellular automaton be denoted as $\delta(x, n)$, where x, being the state of a finite Markov process, is itself describable in terms of the transition probability $P[x_{t+1} = x | x_t = x']$ between x' and x. An equivalent autonomous stochastic cellular automaton can now be constructed from δ and P by defining the next state probability $f(n^*, q^*)$ to be $P[x_{t+1} = x | x_t = x'] D(q - \delta(x', \mathbf{N}))$, where $q^* = (x, q) \in \mathbf{Q}^* = \mathbf{X} \times \mathbf{Q}$, n^* is the neighborhood configuration defined in terms of q^*, and $D(z)$ is the Kronecker delta function.

The case where the inputs are *independent identically distributed* (IID) *random processes* admits additional simplification. In fact, the behavior of the deterministic cellular automaton with IID random inputs can be described by an equivalent autonomous stochastic cellular automaton with the original state set $\mathbf{Q}$. The state transition probability $f(n, q)$ can be obtained from

$$\sum \mathrm{d}x \, D(q - \delta(x, n)) \, P(x),$$

where $P(x)$ is the PDF of the IID x. The converse, that is, given an autonomous stochastic cellular automaton, to find an equivalent deterministic nonautonomous cellular automaton and the corresponding PDF for the IID random inputs $\boldsymbol{x}$, in general has no solution. However, it is possible to find the best approximate solution in the least mean square sense. The details are omitted here.

2.3. Equivalence relations for stochastic automata

The most obvious intuitive definition of the equivalence of two different automata, either deterministic or random, is that they should "behave" the same. Carlyle [30,31] and others have extended the state and automata equivalence relations from the deterministic case to the case of stochastic automata in a mathematically rigorous fashion provided that the automata states are finite. Clearly, as long as the lattice space $\mathbf{U}$ is finite with m being the number of lattice points, it follows that a cellular automaton defined on this lattice can also be considered to be a finite automaton since the total number of global states in this case is just Q^m, where Q is the cardinal number for $\mathbf{Q}$.

However, except for very small lattice, it is in general not very useful to think of a cellular automaton as just another finite automaton. This is evidenced by the fact that even with Q as small as two (i.e. two states per site), a mere 100-site lattice would give a global state count of $2^{100} \approx 10^{30}$, an astronomical number. Thus, even with finite lattice, it is far more useful to consider the corresponding cellular automaton to be an approximation to an infinite state automaton with decomposable rules. For this reason, it is desirable that global equivalence be established through the examination of local equivalence.

To facilitate discussion, a (temporal) sequence of input arrays consisting of the totality of inputs $\boldsymbol{x}$'s spanning the lattice $\mathbf{U}$ is called an *input trajectory*. Similarly, a sequence of output arrays over the entire lattice is called an *output trajectory*, a sequence of global states is called a *state trajectory*, etc. The probability of producing the (finite) output trajectory $\mathbf{O}$ given the input trajectory $\mathbf{I}$ of length n *and* the initial state PDF $\boldsymbol{\Psi}(\mathbf{S}_0) = \boldsymbol{\Psi}_0$ is

$$P_{\boldsymbol{\Psi}_0}(\mathbf{O}|\mathbf{I}) = P(\mathbf{O}|\mathbf{I}, \boldsymbol{\Psi}_0).$$

Definition: Two stochastic cellular automata are said to be equivalent if, and only if, for all input–output trajectory pairs $(\mathbf{O}|\mathbf{I})$ there exist initial distributions $\boldsymbol{\Psi}_0$ and $\boldsymbol{\Psi}_0'$ such that

$$P_{\boldsymbol{\Psi}_0}(\mathbf{O}|\mathbf{I}) = P_{\boldsymbol{\Psi}_0'}(\mathbf{O}|\mathbf{I}).$$

Definition: Two initial distributions $\boldsymbol{\Psi}_0$ and $\boldsymbol{\Psi}_0'$ are said to be equivalent if, for all $(\mathbf{O}|\mathbf{I})$,

$$P_{\boldsymbol{\Psi}_0}(\mathbf{O}|\mathbf{I}) = P_{\boldsymbol{\Psi}_0'}(\mathbf{O}|\mathbf{I}).$$

Furthermore,

Definition: Two stochastic cellular automata are said to be k-equivalent if, and only if, for all input–output trajectory pairs $(\mathbf{O}|\mathbf{I})$ of length k, there exist initial distributions $\boldsymbol{\Psi}_0$ and $\boldsymbol{\Psi}_0'$ such that

$$P_{\boldsymbol{\Psi}_0}(\mathbf{O}|\mathbf{I}) = P_{\boldsymbol{\Psi}_0'}(\mathbf{O}|\mathbf{I}).$$

Definition: Two initial distributions $\boldsymbol{\Psi}_0$ and $\boldsymbol{\Psi}_0'$ are said to be k-equivalent if, for all $(\mathbf{O}|\mathbf{I})$ of length k,

$$P_{\boldsymbol{\Psi}_0}(\mathbf{O}|\mathbf{I}) = P_{\boldsymbol{\Psi}_0'}(\mathbf{O}|\mathbf{I}).$$

Definition: Two automata (or initial distributions) are k-distinguishable when they are not k-equivalent. If two automata (initial distributions) are k-equivalent but $(k+1)$-distinguishable, then they are $(k+1)$-differentiable.

We see that the state distribution $\boldsymbol{\Psi}_t$ as a function of time can be uniquely determined given the initial state distribution $\boldsymbol{\Psi}_0$ together with a known observation of the input and output trajectories $\mathbf{I} = \mathbf{I}_0, \mathbf{I}_1, \cdots, \mathbf{I}_t, \cdots$, $\mathbf{O} = \mathbf{O}_0, \mathbf{O}_1, \cdots, \mathbf{O}_t, \cdots$. It suffices to show that knowing $\mathbf{I}_t$ and $\mathbf{O}_t$ together with the state distribution $\boldsymbol{\Psi}_t$ uniquely determines the next state distribution $\boldsymbol{\Psi}_{t+1}$. It is clear from the definition of the stochastic cellular automata that the state transition function $P(\boldsymbol{\Psi}_{t+1}|\boldsymbol{\Psi}_t, \mathbf{I}_t)$ and the output emission probability function $P(\mathbf{O}|\boldsymbol{\Psi}_t)$ can be derived directly from the local PDFs, $f(\boldsymbol{x}, \boldsymbol{n}, \boldsymbol{q})$ and $g(\boldsymbol{n}, \boldsymbol{y})$. Applying Bayes' theorem:

$$P(\boldsymbol{\Psi}_{t+1}) = \sum_{\boldsymbol{\Psi}_t} P(\boldsymbol{\Psi}_{t+1}|\boldsymbol{\Psi}_t, \mathbf{I}_t) \cdot P(\boldsymbol{\Psi}_t|\mathbf{O}_t)$$

$$= \sum_{\boldsymbol{\Psi}_t} P(\boldsymbol{\Psi}_{t+1}|\boldsymbol{\Psi}_t, \mathbf{I}_t) \cdot \frac{P(\mathbf{O}_t|\boldsymbol{\Psi}_t)}{P(\mathbf{O}_t)} \cdot P(\boldsymbol{\Psi}_t), \quad (1)$$

where $P(\mathbf{O})$ can be computed from $P(\mathbf{O}|\Psi)$ by summing the latter over Ψ. Repeat applications of the above equation enable us to compute the state distribution function $P(\Psi_t)$ from any initial distribution function $P(\Psi_0)$.

The preceding derivation directly implies the following factorization property of the conditional probability function $P_{\Psi_0}(\mathbf{O}|\mathbf{I})$:

Proposition 1: For an input trajectory $\mathbf{I}$ which is the union of two subtrajectories $\mathbf{I}_{0,1}$ and $\mathbf{I}_{1,2}$, namely,

$$\mathbf{I} = \mathbf{I}_{0,1} \bigcup \mathbf{I}_{1,2}$$

and the corresponding output trajectories

$$\mathbf{O} = \mathbf{O}_{0,1} \bigcup \mathbf{O}_{1,2}$$

and for the initial state distribution Ψ_0 as well as the intermediate state distribution function Ψ_{t_1} which is obtained from Ψ_0, $\mathbf{I}_{0,1}$ and $\mathbf{O}_{0,1}$ (using eq. (1)), the following is true:

$$P_{\Psi_0}(\mathbf{O}|\mathbf{I}) = P_{\Psi_0}(\mathbf{O}_{0,1}|\mathbf{I}_{0,1}) \cdot P_{\Psi_{t_1}}(\mathbf{O}_{1,2}|\mathbf{I}_{1,2}).$$

If, for a fixed stochastic cellular automaton, two initial state distributions Ψ_0 and Ψ_0' can be found such that they are k-differentiable, then there exists at least one pair of input–output trajectories of length k for which

$$P_{\Psi_0}(\mathbf{O}_{0,k-1}|\mathbf{I}_{0,k-1}) = P_{\Psi_0'}(\mathbf{O}_{0,k-1}|\mathbf{I}_{0,k-1})$$

but

$$P_{\Psi_0}(\mathbf{O}_{0,k}|\mathbf{I}_{0,k}) \neq P_{\Psi_0'}(\mathbf{O}_{0,k}|\mathbf{I}_{0,k}).$$

By simply applying eq. (1) to the initial distributions Ψ_0 and Ψ_0', two new state distributions Ψ_1 and Ψ_1' are obtained. From proposition 1 and the fact that

$$P_{\Psi_0}(\mathbf{O}_0|\mathbf{I}_0) = P_{\Psi_0'}(\mathbf{O}_0|\mathbf{I}_0)$$

it is straightforward to deduce that

$$P_{\Psi_1}(\mathbf{O}_{1,k-2}|\mathbf{I}_{1,k-2}) = P_{\Psi_1'}(\mathbf{O}_{1,k-2}|\mathbf{I}_{1,k-2})$$

but

$$P_{\Psi_1}(\mathbf{O}_{1,k-1}|\mathbf{I}_{1,k-1}) \neq P_{\Psi_1'}(\mathbf{O}_{1,k-1}|\mathbf{I}_{1,k-1}).$$

We have thus found two new initial state distributions Ψ_1 and Ψ_1' which are $(k-1)$-differentiable.

Continuing this argument leads to the following theorem [30,31]:

Theorem 2: For any given stochastic automaton, if two initial state distributions Ψ_0 and Ψ_0' can be shown to be k-differentiable, then for all integers $l \leq k$, there exists a pair of state distributions Ψ and Ψ' which are l-differentiable.

Suppose now we have two stochastic automata with s and s' global states, respectively. Using a procedure essentially identical to that used in deriving eq. (1), it is easy to see that for a given input trajectory $\mathbf{I}$, the corresponding output trajectories $\mathbf{O}$ and $\mathbf{O}'$ generated by these two automata depend linearly on the initial state distributions Ψ_0 and Ψ_0'. Note also that any arbitrary initial state distribution can be synthesized as a linear mixture of pure states. This, together with the normalization conditions on Ψ_0 and Ψ_0' implies that Ψ_0 and Ψ_0' can be considered to live in a $(s+s'-2)$-dimensional vector space $\mathbf{V}$. The condition of k-equivalence can be reduced to $\mathbf{O} = \mathbf{O}'$. Since both $\mathbf{O}$ and $\mathbf{O}'$ are linear (vector) functions in $\mathbf{V}$, it follows that the complete equality of $\mathbf{O}$ and $\mathbf{O}'$ of length $l = s+s'-1$ is sufficient to guarantee the equivalence of $\mathbf{O}$ and $\mathbf{O}'$ for all l. Thus, we have [30,31]:

Theorem 3: For any two stochastic automata with s and s' global states, and any two initial state distributions Ψ_0 and Ψ_0' of these two automata, respectively, $(s + s' - 1)$-equivalence is a sufficient condition for their equivalence.

While theorem 3 looks impressive, it is actually of little use for the stochastic cellular automata of even moderate size because of the exponentially large number of global states involved. Therefore it would be extremely desirable to be able to establish the equivalence relations on the local level. Unfortunately, this proves to be difficult owing to the fact that any local testing will have to deal with the problem that the dynamics of the cellular automata is inherently global in nature.

2.4. Local equivalence relations for stochastic cellular automata

In order to be able to define *relevant* local equivalence relationships, in the sense that their affirmation has direct bearing to the global equivalence among cellular automata, we have to moderate our goals considerably. First of all, we will ignore the distinction between two stochastic cellular automata if they become identical under an isomorphism between their respective local state spaces $\mathbf{Q}$ and $\mathbf{Q}'$. Furthermore, we will only consider the local equivalence of two stochastic cellular automata having the same lattice structure and the same neighborhood relationship as defined by $\mathbf{N}$.

In order to define local equivalence relationships, we need the definition of *proper neighborhood state configuration* $\boldsymbol{n}'$. Here, $\boldsymbol{n}'$ differs from the neighborhood state configuration $\boldsymbol{n}$ in that the state $\boldsymbol{q}$ of the site $\boldsymbol{u}$ itself is ignored.

Definition: The proper neighborhood state configuration $\boldsymbol{n}'$ is defined to be a member of the proper neighborhood state space $\otimes_{\mathbf{N}-\boldsymbol{u}}\mathbf{Q}$ which is the product of the (local) state spaces $\mathbf{Q}_i$ of all the neighboring sites of $\boldsymbol{u}$ excluding $\boldsymbol{u}$ itself.

We are now in a position to define local trajectories. A local trajectory is a sequence of snap-shots of the local configurations relevant to a particular lattice site. Hence

Definition: A *local input trajectory* $\mathbf{X}$ for $\boldsymbol{u}$ is a temporal sequence of input symbols $\boldsymbol{x}_t$'s received at the lattice site $\boldsymbol{u}$.

Definition: A *local output trajectory* $\mathbf{Y}$ of $\boldsymbol{u}$ is a temporal sequence of output symbols $\boldsymbol{y}_t$'s emitted from the lattice site $\boldsymbol{u}$.

Definition: A *proper neighborhood state trajectory* $\mathbf{NS}$ is a temporal sequence of proper neighborhood state configurations $\boldsymbol{n}'_t$'s associated with the lattice site $\boldsymbol{u}$.

Definition: A *local state trajectory* $\mathbf{S}$ is a temporal sequence of local states $\boldsymbol{q}_t$'s associated with the lattice site $\boldsymbol{u}$.

In complete analogy with the definition of global equivalence relationships, we offer the following definition for the local equivalence:

Definition: Two stochastic cellular automata defined on the same lattice space $\mathbf{U}$ and with the same neighborhood definition are said to be locally equivalent at $\boldsymbol{u}$ if for all local input–output trajectory pairs $(\mathbf{X},\mathbf{Y})$, all possible initial states

$\boldsymbol{q}_0$, and all neighborhood state trajectories $\mathbf{NS}$, then

$$P_{\boldsymbol{q}_0}^{\boldsymbol{u}}(\mathbf{Y}|\mathbf{X},\mathbf{NS}) = P_{\boldsymbol{q}_0}'^{\boldsymbol{u}}(\mathbf{Y}|\mathbf{X},\mathbf{NS}). \tag{2}$$

If the lengths of the trajectories are restricted to k, then we have the definition for k-equivalence.

Note that the proper neighborhood trajectory $\mathbf{NS}$ essentially plays the same role as the input trajectory $\mathbf{X}$ in that they represent the information that $\boldsymbol{u}$ received from the rest of the lattice sites, therefore, to $\boldsymbol{u}$ they do represent external information.

Indeed it is straightforward to show that the counterpart to theorems 2 and 3 for the local case are just as valid. Hence they will be omitted here to avoid redundancy.

To see that local equivalence implies global equivalence, all that is required to be proven is that starting from an arbitrary (global) initial state distribution $\boldsymbol{\Psi}_0$ and the initial input $\mathbf{I}_0$, both cellular automata will produce the same output distributions $\mathbf{O}_0$ and $\mathbf{O}'_0$ as well as the same state distributions $\boldsymbol{\Psi}_1 = \boldsymbol{\Psi}'_1$. The rest follows directly from mathematical induction.

Starting from an initial pure global state $\mathbf{S}_0$, taking into account the statistical independence among the G functions at different lattice sites, it is not hard to show that

$$P(\mathbf{O}) = \otimes_{\mathbf{U}} P_{\boldsymbol{q}}^{\boldsymbol{u}}(\mathbf{Y}|\mathbf{X},\mathbf{NS}) \tag{3}$$

from which the equivalence of the output distributions immediately follows by averaging over the initial global state distribution. Similar consideration leads to the equivalence of the next state distributions.

3. Stochastic cellular automata with adaptive mappings

3.1. Reinforcement learning models

The importance of reinforcement learning in animal learning has long been recognized by psychologists. This in turn, has been taken up by artificial neural net researchers and genetic algorithm/classifier system practitioners interested in incorporating general learning paradigms into artificial intelligent systems. Typically, reinforcement takes the form of reward and punishment, and the system response probabilities get modified in direct proportion to the frequencies at which the rewards/punishments are administered [32–37].

In an artificial neural network, the behavior modification is accomplished by parameter-adjustment. Often, the adjustable parameters in the neural net are called synaptic weights or connection strengths, but their analogy to real biology is tenuous at best. Most neural net research on reinforcement learning to date has been on pattern classification and simple time-delayed feedback control tasks [38]. Except for the works on grammatical inference learning [39,40] mentioned in the section 1, to our best knowledge there has been no significant work to use reinforcement learning techniques to carry out structural learning.

Holland's genetic algorithm-based classifier system [2], on the other hand, has the syntax of a production system and therefore has more built in computational capability than most artificial neural nets. In this regards it is perhaps worth pointing out that a feedforward neural network is simply an adaptive function, whereas a recurrent network has roughly the computational capability of a state-machine. On the other hand, a production system is computationally equivalent to a Turing machine. Thus it is perhaps somewhat disappointing that classifier systems so far have not achieved the kind of success that artificial neural networks are enjoying. In fact, despite the classifier system's more powerful structure, the most successful classifier systems still can only learn direct input–output mapping in the manner of a feedforward neural network.

Stochastic cellular automata are computationally just as capable as the production system that underlies a classifier system. Therefore, one would expect the stochastic CA to be just as hard to train as the classifier system. However, stochastic cellular automata differ from production systems in that, first of all, the former rely almost entirely on a large number of extremely simple (local) rules to collectively perform powerful computations, whereas the latter typically require much more complex rules to execute the same tasks, and

secondly, even though both the stochastic CA and production systems employ discrete set of rules, the stochastic nature of the state transitions and input–output relations allows the stochastic cellular automata to be trained using continuous optimization techniques in addition to discrete ones. Both of these promise a drastic reduction of learning complexity.

In all fairness, classifier system researchers have attempted to reduce the learning complexity by forcing the production rules to be uniform as well as by padding the rules with "don't care" symbols to simulate shorter rules and to increase the chance of a match between the condition part of the production rules and the inputs. However, this tends to make the rule length rather large, and attempts to shorten the length without any other way of compensating for the lost capability generally precludes more complex tasks.

The approach we take in implementing the reinforcement learning in a stochastic cellular automaton is closely parallel to that of the learning stochastic automata [41,42]. The basic idea there is to modify either the state transition probabilities or the state probabilities of the learning automaton in order to appropriately reflect the (stochastic) environmental feedback, such as rewards and/or punishments which represent favorable and/or unfavorable reactions of the environment to the system response.

In its simplest form, a learning automaton is supposed to deal with a random environment in such a fashion that any response of the learning automaton to the environmental inputs is rewarded either with a nonpenalty or a penalty instantly. The random environment E is assumed stationary and is characterized by a set of probabilities $\omega_{\mathbf{E}} = \bigcap_{\boldsymbol{y},\boldsymbol{x}} P[r = 1|\boldsymbol{y},\boldsymbol{x}]$, where r is the reinforcement signal; $r = 1$ is associated with penalty and $r = 0$ with reward, and $P[r = 1|\boldsymbol{y},\boldsymbol{x}]$ is the probability that a given input–output pair will receive punishment from the environment.

It is clear that in this formulation of the random "greedy" environment, the output $\boldsymbol{y}$ of the learning automaton could not possibly be evaluated at the same time step t as the input $\boldsymbol{x}$, since then the output would depend only on the current state $\boldsymbol{s}$ which does not have any knowledge of the input $\boldsymbol{x}$ in general (unless the learning automaton could anticipate the coming of the input signal, which is impossible. Therefore we will slightly modify the usual formulation of the stochastic automata so that the output probability now depends on the next state s_{t+1} information rather than that of the present state. This means that the output is always delayed by one time step, which should cause no problem.

Let us denote by $\sigma^{\boldsymbol{y}}$ the probability that the output of the learning automaton is $\boldsymbol{y}$. The standard measure of performance of such a system in a random environment is the *expectation of penalty*, which is

$$\text{EP} = \sum_{\boldsymbol{y}} \sigma^{\boldsymbol{y}} \cdot P[r = 1 | \boldsymbol{y}, \boldsymbol{x}]. \tag{4}$$

When

$$\text{EP} < \frac{1}{m} \sum_{\boldsymbol{y}} P[r = 1 | \boldsymbol{y}, \boldsymbol{x}], \tag{5}$$

where m is the total number of output symbols, the performance of the automaton is called *expedient*. *Expediency* ε is defined as a measure of the closeness to optimal performance $\text{EP}_{\min} = \min_{\boldsymbol{y}, \boldsymbol{x}} P[r = 1 | \boldsymbol{y}, \boldsymbol{x}]$.

Definition: The *expediency* of an automaton is

$$\varepsilon = \frac{\text{EP} - \text{EP}_{\min}}{\text{EP}_{\min}}.$$

From Bayes' theorem,

$$\sigma^{\boldsymbol{y}} = \sum_{\boldsymbol{s}_t} \sum_{\boldsymbol{s}_{t+1}} P[\boldsymbol{y} | \boldsymbol{s}_{t+1}] \cdot P[\boldsymbol{s}_{t+1} | \boldsymbol{x}, \boldsymbol{s}_t]$$

$$= \sum_{\boldsymbol{s}_t} P[\boldsymbol{y} | \boldsymbol{x}, \boldsymbol{s}_t]. \tag{6}$$

In order to achieve optimal performance, the conditional emission probability $P[\boldsymbol{y} | \boldsymbol{s}]$ of the automaton must be 1 for at least one state $\boldsymbol{s}$ for each output symbol $\boldsymbol{y}$ for which there is at least one $\boldsymbol{x}$ such that $P[r = 1 | \boldsymbol{y}, \boldsymbol{x}]$ is a minimum. That this is necessary can be seen from the fact that if this were not the case, then for some output $\boldsymbol{y}$ for which there exists $\boldsymbol{x}$ such that $P[r = 1 | \boldsymbol{y}, \boldsymbol{x}]$ is the minimum among all output symbols, the conditional probability $P[\boldsymbol{y} | \boldsymbol{s}]$ is $1 - \delta < 1$, and all other states $\boldsymbol{s}'$ with nonzero output probabilities $P[\boldsymbol{y} | \boldsymbol{s}']$ have smaller probabilities than that of

$\boldsymbol{s}$. Now since $P[\boldsymbol{y} | \boldsymbol{s}]$ is less than 1, it follows that for the particular input $\boldsymbol{x}$, the expected penalty $\text{EP}_{\boldsymbol{x}}$ is strictly larger than the optimal expected penalty $\text{EP}_{\min}$. In fact if we denote the minimum difference between $P[r = 1 | \boldsymbol{y}, \boldsymbol{x}]$ and all other $P[r = 1 | \boldsymbol{y}', x]$ to be ζ, then the expected penalty $\text{EP}_{\boldsymbol{x}}$ is at least larger than the optimal expected penalty (i.e., optimal performance) by the amount $\zeta\delta$. Hence,

Proposition 4: A necessary condition for the attainment of optimal performance for the learning automaton $\mathbf{A}$ is that the conditional probability $P[\boldsymbol{y} | \boldsymbol{s}]$ for $\boldsymbol{y}$ must be 1 for at least one state $\boldsymbol{s}$ for each output symbol $\boldsymbol{y}$ for which there is at least one $\boldsymbol{x}$ such that $P[r = 1 | \boldsymbol{y}, \boldsymbol{x}]$ is a minimum.

It is readily obvious from the above proposition that the stochastic output mapping for the learning automaton has to become deterministic at the end of the learning session in order for it to have any chance of achieving optimal level of performance. Even this is not enough to ensure optimal behavior. As a matter of fact, the state transition probabilities $P[\boldsymbol{s}_{t+1} | \boldsymbol{x}, \boldsymbol{s}_t]$ themselves also need to become essentially deterministic for it to happen. To understand what we mean by *essentially deterministic*, we need the following definitions:

Definition: s_1 and s_2 are said to be *equivalent with respect to the output probability* $P[\boldsymbol{y} | \boldsymbol{s}]$ if $P[\boldsymbol{y} | \boldsymbol{s}_1]$ and $P[\boldsymbol{y} | \boldsymbol{s}_2]$ are equal for all $\boldsymbol{y}$'s. The binary relation between s_1 and s_2 is denoted by $R_{P[\boldsymbol{y} | \boldsymbol{s}]}$.

Proposition 5: $R_{P[\boldsymbol{y} | \boldsymbol{s}]}$ is an equivalence relation.

We will write $[a]_R$ for the equivalence class containing an element a with respect to the equivalence relation R.

Definition: Two state transition probabilities $P[v | \boldsymbol{x}, \boldsymbol{s}_t]$ and $Q[v | \boldsymbol{x}, \boldsymbol{s}_t]$ are said to be *equivalent with respect to* $R_{P[\boldsymbol{y} | \boldsymbol{s}]}$ if

$$\sum_{v \in [\boldsymbol{s}]_R} P[v | \boldsymbol{x}, \boldsymbol{s}_t] = \sum_{v \in [\boldsymbol{s}]_R} Q[v | \boldsymbol{x}, \boldsymbol{s}_t].$$

Proposition 6: The set $\mathbf{A}$ of all $\boldsymbol{s}$'s for which the output probability $P[\boldsymbol{y} | \boldsymbol{s}]$ is 1 for some output symbol $\boldsymbol{y}$ is partitioned by the equivalence relation $R_{P[\boldsymbol{y} | \boldsymbol{s}]} = 1 \equiv R_1$.

Definition: The state transition probabilities $P[s_{t+1}|x, s_t]$ are *essentially deterministic* if for every y, there exists one equivalence class $[s]_{R_1}$ for it.

It is apparent from the above definition that *essentially deterministic* simply means deterministic to within an equivalence class. Having deterministic output probabilities $P[y|s]$ together with essentially deterministic state transition probabilities $P[s_{t+1}|x, s_t]$ is clearly sufficient to ensure that the optimal performance is reachable for the learning machine, since all that remains is to modify transition probabilities $P[s^+|x, s]$ in such a way that for any input signal x and current state s, the transition probabilities $P[s^+|x, s]$ belong to the equivalence class of $\min[P[r = 1|y, x]]$. That it is also necessary can be seen from the fact that if $P[s^+|x, s]$ is not essentially deterministic, then

$$\sum_{R_{P[y|s^+]=1}} P[s^+|x, s] = 1 - \delta < 1. \qquad (7)$$

Hence the expected penalty will again exceed that of the optimal performance by at least the amount $\zeta\delta$ (see the discussion which comes immediately after eq. (6)). Thus

Theorem 7: Let P be the property of having output probabilities $P[y|s]$ which satisfy the necessary condition in proposition 4 as well as having state transition probabilities $P[s_{t+1}|x, s_t]$ which are essentially deterministic. Then P is both necessary and sufficient for the attainment of optimal performance by the learning automaton.

One of the interesting and perhaps unanticipated consequences is that optimal learning automata as described above actually do not use the finite state memory that is provided by s's. This follows directly from eq. (7), which states that state transition probabilities $P[s^+|x, s]$ are completely independent of the state information (i.e. s), which also implies that s^+ is not needed either. Hence the *ideal* automaton is not a finite state machine at all, but a deterministic map. How ironic.

3.2. Learning algorithms

Perhaps the simplest possible reinforcement learning algorithm is the following: If $y_t = y^1$ is the output of the automaton at time t and the automaton is at state $s_t = s^1$, and the reinforcement signal r_t is 1 (that is, the automaton has been penalized), then

$$P[y^1|s^1, t + 1] = \lambda P[y^1|s^1, t], \qquad (8)$$

where $0 < \lambda < 1$ is the learning rate for reinforcement. And for all other output symbols $y \neq y^1$, the output probabilities are modified in the following way:

$$P[y|s, t + 1] = \lambda P[y|s, t] + \frac{1 - \lambda}{n_y - 1}, \qquad (9)$$

where n_y is the cardinality of the set of output symbols. Eqs. (8) and (9) preserve the normalization of the conditional probabilities, as they should.

When the reinforcement signal r_t is 0, meaning the automaton has not been penalized by the environment, then no updating of the output probabilities is to be carried out.

There really is nothing fundamentally wrong about eqs. (8) or (9) except for the fact that since the output probabilities $P[y|s, t]$ do not have any explicit dependence on x's, the ultimate performance of the learning automaton depends strongly on the fixed state transition probabilities $P[s^+|x, s]$. If, for example, the transition probabilities are such that all automaton states are equally probable irrespective of the inputs x, then clearly there is no way that the learning automaton can learn anything useful. Therefore, in view of the above argument as well as that of theorem 7, only a more general learning algorithm which will modify both the output probabilities and state transition probabilities can be expected to provide the necessary ingredient for optimality.

The more complete reinforcement learning algorithm includes not only eqs. (8) and (9), but also the following equations for modifying the state transition probabilities. As before, we assume that at time t the automaton is at state $s_t = s^1$, and the input is $x_t = x^0$. In addition, the reinforcement signal $r_t = 1$. Then,

$$P[s^1|x_t, s^0, t+1] = \lambda P[s^1|x_t, s^0, t]. \qquad (10)$$

Again, for all other states $s \neq s^1$, the update equations become

$$P[s|x_t, s^0, t+1] = \lambda P[s|x_t, s^0, t] + \frac{1-\lambda}{n_y - 1}, \qquad (11)$$

where n_y is the cardinality of the set of state symbols. Here again, the conservation of total probabilities is preserved by the above set of equations.

For the case the reinforcement signal r_t is 0, no update of the state transition probabilities will take place. It is also possible to reward both the output probabilities and the state transition probabilities when there is no penalty. In that case the updating equations for $r_t = 0$ become

$$P[y^1|s^1, t+1] = \lambda P[y^1|s^1, t] + (1-\lambda) \qquad (12)$$

for the output probability and

$$P[s^1|x_t, s^0, t+1] = \lambda P[s^1|x_t, s^0, t] + (1-\lambda) \qquad (13)$$

for the state transition probability, assuming that at time t the automaton is at state $s_t = s^0$, the input is $x_t = x^0$, and $y_{t+1} = y^1$ is the output of the automaton at time $t+1$ and the next state of the automaton is $s_t = s^1$.

As we will show in the next subsection, both strategies will converge to the desired behavior of minimizing penalty. The first strategy of penalizing only the offenders (those actions which lead directly to punishment) probably generalize better since it does not give any preference to those actions which were active in the past and did not get penalized. Whereas the second strategy actively rewards those actions which have been activated and did not receive punishment, therefore while it learns faster, it probably would not fare well in unfamiliar surroundings since those actions responsible for handling these regions were never trained.

3.3. Convergence of the reinforcement learning procedures

The reinforcement learning algorithms presented in the preceding subsection all have the follow-

ing general form:

$$P[\mathfrak{R}^1|\mathfrak{R}_t, \mathfrak{S}_t, t+1]$$
$$= (1-\lambda_t)P[\mathfrak{R}^1|\mathfrak{R}_t, \mathfrak{S}_t, t] + \lambda_t \Theta^{r_t}_{\mathfrak{R}^1|\mathfrak{R}_t, \mathfrak{S}_t}, \qquad (14)$$

where $0 \leq \Theta \leq 1$ represents the reward/punishment that a given action (represented by $P[\mathfrak{R}^1|\mathfrak{R}_t, \mathfrak{S}_t, t]$) received from the learning algorithm to reflect the actual penalty/nonpenalty that the (assumed) random environment **E** dished out. To conserve probability, Θ must satisfy

$$\sum_{\mathfrak{R} \in \Omega_{\mathfrak{R}}} \Theta = 1, \qquad (15)$$

where $\Omega_{\mathfrak{R}}$ is the set of all possible symbols of the type $\mathfrak{R}$. It is trivial to show from eqs. (14), (15) that

$$\sum_{\mathfrak{R} \in \Omega_{\mathfrak{R}}} P[\mathfrak{R}^1|\mathfrak{R}_t, \mathfrak{S}_t, t+1] = \sum_{\mathfrak{R} \in \Omega_{\mathfrak{R}}} P[\mathfrak{R}^1|\mathfrak{R}_t, \mathfrak{S}_t, t]$$
$$\qquad (16)$$

as it should.

Since $\Theta^{r_t}_{\mathfrak{R}^1|\mathfrak{R}_t, \mathfrak{S}_t}$ represents the actual changes the reinforcement algorithm makes on the probabilities, they must be related to some instantaneous performance measure of the learning automaton. Because of the inherent stochastic nature of the training procedure, $\Theta^{r_t}_{\mathfrak{R}^1|\mathfrak{R}_t, \mathfrak{S}_t}$ is liable to be extremely noisy. However, as long as the expectation of Θ exists in the time asymptotic limit, then the probability $P[\mathfrak{R}^+|\mathfrak{R}, \mathfrak{S}]$ will tend to that limit in the mean. Convergence in the mean is defined by:

Definition: An infinite sequence $c_1, \ldots, c_k, \ldots$ is said to *converge to c in the mean* if

$$\lim_{k \to \infty} c_k = c.$$

That $P[\mathfrak{R}^+|\mathfrak{R}, \mathfrak{S}]$ does indeed converge to the expectation $E[\Theta]$ in the mean can be seen from the following lemma:

Lemma 8: Consider the following iterative equation:

$$\alpha_{k+1} = \alpha_k + \lambda_k(\beta_k - \alpha_k), \qquad (17)$$

where β_k, $k = 0, 1, 2, \ldots$ is a sequence of random variables whose means converge to β in the limit

$k = \infty$, and the coefficients λ are chosen such that $0 \leq \lambda_k \leq 1$ and

$$\sum_{k=0}^{\infty} \lambda_k = \infty. \tag{18}$$

Then α_k converges to β in the mean.

To prove lemma 8, we need the following

Lemma 9: Let

$$g_k = \sum_{i=l}^{k} \left[\prod_{m=i+1}^{k} (1 - \lambda_m) \right] \lambda_i,$$

where $\lambda_k < 1$, then $g_k = 1$ for all k.

Proof: Clearly, $g_1 = 1$. Assume that $g_j = 1$ for all $j \leq k$, then

$$g_{k+1} = (1 - \lambda_{k+1})\mathbf{g}_k + \lambda_{k+1}$$
$$= 1 - \lambda_{k+1} + \lambda_{k+1}$$
$$= 1. \tag{19}$$

$\square$

Let

$$\zeta_k = E[\alpha_k]$$

and

$$\eta_k = E[\beta_k]$$

be the expectations of α_k and β_k, respectively. Taking the expectation on both sides of eq. (17), we have

$$\zeta_{k+1} = (1 - \lambda_k)\zeta_k + \lambda_k \eta_k. \tag{20}$$

Eq. (20) can be solved to give

$$\zeta_k = \sum_{i=l}^{k-1} \left(\prod_{m=i+1}^{k-1} (1 - \lambda_m) \right) \lambda_i \eta_i + \zeta_l \prod_{m=l}^{k-1} (1 - \lambda_m). \tag{21}$$

Since η_k converges to β, it follows that there exists an k for any $0 < \delta \ll 1$ such that

$$|\eta_k - \beta| < \delta.$$

Hence

$$|\zeta_k - \beta| < \delta + |\zeta_l| \prod_{m=l}^{k-1} (1 - \lambda_m)$$
$$= \delta + |\zeta_l| b_{k,l}, \tag{22}$$

where $\lim_{k \to \infty} b_{k,l} = 0$ because of

$$\sum_{k=0}^{\infty} \lambda_k = \infty.$$

Thus the right-hand side of eq. (22) is seen to approach zero as k approaches infinity. $\square$

While lemma 8 ensures that the probability $P[\Re^+|\Re, \Im]$ will ultimately tend to the asymptotic limit of the expectation of Θ in the mean, it does not guarantee that $P[\Re^+|\Re, \Im]$ itself will approach $\lim_{t \to \infty} E[\Theta_t]$ in the limit. However, if the deviation $\tilde{\Theta}_t = \Theta_t - E[\Theta_t]$ is an IID random variable, i.e., *white noise*, and if λ is allowed to decay to zero slowly as a function of t, namely

$$\lim_{t \to \infty} \lambda_t = 0 \tag{23}$$

then the variance of $P[\Re^+|\Re, \Im]$ will also converge to zero.

Lemma 10: If the learning rate λ_k in eq. (17) satisfies, in addition to eq. (18),

$$\lim_{t \to \infty} \lambda_t = 0 \tag{24}$$

then one has

$$\lim_{k \to \infty} E[\| \tilde{\alpha}_k \|^2] = 0, \tag{25}$$

where

$$\tilde{\alpha}_k = \alpha_k - E[\alpha_k]$$

is the deviation of α_k from its mean value.

Proof: To prove that lemma 10 is indeed true, one first notes that eq. (17) can be solved directly after the average values have been subtracted out on both sides. Denoting $\beta_k - E[\beta_k]$ by ξ_k, one has

$$\tilde{\alpha}_k = \sum_{i=l}^{k-1} \left(\prod_{m=i+1}^{k-1} (1 - \lambda_m) \right) \lambda_i \xi_i + \alpha_l \prod_{m=l}^{k-1} (1 - \lambda_m). \tag{26}$$

Squaring both sides of eq. (26) and taking the expectation, taking into account of the statistical independence of ξ_i's, one arrives at:

$$E[\| \tilde{\alpha}_k \|^2] = \sum_{i=l}^{k-1} \left(\prod_{m=i+1}^{k-1} (1 - \lambda_m)^2 \right) \lambda_i^2 \sigma$$

$$+ E[\| \tilde{\alpha}_l \|^2] \prod_{m=l}^{k-1} (1 - \lambda_m)^2, \qquad (27)$$

where σ is the expectation of the square-norm of ξ_k, and is independent of k, due to the IID assumption.

For any given small δ, let l be chosen such that for all $k > l$, one has $\lambda_k < \delta$. In the limit of infinite k, the second term on the right-hand side of eq. (27) once again tends toward zero. On the other hand, $(1 - \lambda_m)^2$ is bounded from above by $(1 - \lambda_m)$. Hence for sufficiently large k, one finds

$$\begin{aligned} E[\| \tilde{\alpha}_k \|^2] &< \sum_{i=l}^{k-1} \left(\prod_{m=i+1}^{k-1} (1 - \lambda_m) \right) \lambda_i^2 \sigma \\ &< \sum_{i=l}^{k-1} \left(\prod_{m=i+1}^{k-1} (1 - \lambda_m) \right) \lambda_i \lambda_l \sigma \\ &= \lambda_l \sigma. \end{aligned} \qquad (28)$$

The right-hand side of eq. (28) can be made arbitrary small by letting l become arbitrary large.
□

Lemma 10 can be generalized to the case where ξ_k is a colored noise process provided the autocorrelation of ξ_k vanishes with the separation in k sufficiently fast. In general, the longer the autocorrelation time, the slower the variance of α_k will go to zero.

An even more powerful lemma can be obtained if one is willing to impose a slightly stronger condition on the learning rate λ_k:

$$\sum_{k=0}^{\infty} (\lambda_k)^2 < \infty. \qquad (29)$$

In return, the previous assumption about the statistical nature of the random variable ξ_k (white noise process, colored noise process, etc.) can be dispensed with. This powerful lemma can be stated as follows:

Lemma 11: Consider again the following iterative equation:

$$\alpha_{k+1} = \alpha_k + \lambda_k (\beta_k - \alpha_k). \qquad (30)$$

Assume that the learning rate λ_k satisfies eq. (29) in addition to eqs. (18) and (23). The following

condition is assumed to be satisfied by $\chi_k = \beta_k - \alpha_k$:

$$E[\| \chi_k \|^2] \leq \kappa (1 + \| v_k \|^2), \qquad (31)$$

where $v_k = \alpha_k - \beta$ with β the asymptotic mean of β_k, and κ is a positive constant. Last, we require that

$$(1 - \tau) \| v_k \|^2 \geq E[(\beta_k - \beta) v_k], \qquad (32)$$

where $0 < \tau \ll 1$. Then $E[\| v_k \|^2]$ will converge to zero with probability one.

Before we proceed with the proof, let us pause to see if those conditions that we assume are reasonable. Eq. (31) is clearly reasonable since it is satisfied automatically for both large and small v_k values. Therefore, by choosing a suitably large κ value, eq. (31) can always be satisfied. Eq. (32) is likewise trivially satisfied for large v_k values. It can also be justified from the fact that the statistical interdependence between v_k and $(\beta_k - \beta)$ is very weak (and approach zero at least as fast as λ_k) for most statistical models that one can come up with. For example, for white noise, the right-hand side of eq. (32) exactly vanishes, and for colored noise, the cross-correlation goes to zero like λ_k. Thus eq. (32) seems eminently reasonable.

The proof of the lemma is as follows: First note that the infinite product

$$\prod_{m=k}^{\infty} (1 + \lambda_m^2 \kappa)$$

converges owing to the fact that the summation of λ_m^2 is bounded. Subtracting the asymptotic mean β from both sides of eq. (30), squaring and taking conditional expectations (given v_m, $m < k$) over both sides of the resulting equation, one finds

$$\begin{aligned} E[\| v_{k+1} \|^2] &\leq \| v_k \|^2 + \lambda_k^2 \kappa (1 + \| v_k \|^2) \\ &= (1 + \kappa \lambda_k^2) \| v_k \|^2 + \kappa \lambda_k^2, \end{aligned} \qquad (33)$$

where use has been made of the inequality condition eq. (31). Defining the function v_k to be

$$\begin{aligned} v_k = {}& \| v_k \|^2 \prod_{m=k}^{\infty} 1 + \lambda_m^2 \kappa) \\ & + \sum_{m=k}^{\infty} \lambda_m^2 \kappa \prod_{j=m}^{\infty} (1 + \lambda_j^2 \kappa). \end{aligned} \qquad (34)$$

Note that v_k satisfies what essentially is eq. (33) except that the unequal sign in eq. (33) is replaced by an equal sign. Eqs. (33) and (34) can be combined to give the following inequality

$$E[\varrho_{k+1}|v_k, v_{k-1}, \ldots, v_1] \leq v_k. \tag{35}$$

Taking again the conditional expectation values on both sides of eq. (35) (given $\varrho_k, \varrho_{k-1}, \ldots, \varrho_1$), one obtains

$$E[\varrho_{k+1}|\varrho_k, \varrho, \ldots, \varrho_1] \leq \varrho_k. \tag{36}$$

Eq. (36) implies a semi-martingale [43] for ϱ_k. This, together with the fact that ϱ_1 is bounded, implies that the sequence ϱ_k converges with probability one. To show that the sequence ϱ_k indeed converges to zero, we need to show that the variance of ϱ_k converges to zero. Indeed, using the inequality eq. (29) and lemma 10, with λ_k replaced by $\lambda_k^2(\kappa+\tau)$, the convergence of the variance of ϱ_k in the limit of large k immediately follows. Therefore, the assertion that ϱ_k converges to zero with probability one has been demonstrated.

The preceding four lemmas convey an important message, namely, that constant learning rate which most neural net researchers employ in training their systems in general can only guarantee mean convergence, which may or may not be good enough for the particular tasks these networks are designed for. A decreasing learning rate in a sufficiently random environment may be sufficient to ensure mean square convergence. However, if the environment is highly structured (correlated), then the learning rate will have to decrease faster than $1/\sqrt{k}$ in order to ensure that convergence will be achieved with probability one.

In practice, though, it might not be desirable to use learning rate which goes down to zero like $1/\sqrt{k}$ or faster, simply because such a rapid decrease in the learning rate will make the learning system less and less responsive to the ever-changing environment, since any novel feature that occurs after the learning rate has decreased to near zero will be essentially ignored by the learning system. For a stationary (in the statistical sense) environment the rapidly decreasing learning rates are indeed ideal in view of its superior convergence property. But it is probably always wise to design a system for a changing environ-

ment rather than a stationary one.

A hybrid scheme for choosing the learning rate *cooling schedule* so as to combine the advantage of the superior convergence property of the faster cooling scheme and the faster novelty learning of the constant learning rate scheme is to use the frequency of reward/penalty as a switch. Namely, the learning rate will be allowed to cool down as long as the frequency of reward/penalty is either very small or is steadily going down. On the other hand, if the frequency of reward/penalty all of a sudden starts to go up sharply, then the learning rate is allowed to jump back up immediately. This particular hybrid scheme has been tested and the preliminary results seem promising. Unfortunately, the inherent complexity of the scheme has hampered our effort to analyze it mathematically.

It remains to show that with a suitable choice of the reward/penalty signals Θ^{r_t}, we can indeed obtain a convergent learning scheme which will minimize certain performance index $\mathcal{I}$. In this paper, we will concentrate on one such performance measure, i.e., the probability that the learning automaton will receive penalty from the random environment. Hence

$$\mathcal{I} = P[r_t = 1]. \tag{37}$$

Note that we have taken the performance measure to be the instantaneous probability rather than a time-averaged one, because we would like to treat learning as a sequential (or online) process rather than a batch process. This makes the algorithms much simpler to implement and at the same time possibly is also biologically more plausible (Although up to now we had not used biology as a justification for our choice of automaton structure and learning procedure, we do believe that all learning models try, in one way or another, to imitate nature.)

Although there are numerous batch and online optimization schemes which can be used to optimize $\mathcal{I}$, or, more precisely, the time/ensemble average of $\mathcal{I}$, one of the most popular choice is the so called *stochastic gradient descent* method. In our case, the *parameters* we would like to modify are the state transition probabilities and output probabilities. The stochastic gradient descent method basically says that the adjustable parameters, considered together as a vector, should move in the direction of the negative gradient of the objective function to be minimized. Introduce the probability vector $\boldsymbol{P}_t$ at time t

$$\boldsymbol{P}_t = (P[\Re^1|\Re^1, \mathfrak{S}_t, t], \ldots, P[\Re^I|\Re^J, \mathfrak{S}_t, t]), \quad (38)$$

where, as was in eq. (14), $\Re^i$ represents either an internal state of the automaton or an output symbol, $\mathfrak{S}_t$ represents either the current input symbol or no symbol, and I is the total number of states plus output symbols, while J represents the total number of state symbols alone. If various components of $\boldsymbol{P}$ were completely independent, then the gradient of $\mathcal{I}$ is simply $\operatorname{grad}_{\boldsymbol{P}} \mathcal{I}$. However, the components of $\boldsymbol{P}$ all have the meaning of conditional probabilities, as such, they have to satisfy conservation laws of the form:

$$\Lambda \cdot \boldsymbol{P} = \boldsymbol{e}, \quad (39)$$

where Λ is an $n \times I$ partition matrix with a total of I unity elements, i.e., each column of Λ having one and only one unity element and zero elsewhere, and $\boldsymbol{e}$ is an n-component column vector with all unity elements, n being the number of partitions in Λ. Eq. (39) introduces n constraints, thereby reducing the dimensions of $\boldsymbol{P}$ by n. Eq. (39) implies the following differential relations:

$$\Lambda \cdot \mathrm{d}\boldsymbol{P} = 0. \quad (40)$$

Let us construct the following projection operator Π:

$$\Pi = \Lambda'[\Lambda\Lambda']^{-1}\Lambda' \quad (41)$$

where Λ' denotes the transpose of the matrix Λ. It is easily seen that Π acts like an identity matrix when restricted to the subspace $\mho$ spanned by the row vectors (or the transposed row vectors for left multiplication) of Λ and like a null matrix for vectors which are orthogonal to $\mho$. This serves to verify that Π is indeed a projection matrix.

Since the probability vector $\boldsymbol{P}$ cannot vary in the subspace $\mho$ as can be seen from eq. (41), it then follows that the components of the gradient vector $\operatorname{grad}_{\boldsymbol{P}} \mathcal{I}$ which are in $\mho$ will have to be removed before it can be used to update the probabilities. The reduced gradient vector can be obtained from

$$\boldsymbol{D}_{\boldsymbol{P}} \mathcal{I} = \Pi \cdot \operatorname{grad}_{\boldsymbol{P}} \mathcal{I}. \quad (42)$$

Let us take a closer look at the individual component of the reduced gradient vector $\boldsymbol{D}_{\boldsymbol{P}} \mathcal{I}$. First of all, we will look at the component which corresponds to the the state transition probability $P[\Re^i|\Re^j, \mathfrak{S}_t, t]$, where $\Re^i$ and $\Re^i$ represent the ith and jth internal states of the automaton. The unconstrained partial differentiation of $\mathcal{I}$ with respect to $P[i|j, t] = P[\Re^i|\Re^j, \mathfrak{S}_t, t]$ is

$$\frac{\partial P[r_t = 1]}{\partial P[i|j, t]} = P[r_t = 1|i]\, P[j, t]. \quad (43)$$

Eq. (43) has been derived using Bayes' theorem. Assume now that the ith state $\boldsymbol{s}^i$ belongs to the kth partition Ξ_k and that the jth state is only allowed to make transition to any one of the states $\boldsymbol{s}^{l_k}$, of which $\boldsymbol{s}^i$ is one of them. The conservation of the transition probabilities means that the following equation must hold:

$$\sum_{l \in \Xi_k} P[l|j, t + 1] = 1. \quad (44)$$

The above condition is satisfied by the reduced gradient $(\boldsymbol{D}_{\boldsymbol{P}} \mathcal{I})_{i,j}$, which is obtained from eq. (43) by subtracting the average value of the unconstrained gradient components $(\operatorname{grad}_{\boldsymbol{P}} \mathcal{I})_{i,j}$ over Ξ_k from $(\operatorname{grad}_{\boldsymbol{P}} \mathcal{I})_{i,j}$, or

$$(\boldsymbol{D}_{\boldsymbol{P}} \mathcal{I})_{i,j} = P[r_t = 1|i]P[j, t]$$
$$- \frac{1}{n_k} \sum_{l \in \Xi_k} P[r_t = 1|l]\, P[j, t]. \quad (45)$$

By identifying the reduced gradient $(\boldsymbol{D}_{\boldsymbol{P}} \mathcal{I})_{i,j}$ with $\Theta - P[i|j, t]$ in eq. (14), the state transition probability $P[\Re^i|\Re^j, \mathfrak{S}_t]$ can be updated as follows:

$$P[\Re^i|\Re_j, \mathfrak{S}_t, t + 1]$$
$$= P[\Re^i|\Re_j, \mathfrak{S}_t, t] - \lambda_t(\boldsymbol{D}_{\boldsymbol{P}} \mathcal{I})_{i,j,t}. \quad (46)$$

In order to see whether the above learning algorithm converges or not, we first need to make sure that the expectation of the learning increment $(\boldsymbol{D}_{\boldsymbol{P}} \mathcal{I})_{i,j,t}$ is well behaved. It is immediately apparent that when there are ideal transition probabilities, such that if the $P[i|j, t]$ assume those values, then no penalty signal will be issued by the environment, i.e., $P[r_t = 1]$ will be identically zero. Hence in this case, $\Theta_{[i|j,t]}$ becomes $P[i|j, t]$, the ideal solution. When $P[i|j,t]$ is only slightly different from $P^*[i|j,t]$, the environment tends to respond much more favorably to the automaton outputs, dishing out penalties infrequently. According to eqs. (45) and (46), those transitions which led to more punishments than average, i.e.,

$$P[r_t = 1|i]\, P[j,t] \geq \frac{1}{n_k} \sum_{l \in \Xi_k} P[r_t = 1|l]\, P[j,t]$$

$$(47)$$

will have their transition probabilities $P[i|j,t]$ reduced. This evidently will reduce the probability of receiving future penalty. However, it is entirely possible that at some point all adjustable state transition probabilities within any given partition become equal, irrespective of the input signal that the automaton received, even when the probability of penalty is still not zero. When that occurs, the learning automaton is said to be at a suboptimal stationary point. If the suboptimal stationary point has directions along which further descent is possible, then the stationary point is termed unstable. Because of the stochastic nature of the automaton, it is highly unlikely that the automaton can stay in the vicinity of an unstable fixed point for any length of time, if at all. Therefore, the reinforcement learning with gradient descent will almost always converge to a local minimum.

The biggest drawback of the gradient learning algorithm is that the algorithm may not be realizable under certain situations. For example, there is no guaranty that the right-hand side of eq. (45) always has components which are between zero and one. And since the components of $(D_{\boldsymbol{p}}\boldsymbol{\mathcal{I}})_{i,j}$ have the meaning of probabilities, as soon as they become negative or larger than one, they are no longer realizable. An awkward way of solving this problem is to limit the right-hand side to values between zero and one. No doubt this type of scheme might work for certain situations, but to prove that this will actually work seems difficult. Then there is always the implementation issue associated with this kind of scheme.

A more elegant solution is to use the following Baum–Welch-like [44] re-estimation algorithm to update the transition probabilities. This procedure is based on the three propositions given below:

Proposition 12: Suppose $a_i \in [0,1]$ and $b_i \leq \infty$ are such that

$$\sum_i a_i = 1$$

then

$$\log\left(\sum_i a_i b_i\right) \geq \sum_i a_i \log b_i. \qquad (48)$$

Proposition 13: If a_i and b_i obey the same condition as in proposition 12, then the inequality below is true

$$\sum_i a_i \log\left(\frac{b_i}{\sum_j b_j}\right) \leq \sum_i a_i \log\left(\frac{a_i}{\sum_j a_j}\right). \qquad (49)$$

Proposition 12 is well known so its proof will not be reproduced here. Proposition 13 can be derived by partially differentiating the left-hand side of eq. (49) with respect to b_i to obtain

$$\frac{a_i}{b_i} - \frac{1}{\sum_j b_j},$$

which can only be zero when b_i is exactly proportional to a_i. In addition, the second derivatives of the left-hand side of eq. (49) induce the quadratic form

$$\frac{(\sum_i \xi_i)^2}{(\sum_i b_i)^2} - \sum_i \frac{a_i (\xi_i)^2}{(b_i^2)}, \qquad (50)$$

which reduces to

$$\frac{(\sum_i \xi_i^2)}{\sum_i a_i} - \sum_i \frac{(\xi_i)^2}{a_i} \qquad (51)$$

at $b_i = a_i$. In deriving eq. (51) we deliberately left one of the factors $\sum_i a_i$ in the denominator even though $\sum_i a_i$ is exactly one by definition. Eq. (51) can be easily seen to be positive definite by Schwarz inequality. It thus follows that the left-hand side of eq. (49) attains its maximum value only at $b_i = a_i$. $\qquad \square$

Proposition 14: Let a_i and b_i be as previously defined, and $\lambda \in \mathbf{R}^+$, then the following is true

$$\sum_i a_i \log\left(\frac{b_i}{\sum_j b_j}\right) \leq \sum_i a_i \log\left(\frac{b_i + \lambda a_i}{\sum_j (b_j + \lambda a_j)}\right).$$

$$(52)$$

Note that eq. (52) becomes eq. (49) when λ is equal to infinity. Differentiating the right-hand side of eq. (52) with respect to λ, one has

$$\sum_i \frac{a_i^2}{b_i + \lambda_i} - \frac{a_i}{\sum_j (bj + \lambda_j)} \geq 0. \tag{53}$$

The above inequality eq. (53) can be shown to be true by Schwarz inequality provided that the missing factor $(\sum_i a_i)^2$ is supplied to the second term on the left-hand side of eq. (53). It is now possible to prove the following important lemma:

Lemma 15: Consider the positive homogeneous polynomial of degree p and q variables

$$\Theta_q^p = \sum_{1 \leq i_1 < i_2, \cdots, < i_p} M_{i_1, \cdots, i_p \leq q} x_{i_1} x_{i_2} \cdots x_{i_p}, \tag{54}$$

where

$$M_{i_1, \cdots, i_p} = 0, \tag{55}$$

if $i_1, \ldots, i_p$ are not totally distinct, and ≥ 0 otherwise. Let

$$\Delta^{(i)} = x_i \frac{\partial}{\partial x_i} \Theta_q^p.$$

Further, let $0 \leq x_i \leq 1$, $i \leq q$ and

$$\sum_{i \in \Pi_j} c_i^{\Pi_j} x_i = 1 \tag{56}$$

are a set of constraint equations for x_i's. The coefficients $c_i^{\Pi_j}$ in eq. (56) are all positive within the jth partition $\Pi_j = i_{1,j}, \ldots, i_{m_j, j}$, where the partitions Π_j jointly span the set $1, \ldots, q$, and zero outside. Then the following transformation

$$T: x_i \rightarrow \frac{x_i + \lambda \Delta^{(i)}}{\sum_{k \in \Pi_i} c_k^{\Pi_i}(x_k + \lambda \Delta^{(k)})} = z_i \tag{57}$$

is strictly increasing for the polynomial Θ_q^p for any given positive learning rate λ except when $\Delta^{(i)}$ becomes proportional to x_i's, in which case Θ_q^p is invariant with respect to T.

To prove the above lemma, we first note that

$$\log \left(\frac{\sum_i e_i}{\sum_j b_j} \right) \geq \frac{1}{\sum_j b_j} \sum_i b_j [\log(e_i) - \log(b_i)], \tag{58}$$

which can be obtained from eq. (48) by replacing a_i by $b_i / \sum_j b_j$ If we now identify b_i to be the individual monomial $M_{i_1, \ldots, i_p} x_{i_1} x_{i_2} \cdots x_{i_p}$ and e_i to be $M_{i_1, \ldots, i_p} z_{i_1} z_{i_2} \cdots z_{i_p}$ then the left-hand side of eq. (58) becomes

$$\log \left(\frac{\Theta_q^p[z_i]}{\Theta_q^p[x_i]} \right).$$

The right-hand side of eq. (58) has only one term which depends on z_i, namely

$$\sum_{i_1 < i_2 \cdots < i_p} M_{i_1, \ldots, i_p} x_{i_1} x_{i_2} \cdots x_{i_p}$$
$$\times \log(M_{i_1, \ldots, i_p} z_{i_1} z_{i_2} \cdots z_{i_p}).$$

Using the fact that

$$\log(M_{i_1, \ldots, i_p} z_{i_1} z_{i_2} \cdots z_{i_p})$$
$$= \sum_{i_k, k \in 1, \cdots, p} \log z_{i_k} + \log M_{i_1 \ldots i_p},$$

as well as that the coefficient of $\log z_{i_k}$ in the above expression is just

$$\Delta^{i_k} = x_{i_k} \frac{\partial}{\partial x_{i_k}} \Theta_q^p.$$

The above lemma can now be proved by the application of proposition 14. $\square$

In order to apply the above lemma to reinforcement learning of a stochastic automaton, we note that the performance objective function $\mathcal{I} = P[r_t = 1]$ in general can be written as

$$P[r_t = 1]$$
$$= \sum_{(i_1, i_2, \ldots, i_n) \in \Omega} P[r_t = 1 | i_1, t]$$
$$\times P[i_1 | i_2, t] \cdots P[i_{n-1} | i_n, t], \tag{59}$$

where Ω is the set of allowable state transitions $(i_1, i_2, \ldots, i_n)$, i_k is an abbreviation for the state s_{i_k}, and the dependence of the transition probabilities $P[i | j, t]$ on inputs $\boldsymbol{x}_t$ is implicitly assumed so long as it does not lead to any confusion.

Eq. (59) is easily seen to be in the form of eq. (54) for Θ_q^p provided that one identifies the $\boldsymbol{x}_k$ in eq. (54) with the state transition probability $P[i | j, t]$. The reinforcement learning algorithm can now be stated:

$$P[i|j, t+1] = \frac{P[i|j,t] + \lambda \Delta^{(i|j,t)}}{\sum_k (P[k|j,t] + \lambda \Delta^{(k|j,t)})}, \qquad (60)$$

where

$$\Delta^{(i|j,t)} = P[i|j,t] \frac{\partial}{\partial P[i|j,t]} P[r_t = 1] \qquad (61)$$

and the sum $\sum_k$ in eq. (60) is over the allowable state transitions from the given state s_j.

Eq. (60) bears a strong resemblance to the gradient learning algorithm described by eq. (46). The main differences here are that eq. (60) is guaranteed to have values strictly between 0 and 1, and that the new algorithm can converge to a local minimum for any positive learning rate λ, whereas eq. (46) will fail to converge once the learning rate is above a certain threshold. In addition, the stochastic implementation of the non-gradient based algorithm is much more straightforward in a stochastic CA than the gradient based one, as we will see in the next subsection.

3.4. Stochastic implementation of learning algorithms

One of the chief attractions of a stochastic cellular automaton is that only simple processes are performed by the individual node of the automaton. Therefore it is highly desirable to maintain the same simplicity in the design of a learning mechanism for the stochastic CA. Superficially, both the gradient and non-gradient based learning algorithms require some complicated partial differentiations to be carried out. Fortunately, the latter method is amenable to a stochastic implementation in a very natural way, as we shall illustrate below.

The first thing to notice is the fact that the function $\Delta^{(i|j,t)}$ as defined in eq. (61) can be expanded as follows:

$$\Delta^{(i|j,t)} = P[r_t = 1|i,t] P[i|j,t] P[j,t] \qquad (62)$$

by using Bayes' theorem. The above equation clearly demonstrates that $\Delta^{(i|j,t)}$ is nothing but the conditional probability that the automaton is observed to be in the s_i state at time $t + 1$ after having made the transition from s_j state at time

t and is being penalized at time $t + 1$, since this is simply proportional to the frequency at which such events are taking place. Thus the stochastic learning algorithm which realize eqs. (60) and (61) would be one which would decrease by a fixed small amount (depending on the learning rate used) the state transition probability for the particular transition that was made when the system received a penalty message, followed by a renormalization of the transition probabilities which share the same partition Π_k with the destination state of the transition that was penalized. The renormalization is necessary in order to conserve the probabilities.

It is easy to see that with the afore-mentioned learning procedure, the expectation of the state transition probabilities indeed assume the right form. This is because the expectation of the occurrence of a penalty for the particular transition is just $\Delta^{(i|j,t)}$. And subsequent renormalization brings the expected transition into even closer agreement with eq. (60). The agreement is not perfect for any finite λ value because in general the expected value of a non-linear expression is not the same as the non-linear function of the expected value, i.e.,

$$E\left[\frac{a_i + b_i}{\sum_j (a_j + b_j)} \right] \neq \frac{E[a_i + b_i]}{E[\sum_j (a_j + b_j)]}. \qquad (63)$$

However, if the learning rate λ is assigned a cooling schedule such that $\lambda \to 0$ as t approaches ∞, then both sides of eq. (63) will indeed tend toward each other (their difference being of second order in λ) asymptotically. It can be further demonstrated that the above procedure does indeed satisfy the conditions for both lemma 10 and lemma 11 provided the learning rate cooling schedule is chosen properly. Hence,

Theorem 16: The stochastic learning scheme has mean square convergence if the cooling schedule of lemma 10 is used for the learning rate λ, whereas convergence with probability one is obtained if the faster cooling schedule of lemma 11 is used.

The detailed proof of the theorem is straightforward but messy, and since it is intuitively obvious, we will omit the proof here for the sake of clarity.

The fact that the afore-mentioned training procedure can converge to local minimum without the need for explicitly performing complex differential operations is at once surprising and satisfying. The algorithm itself has a close kinship with the well known Baum–Welch (BW) re-estimation formula widely used in the so-called hidden Markov model [44]. However, the BW scheme is strictly batch processing, but for the kind of intelligent tasks we are interested in, the need to store a potentially unlimited amount of training data in order to be able to train up the learning system is both impractical and potentially much too expensive. A batch operation requires that all the training data have to be in before an update of the weights (adjustable parameters) can be computed. Thus if there is a very large data set, it would take essentially forever for the batch scheme to make just the first update. By contrast, the on-line scheme is greedy in the sense that as soon as new information starts to come in, the weights are immediately adjusted to reflect the new knowledge.

Take, for example, one of the problems we have studied as an application of this technique, the problem of how to get the stochastic cellular automaton to learn how to balance an inverted pole, the detail of which will be discussed fully in the other paper [45]. In that case it is not possible to collect data about the dynamics of the pole beforehand and feed those data all at once to the stochastic CA, since the pole has to interact with the CA for the training to be meaningful, and a CA which has not learned how to balance a pole yet will not provide much useful training information. This is a prototypical chicken and egg dilemma for the batch type procedure. On the other hand, for pattern recognition type problems and with a small training set, a batch type (off-line) training scheme can be more efficient computationally since simultaneous presentation of data usually can reduce statistical fluctuation more effectively.

The original form of the Baum–Welch algorithm is not suited for on-line operation for the following reason: It requires that the weights be updated in a most drastic fashion, so much so that the new weights bear essentially no relationships to the ones immediately preceding them. In fact, the Baum-Welch algorithm can be considered to be a special case of our learning algorithm, obtainable by simply letting λ in our equation (60) go to infinity, which, as we have seen, most assuredly would not satisfy even the most lenient conditions for convergence.

For stochastic cellular automata, it is possible to go one step further by eliminating the need to keep the individual state transition probabilities altogether. In their place we can have a separate local rule table for each site, so that the state transition probabilities are explicitly represented by the entire rule table population. By constantly shuffling the individual entries of a given table with the corresponding ones of neighboring tables, we can ensure the statistical homogeneity of the CA rules. Learning can be easily accomplished by simply flipping the particular offending entry bit, i.e., one which was retrieved to determine the local action when the system received a penalty message. To allow the learning rate to vary, the flipping can be done in a probabilistic fashion. For example, if the flipping probability is small, corresponding to a small learning rate, then most of the time the offending bits do not get flipped. If, on the other hand, a larger learning rate is desired, then even the bits which are neighbors to the offending bits are fair game.

The bit-flipping learning scheme does not exactly duplicate the learning algorithm using the state transition probabilities directly. Nevertheless in the limit of small learning rates they become equivalent, and that is sufficient for convergence provided a suitable cooling schedule is used for learning rates. For our actual computer simulations that we will discuss in the next section, we elect not to use a cooling schedule, but instead opted for a constant but small learning rate. We feel (and hope to be able to justify in a later publication) that in a non-stationary environment, a constant learning rate is probably more robust since it better enables the automaton to adapt to the constant changes in the environment.

So far, our discussion of the learning mechanism has been limited to the situation where the actions of the learning system in response to environmental inputs are instantly evaluated by the environment. This so-called *instant gratification* type reward/penalty situation is obviously easier to train than if the reinforcement signals were to have time-delay. In a more typical realistic situation there is always some inevitable delay in the environmental response, mostly because it usually takes a whole sequence of actions in order to get to a point where the performance of the system could be evaluated. A case in point is the chess game, in which the performance of the system really can not be evaluated until the game ends. Chess playing requires a lot of intelligence, one that we are not yet prepared to deal with. Pole balancing is a much lower level task than that of chess playing, yet it shares some of the same characteristics with the latter, especially if we stipulate that penalty will be given only if the task of balancing the pole has failed.

Broadly speaking, there are three kinds of delays; first, the delay may be caused by the inevitable physical processes of transmission and measurement lags; second, the environment may have to wait for the completion of a chain of actions before an evaluation can be made; third, the process that the environment is waiting for may be just a collection of actions, the sequencing of which does not affect the outcome. Of the three, the second kind which requires the exacting science of properly sequencing the actions is most difficult to handle. Both the game of chess and pole balancing belong, to various degree, to the second kind.

4. Concluding remarks

In our second paper in these Proceedings [45], *Adaptive stochastic stochastic cellular automata: applications*, two control problems are examined: balancing a single pole and a double pole. In these two examples, the adaptive stochastic CA has demonstrated its power to control unstable systems. Unlike previous works on the pole-balancing problem [46,47], we made almost no assumptions about the dynamics of unstable systems when we design the adaptive CA controller. Even so, the stochastic CA is able to learn to balance the pole after only a few training cycles. Moreover, the stochastic CA is able to control a more complex unstable system, namely an inverted double pendulum.

Cellular automata have a number of potential advantages over traditional artificial intelligence methods, genetic techniques and neural networks. CA are less problem specific than machine learning techniques, and the brittleness problem is less of an issue. CA are numerically faster than genetic algorithms, which often require extensive replication of the original system. Neural networks have yet to demonstrate the computational ability of finite state machines or pushdown automata. CA show much more promise in this area. Hardware implementation is easy with CA. Finally, the stochastic modification to deterministic CA permits the use of familiar continuous optimization and function analysis techniques.

There are open issues. While stochastic CA have improved on neural network control systems such as the pole-balancing problem discussed in a separated paper, a demonstration of complicated grammatical inference needs to be performed. A demonstration is needed of the integration of a rule-based expert system with a neuron-like emergent system.

References

[1] R.S. Michalski, Carbonell and T. Mitchell, eds., Machine Learning, An Artificial Intelligence Approach Vols. I, II (Kaufmann, Los Altos, CA, 1983).

[2] J.H. Holland, Adaptation in Natural and Artificial Systems (The University of Michigan Press, Ann Arbor, MI, 1975).

[3] J.H. Holland, Genetic algorithms and adaptation, in: Proceedings of the NATO Advanced Research Institute on Adaptive Control of Ill-Defined Systems, eds. O.G. Selfridge, E.L. Rissland and M.A. Arbib (Plenum Press, New York, 1983) pp. 317–333.

[4] J.H. Holland, Genetic algorithms and classifier systems: foundations and future directions, in: Genetic Algorithms and their Applications: Proceedings of the Second International Conference on Genetic Algorithms (1987) pp. 82–89.

[5] J.H. Holland, K.J. Holyoak, R.E. Nisbett and P.R. Thagard, Induction: Processes of Inference, Learning, and Discovery (MIT Press, Cambridge, MA, 1986).

[6] D.E. Goldberg, Genetic Algorithms in Search, Optimization, and Machine Learning (Addison–Wesley, Reading, MA, 1989).

[7] K.J. Holyoak, R.E. Nisbett and P.R. Thagard, Classifier systems, q-morphisms, and induction, in: Genetic Algorithms and Simulated Annealing, ed. L. Davis (Artificial Neural Networks, 1987) pp. 116–128.

[8] J.J. Hopfield, Neurons with graded response have col-

lective computational properties like those of two-state neurons, in: Proc. Natl. Acad. Sci. US 81 (1984) 3088–3092.

[9] T.J. Sejnowski and C.R. Rosenberg, NETtalk: A parallel network that learns to read aloud, The Johns Hopkins University Electric Engineering and Computer Science Technical Report, JHU/EECS-86/01 (1986) p. 32.

[10] D.E. Rumelhart, G.E. Hinton and R.J. Williams, Learning internal representations by error propagation, in: Parallel Distributed Processing: Explorations in the Microstructures of Cognition, Vol. I, eds. D.E. Rumelhart and J.L. McClelland (MIT Press, Cambridge, MA, 1986) pp. 318–362.

[11] D.E. Rumelhart, G.E. Hinton and R.J. Williams, Learning representations by back-propagating errors, Nature 323 (1986) 533–536.

[12] S. I. Amari, Neural theory of association and concept formation, Biol. Cybern. 26 (1977) 175–185.

[13] T. Kohonen, Self-organized formation of topologically correct feature maps, Biol. Cybern. 43 (1982) 59–69.

[14] S. Grossberg, How does a brain build a cognitive code, Psychol. Rev. 87 (1980) 1–51.

[15] K. Fukushima, S. Miyake and T. Ito, Neocognitron: a neural network model for a mechanism of visual pattern recognition, IEEE Trans. Systems, Man Cybern. 13 (1983) 826–834.

[16] L.N. Cooper, A possible organization of animal memory and learning, in: Proceedings of the Nobel Symposium on Collective Properties of Physical Systems, eds. B. Lundquist and S. Lundquist (Academic Press, New York, 1973) pp. 252–264.

[17] F. Rosenblatt, The perceptron: a probabilistic model for information storage and organization in the brain, Psychol. Rev. 65 (1958) 384–408.

[18] M. Minsky and S. Papert, Perceptrons (MIT Press, Cambridge, MA, 1969).

[19] H.D. Block, The perceptron: a model for brain functioning I, Rev. Mod. Phys. 34 (1962) 123–135.

[20] G.C. Bacon, The decomposition of stochastic automata, Inform. Control 7 (1964) 320–339.

[21] A. Paz, Some aspects of probabilistic automata, Inform. Control 9 (1967) 26–60.

[22] M.D. Waltz and K.S. Fu, A heuristic approach to reinforcement learning control systems, IEEE Trans. Autom. Control 10 (1965) 390–398.

[23] V.Y. Krylov, On one automaton that is asymptotically optimal in a random medium, Autom. Remote Control 24 (1963) 889–899.

[24] J. von Neumann, Theory of self-reproducing automata (edited and completed by A.W. Burks) (University of Illinois Press, Champaign, IL, 1966).

[25] A.R. Smith, Simple computational-universal cellular spaces and self-reproduction, in: Conference Record, IEEE 9th Annual Symposium on Switching and Automata Theory (1968) pp. 269–277.

[26] J.W. Thatcher, Universality in the von Neumann cellular model, in: Essays on Cellular Automata, ed. A.W. Burks (University of Illinois Press, Champaign, IL, 1970) pp. 132–186.

[27] E.F. Codd, Propagation, computation and construction in two-dimensional cellular spaces, in: Cellular Automata (Academic Press, New York, 1965);
J. von Neumann, The general and logical theory of automata, in: Cerebral Mechanisms in Behavior: The Hixon Symposium (Wiley, New York, 1951).

[28] S. Wolfram, Cellular automata as models of complexity, Nature 311 (1984) 419–424.

[29] S.E. Levinson, L.R. Rabiner and M.M. Sondhi, An introduction to the application of the theory of probabilistic functions of a Markov process to automatic speech recognition, Bell Syst. Techn. J. 62 (1983) 1035–1074.

[30] J.W. Carlyle, Reduced form for stochastic sequential machines, J. Math. Anal. Appl. 7 (1963) 167–175.

[31] J.W. Carlyle, Equivalent stochastic sequential machines, Technical Report, Ser. 60 I 415, Electronics Research Laboratory, University of California, Berkeley (1961).

[32] O.H. Mowrer, Learning Theory and Behavior (Wiley, New York, 1960).

[33] J. Spragins, Learning without a teacher, IEEE Trans. Inform. Theor. 12 (1966) 223–230.

[34] M.D. Waltz and K.S. Fu, A heuristic approach to reinforcement learning control systems, IEEE Trans. Autom. Control 10 (1965) 390–398.

[35] R.C. Atkinson, G.H. Bower and E.J. Crothers, An Introduction to Mathematical Learning Theory (Wiley, New York, 1965).

[36] R.B. Allen, Adaptive training for connectionist state machines, in: ACM Computer Conference, Louisville (1989).

[37] R.J. William and D. Zisper, A learning algorithm for continually running fully recurrent neural networks, ISC Report (1988) p. 8805.

[38] Y. Matsuyama, Multiple descent cost algorithms for standard pattern self-organization, in: Proceedings of the International Joint Conference on Neural Networks, (1990) pp. I-436–I-439.

[39] G.Z. Sun, H.H. Chen, C.L. Giles, Y.C. Lee and D. Chen, Connectionist pushdown automata that learn context-free grammars, in: Proceedings of the International Joint Conference on Neural Networks (1990) pp. I-557–I-580.

[40] M.I. Jordan, Attractor dynamics and parallelism in a connectionist sequential machine, in: Proceedings of the Cognitive Science Society (Amherst, 1986) pp. 531–546.

[41] K.S. Fu, Stochastic automata as models of learning systems, in: Computer and Information Science II (Academic Press, New York, 1967) pp. 177–191.

[42] B. Chandrasekaran and D.W.C. Shen, On expediency and convergence in variable structure automata, in: Proceedings of the National Electronics Conference 22 (1966) pp. 845–850.

[43] J.L. Doob, Astochastic Processes (Wiley, New York, 1953).

[44] L.R. Rabiner and R.H. Juang, An introduction to hidden Markov models, IEEE ASSP Mag. 3 (1986) pp. 4–16.

[45] S. Qian, Y.C. Lee, R.D. Jones, C.W. Barnes, G.W. Flake, M.K. O'Rourke, K. Lee, H.H. Chen, G.Z. Sun, Y.Q. Zhang, D. Chen, and C.L. Giles, Adaptive stochastic cellular automata: Applications, Physica D 45 (1990) 181–188, these Proceedings.

[46] A.G. Barto, R.S. Sutton and C.W. Anderson, Neuron-like adaptive elements that can solve difficult learning control problems, IEEE Trans. Systems Man Cybern. 13 (1983) 834–846.

[47] M.I. Jordan and R.A. Jacobs, Learning to control an unstable system with forward modeling, preprint, MIT, Cambridge, MA (1990).

Physica D 45 (1990) 181–188
North-Holland

ADAPTIVE STOCHASTIC CELLULAR AUTOMATA: APPLICATIONS

S. QIAN, Y.C. LEE[1]
Center for Nonlinear Studies, Los Alamos National Laboratory, Los Alamos, NM 87545, USA

R.D. JONES, C.W. BARNES, G.W. FLAKE, M.K. O'ROURKE, K. LEE
Los Alamos National Laboratory, Los Alamos, NM 87545, USA

H.H. CHEN, G.Z. SUN, Y.Q. ZHANG, D. CHEN
*Department of Physics and Astronomy and
Institute for Advanced Computer Studies, University of Maryland, College Park, MD 20742, USA*

and

C.L. GILES
NEC Research Center, Princeton, NJ 08540, USA

Received 28 February 1990
Revised manuscript received 1 April 1990

The stochastic learning cellular automata model has been applied to the problem of controlling unstable systems. Two example unstable systems studied are controlled by an adaptive stochastic cellular automata algorithm with an adaptive critic. The reinforcement learning algorithm and the architecture of the stochastic CA controller are presented. Learning to balance a single pole is discussed in detail. Balancing an inverted double pendulum highlights the power of the stochastic CA approach. The stochastic CA model is compared to conventional adaptive control and artificial neural network approaches.

1. Introduction

Learning to control an unstable system has been studied by many authors [1,2]. There are two different kinds of learning paradigms used in adaptive control problems. One is the predictive learning scheme. Predictive learning is also called supervised learning because the learning automaton needs a teacher to evaluate its performance [3]. Predictive learning occurs by error-correction. The other kind of learning paradigm is reinforcement learning or unsupervised learning [4–6]. This learning scheme is evaluated by a critic who sends a reward or punishment signal to the learning au-

tomaton according to its performance. Reinforcement learning is a more realistic learning model in the sense of learning in biological systems. Because there is no need for a detailed model of the controlled process, this method can successfully be applied to complicated control problems. The adaptive stochastic CA algorithm uses the reinforcement learning scheme, and therefore it has the potential to control complicated systems.

The mathematical concept and learning theory of the adaptive CA are studied in the first our two papers in these Proceedings [7], *Adaptive stochastic cellular automata: Theory*. The stochastic CA is generalized from conventional cellular automata which have uniform local interaction rules, to allow for nonuniform local interaction rules which are drawn from some probability dis-

<hr>

[1] Also at: Department of Physics and Astronomy and Institute for Advanced Computer Studies, University of Maryland, College Park, MD 20740, USA.

tribution function. Adaptation and learning are accomplished by evolving the probability distribution function that the stochastic CA's dynamical behavior approaches to the desired behaviors asymptotically. Training of the stochastic CA is done by reinforcement learning, which increases or decreases the populations of the sets of rules according to the fitness of the rules based on environmental feedback. In the control problem, the feedback is simply given by the critic of network performance. Therefore, the learning scheme strongly resembles to the genetic algorithm [8,9]. The major difference is that instead of a whole population of stochastic cellular automata, it is the entire population of logical rules within a single cellular machine that is the subject of the simulated evolutionary procedure. Since a cellular automaton presumably derives much of its power from its structure, the cellular-automaton rule space that the population of rules needs to explore can be considerably smaller.

In this paper, two unstable systems are studied. Our interest in the first problem, the pole-balancing problem, arises from the fact that this problem has been studied by a variety of adaptive networks [1,2] and it conveniently serves as a benchmark. While many adaptive control methods can be successfully applied to the pole-balancing problem, the task the adaptive stochastic CA has achieved is one of the most difficult, because we made no assumption about the dynamics of the pole-balancing problem. In the second problem, the adaptive stochastic CA successfully controls a more difficult unstable system, balancing an inverted double pendulum.

2. The adaptive stochastic CA controller

The control problem we study here is identical to the problem studied by Michie and Chambers [10,11], Barto et al. [1], and Jordan and Jacobs [2] using various adaptive networks. It consists of a cart which can freely move on a linear track. A rigid pole is attached at one end to the top of the cart by means of a hinge so that it can freely swing in the vertical plane defined by it and the track. The objective is to balance the pole in the vertical position by means of a control force to the cart, applied co-linearly with the track. We use the same parameters of Barto et al., a 1 m pole of mass 0.1 kg atop a 1 kg cart.

The cart–pole control problem involves four input variables:

θ angle of the pole with vertical,
$\dot{\theta}$ angular velocity of the pole,
z position of the cart on the track, and
$\dot{z}$ velocity of the cart.

There is one control variable, the control force f_c along the track direction. Both input variables and control force are each coded into a 64-bit string. The number of 1-bits in each string is determined by the value of input and output variables. For example, the number of 1-bits in θ is given by the integer part of

$$64 \, \frac{\theta - \theta_{\min}}{\theta_{\max} - \theta_{\min}} \, .$$

Therefore, the stochastic CA has a vector input $\boldsymbol{x}$ of length four, corresponding to θ, $\dot{\theta}$, z, $\dot{z}$, and a scalar output y, the control force. The binary coding used here is similar to the BOXES system used by Barto et al. [1]. The input state space in which we are interested is evenly divided into $64 \times 64 \times 64 \times 64$ boxes and the input values are obtained by quantization of the inputs into one of 64 possible values. To ensure homogeneity, the 1-bits are randomly distributed in the 64-bit strings.

The CA stencil (neighborhood state configuration function) we used to determine the local state is the neighborhood-n_i rule [12] for the input x_i. In the pole-balancing problem, the nearest four neighbors around each cell ($n_1 = 5$) are used for input x_1 and the nearest two neighbors around each cell ($n_i = 3, i = 2, 3, 4$) are used for all other input variables. The output mapping is obtained by a lookup rule table. To determine the output on jth cell, we first map the input at the jth cell and its neighborhood-n_i state configuration, to state q_j which is labeled by four indices ($I_i, i = 1, \ldots, 4$). The state index I_i for input x_i equals the number of 1-bits in n_i bits around the jth cell. The index I_i varies from 0 (all zero bits) to n_i (all 1 bits), and over all there are $\prod_{i=1}^{4}(n_i + 1)$ different states ($6 \times 4 \times 4 \times 4$ in this case). The four-dimensional state indices I_i can be mapped to a one-dimensional index R_j, the row number of the rule table, as follows:

$$R_j = \sum_{i=0}^{4} M_i I_i,$$

where

$$M_i = \prod_{j=2}^{5-i}(n_j + 1), \quad i = 2, 3, 4$$

and $M_1 = 1$. The output is obtained by copying the bit from the jth column and the R_jth row in the lookup table to the jth cell in the output bit array. After mapping to the output array on all 64 cells, the control force is calculated from the output array y,

$$f_c = f_{\max} \frac{N_y - 32}{32},$$

where $f_{\max}$ is the maximum control force and N_y is the number of 1-bits in y.

While the input 64-bit representation is a quantized representation of input variables in phase space, the state indices I_i label an even more coarsely quantized state in phase space, here the space is divided into only $6 \times 4 \times 4 \times 4$ states. The *next state mapping F* [7] in this problem is simply a mapping which maps the fine quantized input to a very coarsely quantized state in phase space, and the control action is defined on the coarsely quantized grids. For the input $x = \{x_i\}$, indices I_i are calculated for each of 64 cell positions within each 64-bit representation of the x_i. These indices range in value between 0 to $n_i + 1$ with some probability. For instance, an angle close to $\theta_{\max}$ yields a most likely value of 5 for I_1, and an angle close to $\theta_{\min}$ will most likely yield zero. A given input is mapped to 64 different states at each cell, labeled by (I_1, I_2, I_3, I_4) in state space, and states at each cell obey different local rules defined on each column at the rule table. These local rules provide the different control actions, and the average action is applied to the cart–pole system. It is not hard to show that the next state PDF $f(x, n, q)$ for a given input x usually has a peak at one of the states, say $(\bar{I}_1, \bar{I}_2, \bar{I}_3, \bar{I}_4)$. Therefore, in the statistical sense, for an input state which is close to a state labeled by $(\bar{I}_1, \bar{I}_2, \bar{I}_3, \bar{I}_4)$ in the

coarsely quantized phase space, the output is primarily determined by the average output on the row whose row number is calculated from $\bar{I}_i$. The control action defined on a very coarsely quantized space reduces the size of the rule table, and therefore enhances the learning speed.

In the simulation we will discuss below, we use the majority rule to calculate the average action on each row in the rule table. For the jth cell, the output bit is 0 or 1 according to majority rule on the R_jth row, i.e. if there are more ones on row R_j, the output is one and vice versa. Considering that the adaptive learning algorithm will make a stochastic CA converge to a deterministic one, the latter will have only a one or zero on each row in the rule table. The majority rule certainly enhances the converge rate dramatically. In simulation, we find the majority rule greatly improves the learning rate.

The bit-flipping learning algorithm [7] is performed either immediately after receiving the reward or punishment signal, or after some time delay. In the instant gratification reward/punishment reinforcement learning scheme, each bit in the rule table which has been used to determine the local output in the previous time step is rewarded or punished. In fact, the nearest l_r neighbors of those bits are set to the same value according to a given probability when a reward is given. When the controller receives a punishment signal, the activated bits and their l_p nearest neighbors are flipped to the opposite with some probability. By changing l_r, l_p and the bit-flipping probability, we can vary the learning rate. After reinforcement learning, the 64 columns in the rule table are randomly shuffled to guarantee the statistical homogeneity of the rules.

To simulate the motion of the cart–pole system, a fourth-order Runge–Kutta method is used to integrate the equation of motion

$$\ddot{\theta} = \frac{g \sin\theta - \cos\theta\left[(f_c + ml\dot{\theta}\sin\theta)/(M+m)\right]}{l\left[\frac{4}{3} - (m\cos^2\theta)/(M+m)\right]},$$

$$\ddot{z} = \frac{f_c + ml\left(\dot{\theta}^2 \sin\theta - \ddot{\theta}\cos\theta\right)}{M + m}, \tag{1}$$

where the control force f_c is provided by the stochastic CA. In all cases discussed below, we use a time step $\Delta t = 0.002$ s to integrate the equation

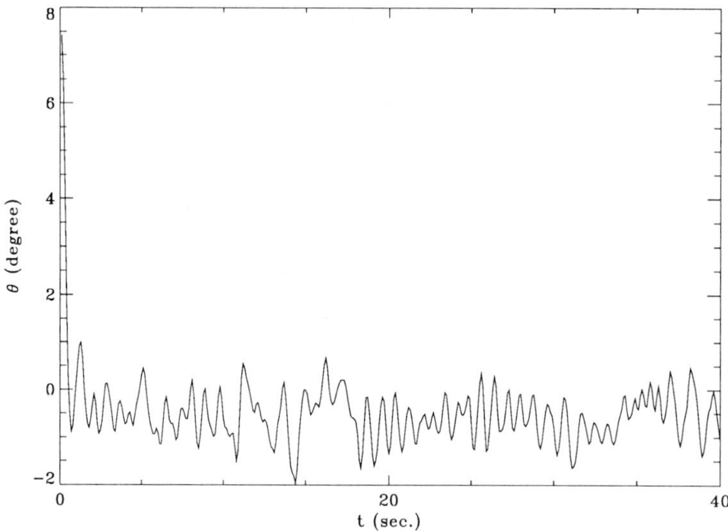

Fig. 1. Time history of angle θ.

of motion, and the reinforcement learning is performed at every 10 time steps (every 0.02 s).

In the first case studied, the stochastic CA only controlled the angle θ. It is a two-input (θ and $\dot{\theta}$) and one-output (force f_c) problem. The reward (punish) signal is given if the cost function $C = \theta^2 + \dot{\theta}^2$ decreases (increases) in each time step, when the angle from vertical is larger than $10°$. When the angle is less than $10°$, no reward/punish signal is given. We choose a small learning rate, $l_p = 0$, $l_r = 0$ and 30% of the time bit-flipping learning is performed. Fig. 1 shows the time history of the angle in a typical try. The initial angle is randomly chosen in the range $-10° < \theta < 10°$ and the angular velocity is chosen to be small. The initial position z and $\dot{z}$ are set to zero. For the case shown in fig. 1, the stochastic CA controller is able to balance the pole in the first attempt. Except for some severe initial conditions, the controller usually succeeds in balancing the pole in a few tries.

In a more difficult learning problem, the stochastic CA adaptive controller is used to balance a pole with a reward/punishment signal which is only given when the pole hits the boundary at $\pm 10°$. In this problem, the activated bits in the rule table are memorized by a second table (memory table). When a bit in the rule table is used for local output mapping, a bit at the same loca-

tion in the memory table is turned on. We wish the adaptive controller to have a stronger memory for a most recent action and faint memory for an action taken long ago. Therefore, we turn the 1-bits in the memory table off with some probability determined by a forgetting rate. When a failure occurs, learning is accomplished by penalizing all the past action indicated by the memory table by flipping the bits in the rule table corresponding to the bits in the memory table which are on. The stochastic CA learning model succeeds in balancing the pole after about one hundred failures.

One may notice that in fig. 1 the pole is stabilized at an angle slightly less than zero. Therefore, when the controller balances the pole, it also constantly pushes the cart in one direction. To control both angle and position, we need a four-input (θ, $\dot{\theta}$, z and $\dot{z}$) and one-output (control force f_c) stochastic CA adaptive controller. A critic then can warn the controller when either pole or cart hit the limit boundaries. This problem is much more difficult because in some regions in phase space, the cart–pole system is not controllable. For instance, when the cart almost reaches the boundary at $z = z_{\mathrm{max}}$ and at the same time the pole has almost reached the boundary at $\theta = \theta_{\mathrm{max}}$, the system is obviously not controllable. Therefore, to control a system with an uncontrollable region, we

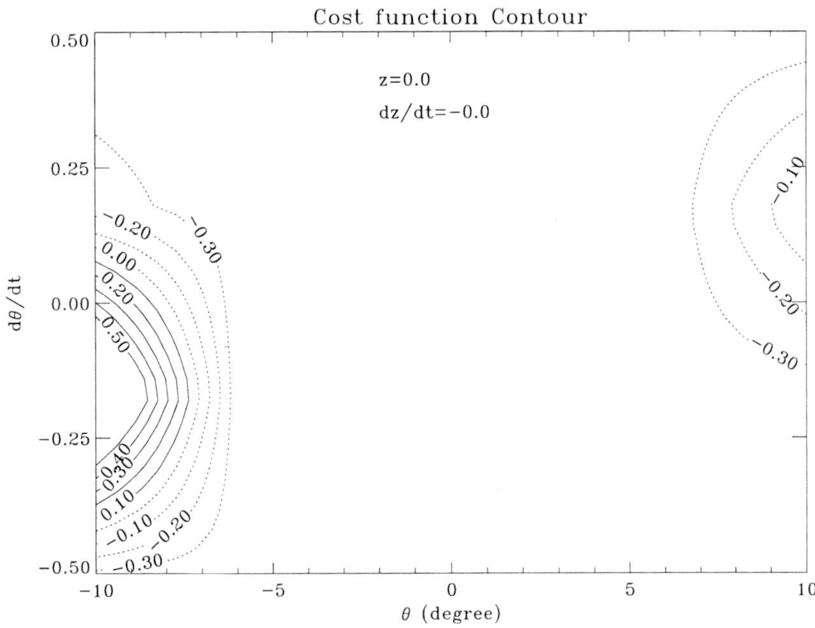

Fig. 2. Contour of cost function when $z = 0$ and $\dot{z} = 0$.

need an adaptive learning machine which can sort out and avoid the uncontrollable region, based on experience.

A stochastic CA learning model is designed to have that ability. We again use a memory table to memorize the previous action in the same way as discussed before. In this case, the memory table is not only used to punish past action, but also is used to construct a cost function. As we discussed before, each row in the rule table can be mapped to some position in phase space, a one bit in the memory table not only means that a local mapping may lead to failure but also draws out a very rough trajectory of the cart–pole system in phase space. Because of the forgetting rate, the positions in phase space immediately before failure are marked by more bits than those traversed earlier. Therefore, from the number of bits at each row in the memory table, we can assign some value to the cost function at that location in phase space. The value of the cost function at each location in the coarsely quantized phase space is then calculated by taking the average after many failures, and the cost function at any point within the coarse grid is calculated by linear interpolation. Fig. 2 shows a contour plot of the cost func-

tion at $z = 0$ and $\dot{z} = 0$. The cost function is calculated after the cart–pole system has hit the boundary several times.

The problem of controlling both angle and position is a four-input (θ, $\dot{\theta}$, z and $\dot{z}$) and one-output (control force F_c) control problem. In all tries, the cart–pole system starts from a random initial condition $-5° < \theta < 5°$, $\dot{\theta} = 0$, $z = 0$ and $\dot{z} = 0$. The cost function is initialized as a constant function and is recalculated each time when either the pole reaches 10° from vertical position or the cart reaches the boundary at $z = \pm 1$ meter. The performance of the stochastic CA is evaluated instantly according to whether the change of the cost function is positive or negative along the cart–pole trajectory. The learning algorithm is performed at each time step when a reward/punishment signal is given. Also when the cart–pole system hits the boundary, the past actions are penalized according to the memory table. After 200 failures, the stochastic CA successfully balances the pole. Fig. 3 shows a successful run where the final angle of the pole oscillates around zero degrees and the position of the cart oscillates around $z = 0.4$ m.

Although our use of the adaptive stochastic CA

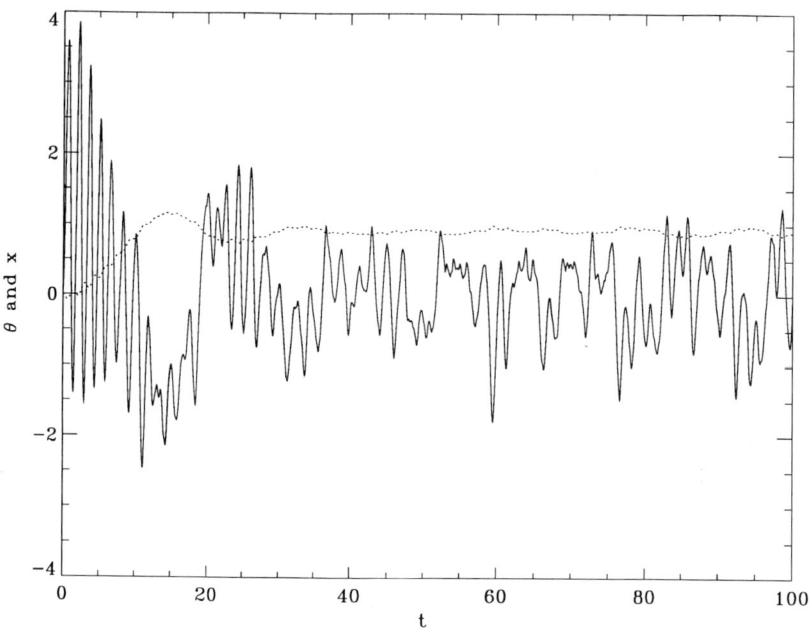

Fig. 3. Time history of the angle θ (solid line) and the position z (dashed line). The case shown is a successful run after 1000 failures. The angle θ is in the unit of degree and position z is in the unit of half pole length (0.5 m).

controller on real problems of interest is still in a preliminary stage, it already shows its power on adaptive learning in the cart–pole control problem. The stochastic CA is able to learn to balance a pole under more difficult conditions than assumed by Barto et al. For instance, the associative search element used by Barto et al. works only if the cart–pole system always starts from $\theta = 0$, $\dot{\theta} = 0$, $z = 0$ and $\dot{z} = 0$. When we train the stochastic CA, the cart–pole system starts from non-zero angles and angular velocities. When the stochastic CA has already learned to balance the pole and the learning is turned off, the controller balances the pole even if the pole starts from an angle every close to the boundary $\pm 10°$. We have not assumed any symmetry of the cart–pole dynamics system in our model. The stochastic CA learns the symmetry when both angle of pole and position of cart are controlled. In the next section, we show a more difficult control problem which the adaptive stochastic CA has successfully solved, balancing an inverted double pendulum.

3. Balancing an inverted double pendulum

The double-pendulum system that we want to control with the stochastic CA is similar to the cart–pole system studied in the previous section. In addition to one pole hinged on the cart, we hinge another pole on the top of the first pole. While balancing the single pole is a problem with one unstable degree of freedom (angle θ), balancing a inverted double pendulum is a problem with two unstable degrees of freedom (angle θ_1 and θ_2). If the double-pole system is still controllable by the force f_c along the track direction, the two poles must oscillate in sufficiently different time scales. To make the system easy to control, the first pole is chosen to be 10 times longer than the second pole; the characteristic time scale is about three to one.

The parameters used in simulation are: cart weight is 1 kg. The first pole weights 0.1 kg and is 10 m long, and the second pole weights 0.1 kg and is 1 m long. In this example, we only control θ_1 and θ_2. The input parameters then are:

θ_1 angle of the first pole with vertical,
$\dot{\theta}_1$ angular velocity of the first pole,
θ_2 angle of the second pole with vertical,
$\dot{\theta}_2$ angular velocity of the second pole.

The initial condition is chosen randomly with $-3° < \theta_1 < 3°$, $-5° < \theta_2 < 5°$, and some small angular velocities. The time step in reinforcement learning is 0.015 s. The input variables are coded into 64-bit string as in the single-pole case. The CA neighborhood state configuration used is neighborhood-two rule for angles and neighborhood-one rule for angular velocities. Therefore, we have about the same size rule table to train as before.

The reinforcement learning scheme is exactly the same as the scheme used in the single-pole case. The stochastic CA is rewarded or penalized according to the adaptive critic, and when either θ_1 or θ_2 reaches $\pm 10°$, a fraction of the previous actions are punished. After about two thousand failures, the adaptive controller is able to balance the inverted double-pendulum system for two hours simulated real time, which is the time limit in the simulation. However, the adaptive controller fails to balance the double-pendulum system for such a long time for all initial conditions, owing to the much more complicated dynamics. After using $1/\sqrt{t}$ cooling (of the learning rate), we are able to train the stochastic CA to control the double-pendulum system for most initial conditions, but often with a very long training time, however.

4. Concluding remarks

The simulation results for controlling single-pole and double-pole systems are still in early stages. More study of the reinforcement learning scheme and network architecture will be required to achieve faster learning speed and a more powerful control ability. For example, the cooling schedule in learning rate, the structure of the cost function and its learning scheme, the resolution in input coding (number of bits in input strings) and the resolution in state space (the number of columns of rule table) have been found to have an effect on the performance of stochastic CA.

From our experience on controlling the single pole, the constraint on position of the cart will not make the problem impossible to learn, but will make learning slower because of the need for a larger rule table. This is because the position z is not an unstable degree of freedom as are the angles θ_1 and θ_2. Since we use a force with only one degree of freedom to control a system with two unstable degrees of freedom, we must use a large ratio of pole lengths to separate the characteristic time scales of the two poles. Our results so far show that a ratio 8 to 1 is still controllable, but a system with a ratio of 6 to 1 cannot be balanced.

We suspect the dependence on initial conditions in the double-pendulum problem is due to very low resolution in state space in the present model. However, the low-resolution model has the advantage of fast learning and better statistics because of the small rule table. However, with a coarse representation it is difficult to make correct responses when the dynamics of the unstable system is very complicated. To maintain the advantages of coarse representation, we may design a controller with a multi-resolution layered hierarchy to handle complicated systems. The first layer with low resolution would learn roughly the correct action very rapidly, the higher-resolution layers would do fine adjustment to cope with complicated situations.

The pole-balancing problem presented in this paper illustrates the power of the adaptive stochastic CA. Compared to previous works on the pole-balancing problem, we made no assumption about the dynamics of the cart–pole system when we built the adaptive network. However, the stochastic CA still can learn to balance the pole after only a moderate number of training cycles. Moreover, the stochastic CA is able to control a much more complex system, such as the inverted double pendulum.

If we consider each row of the rule table as a neuron, the stochastic CA with a 384-row rule table used in the pole-balancing problem is not a very small network. However, because only bit operations are used in the learning algorithm, the learning is quite fast. For instance, 1000 trials only takes about 1 h of CPU time on a Sun SPARC-Station 1. Because the learning in the stochastic CA is a parallel process which involves only simple binary operations, it can be very easily implemented on hardware. We expect that a hardware implemented controller should be very fast and should be able to handle very large networks.

References

[1] A.G. Barto, R.S. Sutton and C.W. Anderson, Neuron-like adaptive elements that can solve difficult learning control problems, IEEE Trans. Systems Man Cybern. 13 (1983) 834–846.

[2] M.I. Jordan and R.A. Jacobs, Learning to control an unstable system with forward modeling, Preprint, MIT, Cambridge, MA (1990).

[3] B. Widrow and S. D. Stearns, Adaptive Signal Processing (Prentice-Hall, Englewood Cliffs, NJ, 1985).

[4] O. H. Mowrer, Learning Theory and Behavior (Wiley, New York, 1960).

[5] J. Spragins, Learning without a teacher, IEEE Trans. Information Theor. 12 (1966) 223–230.

[6] M.D. Waltz and K.S. Fu, A heuristic approach to reinforcement learning control systems, IEEE Trans. Autom. Control 10 (1965) 390–398.

[7] Y.C. Lee, S. Qian, R.D. Jones, C.W. Barnes, G.W. Flake, M.K. O'Rourke, K. Lee, H.H. Chen, G.Z. Sun, Y.Q. Zhang, D. Chen, and C. L. Giles, Adaptive stochastic cellular automata: Theory, Physica D 45 (1990) 159–180; these Proceedings.

[8] J.H. Holland, Adaptation in Natural and Artificial systems (The University of Michigan Press, Ann Arbor, MJ, 1975).

[9] J.H. Holland, Genetic algorithms and adaptation, in: Proceedings of the NATO Advanced Research Institute on Adaptive Control of Ill-Defined Systems, eds. O.G. Selfridge, E.L. Rissland and M.A. Arbib (Plenum Press, New York, 1983) pp. 317–333.

[10] D. Michie, and R.A. Chambers, BOXES: An experiment in adaptive control, in: Machine Intelligence, Vol. 2, eds. E. Dale and D. Michie (Oliver and Boyd, Edinburgh, 1968) pp. 137–152.

[11] D. Michie, and R.A. Chambers, 'Boxes' as a model of pattern formation, in: Towards a Theoretical Biology, Vol. 1. Prolegomena, ed. C.H. Waddington (Edinburgh Univ. Press, Edinburgh, 1968) pp. 206–215.

[12] S. Wolfram, Theory and Applications of Cellular Automata (World Scientific, Singapore, 1986).

Physica D 45 (1990) 189–202
North-Holland

EXTRACTING CELLULAR AUTOMATON RULES
DIRECTLY FROM EXPERIMENTAL DATA

Fred C. RICHARDS, Thomas P. MEYER and Norman H. PACKARD

*Department of Physics and Center for Complex Systems Research, Beckman Institute, University of Illinois,
405 N. Mathews Avenue, Urbana, IL 61801, USA*

Received 7 January 1990
Revised manuscript received 13 March 1990

We outline a method for extracting two-dimensional cellular automaton (CA) rules directly from experimental data. The data consist of discrete-space patterns evolving in discrete time. We employ a learning algorithm, the genetic algorithm, to search efficiently through a space of probabilistic CA rules for a local rule that best reproduces the observed behavior of the data. Included are the results of an analysis of patterns generated by the dendritic solidification of NH_4Br from a supersaturated aqueous solution.

1. Introduction

In this paper we address the problem of constructing models for two-dimensional spatial patterns directly from experimental data. We expand on earlier work reported in ref. [1], in which we dealt in a similar but simpler way with data generated from an artificial model. The data we analyze here are sequential patterns of dendrites formed by NH_4Br as it solidifies from a supersaturated solution. We consider a set of probabilistic cellular automaton (CA) rules as possible models for the data, and we search this space of rules with a learning algorithm – a variant of the genetic algorithm [2] – to find the rule which most accurately predicts the behavior of the evolving patterns. Included are some preliminary results.

The experimental analysis of complex spatial patterns is a problem only recently tackled in the literature [3–6]. Typically, one is presented with a two-dimensional pattern which evolves in time. Examples include satellite images of weather data (clouds, temperature, ozone, etc.) and patterns produced by chemical processes (the circular and spiraling waves of the Belousov–Zhabotinsky reaction [7] and the complex shapes of growing crystals [8]). These patterns often have a complicated global structure.

Recent studies of CA have shown that complicated global patterns can be generated by relatively simple local state-transition rules applied uniformly in space [9] [#1]. CA are also perhaps the simplest class of models to investigate using a digital computer. These two factors are what motivate us to consider CA rules as a class of models for spatially dynamic phenomena: complex spatial effects generated by very simple dynamical rules. When one attempts to model spatial dynamics with a CA rule, however, one assumes that the observed global structure can be regenerated with a rule that is local in both space and time. Since this may not always be the case in nature, we will describe later in this paper techniques which expand the class of CA models to incorporate effects that are more global in both space and time.

A CA rule maps the state of a given site on a discrete lattice to a future state. The future state is some function of the states of the sites in a neighborhood containing the given site. Therefore, in order to specify a CA rule one must specify not only the mapping function but also the structure of this neighborhood, or the structure of the input space. In order to specify the structure of the input

---

[#1] For a overview of CA refer to the collection of articles in ref. [10].

space of a CA rule we will employ the notion of a *template*. A template determines which of the sites in a neighborhood around a given lattice point are used to define an input state for the CA rule. We will let the experimental data determine the mapping function for any given template; the mapping function will be a probability histogram derived from the frequency of occurrence of each input-state/output-state pair found by viewing the sites of the experimental data through the given template.

As we shall see in section 3, the set of possible CA templates can number in the hundreds of millions. In order to find the template yielding a probabilistic CA which best describes the observed spatial dynamics, we must first establish some measure for ranking each template. We must also be able to search efficiently through the space of possible templates for those that rank highest. Both of these issues are discussed in detail in section 4. There we introduce a fitness function by which we can rank each template. The fitness function employs a measure of informational correlation between the input states of a given CA template and the output state. We then describe a variation of the genetic algorithm which is used to search through the space of possible templates for the ones that are most fit. Finally, in section 5 we discuss the results of applying this approach to model the dendritic solidification of NH_4Br.

2. The experimental data

In this section we will describe the dynamical patterns that we are modeling; we begin with a description of the patterns and then describe the experimental apparatus used to generate them.

2.1. Patterns

The data we wish to model consist of sequential images of NH_4Br crystals solidifying in a supersaturated aqueous solution [12,13]. This system exhibits at once many of the problems that will be encountered in trying to model spatial dynamics. One can see from the static image in fig. 1 that these crystals are dendritic in nature, and they generate complex shapes at different length scales.

The dynamic video images show growth occurring on different time scales as well. Also, the observational data do not provide access to all pertinent physical variables. Existing models for this process are based on equations of motion for the diffusion of the latent heat of solidification generated by the growing crystals [8]; but the video images provide no thermal information, just a record of the phase of the system at a given point.

The fourfold symmetry of the dendrites makes it a straightforward task to convert video images of the solidifying crystal to a sequence of two-dimensional square lattices. Because we are interested only in whether a given point in space is solid or liquid, the lattice will consist of binary-valued sites, where a site is zero (or black) if that point in space and time is solid, and a site is one (or white) if that point is in the liquid phase. We can make the further simplifying assumption that if a point is black/solid then it will remain so for the lifetime of the image (i.e. no melting occurs on this time scale).

2.2. Experimental apparatus

The dendritic NH_4Br patterns were generated in the following way: Start with an aqueous solution of NH_4Br saturated at $T_c \approx 45$–$50°C$, heat the solution above T_c, and sandwich a small amount between two pieces of glass. The glass plates – a pair of microscope slides – are separated by strips of mylar, 5 mil thick (13 μm), placed at the outside edges of the plates. These plates are surrounded by a rectangular box made of aluminum blocks (to add thermal inertia), which is

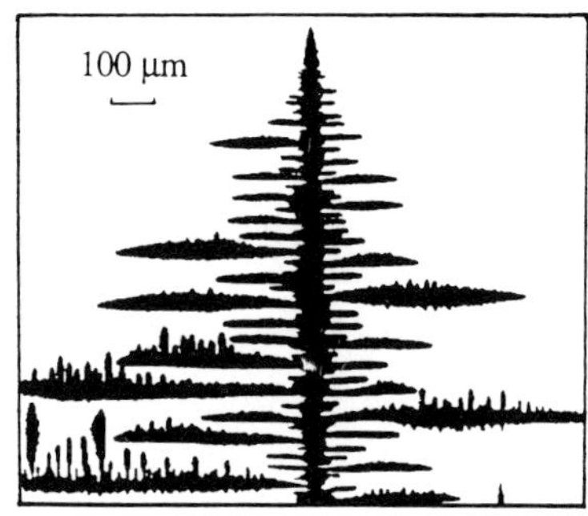

Fig. 1. A typical NH_4Br dendrite formed from a supersaturated aqueous solution.

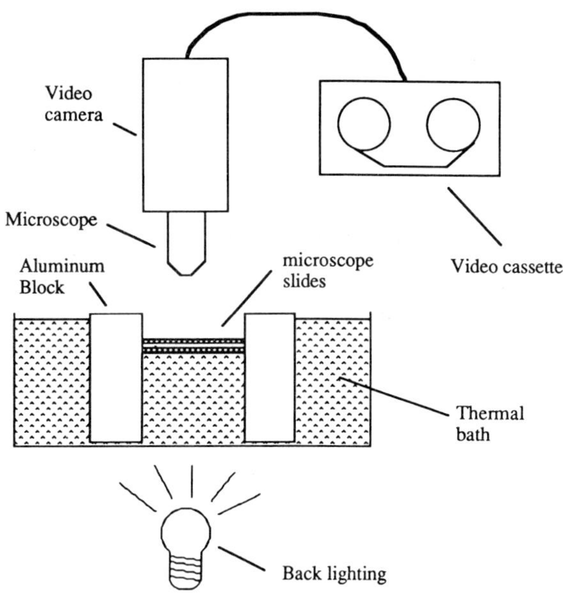

Fig. 2. A cross-section of the experimental apparatus. The supersaturated solution is placed between two microscope slides (separated by 5 mil mylar strips) which are supported in a four-sided aluminum box. The box is immersed in water so that water makes contact with the base slide but not with the cover slide. The image is recorded from above the slides (with backlighting) and saved with a video cassette recorder.

then placed in a water bath so that the bottom plate makes contact with the water bath and the top plate is open to the air (see fig. 2). The heated water bath is then transferred to the microscope stage and allowed to cool below T_c. When the water bath drops to about 5–8°C below T_c growth is initiated, typically by tapping on the cover slide.

The backlighting with the opaque crystals produces images that are primarily black (solid) or white (translucent liquid). The microscope eyepiece is removed and a video camera is mounted directly to the microscope so that the magnified image focuses directly on the video photo tube. We used a microscope lens with $6\times$ magnification for these experiments. The length scale has been indicated in fig. 1. The video images are saved using a standard video cassette recorder.

The video images are digitized using a video frame grabber. The frame grabber accepts video images that comprise 512×512 pixels, and each pixel is an eight-bit, gray-scale value. Operating at full speed, the frame grabber can record roughly 15 frames per second. The pixels of the images are converted to black or white, thus giving us sequential lattices of binary sites. The images are typically processed in the following way before being recorded in computer memory: sixteen pixels around each edge of the image are discarded so that only the central 480-by-480 pixels remain, and then the images are *zoomed* so that only every fourth pixel (in both the horizontal and vertical directions) is recorded, thus producing a 120-by-120 pixel image [#2]. A zoomed pixel represents the state of roughly a 10-by-10 μm region of space.

The image sampling rate depends on the growth rate of the crystals. We adjust the data sampling rate such that the tip velocity of the fastest dendrite is roughly one pixel per time unit. For the analysis presented here, the sampling rate was typically between one and two frames per second. In the following section we will discuss how we attempt to deal with the dynamics on different time scales in our CA model.

3. Nature of the CA rule space

A CA rule takes a pattern of discrete values over a spatial lattice to another (future) pattern, with a local map that depends only on the value of sites in a local region of the lattice [9,10]. In general, a CA rule can be expressed as

$$a_{\mathbf{x}}^{t+1} = \mathbf{\Phi}[\boldsymbol{y}_{\mathbf{x}}^{t}], \qquad (1)$$

where $\boldsymbol{y}_{\mathbf{x}}^{t}$ represents the value of the lattice sites in some neighborhood around site $a_{\mathbf{x}}$ at time t, and $\mathbf{\Phi}$ is a local map which takes $\boldsymbol{y}_{\mathbf{x}}^{t}$ to $a_{\mathbf{x}}$. A two-dimensional CA rule therefore has the form,

$$a_{i,j}^{t+1} = \mathbf{\Phi}[\boldsymbol{y}_{i,j}^{t}], \qquad (2)$$

where the indices, i, j, indicate the spatial position of the site. Since our data have been reduced to lattices of binary-valued sites, $a_{i,j} \in \{0, 1\}$.

[#2] Whether we selected every fourth pixel (in both the horizontal and vertical directions) or averaged over four-by-four pixel neighborhoods and then selected every fourth pixel as representative of the neighborhood did not greatly affect the outcome of the analysis. For this work we simply discard 15 pixels of each four-by-four pixel block and use just one pixel to represent the state of the entire neighborhood.

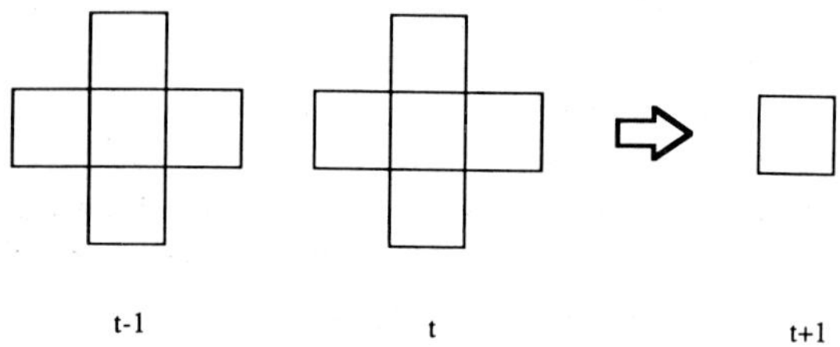

Fig. 3. A two-time-step CA rule. The future value of the site is determined from the present values of the four nearest neighbors and the past values of the four nearest neighbors.

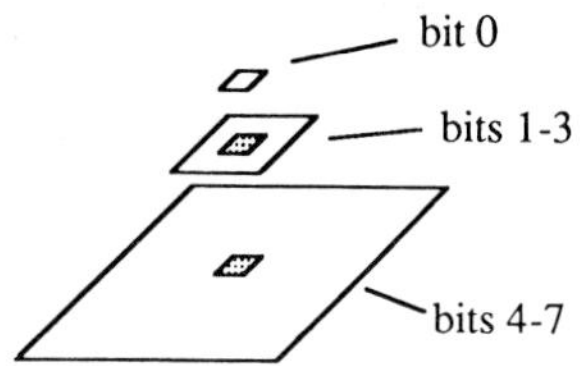

Fig. 4. The pyramid encoding scheme. The least-significant bit (bit 0) at each site contains the binary value of that site (black or white). Bits 1 through 3 are used to store the 3 most-significant bits of the 3×3 site neighborhood around each site. These values are then used to compute a sum for the 9×9 site neighborhood around each site. The 4 most-significant bits of this sum are stored in bits 4 through 7.

We will expand this definition of a CA to include dynamics on different times scales and different length scales. To incorporate the behavior of different time scales in our mapping function, we make the future state of a site depend on not just one past neighborhood configuration but two. In fig. 3 we see one example of such a rule: the site $a_{i,j}^{t+1}$ is a function of the state of the four nearest neighbors at time t (often referred to as the von Neumann neighborhood) as well as the state of the four nearest neighbors at time $t - 1$. Using this two-time-step CA rule we hope to capture dynamics that depends on past *dynamics* as well as past static information, and to capture dynamics that occurs on different time scales. In the context of dendritic solidification, information on past dynamics (i.e. past *change* of phase) may give us indirect information about latent heat of solidification in a given region. By including data from multiple past time steps we may also be able to model fast and slow interfacial growth.

In addition to incorporating information from different time scales in our CA rule, we will also expand our definition of $\mathbf{\Phi}$ to simultaneously incorporate information from different length scales. One noteworthy feature of the experimental data is the complexity of the patterns on different length scales: a dendrite forms a sidebranch, which in turn may form sidebranches, which may also form sidebranches, and so on (see fig. 1). We would like to choose a length scale where the behavior is dominated by the dynamics instead of noise, and yet the dynamics appears to occur on several length scales.

To capture dynamical effects on many length scales we represent the data in a "pyramid" form [11], where each level of the pyramid represents information about a different length scale. At the bottom level of the pyramid we store the site value. The next level contains information about the three-by-three site neighborhood around the site. The next level contains information about the nine-by-nine site neighborhood around the neighborhood, and so on. In practice, the amount of computer memory limits how high and how detailed the data pyramid is.

We now describe the encoding scheme used for the analysis of the NH_4Br data. At each site we retain the binary value, a single bit. We then calculate the three-by-three site sum around each site and save the three most-significant bits of this sum. Using these sums, we then calculate a nine-by-nine site sum with a four-bit resolution (see fig. 4). These sums are then converted to binary values using different thresholds for the different pyramid levels. Our definition of a CA rule has thus been expanded to

$$a_{i,j}^{t+1} = \mathbf{\Phi}[\boldsymbol{y}_{i,j}^{t}, \boldsymbol{y'}_{i,j}^{t}, \boldsymbol{y''}_{i,j}^{t}, \boldsymbol{y}_{i,j}^{t-1}, \boldsymbol{y'}_{i,j}^{t-1}, \boldsymbol{y''}_{i,j}^{t-1}],$$

$$(3)$$

where the different neighborhood configurations, $\boldsymbol{y}$, $\boldsymbol{y'}$ and $\boldsymbol{y''}$, are taken from the different levels of the data pyramid. Now $\boldsymbol{y}_{i,j}^{t}$ represents the values of sites taken from the first level of the pyramid for some neighborhood around $a_{i,j}^{t}$. The state vector $\boldsymbol{y'}_{i,j}^{t}$ represents the values of the sites in some other neighborhood around $a_{i,j}^{t}$, with the site values taken from the second level of the data pyramid. Finally, $\boldsymbol{y''}_{i,j}^{t}$ represents the values of the sites in another neighborhood around the site $a_{i,j}^{t}$, where the site values come from the third level of the pyramid. In other words, the future value of a given site, $a_{i,j}^{t+1}$, depends on neighborhood configurations around the site at time t and $t-1$, and these neighborhood configurations can be derived from representations of the data from many different length scales.

Our expanded definition of a CA rule can now be represented as

$$a_{i,j}^{t+1} = \mathbf{\Phi}[\mathbf{Y}_{i,j}^{t,t-1}], \tag{4}$$

where we now use $\mathbf{Y}$ to indicate that the site values of the neighborhood around $a_{i,j}$ can come from different levels of the data pyramid. In order to find a CA rule which best models a given two-dimensional dynamical pattern we must address two questions: which neighboring sites (in space and time) best determine the future state of a given site, and given an input configuration, how do we determine the future site value (i.e. what is $\mathbf{\Phi}$)? The local mapping function, $\mathbf{\Phi}$, will be determined empirically from the pattern data. Having determined which sites constitute an input state, $\mathbf{\Phi}$ is constructed by collecting a probability histogram. For every input state we record how many times $a_{i,j}^{t+1}$ is 0 (black) and how many times it is 1 (white). This probability histogram defines a probabilistic CA rule. Hence, for any given set of sites defining an input state there will be only one probabilistic CA rule defined by the observational data.

In order to construct a CA rule for the experimental data, we must first determine which sites make up an input state. Although we restrict our sites to binary states (at all levels of the pyramid), we still have many degrees of freedom to an input state. In order to specify a neighborhood around a site $a_{i,j}^t$, we must specify not only the spatial location of the neighboring sites (relative to i and j), but also the temporal location (t or $t-1$) and the pyramid location (level 1, 2, or 3). Neighborhood configurations will be defined by a template. In the following sections we will describe how to determine which sites to include in the template in order to construct the probabilistic CA rule which best models the observed behavior.

4. The genetic algorithm

The expanded definition of a two-dimensional CA rule (4) defines a space of possible models for the observed behavior. The size of this model space is determined by how many lattice sites we actually use in our input states (i.e. the template size). If the largest template has M sites, we can construct up to 2^M different templates that use no more than those M sites. Each of these templates can then be used to generate a CA rule from the experimental data. Clearly, we cannot construct probabilistic CA rules for each template in order to find the one that best describes the pattern dynamics. In order to search this space of models more efficiently we must introduce some set of operators to provide motion in the space of rules. For this we introduce the genetic algorithm, a machine-learning technique put forth originally by John Holland [2].

4.1. Outline

The genetic algorithm is a machine-learning technique used to find an optimal (or at least nearly optimal) solution to a problem given a large set of possible solutions. In general, the algorithm is as follows:

(i) Define a population; in our case the population is a set of CA rule templates. Each template is given a unique genetic label. We will establish the bounds of this template space below.

(ii) Select some small subset of the population at random.

(iii) Assign a "fitness" to each of these members, where the fitness is a measure of how well a member performs at solving the given problem.

(iv) Retain the fittest members and discard the least fit members.

(v) To the labels of the fittest members, apply the transformations operators, appropriately named the genetic operators, in order to produce new labels of other members in the population.

(vi) Assign a fitness to these new members, compare them to the previously selected members, retain the fittest, discard the least fit, and re-iterate the process.

Now we must address the following: (a) how do we represent the CA rules in a symbolic *genetic* form, (b) what do we mean by fitness, and (c) what are the genetic operators? These questions are answered in the following sections.

4.2. The genetic form of CA rules

We now define the concept of a *master template*. One can see from (4) that the future value of a single site $a_{i,j}^{t+1}$ can depend on many variables. On the other hand, it may turn out that the rule which best describes the experimental data depends on only a small subset of these variables. We use the template illustrated in fig. 5 to describe the full set of variables allowed in the input state of a CA rule. The future value of a site, $a_{i,j}^{t+1}$, can depend on any of the eight nearest-neighbor values from time t, or the four nearest-neighbor values from time $t-1$; it can depend on the values of the four nearest three-by-three neighborhood values (taken from the second level of the data pyramid) at either time t or time $t-1$; finally, $a_{i,j}^{t+1}$ may depend on the four nearest nine-by-nine neighborhood values (taken from the third level of the data pyramid) measured at either t or $t-1$. This master template defines a space of CA rules.

We can specify classes of rules within the rule space with a *subtemplate*. A subtemplate is made up of only sites from the master template that are used to determine the input states for a rule. The subtemplates, or rule classes, can then be labeled by 28-bit integers, where each bit corresponds not to a specific site *value*, but to whether or not that site is to be used in determining the input state. If a bit is equal to 1 then that site is included in the subtemplate, otherwise it is not. The 28-bit integer label therefore specifies the form of Y in (4). It will be helpful to think of these integer labels as analogous to gene patterns in biological organisms. Each template has a unique label, and by manipulating the bits of this label we generate a new template. As was stated above, the map which takes $Y_{i,j}^{t,t-1}$ to $a_{i,j}^{t+1}$ is determined by collecting a transition-probability histogram from the experimental data. Thus there is a one-to-one correspondence between a template and a probabilistic CA rule for any given set of experimental data.

Notice that the master template we have selected (see fig. 5) contains 28 sites, and hence we

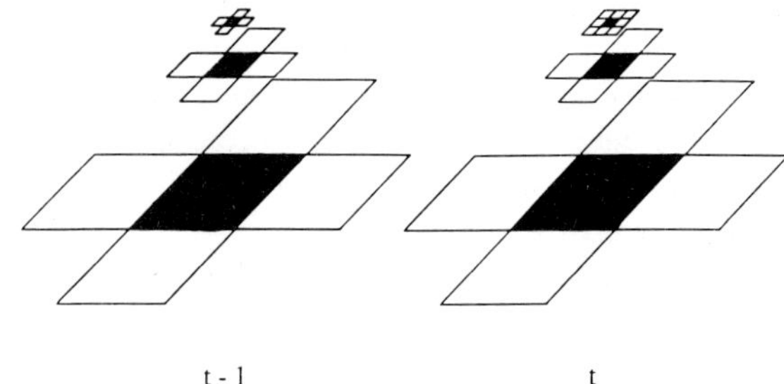

Fig. 5. The master template. Each block represents a binary value. The smallest blocks represent a single site value. The intermediately sized blocks represent the thresholded 3×3 site neighborhood sum, taken from the middle level of the data pyramid. The largest blocks represent the thresholded 9x9 site neighborhood sum. Because we are only considering neighborhoods around interface sites, the blocks colored black are not used.

may construct up to 2^{28} different subtemplates from it. Even given that each subtemplate defines only one probabilistic CA rule for a given set of observational data, there are still over 200 million different rules contained in our rule space. This rule space is too large to evaluate the performance of each individual rule. Instead we use the genetic operators to move us through the rule space in search of the optimal rule for the given experimental data.

4.3. The meaning of "fitness"

The fitness of a CA rule is a measure of how well that rule can regenerate the behavior observed in the experimental data. A fit rule would reproduce global and local features of the spatial pattern dynamics, but it need not necessarily reproduce patterns exactly. Under most circumstances, no CA rule will be able to reproduce exactly the behavior observed in the pattern data because the data provide only incomplete information about the observed phenomenon. For the example of solidifying NH_4Br dendrites, the video images do not tell us about the concentration of the NH_4Br in the solution, nor do they provide direct information about the temperature field around the dendrites, even though both variables may be crucial for a complete description of the physical phenomenon (see ref. [8]). Further, we have reduced what data we do have about the neighborhoods surrounding a given site first to a sum and then to a binary value. By choosing a master template that spans different length scales and different time steps, however, we hope to capture indirectly the effects of such phenomena as temperature and concentration fields and gradients.

Rather than evaluate rules based on their ability to regenerate global spatial structures in the experimental data, we rank our CA rules according to how much information their past and present sites contain on average about the future site value. We can quantitatively measure the informational correlation between the sites in a CA rule at t and $t-1$ and the state of a single site at time $t+1$. To do this we use Shannon's measures of information and mutual information [16].

Given a set of measurements of the state of some system, X, the information contained therein is

$$H(x) \equiv - \sum_i P(x_i) \log P(x_i), \tag{5}$$

where x is a state variable, $P(x_i)$ is the probability of the variable x being in the discrete state i, and the sum is taken over all possible states of x. For a system with N possible states, where we are equally likely to observe each state (probability $= 1/N$), the information associated with the state variable is $\log N$. The base of the logarithm is typically 2, so the units of information are bits.

Mutual information is a measure of how two variables (or symbol sequences) are correlated. The mutual information between two state variables x and y is defined by

$$I(x,y) \equiv \sum_{i,j} P(x_i,y_j) \log \frac{P(x_i,y_j)}{P(x_i)\,P(y_j)}, \tag{6}$$

where $P(x_i, y_j)$ is the joint probability, or the probability of finding the variable x in state i and simultaneously finding y in state j. An alternate definition of mutual information is

$$I(x,y) = H(x) + H(y) - H(x,y), \tag{7}$$

where now $H(x,y)$ is the joint information (i.e. the information gained by measuring x and y simultaneously).

We wish to measure the mutual information between the sites of a rule subtemplate and the future site value. For our application, the future

value of a site is $y \in \{0, 1\}$. We let x represent the state of the sites in the subtemplate of the CA rule. If a rule uses m cells of the master template, and each cell is reduced (using some threshold) to a binary value, the variable x can have as many as 2^m unique values. If the mutual information between x and y is high for a given rule subtemplate, we say that the sites of the rule are strongly correlated with the future value of the central site. The fitness measure of a CA rule should be directly proportional to this mutual information.

We stated above that if a rule uses m cells of the master template then there are 2^m different configurations, or states, of the past neighborhood around a site. Clearly, as the template size m increases, the number of possible states for a template grows exponentially. Since we are approximating probability distributions by measuring the frequency of occurrence of the different states, our data set must also grow exponentially for this approximation to be valid. The mutual information is *over-estimated* if the number of measurements is small compared to the number of possible discrete states [#3]. If we make N pairs of measurements on a system, and there are 2^m equally probable outcomes for each pair (i.e. the data are random), the overestimation of the mutual information will be $2^m/N$ [17]. Because the information generated by a random system is maximum, we use this factor as an upper bound for the overestimation of in the mutual information between any two finite sets of data.

The fittest CA rule should be the rule with the highest informational correlation between the subtemplate sites and the future site, while at the same time the rule should not contain so many sites that the statistics are so poor that the mutual information is artificially high. To account for both of these effects we introduce the following fitness function:

$$F = I(x,y) - \frac{2^m}{N}, \tag{8}$$

where again, m is the number of cells in our rule

[#3] As an intuitive example of this, considering tossing two dice. The statistics of just one toss indicate that the dice are completely correlated. As further tosses are performed, however, this apparent correlation quickly goes away.

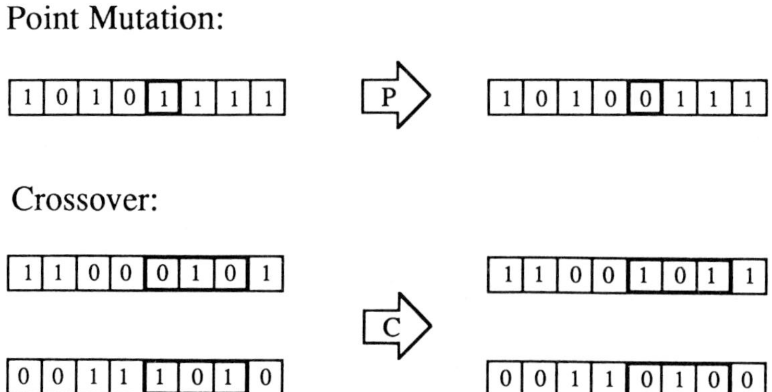

Fig. 6. The genetic operators, point mutation and crossover. Point mutation involves making a small change in just one rule, in this case flipping a bit in the label, which corresponds to adding or deleting a cell from the rule template. The crossover operator takes the labels from two rules and splices them together to form two new rule labels. A bit=1 in a rule label indicates that the corresponding site from the master template is used to construct the rule, a bit=0 means the site is ignored. Notice that these hypothetical labels would represent a master template with only 8 sites.

template, N is the number of experimentally determined data points, y represents the state of the future site variable, and x represents the state of the sites in the subtemplate. We can imagine constructing a CA rule template by starting with one site ($m = 1$) and then adding cells from the master template. As long as each new site provides new information about the state y, the fitness function will increase. It will continue to increase until the number of sites m is so large that the additional mutual information is smaller than the overestimation due to having finite statistics. The fitness of a CA rule is measured by scanning the pattern data with the appropriate subtemplate and collecting probability histograms for $P(x)$, $P(y)$ and $P(x,y)$. These probability histograms are then used to compute $I(x,y)$, which is taken with the template size m and the data-set size N, to calculate F.

4.4. The genetic operators

Having defined the space of possible models for our pattern data and a notion of fitness by which we can rank rules, we must now efficiently search through this space to find the fittest model. In this section we introduce motion operators for the space of CA rule templates. These operators must be closed with respect to the space of rules, and they should transform the rule labels, which are integer representations of the subtemplates, from one value to another.

For our application we consider two unique genetic operators: the point-mutation operator, and the crossover operator. The names and functions of these operations are derived from the analogy to genetic evolution in nature. Point mutation corresponds to changing just one bit in a rule label in order to generate a different rule label, or select a new rule. Remember that these bits correspond to sites in the master template. If a bit is one then that site is used in the input configuration of the given rule, otherwise it is not. Changing the value of a bit is equivalent to adding or taking away a site from a given subtemplate. The crossover operation involves taking the labels of two rules and replacing a segment of one with the corresponding section of the other. With the crossover operation we are combining part of the template from one rule with part of the template from another in the hopes of generating a better rule (see fig. 6).

We can think of the genetic operators as providing motion in the space of CA subtemplates. Imagine that we assign a fitness value to every template, and hence to every probabilistic CA rule, in the space delimited by the master template. The fitness measure defines a landscape in some high dimensional space. Fig. 7 is a schematic representation of a one-dimensional fitness landscape. The point mutation operator produces small changes

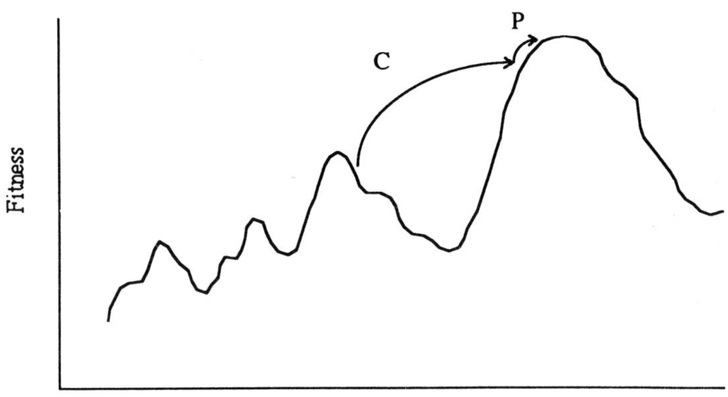

Fig. 7. A one-dimensional sketch of a fitness landscape. The genetic operators generate translations along the horizontal axis: crossover (C) provides large-scale steps (long jumps), and point mutation (P) provides local movement.

in the rule label. Since we only retain changes that result in an increase in the fitness of rules, the corresponding motion on the fitness landscape is a local hill climbing action. The crossover operation, however, can generate rule labels that are quite different from any in the original subpopulation (analogous to the gene pool), and hence this operation provides a long-jump mechanism on the fitness landscape. Using the two operations in combination we hope to efficiently find the region of the global maximum in the fitness landscape and then climb up to that local peak. There is no guarantee, however, that the resulting rule will be the absolute optimum. The success of the genetic algorithm depends on the nature of the fitness landscape. If the landscape is very jagged and has many peaks that are nearly the same height, the algorithm is not as likely to find the fittest rule. If, on the other hand, the fitness landscape is very smooth and contains just one large peak then the probability is much higher that the genetic operators will move the population to this point.

Our description of the genetic algorithm is now complete. We have defined a space of templates and hence a space of probabilistic models for the two-dimensional pattern data using the concept of a master template. The members of this population are specified by subtemplates, and each subtemplate has an integer label which specifies which sites of the master template are used to make up that subtemplate. We have provided a measure of fitness for the members of the population. Finally, we have introduced the genetic operators, point mutation and crossover, which manipulate rule labels to provide motion in the space of templates. By following the steps outlined above, we can move through the given space of two-dimensional probabilistic models (defined uniquely by the subtemplates) in search of the one which best describes the observed behavior of the solidifying NH_4Br dendrites.

5. Results

In ref. [1] we report on the results of applying the genetic algorithm to patterns generated by a one-time-step, twelve-nearest-neighbor CA rule. The master template (i.e. the rule space) used for this study contained only single-site cells and did not make use of any pyramid encoding scheme for the training data. In this section we present some preliminary results obtained for the NH_4Br solidification process.

The results described below were obtained from a data set consisting of approximately 10^3 images of solidifying NH_4Br, sampled at a rate of roughly two frames per second. The sequences of images provided a total of about 8×10^5 data points with which to construct the probability histogram, or

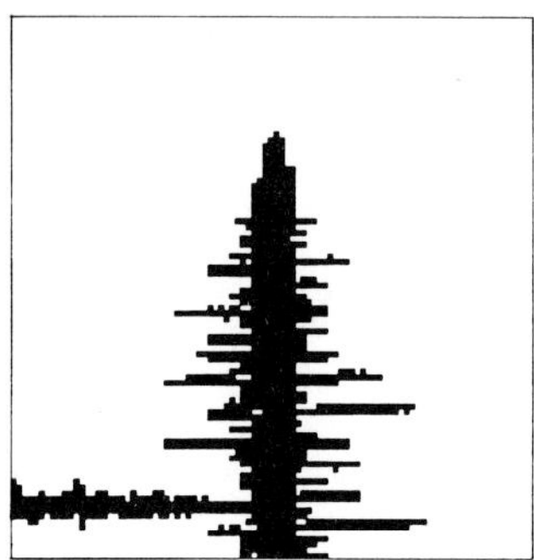

Fig. 8. An NH$_4$Br dendrite as the computer "sees" it. This particular pattern is the latter half of a two-time-step seed, or initial condition for growth.

roughly 8×10^2 interface points per frame [4]. The data were converted to a binary form for the genetic algorithm using several different threshold values for the different levels of the pyramid. In general, the best rules were produced when both of the two thresholds were roughly halfway between their respective minima and maxima.

When the genetic algorithm finished its search of the rule space, the final 10 subtemplates all had roughly the same fitness measure. Each subtemplate contained 13 of the 28 cells of the master template. All but one of the sites of the master template were used in at least one of the fittest subtemplates. In general, the same spatial cell was not selected from both past time steps in any given subtemplate.

In fig. 8 we see a snapshot of a dendrite as it grows from left to right. This pattern (and a previous image not shown) acts as the seed for the learned CA rule. In fig. 9 we show three different patterns derived from the seed in fig. 8. The rule that generated the pattern in fig. 9a was found in one of the early generations of the learning process. Fig. 9a resulted from applying the not-so-fit rule to the initial conditions of fig. 8 for 36 steps of evolution. Fig. 9b was generated by the fittest rule that the genetic algorithm found. Again, the rule used the seed pattern of fig. 8 as an initial condition, and fig. 9b was the resulting pattern after 36 time steps. Fig. 9c is the true shape of

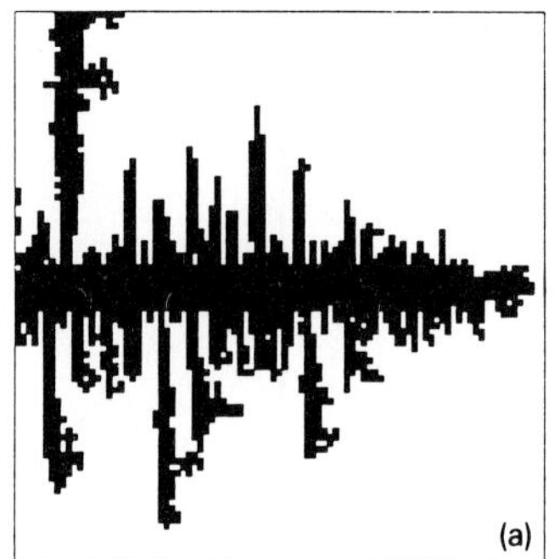

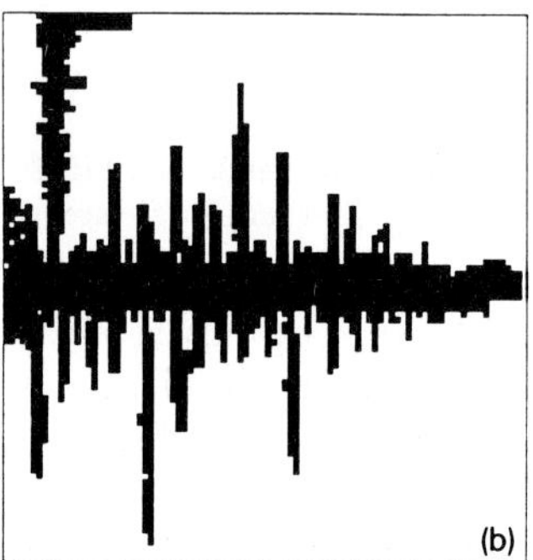

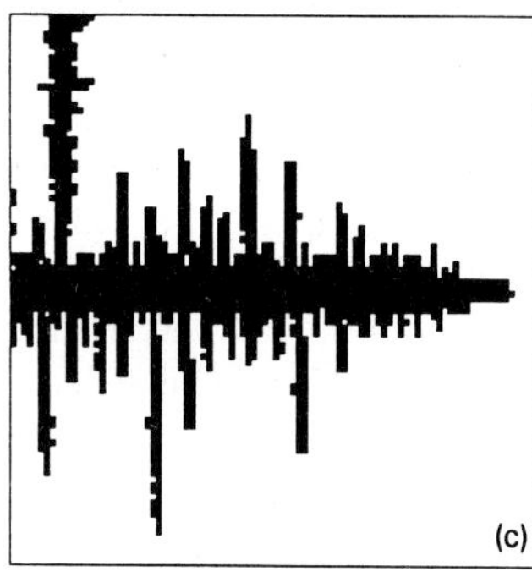

Fig. 9. The growth that follows 36 time steps after the initial conditions shown in the previous figure. (a) Result of an "unfit" CA rule; (b) result of the fittest rule found by the genetic algorithm, and (c) how the real crystal looked after 36 time steps.

the dendrite 36 time steps after the seed pattern was recorded.

In fig. 10 we show the patterns produced by the unfit rule, the fittest learned rule and nature 72 time steps after the seed pattern. Note that figs. 10a and 10b do *not* follow from fig. 9a and 9b respectively, but rather were generated independently from the same seed (fig. 8).

[4] An interfacial site is one for which the site itself is white (liquid) while at least one of its four nearest neighbors is black (solid).

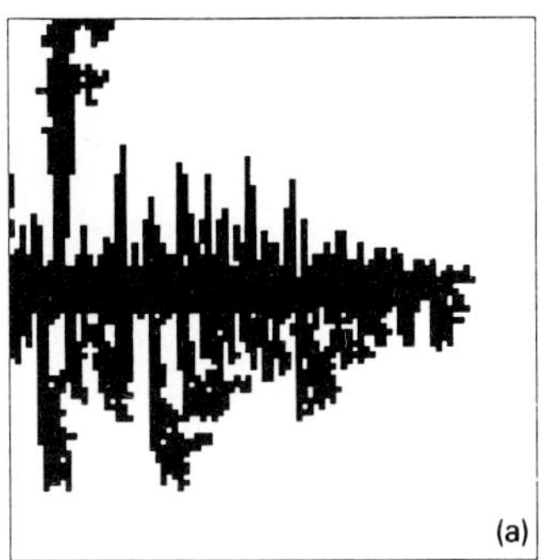

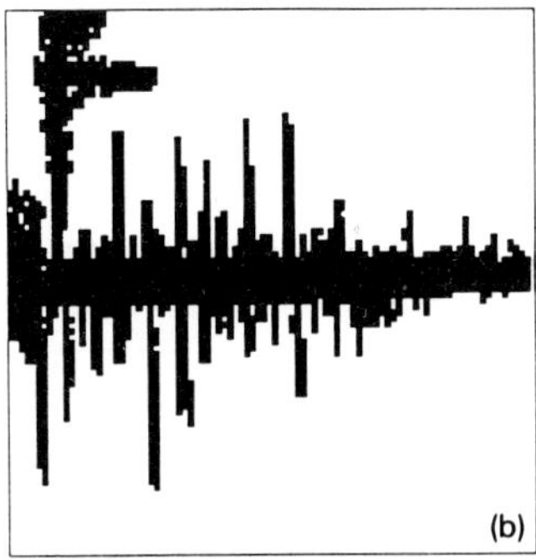

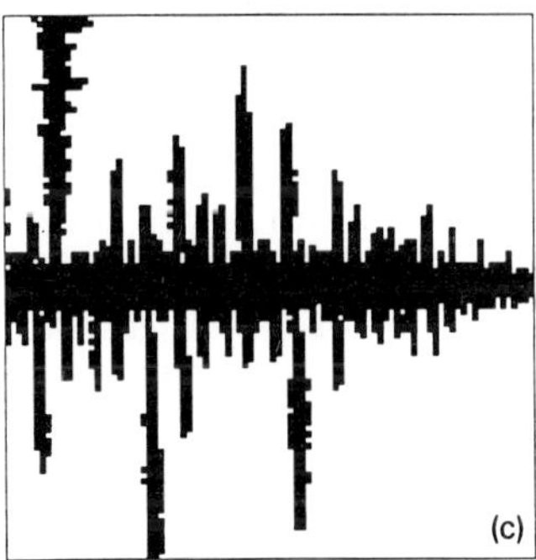

Fig. 10. The growth that follows 72 time steps after the initial conditions shown in fig. 8. (a) was generated by an unfit rule, (b) by the fittest rule found, and (c) is the actual dendrite after 72 time steps.

6. Analysis

The results as illustrated in figs. 9 and 10 indicate that the CA rule selected as optimum is improving over successive iterations of the genetic algorithm (witness fig. 10a versus fig. 10b). We see that although the optimal learned CA rule does not generate the exact same results as does nature, the CA rule is starting to capture some of the salient features of the crystal growth, such as the sidebranching behavior. By comparing the structure of the pattern in fig. 9b with fig. 10b we see that the same rule applied to the same initial conditions does not even regenerate the exact same global structure (compare, for example, the secondary horizontal sidebranch in the upper left corner of fig. 9b, produced after only 36 iterations, not present in fig. 10b after 72 iterations). This sensitive dependence of the global structure to small-scale fluctuations suggests the spatiotemporal dynamics is chaotic, and that no exact quantitative comparison between predictions and observations is possible.

We can see the trend towards improving fitness of the rules even more clearly if we look at fig. 11. Here we plot the fitness of the best rule found by the learning algorithm against the number of generations of rules that have been investigated. Also plotted is the fitness of the best rule found when using just crossover operations and just the point mutation operator. What we typically see for the NH_4Br data is that, except for the initial few generations, the search is expedited most by the mutation operator. One possible conclusion from this is that the fitness landscape has many peaks that are roughly the same height. These peaks are discovered most efficiently with the hill-climbing action of the mutation operator. If the landscape had just one very large peak but many smaller peaks, we would expect to need more crossover operations to find the general region of this global maximum before applying the mutations to take us to the top.

To quantitatively rate the performance of the learned CA rule we should compare the results of the rule's predictions to the actual crystal growth for a single time step. To do this we use a correlation coefficient which takes into account not only the correct positive (growth) and negative (no growth) predictions, but the false positive and false negative predictions as well. The number of times that a 1 is predicted correctly we call p, while the number of times that a 1 is predicted incorrectly is $\tilde{p}$. The number of times that a 0 is predicted correctly we label by n, while the number of incorrect predictions of 0 is $\tilde{n}$. The following correlation coefficient [18] gives us a measure of how well a rule predicted what actually occurred:

$$C = \frac{pn - \tilde{p}\tilde{n}}{[(n + \tilde{n})(n + \tilde{p})(p + \tilde{n})(p + \tilde{p})]^{1/2}} . \tag{9}$$

If the predicted behavior is always correct, $C = 1$. If the predictions are always incorrect, $C = -1$. If our predictions are completely random, $C = 0$. When we compare the predictions of the learned rule (used to generate figs. 9b and 10b) to the actual growth for a single-time-step prediction we find that $C = 0.163$.

In fig. 12 we see how the correlation coefficient varies with the range of our predictions. If we predict the shape of the dendrite (for various initial configurations) just one time step into the future, we find $C = 0.163$. If we use the learned CA rule to predict what the shape of the dendrite will be two steps into the future, we find the correlation between our predictions and reality *rising* to $C = 0.253$. We see from fig. 12 that when we try to predict even further into the future we witness a steady decline in the accuracy of our predictions. It is not until we try to predict the dendrite shape eight time steps into the future, however, that C falls below 0.163.

One possible explanation for the decreasing trend in fig. 12 is that the dendrite pattern dynamics is inherently chaotic. The probabilistic dynamical rule and low-resolution data introduce noise into the system that is amplified over time. We see

Template Evolution

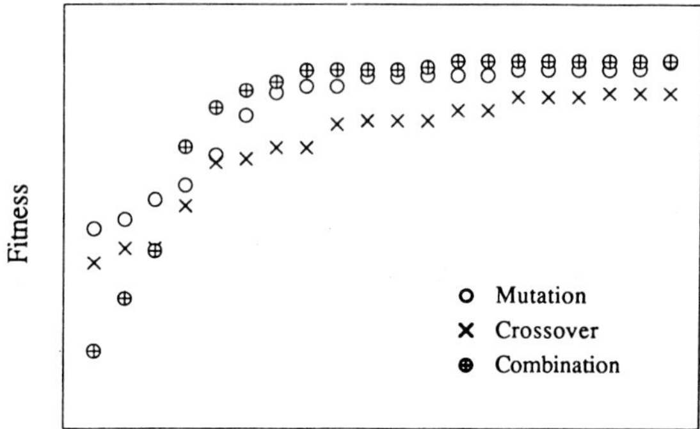

Fig. 11. The evolution of the population fitness. The crosses show the evolution of the population when just the crossover operation is applied. The circles show the fitness of the population when just point mutations are used by the genetic algorithm. The circles-around-crosses show the fitness using a combination of crossover and mutation operations.

a sensitive dependence to initial conditions in the learned rule; the CA rule does not generate the same patterns from the same initial conditions. If the dynamics is inherently chaotic this would also be true for a completely deterministic CA rule.

The initial rise in C seen in fig. 12 we attribute to the inability of the template to capture enough temporal information. Dynamic images of the growing dendrites show the fastest interfacial growth is at the dendrite tips, a small percentage of the total interface. The majority of the interface grows at a much slower rate. We set our sampling rate so that we maintain continuity of the interface at the fastest growing tips. By doing this, however, deterministic dynamics of much of the rest of the interface can only be described probabilistically. To illustrate this we invoke the following intuitive argument. Suppose that we want to model the interface behavior with a single-time-step CA rule. Suppose also that we have an input state which is certain to map to a 1 (solid) at a given point, but only after two time steps. In other words, the growth velocity of that interface configuration is 0.5 sites per time step, and the growth probability for this input state is zero after one step and one after two time steps. If we try to describe this behavior with a single-time-step CA rule, the best we can do is predict solidification with a probability of 0.5. The probability of correctly predicting the no-growth condition after one time step is 0.5. However we know that the rule will generate two successive no-growth predictions with only a 0.25 probability, therefore the probability of correctly having predicted growth after two time steps is 0.75. Our predictions for such an input state would be more accurate on average if we make two sequential predictions before comparing results to the real behavior. We claim that this time-averaged increase in accuracy occurs for our probabilistic CA predictions – for short-time predictions where there is not enough time for the dynamics to amplify the noise – thus giving us the initial upward trend in fig. 12. The implication is that we must better account for the multiple growth rates present in the NH_4Br data if we wish to improve the accuracy of our model predictions. We could do this by either adding another past time step to the master template, or by increasing the temporal separation between the past

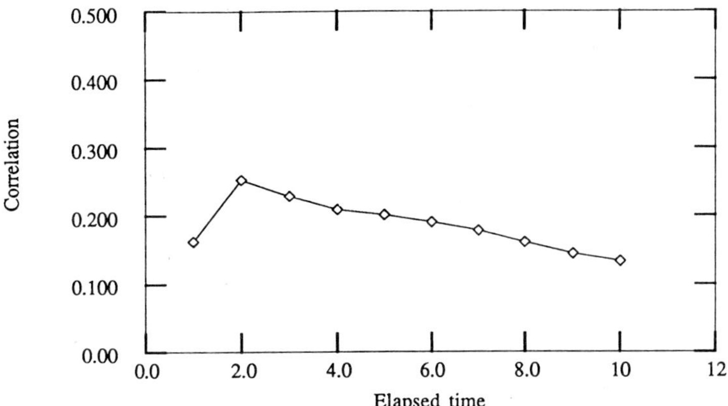

Fig. 12. The correlation between predictions and reality after successive predictions. The first point is the correlation between the predicted growth and real growth after just one time step. The second point is the correlation between real and predicted growth after two time steps of independent growth, and so on.

and present data in the existing master template.

7. Conclusions

We have presented a method for extracting spatial models directly from experimental data. The learning algorithm starts with no *a priori* knowledge about the physical system and builds models that regenerate the observed behavior.

The results shown for dendritic NH_4Br data are encouraging. This is a system that exhibits at once many problems that will be typical of different spatial systems. There is pattern structure on many different length scales and there is dynamics on different time scales (some parts of the interface change rapidly while others change little or none). Information about the physical variables is limited, as is often the case with complex spatial data from physical systems. While we can see whether a point in space is liquid or solid, we know nothing about the solute concentration field or the temperature field in the liquid. In spite of this the learning algorithm was able to find a CA rule which reproduced qualitatively the pattern dynamics of the training set.

The exact relationship between the learned dynamical rule and fundamental equations of motion for solidification is an open question. This question may be in principle impossible to resolve. Though some results may be obtained from the equations of motion [8,15,14], and even tested experimentally [12,13], our data represent patterns on length scales roughly two orders of magnitude larger than the length scales in the equations of motion. The equations of motion are extremely nonlinear, and there is little hope for closed form solutions that capture the full observed complexity. In this sense, it may be impossible to "derive" the observed rules from the fundamental equations of motion. It may be possible to connect certain features of the learned rule to phenomenological models [19], even if derivation from equations of motion is impossible.

We propose that this type of "derivability gap" is the rule, rather than the exception for most complex spatial patterns observed in nature. For such phenomena, it may be impossible to derive models which explain observed spatiotemporal complexities directly from fundamental equations and "first principles". Though perhaps underivable, the dynamical structure extracted by the learning algorithm is undeniable, and represents a new type of progress, perhaps the primary kind of understanding possible for complex patterns.

Acknowledgements

FCR would like to thank Annette Matheny for her assistance with the experimental apparatus. The data were digitized using hardware from the Datacube Corporation, and the learning algorithm was run on a Sun 3/110 workstation with 4 megabytes of memory. This work was supported by the Sloan Foundation, the National Science Foundation Grant No. NSF Phy 86-58062 and the Office of Naval Research, Grant No. N00014-88-K-0293.

References

[1] T.P. Meyer, F.C. Richards and N.H. Packard, Learning algorithm for modeling complex spatial dynamics, Phys. Rev. Lett. 63 (1989) 1735–1738.

[2] J. Holland, Adaptation in Natural and Artificial Systems (Univ. of Michigan Press, Minneapolis, MI, 1975).

[3] J.P. Crutchfield and K. Kaneko, in: Directions in Chaos, ed. Hao Bai-lin (World Scientific, Singapore, 1987).

[4] J.A. Vastano and H.L. Swinney, Information transport in spatiotemporal systems, Phys. Rev. Lett. 60 (1988) 1773–1776.

[5] S. Ciliberto and J.P. Gollub, Pattern competition leads to chaos, Phys. Rev. Lett. 52 (1984) 922–925.

[6] Y. Sawada, A. Dougherty and J.P. Gollub, Dendritic and fractal patterns in electrolytic metal deposits, Phys. Rev. Lett. 56 (1986) 1260–1263.

[7] A.M. Zhabotinsky and A.B. Rovinsky, Mechanism and nonlinear dynamics of an oscillating chemical reaction, J. Stat. Phys. 48 (1987) 959–975.

[8] E. Ben-Jacob, N. Goldenfeld, J.S. Langer and G. Schön, Dynamics of interfacial pattern formation, Phys. Rev. Lett. 51 (1983) 1930–1932.

[9] J. von Neumann, Theory of self-reproducing automata, ed. A.W. Burks (Univ. of Illinois Press, Champaign, IL, 1966).

[10] S. Wolfram, ed., Theory and Applications of Cellular Automata (World Scientific, Singapore, 1986).

[11] P.J. Burt, Computer graphics and image processing, Comput. Graph. Image Processing 16 (1981) 20.

[12] A. Dougherty and J.P. Gollub, Steady-state dendritic growth of NH_4Br solution, Phys. Rev. A 38 (1988) 3043–3053.

[13] A. Dougherty, P.D. Kaplan and J.P. Gollub, Development of side branching in dendritic crystal growth, Phys. Rev. Lett. 58 (1987) 1652–1655.

[14] O. Martin and N. Goldenfeld, Origin of sidebranching in dendritic growth, Phys. Rev. A 35 (1987) 1382–1390.

[15] F. Liu and N. Goldenfeld, Linear stability of needle crystals in the boundary-layer model of dendritic solidification, Phys. Rev. A 38 (1988) 407–417.

[16] C.E. Shannon and W. Weaver, The Mathematical Theory of Communication (Univ. of Illinois Press, Urbana, IL, 1962).

[17] W. Li, Mutual information functions versus correlation functions, Center for Complex Systems Research Technical Report CCSR-89-1, University of Illinois, Urbana (1989).

[18] B.W. Mathews, Biochim. Biophys. Acta 405 (1975) 442–451.

[19] N.H. Packard, Lattice models for solidification and aggregation, in: Science on Form, eds. S. Ishizaka, Y. Kato, R. Takaki and J. Toriwaki (KTK Scientific Publishers, Tokyo, 1986).

CHAPTER 4

CELLULAR AUTOMATA AND THE NATURAL SCIENCES

Physica D 45 (1990) 205–207
North-Holland

WHAT CAN AUTOMATON THEORY TELL US ABOUT THE BRAIN?

Jonathan D. VICTOR

*Department of Neurology, Cornell University Medical College, 1300 York Avenue, New York, NY 10021, USA
and Laboratory of Biophysics, The Rockefeller University, 1230 York Avenue, New York, NY 10021, USA*

Received 13 September 1989
Revised manuscript received 11 January 1990

The potential contribution of the theory of cellular automata to understanding of the brain is considered.

1. Introduction

How the brain works is one of the fundamental problems in biology. Despite dramatic advances in understanding at the molecular and cellular level, a number of basic issues remain. These include, for example, the mechanism for the determination of neural connectivity by genetic and epigenetic factors, and the basis for the difference between the intellectual capacity of the human brain and that of other species. Satisfactory answers require integration of understanding at multiple levels of structure into a coherent whole. Thus, a theoretical framework, as well as experiment, is required. But what can one reasonably expect from theory, and what is its relationship to experimental investigations? With these questions in mind, I would like to outline areas of inquiry in which I believe that theoretical insights will make significant contributions towards understanding the brain.

The best place to begin is to consider the landmark contribution of von Neumann [1]. Von Neumann constructed a cellular automaton capable of universal computation and self-reproduction. This construction demonstrated that a small set of local rules acting on a large repetitive array can result in a structure with very complex behavior. The von Neumann construction thus immediately suggests how an organ with behavior as complex as the brain's can be specified from limited genetic information.

That each unit of the von Neumann automaton had twenty-nine states and four neighbors is inessential to its biologic import; what is important is that the construction could be done at all. The property of universal computation is only indirectly important: we do not need to think of the brain as a universal Turing machine in order to apply the cellular automaton metaphor; universal computation is simply a rigorous way to guarantee that the von Neumann construction has behavior that, all would agree, is complex. The property of self-reproduction is relevant in a similar fashion, in that it is a rigorous way to guarantee a rich behavioral repertoire.

The observation that the cerebral cortex is composed of a large number of local neural assemblies that are iterated throughout its extent, by itself, is not an existence proof that complex behavior may result from a network of simple elements. Von Neumann's construction is necessary to show that such a structure is indeed capable of complex behavior, *without the need to invoke* region-to-region variability, long-range interactions, stochastic components, or mysticism.

Whether the von Neumann metaphor is qualitatively correct is open to question. However, even if we agree that it does capture an essential feature of brain organization, there is more to be done than to turn the metaphor into an allegory: several challenging areas of theoretical inquiry remain.

2. The problem of robustness

One major qualitative difference between the behavior of the von Neumann construction and that of the brain is that of robustness, or stability to perturbation. Alteration of the state of a single unit of the von Neumann machine typically leads to catastrophic failure; malfunction of a single neuron or neural assembly should have no measurable effect. To some extent, the instability of the cellular automaton model may be a consequence of the discretization of states – but how many states are needed to provide for robustness?

Closely related is the problem of robustness of behavior on an imperfect lattice. The cellular automaton metaphor becomes very inattractive if it would require an *exactly* periodic brain. Presumably, neural connections are formed on the basis of a regular overall plan, but with variability at the level of the individual neuron. If the four neighbors of a von Neumann unit were selected at chance from a somewhat larger local neighborhood, the construction would fail catastrophically. Again, this failure is probably a consequence of the small number of neighbors in the von Neumann machine – but how many neighbors are necessary?

Yet another kind of robustness is that of relative insensitivity to global alterations of the transition rule itself. In this regard, the brain does *not* appear to be robust. Rather, special homeostatic mechanisms such as autoregulation of cerebral blood flow act to minimize changes in the metabolic milieu that might induce global changes in neural function. When the homeostatic mechanisms fail in even apparently subtle ways, the result is a gross disturbance of consciousness. It is plausible that the cost of ensuring some measure of insensitivity to mild changes in the transition rule itself is much higher than that of ensuring insensitivity to state change or connectivity – but can this notion be made precise?

Introduction of more states and more neighbors leads to other questions. What is the tradeoff between number of states and number of neighbors in providing a given level of reliability? A many-neighbor (100 or more) many-state (20 or more) automaton is biologically implausible unless there is some regularity to the state space and the transition rule. An "irregular" rule is also very likely to be unstable to perturbation. For a plausible rule, it should be possible to view the states as discrete points in a continuous state space, and the transition rule should be at least piecewise continuous on state space. But what can be said about the topology and dimension of that space, and how complex must the transition rule be?

These abstract questions can be made quite concrete if, say, we consider a von Neumann unit to be a neuron. Does a single variable (such as a transmembrane voltage) suffice to specify a neuron's state, or must additional variables (perhaps concentration of calcium or a trophic factor) be considered? Such factors are crucial for development and synaptic plasticity. But are they required for the moment-by-moment information-processing as well? Questions about the transition rule become questions about the information-processing capacity of a single neuron. Does it suffice to lump all the neighbors' inputs into a few pools which combine additively (e.g. an excitatory pool and an inhibitory pool), or is it necessary to hypothesize more complex interactions among a neuron's inputs? Such interactions, such as presynaptic inhibition, are well known – but are they a requirement, an efficiency, or an inessential byproduct of evolution from a more primitive nervous system?

3. The problem of scaling

To understand the behavior of a model neural network, it often appears necessary to create an explicit computer model. Indeed, if the behavior of the model could be predicted in a simple fashion from its axioms, the model would likely be criticized as not possessing the "emergent" properties that are the essence of an acceptable model for cortex. But even the most ambitious explicit computer simulations are dwarfed by the number of elements in a real cortex. The scale of a computer model (10^4 to 10^5 elements) is about halfway between that of just a single unit and that of a real brain (10^8 to 10^9 elements). We need to know how characteristic lengths and times scale as the number of network elements increase from that of a computer model to that of a real brain. A network whose settling time increases even linearly with network size will have very different behavior when scaled up by four or five orders of magnitude.

More generally, we need to be able to understand what happens to global dynamics (number, dimension, and stability of attractors, for example), and whether additional qualitative properties of the model will "emerge" at a more realistic scale. A thorough understanding of scaling behavior may permit a clearer answer to the question of whether the qualitatively distinct features of human brain function may simply be viewed as consequences of its size, or rather, whether other processes (such as the development of new, specialized brain regions) must be invoked.

4. The problem of model testing

Let us now assume that we have a model in hand, with acceptable robustness, a biologically reasonable transition rule, and scaling behavior within grasp. How, and to what extent, can the model be tested?

Model testing requires two levels of investigation. (i) Does the overall behavior of the model correspond to that of the brain? (ii) Is there a detailed correspondence between parts of the model and parts of the brain? These questions seem inextricably related: it only makes sense to inquire about a detailed correspondence if overall behavior is acceptable, but how do we ask about overall behavior if we do not know what are the model counterparts of biologic observables?

Theory can help in two ways. Firstly, theory may be able to identify certain kinds of observables that are relatively insensitive to hypotheses about detailed correspondence. Possibilities for such observables might include dimensionality and stability of limit trajectories, or, how mutual information at two points in the model scales with separation or time lag. Such ideas might suggest new ways to interpret anatomic, single-cell, and gross-potential data.

Secondly, we would like to know to what extent two models which have different internal descriptions may have similar overall behavior. This is perhaps the most important contribution that theory can make. A relatively minor benefit is that competing models that can only be distinguished on the basis of detailed biologic correspondence will be recognized as such. The major benefit is that such an understanding will identify the critical features of internal structure which *do* affect overall behavior – and thus, define the critical experimental questions.

5. Conclusion

I hope that the questions raised above will serve as a focus for theoretical efforts, and as a framework for the design and interpretation of experiments. The viewpoint implicit in these questions is likely to be considered by experimentalists to be an apology for automaton theory, and by theorists, to be one of dissatisfaction with the state of the art. It is neither. My view is that the questions raised above are difficult but not unanswerable. Progress will be made, but progress will require new theoretical insights and mathematical analysis, and not merely brute-force computer simulations. Von Neumann's work has had a profound impact on neuroscience; if we can continue in his tradition, there will be further rewards.

Acknowledgements

This acknowledges the support, in part, of the McKnight Foundation and NIH grant EY7977.

References

[1]   J. von Neumann, in: Theory of Self-Reproducing Automata, ed. A.W. Burks (University of Illinois, Urbana, IL, 1966).

Physica D 45 (1990) 208–227
North-Holland

SIMULATION OF HIV INFECTION IN ARTIFICIAL IMMUNE SYSTEMS

Hans B. SIEBURG [a,1], J. Allen McCUTCHAN [b], Oliver K. CLAY [a],
Lisa CABALERRO [c] and James J. OSTLUND [d]

[a] *Department of Psychiatry, M-003-H, University of California at San Diego, La Jolla, CA 92093, USA*
[b] *Department of Medicine, H-811-F, UCSD Medical Center, University of California at San Diego, San Diego, CA 92103, USA*
[c] *Bio-Computing Laboratory, The Salk Institute for Biological Studies, P.O. Box 85800, San Diego, CA 92138, USA*
[d] *Office of University Computing, University of Notre Dame, Notre Dame, IN 56556, USA*

Received 29 January 1990
Revised manuscript received 5 March 1990

Infection by the human immunodeficiency virus (HIV) causes a multi-faceted disease process which ultimately leads to severe degenerative conditions in the immune and nervous systems. The complexity of the virus/host-system interaction has brought into sharp focus the need for alternative efforts by which to overcome the limitations of available animal models. This article reports on the dynamics of HIV infection in an artificial immune system (AIS), a novel in silico tool for bio-medical research. Using a method of graphical programming, the HIV/AIS interactions are described at the cellular level and then transferred into the setting of an asynchronous cellular automaton simulation. A specific problem in HIV pathogenesis is addressed: To determine the extent by which the physiological connectivity of a normal B-cell, T-cell, macrophage immune system supports persistence of infection and disease progression to AIDS. Several observations are discussed which will be presented in four categories: (a) the major known manifestations of HIV infection and AIDS; (b) the predictability of latency and sudden progression to disease; (c) the predictability of HIV-dependent alterations of cytokine secretion patterns, and (d) secondary infections, which are found to be a critical element in establishing and maintaining a progressive disease dynamics. The effects of exogenously applied cytokine Interleukin 2 are considered. All results are summarized in a phase-graph model of the global HIV/AIS dynamical system.

1. Introduction

The immune system is a self-regulated anticipatory system [1] serving as part of the host's defense mechanisms. It is composed of large and heterogeneous populations of cells capable of recognizing, processing and deactivating harmful substances termed antigens. The cell populations are functionally interrelated by intricate networks of cooperative processes. Many of these processes are evoked by antigen and subsequently evolve into complex patterns of cellular signaling through interdependent secreting and binding of cytokines. Cytokines are regulatory molecules that are released by cells and affect the movement or metabolism of other cells. During infections or, due to neurological stresses, normal cytokine secretion patterns can change. Experimental and clinical research has indicated that cytokines are effective in tumor-therapy, but, in higher concentrations, can also have toxic effects.

Important immune system cell types are the mononuclear phagocytes, e.g. monocytes and macrophages,

---

[1] Author to whom all correspondence should be addressed.

and the lymphocytes, e.g. B cells and T cells. Macrophages can non-specifically recognize, process and present antigens to activate specific T lymphocyte clones. B cells also process and present antigen to T lymphocytes, but the preceding recognition process is specific. T lymphocytes activated after the recognition of antigen presented in the context of self-molecules on the surface of macrophages or B cells can secrete cytokines that affect the differentiation and proliferation behavior of B cells, T cells, and macrophages. T cells are therefore regarded as pivotal transducer elements that recognize antigen-specific signals and convert them into antigen-nonspecific cytokine signals, which ultimately enhance antigen-specific responses such as the secretion of specific antibody by plasma B cells or the lysis of infected cells by cytotoxic T cells [2]. Important characteristics of the immune system are the capacity to distinguish self from non-self and the long-term preservation of antigen-specific events due to a process defined as immunological memory.

Large subsets of both the macrophages and the T lymphocytes carry a surface marker denoted as the cluster designation 4 (CD4) molecule. Since the human immunodeficiency virus (HIV) is characterized by a selective tropism for human CD4-positive (CD4$^+$) cells [3–5], it has been difficult to find experimental models for the study of AIDS. For ethical reasons, experiments with human tissues are generally confined to in vitro analyses. Chimpanzees, which in some, but not all, respects mimic the human responses to HIV [6], are rare and expensive animal models whose usage is severely complicated by animal rights issues. The recently reported reconstitution of SCID mice by human hematopoietic cells [7,8] may provide a valuable in vivo model for studying the interaction of HIV with immune system cells. Clearly, even this system is restricted to a subset of human cells. Because of this paucity of experimental systems and the complexity of HIV's target system, computer simulation of HIV infection has received increasing attention. The primary focus in this area has been on the development of differential equation models of HIV epidemiology [9–15]. Due to the physiological complexity, largeness and heterogeneity of the immune system, and the complexities of the virus/host-system interactions, it is difficult to apply the same methods for the development of predictive models of HIV pathogenesis [#1].

2. Direct simulation strategies

In the true sense of the word, simulation means the feigning of a particular situation, with the practical purpose of identifying the surprises. Experiments with wind-tunnel models are good examples and indeed many surprising effects regarding the stability of airplanes under different wind stresses have been discovered [16]. Therefore, simulation traditionally studies the interaction of objects (such as down-scaled airplanes) with a certain environment that is experimentally realizable (for example, the wind tunnel). The outcome of a simulation is then always a prediction on a priori grounds: this meaning that surprising effects become known before the actual large-scale experiment has been performed (for example, having a test-pilot fly a prototype airplane).

[#1] Technical note: Modeling the immune system by differential equations means to determine, simultaneously, the rates of change over time $(\mathrm{d}/\mathrm{d}t)\,X_i(t)$ of system variables $X_i(t)$ as a continuously evolving function F_i of all system parameters, i.e. $(\mathrm{d}/\mathrm{d}t)\,X_i(t) = F_i(X_1(t),\ldots,X_n(t))$, where $1 \leq i \leq n$. The variables $X_i(t)$ may denote the concentrations of cell types, factors, antibodies and antigens. Considerable difficulties can arise when this approach is used to describe very complex problem areas, such as HIV infection, where many cell types, clonal diversity, and changes in cellular differentiation and half-lifes, have to be considered. In this case, the number n of variables tends to be large and the functions F_i complicated. Already for small n it is often difficult to determine all F_i, and the predictive value of a differential equation system is proportional to the degree to which the format of these functions can be established. The differential equation approach therefore shares an important problem with all other modeling methods, namely to find a way of minimizing the amount of biological input, of maximizing the predictive output, and at the same time avoid the embarrassment of representing in a shorter form what is known already.

Computers are the experimental environment for simulations that involve a very large number of objects. Two major approaches to the direct computer simulation of large systems are currently used. The connectionist, or network-oriented, simulation strategies are essentially based on assumptions that include linkages between sensors and signals as part of the system implementation. The linkages are maintained, though varying in strength, for the duration of the simulation experiment. The network performance, for example in neural networks [17], can be measured relative to the overall distribution of linkage strengths

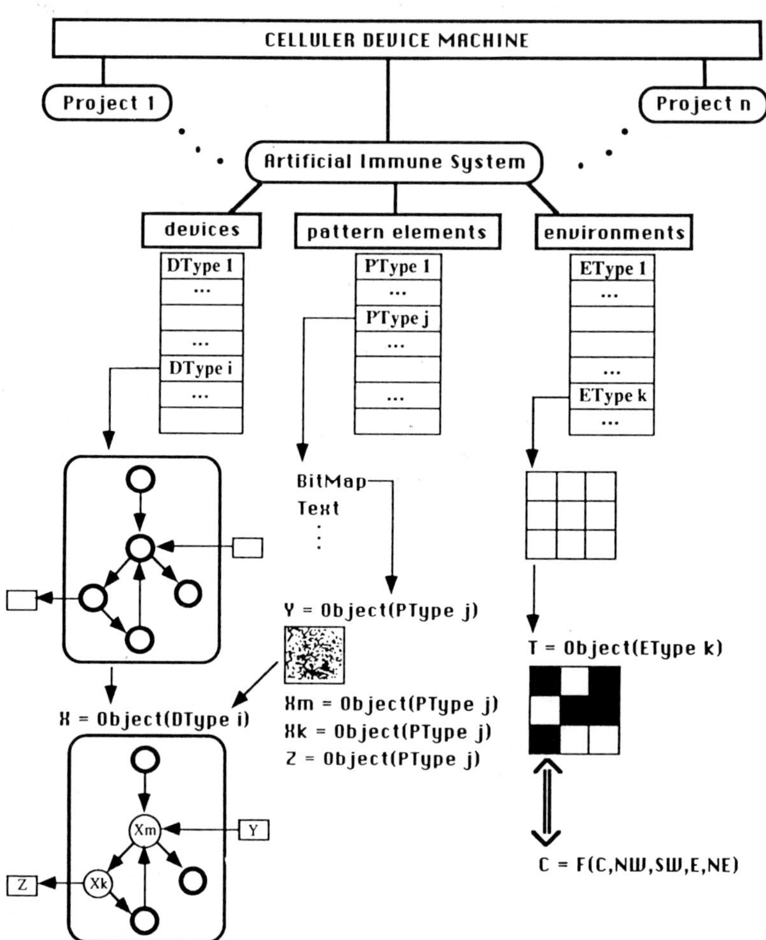

Diagram 1. The cellular device machine (CDM) uses a simple organizational scheme where all the files associated with a particular program are organized into a project. Programs are entered through the graphical project interface, and consist of information about three categories of classes: The "devices" (DEVs) category, the "pattern elements" (PELs) category and the cellular automaton "environments" (ENVs) category. Objects of a particular class are created in two steps. First, one designs a class "structure" (or chooses a structure from an existing set). For example, the transition graph of a particular device (class DType i), a bitmap for a pattern element (class PType j), or the fundamental domain (e.g. square) and a neighborhood template (object of class EType k) for a tessellation geometry of the plane, are structures within their respective categories. Second, one defines "realizations" of the structures. For example, the realization of an object of type DType is achieved by expressing the vertex set of its transition graph in terms of PType objects. Objects of PType or EType are also designed by "filling in the blanks", e.g. the actual drawing of a bitmap, or defining a cellular automation neighborhood rule through mouse-clicks in the neighborhood template.

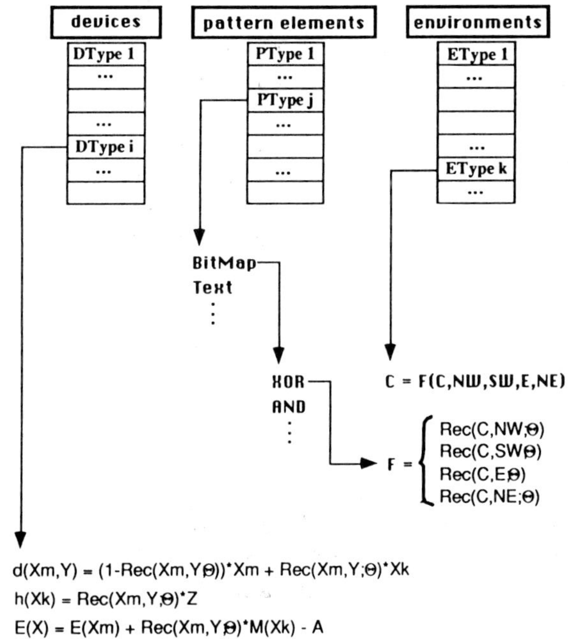

Diagram 2. Only in the case of PType objects it is necessary to make a choice of method(s). Any particular choice of method, e.g. a "bitmap exclusive or" (XOR), will fully determine the pattern element recognition function $Rec(X, Y; \Theta)$. $Rec(X, Y; \Theta)$ is a Boolean function which returns a value 1, iff the number of black pixels of the resulting bitmap $XOR(X, Y)$ is greater than a fixed integer Θ, and a value 0 otherwise. Since Rec is defined for all pairs of PELs, the CDM can use Rec to define a cellular automaton input function F for some Etype k which describes that a center site "C" receives information from specific neighboring sites, e.g. northwest (NW), southwest (SW), east (E), northeast (NE). The methods for objects of DType are a state transition function $d(Xm, Y) = (1 - Rec(Xm, Y; \Theta)) * Xm + Rec(Xm, Y, \Theta) * Xk$, an output function $h(Xk) = Rec(Xm, Y; \Theta) * Z$ and the "eigenlife" function $E(X) = Rec(Xm, Y; \Theta) * M(Xk) - A$. Here the right-hand side of $d(Xm, Y)$ means that the state-pointer continues to point to Xm iff Y is not recognized, but will point to the next state Xk iff Y is recognized through the PType object Xm. The right-hand side of $h(Xk)$ means that the PType object Z will be released following recognition of Y through Xm. The process of secretion requires an output neighborhood (EType). h shows that in a CDM simulation we do not only update center sites with each iteration step, but also a certain pattern of neighborhood sites. $E(X)$ is an integer-valued function which relates the event history of a DType object to time, by mapping events, e.g. the recognition of a certain bit pattern, onto the update cycle of the cellular automaton. The eigenlife function works like an event-driven biological clock, where time is accumulated through a sequence of successfully accomplished state transitions, starting from an initial value assigned at birth, and where time is lost in the form of a fixed universal aging parameter A which is subtracted globally with each iteration cycle. When a non-positive eigenlife value is reached, the respective device will be discontinued, a condition signifying death. Thus, the repeated failure to interact successfully with its environment will lead to the rapid death of a device, whereas successful interactions will expand its life span. $M(Xk)$ is a numerical identifier of the PType object Xk that defines the amount by which the eigenlife will increase should a state transition from Xm to Xk occur.

over time. In contrast, an object-oriented simulation strategy will primarily provide information about sensor and signal structures, but will not define particular relationships between them. Rather, the system components are free to explore, and engage with, their environments, and all communication linkages between objects can appear and disappear as functions of sensor states and available signals. Cellular automata[#2] and classifier systems [20] are important object-oriented simulation methods, where local interaction rules or natural selection criteria are used to explore the formation of micro-environments or other adaptive processes that determine a system's dynamical behavior. Which of the two strategies is chosen, or whether a combination of methods is employed, depends strictly on the application, or investigator preferences.

We previously described a virtual computer, the cellular device machine (CDM), which can be programmed graphically[#3]. Due to this approach, object-oriented simulations of large and heterogeneous systems can be easily implemented and executed even by a non-specialist [21,22]. The CDM consists of two major components: The interface, which contains editors for devices, pattern elements and environments, and the simulator (compare diagrams 1 and 2 for technical notes). The building blocks of any CDM simulation are appropriately defined objects called cell devices. Biologically, a cell device resembles a cell insofar as it consists of a small number of differentiation states which are linked by differentiation pathways. Associated with each differentiation state are particular data structures, called pattern elements (PELTs), which serve as analoga for the binding sites of receptors on a cell surface. In each of its states, a cell device can therefore recognize and process specific signals from its surrounding environment, for example factors for activation or proliferation, which are also represented by PELTs. Upon processing such signals, a cell device undergoes a differentiation state transition that is indicated by the expression of new receptor structures, the secretion of factors, or by cell device division. Therefore, the development of a cell device is event-, rather than time-driven, and a sequence of internal responses describes its history of communication with its environment[#4]. The information flow between cell devices and environmental signals is strictly local.

[#2]   Technical note: Cellular automata can be thought of as large parallel computers whose individual processors are located in the fundamental domains (sites) of a tessalated manifold, e.g. a grid of squares covering a plane. The information flow is coordinated among groups of nearest-neighbor processors according to local interaction rules that are valid for all sites. An important advantage of cellular automata as simulation tools lies in the liberty and simplicity by which the individual processors can be programmed, placed with respect to each other, and be related by local interaction rules. Comparatively little modeling detail is therefore needed to arrive at complex dynamical behaviors that by far exceed that of the original model implementation. For more detailed information on cellular automata and their applications compare ref. [18] or [19].

[#3]   Technical note: Scenarios 1 and 2 illustrate how the method of graphical programming is used to implement the facts and hypotheses of a CDM project. The fundamental instructions are: (i) circles, to indicate state types, (ii) rectangles, to indicate signal types, (iii) thin arrows, to indicate whether a signal is to be secreted or investigated, (iv) fat arrows and attached rectangles, to indicate transitions between two states that are driven by signal-recognition, and (v) round rectangles, to indicate the object data domains that are normally inaccessible to environmental events. Furthermore, the CDM interface contains a pattern editor which can be applied to associate particular properties with a given state-type or signal-type.

[#4]   Technical note: We have expanded the concept of event-dependent development to include a notion of life-span, which we called eigenlife. The eigenlife is an integer-valued function that relates the event history of a cell device to time, by mapping events, e.g. the recognition of a certain bit pattern, onto the iteration cycle T of the simulator. Each cell device is given its own eigenlife function. The way the eigenlife function works is to first assign a characteristic value, denoted $E^{(0)}(\mathbf{X}, T)$, to a cell device $\mathbf{X}$ upon its birth into the simulation environment. This number is then used as the initial value of a discrete function $E^{(t+1)}(\mathbf{X},T) := E^{(t)}(\mathbf{X},T) + \mathrm{Rec}(\mathrm{X}^{(t)}, \mathrm{Y}) \cdot m(\mathrm{X}^{(t+1)}) - d(T)$, which increases upon a successfully accomplished state transition and decreases with respect to the universal cycle T. Here, $E^{(t)}(\mathbf{X}, T)$ denotes the eigenlife value at iteration t starting from the birth of $\mathbf{X}(t = 0)$, $\mathrm{X}^{(t)}$ and $\mathrm{X}^{(t+1)}$ denote the current and future states of $\mathbf{X}$, and $m(\mathrm{X}^{(t+1)})$ is a numerical identifier of the state $\mathrm{X}^{(t+1)}$ that defines the amount by which the eigenlife will have increased at iteration $t + 1$. $\mathrm{Rec}(\mathrm{X}^{(t)}, \mathrm{Y})$ is a Boolean function that indicates the success or failure in handling an event Y. The fixed universal aging parameter $d(T)$ is determined by the speed by which the simulation environment is updated and is subtracted with each iteration in T. When a non-positive eigenlife value is reached, the respective cell device will be discontinued, a condition signifying death. Thus, the repeated failure to interact successfully with its environment will lead to the rapid death of a cell device, whereas successful interactions will expand its life-span.

In previous simulations of the immune system, we have shown that the dynamics of an appropriately defined cell device system can account for many important behavioral aspects such as spontaneous self-organization (reflecting the cooperation of different cell types during an immune response), non-linearities (e.g. due to overlapping immune responses), non-stationarity (reflecting the self-regulation of pathogen-eliminating processes) and memory (as the basis for vaccination), which are generated by the model. The simulator's capacity for a priori predictions of functionally evolved complex behaviors has offered new perspectives for immune system and infectious diseases modelling [21,23]. The present article applies the simulator to address a specific problem of HIV pathogenesis in the context of an artificial immune system (AIS) that we constituted on the CDM. The major AIS-object classes (scenario 2a; [24]) are derived from well established immunological data that describe the developmental and physiological connectivities of a normal B-cell, T-cell, macrophage immune system [2,24–29].

3. Acquired immune deficiency

A considerable body of experimental and clinical data has been collected that describes potential interactions between HIV and its target system. From these we selected a number of interactions which address the following questions: (1) How does the infectious agent establish contact with its target cells? This is answered by the earliest findings about HIV, whereby the HIV envelope glyco-protein gp120 binds to the CD4 surface molecule on helper T cells and macrophages [3,4]. (2) Under which conditions does

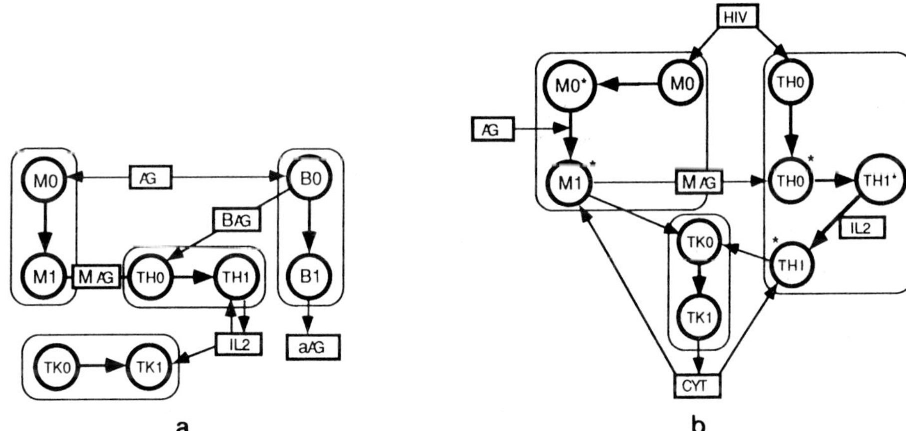

Scenario 1a. The Boolean version of a normal immune system consisting of helper and cytotoxic T cells (TH and TK), and macrophages (M). Antigen (AG) can be taken up in the inducible states M0 and B0. The macrophage cell device and the B cell device can present antigen to a T cell device in state TH0 specific for the antigen's pattern element. Cell–cell contact during antigen presentation (MAG or BAG) leads to the TH0 → TH1 state transition and therefore T cell activation (and the B0 → B1 transition in the BAG case). The activated state TH1 is to represent the expression of IL2 surface receptors as well as the capacity to secrete IL2. Binding of IL2 pattern elements leads to proliferation of activated helper and cytotoxic T cell devices (TK1). Activated B cells in state B1 will secrete antigen-specific antibody (aAG).

Scenario 1b. The "hypothesis graph" formalizing potential interactions between HIV and the T-cell–macrophage component of the normal Boolean immune system depicted in scenario 1a combines in the simplest possible way four features which are considered important for the pathogenesis of AIDS: (1) selective binding of HIV to the CD4 surface receptor, (2) productive infection dependent on the activation of non-productively infected cells (e.g. due to the activation of macrophages by concomitant infections (AG), or the activation of T helper cells during antigen-presentation (MAG)), (3) soluble signals, such IL2, as possible modulators of disease progression, and (4) autoimmune phenomena such as the lytic response (CYT) by cytotoxic T cells against uninfected cells expressing viral envelop protein on their surface (indicated by an asterisk).

the infectious agent enter or exit its target cells? Here, it has been indicated that T-cell activation is required for HIV entry into $CD4^+$ lymphocytes [30] and possibly also for the activation of productive infection in this cell type [31]. (3) By which mechanisms are cells destroyed that have made contact with the infectious agent? This is answered by the finding that cells expressing virus, infected or not, are lysed following recognition by cytotoxic T cells [32,33]. For the purpose of simulation, we have transformed these findings into formal hypotheses that can be tested with regard to a data base of immunological facts, i.e. observations about normal immune system behavior derived from frequently repeated wetlab experiments. We investigated two scenarios to predict the long-term effects of HIV on the immuno-regulatory activity of a normal B-cell, T-cell, macrophage immune system. Scenario 1 is to illustrate in the simplest possible way how the techniques of project implementation on CDM are actually encoded. In this example, we use three device types: TH, for T helper, TK, for T killer, and MP, for macrophage. Each of the device types is capable of two states only: The inducible state, denoted by X0, where $X \in \{TH, TK, MP\}$, and the activated state X1. Different TH or TK clones can be created by assigning different pattern elements to the respective inducible states TH0 and TK0. The transition from state X0 to state X1, which depends on the availability of a pattern element M, is encoded in the form $(X0.M) \rightarrow (X1;N)$, where N represents a pattern element that is secreted once the state X1 is reached. Proliferation steps are denoted as $(X1.M) \rightarrow 2(X1)$. This meaning that the cell device X, when in state X1 and after recognizing the pattern element M, will self-replicate into, for example, two identical copies each of which is then in the same state X1. Thus the "facts-graph" of scenario 1a translates into the following script:

$(M0.AG) \rightarrow (M1; MAG);$
$(M1) \rightarrow (M0);$
$(TH0.MAG) \rightarrow (TH1; IL2);$
$(TH1.IL2) \rightarrow 2(TH1);$
$(TK1.IL2) \rightarrow 2(TK1).$

IL2 denotes the T-cell growth factor Interleukin 2. Without loss of generality, we assume that the IL2 pattern elements are secreted in single bursts. The "hypotheses-graph" in scenario 1b translates into the script:

$(M0.HIV) \rightarrow (M0^*);$
$((M0^*).AG) \rightarrow ((M1)^*; MAG);$
$(TH0.HIV) \rightarrow (TH0)^*;$
$((TH0)^*.MAG) \rightarrow ((TH1^*); IL2);$
$((TH1^*).IL2) \rightarrow 2(TH1)^*;$
$(TK0.(TH1)^*) \rightarrow (TK1; CYT);$
$(TK0.(M1)^*) \rightarrow (TK1; CYT);$
$((TH1)^*.CYT) \rightarrow D;$
$((M1)^*.CYT) \rightarrow D.$

Here, we use $(Xj)^*$, (Xj^*) to describe a cell device, in state Xj, carrying HIV on its surface or integrated into its gene pool. CYT denotes a cytolytic enzyme, and D the empty cell device, respectively.

    The reasons why we can expect surprising, i.e. otherwise difficult to predict, complex behavior from this deterministic script are that in a CDM simulation (a) a large number of objects is distributed across a large cellular automaton space, (b) clonal diversity is generated by initially associating a variety of pattern elements with the labels TH0, TK0, AG, HIV, and (c) the cell device life history is dependent on recognition events (see footnote #4), i.e. the local timing regime of event-driven biological clocks in the cell devices implies asynchronicity in the underlying cellular automaton whose dynamics therefore evolves without implicit global conditions. The dynamics of the CDM simulation of scenario 1 (fig. 1) describes how a set of hypotheses about HIV/immune system interactions performs relative to a set of facts about normal immunity.

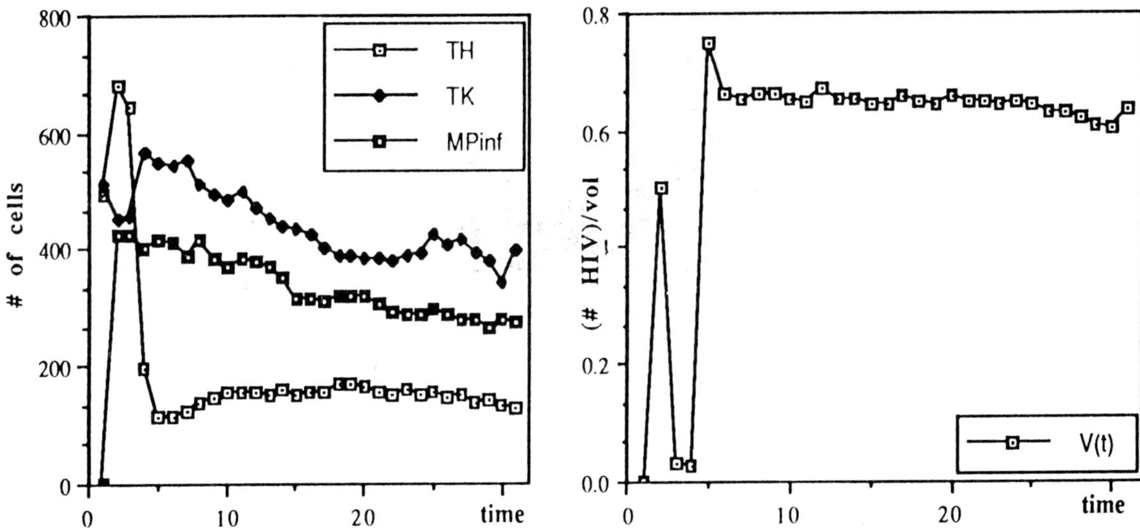

Fig. 1. To study the dynamics of a particular model, we represented scenario 1 in a cellular automaton consisting of 400 sites into which the cell devices were distributed randomly according to the ratio M : TH : TK = 20 : 50 : 30. For each parameter, we recorded its respective concentration function at discrete time points, i.e. Xp(t) := (number of active devices/volume, where X $\in$ {M, TH, T}, t a non-negative integer, p $\in$ {u, u*, e, e*, i}, or Z(t) := number of particles/volume, if Z $\in$ {IL2, HIV}. Here, the volume is defined as the number of sites in the cellular automaton. The symbols u, u* ,e ,e*, i are to denote uninfected, non-productively infected, uninfected expressing viral antigen, non-productively infected expressing viral antigen, and infected cell devices, respectively. In this notation, we considered the fixed initial values Mu(0) = 0.2, THu(0) = 0.5, TKu(0) = 0.3, IL2(0) = 0, V(0) = 0. Due to the cellular automaton representation, all model rules are executed locally and in parallel.

The information we gained from the scenario 1 simulations already describes a qualitatively correct picture of disease progression from a normal status to a T-cell-depleted status (left-hand side of fig. 1), with a persisting reservoir of infected T cells and macrophages describing the intermittant (latent) state. Therefore, scenario 1 is a valuable preview of an interesting hypothesis set. The system dynamics evolves unfortunately too quickly to allow for biologically relevant predictions which are applicable to therapy.

To achieve higher predictive power, we incorporated more biological detail into the description of a normal immune system (scenario 2a), and its interaction with HIV (scenario 2b). Using a simulation protocol based on the physiological connectivity represented by the cell device transition diagrams shown in scenario 2a, and the hypotheses shown in scenario 2b, we were able to investigate the long-term effects of HIV on the immuno-regulatory activity of a normal B-cell, T-cell, macrophage immune system more thoroughly and with a higher degree of therapeutic relevance.

To simulate HIV infection in the CDM environment, we placed T-cell, B-cell and macrophage devices randomly into a two-dimensional cellular automaton (CA) environment. The automaton was implemented as a regular grid of sites with periodic boundary conditions, so that the uppermost and lowermost rows within the grid, and the leftmost and rightmost columns, respectively, become neighbors. We used two kinds of CA rules, one to emulate rapid flow for peripheral blood, and the second to pattern movement in the densely packed lymphoid tissues. The system dynamics was recorded in the form of time series $(X(t))_{1 \leq t \leq T}$, where $X(t)$ either denotes the number of cells of a particular type or in a particular condition, or the titers of cytokines, antigens or antibodies. The initial T-cell, B-cell and macrophage populations were seeded into the simulation environments at the ratio of 70 : 20 : 10 of total cells, which correlates with the ratio normally found in peripheral blood. This allowed us to observe up to 10^4 cell devices at any given time t. We introduced HIV into already running simulations by injecting specific quantities of free

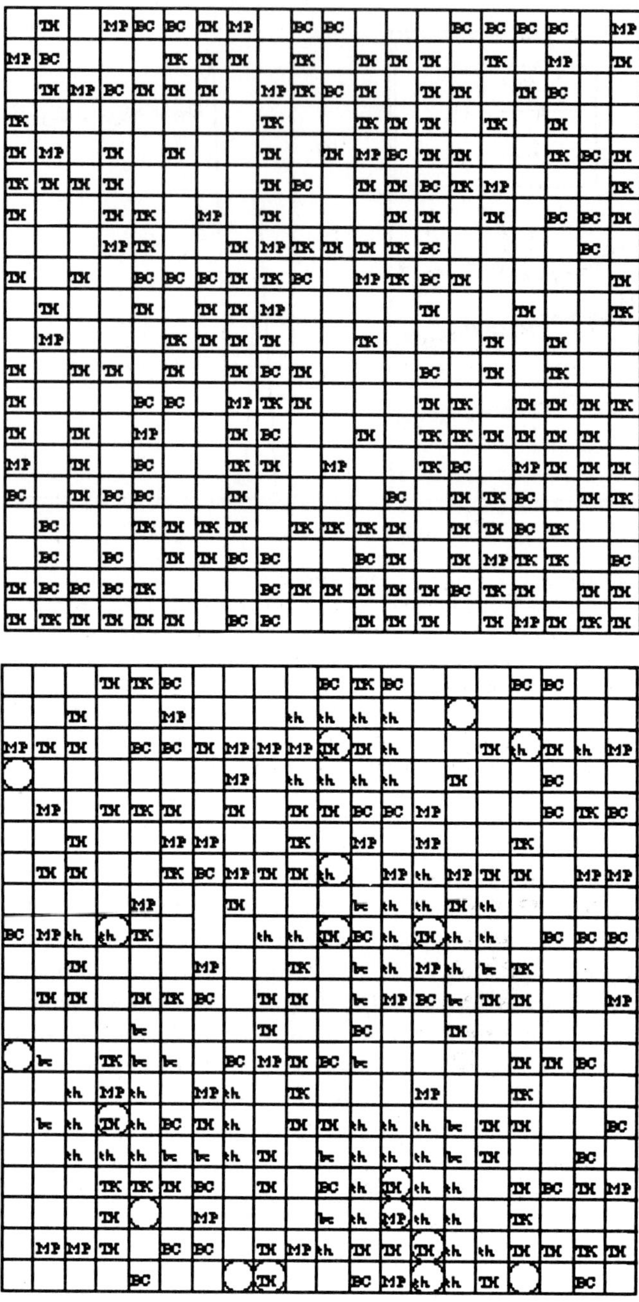

Fig. 2. Before running a program, the CDM simulator will first build the program's project. The building process involves combining the non-numerical object information accumulated by the interface with numerical parameters such as initial cell device population densities and distributions across the cellular automaton space, the amount of pattern elements released by the output function h of a particular device type, or the binding strength Θ. Also, cell devices can be expanded into clones, i.e. a user-definable amount of identical copies are made from a template object. The picture shows a random initial distribution of four cell device types (upper row), denoted TH, TK, MP, and BC on a 20×20 cellular automaton with periodic boundary conditions. During simulation, the initial distribution will change and new patterns marking the response against an antigenic stimulus will form (lower row). Antigen is indicated by circles inside the sites.

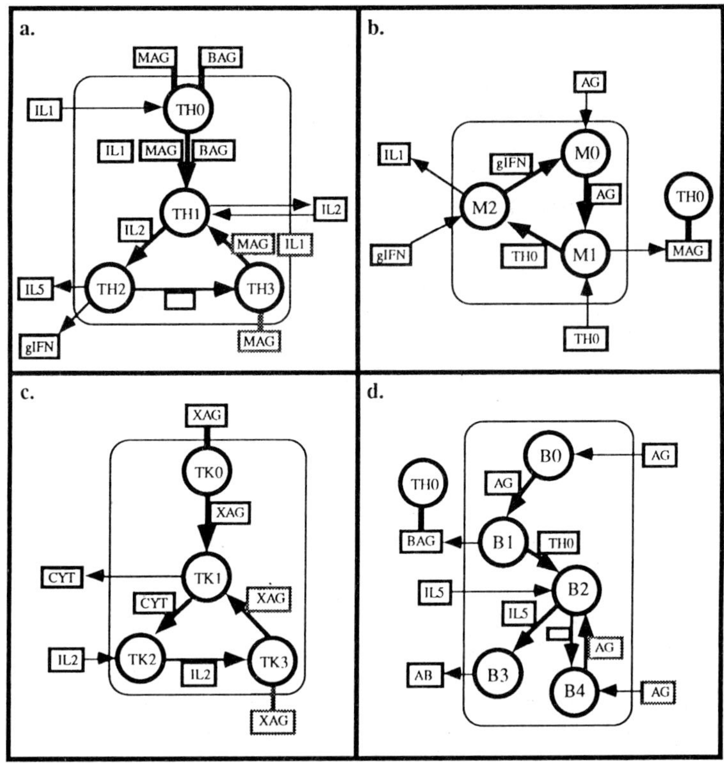

Scenario 2a. The cell device diagrams shown represent a data base of normal immunoregulatory activity in a B cell, T cell, macrophage immune system. This data base was derived from observations about immune system behavior in frequently repeated wetlab experiments. (a) A helper T cell device (TH). Inducible helper T cells (TH0) are able to recognize antigen presented by antigen-presenting macrophages (MAG) or B cells (BAG). The macrophage is then stimulated to secrete Interleukin 1 (IL1). IL1 binds to a specific receptor type on the T cell, which also becomes a highly secretory cell (TH1). Among its soluble products are Interleukin 2 (IL2), Interleukin 5 (IL5) and γ-Interferon (gIFN). IL2 causes T cells to differentiate and proliferate (TH2). The gIFN secretion will further activate nearby macrophages. IL5 causes terminal differentiation in B cells, thus leading to the production of large quantities of specific antibody. Upon secretion of gIFN or IL5, cells in state TH2 are capable of memory (TH3), from where they can be newly stimulated (shaded box). Event-processing may not be required for the state transition to memory (empty box). (b) A macrophage cell device (MP). Antigen (AG) can non-specifically bind to the macrophage surface (M0). It is internalized, degraded and subsequently presented on the cell surface (M1) in the context of self (MAG). Antigen presentation sets the stage for interaction with inducible or memory helper T lymphocytes (TH0 or TH3). Recognition of the self/antigen complex will lead to tight binding between macrophages and lymphocyte. This binding stimulates the macrophage to secrete Interleukin 1 (IL1) in state M2, which in turn promotes the secretion of the T cell cytokine γ-Interferon (gIFN). (c) A cytotoxic T cell device (TK). Inducible cytotoxic T lymphocytes (TK0) bind their targets (X) while specifically recognizing antigen in the context of self determinants (XAG). The T cells involved are generally CD8[+] (in humans) and major histocompatibility complex (MHC) class I restricted. After binding (TK1), the target cell is lysed (CYT). Following lysis of the target cell (TK2) cytotoxic T cells may proliferate (as indicated by the IL2-dependent state transition), and may go on (as indicated by the memory state TK3) to kill more target cells (shaded box). (d) A B cell device (BC). Mature B cells (B0) recognize antigen (AG) specifically. They become activated (B1) and present antigen (BAG). Specific T cells (TH0) are able to recognize the complex, and a binding between the two lymphocytes ensues. This binding stimulates the B cell to express surface receptors for T-cell-derived factors (B2) such as Interleukin 5 (IL5). IL5 promotes terminal B cell differentiation to a plasma cell (B3). Plasma cells are capable of producing large amounts of specific antibody (AB). If IL5 is not available, B cells in state B2 are capable of memory (B4), from where they can be again stimulated by antigen to launch a secondary response (shaded box).

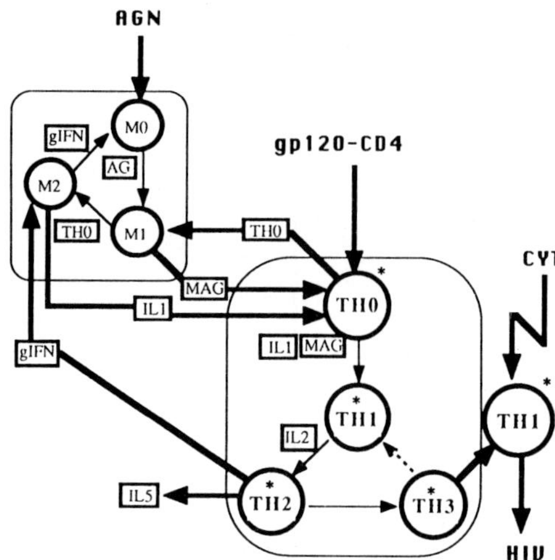

Scenario 2b. This schematic represents events that might occur during a hypothetical interaction between HIV and a CD4$^+$ T cell. The fat arrows reflect the cell–cell interactions and the exchange of cytokine signals during a normal immune response against antigen AGN. Prior to this response, we imagine that HIV particles have attached themselves to their target receptors (gp120-CD4) and are now residing on the cell surface (indicated by the asterisk). Following the interaction with the antigen-presenting macrophage (MAG), the receptor–ligand complexes are incorporated into the cell interior and the T cell becomes latently infected (TH1*, TH2*, TH3*). Maybe quite some time later, we imagine that the T cell becomes reactivated from the memory state (TH3*), upon which the infection turns productive. As a result, virus particles are again expressed on the cell surface (mimiced for lack of animation by a new state on the outside of the cell device). Finally, this unlucky cell device is recognized by a cytotoxic T cell device in the vicinity and is eliminated (lightning bolt). The lytic action liberates virus particles (HIV), which may succeed in infecting more cells. During a running simulation many kinds of interaction patterns can form, which are likely to be different from the one presented here.

or T-cell-device-associated virus[5]. Experimental studies have shown that both forms can, for example, be found in seminal fluid [34].

We classified the evolving computer disease as AIDS for several reasons. First, a lasting depletion of CD4$^+$ T cells occurs at one point during infection (fig. 3). Second, the depletion event is preceded by a long latency period as can be deduced from the ratio $(t_{1,\text{sim}} - t_{0,\text{sim}}) : (t_{2,\text{sim}} - t_{1,\text{sim}}) = 14 : 1$. Here, $t_{0,\text{sim}}, t_{1,\text{sim}}, t_{2,\text{sim}}$ denote the time points of virus introduction, the onset of T-cell depletion, and the onset of immune deficiency, respectively. Third, the initial humoral immune response to the infectious agent, i.e. sero-conversion, occurs late after the agent is already well established in the system. The virus titer generally remains low, except for episodical flare-ups following the elimination of large numbers of infected cells by cytotoxic T cells (fig. 4). The above findings indicate that the hypotheses we used to define the interaction between HIV and the AIS are sufficient to recreate the major known immunological manifestations of AIDS.

The artificial immune system (AIS) simulations have carried us in several ways beyond clinically established findings. For example, we find that progression to AIDS in AIS occurs in a sudden transition. Also, we find that the infectious agent persists in two reservoirs of latently infected cells, the larger of which is the macrophage reservoir. Furthermore, we observed that secondary infections are a critical element in triggering productive infection in latently infected cells. This result may help to explain why progressively more T cells than macrophages are eliminated by cytotoxic T cells among HIV's two target cell populations.

---

[5] Technical note: The cellular device machine simulator provides a special "syringe tool" which allows the user to place selectable quantities of new entities anywhere into the simulation environment anytime during a running simulation.

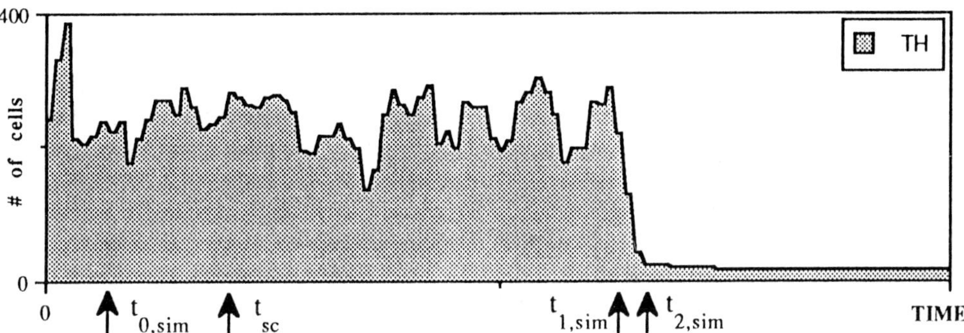

Fig. 3. The acid test of AIDS simulations: T-cell depletion occurs after infection with a CD4-tropic, cytolytic virus. The time series depicts the dynamics of the T cell population shortly before and at the time point of infection ($t_{0,\text{sim}}$), during latency and following the onset of T cell depletion ($t_{1,\text{sim}}$) and disease ($t_{2,\text{sim}}$). t_{sc} marks the time point of sero-conversion. The depletion of $CD4^+$ T cell devices is an example for adaptive dynamical behavior that occurs during simulation runs under different initial AIS statuses and was not part of the original simulation protocols. All concentrations are calculated as the number of objects, in this case $CD4^+$ T cell devices, divided by the size of the cellular automaton, i.e. the number of its sites. The fluctuations in the population size prior to infection reflect the event dependence of the cell device life-spans (see footnote #4). Time (TIME) is represented in terms of the update cycle of the simulator.

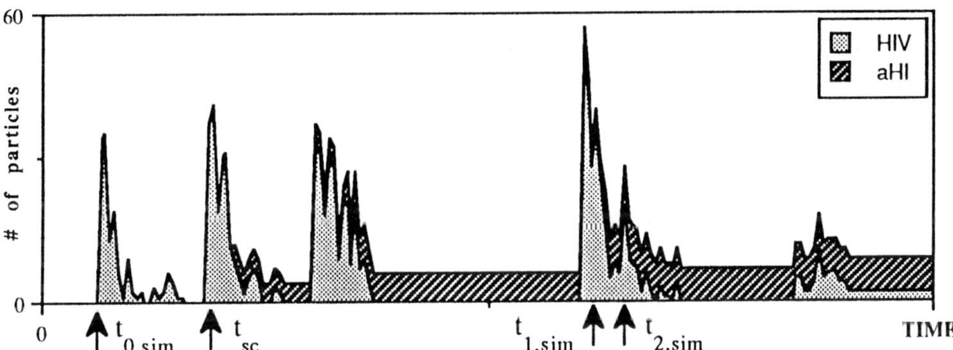

Fig. 4. The time series depict the HIV and anti-HIV (aHI) titers as they occur during the simulation experiment discussed in fig. 3, where we injected free virus ($t_{0,\text{sim}}$). The bursts are due to the onset of productive infection in activated T helper cells and macrophages, and the lysis of such cells by cytotoxic T cells. During these periods of enhanced cytotoxic T cell activity, we also observed the elimination of "by-stander cells" which carry virus particles on their surface, but are not infected. After each burst, the virus titer declines more or less rapidly as a result of virus particles binding to $CD4^+$ cells or virus-specific antibody, or due to uptake by antigen-presenting cells. We generally assumed a long half-life for antibodies, which explains the relative stability of the aHI curve between bursts. Biologically, this may mean that no or only small fluctuations can be observed. Mathematically, the peaks and valleys represent a sequence of transitions between global dynamical states as the system progresses towards disease.

4. Latency and transition to disease

It is commonly believed that progression to AIDS occurs gradually [35]. In contrast, the T-helper-cell time series (fig. 3) shows that the progression from infection to disease in AIS is the result of a rapid event similar to a phase transition. This finding was recently confirmed in a UCSD study (conducted by J.A.M.) of 22 men progressing to AIDS. Here, the period ($t_{2,\text{cli}} - t_{1,\text{cli}}$) was estimated to 8–10 months. This value divided by the period ($t_{2,\text{sim}} - t_{1,\text{sim}}$) obtained from fig. 3, can be applied to calculate a "clinical-to-simulation" scaling factor c of 4.25 weeks for each simulator iteration cycle. Using this scaling factor, we arrived at a latency period $P_{\text{lat}} = c(t_{1,\text{sim}} - t_{0,\text{sim}})$ of 9.41 years in this particular example. Allowing for different initial immune system statuses through systematic variation of the initial sizes and clonal distributions of the T-cell, B-cell and macrophage populations, we found a latency variability of $6 < P_{\text{lat}} < 11$ years over 46 independent simulation studies. These values are in good agreement with known or estimated disease progression rates [36].

We have investigated the effects of presenting the host with the virus in two forms: cell-associated and free. Our data show that if HIV is given in free form, many virus particles are taken up by antigen-presenting cells, or bind to the CD4 receptor on the surface of macrophages and T cells. Antigen presentation by macrophages and B cells activates specific T cells and leads to increase in the size of the pools of latently infected macrophages and T cells (fig. 5). Surprisingly, we observed that T-cell help to B cells is less pronounced than T-cell help to macrophages, a finding which might explain why the anti-HIV humoral immune response is late to develop. When we presented the system repeatedly with T-cell-associated virus (see fig. 6), but otherwise kept the environment antigen-free, we saw that the latently infected T cells died out on reaching their normal life-spans (see footnote #4) without virus particles being produced and released. We concluded that additional factors, or at least the continuing presence of free virus particles, are needed to maintain the latent reservoirs.

In order to identify mechanisms that help to replenish the naturally diminishing reservoirs of infected cells, we injected a latently infected system with a second antigen. The antigen was processed and presented normally to specific T cells. Following activation of latently infected T cells during antigen-presentation, a burst of productive infection emerges (fig. 6). Once productive infection has occurred, the system showed behaviors similar to those we described in the previous experiment, where we injected only free virus. These findings show that secondary antigens are necessary to drive the spontaneous outbreak of productive infection, a dynamical process that promotes the persistence and spread of HIV in the

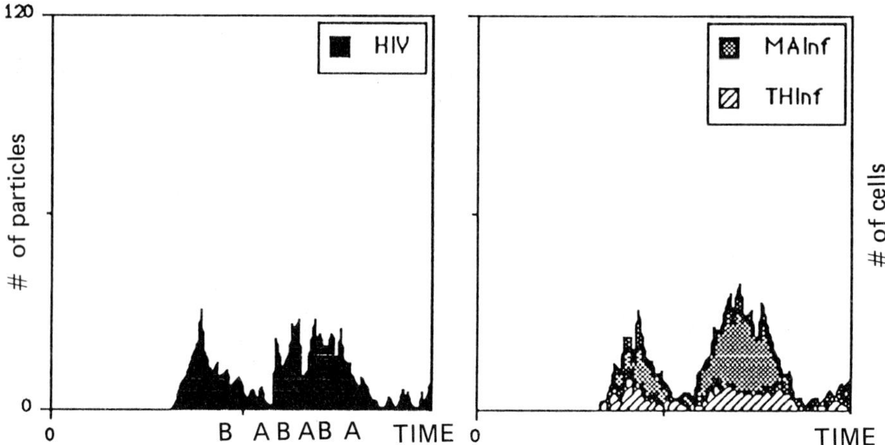

Fig. 5. In these time series we recorded the concentrations of HIV particles (left-hand side) and the sizes of the newly arising reservoirs of latently infected macrophages and CD4$^+$ T cells (right-hand side) when virus appears in free form. The data show that the system tends to alternate between two global dynamical states, namely one where the virus titer is low (A) and another, where the virus titer is high (B). This general behavior is a pattern that we were able to observe in most simulations. It is reflected, although with a different phase, in the fluctuations of the reservoirs of latently infected cells. The elapse time before a transition to a new state occurs can be used to calculate disease progression rates. A second pattern that arises in nearly all simulations, is that the macrophage reservoir is markedly larger than the T cell reservoir. This particular behavior is an example of an adaptive process that occurs during a simulation and was not part of the original system implementation.

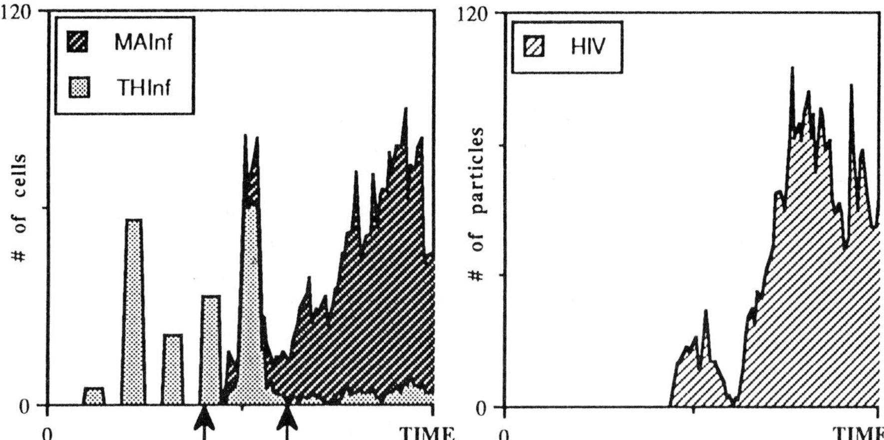

Fig. 6. This experiment shows that secondary infections are necessary to kick off productive HIV infection. We presented the host several times with HIV in T cell associated form, but otherwise kept the environment antigen-free, i.e. clear of HIV particles and other infectious agents. We observed that the latently infected cells die out within their normal life span. The first arrow indicates the time point when we introduced a second antigen (AGN). The time series shown in the left-hand diagram represent the sizes of the reservoirs of latently infected macrophages and CD4$^+$ T cells that evolve following the eruption of productive infection. Productive infection is evidenced by the increasing HIV titer shown in the right-hand diagram. The macrophage reservoir is again larger than the T cell reservoir and stays that way throughout the observation period. In the experiment shown, the hypothetical host progresses rapidly to disease (indicated by the second arrow) with a scaled latency period of 5.88 years.

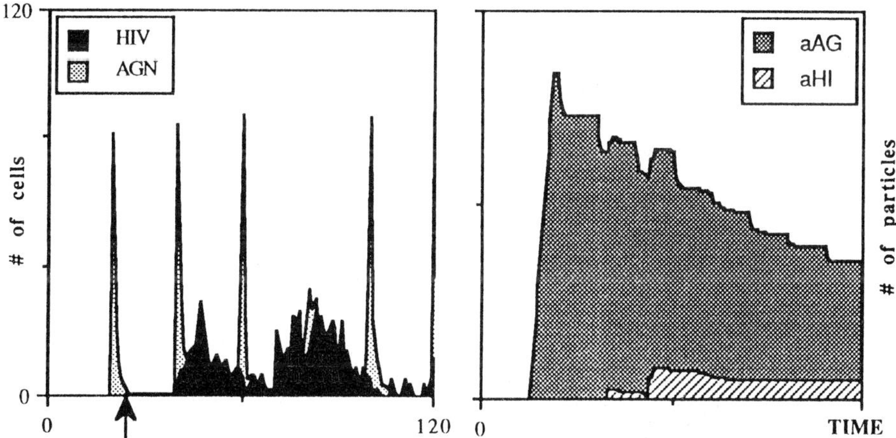

Fig. 7. In this experiment we investigated possible dynamical couplings between two, on the outset independent, infections. Plotted in the left-hand diagram are the respective antigen concentrations and, in the right-hand diagram, the respective antibody titers. The second agent (AGN) is assumed to exist latently in some unnamed cell type, from where it will surface in regular intervals. We observe that the first outbreak of AGN in responded to with a strong antibody response (the time point of sero-conversion is indicated by the arrow) that clears the agent, and which can be considered as a control example for a normal humoral immune response in AIS. Compared to this normal control, the anti-HIV (aHI) humoral response is weak. Three observations indicate that a coupling has occurred between the two agents. First, shortly after the second outbreak of AGN, productive HIV infection erupts. Second, the antibody response declines after HIV has manifested although we found a well developed pool of specific memory B cells. Third, the fourth outbreak of AGN is markedly weaker than those before and those after, and its clearance takes more time (not visible in the diagram).

AIS. The experiment also shows that these spontaneous outbreaks can critically influence latency and progression to disease.

It has been widely speculated how concomitant infections can influence disease progression [35]. Such infections might arise through newly encountered external agents, or through the outbreak of internal latent antigens other than HIV, or by mutations of HIV. Several pathogenic strains of HIV have so far been discovered that can coexist in one host [35,37]. To understand the dynamical effects caused by multiple infections, we conducted a series of preliminary experiments that aim at identifying non-linearities that might arise from overlapping immune responses.

In the example shown (fig. 7), we consider a host carrying HIV in T-cell-associated form as well as a different, also latent, antigen denoted AGN. We assumed that AGN materializes in bursts that occur periodically. We recorded the titers of both AGN and HIV, the numbers of newly infected macrophages and T cells, and the humoral immune responses launched against both agents. We observed that, during the first outbreak of AGN, a strong humoral immune response is launched. This response declines over the course of the observation period. This suggests a weakening of specific humoral responsiveness, since we found a sizable pool of memory B cells. During the second outbreak of AGN, productive HIV infection occurs. Again we see the pattern of a slowly developing anti-HIV response that is noticeably weaker than a normal response (for which we use anti-AGN as an example). The fourth outbreak of AGN is weaker than those before, but of longer duration, and is preceded by a burst of productive HIV infection. These findings show that two latent antigens can successfully enhance or suppress each other's acute phases and humoral responses. The resulting non-linear coupling between the two agents may render the outcome of the responses against each single agent unpredictable. By assigning more properties to describe AGN, we have started to investigate the implications of these findings for different therapy strategies, a project that will allow us to identify specific non-linear behaviors and thereby reduce uncertainty in therapy design.

5. CytoTalk

One of the many difficult problems in AIDS is to describe the way in which the cytokine immunoregulators influence disease progression. The study of cytokine secretion patterns in the context of HIV infection is important for several reasons. For example, cytokine-dependent cell activation may lead to the conversion of latent or chronic infection to an acute, productive one [31,37]. Productive infection causes intermittent bursts of virus particles, which can lead to the spread of virus to other $CD4^+$ cells and their destruction. Also, viral proteins bearing similarities to cytokines may unspecifically activate regulatory functions in uninfected cells. The resulting distortions of normal immunoregulatory patterns may influence the immune response to HIV and other pathogens. Furthermore, it is likely that under as yet to be identified disease conditions non-physiologic cytokine concentrations are reached. This can lead to lesions in surrounding tissues, or the unspecific elimination of healthy immune system cells, infected or not, at the place of action.

We have investigated the secretion patterns of Interleukins 1, 2 and 5 (IL1, IL2, IL5) and Interferon gamma (γ-IFN) with the goal of identifying periods during the course of disease in which increased levels of these immunoregulators can occur. From our data we derive a clear immuno-regulatory profile of HIV infection in the B-cell, T-cell, macrophage AIS. Fig. 8 shows, for example, several episodes of enhanced IL1 and γ-IFN secretion that were accompanied by substantial increases in the number of cytotoxic T cells and decreasing numbers of helper T cells. The secretion of cytokines IL1 and γ-IFN result from the uptake of viral particles by antigen-presenting cells and cell–cell contact activation of specific helper T cells during antigen presentation. The amount of detectable IL2 is noticeably smaller than that of the other cytokines. Similar observations were reported in earlier experimental studies [38,39], but were attributed to impairments at the level of IL2-producing cells. In the simulations, this effect does not originate from impairments at the level of IL2-producing cells; instead, it is due to the presence of many

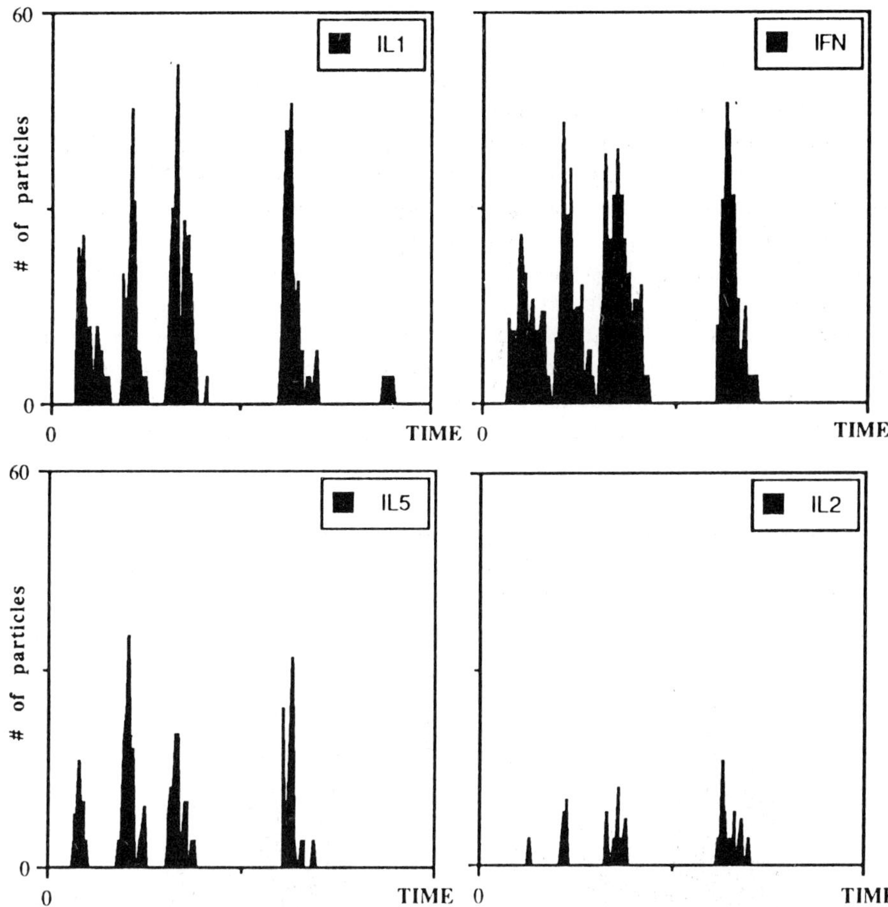

Fig. 8. Shown in these diagrams are the concentrations, over time, of the four regulatory factors that we used to represent the physiological connectivity of the B-cell, T-cell, macrophage artificial immune system during the simulation experiment described in figs. 3 and 4. Mathematically, the spike trains represent a sequence of non-stationary events involving cell device activation, differentiation, and proliferation that may be interpreted, in a more general context, as the temporary allocation and reallocation of communication links between cells in close proximity. In the AIS, it is possible to track these events, which allowed to obtain information about how different infections are translated into the language of immune system physiology. Interpreted biologically, the peaks and valleys represent periods of cytokine secretion at different levels. This information may be important for therapy or vaccine design.

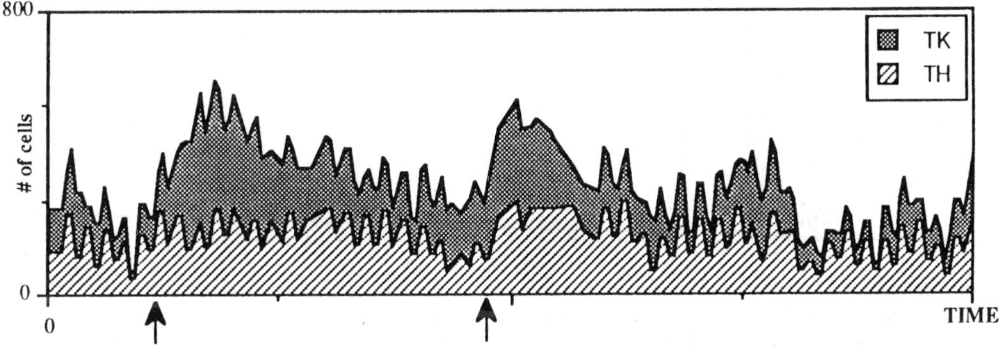

Fig. 9. This chart shows the sizes, over time, of the T helper (TH) and T killer (TK) cell populations when we injected the infected system with exogenous Interleukin 2 (IL2). The first (second) time point of the injection is indicated by the first (second) arrow. The time series evidences a strong proliferation in the TK cell population triggered by the growth factor, whereas the TH cell population remains unaffected.

IL2-responsive cells in close proximity to IL2-producing cells. To identify the type of the responsive cells, we investigated the effects of exogenously applied IL2 on the overall population of T cells.

Therapeutic applications of cytokines have been frequently discussed and used in the treatment of cancers [40] and have recently also been applied in the treatment of AIDS [38,41]. When we tested the effect of exogenous IL2 in our simulation experiments, we found that this T-cell growth factor selectively boosts the cytotoxic T-cell (CD8$^+$) population (fig. 9). This shows that many CD8$^+$ T cells have been activated and are IL2 responsive. As indicated by the IL5 and γ-IFN secretion pattern obtained from a different experiment (fig. 8), many CD4$^+$ T cells have been properly activated during antigen presentation and subsequently achieved their normal effector functions. We noticed that few memory helper T cells survive following secondary activation. This indicates that, following the periods of active cytokine secretion, many CD4$^+$ T cells are in the resting inducible or memory states and are therefore unresponsive to IL2.

Since the booster effect of IL2 is short-lived, these findings may suggest the possibility of using IL2

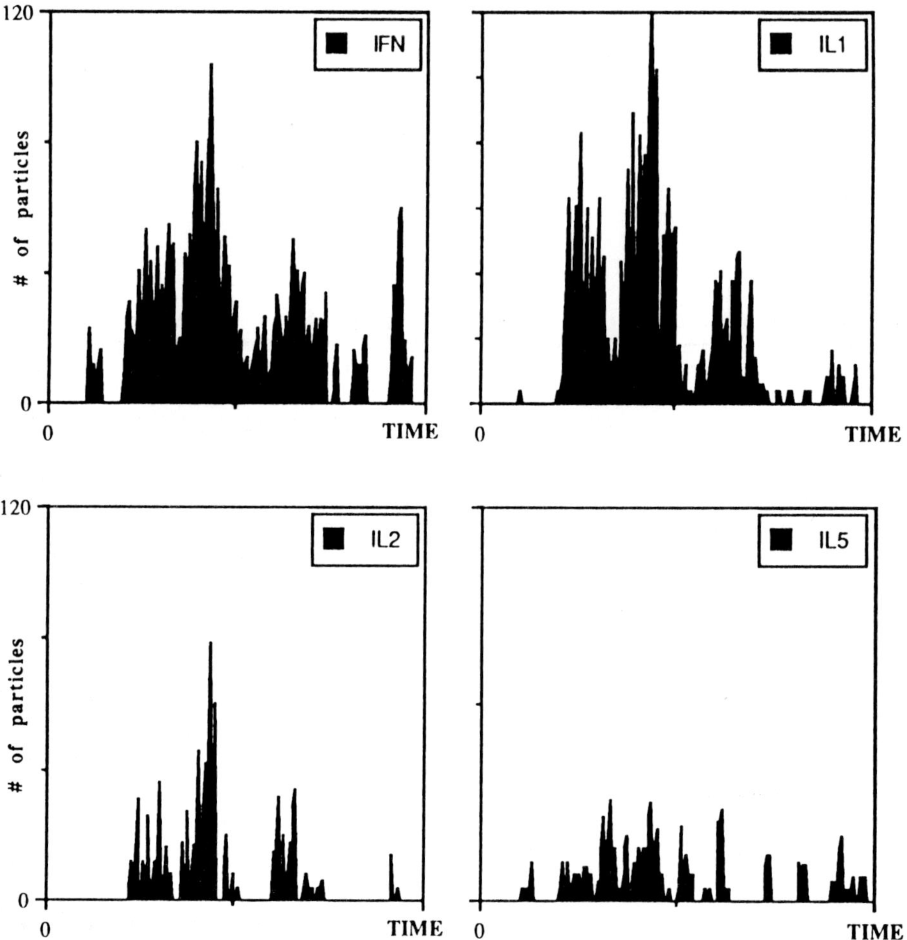

Fig. 10. Shown in these diagrams are the concentrations, over time, of the four regulatory factors of the B cell, T cell, macrophage artificial immune system during the simulation experiments described in figs. 6 and 7. When compared to the regulatory profiles shown in fig. 8, the profiles here indicate an increasing disorder in the coordination between individual immune responses which may ultimately reflect a growing potential for chaotic behavior such as heterogeneous antibody responses that have enhancing, suppressing or no effect on HIV.

in order to rapidly amplify the cellular immune response against HIV at times when many latently infected cells have been activated. This approach to controlling disease dynamics by targeted growth factor applications could be effective, if there were to exist a controlled approach to activating latently infected cells that could be safely used in conjunction with HIV-neutralizing antibody or immunoadhesins [42]. We are currently investigating the systemic implications to this combined strategy.

Uncontrolled cytokine regulation has been pointed out as a possible cause for the lack of natural lymphocyte replenishment from bone marrow derived pluri-potent stem cells [36,37], thus indirectly contributing to the global depletion of CD4$^+$ T cells. We conducted a series of experiments to explore the effects of multiple infections on the IL1, IL2, IL5 and γ-IFN secretion patterns (fig. 10). When compared to the distinct burst patterns that we saw during a single infection, we find that each cytokine is secreted

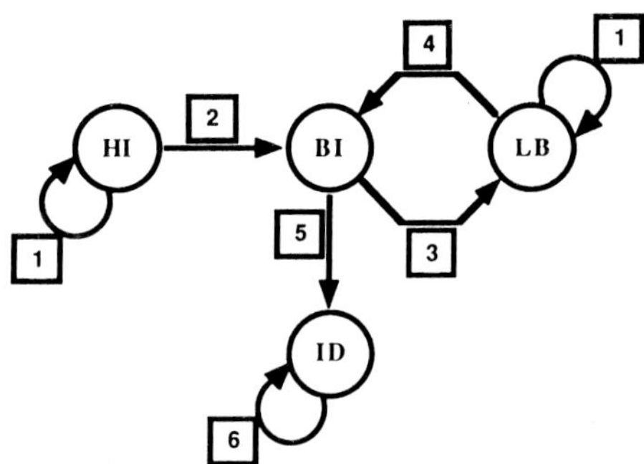

Fig. 11. Phase graph of the global HIV/AIS dynamical system. By exposing the system to different experimental conditions involving the single HIV antigen in free or cell-associated form, a combination of HIV and another antigen, or exogenous applications of IL2, we were able to identify a preliminary four-stage model of HIV pathogenesis which has one entry state (HI), a loop (BI) $\rightarrow$ (LB) $\rightarrow$ (BI), and a terminal (ID). The vertices define the following global dynamical states: (HI) the host's initial immune system status by the time of virus introduction, (BI) a bifurcation state from where the system can reach one of two other states, namely the latency basin (LB), and the immune deficiency state (ID). The edges, as depicted by the transitions between the dynamical states, derive as follows. We showed that a host will remain, as indicated by the self-transition (1), in the normal, broadly responsive and functionally diverse immune system status (HI) as long as HIV exists only in the cell-associated form, and the environment is kept antigen-free. A transition, marked (2), to the next following state (BI) will however occur, if HIV manifests in free form, or if other infectious agents materialize. When (BI) is entered for the first time, we find increased levels of virus titer. Our data show that a rapid transition, marked (3), into (LB) will occur due to the binding of HIV to CD4$^+$ cells or due to the uptake of virus particles by antigen-presenting cells. When (BI) is entered a second or more times, we find, in addition to the elevated virus titer, increased cytokine concentrations (except for IL2), detectable anti-HIV antibody titer, reservoirs of latently infected macrophages and T cells that the immune system is unable to clear, and an increase in the population of cytotoxic T cells. (LB) is characterized by well-established reservoirs of latently infected macrophages and CD4$^+$ T cells and low virus titer. Since the latent form is favorable to HIV's natural lifecycle, one may think of (LB) as a dynamically attracting state, hence the term basin. If (LB) is entered a second or more times, we are likely to find newly acquired patterns, e.g. diminished IL2 responsiveness in CD4$^+$ T cells, sero-conversion, and increased IL2 responsiveness in the CD8$^+$ cell population. Our simulations showed that the host remains in the latency basin unless other infections materialize, at which point a transition, labelled (4), into the bifurcation state (BI) occurs. Arrived again at (BI), the immune system status is marked with uncertainty, since a variety of immunological forces (e.g. cytokines, cytotoxic T cells, or abundant, but unspecific, antibodies) may interact chaotically, and therefore make it difficult to predict the future destination state: The system may progress towards immune deficiency, i.e. enter (ID), or go into remission, i.e. enter (LB). The actual causes for the transition from (BI) to (ID), labelled (5), are presently unknown. Empirical studies have shown that the immunedeficiency state can be maintained for varying periods of time, a possibility that is indicated by the self-transition (6). The global dynamical model discussed here can be used to calculate the rate of disease progression in terms of the number of times the system traverses the loop (BI) $\rightarrow$ (LB) $\rightarrow$ (BI) and the speed at which individual transitions occur. This indicates the possibility of using phasegraphs to predict epidemiology by systematically modifying initial immune system statuses and the state-transition conditions. Cell devices can again be used to implement this approach.

in higher concentrations and also almost continually. The reason for this is that the outbreaks of infection overlap between the two agents in such a way as to mimic a persistently acute infection. The resulting amplification or diminution of particular cytokines might seriously obstruct their controlling effect over the immune response against either agent.

6. Concluding remarks

In this paper we described a new method which allows the simulation of large and heterogeneous cellular systems. We have applied this method to create a variety of scenarios, which enables us to predict dynamical events at the physiologic and population levels of a B-cell, T-cell, macrophage immune system during HIV infection. In particular, we were able to show that immunological stresses, such as concomitant infections, are a critical driving force of disease progression. This finding shows that the set of rules we applied to implement the interaction between the infectious agent and the artificial immune system, is minimal, and sufficient to lead to disease. We also established systemic patterns, e.g. the predominance of the macrophage reservoir, increased IL2 responsiveness of the cytotoxic T-cell population, higher productions of IL1 and γ-IFN than IL2 and IL5, that might help to explain the surprising asymmetry that is the long-term effect of HIV on its two equally susceptible target cell populations. The different aspects of our findings are summarized in a preliminary model of HIV pathogenesis (fig. 11).

Due to its flexibility, its capacity to provide *a priori* predictions that can give rise to further experimental studies, and its expandibility, the cellular device system approach will allow us to address a wide spectrum of immunological and also neuro-immunological complications of HIV infection in the future. Clearly, all our findings apply to the class of pathogens for which we chose HIV as a representative and may therefore contribute to the more general understanding of retroviral infections and therapy design.

Acknowledgements

HBS would like to thank Drs. J. Salk, C.E. Müller-Sieburg, D. Mosier, D. Wilson, R. Olshen and J. Rice for many enlightening discussions and helpful suggestions. This work is supported by NIMH grant 1 R29 MH45688-01 to HBS.

References

[1] M. Cohn, Anticipatory mechanisms of individuals, in: Control Processes in Multicellular Organisms, Ciba Foundation Symposium, eds. G.E.W. Wolstenholme and J. Knight (Churchill, London, 1970).

[2] A. Miyajima et al., FASEB J. 2 (1988) 2462.

[3] A.G. Dalgleish, P.C.L. Beverly, P.R. Clapham, D.H. Crawford, M.F. Greaves, and R.A. Weis, Nature 312 (1984) 763.

[4] D. Klatzmann, E. Champagne, S. Chamaret, J. Gruest, D. Guetard, T. Hercend, J. Gluckman and L. Montagnier, Nature 312 (1984) 767.

[5] Q.J. Sattenau and R.A. Weiss, Cell 52 (1988) 631.

[6] A.M. Prince, J. Moor-Jankowski, J.W. Eichberg, H. Schellekens, R.F. Mauler, M. Girard and J. Goodall, Nature 333 (1988) 513.

[7] J.M. McCune, R. Namikawa, H. Kaneshima, L.D. Shultz, M. Liebermann and I.L. Weismann, Science 241 (1988) 1632.

[8] D.E. Mosier, R.J. Gulizia, S.M. Baird and D.B. Wilson, Nature 335 (1988) 256.

[9] R.M. May and R.M. Anderson, Nature 326 (1987) 137.

[10] R.M. Anderson and R.M. May, Nature 333 (1988) 514.

[11] P.J. Denning, Am. Sci. 76 (1988) 552.

[12] P.J. Denning, Am. Sci. 75 (1987) 347.

[13] J. Glimm, SIAM News 21 (5) (1988) 1.

[14] S.P. Layne, T.G. Marr, St.A. Colgate, J.M. Hyman and E.A. Stanley, Nature 333 (1988) 511.

[15] A.S. Perelson, ed., Theoretical Immunology, SFI Series in the Sciences of Complexity, Nos. 2, 3 (Addison–Wesley, New York, 1988).

[16] A. Jameson, Science 245 (1989) 361.

[17] T. Kohonen, Neural Networks 1 (1988) 3;
S. Grossberg, Neural Networks 1 (1988) 17.

[18] S. Wolfram, Theory and Applications of Cellular Automata, Advanced Series on Complex Systems, Vol. 1 (World Scientific, Singapore, 1986).

[19] T. Toffoli and N. Margolus, Cellular Automata Machines (MIT Press, Cambridge, MA, 1987).

[20] J.H. Holland, Physica D 22 (1986) 307.

[21] H.B. Sieburg, A logical dynamic systems approach to the regulation of antigen-driven lymphocyte stimulation, in: Theoretical Immunology, SFI Series in the Sciences of Complexity, Nos. 2, 3, ed. A.S. Perelson (Addison–Wesley, New York, 1988).

[22] H.B. Sieburg, The cellular device machine: point of departure for large-scale simulations of complex biological systems, Int. J. Comput. Math. Appl., in press.

[23] H.B. Sieburg and C.E. Müller-Sieburg, Nucl. Phys. B. (Proc. Suppl.) 2 (1987) 615.

[24] B.D. Jameson and R. Ahmed, J. Exp. Med. 169 (1989).

[25] I.M. Roitt et al., Immunology, 2nd Ed. (Gower Medical Publishing, London, New York, 1989).

[26] E.R. Unanue and P.M. Allen, Hospital Practice (15 April 1987) 87.

[27] D.C. Linch, D.L. Wallace and O. Flynn, Immunol. Rev. 95 (1987) 137.

[28] E.F. Braakman, T.M. Rotteveel, G. van Bleek, G.A. van Seventer and C.J. Lucas, Immunol. Today 8 (1988) 265.

[29] A. Alcover, D. Ramarli, N.E. Richardson, H. Chang and E.L. Reinherz, Immunol. Rev. 95 (1987) 5.

[30] S.D. Gowda et al., J. Immunol. 142 (1989) 773.

[31] J.S. McDougal, A. Mawle, S.P. Cort, J.K.A. Nicholson, G.D. Cross, J.A. Scheppler, D. Hicks and J. Sligh, J. Immunol. 135 (1985) 3151.

[32] R.N. Germain, Cell 54 (1988) 441.

[33] H.K. Lyerly, T.J. Matthews, A.J. Langlois, D.P. Bolognesi and K.J. Weinhold, Proc. Natl. Acad. Sci. US 84 (1988) 460.

[34] D.E. Lewis et al., FASEB J. 2 (1988) 251.

[35] J.A. Levy, Nature 333 (1988) 519.

[36] A.R. Moss et al., British Medical J. 290 (1988) 745.

[37] A.S. Fauci et al., Science 239 (1988) 617;
Th.M. Folks et al., Proc. Natl. Acad. Sci. US 86 (1989) 2365.

[38] J. Alcover-Varela, D. Alarcon-Segovia and C. Abud-Mendoza, Clin. Exp. Immunol. 60 (1985) 31.

[39] S. Gupta et al., J. Clin. Lab. Immunol. 22 (1987) 113.

[40] M.T. Loetze and S.A. Rosenberg, CA-A, Cancer J. Clin. 38 (2) (1988) 69.

[41] A.S. Fauci et al., Proc. Natl. Acad. Sci. US 83 (1986) 9278.

[42] D.J. Capon et al., Nature 337 (1989) 525;
A. Traunecker et al., Nature 339 (1989) 68.

Physica D 45 (1990) 229–253
North-Holland

INVERTIBLE CELLULAR AUTOMATA: A REVIEW

Tommaso TOFFOLI and Norman H. MARGOLUS
MIT Laboratory for Computer Science, Cambridge, MA 02139, USA

Received 15 January 1990

In the light of recent developments in the theory of invertible cellular automata, we attempt to give a unified presentation of the subject and discuss its relevance to computer science and mathematical physics.

1. Introduction

1.1. Preliminaries

One of the goals of computer science is to provide abstract models of concrete computers, i.e., of whatever computing apparatus can ultimately be built out of the physical world. In this, one seeks *expressiveness* (whatever aspects of the computer are deemed relevant should be captured by the model) and *accuracy* (whatever one can prove about the model should be true about the computer). *Invertible cellular automata* (ICA) are an important development in this direction, of significance comparable to the introduction of Turing machines (and similar paradigms of effective computation) in the late '30s.

Turing machines represent a conscious effort [81] to capture, in axiomatic form, those aspects of physical reality that are most relevant to computation. In this, they are quite unlike Cantor's transfinite sets or similar inventions of the romantic era, and much more like Euclid's principles, which were meant to describe the geometry of our physical world (and one should not forget that, for the Greeks, computation was synonymous with geometrical construction).

Cellular automata are more expressive than Turing machines, insofar as they provide explicit means for modeling *parallel* computation on a spacetime background. However, both classes of models are indifferent to one fundamental aspect of physics, namely *microscopic reversibility*, and

thus help create the illusion that, computationally speaking, one lives in a fairy world where the second principle of thermodynamics is not enforced, and "perpetual computation" (in the same sense as "perpetual motion") is possible. By explicitly coming to terms with this aspect of reality, *invertible cellular automata* provide a modeling environment that is more accurate and, in the end, more productive.

Only a few years ago, what was known about ICA could be summarized in a few lines – and was not very exciting either. Today, one can tell a more interesting story, and we shall try to do so in this paper.

Our involvement with ICA represents the convergence of several research trails, including
- Concurrent computation in networks having uniform structure.
- Reversible (or "information preserving") computing processes.
- Fundamental connections between physics and computation.
- Foundations of relativity.
- Quantum computation.
- Fine-grained modeling of physical systems.
- High-performance simulation of cellular automata.
- Data encryption.

1.2. An apology

We have many things to say, and we have a quite varied audience in mind. We want to make sure that the readers perceive the essential issues before plunging into endless technicalities. For these reasons, we shall follow an informal approach when this seems to enhance clarity, and we shall often present a specific example rather than the most general case.

2. Invertible cellular automata

2.1. Cellular automata

Cellular automata are abstract dynamical systems that play a role in discrete mathematics comparable to that played by partial differential equations in the mathematics of the continuum. In terms of structure as well as applications, they are the computer scientist's counterpart to the physicist's concept of a 'field' governed by 'field equations'. It is not surprising that they have been reinvented innumerable times under different names and within different disciplines. The canonical attribution is to Ulam and von Neumann [82,84] (circa 1950).[1]

Concise formal definitions are given in section 2.4 (for a more complete formal treatment, see refs. [68,79]). Intuitively, a *cellular automaton* is an indefinitely extended network of trivially small, identical, uniformly interconnected, and synchronously clocked digital computers.

More specifically, we start with an indefinitely extended n-dimensional lattice which represents "space" (typically, n equals 1, 2, or 3 in physical modeling applications). To each site of this lattice, or *cell*, there is associated a state variable, called the *cell state*, ranging over a finite set called the *state alphabet* (typically, the cell state consists of just a few bits' worth of data).

"Time" advances in discrete steps; the dynamics is given by an explicit recipe, called the *local map*, which is used at every time step by each cell to determine its new state from the current state of certain cells in its vicinity.

The local map itself is the composition of two operators, namely, the *neighborhood*, which specifies *which* cells affect the given cell, and the *table*,

which specifies *how* those cells affect it. In more detail:

(i) The *neighborhood* lists the relative positions, with respect to the generic cell (in this context also called *center cell*), of a finite number of cells called the cell's *neighbors*. (The neighbors of a cell need not coincide with the cell's "first neighbors" in the array. They may include the cell itself or cells that are several sites away, while cells in between may be skipped. All that is required is that they be finite in number, and be arranged in the same spatial pattern with respect to each cell.)

(ii) The *table* takes the states of a cell's neighbors as arguments, and returns as a result the cell's corresponding new state.

Thus, a cellular automaton's laws are *local* (no action-at-a-distance) and *uniform* (the same rule applies to all sites at all times); in this respect, they reflect two fundamental aspects of physics. Moreover, the system's laws are *finitary*, that is, by means of the local map one can explicitly construct *in an exact way* the forward evolution of an arbitrarily large portion of a cellular automaton through an arbitrary length of time, all by finite means. It is such strong effectiveness built in their definition that makes dynamical systems based on cellular automata appealing to the computer scientist. In continuous dynamical systems, such as those defined by differential equations, the state variables range over an uncountable set, and one has to accept a weaker standard for "effectiveness"; i.e., a specification of the dynamics is accepted as effective if the state of any finite portion of the system at any future time can be computed with an *arbitrarily small error* by finite means. In cellular automata, we demand and obtain zero error.

In this sense, cellular automata present themselves as a finitary alternative to the methods of the calculus in the modeling of spatially extended systems.

2.2. Invertibility

An assignment of states to all cells, i.e., a state for the entire cellular automaton, is called a *configuration*. By applying the local map to every

---

[1] At about the same time but quite independently, Zuse [91] proposed structures, intended as digital models of mechanics, that are essentially cellular automata.

site of the array, from any configuration q one obtains a new configuration q', called its *successor*. Thus, the local map defines a transformation $q \overset{\tau}{\mapsto} q'$, called the *global map*, on the set of configurations.

A cellular automaton is *invertible* if its global map is invertible, i.e., if every configuration – which, by definition, has exactly one successor – also has exactly one predecessor.

In the context of dynamical systems, invertibility coincides with what the physicists call 'microscopic reversibility'. This should not be confused with 'invariance under time reversal', which is a stronger property.[#2]

Initially, cellular automata were used chiefly as "toy models" for phenomenology associated with dissipative (i.e., macroscopically irreversible) processes; typical topics were biological organization [40,23], self-reproduction [84], chemical reactions, and visual pattern processing [59]. Since the interest was more in exploring the *consequences* of irreversible behavior rather than its *origins*, it was not only harmless but actually expedient (given the severely limited computing resources) to use models where irreversibility happened to be built-in.[#3] Thus, it is not surprising that no need was felt for ICA.

It should be noted that the issue of invertibility wasn't even present in the minds of most cellular automata investigators. To the few to whom it was, it wasn't at all clear whether ICA could actually lend themselves to the modeling of microscopically reversible physics.

The perceived difficulties were of two kinds. On one hand, there were no practical procedures known for constructing nontrivial ICA; on the other, it was suspected and argued that ICA did not have adequate computing capabilities.

Both of these difficulties have now been amply removed. Indeed, ICA have become an important tool of computational physics in applications where the explicit modeling of reversible phenomena is concerned. Moreover, they are playing an increasingly important role as conceptual tools of theoretical physics.

2.3. Historical notes

As already mentioned, cellular automata were initially treated as some sort of conceptual erector set – a plaything for interdisciplinary biologists and computer scientists – and drew little attention from professional mathematicians. This may explain why the issue of invertibility – which in mathematical systems theory is one of obvious priority – was slow to be explicitly recognized by the cellular automata community.

In 1962, Moore [51] asked whether there could exist "Garden of Eden" configurations – i.e., configurations that do not have a predecessor – and proved that, under certain conditions, if a configuration has more than one predecessor then there must be a configuration that has none [51]. The converse was proved by Myhill [53] in 1963.

Moore's and Myhill's results originated a lengthy "Garden of Eden" debate (see references in ref. [61]), which brought to light a number of subtle issues somehow related to invertibility. But invertibility was explicitly addressed only in 1972, in seminal papers by Richardson [60] and Amoroso and Patt [2].[#4]

After that, theoretical work on invertibility in cellular automata proliferated (see refs. [3,61,54] and [46–48,90,35]). In spite of that work, however, for many years the most interesting ICA actually exhibited remained an extremely simple-minded one (the longest orbit is of period two!), discovered by Patt through brute-force enumeration [56]. ICA continued to "appear to be quite rare" [2].

Not only rare, but also simple-minded! On the

[#2] Let $\tau \colon Q \to Q$ be an invertible dynamical system. A bijective mapping $\phi \colon Q \to Q$ obeying appropriate regularity properties (e.g., continuity, translation invariance, etc., depending on the context) is called a *time-reversal* operator if $\tau^{-t} = \phi^{-1}\tau^t\phi$; the system is *invariant under time reversal* if it admits of such an operator. Thus, a time-reversal invariant system is not only invertible but also *isomorphic to its inverse*. Hamiltonian mechanics has the well-known time-reversal operator $\phi \colon \langle q, p \rangle \mapsto \langle q, -p \rangle$.

[#3] This interest is not abating; see, for instance, refs. [55,7,8,10].

[#4] Unbeknownst to those authors, systems that are in essence one-dimensional cellular automata had already been studied in an abstract mathematical context by Hedlund and associates as early as 1963 [30,31]; both Richardson's results on invertibility (section 4.3) and Patt's search for ICA (section 5.3) had been anticipated by Hedlund's school.

basis of various kinds of circumstantial evidence (cf., e.g., ref. [63]), Burks conjectured that an ICA cannot be computation-universal [11] (note that that was at a time when computation universality was being turned up under almost every stone), and soon Aladyev [1] appeared to have proved Burk's conjecture.

Finally, except for the one-dimensional case [2], no one even knew of a systematic procedure for telling whether or not a cellular automaton was invertible.

In summary, for a long time ICA seemed to lack any appeal or promise.

Following fundamental results by Bennett on invertible Turing machines [6], in 1976 one of us (Toffoli) proved the existence of ICA that are computation- and construction-universal [67], and noted the relevance of this, in principle, to the modeling of physics [68].

In the same year, and unnoticed by the (then meager) cellular automata establishment, Pomeau [26] discussed, as a model for hydrodynamics, a "lattice gas" that is in fact an ICA and a useful stylization of certain microscopically reversible physical interactions.

Independently, Fredkin had been studying invertible recurrences as models of dynamical behavior; had arrived at techniques for synthesizing arbitrary sequential behavior out of invertible Boolean primitives [19]; and had studied a class of ICA (see section 5.4) that displayed some analogy with Lagrangian mechanics.

Rapid, synergistic developments finally started taking place in the early '80s.

In 1981, a conference on "Physics and Computation" [20] explicitly addressed the theme of *fundamental* connections between physics and computer science (rather than more incidental ones, such as computer programs for the numerical integration of differential equations). Ideas such as "virtually nondissipative computation" [19] and "quantum computation" [5,18] started gaining legitimacy.

The links between a number of physicists and computer scientists interested in these themes were tightened by a follow-up workshop on Moskito Island (1982), where an early prototype of cellular automata machine was demonstrated by us.

Wolfram's 1983–1986 sortie into the cellular au-

tomata arena [85–89], stimulated by that workshop, was in turn a determining factor in introducing a generation of mathematical physicists to the cellular-automaton paradigm.

Inspired by Fredkin's "billiard-ball" model of computation [19], Margolus arrived in 1983 at a very simple computation-universal ICA[41] that is suggestive of how a computer could in principle be built out of microscopic mechanics. At about the same time, Vichniac [83] and Creutz [13] pioneered the use of cellular automata for the microcanonical modeling of Ising spin systems.

The introduction of dedicated cellular automata machines [71] encouraged much new experimental work on ICA, and stimulated further theoretical developments.

For instance, according to Pomeau, seeing (at a second Moskito Island workshop, in 1984) his lattice-gas model running on one of these machines made him realize that what had been conceived primarily as a *conceptual* model could indeed be turned, by using suitable hardware, into a *computationally accessible* model. This stimulated his interest in finding lattice-gas rules that would provide better models of fluids. In the past few years, lattice-gas hydrodynamics has grown into a substantial scientific business (see section 5.6).

In turn, the growing interest in fine-grained models of physics and in their potential applications to important practical problems created the need for computers capable of handling large models of this kind much more efficiently than conventional scientific computers [45]. A second generation of cellular automata machines, whose development is almost complete [44,78], reflects in its architecture the objective to efficiently support the simulation of ICA– which are likely to constitute a major portion of its fare.

2.4. Terminology

The present section complements with precise definitions and notation the informal terminology introduced in sections 2.1–2.2. Refer to refs. [31,72] for more abstract, but equivalent, definitions given in terms of continuity in the Cantor-set topology.

Space. Let $S = \mathbf{Z}^n$ denote the Abelian group of translations of an n-dimensional lattice I onto itself. It will be convenient to call the elements of S *displacements*. The sum and the difference of two displacements are again displacements. The application of a displacement $s \in S$ to a site $i \in I$ yields a new site denoted by $i + s$. The *difference* $i' - i$ between two sites is the displacement s such that $i' = i + s$.

Interconnection. A *neighborhood* is a finite set of displacements (i.e., a subset of S). Typically, a neighborhood X is applied as an operator to a site i, yielding a set of sites. That is, the X-*neighborhood* (or simply the *neighborhood*, when X is understood) of i is the set $i + X = \{i + x \mid x \in X\}$; the elements of $i + X$ are the *neighbors* of i, and are naturally indexed by the elements of X.

The *radius* of X is the length of its longest element.[#5]

State. Given a lattice I and a nonempty state alphabet A, the set Q of *configurations* (of A over I) is the Cartesian product of copies of A indexed by the set I, i.e., $Q = A^I$. The ith component, in this product, of a configuration q is called the state of site i in configuration q, and is denoted as usual by q_i. Note that $q_i \in A$ can be thought of as the result of applying to configuration $q \in A^I$ the projection operator $[i]$ associated with the ith coordinate of the Cartesian product, i.e., $q_i = [i]q$.

More generally, a *neighborhood projection* operator $[i + X]$ will extract from a configuration q the collective state of the neighbors of i, denoted by $[i + X]q$ or q_{i+X}. Note that $q_{i+X} \in A^X$.

Dynamics. A *local map* is a pair $\lambda = \langle X, f \rangle$, where X is a neighborhood and f a *table*, i.e., a function of the form $f: A^X \to A$. The table f can be applied to any site i of a given configuration q through the agency of the neighborhood projection operator, which will extract from q and supply to f the correct set of arguments. Let q_i' be the result of this application, i.e.,

$$q_i' = f q_{i+X} \ (= f[i + X]q). \tag{1}$$

The symbol q_i' can be interpreted as the state at site i of a new configuration q'. The relation $q' =$

τq defines a new function $\tau: Q \to Q$, called the *global map* induced by the local map λ.

Note that the local map can be thought of as a function $\lambda: \langle Q, I \rangle \to A$ defined (cf. (1)) by

$$\lambda(q, i) = \langle X, f \rangle(q, i) = f q_{i+X}. \tag{2}$$

The sequence of configurations obtained from an initial configuration q^0 by iterating the map τ will be denoted by

$$q^0, q^1, q^2, \ldots \tag{3}$$

where the superscript represents the sequence index rather than an exponent.

A table f, formally given as a function of k arguments (k is the size of the neighborhood X), may happen to depend vacuously on some of these arguments. In that case, one can maximally reduce neighborhood and table in an obvious way, yielding the *effective* neighborhood and the corresponding table – which together make up the *reduced* local map. Unless otherwise noted, we shall tacitly assume that local maps are given in reduced form.

3. Universality

Little needs to be added here on the universality theme.

In ref. [67], the computation universality of ICA was proved by showing that every computation-universal cellular automaton can be embedded in an invertible one having one more dimension. This left open the question of whether *one*-dimensional ICA could be computation-universal. A positive answer was recently given by Morita and Harao [50].

The constructions of refs. [67,50] are more concerned with existence than with efficiency. More direct constructions can be more instructive as well as more efficient. Indeed, if one wants to build a general-purpose computing structure within a cellular automaton, the most practical approach is to start with a local map that directly supports logic gates and wires, and then build the appropriate logic circuits out of these primitives [4,77]. As explained in refs. [70,19], in an *invertible* cellular automaton the gates will have to be invertible; because of this constraint, a complete, self-contained

---

[#5] For our purposes, the Euclidean metric will do, even though it is an overkill.

logic design will have to explicitly provide, besides circuitry for the desired logic functions, additional circuitry for functions (analogous to energy supply and heat removal in ordinary computers) concerned with entropy balance. This issue, which had been bypassed in ref. [67], is directly addressed by ICA such as the BBM model devised by Margolus [41].

4. Decidability

One of the first questions to come to mind is, of course, "How does one tell if a cellular automaton is invertible?" We shall now present the essential aspects of this question; further details will be given in the following sections.

4.1. A parable

FOR SALE: Local map λ of invertible cellular automaton SPRIZE. Interesting behavior, lots of fun. $29.95. Call John at ×3194.

This ad catches my attention. I already have a bunch of cellular automata at home, that I can run on my personal computer, but this one claims to be invertible: its global map τ has an inverse τ^{-1}, and by running τ^{-1} I can watch the automaton go backwards in time! I send my check. Four days later I receive a diskette containing a data file SPRIZE.LOC (which, I presume, tabulates the local map λ of SPRIZE). I bring up my cellular-automaton simulation program, load SPRIZE.LOC, initialize the screen with a blob of random junk on a clear background, and hit the RUN key. The blob starts churning – perhaps it's spreading a bit – yes, it's spreading, but very slowly – rather boring, I'd say. Wait – look at that long caterpillar crawling up the screen! How the heck did it get started?

Can I go backwards in time and see exactly how the caterpillar emerged out of the random blob? I look in the diskette directory, and I find a second file labeled -SPRIZE.LOC. That must be it – the local map $\overline{\lambda}$ of SPRIZE's inverse! I load it. There goes the caterpillar crawling backwards, curling up into some sort of cocoon at the edge of the blob, and finally dissolving into randomness! I make a few more experiments. Indeed, the cel-

lular automaton SPRIZE defined by the local map λ is invertible and $\overline{\lambda}$ is the local map of its inverse.

That's a happy conclusion – but the story could have ended differently.

Suppose I didn't find a file -SPRIZE.LOC. What good is knowing that SPRIZE is invertible if I don't have an effective way to run its inverse? Worse yet, in this situation how can I be sure that John's ad was truthful – that SPRIZE *is* invertible? Can I prove it? Or perhaps disprove it and claim a refund?

Assuming that SPRIZE is after all invertible, can I find $\overline{\lambda}$ by myself, starting from λ? How long will that take? By definition, $\overline{\lambda}$ is a finite object, and I can sequentially generate all possible candidates. But how would I recognize the right one? And is the very existence of $\overline{\lambda}$ guaranteed? In other words, if the global map τ has a local description, does it follow that also its inverse τ^{-1} has a local description – that the inverse of a cellular automaton is again a cellular automaton?

4.2. The finitary connection

To summarize, for any given cellular automaton we would like to know whether or not it is invertible; if it is, we would like to have a finite recipe also for its backward evolution, i.e., the *inverse* local map – as contrasted to the *direct* local map of the forward process.

For the sake of comparison, let's look at a dynamical system defined by a differential equation. Consider, for example, the evolution of the temperature distribution $q(x,t)$ along a metal bar, according to the "heat" equation

$$\frac{\partial q}{\partial t} = \frac{\partial^2 q}{\partial x^2}. \tag{4}$$

This equation can be thought of as a local recipe for an "infinitesimal" forward step,

$$q(t+\mathrm{d}t)|_x = q(t)|_x + \left.\frac{\partial^2 q}{\partial x^2}\right|_x \mathrm{d}t \tag{5}$$

and by just turning +'s into −'s one immediately obtains a local recipe for an infinitesimal backward step. Thus, whenever the forward evolution is defined, so is the backward one, and a local inverse recipe is known as soon as a direct one is. In conclusion, with differential equations there is an immediate connection between direct and inverse local recipes. (However, one should keep in mind that, as noted in section 2, these local recipes are of a less effective kind than those of cellular automata.)

On the other hand, with ICA, the only connection we have in principle between direct and inverse local maps is through a *non-finitary* route, as is illustrated by the following diagram:

$$
\begin{array}{ccc}
\textit{finitary} & & \textit{non-finitary} \\
\textit{constructs} & & \textit{constructs} \\
\text{direct local map} & \longrightarrow & \text{direct global map} \quad (6) \\
 & & \downarrow \\
\text{inverse local map} & \longleftarrow & \text{inverse global map}
\end{array}
$$

Under what conditions can we establish a *finitary* route from a direct local map to an inverse one?

4.3. The fundamental lemmas

Whatever the technical difficulties in charting such a route, it is important to know that the endpoint exists and is recognizable, as shown by the following two lemmas.

Lemma 4.1 (Richardson [60]) *If a cellular automaton is invertible, then its inverse is a cellular automaton.*

This is a fundamental result. It tells us that if the global process described by a local map is invertible, then also the inverse global process has a local map, i.e., it can be described in local terms. On the other hand, the proof (which is based on topological arguments of a general nature) doesn't give any explicit method for constructing the inverse local map.

Lemma 4.2 *There is an effective procedure for deciding, for any two local maps λ and λ' defined on the same set of configurations, whether the corresponding global maps τ and τ' are the inverses of one another.*

Proof. The composition $\lambda'' = \lambda'\lambda$ is a new map operating on the neighborhood consisting of the neighbors (according to λ) of the neighbors (according to λ') of the generic cell. If for any value

a of this cell the map λ'' returns the value a, independently of the values of the other neighbors, then $\tau'\tau$ is obviously the identity function. If a value different from a is returned from some choice of values for the other neighbors, then $\tau'\tau$ obviously differs from the identity. $\square$

The construction in the above proof provides a way to verify whether an alleged inverse local map λ' of a direct local map λ is indeed such an inverse. This test is our touchstone for certifying that a cellular automaton is invertible. Whether the candidate λ' was generated by a rigorous construction, suggested by heuristic methods, dictated by an oracle, encountered on a search, or even arrived at by faulty arguments, if it passes the test it can be accepted without further question.

Conceivably, it might be possible to prove the invertibility of a cellular automaton without exhibiting a local map for its inverse. However, we don't know of any cases where this has been done. For all of the ICA known today, the inverse local map comes "bundled", as it were, with the direct one.

4.4. The fundamental theorems

From lemma 4.2 one immediately obtains

Theorem 4.3 *The class of invertible cellular automata is recursively enumerable*

Proof. Sequentially generate all local maps, say, in order of increasing complexity (measured in terms of state-alphabet size and neighborhood radius). For each item λ in this sequence, start a new enumeration of all local maps, and for each item λ' of this enumeration test whether λ' is the inverse local map of λ. Each match yields an ICA, and every ICA will eventually turn up in the course of this double enumeration. $\square$

In other words, invertibility in cellular automata is at least *semi*decidable: if a specific cellular automaton is invertible, the above procedure will positively let us know; however, if it is not invertible, at no moment in time will we be positively told of that.

How about *full* decidability? One partial result was already mentioned in section 2.3, namely,

Theorem 4.4 (Amoroso and Patt [2]) *There is an effective procedure for deciding whether or not an arbitrary one-dimensional cellular automaton, given in terms of a local map, is invertible.*

Amoroso and Patt thought that the techniques employed by them were "in principle extendable to arrays of higher dimension". Since, however, these techniques were "difficult to manage beyond dimension one", they expected that "generalizations of their results to higher dimensions" would "most likely require a different approach".

Since then, for almost twenty years a quest for these "generalizations" to more than one dimension went on with little success. (Invertibility and related properties for the one-dimensional case were revisited in [54,87,14,29].) Many equivalent characterizations of ICA were given [90,47,48,35], but none that offered a finitary handle on invertibility.

Finally, quite recently, Kari proved that

Theorem 4.5 (Kari [38,39]) *There is no effective procedure for deciding whether or not an arbitrary two-dimensional cellular automaton, given in terms of a local map, is invertible.*

His proof is by reduction to the undecidability of the tiling problem, and depends on the availability of a set of tiles having certain properties. He exhibits such a set, but the construction runs over fifteen pages; given the importance of the result, we hope that a shorter proof will be found soon.

Thus, the invertibility of a cellular automaton is, in general, undecidable. This has important implications, some of which we intend to discuss (section 8). But, lest the readers feel that they are groping completely in the dark, let us first balance the negative result of theorem 4.5 with a body of positive results concerning ICA.

5. Ways to make invertible cellular automata

Suppose one wanted to get a general feeling for what kinds of behavior are possible with ICA. One could start with a good assortment of these automata, and study a number of cases in detail.

The problem is, how does one put together such an assortment?

It turns out that of all cellular automata the invertible ones constitute a vanishingly small subclass [62]. Moreover, as we have seen in section 4, there is no effective procedure for determining whether or not an arbitrary cellular automaton (as specified by a local map) is invertible. Finally, even if one were willing to fall back on a brute-force search (theorem 4.3), a long search time would generate only a few items, and even those would be for the most part quite uninteresting.

In our experience, rather than spend an inordinate amount of resources on a blind search for these objects that are rare, hard-to-recognize, and more often than not quite plain, it is more rewarding to attempt the direct synthesis of special cases having certain desirable features. In this section, we shall discuss a number of synthesis techniques which yield a rich variety of ICA.

5.1. Trivial cases

Let us consider the following two cases:

(a) *Each cell is allowed to look only at itself as a neighbor.*   Then the cellular automaton reduces to a collection of finite, isolated systems – one per cell.

(b) *Each cell is instructed to copy, say, its left neighbor.*   Then the whole configuration will shift one cell to the *right* at each time step. This 'uniform shift' behavior can be factored out of the dynamics by a simple coordinate transformation of the form $x \mapsto x - t$. After this transformation, the global map reduces to the identity, and again the cellular automaton collapses into a collection of finite, noninteracting systems.

A cellular automaton whose neighborhood consists of at most one cell, as in (a) or (b) above, is called *trivial* [2].

Clearly, a trivial cellular automaton is invertible if and only if its effective neighborhood X consists of *exactly one* element, and its table, of the format $A \xrightarrow{f} A$, is invertible, i.e., is a *permutation* of the state alphabet A.

5.2. The incredible shrinking neighborhood

It would be nice if the invertibility of a cellular automaton could always be reduced, as in the previous section, to the invertibility of its table. Unfortunately, when the neighborhood X consists of more than one element, the table f as such cannot be invertible, since the two sets that appear in $A^X \xrightarrow{f} A$ have different cardinalities.

Let us concentrate on this point. From the viewpoint of the global map, $q^{\text{old}} \xmapsto{\tau} q^{\text{new}}$, the configuration q^{old} contains all the information needed to construct q^{new} (via τ); if the automaton is invertible, also q^{new} contains all the information needed to construct q^{old} (via τ^{-1}). This symmetry between the two directions of time is not preserved, in general, when the same dynamics is expressed by a local map $q^{\text{old}}_{i+X} \xmapsto{f} q^{\text{new}}_i$; in fact, while the new state of a cell is completely determined by the old state of its neighbors, the latter is not completely determined by the former – *even if the cellular automaton is invertible.*[#6]

Yet, the only way we know to construct ICA is to somehow manage to overcome the above difficulty of format, and in the end express the local map in terms of permutations of the state alphabet. Different ways of doing this are presented in the next three sections; here we'll try to capture the flavor of this approach.

In a trivial ICA (cf. previous section), a new configuration is obtained from an old one by applying a given permutation of the state alphabet, π, to each cell, i.e., $q^{\text{old}}_i \xmapsto{\pi} q^{\text{new}}_i$; of course, π itself is independent of i. Consider now the set Π of *all* permutations of the state alphabet, and make a dynamical system where each site i uses a permutation $\pi^t_i \in \Pi$ that may be *different* from site to site and from moment to moment, i.e.,

$$q^{t+1}_i = \pi^t_i q^t_i. \tag{7}$$

This system is not a cellular automaton, because its dynamics is space- and time-dependent, but is still invertible. In fact, since the π^t_i are assigned once and for all for each i and t, the inverse system is explicitly given by

$$q^t_i = (\pi^t_i)^{-1} q^{t+1}_i. \tag{8}$$

Now, the trick to restore space- and time-invariance is to make the choice of π^t_i depend on i and t not directly, but only indirectly, as a function of the "landscape" that can be seen from site i at time t (i.e., the state of the neighborhood not including the center cell itself). Now we are back to a cellular automaton, but, unless we are careful, we may lose invertibility, since the landscape itself will in general change under the action of the local map. All that is left to do is make sure of the following: If the "old" landscape of site i told us (by means of a definite recipe ρ) to use permutation π^t_i at site i and time t while going forward in time, then the "new" landscape of site i must be able to point (by means of a matching recipe $\bar{\rho}$) at the same permutation π^t_i (which we will then invert) when going backwards in time.

In section 5.3, this is done by making sure that the relevant landscape does not change at all, so that π^t_i does not in fact depend on t. In section 5.4, only half of the state variables are allowed to change at each step, while the other half, which does not change, provides a landscape that is recognizable from either direction of time-travel; the two halves exchange roles after each step. Finally, in section 5.5, center cell and landscape are fused into an indivisible block of cells, and the dynamics is effectively given by a permutation that acts on the entire block rather than on a single cell.

5.3. Conserved-landscape permutations

In 1971, Patt [56] conducted a search for nontrivial ICA, restricting himself to one dimension (where a decision procedure is known), two states per cell, and contiguous neighbors. He found none for neighborhoods of size 2 and 3. His search stopped at size 4, where he found exactly eight cases (out of 65,536). All the eight cases are variants (obtained by reflection or complementation) of a single cellular automaton, described by the following local map (where the center cell is underscored):

$$
\begin{array}{llll}
0\underline{0}00 \mapsto \underline{0} & 0\underline{1}00 \mapsto \underline{1} & 1\underline{0}00 \mapsto \underline{0} & 1\underline{1}00 \mapsto \underline{1} \\
0\underline{0}01 \mapsto \underline{0} & 0\underline{1}01 \mapsto \underline{1} & 1\underline{0}01 \mapsto \underline{0} & 1\underline{1}01 \mapsto \underline{1} \\
0\underline{0}10 \mapsto \underline{1}* & 0\underline{1}10 \mapsto \underline{0}* & 1\underline{0}10 \mapsto \underline{0} & 1\underline{1}10 \mapsto \underline{1} \\
0\underline{0}11 \mapsto \underline{0} & 0\underline{1}11 \mapsto \underline{1} & 1\underline{0}11 \mapsto \underline{0} & 1\underline{1}11 \mapsto \underline{1}.
\end{array}
\tag{9}
$$

[#6] Intuitively, the neighbors of site i will also affect sites other than i; in turn, when time is made to flow backward, they may be affected by sites other than i.

What makes this cellular automaton invertible? Note that the local map specifies 'no change' except for those two entries (starred in the table) where the center cell is surrounded by the landscape $0\bullet10$ – in which case the cell itself invariably "flips", i.e., complements its state. What property of this landscape is crucial to the automaton's invertibility? [7]

It turns out that, with local map (9), changing the state of the center cell in the landscape $0\bullet10$ cannot lead to the creation or the destruction of other occurrences of that landscape – the landscape is *conserved*. This is easier to verify if the local map, of the form

$$q_i^{\text{new}} = \pi_i q_i^{\text{old}}, \tag{10}$$

is explicitly tabulated as follows ('−' denotes a "don't care" argument)

landscape(i)	π_i
$-\ \bullet\ -\ 1$	'no change'
$0\ \bullet\ 1\ 0$	'flip'
$-\ \bullet\ 0\ -$	'no change'
$1\ \bullet\ -\ -$	'no change'

In these circumstances, if one attempted to "undo" one step of the dynamics, one would know exactly *which* cells have just been changed (because wherever a change was permitted the corresponding landscape was conserved), and *how* to undo the changes (because the effect of a 'flip' can be reversed by flipping again). In fact, the inverse local map for this dynamics is identical to the direct one.

Note that a conserved landscape prevents most of the cells of a configuration from ever changing state; at a higher hierarchical level, those cells can be regarded as structural parameters of the machinery rather than state variables. The remaining cells, which constitute the effective state variables, are decoupled from each other, leading to a situation where invertibility is determined in a trivial way, much as in the previous section.

---

[7] Notice that when the flip is conditioned by a different landscape, the resulting cellular automaton is not, in general, invertible. For instance, with the rule where the center cell flips in the landscape $1\bullet00$, the configuration $\ldots00000100000\ldots$ has no predecessors.

Invertibility in conserved-landscape cellular automata is not limited to trivial cases. For example, by making use of several conserved landscapes that partially overlap one another, one can selectively retain some coupling between cells; in particular, one can construct ICA that simulate any second-order ICA (cf. section 5.4).

5.4. Second-order cellular automata

We shall give a simple method for obtaining, starting from an arbitrary cellular automaton, a new one that is invertible and has a neighborhood at least as large as that of the original – and thus is nontrivial if the original was nontrivial.

To paraphrase Zeno, if we cut a single frame out of the movie of a flying bullet, we have no way of knowing what the bullet is doing. However, if we are given two consecutive frames, then we can figure out the bullet's trajectory. That is, from these two frames, interpreted as the bullet's "past" and "present" positions, we can construct a third frame giving the bullet's "future" position; this procedure can be iterated.

The laws of Newtonian mechanics happen to be such that, if for some reason the two frames got exchanged, we would end up figuring the bullet's trajectory *in reverse*. The present approach to invertibility in cellular automata, suggested by Ed Fredkin of MIT, is based on the above mechanical metaphor.

Let us start with a dynamical system in which the sequence of configurations that make up a trajectory is given by a relation of the form

$$q^{t+1} = \tau q^t. \tag{11}$$

For the moment, we can think of the q^t as real variables. In general, (11) gives rise to a noninvertible dynamics (i.e., there may be no way, or no unique way, to extend the sequence backwards). For the dynamics to be invertible, τ itself must be invertible.

Now, let us consider a new system, defined by the relation

$$q^{t+1} = \tau q^t - q^{t-1}. \tag{12}$$

This is an example of *second-order* system, where the "next" configuration is a function of both the

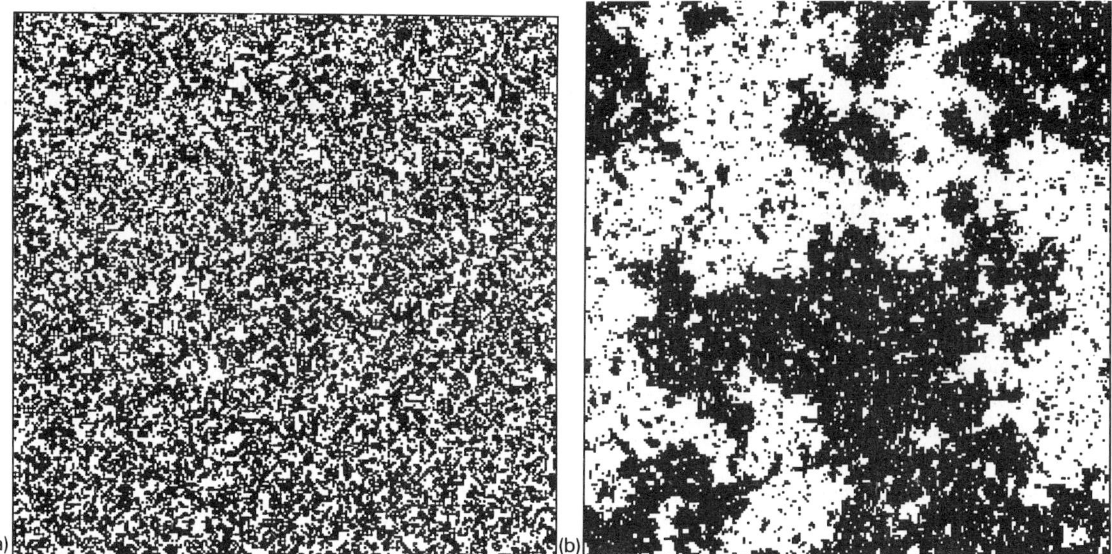

Fig. 1. Equilibrium configurations of **Q2R** above the critical energy (left) and at the critical energy (right).

"current" and the "previous" one – and thus it takes a *pair* of consecutive configurations to completely determine the forward trajectory. In general, second-order relations give rise to noninvertible dynamics. However, a relation of the specific form (12) guarantees the invertibility of the dynamics for an *arbitrary* τ. In fact, by solving (12) with respect to q^{t-1}, one obtains the relation

$$q^{t-1} = \tau q^t - q^{t+1}; \tag{13}$$

that is, a pair of consecutive configurations suffice to determine in a unique way also the *backward* trajectory.

The above considerations can immediately be applied to cellular automata (see ref. [77, ch. 6] for an intuitive presentation). In equation (11), let the q^t be configurations of a cellular automaton, and τ a global map. The local map will be of the form

$$q_i^{t+1} = f q_{i+X}^t. \tag{14}$$

We shall identify the r elements of the state alphabet A with the integers $0, 1, \ldots, r - 1$. Then (12), with "$-$" denoting the *difference* [#8] mod r

[#8] Of course, "$-$" as an operator on configurations is induced from "$-$" as an operator on single cells, as used for instance in (15) below, by applying it in parallel to all sites.

between two configurations, defines a particular *second-order* cellular automaton whose local map is

$$q_i^{t+1} = f q_{i+X}^t - q_i^{t-1}. \tag{15}$$

Note that the above equation can be rewritten as

$$q_i^{t+1} = \pi_{q_{i+X}} q_i^{t-1}, \tag{16}$$

where the permutation π that turns q_i^{t-1} into q_i^{t+1} changes, from site to site, as a function of the "landscape" q_{i+X}.

With this method, from any ordinary cellular automaton with state alphabet A and global map τ one immediately obtains from (15) a second-order cellular automaton that is invertible. The latter, in turn, can always be written as an *ordinary* cellular automaton, with state alphabet $A \times A$, configurations of the form $\langle q^{t-1}, q^t \rangle$ and global map of the form

$$\langle q^{t-1}, q^t \rangle \mapsto \langle q^t, \tau q^t - q^{t-1} \rangle. \tag{17}$$

Thus, in spite of the great "rarity" of ICA, the ones we can construct are at least "as many" as the noninvertible ones!

Second-order recurrences in which the individual state variables are real numbers (rather than bits) are, of course, routinely used in constructing finite-difference schemes for Lagrangian systems, and recurrences of the form (17), in particular, for obtaining invertible dynamics (cf. ref. [12]). Note, however, that – no matter whether the state variables are real numbers or symbols from a finite state alphabet – a permutation operator of the form $\pi_{q_{i+x}}$, as in (16), is much more general than a difference operator of the form '$fq^t_{i+X}-$', as in (15).

We shall briefly present two examples of second-order ICA that combine richness of behavior with economy of means. [#9]

The Q2R rule, introduced by Gérard Vichniac [83] (for more detail, see ref. [77, ch. 17]), is the simplest microcanonical model of a two-dimensional *Ising spin system*. The two elements of the state alphabet $A = \{\uparrow, \downarrow\}$ can be thought of as the two orientations of a spin-$\frac{1}{2}$ particle tied to each lattice site. The neighborhood consists of the four "first neighbors" of a cell (in the four directions of the compass). The operator π in (16) specifies 'flip' if two of the four neighbors are spin-up and two spin-down, and 'no change' in all other cases. If one assigns one unit of potential energy to each occurrence of an antiparallel bond ($\uparrow\downarrow$) between a cell and one of its neighbors, the above rule is equivalent to flipping a spin whenever this operation is energetically indifferent. Thus, the state of the entire spin array moves, in phase space, along a surface of constant energy. [#10] Innumerable variants and generalizations of this basic model can, of course, be devised (cf. ref. [77]). But even in this bare form the model is adequate for illustrating the richness and the theoretical challenges of critical phenomena theory (symmetry breaking, phase transitions, long-range correlations, etc.). Fig. 1

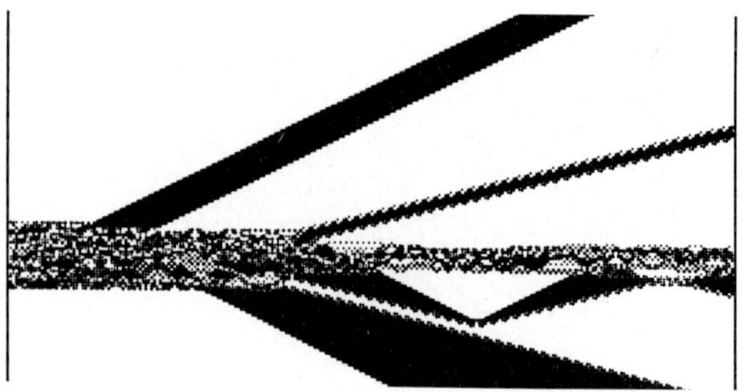

Fig. 2. A spacetime history from the **SCARVES** rule. Time progresses righwards.

illustrates the onset of phase separation as one crosses the critical temperature.

The **SCARVES** rule [77, p. 97], introduced and extensively studied by Charles Bennett, is a one-dimensional system; it supports a variety of "elementary particles" that travel at different speeds and undergo various types of interaction, as illustrated in fig. 2. The four neighbors are now the two "first neighbors" and the two "second neighbors" in the one-dimensional array; except for that, the operator π can be described by the same words as that for Q2R.

Other simple examples of second-order cellular automata relevant to statistical mechanics are discussed in ref. [77] and in refs. [64,66]; the analogy with Lagrangian and Hamiltonian mechanics is explored further in refs. [43,37].

5.5. Partitioning cellular automata

The advantage of trivial cellular automata is that the domain and the range of the table are the same set, so that invertibility of the table implies invertibility of the dynamics; the disadvantage, of course, is that each cell is its only neighbor, so that there is no communication between cells.

Following refs. [41,43], let us try to keep the advantage and remove the disadvantage. At time 0, let us cut up space into finite regions, obtaining a partition p_0 of the set of sites (in fig. 3, the thick lines delimit regions consisting of four-cell squares). Let us introduce a new kind of local map that takes as input the contents of a region and produces as output the new state of the *whole* region (rather than of a single cell). Such a map f_0 allows information to be exchanged between cells

---

[#9]   Another example is the earlier cellular-automaton realization (2 dimensions, 3 states, 9 neighbors) of the "billiard ball" model of computation [19], discussed by Margolus in ref. [41, appendix A]. The encouraging results obtained through this construction eventually lead to to a more compact realization of the billiard-ball model via the partitioning technique (section 5.5).

[#10] Q2R actually consists of two intermeshed but independent sublattices, one evolving in the "white" squares an the other in the "black" squares of a spacetime checkerboard, and each separately conserving energy. In ref. [77], we discuss a number of ways to overcome this redundancy by using cellular automata of slightly different formats.

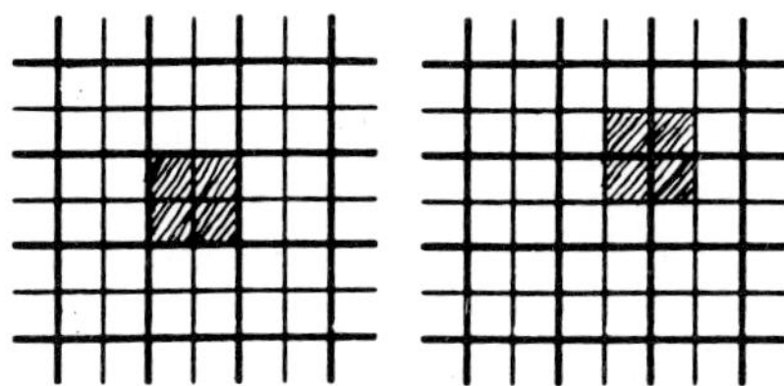

Fig. 3. Even (thick lines) and odd (thin lines) partitions of a two-dimensional array into four-cell blocks. One block in each partition is shown shaded.

of a region, but not across region boundaries. If f_0 is invertible (note that the format of f_0 is "four cells to four cells"), the corresponding global map τ_0 is also invertible. If we kept iterating τ_0, information would remain locked up within each region. Instead, at the next step let us use a *different* partition, P_1 (thin lines in fig. 3), and a local map f_1 of the same kind as f_0, but acting on the regions of P_1. Again, if f_1 is invertible so is the corresponding global map τ_1. The regions of P_1 straddle the boundaries between regions of P_0, and so information that at step 0 had been dammed up within a region of P_0 may at step 1 spill over adjacent regions. We shall keep alternating between τ_0 and τ_1, respectively at even and odd time steps.

We have thus achieved our two main goals, namely, we have constructed a structure, called a *partitioning* cellular automaton (generalizations to larger regions and longer time step cycles are obvious), in which (a) global invertibility derives in a straightforward way from local invertibility, and (b) information can be transmitted over any distance (i.e., the system is an interacting whole rather than a collection of isolated subsystems).[#11]

The partitioning technique is particularly useful for constructing systems consisting of moving

[#11] The definition of partitioning cellular automata introduces a minor departure from uniformity in space and time; but the full uniformity of an *ordinary* cellular automaton can easily be restored; in the present example, one would consider "super-cells" (each consisting of the four cells that make up a region of partition P_0) and "super-steps" (consisting of the composition of two consecutive steps) – yielding a global map of the form $\tau = \tau_1\tau_0$. Quite generally, a wide class of finitary rules that have a periodic spacetime structure can be recast as ordinary cellular automata.

particles. To "move" a particle in a cellular automaton one must actually *erase* the particle from its current site i and *create* a new copy of it on an adjacent site j. These two operations must be carefully matched, lest particles vanish or multiply; unfortunately, in ordinary cellular automata the information available to the local map when acting on site i (i.e., the state of the neighborhood of i) is different from that on site j, and that makes it difficult for two distinct applications of the local map to carry out the two halves of the *same* decision ('move' or 'not move'). Special handshakes must be devised.[#12] With partitioning cellular automata, the two halves of a 'move' operation can be made to fall within the scope of the same neighborhood, and thus coordination is trivial. This feature has an immediate application in lattice gases (section 5.6) and similar interacting particle systems.

In the same way as one insures the conservation of the number of particles, one can insure the conservation of other quantities of physical interest (variables that represent momentum, charge, etc.).

5.6. Lattice gases

Intuitively, a *lattice gas* is a system of particles that move in discrete directions at discrete speeds, and undergo discrete interactions. It will be clear in a moment that a lattice gas is but a special format of cellular automaton; however, it will be useful to start with a example in which the more picturesque terminology of continuous motion is retained.

In the HPP lattice gas [26], identical particles move at unit speed on a two-dimensional orthogonal lattice, in one of the four possible directions. Isolated particles move in straight lines. When two particles coming from opposite directions meet, the pair is "annihilated" and a new pair, traveling at right angles to the original one, is "created" (fig. 4a). In all other cases, i.e., when two particles cross one another's paths at right angles (fig. 4b) or when more than two particles meet, all particles just continue straight on their paths.

[#12] The "firing-squad" problem, of which this is a special case, dates back to the origins of cellular automata [52].

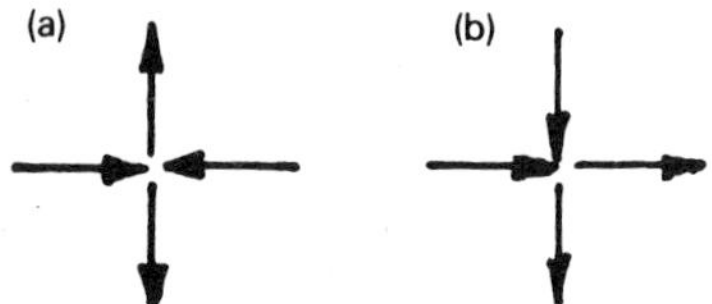

Fig. 4. In the HPP gas, particles colliding head-on are scattered at right angles (a), while particles crossing one another's paths go through unaffected (b).

As soon as the numbers involved become large enough for averages to be meaningful – say, averages over spacetime volume elements containing thousands of particles and involving thousands of collisions – a definite continuum dynamics emerges. And, in the present example, it is a rudimentary *fluid* dynamics, with quantities recognizably playing the roles of density, pressure, flow velocity, viscosity, speed of sound, etc. Fig. 5 illustrates sound-wave propagation in this model. Note that, even though the microscopic interactions only display a more limited form of rotational symmetry (namely, invariance for quarter-turn ro-

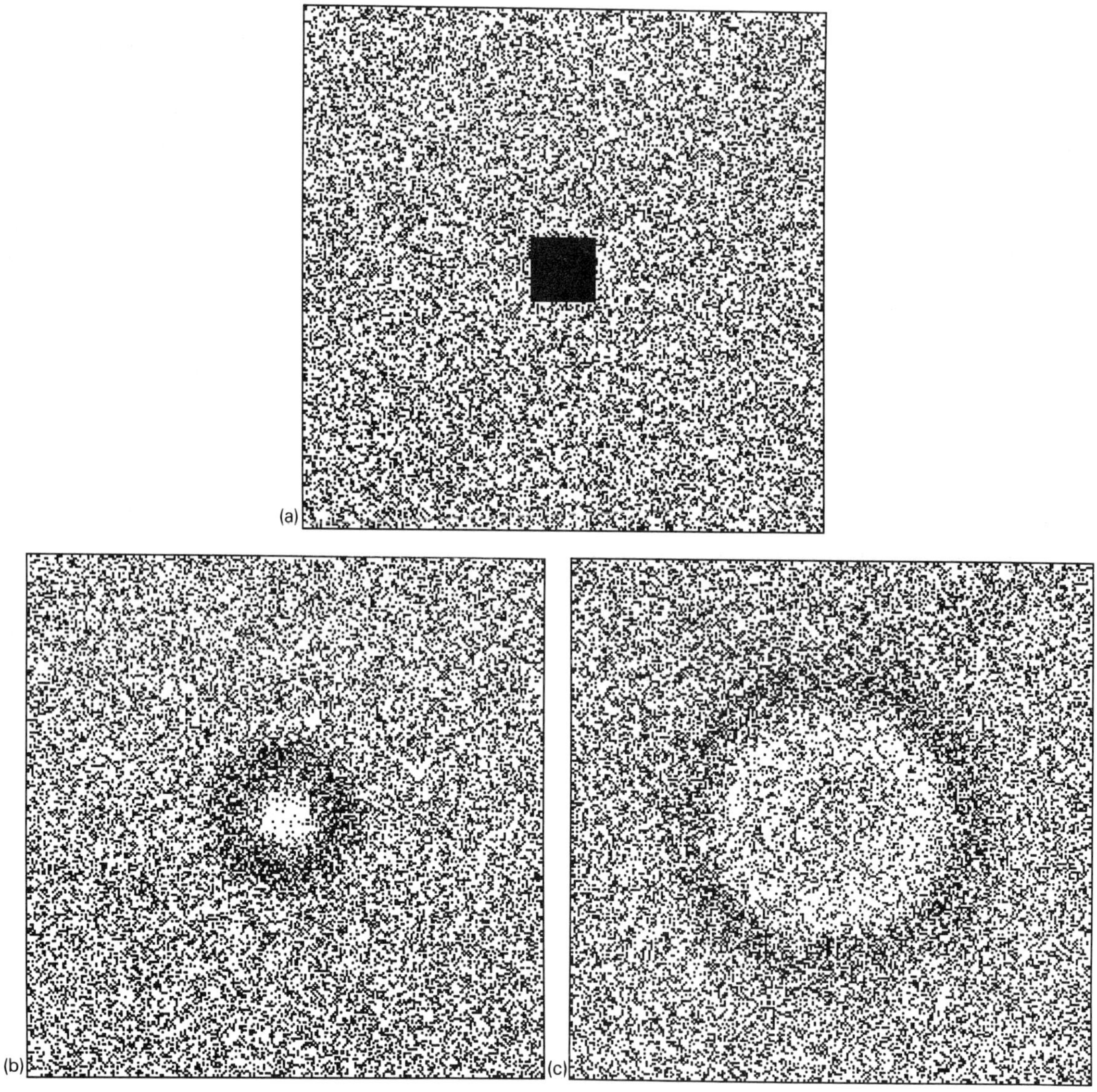

Fig. 5. Wave propagation in the HPP lattice gas. Note the emergence of circular symmetry.

tations), the speed of sound is fully isotropic.

Unlike sound speed, sound attenuation is *not* isotropic in the HPP model. It turns out that, besides conserving energy and momentum (see section 6.1), HPP separately conserves the horizontal component of momentum on each horizontal row and the vertical component on each vertical column. These *spurious* conservations (they have no counterpart in ordinary physics) lead to significant departures from the behavior one would expect from a physical fluid.

The slightly more complicated FHP lattice gas model [21] – which uses six rather than four particle directions (always in two dimensions) – gives, in an appropriate macroscopic limit, a fluid obeying the well-known *Navier–Stokes* equation, and which is thus suitable for modeling actual hydrodynamics (see ref. [27] for a tutorial). Recently, analogous results for three-dimensional models have been obtained by a number of researchers [22].

Lattice gases are rapidly beginning to encroach into modeling niches dominated until recently by differential equations [16,49]; their success in this role is chiefly due to the ease with which they can be made to satisfy local continuity equations, [#13] as discussed below and, more leisurely, in refs. [72,75]).

Let us consider the spacetime texture induced by a lattice gas such as HPP. [#14] In fig. 6, the arcs represent the spacetime paths available to the particles (a, b, c, d denote the four possible directions of motion) and the nodes (labeled f) represent the available collision sites; the entire structure is iterated in space and time, yielding a body-centered cubic lattice. Thus, we have a spacetime diagram (the arcs are *signals* and the nodes *events*) of a kind that is routinely used in illustrating special-relativity arguments. From this diagram, a particular history is obtained by assigning, as initial conditions, a definite occupation state ('particle'

[#13] In this traditional but unfortunate term, 'continuity' does not refer to the continuum, in opposition to 'discreteness'; rather, it refers to "continuity of existence" – detailed balance, in other words, as in Kirkhhoff's laws.

[#14] Much as in the case of **Q2R** (cf. footnote #10), also in HPP the lattice splits into two intermeshed but independent sublattices. Here we shall consider only one of these sublattices.

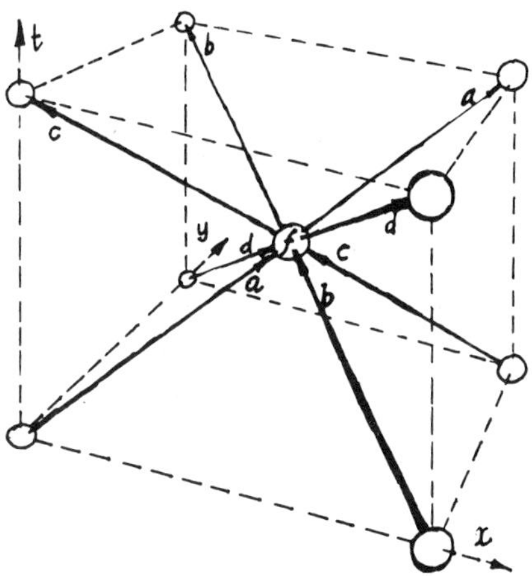

Fig. 6. Spacetime layout of the HPP gas.

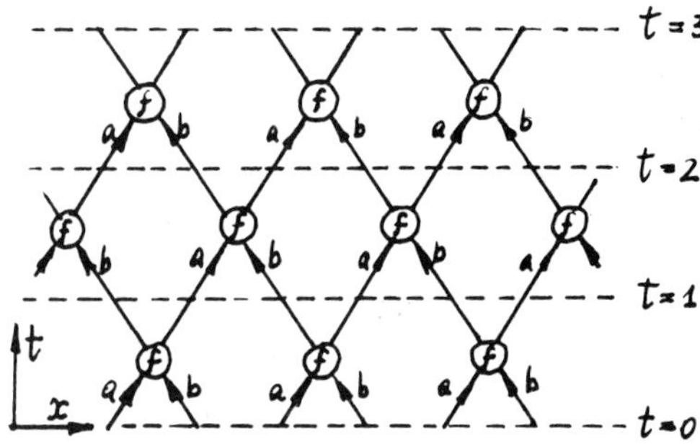

Fig. 7. Evenly spaced spacelike surfaces of the form t=constant in a lattice-gas spacetime diagram.

or 'no particle', which may be denoted by '0' and '1' respectively) to each arc that crosses a given spacelike surface, and then extending the state assignment timewards, using the following table at each event

in	out		in	out
abcd	abcd		abcd	abcd
0000	0000		1000	1000
0001	0001		1001	1001
0010	0010	* 1010	0101
0011	0011		1011	1011
0100	0100		1100	1100
* 0101	1010		1101	1101
0110	0110		1110	1110
0111	0111		1111	1111

$$(18)$$

The table itself represents the collision rule given in words above. Note that in only two cases (marked with an asterisk) out of sixteen does an interaction take place; in all other cases, each of the four signals proceeds undisturbed.

Since the function f represented by this table is invertible, the spacetime history may be extended backwards as well as forwards in time.

If one now considers surfaces of the form $t = $ constant, drawn midway between rows of events, as in fig. 7 (where, for clarity, only one spatial dimension is indicated), it is clear that the collective state of the signals that traverse each such surface can be thought of as the configuration of a cellular automaton. If one groups together the four signals that converge on the same event (fig. 4), and notes the regular alternation, at odd and even times, of these groupings (cf. fig. 7), it will be evident (cf. fig. 3) that the updating scheme of fig. 4 is isomorphic to that described in section 5.5.

In other words, lattice gases and partitioning cellular automata are formally the same thing. Nonetheless, the tendency is to reserve the term 'lattice gas' for cellular automata in which the 'gas' metaphor can be defended [33]. Typically, one has a distinguished "vacuum" state ('0' in table (18)) on whose background isolated "particles" ('1' in table (18)) travel with inertial motion.

Figs. 6 and 7 make it clear not only that (a) the global dynamics is invertible if f is invertible, but also that (b) any additive quantity carried by individual signals (occupation number, momentum, etc.) is globally conserved if it is conserved by f. For example, in ref. [9] we show how certain second-order ICA can be rewritten in a straightforward way as lattice-gas ICA; conserved quantities (such as bond energy) that in the second-order format it took some effort to discover [58] are immediately visible in the lattice-gas format.

6. Physical modeling

The question that we are most often asked about cellular automata is the following.

"I've been shown cellular automata that make surprisingly good models of, say, hydrodynamics, heat conduction, wave scattering, flow through porous media, nucleation, dendritic growth, phase separation, etc. But I'm left with the impression that these are all *ad hoc* models, arrived at by some sort of magic."

"I'm a scientist, not a magician. Are there well-established *correspondence rules* that I can use to translate features of the system I want to model into specifications for an adequate cellular-automaton model of it?"

Physical modeling with cellular automata is a young discipline. Similar questions were certainly asked when differential equations were new – and, despite three centuries of accumulated experience, modeling with differential equations still requires a bit of magic. Mainly, we get new models by dressing up old models, or by microdynamical analogy. As it stands, a "rational" or "analytical" mechanics of cellular automata is beginning to take shape (cf., e.g., refs. [24,25]). Ironically, the most mature aspects of this understanding concern the modeling of *continuum* phenomena. In the rest of this section, we'll try to convey at least a feeling for the issues involved.

6.1. HPP versus Navier–Stokes

How does a ridiculously simple interaction such as that of fig. 4a (represented, in the two possible spatial orientations, by the starred entries in table (18)) possibly come close to capturing the richness of the Navier–Stokes equation?

Let us attempt to sketch an answer, using drastically simplified arguments.

By inspection of table (18), one concludes that (a′) the dynamics specified by this local map is *invertible*; (b′) *energy*, represented simply by particle count, is *conserved* by each event; (c′) the two components of *momentum*, obtained by separately counting (with a + or − sign depending on the sense) particles traveling in the two orthogonal directions, are similarly *conserved*; (d′) the dynamics is *rotationally invariant* for quarter-turn rotations; and (e′) there is *some coupling* between the two spatial *directions*.

By inspecting the Navier–Stokes equation, one concludes that this equation specifies that (a) the dynamics is *invertible*; (b) that *energy* and (c) *momentum* are *conserved* by contact processes (no action-at-a-distance); (d) the dynamics is *rotationally invariant* for continuous rotations; (e) pressure is a *scalar* quantity, i.e., is independent of *direction*; and (f) *nothing else* that doesn't already follow from (a)–(e).

In the cellular-automaton model, the moment one puts on *macroscopic* glasses, as it were, and is only able to see details that are much coarser than the pitch of the lattice, the discreteness of energy, momentum, and position washes out, so that points (a'), (b'), and (c') above closely match (a), (b), and (c).

As for point (d'), it is well known that 90° rotational invariance is sufficient to yield full isotropy for *diffusive* phenomena.[15] However, in a lattice gas not too far from equilibrium (cf. ref. [73]) the chief transport mechanisms are both diffusive transport and wavelike transport, and for the full isotropy of *wavelike* phenomena it turns out that one needs nondiffusive transport of horizontal momentum in the vertical direction and vice versa – which is hard to achieve with 90° invariance and is much easier to achieve with six particle directions and 60° invariance (cf., e.g., ref. [88]). This is why the fluid modeled by the HPP gas significantly departs from the Navier–Stokes equation (cf. section 5.6), and why a model such as FHP manages to achieve the goal.

As for point (e'), by considerations similar to the above one can show that *almost any* coupling between the x and y directions leads to equalization of pressure in all directions.

The only philosophical difficulty lies with the 'nothing else' of point (f). Like sorcerer's apprentices, we commanded HPP to conserve a certain few quantities. HPP complied; but it didn't warn us that it had taken the initiative to conserve an infinity of other quantities as well (section 5.6). And these spurious conservations undermined our modeling efforts.

How do we know that something like this will not happen to us the next time? How can one tell that a model has no spurious invariants [15]? (For that matter, how does one know that a certain unanticipated invariance is 'spurious'? may it not turn out to be a "feature" rather than a "bug"?[16]) A general method for keeping spurious constraints out of a cellular-automaton model is suggested by Jaynes's maximum entropy principle [36]. The idea is to make the local map *maximally random* with respect to any features that are not explicitly demanded by the modeling context.[17] Unfortunately, to guarantee a closer and closer approximation to this ideal of maximal randomness one may have to introduce larger and larger state alphabets and neighborhoods, at a corresponding sacrifice in simulation efficiency. In the end, one must know where to draw the line between accuracy and efficiency.

6.2. Invertibility, symmetries, and conserved quantities

An invertible cellular automaton shares certain important traits with physical systems even when it does not manifestly support waves, particles, energy, momentum, and similar symptoms of "physicalness".

Even in dynamical systems that have very little structure one can show some connections between symmetries and conservation laws.[18] However, such connections manifest themselves with special strength in Hamiltonian systems, where, according to Noether's theorem, to each continuous one-parameter group of transformations that commutes with the dynamics there is associated a real-valued, conserved quantity. Thus, energy is associated with time-invariance of the dynamics, the three components of momentum with translation invariance, etc. These quantities are functions of

[15] Consider the set of all possible algorithms telling an individual how to move, one block at a time, North, South, East, or West on an orthogonal street grid (at each step, the choice may take into account local landscape features, the actions of neighboring individuals, the weather, etc.). *Almost all* such algorithms yield an exploration pattern that is close to a superposition of independent random walks in the two orthogonal directions. In the limit as the binomial distribution goes over to the Gaussian distribution, such a pattern displays full rotational invariance (i.e., $\exp(x^2)\exp(y^2) = \exp(x^2 + y^2) = \exp(r^2)$) – and this in spite of the fact that the microscopic rule can at best have quarter-turn invariance.

[16] For instance, certain constraints that in lattice gases lead to a departure from Galilean invariance [28] may actually bias the system toward *Lorentz* invariance [75] (also cf. ref. [65]).

[17] Hénon's strategy for rule optimization in lattice gas hydrodynamics [32] is somewhat related to this approach.

[18] Given any group of transformations that commutes with the dynamics, each configuration belongs to a definite symmetry class (a normal subgroup of the group itself); all the points of an orbit belong to the same symmetry class, which is thus a *conserved quantity*.

the system's state as such (i.e., its "mechanical" or "microscopic" state); one can literally speak of, say, the *energy content* of a certain state.

The Hamiltonian structure itself, i.e., invariance with respect to canonical transformations, leads to the conservation of an additional quantity, called *(fine-grained) entropy*, which is defined not on individual mechanical states but on *statistical* states (probability distributions on the set of mechanical states); a special case of this conservation is the well-known "incompressibility" of volume in phase space (Liouville's theorem). In other words, (a) the Hamiltonian structure provides an unequivocal way of measuring the *information* content [17, ch. II] of a statistical state, and (b) the quantity thus measured is *conserved* by the dynamics.

Continuum dynamical systems that do not have a Hamiltonian structure lack, in general, an intrinsic "yardstick" for measuring the information content of a (statistical) state, and thus do not come with a built-in statistical mechanics. Cellular automata, on the other hand, owing to their local finiteness (the lattice is discrete, and state alphabet and neighborhood are finite), do possess a natural information measure.[19] And the foremost property of *invertible* cellular automata is, of course, that they are *information-conserving*.

Because of this "information-losslessness" (ach!), ICA automatically obey the second principle of thermodynamics and, more generally, display a full-featured statistical mechanics analogous to that of Hamiltonian systems. As additional structure is introduced (for instance, particle conservation), macroscopic mechanical features such as elasticity, inertia, etc. naturally emerge out of statistics itself. In sum, once we make sure that it is conserved, information has an irresistible tendency to take on a strikingly tangible aspect (cf. ref. [73]) – to *materialize* itself.

For the above reasons, and because they lend themselves to very efficient computer simulations, ICA are an ideal medium for the qualitative study of the connections between microscopic mechanics and statistical mechanics on one hand (cf. refs. [43,77,85,13]), and between statistical mechanics and macroscopic mechanics on the other (cf. ref.

[88]). They are also suitable for the the modeling of an increasingly important type of generalized mechanical activity, namely *computation* [43].

In physics, additive invariants, whether represented by mechanical quantities such as energy or statistical quantities such as entropy, bear a major responsibility for the emergence of nontrivial macroscopic properties. Intuitively, one may expect that almost every detail of the microscopic interactions will be washed out by macroscopic averaging; only features that are supported by a definite conspiracy (such as a particular symmetry or conservation law) will bubble up all the way to the macroscopic surface and emerge there as recognizable laws. The study of invariants in ICA has barely started. Here we can only refer the reader to a few literature items, such as refs. [58,43,15,66]. It must be noted that in general the problem of *discovering* the invariants of a cellular automaton is presumably of a difficulty comparable to that of deciding its invertibility (cf. section 4). Much as for invertibility, it is much easier to directly *synthesize* a cellular automaton having certain desired invariants than to figure out the invariants of a given one.

6.3. Modeling first principles?

As we hinted in sections 5.6–6.2, ICA are quite successful at explaining complex macroscopic phenomenology in terms of the collective behavior of an enormous number of very simple subsystems. So much so, that one begins to wonder whether aspects of physics that are usually regarded as primitive (such as the principles of analytical mechanics) are not, after all, similarly derivable as emergent aspects of an extremely fine-grained underlying dynamical structure.

In this role, ICA have been used with some success as fine-grained models of basic aspects of special and general relativity [75,80,65] and of quantum mechanics [34]

7. Decidability, revisited

Now that we have some concrete examples of ICA in mind, let us pick up the trail that we left off at the end of section 4.4.

---

[19] This is the *uniform* measure, which gives equal weight to all configurations.

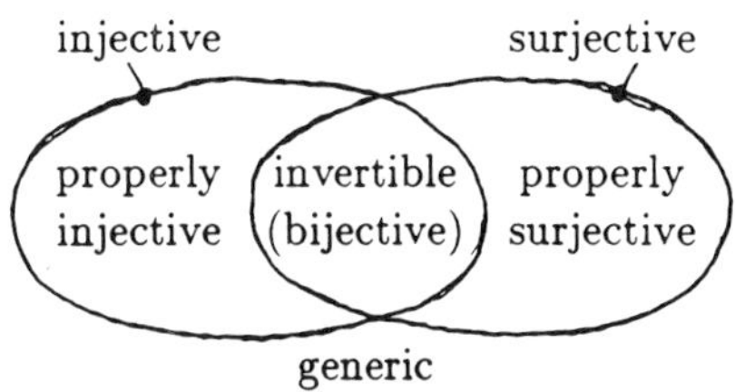

Fig. 8. Venn diagram for surjectivity and injectivity.

cellular automata:

invertible (r.e)	properly surjective	nonsurjective (r.e.)

Fig. 9. Surjectivity and injectivity in cellular automata; 'r.e.' denotes a recursively enumerable set.

A dynamical system is *surjective* if every state has at least one predecessor; *injective*, if it has at most one predecessor. If it is both surjective and injective it is, of course, invertible (or *bijective*). It will be convenient to call *properly surjective* a surjective system that is not invertible, and similarly *properly injective* an injective system that is not invertible. The situation is summed up in fig. 8.

For cellular automata, the 'properly injective' lobe in fig. 8 is empty, i.e., injectivity is equivalent to invertibility (Richardson [60]), leaving us with the simpler situation of fig. 9.

As we have seen in section 4.4, the left set of fig. 9 (i.e., the class of invertible cellular automata) is recursively enumerable (theorem 4.3). The right set too is recursively enumerable, according to the following

Theorem 7.1 *The class of nonsurjective cellular automata is recursively enumerable.*

Proof. According to Myhill's theorem [53] (cf. section 2.3), if a configuration has no predecessors there must exist a configuration having two predecessors that are identical except over a finite area. If such a pair exists, brute-force enumeration (apply the local map to larger and larger finite areas) will eventually turn it up. □

If also the middle set in fig. 9, i.e., the set of properly surjective (PS) cellular automata, were recursively enumerable, then all three sets would be fully *recursive* (i.e., the corresponding predicates would be decidable). From theorem 4.5, we

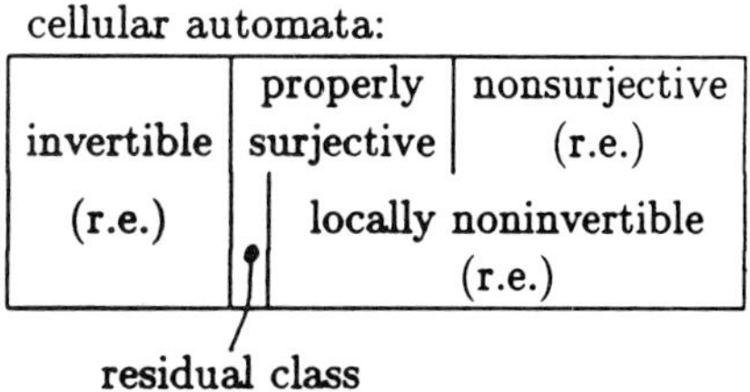

cellular automata:

invertible (r.e.)	properly surjective	nonsurjective (r.e.)
		locally noninvertible (r.e.)

residual class

Fig. 10. Between invertibility and local noninvertibility, which are both semidecidable, there is a no man's land that is not semidecidable.

must conclude that the middle set is not even recursively enumerable. We shall examine the make-up of this PS set in more detail.

Let us apply the local map of a cellular automaton on a *finite*, wrapped-around space array (which is equivalent to imposing periodic boundary conditions along each dimension). [20] The resulting *finite cellular automaton* will be invertible if the original was invertible, and nonsurjective if the original was nonsurjective. If, however, the original was PS, then the resulting cellular automaton will be forced to abandon 'PS-ness' and "choose" between being invertible and being nonsurjective (by a simple counting argument, the PS middle ground is forbidden); note that either choice may occur with the *same* local map, for different sizes of the finite space.

Those cellular automata that become nonsurjective when thus forced on a finite space (at least for some space size), together with those that were nonsurjective to begin with, will be called *locally noninvertible* (they are exactly those cellular automata whose noninvertibility can be verified by local arguments). Local noninvertibility is semidecidable (the proof is similar to that of theorem 7.1). What is left (fig. 10) is a residual class of cellular automata having rather counterintuitive behavior; that is, they

— are invertible on *all* finite spaces, but
— become noninvertible (specifically, properly surjective) on the infinite array.

Because of theorem 4.5, this residual class cannot be empty; on the other hand, no instance of

[20] To simplify the statement of some of the assertions below, we shall consider only finite spaces that are large enough to keep all of the neighbors of a cell distinct.

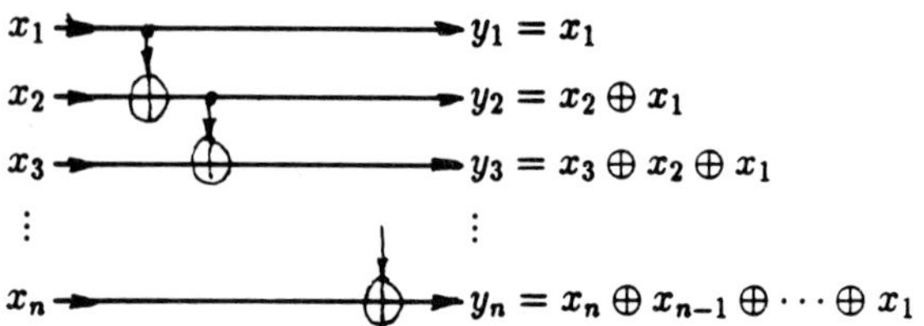

Fig. 11. An invertible function with an asymmetric dependency pattern.

this class has, to our knowledge, ever been exhibited.

Let us go back to the enumeration in the proof of theorem 4.3. If from λ we could determine an upper bound to the neighborhood radius of its hypothetical inverse, then we would, eventually, positively know whether or not this inverse exists. From theorem 4.5, we must conclude that this is not the case; in particular, for any n there must be ICA for which the radius of the *inverse neighborhood* (the neighborhood of the inverse) is at least n times larger than the radius of the *direct* neighborhood.

We shall show how to construct such an ICA, for any n (note that in this example n, though arbitrarily large, is a known function of the size of the state alphabet, so that we *do* have an upper bound for the neighborhood radius). Consider the function with n binary inputs $x_1, \ldots, x_n$ and n binary outputs $y_1, \ldots, y_n$ defined by

$$
\begin{aligned}
y_1 &= x_1 \\
y_2 &= x_2 \oplus x_1 & (&= x_2 \oplus y_1), \\
y_3 &= x_3 \oplus x_2 \oplus x_1 & (&= x_3 \oplus y_2), \\
&\;\cdots \\
y_n &= x_n \oplus x_{n-1} \oplus \cdots \oplus x_1 & (&= x_n \oplus y_{n-1}),
\end{aligned}
$$

$$(19)$$

where $\oplus$ denotes Boolean exclusive-OR. This function, pictorially shown in fig. 11, is invertible, and its inverse is given by

$$
\begin{aligned}
x_1 &= y_1, \\
x_2 &= y_2 \oplus y_1, \\
x_3 &= y_3 \oplus y_2, \\
&\;\cdots \\
x_n &= y_n \oplus y_{n-1}.
\end{aligned}
$$

$$(20)$$

Note the asymmetric dependency pattern: a y may

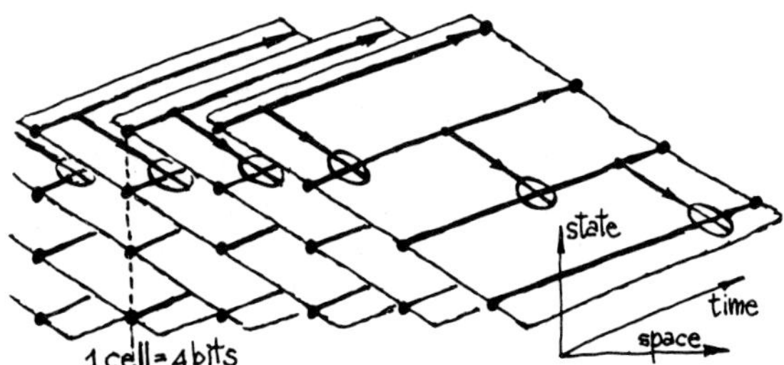

Fig. 12. Infinite juxtaposition of copies of function (19) yields a one-dimensional ICA having n bits per cell. Here $n = 4$ for definiteness.

depend on up to n of the x's, while each x depends on no more than *two* of the y's.

From (19), by using a step-and-repeat construction (fig. 12), we obtain a one-dimensional cellular automaton having n bits per cell. This cellular automaton is invertible; going forward in time, the new state of each cell depends on the current state of itself and the $n-1$ cells to its left; going backwards, each cell depends only on itself and the cell on its left. The neighborhood radii for the direct local map λ and the inverse local map $\overline{\lambda}$ are, respectively, n and 1.

This cellular automaton is an invertible system governed by local and uniform laws, but the cause-and-effect tree (the "light cone") in one direction of time is much wider than in the opposite direction. [21] Such behavior doesn't seem to have any counterpart in physics.

8. Structural invertibility

The usual definition of a cellular automaton, as given in section 2.1, is a *structural* one; i.e., one explicitly tells how a cellular automaton *is made*. [22] The structure in question is a *uniform sequential*

[21] See ref. [57] for a related discussion on different aspects of causality.

[22] One can think of the overall state of the system as a point in some abstract space, and of a configuration as a way of expressing this point in a convenient coordinate system (each site represents a coordinate); the local map tells how to update the state-component associated with each coordinate. In section 2.4 we mentioned the possibility of an equivalent *functional* definition, where the global map is characterized in terms of what it *does* to the state points themselves, without making recourse to coordinates.

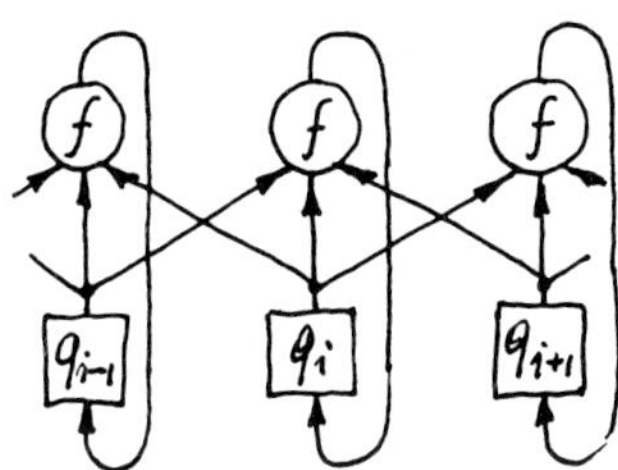

Fig. 13. A cellular automaton as a sequential network. We illustrate the case of one dimension and neighborhood $X = \langle -1, 0, +1 \rangle$. The table is denoted by f.

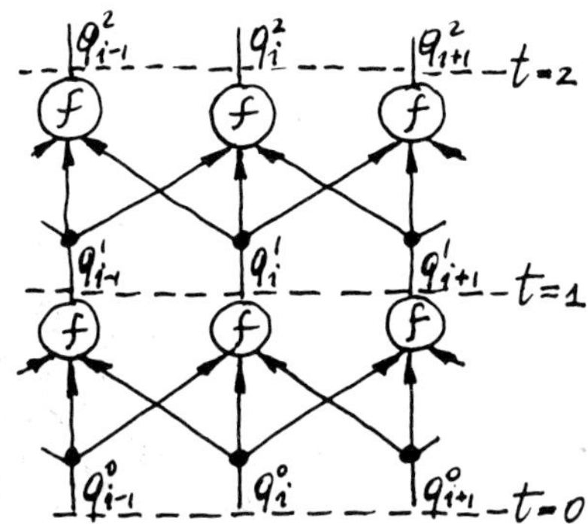

Fig. 14. The cellular automaton of fig. 13, re-expressed as a combinational network. The f nodes denote occurrences of the table; the black dots, occurrences of fan-out.

network: the state q_i of a cell (a square in fig. 13) is realized by a collection of flip-flops or similar memory elements, and the table f (a circle in the same figure) by a composition of NAND gates or similar logic elements. Both kinds of element have a straightforward implementation in a number of technologies.

Since a flip-flop is simply a digital sample-and-hold device, used for regulating the traffic in and out of the gates,[23] it will be conceptually simpler to represent the design of a cellular automaton as a uniform *combinational* network, where each usage of a table is explicitly represented by a separate node, as in fig. 14a.

Now, if we were given the design of a deterministic physical machine A whose behavior happens to be invertible (i.e., for any possible state there is a unique way one can arrive at it), we would tend to think that

(i) One can construct, out of the same technology, a machine $\overline{A}$ that has the same orbit structure as A but follows each orbit in reverse.

(ii) From the detailed design plans of machine A one should be able to arrive in a straightforward way at the design plans for machine $\overline{A}$.

Both points are certainly true if the "technology" in question is Newtonian mechanics (think of the sun and the planets). In fact, since this mechanics is time-reversal invariant (cf. footnote #2), we can use the *same* machinery ($\overline{A} = A$), and just reverse the direction of each particle in order to traverse orbits backwards.

Lemma 4.1 tells us that point (i) above is true also for more practical technologies, where non-Newtonian devices such as amplifiers, latches, dampers, and other dissipative devices are available. Note that the usual way cellular automata are specified tacitly implies the availability of such a dissipative technology. In fact, the f nodes in fig. 13 are typically noninvertible because they are *noninjective*, and the fan-out nodes noninvertible because they are *nonsurjective*; as explained in [19], nonsurjective computing primitives imply the availability of a *power supply*, and noninjective ones imply a *heat sink*.

However, theorem 4.5 says that, in this dissipative context, point (ii) is no longer tenable. That is, from design plans of the type represented by fig. 14 we do not know, in general, how to make similar design plans for a new machine that, if the original machine was invertible, will have exactly the inverse behavior. Thus, to rescue point (ii) we have to turn our attention to design plans that do not imply dissipative primitives. For *finite* invertible automata this is always possible [70,19] (see also [69] for the analogous problem in continuous systems); is it possible for cellular automata?

Now that the job of motivating it is done, let us reword the above question. Given an arbitrary ICA, which can always be thought of as realized by a uniform combinational network of the type of fig. 14, is it always possible to give an alternative realization of it by means of a uniform combinational network in which (a) all nodes stand for *invertible* functions, and (b) no extra state variables are introduced? We shall call such a realization

---

[23] Each gate is used over and over, at each time step evaluating the same function on a new set of arguments.

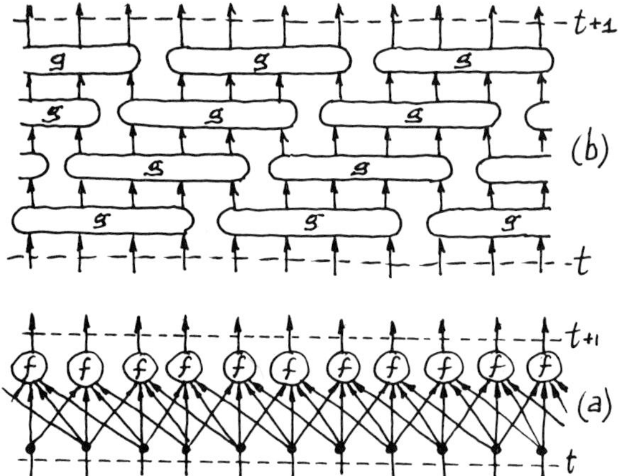

Fig. 15. (a) Combinational network representation of the ica defined by table (9). (b) The same ICA, in a version that exhibits structural invertibility; the nodes are all invertible Boolean functions.

structurally invertible;[24] we shall also call *structurally invertible* an ICA for which a structurally invertible realization exists.

It is clear that from a structurally invertible realization one immediately obtain design plans for the inverse automaton.

We have succeeded in exhibiting structural invertible realizations for every ICA for which we tried (here we shall give only one example, in fig. 15), and apparently all ICA we are aware of are structurally invertible. From the above considerations, we are led to the following

Conjecture 8.1 *All invertible cellular automata are structurally invertible, i.e., can be (isomorphically) expressed in spacetime as a uniform composition of finite invertible logic primitives.*

In other words, we conjecture that there are no cellular automata for which invertibility (a *functional* property) cannot be explained in terms of structural invertibility, which is, of course, a *structural* property.

For an example, fig. 15 shows a structurally invertible design for the conserved-landscape ICA discussed in section 5.3, and contrasts it with the

conventional design. The function f is given by table (9); g is an invertible Boolean function with four inputs and four outputs obtained in a straighforward way from f as follows:

$$y_1 = x_1,$$
$$y_2 = f(x_1, x_2, x_3, x_4) = (\overline{x}_1 \cdot x_3 \cdot \overline{x}_4) \oplus x_2,$$
$$y_3 = x_3,$$
$$y_4 = x_4, \tag{21}$$

where a dot denotes Boolean AND, a bar denotes Boolean complement, and the order (1, 2, 3, 4) of inputs and outputs corresponds to left-to-right in fig. 15.

Note that, even though the new network is uniform, its spatial "pitch" is coarser than that of the original network; in other words, in fig. 15b we have a structure whose *function* is left invariant by shifting, say, one position to the left, but whose *structure* needs to be shifted *four* positions before it again coincides with itself![25] We conjecture that, for a structurally invertible realization, recourse to a network having a *coarser* pitch than the original one is, in general, unavoidable.

9. Conclusions

We have presented invertible cellular automata from a number of angles, trying to illustrate the concrete motivations of many theoretical questions.

This paper also constitutes an original contribution to the study of the mechanisms and pathways of *causality* in distributed reversible systems.

Acknowledgements

This research was supported in part by the Defense Advanced Research Projects Agency (N00014-89-J-1988), and in part by the National Science Foundation (8618002-IRI).

---

[24] Condition (b) simply asks for an *isomorphic* realization. Realizations that satisfy (a) but make use of extra state variables can be obtained by a straightforward procedure, as shown in ref. [35].

[25] More formally, the uniformity group of the structural description is a proper subgroup of that of the functional behavior.

References

[1] V. Aladyev, Computability in homogeneous structures, Izv. Akad. Nauk. Estonian SSR, Fiz.-Mat. 21 (1972) 80–83.

[2] S. Amoroso and Y. N. Patt, Decision procedures for surjectivity and injectivity of parallel maps for tessellation structures, J. Comp. Syst. Sci. 6 (1972) 448–464.

[3] S. Amoroso et al., Some clarifications of the concept of a Garden-of-Eden configuration, J. Comp. Syst. Sci. 10 (1975) 77–82.

[4] E. Banks, Information processing and transmission in cellular automata, Tech. Rep. MAC TR-81, MIT Project MAC (1971).

[5] P. Benioff, Quantum mechanical Hamiltonian models of discrete processes that erase their own histories: applications to Turing machines, Int. J. Theor. Phys. 21 (1982) 177–201.

[6] C. Bennett, Logical reversibility of computation, IBM J. Res. Develop. 6 (1973) 525–532.

[7] C. Bennett and G. Grinstein, Role of irreversibility in stabilizing complex and nonenergodic behavior in locally interacting discrete systems, Phys. Rev. Lett. 55 (1985) 657–660.

[8] C. Bennett, On the nature and origin of complexity in discrete, homogeneous, locally-interacting systems, Found. Phys. 16 (1986) 585–592.

[9] C. Bennett, N. Margolus and T. Toffoli, Bond-energy variables for Ising spin-glass dynamics, Phys. Rev. B 37 (1988) 2254.

[10] C. Bennett, T. Toffoli and S. Wolfram, eds., Cellular Automata '86, Tech. Memo MIT/LCS/TM-317, MIT Lab. for Comp. Sci. (December 1986).

[11] A. Burks, ed., Essays on Cellular Automata (University of Illinois Press, Champaign, IL, 1970).

[12] R. Courant, et al., On the partial difference equations of mathematical physics, Math. Ann. 100 (1928) 32–74 (in German), republished in English translation in IBM J. (March 1967) 215–238.

[13] M. Creutz, Deterministic Ising dynamics, Ann. Phys. 167 (1986) 62–76.

[14] K. Culik, On invertible cellular automata, Complex Systems 1 (1987) 1035–1044.

[15] Invariants in lattice gas models, in: Discrete Kinetic Theory, Lattice Gas Dynamics, and Foundations of Hydrodynamics, ed. R. Monaco (World Scientific, Singapore, 1989) pp. 102–113.

[16] G. Doolen et al., eds., Lattice-Gas Methods for Partial Differential Equations (Addison–Wesley, New York, 1990).

[17] H. Everett, The theory of the universal wave function, in: The Many-World Interpretation of Quantum Mechanics, eds. B. DeWitt and N. Graham (Princeton Univ. Press, Princeton, NJ, 1973).

[18] R. Feynman, Quantum-mechanical computers, Opt. News 11 (Feb. 1985) 11–20, reprinted in Found. Phys. 16 (1986) 507–531.

[19] E. Fredkin and T. Toffoli, Conservative logic, Int. J. Theor. Phys. 21 (1982) 219–253.

[20] E. Fredkin, R. Landauer and T. Toffoli, eds., Proceedings of a Conference on Physics and Computation, Int. J. Theor. Phys. 21(3/4), 21(6/7), 21(12) (1982).

[21] U. Frisch, B. Hasslacher and Y. Pomeau, Lattice-gas automata for the Navier–Stokes equation, Phys. Rev. Lett. 56 (1986) 1505–1508.

[22] U. Frisch et al., Lattice gas hydrodynamics in two and three dimensions, in: Lattice-Gas Methods for Partial Differential Equations, eds. G. Doolen et al. (Addison–Wesley, New York, 1990) pp. 77–135.

[23] M. Gardner, The fantastic combinations of John Conway's new solitaire game 'Life', Sci. Am. 223 (April 1970) 120–123.

[24] H. Gutowitz, J.D. Victor and B.W. Knight, Local structure theory for cellular automata, Physica D 28 (1987) 18–48.

[25] H.A. Gutowitz, A hierarchical classification of cellular automata, Physica D 45 (1990) 136–156, these Proceedings.

[26] J. Hardy, O. de Pazzis and Y. Pomeau, Molecular dynamics of a classical lattice gas: transport properties and time correlation functions, Phys. Rev. A 13 (1976) 1949–1960.

[27] B. Hasslacher, Discrete fluids, Los Alamos Science, Special Issue No. 15 (1987) 175–200, 211–217.

[28] F. Hayot, The effect of Galilean non-invariance in lattice gas automaton one-dimensional flow, Complex Systems 1 (1987) 753–761.

[29] T. Head, One-dimensional cellular automata: injectivity from unambiguity, to appear in Complex Systems 3 (1989).

[30] G.A. Hedlund, K. I. Appel and L. R. Welch, All onto functions of span less than or equal to five, Communications Research Division, working paper (July 1963).

[31] G.A. Hedlund, Endomorphism and automorphism of the shift dynamical system, Math. Syst. Theory 3 (1969) 51–59.

[32] M. Hénon, Optimization of collision rules in the FCHC lattice gases and addition of rest particles, in: Discrete Kinetic Theory, Lattice Gas Dynamics, and Foundations of Hydrodynamics, ed. R. Monaco (World Scientific, Singapore, 1989) pp. 146–159.

[33] M. Hénon, On the relation between lattice gases and cellular automata, in: Discrete Kinetic Theory, Lattice Gas Dynamics, and Foundations of Hydrodynamics, ed. R. Monaco (World Scientific, Singapore, 1989) pp. 160–161.

[34] H. Hrgovčić, personal communication.

[35] G. Jacopini and G. Sontacchi, Reversible parallel computation: an evolving space model, to appear in Theor. Comp. Sci. 75 (1990).

[36] Jaynes, Information theory and statistical mechanics, Phys. Rev. 106 (1957) 620–630; 108 (1957) 171–190.

[37] K. Kaneko, Symplectic cellular automata, Phys. Lett. A 129 (1988) 9–16.

[38] J. Kari, Reversibility and surjectivity of cellular au-

tomata, Licentiate's Thesis, University of Turku, Finland (May 1989).

[39] J. Kari, Reversibility of 2D cellular automata is undecidable, Physica D 45 (1990) 379–385, these Proceedings.

[40] A. Lindenmayer and G. Rozenberg, eds., Automata, Language, and Development (North-Holland, Amsterdam, 1976).

[41] N. Margolus, Physics-like models of computation, Physica D 10 (1984) 81–95.

[42] N. Margolus, Quantum computation, Ann. NY Acad. Sci. 480 (1986) 487–497.

[43] N. Margolus, Physics and Computation, Ph. D. Thesis, Tech. Rep. MIT/LCS/TR-415, MIT Lab. for Comp. Sci. (March 1988).

[44] N. Margolus and T. Toffoli, Cellular Automata Machines, in: Lattice-Gas Methods for Partial Differential Equations, eds. G. Doolen et al. (Addison–Wesley, New York, 1990) pp. 219–248.

[45] N. Margolus, T. Toffoli and G. Vichniac, Cellular-automata supercomputers for fluid dynamics modeling, Phys. Rev. Lett. 56 (1986) 1694–1696.

[46] A. Maruoka and M. Kimura, Conditions for injectivity of global maps for tessellation automata, Info. Control 32 (1976) 158–162.

[47] A. Maruoka and M. Kimura, Injectivity and surjectivity of parallel maps for cellular automata, J. Comp. Syst. Sci. 18 (1979) 47–64.

[48] A. Maruoka and M. Kimura, Strong surjectivity is equivalent to C-injectivity, Theor. Comp. Sci. 18 (1982) 269–277.

[49] R. Monaco, ed., Discrete Kinetic Theory, Lattice Gas Dynamics, and Foundations of Hydrodynamics (World Scientific, Singapore, 1989).

[50] K. Morita and M. Harao, Computation universality of one-dimensional reversible (injective) cellular automata, Trans. IEICE E 72 (1989) 758–762.

[51] E. Moore, Machine models of self-reproduction, Proc. Symp. Appl. Math. 14 (1962) 17–33; reprinted in: Essays on Cellular Automata, ed. A. Burks (University of Illinois Press, Champaign, IL, 1970) pp. 187–203.

[52] E. Moore, The firing squad synchronization problem, in: Sequential Machines, ed. E. F. Moore (Addison–Wesley, New York, 1964) pp. 213–214.

[53] J. Myhill, The converse of Moore's garden-of-Eden theorem, Proc. Am. Math. Soc. 14 (1963) 658–686; reprinted in: Essays on Cellular Automata, ed. A. Burks (University of Illinois Press, Champaign, IL, 1970) pp. 204–205.

[54] M. Nasu, Local maps inducing surjective global maps of one-dimensional tessellation automata, Math. Syst. Theor. 11 (1978) 327–351.

[55] N. Packard and S. Wolfram, Two-dimensional cellular automata, J. Stat. Phys. 38 (1985) 901–946.

[56] Y.N. Patt, Injections of neighborhood size three and four on the set of configurations from the infinite one-dimensional tessellation automata of two-state cells (unpublished report, cited in ref. [2]), ECON-N1-P-1,

Ft. Monmouth, NJ 07703 (1971).

[57] C.A. Petri, State-transition structures in physics and in computation, in: Proceedings of a Conference on Physics and Computation, eds. E. Fredkin, R. Landauer and T. Toffoli, Int. J. Theor. Phys. 21 (1982) 979–992.

[58] Y. Pomeau, Invariant in cellular automata, J. Phys. A 17 (1984) L415–L418.

[59] K. Preston and M. Duff, Modern Cellular Automata (Plenum Press, New York, 1984).

[60] D. Richardson, Tessellation with local transformations, J. Comp. Syst. Sci. 6 (1972) 373–388.

[61] T. Sato and N. Nonda, Certain relations between properties of maps of tessellation automata, J. Comp. Syst. Sci. 15 (1977) 121–145.

[62] M. Sears, The automorphisms of the shift dynamical systems are relatively sparse, Math. Syst. Theory 5 (1971) 228–231.

[63] A. Smith, Cellular automata theory, Tech. Rep. 2, Stanford Electronic Lab., Stanford Univ. (1969).

[64] S. Takesue, Reversible cellular automata and statistical mechanics, Phys. Rev. Lett. 59 (1987) 2499–2502.

[65] M.A. Smith, Representations of geometrical and topological quantities in cellular automata, Physica D 45 (1990) 271–277, these Proceedings.

[66] S. Takesue, Relaxation properties of elementary reversible cellular automata, Physica D 45 (1990) 278–284, these Proceedings.

[67] T. Toffoli, Computation and construction universality of reversible cellular automata, J. Comp. Syst. Sci. 15 (1977) 213–231.

[68] T. Toffoli, Cellular automata mechanics, Tech. Rep. 208, Comp. Comm. Sci. Dept., The University of Michigan (1977).

[69] T. Toffoli, Bicontinuous extension of reversible combinatorial functions, Math. Syst. Theory 14 (1981) 13–23.

[70] T. Toffoli, in: Reversible Computing, Automata, Languages and Programming, eds. de Bakker and van Leeuwen (Springer, Berlin, 1980) pp. 632–644.

[71] T. Toffoli, CAM: A high-performance cellular-automaton machine, Physica D 10 (1984) 195–204.

[72] T. Toffoli, Cellular automata as an alternative to (rather than an approximation of) differential equations in modeling physics, Physica D 10 (1984) 117–127.

[73] T. Toffoli, Information transport obeying the continuity equation, IBM J. Res. Develop. 32(1) (January 1988) 29–36.

[74] T. Toffoli, Four topics in lattice gases: ergodicity; relativity; information flow; and rule compression for parallel lattice-gas machines, in: Discrete Kinetic Theory, Lattice Gas Dynamics, and Foundations of Hydrodynamics, ed. R. Monaco (World Scientific, Singapore, 1989) pp. 343–354.

[75] T. Toffoli, How cheap can mechanics' first principles be?, to appear in: Complexity, Entropy, and the Physics of Information, ed. W.H. Zurek (Addison–

Wesley, Reading, MA, 1990).

[76] T. Toffoli, Frontiers in computing, in: Information Processing 89, ed. G.X. Ritter (North-Holland, Amsterdam, 1989) p. 1.

[77] T. Toffoli and N. Margolus, Cellular Automata Machines – A New Environment for Modeling (MIT Press, Cambridge, MA, 1987).

[78] T. Toffoli and N. Margolus, Programmable matter, in: Lattice Gas Methods for PDEs, ed. G. Doolen, Physica D 46 (1990), special issue, to appear.

[79] T. Toffoli and N. Margolus, Invertible cellular automata, occasionaly updated technical memo (MIT Lab. for Comp. Sci.), eventually to appear in book form.

[80] G. 't Hooft, A two-dimensional model with discrete general coordinate-invariance, preprint.

[81] A. Turing, On computable numbers, with an application to the Entscheidungsproblem, Proc. London Math. Soc. Ser. 2, 42 (1936) 230–265.

[82] S. Ulam, Random processes and transformations, Proc. Int. Congr. Mathem. (held in 1950) 2 (1952) 264–275.

[83] G. Vichniac, Simulating physics with cellular automata, Physica D 10 (1984) 96–115.

[84] J. von Neumann, Theory of Self-Reproducing Automata, edited and completed by A. Burks (University of Illinois Press, Champaign, IL, 1966).

[85] S. Wolfram, Statistical mechanics of cellular automata, Rev. Mod. Phys. 55 (1983) 601.

[86] S. Wolfram, Universality and complexity in cellular automata, Physica D 10 (1984) 1–35.

[87] S. Wolfram, Computation theory of cellular automata, Commun. Math. Phys. 96 (1984) 15–57.

[88] S. Wolfram, Cellular automaton fluids 1: basic theory, J. Stat. Phys. 45 (1986) 471–526.

[89] S. Wolfram, Theory and Applications of Cellular Automata (World Scientific, Singapore, 1986).

[90] T. Yaku, Inverse and injectivity of parallel relations induced by cellular automata, Proc. Am. Math. Soc. 58 (1976) 216–220.

[91] K. Zuse, Rechnender Raum (Vieweg, Braunschweig, 1969); translated as Calculating Space, Tech. Transl. AZT-70-164-GEMIT, MIT Project MAC (1970).

Physica D 45 (1990) 254–270
North-Holland

DIGITAL MECHANICS

AN INFORMATIONAL PROCESS BASED ON REVERSIBLE UNIVERSAL CELLULAR AUTOMATA

Edward FREDKIN

Department of Physics, Boston University, Boston, MA 02215, USA

Received 16 January 1990
Revised manuscript received 2 April 1990

This paper is written from the perspective of a computer scientist and addressed to theoretical physicists. As a consequence it may seem somewhat unusual, but rest assured, everything is really quite simple! The point of this paper is that the study of certain phenomena from computer science suggests that there are computer systems (cellular automata) that may be appropriate as models for microscopic physical phenomena. Cellular automata are now being used to model varied physical phenomena normally modelled by wave equations, fluid dynamics, Ising models, etc. We hypothesize that there will be found a single cellular automaton rule that models all of microscopic physics; and models it exactly. We call this field DM, for digital mechanics.

1. Cellular automaton models

It always bothers me that, according to the laws as we understand them today, it takes a computing machine an infinite number of logical operations to figure out what goes on in no matter how tiny a region of space and no matter how tiny a region of time. How can all that be going on in that tiny space? Why should it take an infinite amount of logic to figure out what one tiny piece of space–time is going to do?

– Richard Feynman

A cellular automaton (CA) is a kind of digital computer. It is discrete and deterministic. It is made up of cells like the points in a lattice or like the squares of a checkerboard. It is called "automaton" because it follows a simple digital (discrete) rule. A well known CA is the *Game of Life* [2] invented by John Conway. The rule is as follows: given a rectangular 2D lattice (like a checkerboard), with two states per cell (up or down, + or −, 1 or 0; we will use 1 or 0), the rule states that at each tick of a clock, the value at each cell becomes a one if exactly 3 of its nearest 8 neighbors are ones, it will remain a one if 2 or 3 neighbors are ones, and it will become a zero in all other cases.

Our thesis is that some CA model may be, in effect, programmed to act like physics. We call

such models digital mechanics (DM). In short, DM is a discrete and deterministic modelling system which we propose to use (instead of differential equations, for example) for modelling phenomena in physics. We are driven in this direction by many heuristics; primarily by the concept, borrowed (and extended) from automata theory, of the universal machine [3]. Any ordinary commercial computer would be a universal machine, except for the fact that it does not have an infinite memory. In this paper, we shall extend the meaning of the term "universal machine" to include the universal cellular automaton (UCA) or other kinds of general purpose computers that have large but finite memories; this would include any commercial computer. A universal machine can exactly mimic the behavior of any other finite computer, provided its memory is just a very little bit larger than the target machine.

For example, let us imagine a Cray YMP, with a total of 256 MB (256 MegaBytes) of RAM (random access memory and other memory, such as registers, but no disks or other auxiliary storage) solving a large, complex problem. An Apple Macintosh with 1 MB of RAM and a 256 MB hard disk can solve the same problem by exactly mimicking the behavior of the Cray. The 1 MB of RAM would be used to contain the simulation program and the 256 MB of disk would represent the state of the memory of the Cray. Simulating a Cray with 16 GB (GigaBytes) of RAM would mean that the Mac's disk would also have to have 16 GB, but the 1 MB of RAM would still be enough (technically, a few hundred more bytes might be needed). Of course the Macintosh would be seen by us to be very much slower than the Cray, but as seen from within (the evolution of the information in the memory) the two processes would be absolutely identical.

A sufficiently large universal [#1] cellular automaton (UCA) can, by definition, exhibit any (and every) kind of mechanistic, locally finite behavior. What this means is that if you can imagine a process that could take place in a particular CA of any degree of complexity, then the same process can also be done by any CA that happens to be *universal* even though the universal CA is governed by a drastically simpler rule than the complex CA. By "complexity" we mean some combination of: a large number of states per cell, a complex CA rule, neighborhood (spatial connectivity and dimensionality), boundary condition or initial condition. If any particular CA is a good model of microscopic processes in physics, then every UCA could be programmed to exhibit the isomorphic behavior (after a space–time mapping). While it is true that there is no need to match neighborhoods, a UCA with a lower dimensionality than the CA it is modelling will spend almost all of its resources compensating for that deficiency.

A reversible universal cellular automaton (RUCA), belongs to a subclass of UCA. RUCA have properties that are both counter intuitive and very suggestive of being useful as a more direct substrate for an informational process that might be an exact model of microscopic processes in physics. Because RUCA are microscopically reversible, information and other quantities of the process are conserved in a very natural way. This means that it is possible to map properties of a RUCA information process onto properties of

physics, such as energy or angular momentum, and have those properties conserved as a consequence of the reversible rule. RUCA also exhibit unusual and counter-intuitive behavior that is a consequence of perfect reversibility combined with extreme quantization. RUCA reversibility is very different than the intuitive notion of microscopic reversibility that relies on continuity to ensure that no effect gets lost no matter how infinitesimal it becomes.

DM is a model of the basic phenomena of nature that uses the cellular automaton paradigm. What we will do in this paper is to suggest how various physical phenomena might be recast in the DM genre by mapping certain properties of physics onto the collection of kinds of behaviors that we have either seen in RUCA or which our experience leads us to believe can be found in RUCA. In this paper, we will always be assuming that our world is a large but finite system; finite in the amount of information in a finite volume of space–time, and finite in the total volume of space–time. We call that assumption "finite nature". The main implication of finite nature is that a finite volume of space–time can be exactly represented by a finite number of bits. DM like systems may be useful for creating approximate models of continuous phenomena, but if finite nature is true, then DM can be an exact model.

2. A short history

The game isn't over until it's over .

– Yogi Berra

Ulam [4] and von Neumann [5] gave birth to the field of cellular automaton models and of universal cellular automata in the late 40's. Then it slowly withered away. For many years, no one knew of a simple and symmetric UCA, or believed that there could be a reversible CA that was also universal. The intuitive reasoning was simple; first, universality seemed tied to an irreducible amount of ad hoc complexity, as in a Turing machine and second, a reversible CA seems by its nature to be very constrained in what it can do while a UCA can do anything. By the 60's and early 70's, very few people were interested in this field. In 1975,

---

[#1] In this context, by universal we mean capable of supporting the operation of a finite universal computer, such as an Apple Macintosh. Obviously physics allows this, and every sufficiently large RUCA or UCA also allows this.

a leading computer science theoretician at MIT made the following pointed remark – "You can tell that a person is out of it if he doesn't realize that cellular automata is a dead field." However, slow but steady progress took place, starting with the discovery of the XOR rule in 1962, and followed by Zuse [6], Banks [7], Toffoli [8] and Margolus [9,10]. More recently many others have plunged into this greatly revitalized field, helped to make it more of a science, and discovered many simple, highly symmetric, useful, interesting and provocative UCA and RUCA. In particular, Wolfram carefully explored the one-dimensional cellular automaton [11,12] and wrote an important series of papers that put many of the existing observations about cellular automata onto a firmer basis [13]. CAs were prematurely pronounced dead.

Our approach, since 1958 when this work began, has been to enlarge the numbers and kinds of physics-like properties and behaviors that we can demonstrate within a computer model, while looking for fundamental principles that could guide our search. In nature, for example, we have 3 plus 1 space–time, particles, fields, conserved quantities, charge, computation universality, spin, etc. CAs can deal naturally with Euclidean space–time, but finding CA systems that model aspects of the real physical world is work in progress. What is true is that most serious attempts to find a way to incorporate a property of nature into the CA environment have met with progress; today we see ways of incorporating dozens of characteristics of the real world into CA models. At one time, it seemed unlikely that a simple and symmetric universal rule could be found. In the 60's, Marvin Minsky challenged the author to find a way to demonstrate any form of spherical propagation.

Fortunately, ways around many of the apparent limitations of CA models have been found as more people have begun working in this field. When it appears to be true that some essential property of physics cannot be modelled by CA, then many observers feel justified in not looking further. For example, it once seemed that a CA could neither model Lorentz invariance nor be quantum-mechanically correct. Since then we have learned that it is naive to state what a UCA cannot do! Given finite nature, universality has the power to

overcome every objection; we just have to figure out how to do it. Unfortunately, knowing it can be done is very far from doing it.

Those criticals of DM often state "A UCA will never be able to do X", where X is some part of physics. Normally, it is that the speaker, quite reasonably, does not subscribe to finite nature. Sometimes the speaker has tried to think of a way for a UCA to do X and has come up empty handed. Sometimes we do not believe that X is a property of nature. It is hard to imagine a rigorous proof about what kind of finite problem cannot be solved by a finite UCA. Of course, for DM to make sense we must assume finite nature. Finite nature may or may not be true, but surely the assumption is not known to be false. If finite nature turns out to be false then DM is irrelevant as an exact model, but it might be useful as an approximate model (the way computers are presently used to model physics).

What is most interesting is not what DM is today, but the steady progress that the field has made in overcoming a series of limitations that each appeared insurmountable in its day.

3. The computer science approach

Do not become attached to the things you like, do not maintain aversion to the things you dislike. Sorrow, fear and bondage come from one's likes and dislikes.

           – Buddha

The computer science approach to modelling physics with CA is qualitatively different from either theoretical or experimental physics, or from the kinds of abstract mathematical work that so often leads to progress in physics. The problem is that the study of cellular automata is both a theoretical and an experimental science. However, the experiments, which often produce results we did not anticipate, are not like physics experiments. They are the kind of experiments that never existed before the age of the computer. While there are not yet any formal methods for creating CA models with properties one is looking for, practitioners of this arcane field have developed skills over the years that make searches more efficient. The primary skill is the ability to understand the kinds of behaviors that one can expect to find as a consequence of certain kinds of CA rules, structures and initial conditions. In suggesting CA models for DM we are guided by both basic principles and aesthetic considerations. It is only aesthetics (including economy and simplicity) and initial conditions that separates one UCA model from another.

4. Computational universality

The first principle of DM, dating from 1958, was that the model had to be computation universal. If microscopic physics (assuming finite nature) was not universal, then it would be tautologically true that the macroscopic construction of an ordinary computer would not be possible; but nature allows us to construct computers! Universality is the most important property. Strangely enough, it is not only necessary, but as previously mentioned, it is sufficient. It is even feasible to use a non-reversible model to model a reversible process; it is just aesthetically obnoxious. When we say that any UCA can model the behavior of any target CA, we are normally thinking of those cases where the UCA doing the modelling has the same dimensionality as the target system. In that case *tiles* made up of compact and tile-able (such as a triangle or square in 2D or a cube in 3D) sets of cells, set to initial patterns of states, can mimic the behavior of any particular kind of cell. To mimic the behavior of a particular cellular automaton, one would have to wallpaper the entire space with states that correspond to the appropriate tile patterns, as part of the initial conditions. Of course, the simulation would proceed slowly, use more space, and require the identification of the particular cells within the tiles that at certain time steps (such as "at every 1296th time step") represent the state of the simulated cellular automaton.

5. Finding physics-like behavior

We have tried to imagine the kinds of cellular automaton behavior that would be necessary to model various aspects of physics, and we have sought to find examples of such behavior. A simple example was the successful search for a CA that exhibited spherical propagation of any sort, which was undertaken and completed successfully in 1969. Not long afterwards we found a new approach to creating simple and symmetric UCA. This work culminated in Bank's dissertation [7]. In the early 70's we began to look for reversible models of computation. We recognized the desirability of finding reversible, universal systems. Reversibility makes it easy and natural to create systems that can conserve quantities (such as energy or angular momentum) exactly, despite the fact that translational invariance or angular isotropy cannot be microscopically exact in a discrete cellular space. We then found a path to solving the problem of reversible computation in general with the Fredkin gate and conservative logic [14]. Bennett [15] and Toffoli [8] found other means of solving the same problem. That led to the billiard ball model [14] as a Newtonian mechanics realization of conservative logic, which led to the Margolus rule [9], which is the basis of many interesting applications of cellular automata to physics.

As the armamentarium of known kinds of behavior increased, it became possible to hypothesize new behavior that we could expect to find, in order to match such hypothetical behavior against the needs of modelling physics. The computational approach started with a struggle to see if objects (like particles) could be created, made to persist, move and interact. Once John Conway showed us that possibility (the Game of Life) [2], we continued by looking for ways to represent other basic and simple properties of nature; of space, time, matter and energy. We ask:

"What are the ways in which we can get the right kind of conservation laws?"

"Where will the information that represents a particle's momentum be, and how will it be encoded?"

"How will the RUCA rule support both superposition and interaction?"

"How might the groups that define particle families arise from the lattice neighborhood and the nature of a multi-phase clock?"

"What aspect of the fundamental properties of the RUCA might be related to charge, spin, color, energy, momentum, ...?"

"What is the significance in DM of Planck's constant, speed of light, unit of charge, particle masses, CPT symmetry, ...?"

It is natural to wonder how a DM model of nature, based on a Cartesian lattice, can be Lorentz invariant (be relativistically correct). The space of the DM is certainly not the space of the RUCA. Consider the space in which the space shuttle simulation flies; it is very different than the space in which the computer is (the computer doing the simulation of the space shuttle). We now know of reasonable approaches that allow the physics of DM to be Lorentz invariant even though the process runs on a Cartesian lattice. The lattice remains Cartesian, but measurements made from within the DM model by sending simulated photons back and forth to make measurements can be expected to give relativistically correct results.

6. Bridging the gap

By convention there is color, by convention there is sweetness, by convention bitterness, but in realty there are atoms and space.

– Democritus

Of course it is conceivable that one could be lucky; simply find the magic equation and in one fell swoop recreate all of physics in a new genre. We, on the other hand, have felt a need to understand how more of the simple things could be done before attempting to do everything. In summary, we have worked to knock down, one at a time, the bugaboos that made CAs seem inappropriate, while finding ways to model more and more of the properties of physics. Someday we hope to see it all put together with a grand equation or rule that is the answer to everything, but this process may more closely resemble the progress of quantum mechanics as opposed to the theory of relativity. Once we do have *the rule* we will be able to work with it in two ways: analytically, as with any mathematical equation, and through simulation, where we use computers or specially built cellular automaton machines. The usefulness of simulation will depend on the scale; if it is at Planck's length,

then we will be unable to compute much. If however, the scale is closer to a fermi, then we may be able to compute a lot. Surprisingly, the rule would be the only parameter of this model of physics. We assume that the rule includes the specification of the neighborhood. To be complete, everything in our universe (the exact present state) would be a consequence of the rule, the size and shape of the cellular space, the boundary conditions (which can be eliminated if the space wraps; e.g., if cell coordinates are treated modulo some number), the initial conditions and the time (which would be an integer, the number of time steps from the initial conditions to the present). From the rule it should be possible in principal to compute every constant of physics and answer every question posed at the appropriate scale. The reason is that given the rule, we need only run a simulation and watch it to determine the values of whatever we are unable to calculate analytically. To be practical (meaning that we do not plan on spending much more money on CA machines then we are planning to spend on accelerators) the scale of the rule (of the CA) must not be too far below the scale of the phenomena we wish to model.

7. Scaling, anti-scaling and common sense

The whole of science is nothing more than a refinement of everyday thinking.

– Albert Einstein

DM has one set of properties that scale and another that do not scale. The informational processes that look like field equations must be able to scale because there is not always an intrinsic size associated with the process. The informational process that constitutes a particle must not scale, because there is only a very small number of sizes associated with a particular particle. An electron-like particle comes in only a few masses and charges. These quantities do not scale because they would be a direct consequence of the rule and the lattice size. If you could erase physics from the mind of a computer scientist, he or she would be hard pressed to imagine how a particle could emerge from a programmed model of physics if there were no fundamental unit of length, nor could he or she imagine how uniform motion could occur without a fixed reference lattice. Finite nature would mean that our world is an informational process – there must be bits that represent things and processes that make the bits do what we perceive of as the laws of physics. This is true, because the concept of computational universality guarantees that if what is at the bottom is finite, then it can be exactly modelled by any universal machine. Finite nature does not just hint that the informational aspects of physics are important, it insists that the informational aspects are all there is to physics at the most microscopic level. What is speculative is that a $(3+1)$-dimensional RUCA can be a good model of physics even if finite nature is true.

A scientist steeped in the field of computation who happens to believe in finite nature begins to understand certain laws of nature that have not yet been formalized. For example, "What cannot be programmed, given the necessary resources, cannot be physics." What this means is that if we can prove that we cannot program any kind of computer to model telepathy in accordance with the laws of physics, then telepathy cannot be part of physics. No physics experiment can, in general, give an answer to the halting problem [3]. We believe that we cannot program rectilinear motion without a fixed reference frame. We cannot program particles without a unit of length. We cannot program angular orientation (and consequently angular momentum) without coordinate axes. In short, as an informational scientist it is not that we see DM as a possible model of physics, rather it is that we see no way to model physics without the incorporation of much of what is in DM. In other words, to a programmer who believes in finite nature, physics cannot be imagined in microscopic detail without its having most of the characteristics of DM; unless one resorts to magic. A programmer who does not believe in finite nature, knows that he can only model physics on a computer by making the same kinds of assumptions that DM makes. The author believes that if finite nature is false, then there are not yet enough ideas in contemporary physics to adequately explain from first principles such questions as conservation of momentum or small number phenomena ranging from group theory to spin or

charge quantization. Unfortunately, explanations of the most fundamental aspects of physics either do not exist or they are tautological in nature.

8. DM and RUCA

We must carefully distinguish the RUCA from DM, the informational process that may be running in the RUCA. This is similar to distinguishing a chess board, the chess men and a book of the rules from a game of chess. One is the physical representation of the state of the system and of the rules; the other is an informational process that is identically the same whether it takes place on a real chess board or in a computer memory. For each RUCA, there are one or more such informational processes. DM is simply an informational process that runs on a RUCA and that in one way or another should be able to model nature. Our common sense world (intuitive physics) and RUCA share certain properties in common: they are both universal, locally Euclidian, locally connected, partially described by small integers and deterministic. In common with modern theoretical physics, RUCA are microscopically reversible, have certain symmetries and conserved quantities. In other ways RUCA seem very different than current concepts of theoretical physics, they are: globally Cartesian, simply deterministic, nonisotropic, have a fixed reference frame and are discrete in all regards.

Nature and DM have many more things in common than physics and RUCA, in that DM can be relativistically correct [16], apparently nondeterministic [17], asymptotically isotropic [18], perform as though there is no fixed reference frame, model apparently continuous phenomena with great or perfect accuracy, produce a stable of particles as easily as Conway's Game of Life produces gliders, puffer engines, and other stable Life forms. We will show that DM systems with a local rule nevertheless have the property that the state of a cell at a particular time can be, in an unusual way, functionally related to the state of distant cells that are outside the conventional light cone of physics. This makes us optimistic about the possibility that DM may be capable of using mechanistic, deterministic and local rules as a substrate and yet produce behavior that obeys the laws of QM. In short, we cannot find good reason to believe that DM is incapable of manifesting all phenomena associated with fundamental particles and processes in physics.

9. Modelling macroscopic phenomena

It has been discovered through experimentation that cellular automata can be constructed to behave in ways similar to higher-order physical systems; in other words, they are useful for simulating physical phenomena including hydrodynamics [18], percolation, nucleation, Ising dynamics [19], the movement of sand on the beach, and other phenomena [1]. We distinguish such systems from DM, which is intended only as a direct model of the most microscopic physics. We do not anticipate that direct DM models will be useful for anything other than such microscopic models, but the principles of DM may be more widely applicable.

We will discuss many properties of DM in the context of models of physics. Rather than covering the spectrum of possible RUCA and DM systems, we will describe a hypothetical yet plausible DM that combines properties that we have seen in various other DM systems; we will match up those properties with corresponding properties of physics.

10. DM-4, the nature of a rule

So I have often made the hypothesis that ultimately physics will not require a mathematical statement, that in the end the machinery will be revealed, and the laws will turn out to be simple, like the chequer board with all its apparent complexities.

– Richard Feynman

DM-4 is a hypothetical member of a class of typical DM systems, some of which may be suitable for modelling physics. The purpose of this example is not to put forward a candidate for modelling physics, but rather to impart to the reader the flavor of what such a candidate might be like. We will first give a general definition of the kind of RUCA which runs DM-4.

The RUCA for DM-4 is a four-dimensional space–time lattice. In spatial extent, x, y and z, the RUCA can be any size, according to resources. If programmed on a Macintosh, a reasonable size might be $64 \times 64 \times 64$, for a Cray, one might do $256 \times 256 \times 256$, and a specially constructed cellular automaton machine such as the proposed CAM-8 might handle more than $1024 \times 1024 \times 1024$ [20]. Modelling the whole universe would require a very large but finite array which would have to exist somewhere else. While our universe is exactly large enough, it is busy doing its own thing. That large but finite array would have to exist in some other universe.

In the time dimension, the system has a depth of 2. This is because it is a second-order system, which requires both the *present* and the *immediate past* in order to calculate the *future* . There are many different kinds of rules that are possible, but we have chosen one kind for concreteness. The RUCA is a synchronous and deterministic system that operates according to a simple rule. There is a clock that governs the timing of the system, and the clock is a 6-phase clock; instead of going "...tick, tick, tick..." the clock goes "...tick, tock, tack, toock, teck, tuck, tick, tock, tack, toock, teck, tuck, tick...". Various phases of the clock are associated with the fixed coordinate system. There are six directions, north and south, east and west, and up and down. Each phase of the clock is associated with a single direction, in the order east, down, south, west, up, north. This sequence is abbreviated EDSWUN. It is easy to design such systems so that certain properties are either present or absent. However, there is not yet any magical method for designing just what you might want; you have to try it out and see if it does what you want.

The total operation of DM-4 is determined by the geometry and connectedness of the RUCA, the rule, and the initial state. More accurately, the complete time evolution is determined, yet in general it is unknown and unknowable in advance. This is an interesting consequence of such systems; while they are simple and deterministic, there is no shortcut to determining their future; every step must be carried out (in general). This means that there is no possibility of a *superman* within the system, who could predict any exact future state before it happens. The amount of work to predict the future exactly for any non-trivial DM model is always proportional to the space–time volume. "unknowable determinism" is a good description of what is happening in DM models.

When a RUCA is set to an initial state and put into motion, it is nearly always full of surprises. Of course, a poor choice of rules or initial conditions can quickly result in pure chaos or perfect order, neither of which seems as interesting as the right combination of chaos and order and the right combination of interaction and superposition. We need order to allow for the kind of macroscopic determinism that is embodied in Newton's laws. We need chaos in order to supply the kind of randomness that is a part of QM. We need superposition to allow particles and fields to persist for long periods of time. We need interaction in order to make QED work. We have observed RUCA that simultaneously exhibit order and chaos, interaction and superposition. Many rules have structures that persist, objects that travel in various directions and at various speeds, and interesting interactions when objects collide. One of the most famous of such systems is Conway's Game of Life [2], a simple two-dimensional system that is full of interesting objects. However, Life, in its basic form, is definitely not a RUCA; it can be universal but it is not reversible.

11. The basic dimensional units

The RUCA that runs a DM model can be characterized by three basic dimensional units from which others can be synthesized. They are: the digit-transition, length and time; D, L and T. The L and T are similar but yet slightly different than the L and T of ordinary physics. First of all, T can be thought of as an integer, and counting! Whether T is odd or even is important, and affects everything that happens everywhere in the RUCA. In DM-4 the value of T mod 6 is important. The RUCA T is not Lorentz invariant, and while it is locally similar to the time of physics, is not the same thing. The RUCA unit of time is one 6-phase clock cycle.

12. Numbers in the RUCA

The unit of length, L, in the RUCA is the distance from one cell to its nearest neighbor. Values of L are always numbers that have finite representations. All effective values (or numbers) within a DM system are capable of having a finite representation. They may be real or complex. This is a simple consequence of the fact that the RUCA is finite; there is not enough information in a finite RUCA to represent even one arbitrary real number (or even a rational number with too large a numerator or denominator). Many real numbers have a finite description, such as "pi" or "$2^{1/2}$". Such descriptions of real numbers can exist in a DM system as a consequence of a process which can be shown to asymptotically approach the representation of the real number. In a DM, the representation of a number might reach an exact limit in a finite number of steps if the limit has a finite representation, or it might continue to do the digital equivalent of an asymptotic approach to a value with an infinite representation, never reaching the limit.

The contemporary way that Achilles catches the tortoise is to use calculus, which involves an infinite number of time steps in a finite amount of time. This is not possible in DM. Instead, in DM Achilles chases the tortoise on a chess board where Achilles gets to move to the next square every 3 ticks of the clock, while the tortoise only gets to move every 7 ticks of the clock. In DM, as in the real world, Achilles catches the tortoise. We see the flaw in the calculus solution to Zeno's paradox as ignoring the possibility that the informational work to calculate whether to take a step (or which way to take a step) is not proportional to the distance to the tortoise. This means that it is not clear how the computational task in making an infinite number of steps can be done in finite time.

13. Asymptotic isotropy

While DM can be asymptotically isotropic, it cannot be microscopically isotropic. The RUCA has a preferred and absolute coordinate system and is definitely non-isotropic. The same is true for the DM system on the very microscopic level; on the other hand there is no doubt that a DM system can be asymptotically isotropic. Nearly all trace of the preferred space–time coordinate system, its anisotropy, its absolute reference frame and its absolute lengths and times can be totally washed out so as to become relativistically correct as the scale of events moves away from the most microscopic.

One way to see why asymptotic isotropy can be a consequence of a basic conservation law in the DM system is as follows. We know that conservation laws can be a consequence of certain symmetries. For example, translational invariance leads to conservation of momentum. This concept can be thought of as a two-way street; conservation of momentum implies translational invariance. It is easy to create a cellular automaton rule where the quantity associated with angular momentum is conserved as a consequence of the most basic operation of the rule. This microscopic conservation of angular momentum should allow us to derive a version of angular isotropy that is true at some scale above the most microscopic.

14. The digit-transition

Suppose that physics, or rather nature, is considered analogous to a great chess game with millions of pieces in it, and we are trying to discover the laws by which the pieces move.

– Richard Feynman

The most novel dimensional unit is the digit-transition. We are assuming that the information in each cell of the RUCA is 2-state or 3-state; one bit (for 2-state) or one trit (for 3-state) per cell. We will use the word "digit" to mean a bit or a trit. Of course such systems are possible with any number of states per cell. Each cell contains just one digit, therefore a digit-transition must take place over space, time or space–time. Let us assume that the digit is a bit. The digit-transition would then be either a 1 to 0 transition or a 0 to 1 transition. It could be temporal or spatial. If a given cell changes from a 1 to a 0 in a particular time step then we have a temporal transition. Whether or not it changes during an odd

or even time step is related to other derived units such as charge. The frequency of temporal digit-transitions is related to energy. If a cell is a 1 while at the same time its neighbor is a 0, then we have a spatial digit-transition. Spatial digit-transitions also have phase, and the frequency of spatial digit-transitions is related to momentum.

All the other units of the physics of DM are derived from D, L and T. In DM, the dimensions of the RUCA's digit-transition is the same as for Planck's constant. A series of digit-transitions can have angular momentum and a digit transition's numerical value is very simply related to Planck's constant.

The principle of *least action* takes on new significance in DM. Since action and the digit-transition are the same unit, least action seems related to the least amount of information in a process.

As we have begun to understand DM, we have realized that there must be certain universal laws that govern the behavior of such systems. For example, "There can be no information without a means of its representation." This means that if a particle is moving with a particular velocity, it must be true that there is an interpretation of a conglomeration of bits of information in the system that represent the information that describes the particle's velocity. This information may be spread out over a considerable volume of space–time, superimposed onto a great deal of other unrelated information. "For a process to evolve, there must be a means of interpreting and transforming the representational information." For example, if a particle is to *move*, then the information that represents its velocity must interact with the coordinate system (be, in effect, processed by the RUCA) to produce the change in position over a series of time steps that is in accord with the *velocity*. If a field is accelerating a particle, this means that the information that represents the the state of the field must interact with the information that represents the velocity of the particle so as to change the information that represents the velocity of the particle. Of course we may find that the information that represents the state of the field is carried by a digital version of a boson.

Certain quantities must be conserved. There are some things loosely called "information" and in DM the thing called "information" is conserved.

Depending on both the rule and the initial conditions, there will be other conserved quantities that would correspond to energy, momentum, angular momentum, charge, etc.

Every RUCA transits from one state to another along a closed trajectory. The trajectory consists of a finite sequence (ring) of states of the total system. Each state has one successor state (because the RUCA is deterministic) and one predecessor state (because the RUCA is reversible) The length of the trajectory is finite (because the RUCA is finite). We will show a physically impractical but mathematically correct method of determining a unique quantity for each trajectory. Since this quantity can be determined for each trajectory and because it is different for each different trajectory, it is a unique conserved quantity. This means that there is a one-to-one mapping between trajectories and distinct sets of conserved quantities. In other words for a given RUCA that is in a particular state, there is a value of a conserved quantity for the trajectory associated with the state that is different than that value for any state that is on a different trajectory. This can be proven as follows.

Assume that the RUCA is on a trajectory of length T. Assume that the number of cells in the x direction is X, and similarly Y for y and Z for z. For each of the XYZ values of the set $[x, y, z, t]$, we map the state of every cell, $C_{x,y,z}$ into a number which we will store in a two-dimensional array $D_{t,s}$ by scanning in a regular way. The digits of all of the cells become the digits of an ordinary integer. Given that we first scan through the entire space with the t subscript equal to 0, then with it equal to 1, continuing until t equals the length of the trajectory, there will be $48XYZ$ different ways to generate the one-dimensional array. The subscript s in $D_{t,s}$ ranges over the $48XYZ$ ways of scanning. This is because we can start at any of the XYZ cells and we can scan through the cells in any of 48 orders, $+x, +y, +z; +x, +y, -z; +x, +z, +y, \ldots$, 8 ways of distributing the signs and 6 different orders for x, y and z. We will do them all. We then consider each of the $D_{t,s}$ as an integer. Every different D is unique to the particular trajectory. Of all of the D's produced, we choose the smallest value V (it is allright if there are many instances of that value).

V is a unique quantity of the trajectory, so we may say that it is a conserved quantity. This approach generates the same value for the conserved quantity V for distinct trajectories that differ from each other only by displacements, rotations (through angles relating the coordinate axes) or reflections. Every different trajectory has a different value for V. The ergodic hypothesis for DM is different because of V. A DM system will pass through every state in the trajectory it is on, but there will most likely be many other distinct trajectories that are identical with respect to all high level descriptions, such as the values of other conserved physical quantities, but which differ only in the value of V.

15. Memory and communication

Memory and communication, must be the same physical or informational process, differing only by a coordinate transformation.

We can define a fundamental act of communication as follows:

$$R_{x',y',z',t'} = S_{x,y,z,t} \; ; \text{ we } Receive, \text{ at } x', y', z', t'$$
information *Sent* earlier at x, y, z, t.

We can define a fundamental act of memory as follows:

$$R_{x',y',z',t'} = S_{x,y,z,t} \; ; \text{ we } Read, \text{ at } x', y', z', t'$$
information *Stored* earlier at x, y, z, t. Normally, $x', y', z' = x, y, z$.

It should be clear that we can always transform memory into communication by means of a coordinate transformation, and similarly for transforming communication into memory. We normally think of memory as involving a system where we store information in a given place and later read it out at the same place. From a moving coordinate system of a different observer, it will not necessarily be the same place. For example, while the pilot thinks he is storing data in his navigation computer about flying from Boston to Paris, someone on the ground will see that he is communicating that information from Boston to Paris. A visitor to our solar system may think that everything saved on tape for 6 months is actually an attempt to communicate it to the other side of the sun. When we write on a floppy disk it is usually memory, but when we buy a program on a floppy disk, it serves as communication. Thus it is only our intent that separates memory and communication. From the point of view of physics, they must be seen as one and the same. If two processes are identical except for a coordinate transformation, they must be the same process. Within a DM, there is only one process, and it serves as memory and communication. Computation is simply conditional memory–communication. This is clearly demonstrated by the conservative logic gate, which is a universal computation element. Anything may be computed if you have enough reversible devices (computer designers would call them "gates") that communicate information from A to D if C is a 0, and from B to D if C is a 1, along with ways to interconnect them.

In DM, there is an information cone that is loosely equivalent to our current idea of the light cone of ordinary physics. However, the information cone does not have the intuitive kind of local causality we sometimes attribute to the light cone. In DM it is simply incorrect to say that only events within the information cone of the past can influence an event in the present. This is surprising, but really has to do with the nature of DM. The state of a particular cell is absolutely determined by the state of its immediate neighbors in space–time.

$$C_{x,t} = F_1(C_{x-1,t}, C_{x+1,t}, C_{x,t-1}, C_{x,t+1}).$$

F can be an unusual function where the equation may be also written as follows:

$$C_{x,t+1} = F_2(C_{x-1,t}, C_{x+1,t}, C_{x,t-1}, C_{x,t}),$$

computing the future;

$$C_{x,t-1} = F_3(C_{x-1,t}, C_{x+1,t}, C_{x,t+1}, C_{x,t}),$$

computing the past;

$$C_{x+1,t} = F_4(C_{x-1,t}, C_{x,t}, C_{x,t-1}, C_{x,t+1}),$$

computing right!;

$$C_{x-1,t} = F_5(C_{x+1,t}, C_{x,t}, C_{x,t-1}, C_{x,t+1}),$$

computing left!

There must be an information cone similar to a light cone. The rub is that it might stretch out in 8 different directions. The function F for all DM systems must have an F_2 and an F_3. In many interesting DM models all eight versions of F exist. In any DM model, because the system is totally deterministic and reversible, no information ever gets lost despite its extreme quantization. Also, in every reversible CA, the space–time neighbors of a cell always include both the past and the future! As we shall see, this means that there are reasonable DM models where every cell is a deterministic function of cells through a region of space, the information cone, that includes parts of space not in the obvious light cone. Thus the language we use to describe situations where the concept of a light cone is used, may make little sense in DM. For example, consider the following statement, "Suppose something happens at x, y, z, t, will it affect things at x', y', z', t' ?" This question poses many problems for DM. First, because of the nature of the reversible, deterministic system it is hard to understand exactly what is meant by "…something happens at x, y, z, t… ". Second, it is not as simple as in ordinary, intuitive physics where we might observe that x, y, z, t is outside of the x', y', z', t' light cone. We have to ask "is x, y, z, t outside of the x', y', z', t' information cone; and that is another question.

16. Digital determinism

The Lord God is subtle, but malicious he is not.

– Albert Einstein

What most people imagine is that something random, unexpected, or a consequence of free will can happen. In DM, the whole universe has spent its whole history arranging for whatever is going to happen in each and every cell. No digit can just change; everything, everywhere at every time, past present and future, has operated, does operate and will operate in perfect lockstep to cause the state of every cell to be what it is. The present is no more a consequence of the past than it is a consequence of the future. A location is no more the immediate consequence of its local neighborhood than it is a consequence of very distant points in space–time. Consider a given event: if we have partially determined [#2] certain states in its past and arrange that in its future certain other states will be determined, those states that will be determined in the future will have an effect on the event! These concepts take some getting used to because they are so foreign to our experience; they are very counter intuitive. But wonder of wonders, this very property of DM can allow for mechanistic models of the mysterious events of QM; certainly including the paradoxical statistical correlations between separated events like those in the 1935 EPR gedanken experiment.

The only way that an event can be completely determined from the past would be if we could determine every digit in that event's nearby past information cone. This might be only a few bits. However, to determine those bits, we would have to consider each of them as events and determine each of those events through controlling each of the bits in those event's nearby past information cone. In short, the past can only be totally determined for an event if the total state of all of space–time is determined. We cannot totally determine the past but we can partially determine the past. In addition, we can, through purposeful activity, partially determine the future of an event. That future can definitely have what must be seen as a causal effect on the present!

17. Causality in DM

We need to develop a new model of causality that will fit into the mathematical results indicating the functional relationship between the state of a point and states throughout all of space–time. In a DM model, the state of a cell is a digit. Whether or not it has a functional relationship with a particular other cell depends not just on the spatio-temporal relationship, but also on the state of other cells. Unlike ordinary physics, where effect can decrease with distance, in DM the probability can decrease with distance, but the effect cannot! The question can best be framed by asking

whether causing a particular cell to turn on and off will cause another cell at another place and time to blink. The answer depends on the states of lots of other cells. Strangely enough, you would have to be God to arbitrarily turn on or off just one particular cell, and the result would clearly be that the probability that any particular cell in the information cone would be affected is a number close to $1/2$. In other words there is no action that has substantially more of an effect than arbitrarily changing one cell.

Let us define an atom of causality. We will say that $d_{s',t'}$ is a cause of $d_{s,t}$ if we can write the following equation:

$$d_{s,t} = d_{s',t'} v_{s,t,s',t'} + r_{s,t}(1 - v_{s,t,s',t'}),$$

$v_{s,t,s',t'}$ is a variable that has the value 1 or 0, and that takes into account everything in all of space–time, aside from $d_{s',t'}$ that could affect the state of $d_{s,t}$. So long as $v = 1$, $d_{s,t} = d_{s',t'}$. When $v = 0$, $d_{s',t'}$ is unable to affect $d_{s,t}$ and the local neighborhood rule $r_{s,t}$ applies. In a DM model there is a v for d's spread outside of the normal light cone. When $v_{s,t,s',t'}$ cannot be one, then there is no causal relationship between $d_{s,t}$ and $d_{s',t'}$, even if $d_{s',t'}$ is in the normal light cone. Another way to explain this is that the steps we take to set up an experiment to determine a particular event set other causal chains into action. Since in DM there is no such thing as a locally determined random number, every particular event is related to what caused it. This means that the information cone from that event is not just focused on the one space–time point of the event, but instead it emanates from the diffuse volume in the space–time information cone that precedes the event. In particular, if that past includes setting out instrumentation to measure the results of the experiment, then the steps taken to set up the measuring apparatus end up affecting the event. A shortcut to seeing the effect is to look at the experiment running backwards, which must be equivalent to looking at it running forwards because of reversibility. The catch is to deal properly with phenomena such as amplification and dissipation in microscopically reversible systems.

[#2] By "determined" we mean that we have made experimental arrangements to guarantee that a given event will happen.

18. Quantum mechanics and common sense

The results of the prior section have very interesting consequences that transcend DM. It should be possible to make a continuum model of the same result, by using an integral that has a function (or variable) that switches back and forth between two states, infinitely often. An equation of physics that is microscopically reversible mandates a different concept of causality than what we are used to. Any such system that has been allowed to evolve from a distant initial state, has causality in all directions! This means that we cannot construct an experiment with one part causally isolated from another. This difference in the concept of causality leads us to hope that one might not need to separate the results of quantum mechanics from simple common sense.

19. The EPR experiment

Quantum mechanics is not a theory about reality, it is a prescription for making the best possible predictions about the future if we have certain information about the past.

– Gerard 't Hooft

We can now see the EPR [21] gedanken experiment in a new light. What happens when measurements are made cannot be separated by mere distance. Certain effects cannot be attenuated by distance. Of course everyone already knows that. What is new is that the mystery of quantum mechanics may be explainable as the consequence of a simple common sense, mechanistic, deterministic and local model! This implies a very funny exception to the laws of quantum mechanics. If we could construct not only our experimental apparatus, but if we could actually bring into being the entire space containing the apparatus – de novo – and if we could conduct our experiment before the system could evolve into a consistent state (which would happen at the speed of light), then we would get a result which would disagree with QM. Of course, such experiments are not easy to do in this world! On the other hand, certain effects may depend on the fact that some part of an apparatus exist in another part's future light

cone. This would mean that experiments that are set up to alter the future might give different results in the present than experiments that leave the future alone.

20. Randomness

I shall never believe that God plays dice with the world.

– Albert Einstein

We must digress to have a discussion about randomness. To a high-level observer, much of what goes on in DM appears random. Certain structures last for a random amount of time before disintegrating into other structures. The way in which disintegrations take place seems to also happen at random. Yet we have said that DM is a deterministic system! The answer lies in the relationship between superposition and interaction. By far the most common kinds of local events involve a superposition principle, so that the flow of information proceeds outward in every direction, virtually unimpeded by whatever is going on in the space that it traverses. In a large RUCA, where every digit can be written as a function over most of space–time, we can see why each particular digit seems random. Digits throughout most of space–time have a causal relationship to a digit that is here and now. Normally this randomness is not noticed by processes because of superposition; however, every so often one of these random configurations of digits interacts with some process, and determines one of several outcomes. It seems random, it mimics randomness, but it is not random. It is not even orthogonal or independent. Yet it can produce statistical results one would expect of random processes. It is interesting to note that from a computational point of view, a locally determined truly random number is very computationally expensive. If it is truly random, then it is like a real number, and might need unlimited computational resources just to compute one such number. In DM, we get the benefit without the cost; and we can expect a mechanistic explanation of statistical correlations of distant events.

21. Amplitudes

The hardest thing to understand is why we can understand anything at all.

— Albert Einstein

In quantum mechanics, we can calculate the amplitudes and consequent probabilities very accurately. In DM, the automaton calculates with the amplitudes; the result is that the events occur with the proper probabilities. This calculation goes through all of the possible events. The information associated with a particular event in DM moves through all of the space–time for which there are amplitudes for the event. The DM can calculate the result with statistical precision even though the information has not had the time to get from A to B, because the information at A and B is computed by the same coordinated process. It is not just like synchronized clocks, rather it is like synchronized and intermeshed computers! It is easy to make models of QM using systems with one particle. Two or more particles constitute another story. In DM, everything is, in some sense, involved in a computation with everything else. What must be done is to find a DM rule that results in having done the computation that nature does. But this seems quite possible, because of the peculiar nature of computing with digits as opposed to continuous variables, when the computation is reversible.

22. Correspondences between DM and physics

That theory is worthless. It isn't even wrong.

— Wolfgang Pauli

It is interesting to make an assignment of properties of DM to properties of the real world. The list in table 1 is based on our experiences in testing many CA systems. We have chosen, often based on little evidence, reasonable correspondences found in various CA. While there is no special reason to believe that all of the things that are going to be suggested can hang together and be mutually consistent, nevertheless there is something to be learned from the effort. In particular, the list in

table 1 should convey the flavor of how DM might make use of the parameters it has in order to produce the particles and laws of physics. In DM-4, there are just three fundamental units: D for the digit-transition, also equal to Planck's constant, L for the unit of length, which is the cell to cell distance, and T for the unit of time, one cycle of the 6-phase clock.

23. Tests of the DM theory

One should no more rack one's brain about the problem of whether something one cannot know anything about exists all the same, than about the ancient question of how many angels are able to sit on the point of a needle.

— O. Stern

If we conjecture that our universe operates according to the rules of digital mechanics, then there would be certain consequences that would allow us to verify that fact. It is not clear what would be measurable by means of experiment, but we take the attitude that what is not expressly forbidden is possible. In DM, much of what is thought of as forbidden by relativity or by quantum mechanics is no longer forbidden. This opens the door to new experiments that would give results that would verify the DM conjecture. We do not expect these results to be at odds with either current experimental evidence or with the mathematical formulas of relativity and quantum mechanics. What will undoubtedly be different is the philosophical implications we ascribe to present theories. For example, relativity seems to imply that there is no fixed metric. However, the discovery of a fixed metric would not destroy the mathematical theory of relativity, it would merely change our philosophical perspective. Another example concerns randomness in quantum mechanics. If we were to discover that the apparent randomness was really due to a deterministic hidden variable model, quantum mechanics would not suddenly stop working. DM has in it, at this stage, something for everybody to object to. However, one should be careful to note whether the objection is that DM will be at variance with experimental observation, and the mathematical equations of physics, or with more generalized theories,

Table 1
Correspondence between digital mechanics and physics.

Physics	Digital mechanics
digit-transition	D
length	L, the cell to cell distance
time	T, one cycle of the CA clock
energy	D/T, one digit-transition per unit time
momentum	D/L, one digit-transition per unit distance
mass	DT/L^2
angular momentum	D
action	D
Other relationships are:	
charge $(+$ or $-)$	space–time parity of D, even or odd
charge quantization	stable D orbits in 3-space
color	structure orientation: N–S, E–W, U–D
2-state system (spin)	actually, measuring one bit!
conservation laws	conservation of information
isotropy	asymptotic isotropy
continuity	discreteness
infinitesimals	the digit, units of length and time
infinities	large but finite
special relativity	asymptotic special relativity
general relativity	consequence of the DM process
measurable acceleration	measurable velocity
measurable rotation	measurable angular orientation
group theory properties	consequence of RUCA symmetries
particle masses	stable structures in the RUCA
too many parameters	the rule, and the initial conditions
why is there anything?	answer: unknowable determinism
complex amplitudes	2-phase clock, time dimension depth 2
spin 1/2	smallest D orbit
isotopic spin	projection of D orbits that represent charge
form of the photon	each particle is a digital machine
form of the electron	where its spin, momentum, energy
form of the quark	charge, color etc. is represented
form of the gluon	by information or by a particular
form of the neutrino	information process.

English, Danish or other natural language explanations, models, heuristics, paradigms or philosophies.

24. Particular tests

For certain microscopic events, we would expect to see angular anisotropy. To detect this it would be necessary to know, when collecting and analyzing data, the angular orientation of the experimental apparatus in relation to the "fixed" stars. DM predicts that the structure of a quark would have a particular orientation with respect to the fixed coordinate axes. As a consequence, a particularly energetic quark–antiquark jet might exhibit angular anisotropy at a microscopic level. This could be detected by collecting data on such jets, and, taking into consideration the geographical site and the side real time, transforming the angular orientation from laboratory coordinates to astronomical coordinates. At a sufficiently microscopic level, the statistics of the angular distribution of such events should reveal a basic anisotropy, and would allow the determination of the preferred coordinate axes.

If L and T are large enough, then a crystal (as is used in a very accurate oscillator) may change its electrical impedance when the combination of its own lattice spacing, spatial orientation, absolute velocity vector through the DM lattice and driven frequency, fall into phase with the DM lattice spacing, orientation and natural frequency, k/T. In doing this experiment, the crystal would be driven at a frequency far from its natural frequency. It might be possible to drive the crystal by illuminating it with light of the proper frequency. This would add another variable, the orientation of the light beam. By "light" we mean any electromagnetic radiation. Once the major parameters of the DM lattice are known, then such a device could be used to measure absolute angular orientation and velocity through the lattice.

A detectable quantization of energy levels in particles might occur as the energy becomes very high. Of course, this would have to be measured in the frame of the preferred coordinate system. This means that one would have to correct for the changing velocity due to both the earth's rotation and orbit. In other words, the search for such quantization requires taking out the earth's local variable motion, and then correcting for the presently unknown average motion. This may not be very difficult.

If it proves possible to arrange to detect the absolute velocity of some experimental apparatus relative to the lattice substrate, it is clear that being able to do so violates no experimentally determined fact; all we have done in that area is to have tried and failed. The author believes that the connection between that failure and the obviously correct theory of relativity is simply heuristic and not compelling.

The theory would have to (in principle) be able to predict all of the parameters of physics without any further need for any experiments other than informational experiments. It is quite likely that the computations required may be too large to be practical if they have to be done as Monte Carlo experiments.

#3   The conservative logic gate is commonly known as the Fredkin gate. See ref. [14].

25. Quantum mechanics and Newtonian mechanics

It is only in the quantum theory that Newton's differential method becomes inadequate, and indeed strict causality fails us. But the last word has not yet been said. May the spirit of Newtons's method give the power to restore unison between physical reality and the profoundest characteristic of Newton's teaching – strict causality.

– Albert Einstein

Finally, a note to end on. There is a way to model or build computers, cellular automata and RUCA based on reversible logic, such as the Fredkin gate #3 . It is also true that there is a model of conservative logic based on idealized Newtonian mechanics, called the billiard ball model (BBM). Any RUCA and consequently any DM system can be exactly simulated or implemented by a BBM. This means that DM is a Newtonian system (highly constrained, but definitely Newtonian). We are proposing DM as a paradigm for fundamental interactions in physics. To make a long, long story very, very short, we believe that QM can be modelled by Newtonian mechanics; but not in a way that would easily have occurred either to Newton or to the average quantum mechanic.

References

[1] T. Toffoli and N. Margolus, Cellular Automata Machines – A New Environment for Modeling (MIT Press, Cambridge, MA, 1987).

[2] M. Gardner, The fantastic combinations of John Conway's new solitaire game of 'Life', Sci. Am. 223 (April 1970) 120–123.

[3] M. Minsky, Computation, Finite and Infinite Machines (Prentice Hall, Englewood Cliffs, NJ, 1967)

[4] S. Ulam, Random Processes and Transformations, Proceedings of the International Congress on Mathematics, 1950, Vol. 2 (1952) pp. 264–275.

[5] J. von Neumann, Theory of Self-Reproducing Automata, edited and completed by A. Burks (University of Illinois Press, Champaign, IL, 1966).

[6] K. Zuse, Rechnender Raum (Vieweg, Braunschweig, 1969); translated as "Calculating Space," AZT-70-164- GEMIT MIT Project MAC (1970).

[7] E. Banks, Information processing and transmission in cellular automata, Doctoral Dissertation and Technical Report MAC TR-81, MIT Project MAC (1971).

[8] T. Toffoli, Cellular automata mechanics, Technical Report 208, Computer and Communication Sciences Department, University of Michigan (1977).

[9] N. Margolus, Physics-like models of computation, Physica D 10 (1984) 81–95.

[10] N. Margolus, Physics and computation, Technical Report 416, Laboratory for Computer Science, MIT (1987).

[11] S. Wolfram, Statistical mechanics of cellular automata, Rev. Mod. Phys. 55 (1983) 601–644.

[12] S. Wolfram, Universality and complexity in cellular automata, Physica D 10 (1984) 1–35.

[13] S. Wolfram, Computation theory of cellular automata, Commun. Math. Phys. 96 (1984) 15–57.

[14] E. Fredkin and T. Toffoli, Conservative logic, Int. J. Theor. Phys. 21 (1982) 219–253.

[15] C. Bennett, Logical reversibility of computation, IBM J. Res. Devel. 6 (1973)

[16] T. Toffoli, Four topics in lattice gases: ergodicity; relativity; information flow; and rule compression for parallel lattice-gas machines, in: Discrete Kinetic Theory, Lattice Gas Dynamics and Foundations of Hydrodynamics, ed. R. Monaco (World Scientific, Singapore, 1989) pp. 343–354.

[17] S. Wolfram, Random-sequence generation by cellular automata, Adv. Appl. Math. 7 (1986) 123–169.

[18] U. Frisch, B. Hasslacher and Y. Pomeau, Lattice-gas automata for the Navier–Stokes equation, Phys. Rev. Lett. 56 (1986) 1505–1508.

[19] M. Creutz, Deterministic Ising dynamics, Ann. Phys. 167 (1986) 62–76.

[20] N. Margolus and T. Toffoli, Cellular automata machines, in: Lattice Gas Methods for Partial Differential Equations, eds. G. Doolen et al. (Addison–Wesley, Reading, MA, 1990) pp. 219–248.

[21] A. Einstein, B. Podolsky and N. Rosen, Phys. Rev. 47 (1935) 777.

Physica D 45 (1990) 271–277
North-Holland

REPRESENTATIONS OF GEOMETRICAL AND TOPOLOGICAL QUANTITIES IN CELLULAR AUTOMATA

Mark A. SMITH

*Department of Physics, Massachusetts Institute of Technology, Laboratory for Computer Science,
Cambridge, MA 02139, USA*

Received 4 February 1990
Revised manuscript received 28 February 1990

This paper addresses the problem of finding novel representations of analogs of physical field variables in cellular automata. Several examples illustrate how this might be done: (1) a manifestly covariant vector law for diffusion, (2) a model for potential energy and exact one-forms, (3) a counting algorithm using a winding number density, and (4) fields which support topological charges. In order to combine the advantages of cellular automata and continuum analysis, it is necessary to have representations which involve more than one cell.

1. Introduction

Cellular automata have proved useful for physical modeling [1]. This is because their physical structure and computational power give them the ability to simulate the complex, nonlinear behavior found in many spatially extended systems. Efforts are under way to extend the physical modeling power of cellular automata to ever more general domains. Toffoli [2] motivates this program with a representation of scalar quantities as well as a general discussion of how cellular automata can be viewed as an alternative mathematical structure for physics.

The fundamental laws of physics do not depend on a particular coordinate system nor on the choice of a basis. Similarly, the properties of a cellular automata model of a field should not depend on the way the model is coordinatized. The following sections present several examples of coordinate-free cellular automata representations of fields which occur in physics. They were implemented using CAM-6 [3], and the configurations below come from the resulting 256×256 images.

2. Vectors and group invariances

This section presents a relativistic diffusion model which originated from a study of the sense in which cellular automata display *Lorentz invariance* [4]. The model provides a natural representation of two-dimensional (spacetime) *vector* fields. This model has interesting properties and has a wide range of applications [5]. Relativistic models of diffusion have been developed before [6], and a similar model has been used as an example of quantum automata [7].

The continuum model describes one-dimensional probability densities, $\rho^+(x,t)$ and $\rho^-(x,t)$, for finding a particle moving in the positive and negative directions respectively. The light-cone coordinates, $x^\pm = (t \pm x)/\sqrt{2}$, will turn out to be useful. Consider an ensemble of systems, each of which contains a single marked particle which bounces back and forth at unit speed due to collisions with a background of similar particles as shown in fig. 1. The background particles give well-defined mean free paths, $\lambda_+(x^+)$ and $\lambda_-(x^-)$, for the marked particle to reverse direction. This dynamics can be thought of as a random walk with inertia or as the diffusion of a massless particle.

One can also define a conserved two-current, $J^\mu = (\rho, j)$, where $\rho = \rho^+ + \rho^-$ and $j = \rho^+ - \rho^-$.

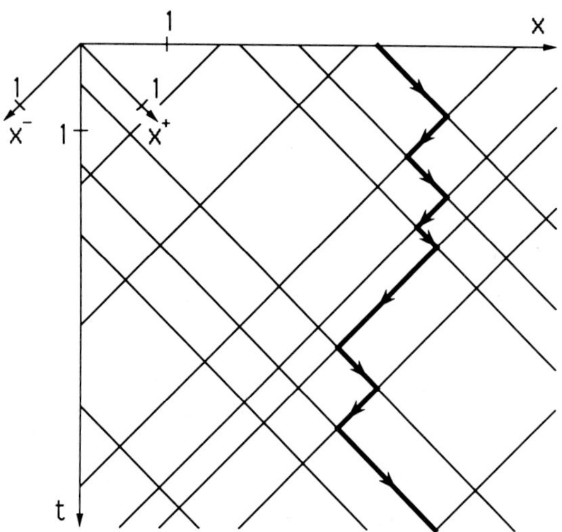

Fig. 1. A spacetime diagram showing the paths of the particles in the relativistic diffusion model. A typical trajectory for a marked particle undergoing collisions with the background gas is emphasized.

The transport equations for the densities are then

$$\frac{\partial \rho^+}{\partial t} + \frac{\partial \rho^+}{\partial x} = -\frac{\rho^+}{\lambda_+} + \frac{\rho^-}{\lambda_-}, \tag{1}$$

$$\frac{\partial \rho^-}{\partial t} - \frac{\partial \rho^-}{\partial x} = -\frac{\rho^-}{\lambda_-} + \frac{\rho^+}{\lambda_+}. \tag{2}$$

Adding these equations gives

$$\frac{\partial \rho}{\partial t} + \frac{\partial j}{\partial x} = 0, \tag{3}$$

and subtracting them gives

$$\frac{\partial j}{\partial t} + \frac{\partial \rho}{\partial x} = -j\left(\frac{1}{\lambda_+} + \frac{1}{\lambda_-}\right) + \rho\left(\frac{1}{\lambda_-} - \frac{1}{\lambda_+}\right). \tag{4}$$

The condition that the background particles stream at unit speed is

$$\frac{\partial}{\partial x^+}\left(\frac{1}{\lambda_-}\right) = 0, \quad \text{and} \quad \frac{\partial}{\partial x^-}\left(\frac{1}{\lambda_+}\right) = 0. \tag{5}$$

If one defines

$$\sigma^t = \frac{1}{\lambda_+} + \frac{1}{\lambda_-}, \quad \text{and} \quad \sigma^x = \frac{1}{\lambda_-} - \frac{1}{\lambda_+}, \tag{6}$$

then eqs. (3)–(5) can be rewritten in manifestly covariant form as

$$\partial_\mu J^\mu = 0, \tag{7}$$

$$(\partial_\mu + \sigma_\mu)\varepsilon^{\mu\nu} J_\nu = 0, \tag{8}$$

$$\partial_\mu \sigma^\mu = 0, \tag{9}$$

$$\partial_\mu \varepsilon^{\mu\nu} \sigma_\nu = 0. \tag{10}$$

The parameter σ^μ is essentially the two-momentum density of the background particles and gives a proportional cross section for collisions. Writing the equations in this form proves that the model is Lorentz invariant.

This model is also *conformally invariant*. This is mentioned because conformal invariance has profound consequences for physics. In 1+1 dimensions, conformal invariance means that the light-cone axes can be locally scaled by any amount $(x^\pm \to f^\pm(x^\pm))$, and the resulting evolution is still described by the original equations. The fact that the above model is conformally invariant is easy to see from fig. 1, since any change in the spacing of the diagonal lines yields a similar picture.

In the case of a uniform background with no net drift ($\lambda_\pm = \lambda$), eqs. (7)–(10) reduce to the telegrapher's equation, which has been studied extensively [8]. For the initial conditions $j(x,0) = \rho(x,0) = \delta(x)$ the solution inside the light cone (i.e. the region $t^2 - x^2 > 0$) is

$$\rho^+(x,t) = \frac{e^{-t/\lambda}}{2\lambda}\sqrt{\frac{t+x}{t-x}}\,I_1\left(\frac{\sqrt{t^2 - x^2}}{\lambda}\right)$$
$$+ e^{-t/\lambda}\delta(t-x), \tag{11}$$

$$\rho^-(x,t) = \frac{e^{-t/\lambda}}{2\lambda}I_0\left(\frac{\sqrt{t^2 - x^2}}{\lambda}\right). \tag{12}$$

Outside the light cone, $\rho^\pm = 0$. The total density, $\rho(x,t)$, for $\lambda = 32$ is plotted in fig. 2. Note how the delta function decays exponentially as it moves along the x^+ axis and how the rest diffuses. Eqs. (11) and (12) can be used to find the solution for any other initial conditions. First, they can be reflected around $x = 0$ to give the solution starting from a left-moving pulse. Second, the light-cone axes can be scaled to change $\lambda \to \lambda_\pm(x^\pm)$ corresponding to an arbitrary background gas. Finally, solutions starting from different values of x can be

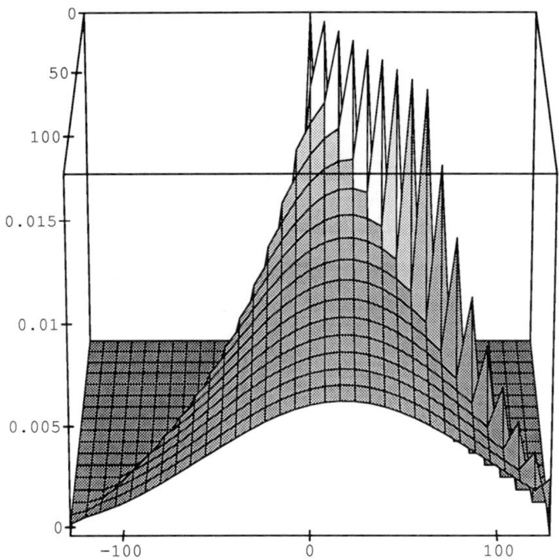

Fig. 2. Solution of the continuum equations in the relativistic diffusion model starting from a localized pulse moving to the right. The t-axis runs from 0 to 128, and the x-axis runs from -128 to 128.

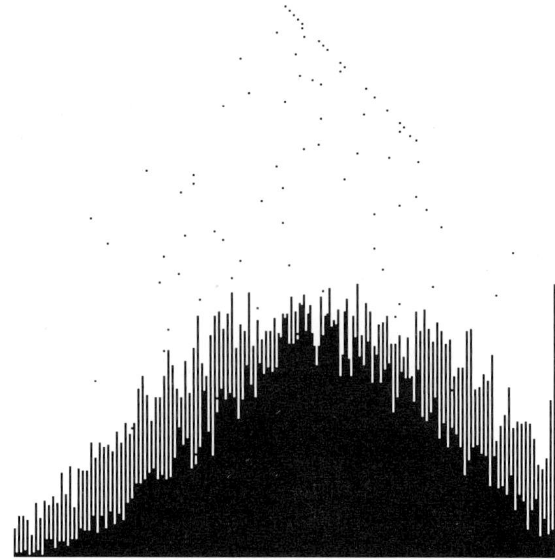

Fig. 3. Histogram of 17 500 particles in the relativistic diffusion cellular automata model for $t = 128$. Particles execute a random walk in the upper half and fall into one of the 256 bins in the lower half.

combined to give any $\rho^+(x,0)$ and $\rho^-(x,0)$.

Fig. 3 shows the result of a cellular automata simulation with $\lambda = 32$. In this case, a particle is restricted to a discrete set of paths, but for $\lambda \gg 1$, the continuum distribution of free path lengths (a decaying exponential) is well approximated by the actual discrete geometric progression. The odd columns contain $\rho^-(x,128)$, and the even columns contain $\rho^+(x,128)$. The separation of these components is quite apparent in the figure and is reflected by eqs. (11) and (12). The rightmost bin, $\rho^+(128,128)$, is full, so the delta function it represents has been truncated. There is a slight excess of particles in the neighboring bin, $\rho^+(126,128)$, because of spurious correlations in the noise source.

3. Potentials and one-forms

One-forms are geometrical quantities very much like vectors. Like vectors, they are rigorously described in terms of infinitesimals of space. Unlike vectors, they cannot always be pictured as arrows. A better picture of a one-form is that of a separate contour map at each point. If the maps fit

together to make a complete contour map, then the one-form field is said to be *exact*. In this case, it can be written as a *gradient*, dV. The scalar field of which it is the gradient is a *potential*, V. Potentials are important in physics because they create forces; in this case, $F = -\mathrm{d}V$.

Contours are unbroken lines (or hypersurfaces in higher dimensions) that do not cross and never end. They can be represented in any way which has these properties and consistently specifies which side of each contour is higher. The magnitude of the field is inversely proportional to the "distance" between contours. The implementation used in fig. 4 encodes the potential very efficiently by digitizing it and then only storing the lowest bit. In this case, it is a harmonic potential, $\frac{1}{2}m\omega^2 r^2$, in real space or alternatively, the classical kinetic energy, $p^2/2m$, in momentum space. The entire potential can be recovered (up to a constant) by integration. Note that the overall slope of the potential is only discernable by looking at a relatively large region. This illustrates the multiple-cell character of this representation.

An important physical application of this representation is to make a reversible lattice gas spontaneously become denser in a particular region of

Fig. 4. Contours representing the gradient of a harmonic potential well. The special encoding of the potential accounts for the width of the steps being nonmonotonic.

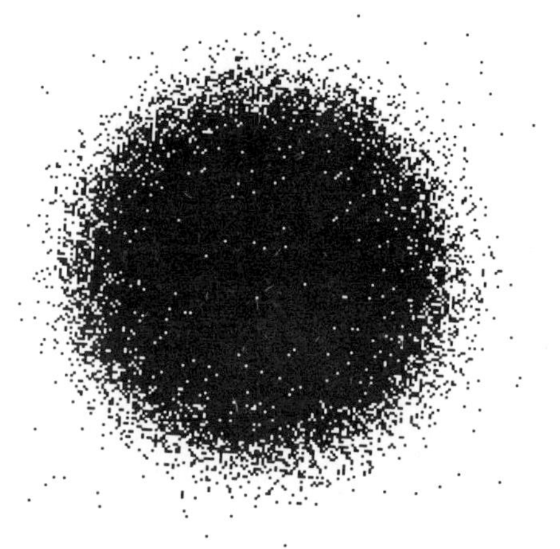

Fig. 5. The Fermi sea formed by a lattice gas in a harmonic potential well. This can be viewed as a spatial density or as a distribution in momentum space. The Fermi energy (compared to the contours above) is approximately 4, and the temperature is about 2.

space by effectively reproducing an external potential [9]. An *isolated* system containing a uniformly distributed lattice gas cannot reversibly self-organize in any way because it is already in a state of maximum entropy. This would be tantamount to perfect compression of N bits of information into fewer than N bits.

However, one can get around this restriction by coupling the system to a heat bath which serves to conserve the microscopic information of the entire system while its macroscopic entropy increases. The heat bath consists of any reversible system that has conserved tokens which can stand for energy. The coupling manifests itself when a particle of the gas tries to cross a contour of the potential. The particle is allowed to go up (down) a contour if the corresponding cell in the heat bath is occupied (empty). In these cases, it may proceed by exchanging energy with the heat bath; otherwise, it must bounce back.

A lattice gas in a potential well distributes itself according to the laws of statistical mechanics – it has a well defined entropy, temperature, free energy, chemical potential, etc. The force which causes it to fall into the well is entirely statistical: the system merely wanders freely and evenly through the space defined by its conserved quantities, but by far the bulk of this space lies where the density of the gas in the well is higher. The result is shown in fig. 5.

A lattice gas has an exclusion principle built in because no two particles can be in the same state; therefore, they are considered to be *fermions*, and they obey *Fermi statistics*. All matter is ultimately made out of fermions, and the exclusion is physically very important because it keeps matter from collapsing. Similarly, the gas in fig. 5 cannot be compressed beyond about the fourth contour (this limit is called the *Fermi energy* and is roughly the same as the *chemical potential*). The density falls from one to zero over about two contours (this is the *temperature*). A collection of particles which follows such a *Fermi distribution* is called a *Fermi sea*. The Fermi distribution is only approximate, since the system is not fully ergodic and has a finite size. Corrections to the distribution have been calculated and measured and are described elsewhere [10].

4. Winding number

Winding number refers to any topological quantity which is given by the number of complete turns made by mappings between a circle and a plane. An archetype to keep in mind is provided by the function of a complex variable, $w = f(z) = z^n$. This maps the circle, $z = r\,e^{i\theta}, 0 \leq \theta < 2\pi$, around the point $w = 0$ a total of n times. The terminology comes by analogy with the principle of the argument in complex analysis, which gives the number of zeros minus the number of poles (of a meromorphic function) surrounded by a contour. This number is the number of times the image of the contour winds around the origin.

4.1. Domain counting

A winding number can be associated with a set of domains in two dimensions (see fig. 6). It is the net number of turns made by following all the boundaries (taken with the same orientation) of all the domains. The orientation of a boundary is determined by whether the interior of a domain is to the right or to the left of the boundary as it is traversed. This number is the number of domains minus the number of holes in the domains and can be shown to be the same as the Euler number of the set. See ref. [11] for a more complete mathematical treatment of these topics.

Fig. 6. Magnetic domains (in black) taken from an Ising model with 64K spins. Each black cell can be thought of as a particle. The winding number is 1357, which is somewhat less than the actual number of domains. The average number of particles per domain is approximately 7.9.

In the discrete case (as for cellular automata), the plane is tiled with cells. A connected group of occupied cells (or tiles) constitutes a domain. If the sides of the tiles are straight, the boundaries of the domains only turn where three or more cells meet at a point (the set of all such points form the vertex set of the lattice *dual* to the original lattice). The net number of turns is simply the sum of these angles ($\times$cells/360°) taken over all the cells of the dual lattice. Hence it is possible to define a density whose integral gives the winding number. This connection between a topological number and a differential quantity is surprising and important.

The *winding number density* on the dual lattice is defined to be zero unless there is a boundary passing though the vertex. In this case, it is 180° minus the angle subtended by the cells constituting the interior of the cluster. This is just the angle that the boundary bends through. On a square lattice, the winding number density is assigned as shown in fig. 7. Note that this quantity depends on correlations in a group of cells, and that the groups overlap.

The winding number density is related to certain many-body potentials in deterministic Ising models. These potentials have been used to bias the curvature of the phase boundary in a model of droplet growth [12]. The winding number then provides an estimate of the number of droplets. In the low-density limit, the domains become simply connected (no holes), and the correspondence between winding number and the number of domains becomes exact. In the case of droplet growth, it

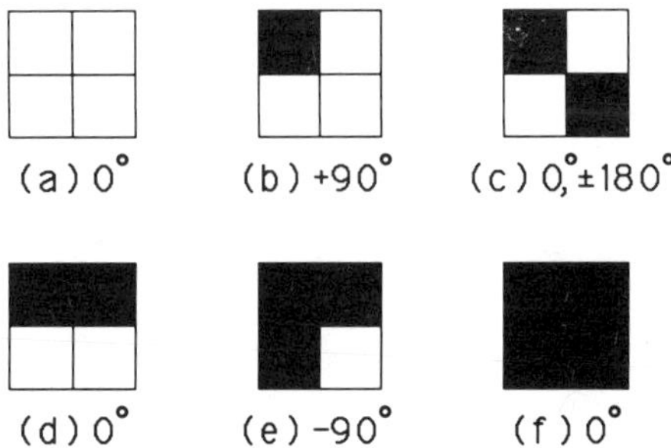

Fig. 7. Winding number density (in degrees per cell) for a square lattice. Rotated versions of these blocks are equivalent. Case (c) has two boundary lines going through the center and can be counted according to how one wants to define a domain.

can be used to find an estimate of the average droplet size as a function of time.

In order to find the average domain size, one needs to count the number of occurrences of cases (b), (c), and (e) in fig. 7 as well as the total number of occupied cells. Fig. 6 has 7380 90° angles, 2310 −90° angles, 179 180° angles, and a total of 10 749 particles. The built-in counter in CAM-6 makes doing these counts over the 64K cells very fast. Finding and tracing out domains one by one requires a relatively slow, complicated procedure, whereas finding the average domain size on CAM-6 only takes four steps or $\frac{1}{15}$ of a second; hence, this is a practical method to use while doing real-time simulations.

4.2. Topological defects

Another form of winding number is illustrated by fig. 8. The cell states of this automaton have a definite cyclic order. A configuration is *continuous* if the states of adjacent cells differ by at most one with respect to this order. The winding number of a closed loop of cells is the net number of times the cell states go through the cycle as the loop

Fig. 8. Configuration from a cellular automata model of an oscillatory chemical reaction. The cells assume eight different values, and the values change gradually between neighboring cells. The centers of the spirals are topological defects.

is traversed in a counterclockwise direction. Any loop with a nonzero winding number is said to contain a defect (see the illustration in ref. [13]). The defects have a topological character because the winding number of a loop cannot be changed by modifying the state of one cell at a time while still maintaining continuity. The centers of the spirals in fig. 8 are the topological defects.

Analogous topological objects are important in several disciplines. The mathematical study of these objects falls under the field of algebraic topology. Their physical relevance stems from the fact that, depending on the dynamical laws and the overall topology of spacetime, most fields can form "knots" – continuity and energy conservation forbid such configurations of fields from being untied. Two physical instances of topological charges which have been postulated to exist (but not definitely observed) are magnetic monopoles and cosmic strings. The dynamical effects of topological objects range from cardiac arrhythmias to extinction in certain predator–prey models.

Many cellular automata models [3] have behavior in which "activity" in a cell spreads to nearby cells and is then suppressed. These include the game of life, neural models, predator–prey models [14], and chemical systems [15]. These rules generate spiral patterns (in a generalized sense) similar to those in fig. 8. One model in particular [16] has definite configurations which are the growth centers for defects. Again, these structures reflect organization over a relatively large region.

5. Conclusions

The above examples show that it is possible to find representations of physical quantities in cellular automata which respect the geometrical and topological foundations of physics. Some people have the misconception that cellular automata models of fields must contain discretized approximations of the variables in individual cells. But the representations given here are distributed over *several* cells and need not be interpreted as averages over those cells – they can also have an interpretation as a single correlated pattern. These alternatives illustrate how cellular automata have a distinct character and should not be viewed as necessarily inferior substitutes for the continuum.

This paper is a step in the development of a catalog of techniques for physical modeling with cellular automata.

Besides yielding practical algorithms, building discrete models of physics will give us the opportunity to reevaluate our understanding of physics and computation.

Acknowledgements

I would like to thank Tommaso Toffoli and the Information Mechanics Group in the MIT Laboratory for Computer Science for inspiring this work.

Support was provided in part by the National Science Foundation, Grant No. 8618002-IRI, and in part by the Defense Advanced Research Projects Agency, Grant No. N00014-89-J-1988.

References

[1] T. Toffoli and N. Margolus, Cellular Automata Machines: A New Environment for Modeling (MIT Press, Cambridge MA, 1986).

[2] T. Toffoli, Cellular automata as an alternative to (rather than an approximation of) differential equations in modeling physics, Physica D 10 (1984) 117–127.

[3] Information Mechanics Group, MIT Laboratory for Computer Science, CAM-6 hardware and software (Systems Concepts, San Francisco, 1987).

[4] T. Toffoli, Four topics in lattice gases: ergodicity; relativity; information flow; and rule compression for parallel lattice-gas machines, in: Discrete Kinetic Theory, Lattice Gas Dynamics and Foundations of Hydrodynamics, ed. R. Monaco (World Scientific, Singapore, 1989) pp. 343–354.

[5] M. A. Smith, T. Toffoli, H. Hrgovčić and N.H. Margolus, A conformally invariant cellular automaton, J. Phys. A, to be submitted for publication.

[6] D.C. Kelly, Diffusion: a relativistic appraisal, Am. J. Phys. 36 (1968) 585–591.

[7] G. 't Hooft, Equivalence relations between deterministic and quantum mechanical systems, J. Stat. Phys. 53 (1988) 323–344.

[8] S. Goldstein, On diffusion by discontinuous movements, and on the telegraph equation, Quart. J. Mech. Appl. Math. 4 (1951) 129–156.

[9] M.A. Smith, A Fermi lattice gas in a potential well, J. Stat. Phys., to be submitted for publication.

[10] M.A. Smith, Broken ergodicity and finite size effects in the Fermi lattice gas model, MIT Laboratory for Computer Science Technical Memorandum, in preparation.

[11] S.B. Gray, Local properties of binary images in two dimensions, IEEE Trans. Computers C-20 (1971) 551–561.

[12] T. Toffoli and N.H. Margolus, CAM-6 experiment, private communication (1988).

[13] A.K. Dewdney, Sci. Am. (August 1989) 103.

[14] R. Fisch, Cyclic cellular automata and related processes, Physica D 45 (1990) 19–25, these Proceedings.

[15] P. Tomayo and H. Hartman, Cellular automata, reaction–diffusion systems and the origin of life, in: Artificial Life, SFI Studies in the Sciences of Complexity, ed. C. Langton (Addison–Wesley, New York, 1988) pp. 105–124.

[16] J.M. Greenberg and S.P. Hastings, Spatial patterns for discrete models of diffusion in excitable media, SIAM J. Appl. Math. 34 (1978) 515–523.

Physica D 45 (1990) 278–284
North-Holland

RELAXATION PROPERTIES
OF ELEMENTARY REVERSIBLE CELLULAR AUTOMATA

Shinji TAKESUE

Faculty of Science, Gakushuin University, 1-5-1 Mejiro, Toshima-ku, Tokyo 171, Japan [1]

Received 24 January 1990
Revised manuscript received 28 February 1990

The relaxation properties of a number of elementary reversible cellular automata are studied. Different rules have different types of relaxation behavior. Possible relationships between the thermodynamic behavior and phase space structure are discussed.

1. Introduction

Reversible cellular automata provide a very efficient means to examine the basis of various concepts and methods in statistical mechanics.

Statistical mechanics, founded by Gibbs, can be applied to the dynamical systems where the volume of phase space is preserved and energy is constant. Most successful applications, of course, have been made to Hamiltonian systems. In Hamiltonian systems, the former condition is satisfied due to Liouville's theorem and energy is Hamiltonian itself.

Because cellular automata have discrete dynamical variables, the cardinality of phase space is finite in a system with a finite number of sites. Therefore, reversible dynamics leads to the preservation of phase space volume. Moreover, some rules have additive invariants [1,2]. Unlike Hamiltonians, these additive invariants do not completely govern the time evolution of the system. Still, one can calculate thermodynamic quantities with those quantities by a standard statistical mechanics argument. That is, a partition function can be calculated from the invariants. Thus, such systems can be used as models of statistical mechanics.

An example of the above type of cellular au-

tomata is Creutz's deterministic Ising dynamics [3]. Creutz added variables representing kinetic energy to the usual two-dimensional Ising Hamiltonian and attached to the model a dynamics which conserved the total energy. This model serves as a microcanonical simulation, where temperature is calculated via the mean value of the kinetic energy at a site. Good agreement is obtained between simulations of this model and exact solution of the two-dimensional Ising model. A simple version of deterministic Ising dynamics called Q2R also agrees with theory, at least in the high temperature region [4]. In addition, lattice-gas automata also belong to this category of cellular automata. Due to the parallel nature of cellular automata, these models can be simulated more efficiently on parallel machines than some types of Monte Carlo simulation.

All of these cellular automata satisfy symmetries and conservation laws characteristic of the physical systems modeled. Studies of these models always assume, but do not prove, the satisfaction of the ergodic-theoretical conditions required to guarantee thermodynamic behavior. Hence, it is worthwhile to ask what rules can realize thermodynamic behavior. To answer this question, a systematic study of a family of simple reversible cellular automata was undertaken. Since cellular automata are particularly simple in structure, such a study is expected to shed new light on the foun-

[1] Present address: Department of Mathematics, Rutgers University, New Brunswick, NJ 08903, USA.

dations of statistical mechanics.

Elementary reversible cellular automata (ERCA) are probably the simplest family of non-trivial reversible cellular automata. Their time evolution rules have the following form:

$$\sigma_i^{t+1} = f(\sigma_{i-1}^t, \sigma_i^t, \sigma_{i+1}^t) \text{ XOR } \hat{\sigma}_i^t, \tag{1a}$$

$$\hat{\sigma}_i^{t+1} = \sigma_i^t, \tag{1b}$$

where σ_i^t and $\hat{\sigma}_i^t$ denote the state of site i at time t, each of which takes values in the set $\{0,1\}$, XOR is the "exclusive OR" operation, namely $\mu \text{ XOR } \nu = \mu + \nu - 2\mu\nu$, and $f: \{0,1\}^3 \to \{0,1\}$ is a Boolean function of three variables. This function f determines each rule which is represented by the number $\sum_{\lambda,\mu,\nu} 2^{4\lambda+2\mu+\nu} f(\lambda,\mu,\nu)$ with an "R" appended to it. For example, if $f(\lambda,\mu,\nu) = \lambda \text{ XOR } \nu$, the rule is called 90R. The reversibility of the dynamics is evident when the time reversal evolution is explicitly written as

$$\sigma_i^{t-1} = \hat{\sigma}_i^t, \tag{2a}$$

$$\hat{\sigma}_i^{t-1} = f(\hat{\sigma}_{i-1}^t, \hat{\sigma}_i^t, \hat{\sigma}_{i+1}^t) \text{ XOR } \sigma_i^t. \tag{2b}$$

This equation has the same form as eq. (1). That is, ERCA are not only reversible but also time reversal invariant as is the case for usual mechanical systems.

In a previous paper [2], the ERCA rules which possessed an additive conserved quantity of the type

$$\Phi = \sum_i F(\sigma_i, \sigma_{i+1}, \hat{\sigma}_i, \hat{\sigma}_{i+1}) \tag{3}$$

were determined. In summary, about half of the rules have the above type of additive conserved quantities. In many such rules, however, there are conserved quantities that are *localized* (i.e. they are bound to a particular site or group of sites). In these rules, statistical mechanics fails since energy propagation is inhibited. There remain seven rules which have one or more additive invariants but no local conservation laws. Only these rules qualify as candidate thermodynamic models. These rules are listed in table 1 along with their additive invariants. As is seen from table 1, there appear to be only four kinds of additive conserved quantities for these rules. They are named energy (a) through (d) as in the table. A rule can conserve

more than one energy and conversely an energy may be invariant under more than one rule. In fact, the rules 90R, 91R, and 123R each have two kinds of energy functions.

Consider a segment in a chain of an ERCA as a subsystem. According to statistical mechanics, if the subsystem is much smaller than the full system (but still contains a sufficient number of sites), then the rest of the system can be regarded as a heat bath for the subsystem. In this case, a canonical distribution of subsystem energies should be realized.

In ref. [6], it was shown numerically that two of these rules, 26R and 90R, do indeed yield a canonical equilibrium distribution of subsystem energy. The temperature of the system was measured via the distribution of subsystem energy. The measured temperature agreed with a value obtained theoretically using ensemble theory. In the simulation, however, the thermodynamic limit should be suitably taken according to the rule: For rule 26R, the number of sites N does not have to be large as compared with the number of iterations T, while in rule 90R, N must be as large as T. This is because any orbit has a period of at most $2N$ under rule 90R. Other rules in table 1 also realize the canonical equilibrium distribution of subsystem energy [7].

In the above experiment, initial conditions were randomly chosen under a given value of energy. Thus, the system was considered as in an equilibrium state from the beginning. However, in order for a state to be truly in equilibrium, the canonical distribution is insufficient and relaxation to the state must also be observed.

In this paper, the relaxation properties of these rules will be discussed in detail. Numerical simulations will be used to show how the type of relaxation behavior depends on the cellular automaton rule. Possible causes for these differences in behavior will be discussed.

2. Relaxation experiment

These experiments are motivated by a textbook example of relaxation to equilibrium. Consider a chamber divided into halves by a partition. Initially, only one half of the chamber is filled with

Table 1
Rules with additive invariants. All the rules that have additive conserved quantities but no local conservation laws are shown, together with energy function $F(\alpha, \beta, \hat{\alpha}, \hat{\beta})$ defined by eq. (3).

Rules	Energy function $F(\alpha, \beta, \hat{\alpha}, \hat{\beta})$
26R	(a): $(\alpha - \hat{\beta})^2 + (\hat{\alpha} - \beta)^2$
90R	(a): $(\alpha - \hat{\beta})^2 + (\hat{\alpha} - \beta)^2$, (d): $\alpha\hat{\beta} - \hat{\alpha}\beta$ [or $(\hat{\alpha} - \beta)^2 - (\alpha - \hat{\beta})^2$]
91R, 123R	(b): $1 + \alpha\hat{\alpha} + \beta\hat{\beta} - [1 - 2(1 - \alpha)(1 - \hat{\beta})][1 - 2(1 - \hat{\alpha})(1 - \beta)]$, (d): $\alpha\hat{\beta} - \hat{\alpha}\beta$ [or $(\hat{\alpha} - \beta)^2 - (\alpha - \hat{\beta})^2$]
77R	(c): $\alpha\hat{\beta}(1 - 2\hat{\alpha} - 2\beta) - \hat{\alpha}\beta(1 - 2\alpha - 2\hat{\beta})$
94R, 95R	(d): $\alpha\hat{\beta} - \hat{\alpha}\beta$ [or $(\hat{\alpha} - \beta)^2 - (\alpha - \hat{\beta})^2$]

air, while the other half is empty. What will happen if, at time 0, the partition is removed? One anticipates that the air will spread rapidly and eventually homogeneously fill the entire chamber. This change is irreversible: the air effectively never spontaneously shrinks back into half of the chamber. This is what we mean by relaxation.

Analogous relaxation was tested in numerical simulations of four rules from table 1, 26R, 90R, 91R, and 123R. An initial condition was randomly chosen under the constraint that energy of the initial configuration is completely concentrated in one half of a cyclic chain, while the other half contains no energy. The time evolution of the distribution of energy was then observed. The experiment was repeated for various values of energy.

Fig. 1 illustrates the temporal variation of the energy values in the half initially filled with energy. Here "energy" is the additive invariant (a) in table 1 for rules 26R and 90R, and (b) for rules 91R and 123R. Other invariants (if they exist) were ignored. For rules with the same type of invariant, the same initial configuration was used. The number of sites was 1000 and the number of iterations was 20 000 for each rule. Each point in the figure represents the energy averaged over 40 time steps.

As is seen from fig. 1b, rule 90R does not relax. The oscillation is exactly periodic and never damped. In fact, rule 90R behaves as a noninteracting ideal gas which consists of particles with velocities ± 1 [6]. Energy (a) represents the number of particles and energy (d) represents the total momentum. Thus, when applied to a finite system, rule 90R does not have good ergodic-theoretic properties.

As mentioned in section 1, however, rule 90R realizes the canonical distribution of subsystem's

energy in the limit of large systems. This is not the contradiction it appears to be. The initial conditions and the ways limits are taken are different in the two cases. In the first case, the initial condition was randomly chosen, though with a fixed total energy. In this case, the system can be considered to be in an equilibrium state from the beginning. In the second case, however, the system is initially in a nonequilibrium state. Further, in the first case, the system size and simulation time were simultaneously increased, while in the second case the system size was held constant while time was increased. The former corresponds to ergodicity of infinite systems, and the latter corresponds to ergodicity of finite systems. It has been mathematically proved that noninteracting ideal gas has Bernoulli property (the top of the hierarchy of ergodic-theoretical properties) [8].

The other three rules do show relaxation, however there exist certain differences among these rules. These will now be discussed.

Fig. 1a illustrates overdamped relaxation in rule 26R. The energy value in the half initially filled with energy rapidly decreases until it approaches the equilibrium value. However, the equilibrium value is not easily reached. In fact, it is known that this rule yields normal thermal conductivity [9]. Therefore, the coarse-grained energy distribution at position x at time t, $\phi(x, t)$, should follow the diffusion equation

$$\frac{\partial}{\partial t}\phi(x, t) = D\frac{\partial^2}{\partial x^2}\phi(x, t), \tag{4}$$

at least near equilibrium. The observed behavior is consistent with this equation.

Rules 91R and 123R exhibit damping oscillations as displayed in figs. 1c and 1d. These two figures are very similar. For both cases, the pe-

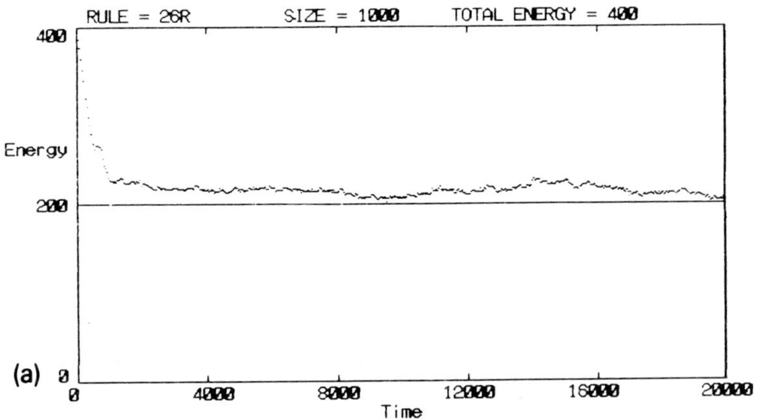

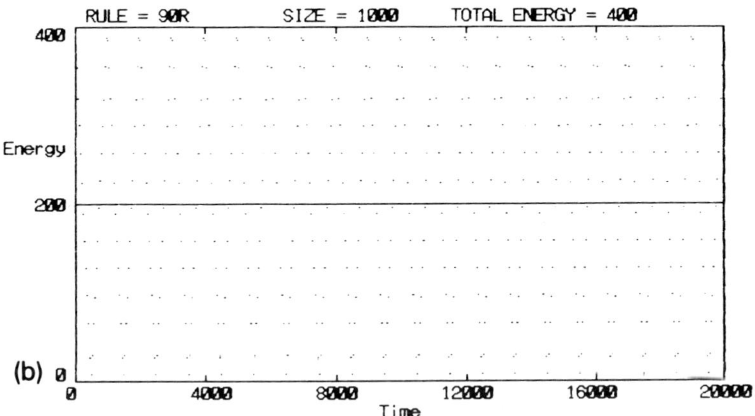

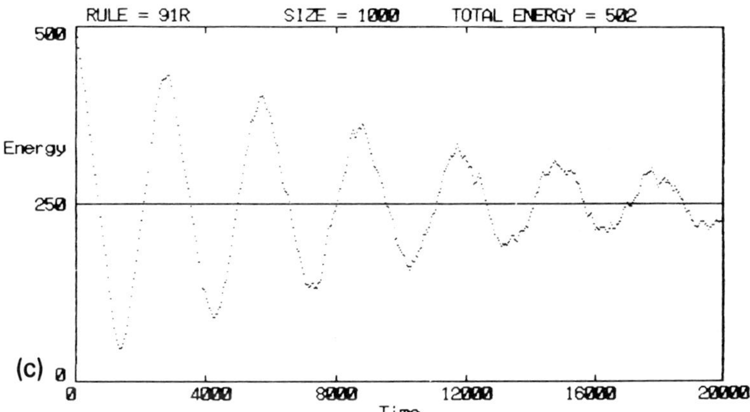

Fig. 1. Temporal variation of energy values in the half of the system initially filled with energy in rules (a) 26R, (b) 90R, (c) 91R, and (d) 123R. Each value is an average over 40 time steps.

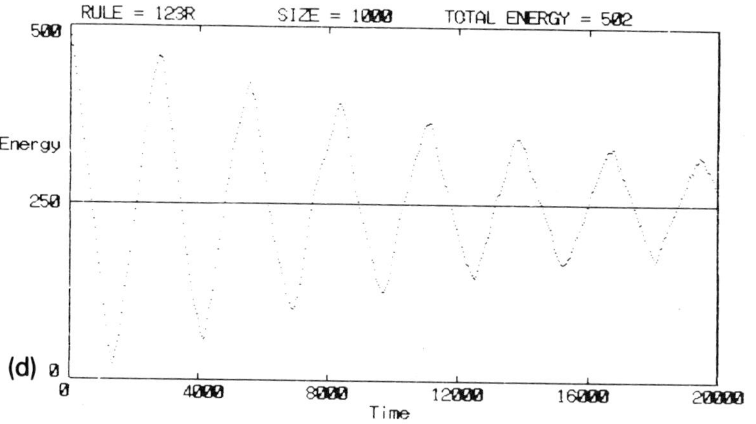

Fig. 1. Continued.

riods of the oscillations are slightly larger than $3N$, and the relaxation times are on the order of 10^5. However, while the oscillation in rule 123R remains undistorted even for large T, that of rule 91R becomes distorted.

Computation of the distribution of subsystem energy clearly reveals the difference between rules 91R and 123R. In the previous paper [6], initial conditions were chosen at random so that the system can be thought of as always in an equilibrium state. Here non-equilibrium initial conditions are used, and it is determined whether equilibrium states are reached, and if so, whether these states follow a canonical distribution.

The energy distribution of a subsystem, $P(E)$, is a canonical distribution if it can be written as

$$P(E) \propto D(E)\,e^{-\beta E}, \tag{5}$$

where $D(E)$ is the density of states, namely the proportion of such configurations that have energy E in all the possible configurations of the subsystem, and β is the inverse temperature. The density of states was obtained exactly through calculation of energy values for all the possible configurations of the subsystem. That is, when the subsystem contains n sites, $D(E)$ is the number of the configurations with energy E divided by 4^n, which is the total number of subsystem configurations. The energy distribution $P(E)$ was numerically calculated by taking time average over every 10^6 time steps in a long run. Then, the linearity of $\ln[P(E)/D(E)]$ versus E was checked.

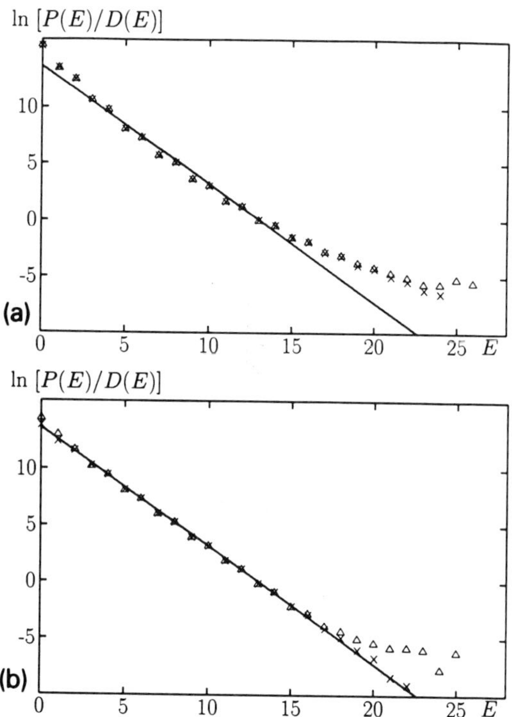

Fig. 2. Subsystem energy distribution function for rules (a) 91R and (b) 123R. The system size is 1000 and the subsystem size is 15. The total number of time steps is 10^7. The initial condition is the same for both rules. Triangles: first 10^6 iterations for each rule. Crosses: tenth 10^6 iterations for (a) and second 10^6 iterations for (b).

Under both rules the energy distribution converged within the simulation time. However, the eventual distributions in the two cases are widely

Table 2
Numbers of periodic orbits in periodic chain with size N

| Rule | N | | | | | | | | |
---	5	6	7	8	9	10	11	12	13
26R	54	200	240	898	918	3722	3518	15066	12870
90R	156	700	1758	8230	21853	104968	285978	1.4×10^6	2.6×10^6
91R	90	296	518	1960	2984	12284	15266	73422	76026
123R	68	216	354	1138	2342	6368	8948	40854	44188

different from each other. Figs. 2a and 2b display the result of the simulation with 10^7 time steps for each rule. It is found that rule 91R does *not* lead to the canonical distribution whereas rule 123R does. However, the shape of the eventual distribution function does not depend on the position of a subsystem. Thus, some relaxation occurs even in rule 91R, but this is not relaxation to the canonical distribution. In addition, rule 26R shows the behavior similar to rule 123R and rule 90R similar to rule 91R in the same kind of experiments.

It should be remarked that this behavior is not caused by the existence of energy (d) in table 1. In general, a system with two additive invariants has two kinds of temperature. One might guess that the other temperature affects the thermodynamic behavior. However, in the present case, energy (b) and energy (d) have different symmetries with respect to the left–right inversion. Thus, if one sets initial conditions according to some distribution of energy (a) only, the temperature corresponding to energy (d) should be infinite. This is assured by the fact that rule 123R shows the relaxation despite the same initial condition as in the case of rule 91R.

One possible origin of the curious behavior of rule 91R is the existence of energy with interaction among more than two sites. If this type of invariant exists, it may affect the relaxation behavior. Actually, it has recently been found that rule 91R has an additive invariant written as

$$\Psi = \sum_i G(\sigma_i, \sigma_{i+1}, \sigma_{i+2}, \hat{\sigma}_i, \hat{\sigma}_{i+1}, \hat{\sigma}_{i+2}), \qquad (6)$$

while rule 123R does not [10]. Whether or not this quantity really causes the behavior shown in fig. 2 is now under investigation.

3. Summary and discussion

In this paper, the relaxation of an initial probability distribution to an equilibrium distribution under four elementary reversible cellular automata (ERCA) was investigated. Numerical experiments revealed that each rule exhibits behavior distinct from the others. Rule 90R does not relax. Rule 26R relaxes with overdamping. Rules 91R and 123R show similar damped oscillations but the equilibrium subsystem energy distributions for the two rules are quite different.

Can one develop a unified understanding of these differences? Ergodic theoretical properties such as mixing do not seem to be applicable to ERCA. Indeed, these rules are not ergodic [2]. On a cyclic chain, every orbit of an ERCA is a periodic orbit. In a relatively small system, the number of periodic orbits can be calculated. Table 2 displays the numbers in the four rules with system size 5 through 13. The number of periodic orbits increases approximately exponentially with system size for all of these rules. On the other hand, the number of possible values of energy grows only linearly with the system size. This means that a huge number of orbits coexist on an energy surface, that is, ergodicity is broken.

One may ask then, why has thermodynamic behavior been observed at all? This question is equivalent to asking what type of conserved quantities influence the thermodynamic behavior of the system. If one labels the periodic orbits in a finite chain, the labels become invariants. I call this type of invariant *ultimate*, because such invariants classify orbits completely. The ultimate invariants necessarily exist for systems on a periodic chain. However, it is almost impossible to write down such invariants unless one knows the phase space structure completely. Some of these invariants may be written as additive invariants or as local conservation laws. However, most of these invariants probably cannot be expressed in a systematic manner. It is questionable even whether a relationship exists between the invariants for different N's. Such unsystematic quantities may not cause perceptibly deviation from thermodynamic behavior. Qualitative and quantitative tests of this hypothesis will be made in the future.

Finally, a comment on a possible connection between reversible cellular automata and continuum dynamical systems. The existence of additive conserved quantities is similar to the problem of the analyticity of integrals in continuum dynamical systems. Without the restriction of analyticity, integrals of motion necessarily exist. This is equivalent to labeling orbits. The integrability of dynamical systems is determined by the presence of analytic integrals. In addition, that rules without additive or locally conserved quantities show similar bit patterns for almost all initial conditions suggests a similarity to dynamical chaos. This parallel between reversible cellular automata and continuum dynamical systems should be investigated further.

Acknowledgements

I would like to thank Dr. Howard Gutowitz for his kind help with improvement of the manuscript.

References

[1] Y. Pomeau, Invariant in cellular automata, J. Phys. A 17 (1984) L415–L418.

[2] S. Takesue, Ergodic properties and thermodynamic behavior of elementary reversible cellular automata. I. Basic properties, J. Stat. Phys. 56 (1989) 371–402.

[3] M. Creutz, Deterministic Ising dynamics, Ann. Phys. 167 (1986) 62–72.

[4] G.Y. Vichniac, Simulating physics with cellular automata, Physica D 10 (1984) 96–116.

[5] H.J. Herrmann, Fast algorithm for the simulation of Ising models, J. Stat. Phys. 45 (1986) 145–151.

[6] S. Takesue, Reversible cellular automata and statistical mechanics, Phys. Rev. Lett. 59 (1987) 2499–2502.

[7] S. Takesue, to be published.

[8] S. Goldstein, J.L. Lebowitz and M. Aizenman, Ergodic properties of infinite systems, in: Dynamical Systems, Theory and Applications, ed. J. Moser (Springer, Berlin, 1975) pp. 112–143.

[9] S. Takesue, Fourier's law and the Green-Kubo formula in a cellular-automaton model, Phys. Rev. Lett. 64 (1990) 252–255

[10] T. Hattori and S. Takesue, to be published.

Physica D 45 (1990) 285–292
North-Holland

A COMPARISON OF SPIN EXCHANGE AND CELLULAR AUTOMATON MODELS FOR DIFFUSION-CONTROLLED REACTIONS

A. CANNING and M. DROZ

*Département de Physique Théorique, Université de Genève,
24 Quai Ernest-Ansermet, CH-1211 Geneva 4, Switzerland*

Received 15 January 1990
Revised manuscript received 9 February 1990

Two different lattice models for diffusion-controlled reaction are studied. The first model is based on the Kawasaki spin exchange model and the second is a cellular automaton algorithm based on the Margolus block. Both models have traps present on the lattice which react with the particles diffusing from a source to a sink. Two cases should be distinguished; in the first one, the traps are moving on a time scale much faster than the diffusion (this is the annealed version of the model), in the second case the traps are fixed (this is the quenched version). Theoretical results for the annealed version are presented along with computer simulations (restricted to small trap concentrations) for the quenched models. It is shown that these models are a viable alternative to standard simulations for this class of problem.

1. Introduction

The use of cellular automata to model physical systems increased greatly after the important paper by Frisch, Hasslacher and Pomeau [1] which presented the now famous FHP algorithm for modeling hydrodynamical flow. In our work we will study a lattice gas and a cellular automaton model for a much simpler class of problems, diffusion-controlled reactions. These processes are described by much simpler differential equations than the Navier–Stokes equations for hydrodynamics. As we shall show later in this paper it is therefore much easier, for certain situations, to show the convergence of our discrete models to solutions of diffusion-controlled reaction equations than is the case for CA models of hydrodynamics.

The first theoretical studies of diffusion-controlled reactions were carried out as far back as 1916 by Smoluchowski [2]. The basic model considered consisted of two types of particles, the diffusing species and the traps. When the diffusing species meets a trap there is an instantaneous reaction and the diffusing species vanishes. The state of the trap remains unchanged by the reaction. A typical application of a model of this type is in combustion where the diffusing species is the oxidant and the traps are the fuel droplets. The behavior for a single particle has been well studied (see ref. [3] for a review) but in this work we will look at the diffusion and steady states of systems with many particles.

In this work two different types of behavior for the traps will be considered: the quenched case, where the traps are at fixed positions during the diffusion and the annealed case where the traps are considered to move on a time scale much faster than the diffusing species. In the latter case at each time step of the diffusion process we have a probability at all positions that there will be a trap. In lattice gas language the annealed model is the Boltzmann approximation for the quenched model.

When a gas or liquid diffuses into a medium with no traps present the process is well described by the diffusion equation

$$\partial P/\partial t = \nabla \cdot (D\nabla P), \tag{1}$$

where $P(\boldsymbol{r}, t)$ is the average probability of finding a particle at point $\boldsymbol{r}$ and time t and D is the diffusion coefficient for the medium. In certain diffusion processes D can depend on position as well as on the properties of the medium in which the diffusion is taking place but in the models studied in this paper D was found not to be a function of position. This diffusion equation can be derived from kinetic theory. The traps, if fixed in position, can be introduced as boundary conditions under which the above equation must be solved. If, for example, we had spherical traps we would have to solve the above equation with the boundary conditions that P is zero inside and at the boundaries of all the spherical traps. The effect of the traps (if statistically distributed) on the diffusion can also be represented on the macroscopic scale by an appropriate sink term added to the right-hand side of the diffusion equation. This gives for the diffusion equation

$$\frac{\partial P}{\partial t} = \nabla \cdot [D(c)\nabla P] - k(c)cP, \tag{2}$$

where c is the trap concentration and $k(c)$ is called the rate coefficient which depends on c. The diffusion coefficient $D(c)$ is now also a function of c. The present authors are not aware of any exact analytical results for $D(c)$ and $k(c)$ for all values of c for any realistic models, but approximate techniques by Felderhof and Deutch [4] give a power series in c for D and k for randomly distributed fixed spherical traps. Their technique involves exploiting the analogy between the diffusion problem with fixed spherical traps and the electrostatic problem of finding the potential due to a collection of ideally conducting grounded spheres in the presence of a given external potential. Their approximations involve neglecting the multipole moments of the trap distribution higher than the dipole moment. It is not clear how to do a similar series expansion for the discrete models studied in this paper. It should be noted in their model [2,4] that if the concentration of traps, c, is sufficiently small then the traps can be assumed to act independently. In this case the diffusion and rate coefficients, D and k, are found to be independent of c.

Since an exact analytical solution of the diffusion-controlled reaction equation (2) for all c values for realistic models is not known and to determine numerical solutions of these equations is very time consuming, the object of this paper is to study a lattice gas and a cellular automaton model which reproduce diffusion-controlled reaction as described by these differential equations. The faster parallel automaton algorithm will also be compared and contrasted with the slower serial spin exchange lattice gas algorithm. The results presented in this paper for the automaton are an extension of the work in refs. [5,6] for the cellular automaton model with no traps. This work showed that the model could efficiently and accurately describe random diffusion without traps. In this paper we show that these models, in certain cases, can reproduce differential equations for a diffusion-controlled reaction process.

In the computer simulations presented in this paper for fixed traps, due to restrictions on computer power and time, we could only work at small values of c ($c < 0.016$) which were found to give results corresponding to the regime where D and k do not depend on c. This will be discussed in more detail in section 2, which also describes the spin exchange algorithm along with some analytical and computational results for the model, and section 3, which describes the cellular automaton algorithm and corresponding results. We will also see in sections 2 and 3 that the annealed case can be solved exactly for the spin exchange model and, with some restrictions, for the cellular automaton model to yield a diffusion equation of the form of eq. (2). Section 4 in this paper contains conclusions and a discussion of the results presented in the previous two sections.

2. Kawasaki spin exchange model

The first model of diffusion to be studied is usually referred to as the Kawasaki [7] spin exchange model at infinite temperature. In this model in two dimensions for example, particles sit at sites on a square lattice and diffuse by choosing at random, one of their four nearest neighbors and then swapping sites with it. The exact algorithm used in this paper consisted of the following steps:

(a) Choose a site at random.

(b) Choose one of its four neighbors at random.

(c) Swap the two sites.

The process is then repeated for a given number of steps. A temperature can be introduced into the system giving a probability that the two sites chosen will not swap. All the simulations presented in this paper were carried out at infinite temperature, which means the swap was always made. It should be noted that this is a very simplified model of a real gas which would typically have an attractive long-range force between the particles and also a repulsive short-range force.

The boundary conditions of the system are that at one end of the system there is a source which immediately replaces the particles diffusing into the system while at the other end there is a sink which continually removes the arriving particles. Traps which absorb the particles are then introduced at random sites on the lattice. The two cases for the behavior of the traps, annealed and quenched, will be studied separately in the next two sections.

2.1. Annealed version

This case can be studied analytically and for simplicity we will look at the two-dimensional case again. Consider a square lattice with each lattice site denoted by the two coordinates (x, y) with the source at $y = 0$ and the sink at $y = L$. The master equation describing the time evolution of this system is

$$P_{x,y}^{t+\delta t} = \tfrac{1}{4}(1 - c)(P_{x+\delta x,y}^{t} + P_{x-\delta x,y}^{t}$$
$$+ P_{x,y+\delta y}^{t} + P_{x,y-\delta y}^{t}), \qquad (3)$$

where c is the concentration of the traps and $P_{x,y}^{t}$ is the probability that there is a particle at site (x, y) at time t. The terms δx and δy are the lattice spacing in the x and y directions, which we choose equal in all our models. The only difference between this master equation and the one for a system with no traps is the $(1 - c)$ factor. This factor reflects the fact that at every site there is a probability c that a trap will be present and hence that the particle will vanish. If we now suppose that in the continuum limit the lattice spacing gets smaller and smaller and the time interval between updates also gets smaller, then eq. (3) becomes

$$\frac{\partial P}{\partial t} = \tfrac{1}{4}(1 - c)\frac{(\delta x)^2}{\delta t}\nabla^2 P - \frac{c}{\delta t}P, \qquad (4)$$

where P is now a continuous function describing the diffusion process. In this equation we have neglected terms of higher order than $(\delta x)^2$ and δt. We expect these approximations to be valid if the lattice we are working with is sufficiently large. This equation is therefore of the same form as the diffusion reaction equation (2) with the diffusion coefficient $D(c) = \tfrac{1}{4}(1 - c)(\delta x)^2/\delta t$ and the rate coefficient $k = 1/(\delta t)$ independent of c.

Choosing $P(x, 0) = 1$ (the source) and $P(x, \infty) = 0$ (the sink) gives for the stationary solution $(\partial P/\partial t = 0)$

$$P(y) = \exp\left(-\frac{2}{\delta x}\sqrt{\frac{c}{1 - c}}\, y\right). \qquad (5)$$

This is an exponential decay of the form $e^{-K(c)y}$ where we shall refer to $K(c)$ as the decay coefficient.

It should be noted at this point that the lattice spacing δx appears explicitly in this expression. This reflects the fact that the traps sit on lattice sites and therefore their effect on the diffusion must depend on the lattice spacing. Since we have a theoretical result for this model which we expect to be valid in the limit of large lattice size, this model represents an opportunity to compare computational and theoretical results for the same system. For the other models studied in this paper we have no good analytical results, so this model gives us the opportunity to test the following assumptions about our analytical and computational results:

(a) The higher-order derivative terms in the diffusion equation are negligible.

(b) The lattice we are working with is sufficiently large to approximate the continuous limit.

(c) The model is in a stationary state when we take the measurements.

(d) The time between measurements is long enough to give statistically independent results.

If these assumptions are valid for this model we can expect them to be valid for the other models studied in this paper.

In our computer simulations the system size studied was 128×130. The starting state for the system was chosen to be a linear distribution of particles starting with $P = 1$ at the source and dropping to $P = 0$ at the sink. This was expected to give a faster approach to a steady state than starting the system with no particles present. The system was then run for 3×10^8 spin updates after which the system was considered to be in a stationary state. Ten measurements of P were then taken every 3×10^7 spin updates. For each value of c ten simulations of this type were carried out. Only small c could be studied with our computer simulations as large c leads to a very sharp drop off in the diffusion profile giving only a few points from which an exponential fit must be made. With larger system sizes we would expect the errors to be smaller, but the number of points from which the exponential fit must be taken would not necessarily be much larger, since the decrease in P, which depends on $K(c)$, is related to the number of lattice points and not the system size. The decay coefficient, if defined in (lattice units)$^{-1}$, is independent of the lattice size. However, we do expect that with a larger system and hence smaller errors a few more points could be extracted from the noise at small values of P. For large values of c the best shape of system to work with would be one that is long in the x direction and quite short in the y direction. Most of the simulations presented in this paper were carried out on a work station, so we were limited in the size of system we could study. However, a few of the simulations were carried out on a Cray.

Fig. 1 shows the results of computer simulations on a 128×130 size system. The source and sink were placed at row 1 and row 130 respectively. For each value of trap concentration studied a good exponential profile was always found. As can be seen from fig. 1 there is an extremely close correlation between the theoretical results and the computer simulations, which suggests that assumptions (a) to (d) mentioned in the last paragraph are well satisfied. We therefore expect these errors to be negligible in our computer simulations of the other diffusion models, so we will not discuss them further.

2.2. Quenched version

This represents a much more difficult problem than the annealed case although in one dimension

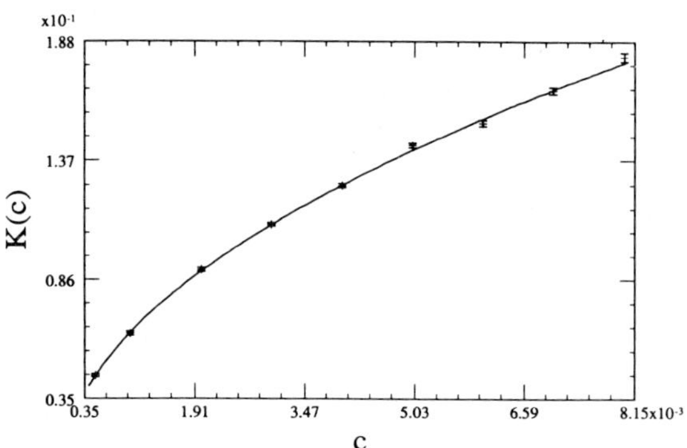

Fig. 1. $K(c)$, the decay coefficient at different values of c (the trap concentration) for the annealed spin exchange model. The points with error bars are the results of computer simulations, the errors being calculated from the sample to sample fluctuations. The solid line is the theoretical result (see eq. (5)).

it is possible to find an exact theoretical solution. The one-dimensional system represents a very special case as the diffusion will be stopped at the first trap encountered, which will then play the same role as the sink in a system with no traps. In a system with no traps the steady state diffusion profile is linear as discussed in ref. [5]. The quenched averaging (averaging over all the possible different trap distributions) for this system can therefore be performed as each trap distribution gives rise to a linear profile with a different gradient depending on where the first trap is encountered. The probability of the first trap being in a given position can be easily calculated. Thus, after quenched averaging the stationary diffusion profile in one dimension is given by

$$P(y) = (1 - c)^y + cy \left(\ln c + \sum_{l=1}^{y-1} \frac{(1 - c)^l}{l} \right) . \quad (6)$$

We do not expect this solution for the one-dimensional case to tell us anything about the solutions in higher dimensions were the particles can diffuse around the traps.

For the two-dimensional system the time evolution at each point is described by the set of equations

$$P_{x,y}^{t+1} = \tfrac{1}{4}(1 - \xi_{x,y})(P_{x+1,y}^t + P_{x-1,y}^t + P_{x,y+1}^t + P_{x,y-1}^t), \quad (7)$$

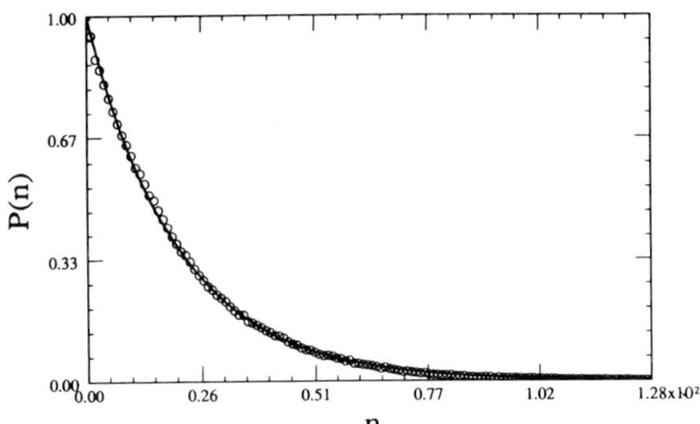

Fig. 2. Steady state diffusion profile for the spin exchange quenched model with $c = 0.002$. n is the number of lattice points from the source. The circles are the results from computer simulations and the solid line is a least-squares exponential fit.

where $\xi_{x,y}$ is a random variable taking the values 1 or 0 with a probability equal to the concentration of the traps. Therefore, for this system we have an infinite set of coupled equations each of which can take one of two different forms. It is not clear how the quenched averaging over all possible trap distributions can be performed to obtain the required master equation and corresponding differential equation. Computer simulations on a 128×130 two-dimensional system were also performed for this model. For each c value these results were averaged over twenty different distributions of the traps and, for each trap distribution, averages were taken over ten statistically independent time points. All the steady state diffusion profiles for different trap concentrations were found to be of exponential form. Fig. 2 shows a typical example for $c = 0.002$. Simulations were carried out for c values from 0.001 to 0.016. Above $c = 0.016$, for reasons discussed in section 1, it was not possible to fit an exponential to the small number of data points we obtained. A least-squares fit on these results gave the decay coefficient $K(c)\,\delta x = (1.60 \pm 0.04)c^{0.511 \pm 0.015}$. This is approximately the same form as the annealed case for small c, with the 2 in the decay length replaced by 1.60 (see eq. (5)). We may expect the annealed case to decay quicker than the quenched case for the same c value, as the number of particles captured by the traps in the quenched

case will depend on the local density of particles at the traps, which will be much lower than the average density. In the case of the annealed model the density is uniform across the system and therefore the traps are always removing particles from sites which have the average particle density.

It may be expected that, like the annealed case, the quenched diffusion process is described by an equation of the form of (2). In fact, in the limit $c = 0$ the two processes must give the same results. Also a differential equation of the form of (2) with D and k independent of c will yield a solution of the type we have found ($K(c) \propto \sqrt{c}$). At small values of c like those the simulations were performed at, we may expect the distance between traps to be sufficiently large for the traps to act independently [2,4]. This means that two traps do not compete for the same particle and the trap term in the differential equation will be linear in c and hence k will not be a function of c. Since our results are a solution of the diffusion equation with k independent of c this suggests that our simulations are in this regime of small enough c such that the traps act independently.

3. Cellular automaton model

The main problem in designing a parallel cellular automaton algorithm is to avoid collisions and overlaps. The following CA algorithm, which is based on the Margolus neighborhood [7], avoids this problem. Like the spin exchange model this CA model also sits on a square lattice with at most one particle per site. The lattice is subdivided into blocks each of which are 2×2 sites lettered a to d. The diffusion process is then carried out in two steps. In the first step the blocks are rotated either $\pm\pi/2$ with a probability $p < 1/2$ for each rotation leaving a probability $p_0 = 1 - 2p$ that the block remains unchanged. Secondly, the block sublattice is shifted by one lattice unit in both directions and then the rotation operation is repeated (see fig. 3). We will refer to the combination of these two steps as a Margolus step. The whole process is then repeated so particles can propagate through the system.

The time-dependent properties, stationary profile and fractal dimension of the diffusion front

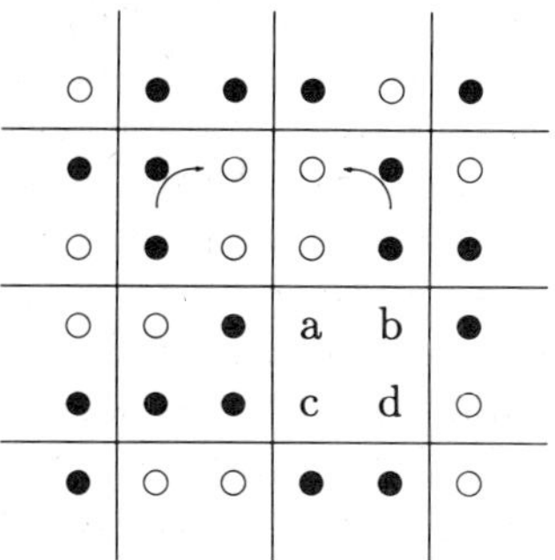

Fig. 3. The Margolus neighborhood. The arrows indicate the two possible motions of the block which generate a diffusion step. At each time step the blocks are shifted by one elementary step along the diagonal. In the text the four different positions in the block are referred to by the letters a, b, c and d.

for the trap free version of this model have been studied by Chopard et al. [5,6]. They found these properties of the model to be very similar to theoretical results for the spin exchange model. It is our objective in this paper to extend their work to the more complicated problem of diffusion with traps.

3.1. Annealed version

By a similar method to the spin exchange model a differential equation describing the diffusion process can be derived for this model, i.e.

$$\frac{\partial P}{\partial t} = \frac{1}{2}p(1-c)^2\frac{(\delta x)^2}{\delta t}\nabla^2 P - \frac{2c-c^2}{\delta t}P$$
$$+ \frac{1}{4}(1-c)^2 pp_0\frac{\delta x}{\delta t}\nabla J, \qquad (8)$$

where P is now

$$P(x,y,t) = \frac{1}{4}[\langle a(x,y,t)\rangle + \langle b(x,y,t)\rangle$$
$$+ \langle c(x,y,t)\rangle + \langle d(x,y,t)\rangle], \qquad (9)$$

the local normalized probability density in a Margolus block positioned at (x,y). $\langle a\rangle$ is the probability of finding a particle at site a in the Margolus block. J is related to the internal fluctuations in the density of the block and is given by

$$J(x,y,t) = \langle a+c-b-d\rangle - \langle a+b-c-d\rangle. (10)$$

We will not go into the derivation of eq. (8) in this paper but for the case of no traps the details are given in ref. [6]. In all our simulations of this system we worked with systems having homogeneous boundary conditions in the x direction giving homogeneous results for the diffusion profiles in the x direction. This means that $\langle a\rangle = \langle b\rangle$ and $\langle c\rangle = \langle d\rangle$ in the same block, so the first term in the above expression was always equal to zero.

Eq. (8) is equivalent to a diffusion reaction equation of the form (2) with the exception of the ∇J term. However, if we choose $p = 1/2, p_0 = 0$ this term becomes zero and the equation reduces to a diffusion equation of the correct form. It should be noted that with the choice $p_0 = 0$ we have two decoupled systems whose particles move on either the black or white squares of a checker board. A particle starting at a lattice site corresponding to a white square will always remain on a white square after each Margolus step; the same is true for a particle on a black square.

Computer simulations of the same type as the spin exchange model were also carried out for this model with $p_0 \neq 0$, and again the diffusion profile was found to be exponential with the decay coefficient proportional to $\sqrt{c}$. It was also found from the simulations that a relationship of the form

$$\nabla J = g(p_0)\,\nabla^2 P\,\delta x \qquad (11)$$

holds for J. This can be easily understood when we look again at the case $p_0 = 0$. In this case in each iteration step the Margolus block is always rotated clockwise or anticlockwise with probability one half. This means that in a system homogeneous in the x direction after the two iteration steps which make up the Margolus step, the density in positions c and d in a block will always be equal to the density in positions a and b in the block below. Therefore, fluctuations in the density of particles always occur within a Margolus block and not between the edges of subsequent blocks. J is thus directly related to the first derivative in space of the probability density. Actually $J = -2\nabla P\,\delta x$ (i.e. $g(0) = 2$) where the factor 2 comes from the fact that $-J$ lies between 0 and 2, whereas P was normalized to lie between 0 and 1. For values of $p_0 > 0$ some of the density fluctuation occurs between the edges of the blocks and so $g(p_0)$ (see eq. (11)) is less than 2.

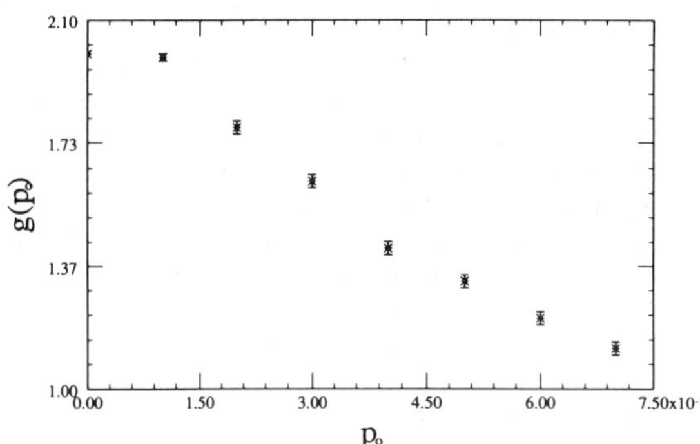

Fig. 4. Plot of the scaling function $g(p_0)$ which gives the relation between J and P for the annealed automaton model. The errors are calculated from the sample to sample fluctuations of systems with different trap distributions.

Putting eq. (11) into (8) gives us a diffusion-controlled reaction equation of the form of (2) whose steady state solution for small c is given by

$$P = \exp\left(-\frac{2}{\delta x}\sqrt{\frac{c}{p[1 - \frac{1}{2}p_0 g(p_0)]}}\, y\right) . \qquad (12)$$

The form of $g(p_0)$ which was derived from computer simulations is shown in fig. 4. As can be seen from this figure the result $g(0) - 2$ also holds for small values of p_0 up to approximately 0.1.

3.2. Quenched version

Again the same simulations as were carried out for the spin exchange model were performed on this automaton model with values of c between 0.001 and 0.016. This model was also found to give exponential steady state profiles with a decay coefficient of the form $K(c) \propto \sqrt{c}$ where the constant of proportionality again varied with p_0. Like the spin exchange model this is again a solution of the diffusion reaction equation (2) with D and k independent of c. This again suggests that we are in the regime where the traps act independently of each other. Again the decay coefficient for this model was always found to be less than the annealed model.

4. Conclusions

We have shown in this work that the spin exchange model and the parallel cellular automaton model can, in the case of the annealed model, reproduce diffusion-controlled reaction type equations of the form of eq. (2). It should be noted however, that in the case of the automaton model it was with two-dimensional systems homogeneous in the x direction that the spurious J term in the differential equation lead only to a rescaling of the $\nabla^2 P$ term (see eq. (8)). We did not study any other system geometries. We have also shown by computer simulations that in the case of the quenched models for small c the steady state diffusion profiles are solutions of eq. (2), the diffusion and rate coefficients being independent of c.

Overall the automaton model has slightly richer properties than the spin exchange model in that the decay coefficient of the diffusion profile is a function of the probability of rotation of the Margolus block. The automaton model also has the potential to be a much faster algorithm if implemented on an appropriate machine since it is totally parallel in nature.

It would be interesting to investigate the properties of the diffusion profiles for the quenched models at higher values of c although we expect this to be computationally very difficult. At higher values of c than our simulations the traps will be close enough together to compete for the same particles and therefore not act in an independent way. We believe that this will lead to values of the rate and diffusion coefficients which depend on c [4]. At very high values of c, above the percolation threshold (this is the trap concentration at which an infinite cluster of traps will span the whole system), the diffusing species would be localized in some area of the system around the source. At this stage the properties of the quenched model would be significantly different from the annealed model. For example, in the case of the spin exchange model, the steady state diffusion profile for the annealed version would still be of exponential form at high values of c.

Acknowledgements

This work was in part supported by the Swiss National Science Foundation.

References

[1] U. Frisch, B. Hasslacher and Y. Pomeau, Phys. Rev. Lett. 56 (1986) 1505.

[2] M.V. Smoluchowski, Phys. Z. 17 (1916) 557, 585.

[3] J.W. Haus and K.W. Kehr, Phys. Rep. 150 (1987) 263–406.

[4] B.U. Felderhof and J.M. Deutch, J. Chem. Phys. 64 (1976) 4551.

[5] B. Chopard, M. Droz and M. Kolb, J. Phys. A 22 (1989) 1609.

[6] B. Chopard and M. Droz, in: Proceedings of the Winter School, Cellular Automata and Modeling of Complex Physical Systems, 1989, Les Houches, eds. P. Manneville, N. Boccara, G.Y. Vichniac and R. Bidaux (Springer, Berlin, 1989) p. 130.

[7] K. Kawasaki, in: Phase Transitions and Critical Phenomena, eds. C. Domb and M. S. Green, Vol. 2 (Academic Press, New York) p. 443;
K. Kawasaki, in: Phase Transitions and Critical Phenomena, eds. C. Domb and M.S. Green, Vol. 5 (Academic Press, New York) p. 165.

[8] T. Toffoli and N. Margolus, Cellular Automata Machine: A New Environment for Modeling (MIT Press, Cambridge, MA, 1987).

Physica D 45 (1990) 293–306
North-Holland

REVERSIBLE CELLULAR AUTOMATA AND CHEMICAL TURBULENCE

H. HARTMAN

Computer Science Department, University of California, Berkeley, CA 94720, USA

and

P. TAMAYO

Physics Department and Center for Polymer Studies, Boston University, Boston, MA 02215, USA

Received 26 January 1990
Revised manuscript received 21 March 1990

We present a class of reversible cellular automata which display soliton-mediated phase transitions and chemical turbulence. This class is defined as the reversible extension of simple reaction–diffusion models of excitable media such as the Schlögl and Greenberg–Hastings models. The phenomenology of these models is discussed in terms of dimensionality, neighborhood and number of states. Reversibility imposes constraints and introduces conservation laws that give rise to invariant quantities.

1. Introduction

Chemical turbulence can be defined as the study of diffusion-coupled fields of similar limit cycle oscillators. The phenomena of chemical turbulence exhibits chaotic behavior not only in time but also in space. This field has been the major topic of a book by Y. Kuramoto [1] titled Chemical Oscillations, Waves, and Turbulence. The major experimental motivation in the study of these reaction and diffusion systems was the discovery of the Belousov–Zhabotinsky reaction, which can spontaneously produce chemical waves and chemical turbulence. The equations which govern the behavior of this system were those of the reaction–diffusion type studied by Turing. Turing was the first to realize the importance of these systems in the understanding of pattern formation in embryology [2]. The nonlinear partial differential equations which model these systems play a significant role in mathematical biology [3]. The Belousov–Zhabotinsky reaction can also be modeled as a "diffusion-coupled field of similar limit cycle oscillators." This allows one to exploit the analogy

between reaction–diffusion systems and the statistical mechanics of cooperative systems. Thus the main feature of reaction–diffusion systems, not shared by fluid dynamical systems, is "that the total system can be viewed as an assembly of a large number of identical local systems which are coupled (i.e. diffusion coupled) to each other [1]." The power spectra of chemical turbulence is quite different from that of hydrodynamic turbulence [4].

From hydrodynamics to plasma physics, physicists are now using cellular automata to explore the nonlinear partial differential equations of the physical world. For introductions to the field see the book by Toffoli and Margolus [5], the collection of articles edited by Wolfram [6] or the proceedings of several cellular automata workshops [7]. Cellular automata have also been used to study reaction–diffusion systems. In particular the Belousov–Zhabotinsky system has been modeled by Greenberg, Hastings and coworkers [8–10]. We review their work in the first sections of this paper. Their model involves a three-state cyclic cell and a two-dimensional lattice. Oono and Kohmoto [11] have used a cellular automata model closely related to the Greenberg–Hastings model to study chemical turbulence in one dimension. The major difference between their model and the Greenberg–Hastings model is the non-existence of a quiescent state. The Oono–Kohmoto model had an automaton which cycled through three states and was diffusively coupled to its nearest neighbors. By varying the coupling term the system underwent a series of phase transitions labeled a periodic phase, a turbulent phase, and a soliton-like phase. This model was the "simplest nontrivial model of chemical turbulence (or coupled limit cycles) which can exhibit many phenomena that we can observe in nonlinear partial differential equations; turbulence, solitons, intermittency, etc. [4]" In one-dimension we have found that by making the Greenberg–Hastings cellular automata model reversible using the Fredkin method we get a variety of phenomena such as solitons and soliton-mediated approach to the turbulent state. This is discussed in section 7.

In section 6 we discuss a two-dimensional reversible Greenberg–Hastings cellular automaton which becomes turbulent from very simple initial conditions. Since in the Greenberg–Hastings model each cell cycles throughout three states, the question arises as to what happens in cellular automata whose cell state table cycles throughout two states (Schlögl model). In the non-reversible case not much of interest is observed. However in the reversible case in two dimensions the system evolves into a turbulent state from simple initial conditions exactly like the reversible Greenberg–Hastings case. Furthermore making the OR function reversible (REV-OR) also gives rise to turbulence in two dimensions. Thus in two dimensions all these reversible systems have a similar behavior. In one dimension these systems differ from one another and it is only the reversible Greenberg–Hastings case which gives rise to soliton-mediated turbulence.

In recent years an attempt has been made to classify one-dimensional cellular automata (two states per cell) which are either reversible or non-reversible [5,12–16]. We have found these attempts useful in interpreting our results in one dimension. Wolfram [6] has classified (non-reversible) one-dimensional cellular automata according to the possible evolutions of a chain initially filled at random with 0's and 1's. The classification involves four classes:

(I) The chain becomes homogeneous (say all 0's).

(II) Appearance of simple localized time-periodic structures.

(III) Evolution leads to chaotic behavior.

(IV) Evolution leads to complex localized structures.

An exhaustive study of the simplest one-dimensional cellular automata (two states per cell), dependence on nearest neighbors only, do not exhibit class IV behavior. When these simple cellular automata are made reversible the Wolfram classification breaks down. Takesue [16] has classified these simple reversible cellular automata by searching for additive conserved quantities and local conservation laws (i.e. wall-like structures). There are three types:

(I) No additive conserved quantities exist.

(II) Additive conserved quantities exist but no walls.

(III) Additive quantities and walls exist.
Another classification of one-dimensional reversible cellular automata was made by Cosenza and Neri [15]. They set up suitable initial conditions which led to periodic structures or phases and then studied the evolution of perturbations. Three classes were found:

(1) The phases were stable (i.e. the perturbations propagated without proliferating).

(2) The phases were not stable and a cone of chaotic structure expands over the original phase.

(3) The phases are not stable and the perturbation generates dislocations that propagate and proliferate.

Additive conserved quantities, walls and the instability of periodic structures are to be found in reversible two- and three-state cyclic automata. These results will be discussed in section 7.

The importance of discrete reaction–diffusion models has been emphasized in the literature in connection to a variety of fields: heterogeneous catalysis [17], non-equilibrium phase transitions [18], clay theories of the origin of life [19–21], models for artificial life [22,23], spatio-temporal chaos [14,24], embryology [25,26] and the development of organized structures [27,28].

2. Reaction–diffusion systems

Thirty years ago Turing began to analyze, from a chemical point of view, the problems of embryology [2]. He introduced a model in which a system of reacting and diffusing chemicals would under certain conditions break the initial symmetry of the system [27]. The development of structures would result from non-homogeneous patterns of chemical substances, called morphogens.

Turing's model for morphogenesis had a large impact on physical chemistry after the Belousov–Zhabotinsky reaction was discovered. This reaction, which is a typical example of Turing's reaction–diffusion, usually involves a mixture of malonic acid, potassium bromate, ceric sulfate and an indicator. The mixture reacts exothermically at room temperature, spontaneously producing chemical waves.

Usually continuum models of reaction–diffusion systems are described by equations of the form

$$\frac{\partial \boldsymbol{u}}{\partial t} = \boldsymbol{\Psi}(\boldsymbol{u}) + \boldsymbol{\Gamma}\nabla^2\boldsymbol{u}, \tag{1}$$

$\boldsymbol{u}(x,t)$ is a vector which represents the state of the system. $\boldsymbol{\Psi}(\boldsymbol{u})$ is the reaction term which describes the kinetics of the medium. The diffusion term is given as the product of the matrix $\boldsymbol{\Gamma}$ and the Laplacian acting on $\boldsymbol{u}$. For a spatially homogeneous state the system evolves obeying

$$\frac{\partial \boldsymbol{u}}{\partial t} = \boldsymbol{\Psi}(\boldsymbol{u}). \tag{2}$$

In practice the interesting models are strongly non-linear and too complicated to be solved analytically, therefore a numerical approach is used. A cellular automata model for the Belousov–Zhabotinsky reaction and diffusion system was proposed by Greenberg, Hastings and coworkers [8–10], based in an earlier excitable media model by Wiener and Rosenblueth [29].

3. The Greenberg–Hastings model

Let us assume that each cell can take one of three states: resting or quiescent (0), active (1) or refractory (2). The cell $a_{i,j}$ has four neighbors in the two-dimensional von Neumann neighborhood:

$a_{i-1,j}$, $a_{i+1,j}$, $a_{i,j-1}$ and $a_{i,j+1}$, where i and j are space indices and t is the time index. The local function to compute the new state $a_{i,j}^{t+1}$ is a function of the neighbors and previous state and is given by

$$a_{i,j}^{t+1} = R(a_{i,j}^t) + D(a_{i-1,j}^t, a_{i+1,j}^t, a_{i,j-1}^t, a_{i,j+1}^t),$$

$$\tag{3}$$

R and D correspond to the reaction and diffusion terms respectively. They are defined by

$$
\begin{aligned}
R &= 2, \quad \text{if } a_{i,j}^t = 1,\\
&= 0, \quad \text{otherwise},\\
D &= N, \quad \text{if } a_{i,j}^t = 0,\\
&= 0, \quad \text{otherwise}, \tag{4}
\end{aligned}
$$

where N is a Boolean function which takes the value 1 if at least one of the neighbors is active, and 0 otherwise:

$$
\begin{aligned}
N = (a_{i-1,j}^t = 1) \quad &\text{or} \quad (a_{i+1,j}^t = 1) \quad \text{or}\\
(a_{i,j-1}^t = 1) \quad &\text{or} \quad (a_{i,j+1}^t = 1). \tag{5}
\end{aligned}
$$

The kinetics implemented by this function is as follows: the state 0 is the quiescent state in which the cell will remain until it is activated. The active state 1 is produced by the existence of one or more active neighbors. When a cell has been activated it will become refractory (2) at the next time step. A refractory cell is not only insensitive to the neighborhood but is also unable to activate its neighbors. After the refractory state a cell will always return to the quiescent state. The reaction term assures that the state will cycle to its successor when the cell is active or refractory and the diffusion term is a Boolean function which assures excitation, taking the value 1, if at least one of the neighbors is in the active state and 0 otherwise.

Fig. 1 shows the state diagram which describes the mechanics for a single cell. Fig. 2 shows the evolution of this rule from simple initial conditions consisting of a short row of active cells on top of a row of refractory ones in a quiescent background. The spirals develop due to the fact that the tips of the initial row act as singularities. The waves will spread unperturbed over the quiescent background.

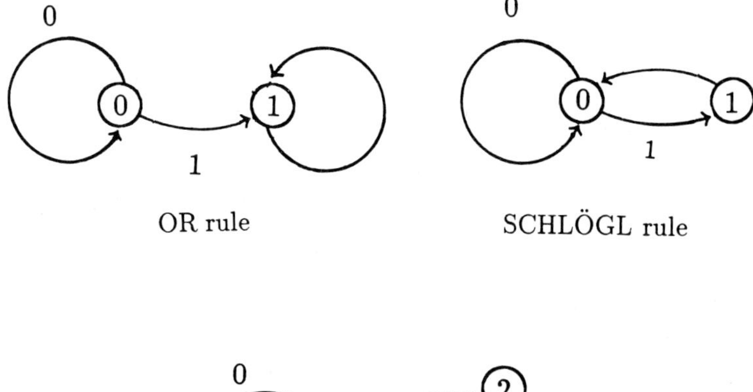

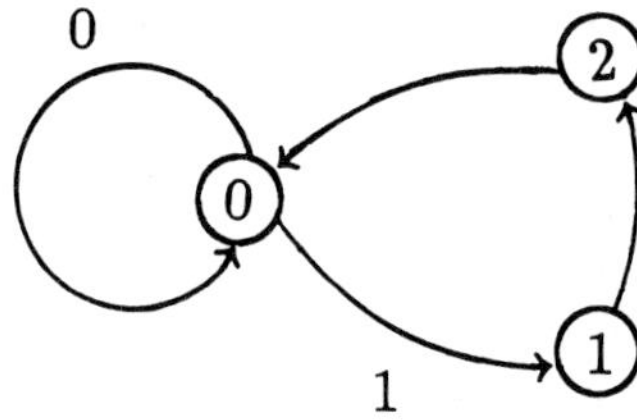

Fig. 1. Single-cell state diagrams for the three rules: OR, Schlögl and Greenberg–Hastings. 0 represents the resting state, 1 the active state and 2 the refractory state. The numbers on the arcs indicate transitions in which the state of the neighborhood is relevant: 1 indicates that at least one of the neighbors is active, and 0 indicates the absence of active neighbors.

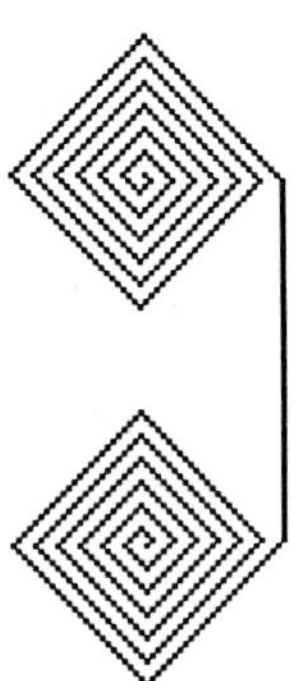

Fig. 2. Evolution of spirals in the Greenberg–Hastings rule. Only active (black) and resting (white) cells are shown. The initial condition consisted of a row of active cells on top of a row of refractory ones in a quiescent background.

The initial configurations which give rise to the organized patterns can be characterized by a "winding" or topological number as was shown by Greenberg et al. [8–10]. The basic "singularities" at the center of the spirals are two-dimensional configurations of the form: $\binom{01}{22}$, $\binom{20}{11}$ or $\binom{12}{00}$ for simple (long-wavelength) singularities and $\binom{102}{201}$ for complex (short-wavelength) singularities. The

arms of the spiral are made of activation wavefronts with refractory tails, which propagate over regions of quiescent cells. Starting from random initial conditions, few active and refractory cells in a background of quiescent cells, singularities would form (see fig. 3). Two singularities will in some cases come close together producing a more complex singularity which is able to generate waves of shorter wavelength. These waves propagate and take over the long-wavelength waves produced by the simple singularities. Each singularity defines a domain which is covered by traveling waves. When traveling waves from different singularities meet, a stationary boundary forms. Fig. 3 displays four of these expanding domains of short-wavelength waves which will destroy the simple singularities.

After some time the system enters a steady state in which the waves propagate steadily from the singularities to the boundaries.

The model can be extended to allow additional active and/or refractory states, basically keeping the same type of reaction and diffusion terms. These extensions show the same basic phenome-

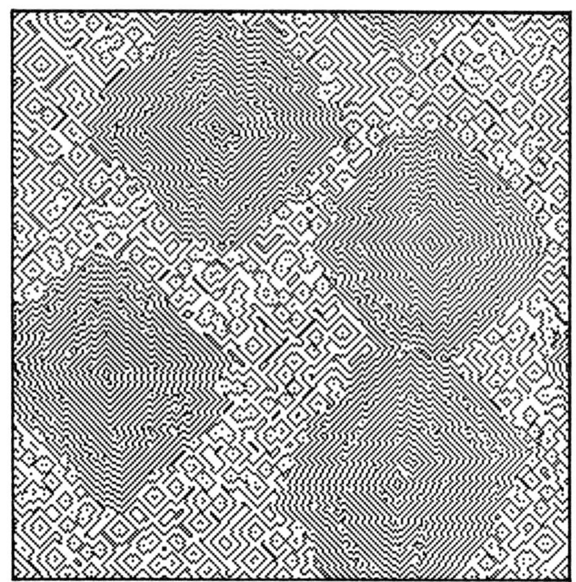

Fig. 3. Evolution of the Greenberg–Hastings model after 200 time steps from a random initial condition (3% active and 3% refractory cells). Only active (black) and resting (white) cells are shown. Many simple singularities can be seen in addition to four complex singularities.

nology as the three-state model. The structure of the singularities in three dimensions is a very intriguing subject, which has been extensively studied by A. Winfree [30]. Other similar models and additional aspects of this model have been studied by Allouche and Reder [31], Winfree [30]. Boon and Noullez [25], and Kapral and Oppo [32].

4. The Schlögl model

In this cellular automaton each cell can take one of two states: quiescent 0 or active 1. This model is like the Greenberg–Hastings model without the refractory state. After a cell has been activated it will change to state 0 at the next time step. In fig. 1 we show the state diagram which describes the mechanics for a single cell. If we start from an initial condition of a single active cell on a quiescent background, a checkerboard pattern propagates out from the initial activated cell. None of the complex behavior of the three-state case is found in the Schlögl model.

The simplest case of diffusion or growth of a

pattern is realized by the OR function. Again we have a two-state cell but now if the cell is activated by a neighboring cell it will remain in the 1 state forever (see fig. 1). From an initial seed and if the neighborhood is the four nearest neighbors (plus the center cell itself), a diamond pattern of activated cells will grow at a uniform rate. If the neighborhood is a Moore neighborhood then the pattern will be a square. The reason for discussing these two rather dull rules is that they give rise to many interesting phenomena when they are made reversible.

5. Reversible cellular automata

A general method for turning a cellular automaton into a reversible one (Fredkin's method) is to subtract (modulus the number of states q) the past value of a site from the value obtained by applying the rule F to the present neighborhood of the site:

$$a_i^{t+1} = F(\text{neighbors of } a_i^t) - a_i^{t-1} \bmod q. \qquad (6)$$

In the case where the states of a site are three-valued the subtraction is carried out mod 3. In the case where the states are two-valued the subtraction is carried out mod 2, which is the same as addition mod 2 or exclusive OR. Applying this transformation to the models described before we obtain three reversible rules: reversible OR (REV-OR), reversible Schlögl (R-SCHLÖGL) and reversible Greenberg–Hastings (R-GREEN). Their cell transition state diagrams are shown in fig. 4.

6. The reversible Greenberg–Hastings model

Based on the transformation described above the rule for the R-GREEN rule is as follows:

$$a_{i,j}^{t+1} = R(a_{i,j}^t) + D(a_{i-1,j}^t, a_{i+1,j}^t, a_{i,j-1}^t, a_{i,j+1}^t)$$
$$- a_{i,j}^{t-1} \bmod 3. \qquad (7)$$

The transformation changes the structure of the function adding a second-order dependence in time. The effect is such that the information

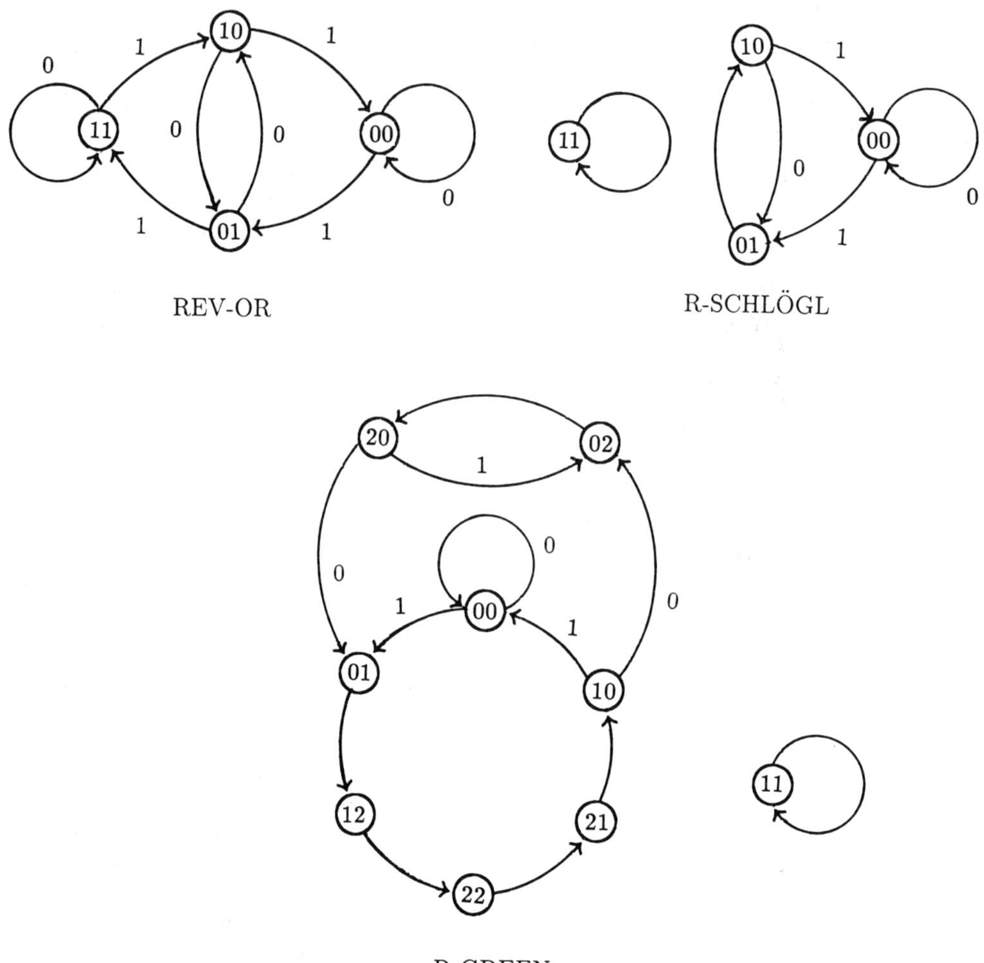

Fig. 4. Single-cell state diagrams for the three reversible rules: REV-OR, SCHLÖGL and R-GREEN. Inside the circles the state of the cell, at present and in the past, is represented by two numbers: (*past present*). As in fig. 1, 0 represents the resting state, 1 the active state and 2 the refractory state. The numbers on the arcs indicate transitions in which the state of the neighborhood is relevant: 1 indicates that at least one of the neighbors is active, and 0 indicates the absence of active neighbors.

stored in the initial conditions will be preserved by the dynamics for all future time. A reversible cellular automaton has as many conserved quantities as there are cells [13,33]. The system cannot forget the initial conditions which are always encoded in the configuration. Dynamical trajectories cannot enter attractors and then form closed orbits. If we run a finite system we will eventually exhaust the possible configurations in the orbit and we will return back to the initial condition. However, the recurrence time could be extremely long. The state description for this rule (see fig. 4)

now includes the state of the cell at time t as well as the state at time $t - 1$.

Depending on the initial conditions the system evolves basically into three different regimes:

    (i) A regular regime with short recurrence time.

    (ii) A chemical-turbulent regime.

    (iii) A disordered random regime.

To obtain the regular regime, with the von Neumann's neighborhood, the initial state should consist of a single active cell, two active cells or a whole line of active cells on a line. In a regular regime wavefronts travel unperturbed in a quies-

Fig. 5. Sequence in the evolution of the reversible Greenberg–Hastings model. Only active and resting cells are shown. (a) $t = 0$. Initial conditions consisting of three active cells at coordinates $(0, 0)$, $(33, 0)$, $(32, -33)$. (b) $t = 12$. The wavefronts propagate. (c) $t = 100$. The wavefronts interact. (d) $t = 1000$. Many new wavefronts are produced by interactions; notice the irregularities along the wavefronts. (e) $t = 1400$. The system is completely populated by wavefronts that form wide rings. (f) $t = 1700$. The rings form around fuzzy regions that act as sources and sinks. (g) $t = 2000$. A central source can now be seen. All the wavefronts have disappeared. (h) $t = 3000$. The rings become wider and more diffused. (i) $t = 5000$. The rings are completely diffused and the system is composed of large patches of quiescent, active and refractory cells.

cent background.

If the initial state consists of a single active cell next to a refractory one the system will develop into a disordered random regime. This random regime displays a fully mixed pattern of quiescent, refractory and active cells. There is no apparent structure or pattern.

The most interesting is naturally the turbulent regime, which is produced by certain initial conditions, for example three active cells in an asymmetric triangle (see fig. 5).

The key feature seems to be the spatial distance between them. They must be separated by an even number of cells. The turbulent regime shows new emergent structures. We have quasi-circular rings emanating from complex regions. A pattern of sources and sinks are set up. Fig. 5 shows a complete sequence in the evolution of the turbulent regime from three asymmetric active cells. The rings grow in size and the waves travel at super-luminal speed, faster than one cell per time step. The reason is that they are not real propagating structures but coherent phases formed by the oscillation of rings of cells. When the Moore neighborhood is used. The simplest initial configuration to develop a turbulent regime is now two active cells. The essential feature is the separation between the active cells. For even/odd separation the rings will/will not form (see fig. 6).

In the case of REV-OR or R-SCHLÖGL there is no random regime, however, the same approach to turbulence is seen in both cases: the same dependence on neighborhood and the same initial conditions. The periodic boundary conditions enter into the considerations on how these systems go to turbulence making the analysis difficult. In all cases from the initial conditions, a system of alternating sources and sinks are set up leading to the turbulence. Due to the richness of phenomena in two dimensions a complete analysis is difficult. We therefore concentrate on the one-dimensional case.

7. One-dimensional cellular automata

When we go from two dimensions to one dimension the three systems behave differently: (1) REV-OR has an additive conserved quantity but no walls. (2) R-SCHLÖGL has walls but we have not been able to find an additive conserved quantity. (3) R-GREEN has walls and again we have not found an additive conserved quantity. Thus by using the Takesue classification REV-OR differs

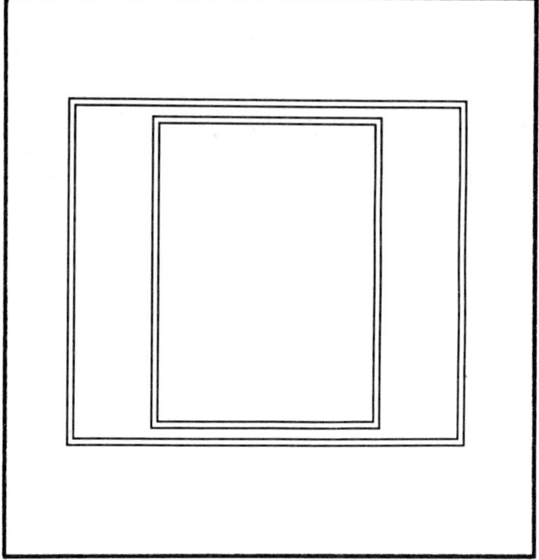

Fig. 6. Reversible Greenberg–Hastings model in the Moore neighborhood. Only active and resting cells are shown. The two pictures show the state of the system after 1500 steps. The initial conditions were two active cells separated by an even number of cells on the right and by an odd number of cells on the left. One system attains the regular regime while the other attains the turbulent one.

from R-SCHLÖGL and R-GREEN. It is when we set up a periodic phase and examine the evolution of simple perturbations, that we can distinguish the behavior of R-SCHLÖGL and R-GREEN.

The one-dimensional non-reversible Greenberg–Hastings model does not show any complex behavior. Starting from an initial state of one active cell, two traveling outgoing waves result. The traveling structures are an active cell followed by a refractory one. When two such structures collide they annihilate each other. An initial state of an active cell followed by a refractory cell gives rise to a single wavefront. Thus starting from any random initial conditions of active and refractory cells the traveling waves will annihilate each other in pairs, leaving the surviving waves to travel endlessly. Similarly the OR and Schlögl rules in one dimension do not show any complex behavior.

7.1. The reversible OR (REV-OR) rule

Applying the analysis of Margolus [33] we were able to find an additive invariant for the REV-OR rule. For the one-dimensional model, the invariant takes the form

$$\Phi^t = 2 \sum_i \left[2a_i^{t-1}(a_{i+1}^t - a_{i-1}^t) - (a_{i+1}^t - a_{i-1}^t) \right]. \tag{8}$$

The details of the calculation can be found in the appendix. Similar invariants have been found for the Q2R rule by Pomeau [34], for some one-dimensional rules by Takesue [16] and for many reversible rules with appropriate symmetries by Margolus [33].

In fig. 7 we see the space–time behavior of a single activated cell. Two outgoing solitons result which collide due to the periodic boundary conditions. We can set up a periodic phase by an initial condition of a line of activated cells. When the periodic phase is perturbed two traveling dislocations result which go through each other when they collide.

7.2. The reversible Schlögl (R-SCHLÖGL) rule

The behavior of this cellular automaton with respect to the initial condition of a single activated cell and to a line of activated cells is similar to the

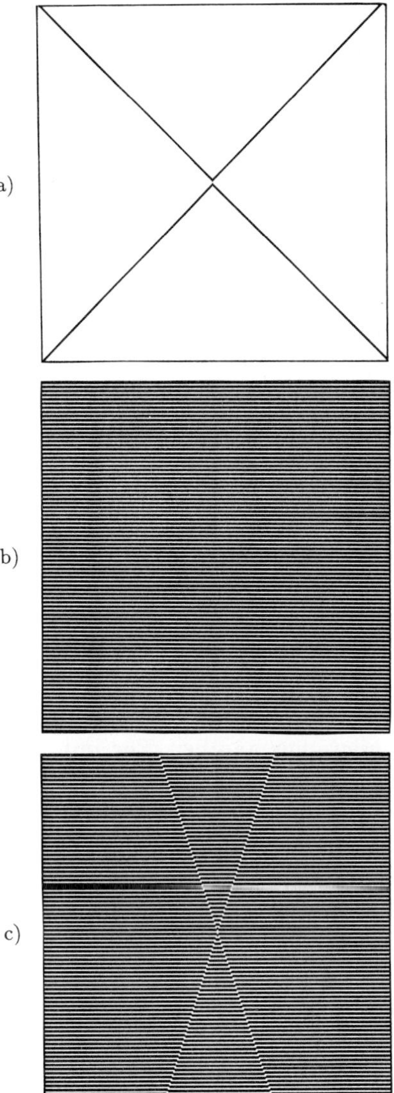

Fig. 7. Space–time behavior of the REV-OR rule from different initial conditions. The time axis runs down vertically. (a) A single active cell. (b) A line of active cells which generates a periodic phase background. (c) A line of active cells with one "hole" (quiescent cell) produces two traveling dislocations.

REV-OR case. Also the behavior of a perturbation in the periodic phase is similar. The appearance of a wall is due to the isolation of the 11 state in the state diagram for state transitions. If the initial state is 11 a wall is set up and also a stream of solitons which interact leading to the appearance of a sink which grows in size until it touches the wall which generates a new source. This alternation between source and sink generates an orderly succession of periodic phases. A single activated cell between two walls leads to the generation of two traveling dislocations which collide with each other without generating new dislocations.

7.3. The reversible Greenberg–Hastings (R-GREEN) rule

The initial state must not only specify the present state of a cell but must also specify its previous state. The state of a cell will be labeled as follows $\left(^{present}_{past}\right)$. The simplest initial condition is a single active cell in the present which evolves into two outgoing waves of the following structure $\left(^{1221000\bar{1}221}_{01221012210}\right)$. Given two active cells two outgoing waves result which collide and give different types of collisions depending upon whether the initial separation was even or odd. In both cases the wavefronts are reconstituted. Thus single active cells give rise to two outgoing waves. However unlike the non-reversible case colliding waves do not annihilate one another. The initial condition where there is a 1 in the present as well as a 1 in the past $\left(^1_1\right)$ will give rise to a train of waves and a wall. The "wall" is formed because as can be seen from the state diagram a cell which is active in the present and in the past $\left(^1_1\right)$ will remain active for all future times. If two of these walls $\left(^1_1\right)$ are placed in the initial condition and are separated by some distance from each other the system separates into two subsystems which do not interact with each other. The $\left(^1_1\right)$ configuration also generates a steady train of wavefronts which interacts with the train of wavefronts generated at the other $\left(^1_1\right)$ configuration. These interactions generate a regular pattern (see fig. 8). However, if in addition to the walls we put a single active cell in the initial conditions, then the wavefronts generated at the walls will interact with the wavefront generated by the active cell and in each collision phase shifts will be produced. After some time the system evolves into a turbulent phase (fig. 8).

In a previous paper we have calculated the two-point correlation functions and the power spectra of these turbulent patterns [23]. Our results agree with those of Oono and Yeung [4]. There are many more initial configurations such as $\left(^2_2\right)$, $\left(^1_2\right)$, $\left(^2_1\right)$ which in pairs lead either to regular patterns or turbulence. Since these systems are reversible there must exist conservation laws which we have not yet found. The invariants could be non-local and involve topological quantities such as winding numbers.

The analysis of this rule is simplified if one sets up a periodic phase background by setting up the initial condition of a line of $\left(^{\cdots22222\cdots}_{\cdots22222\cdots}\right)$. A simple "hole" perturbation $\left(^{\cdots22022\cdots}_{\cdots22222\cdots}\right)$ leads to two traveling dislocations. However if two or more perturbations are separated from each other, the system evolves into a turbulent state by means of a soliton-generated turbulence (see fig. 9). The number of moving dislocations is conserved. The turbulence is generated by the collisions between these moving dislocations. Some of these collisions lead to dislocation-mediated phase transitions and it is the boundaries between phases which ultimately causes the observed disorder. The major observation here is that a new conserved quantity has been found, i.e. the number of moving dislocations. We have calculated the Walsh transform coefficients of the regular and turbulent conditions. The Walsh transform is a transformation analogous to the Fourier transform but with a kernel made up of orthogonal rectangular waveforms [35]. Walsh transforms are better suited than Fourier transforms to deal with Boolean or discrete functions. The Walsh coefficients (in sequency order) for regular and turbulent regimes are shown in fig. 10. It is clear from the plots that the regular regime is made of a few Walsh modes (notice the regularity of the pattern), in contrast the Walsh coefficients for the turbulent regime show a complicated structure that implies many Walsh modes.

8. Conclusions

Chemical turbulence is a new field and a new challenge which is being studied using many different techniques. The use of reversible cellular automata is a novel way to study this most complex phenomena. The major conclusions are: Spatial dimensions are important as the turbulent state was reached by REV-OR, R-SCHLÖGL and R-GREEN in two dimensions but it was only

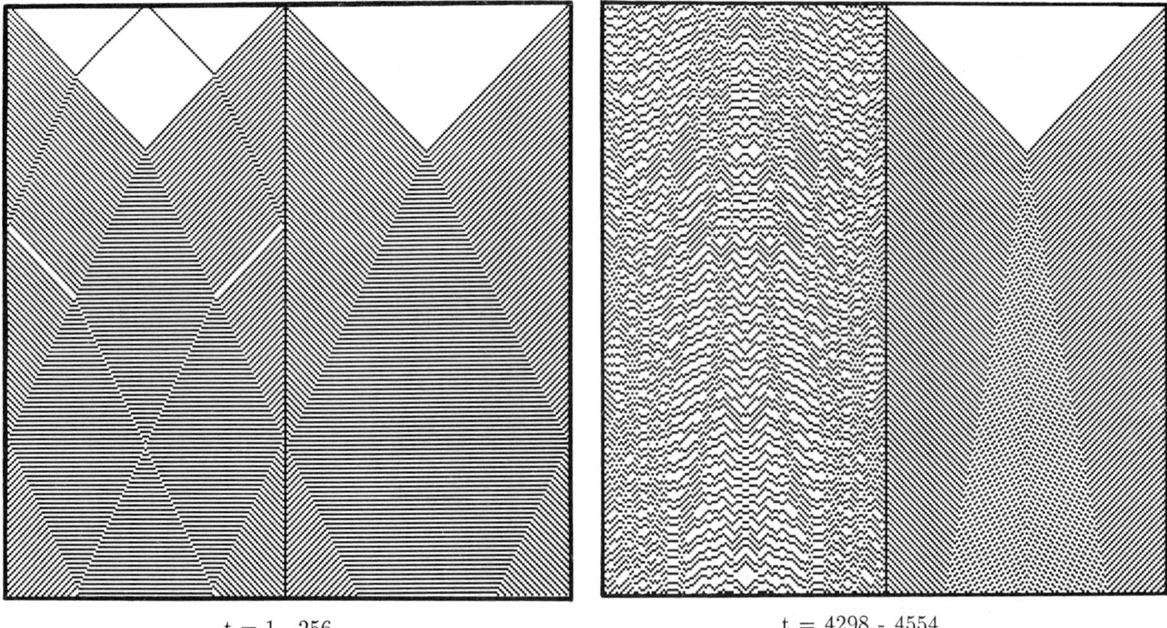

Fig. 8. Two space–time snapshots of the reversible Greenberg–Hastings one-dimensional system (128 cell long) at the times shown below the pictures. The time axis runs down vertically and only quiescent and active cells are shown. Two "walls", one at the center and one at the border, divide the system into two dynamically independent 64-cell-long subsystems. The subsystem on the right-hand side shows a short period regular regime consisting of vacua of different structures. The subsystem on the left-hand side has an additional active cell placed at the center and after few hundred steps develops chemical turbulence.

R-GREEN which became turbulent in one dimension. The neighborhood was important. We needed three activated cells in an asymmetric triangular configuration as the minimal initial configuration if the von Neumann neighborhood was used, but only two activated cells if the Moore neighborhood was used. The dependence on separation between the activated cells clearly involved the periodic boundary conditions.

The tools for studying the one-dimensional case are better but still not adequate. We do have an additive conserved quantity for the REV-OR. We can study the effect of walls in the case of the R-SCHLÖGL and R-GREEN. In both cases we obtain soliton-mediated phase transitions. If we add an active cell between two walls we get soliton-mediated approach to turbulence only in the case of R-GREEN. In the simpler case of a single periodic phase (set up by a line of activated cells), both REV-OR and R-SCHLÖGL gave rise to moving dislocations from an initial perturbation. These dislocations pass through each other

(soliton-like). In the case of the R-GREEN the dislocations interact causing dislocation-mediated phase transitions which create stationary dislocations between the phases. These interactions eventually disorder the system. What is of great interest is the conservation of the moving dislocations.

Thus in the classification of Cosenza and Neri [15], REV-OR and R-SCHLÖGL belong to class 1 (dislocations propagate without proliferation), and R-GREEN belongs to class 3 (dislocations propagate and proliferate). The behavior of the dislocations in the periodic phase of R-GREEN has analogies to the localized structures found in the class IV automata of Wolfram [6]. The original non-reversible Greenberg-Hastings model has many interesting phenomena; however the existence of a stable quiescent state does not permit a turbulent phase to develop in either one or two dimensions.

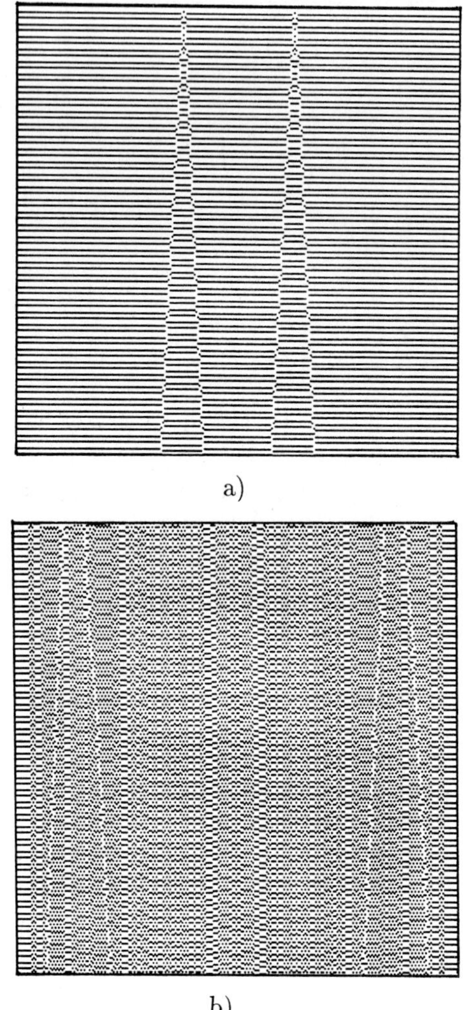

a)

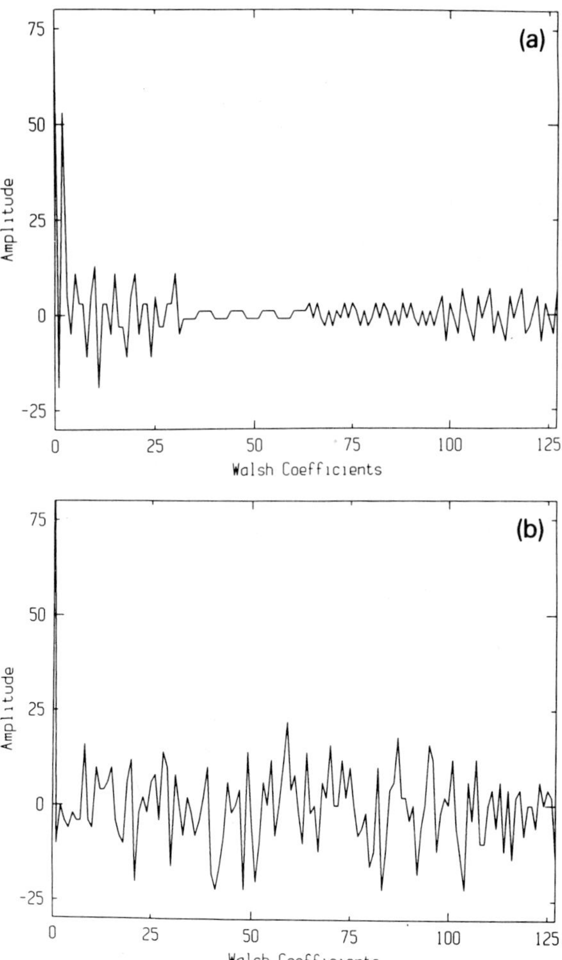

b)

Fig. 9. Two space–time snapshots for the reversible Greenberg–Hastings one-dimensional rule. The time axis runs down vertically and only quiescent and active cells are shown. (a) The first 256 steps starting from two "hole" perturbations of the form $\left(\begin{smallmatrix}...22022...\\...22222...\end{smallmatrix}\right)$. They produce four moving dislocations. (b) The state of the system after 5×10^5 time steps, where the turbulent regime can be seen. The number of moving dislocations is conserved (there are still four of them). The turbulence is generated by the collisions between these moving dislocations and it is the boundaries between phases which ultimately cause the observed disorder.

Acknowledgements

We are grateful for useful discussions to C. Bennett, J. Leao, C. Langton, Y. Pomeau, E. Lorenz, S. Redner, G. Vichniac, W. Klein, R. Brower, E. Fredkin, M. Kolb, R. Livi, S. Ruffo, A. Coniglio, L. De Arcangelis, N. Jan, P. Grassberger, P. Plath, G. Martinez, D. Stauffer, H. Herrmann and H. Gutowitz. Norm Margolus contributed considerably to our understanding of invariant quantities in reversible rules. We also appreciate the ideas and encouragement given to us by the MIT Information Mechanics Group: T. Toffoli, N. Margolus and D. Zaig, and for giving us access to a CAM-6 cellular automaton simulator. This work was supported in part by a grant from the Office of Naval Research.

Fig. 10. Walsh coefficients for two one-dimensional spatial configurations of the reversible Greenberg–Hastings rule. (a) Corresponds to a regular regime like the one shown in fig. 9a. (b) Corresponds to a turbulent regime like the one shown in fig. 9b.

Appendix A

We will show how to obtain a time-invariant quantity for the REV-OR rule. The basic method used is that of Margolus [33]. We will concentrate in the one-dimensional REV-OR rule but the arguments can be applied to higher dimensions and other rules.

Many reversible rules can be casted in the form

$$a_i^{t+1} = F(\text{neighbors of } a_i^t) - a_i^{t-1} \bmod 2 , \qquad (A.1)$$

where F is an arbitrary function of the neighborhood of a_i^t at time t. In our case F is the OR rule.

Now we will try to form a local product $(1 - F)V = 0$, defining V as some linear symmetric function of the neighborhood. This local product must be always zero. If F is the OR rule (a_{i+1}^t or a_i^t or a_{i-1}^t) then one simple choice for V is $(a_{i+1}^t \text{ xor } a_{i-1}^t)$. Then the product $(1 - F)V$ is always zero because V is zero for all cases in which $(1 - F) \neq 0$. To simplify the derivation, we will change the variables from $a_i^t \in \{0,1\}$ to spin-like variables $x_i^t \in \{-1,1\}$; the conversion is simply $x_i^t = 2a_i^t - 1$. By using eq. (1) F can be expressed as

$$F - a_i^{t+1} + a_i^{t-1} \bmod 2 = (a_i^{t+1} \text{ xor } a_i^{t-1})$$
$$= \tfrac{1}{2}(1 - x_i^{t+1} x_i^{t-1}), \qquad (A.2)$$

and V is

$$V = (a_{i+1}^t \text{ xor } a_{i-1}^t)$$
$$= \tfrac{1}{2}(1 - x_{i+1}^t x_{i-1}^t), \qquad (A.3)$$

then, the local product reads,

$$(1 - F)V = \tfrac{1}{4}(1 + x_i^{t+1} x_i^{t-1})(1 - x_{i+1}^t x_{i-1}^t)$$
$$= 0. \qquad (A.4)$$

Multiplying this expression by $x_i^{t+1} x_{i-1}^t$ and dropping the constant $\tfrac{1}{4}$ we obtain,

$$x_i^{t+1} x_{i-1}^t - x_i^{t+1} x_{i-1}^t x_{i+1}^t x_{i-1}^t$$
$$+ x_i^{t+1} x_{i-1}^t x_i^{t+1} x_i^{t-1}$$
$$- x_i^{t+1} x_{i-1}^t x_i^{t+1} x_i^{t-1} x_{i+1}^t x_{i-1}^t = 0. \qquad (A.5)$$

Due to the fact that x variables are always equal to ± 1, the products of repeated variables reduce

to 1 (for example $x_{i-1}^t x_{i-1}^t = 1$), and the equation simplifies to

$$x_i^{t+1} x_{i-1}^t - x_i^{t+1} x_{i+1}^t$$
$$= -x_{i-1}^t x_i^{t-1} + x_i^{t-1} x_{i+1}^t. \qquad (A.6)$$

Summing over all indices i we obtain

$$\sum_i x_i^{t+1} x_{i-1}^t - \sum_i x_{i+1}^t x_i^{t+1}$$
$$= \sum_i x_{i+1}^t x_i^{t-1} - \sum_i x_i^{t-1} x_{i-1}^t. \qquad (A.7)$$

Now changing the indices, the sum over $x_i^{t+1} x_{i-1}^t$ is replaced by a sum over $x_i^t x_{i+1}^{t+1}$. These two sums give the same result because they sum over the same "bonds" or "links". In a similar way the sum over $x_{i+1}^t x_i^{t+1}$ is replaced by a sum over $x_i^t x_{i-1}^{t+1}$. We have then

$$\sum_i x_i^t (x_{i+1}^{t+1} - x_{i-1}^{t+1})$$
$$= \sum_i x_i^{t-1} (x_{i+1}^t - x_{i-1}^t). \qquad (A.8)$$

Now we can see that this equation implies the existence of a time-invariant quantity Φ^t:

$$\Phi^{t+1} = \sum_i x_i^t (x_{i+1}^{t+1} - x_{i-1}^{t+1}),$$
$$\Phi^t = \sum_i x_i^{t-1} (x_{i+1}^t - x_{i-1}^t). \qquad (A.9)$$

In terms of the original Boolean variables the invariant is

$$\Phi^t = 2 \sum_i [2a_i^{t-1}(a_{i+1}^t - a_{i-1}^t) - (a_{i+1}^t - a_{i-1}^t)].$$
$$(A.10)$$

References

[1] Y. Kuramoto, Chemical Oscillations, Waves and Turbulence (Springer, Berlin, 1984).

[2] A.M. Turing, The chemical basis of morphogenesis, Philos. Trans. R. Soc. London Ser. B 237 (1952) 37–72.

[3] J.D. Murray, Mathematical Biology (Springer, Berlin, 1989).

[4] Y. Oono and C. Yeung, A cell dynamical system model of chemical turbulence, J. Stat. Phys. 48 (1987) 593–644.

[5] T. Toffoli and N. Margolus, Cellular Automata Machines (MIT Press, Cambridge, MA, 1987).

[6] S. Wolfram, ed., Theory and Applications of Cellular Automata (World Scientific, Singapore 1986).

[7] D. Farmer, T. Toffoli and S. Wolfram, eds., Proceedings of the Cellular Automata Workshop, published in: Physica 10 D (1984); Proceedings of the CA86 Conference, published in: Complex Systems 1 (1 & 2) (1987).

[8] J.M. Greenberg and S.P. Hastings, Spatial patterns for discrete models of diffusion in excitable media, Siam J. Appl. Math. 34 (1978) 515.

[9] J.M. Greenberg, B.D. Hassard and S.P. Hastings, Pattern formation and periodic structures in systems modeled by reaction–diffusion equations, Bull. Am. Math. Soc. 84 (1978) 1296.

[10] J.M. Greenberg, C. Greene and S. Hastings, A combinatorial problem arising in the study of reaction–diffusion equations, Siam J. Alg. Disc. Meth. 1 (1980) 34.

[11] Y. Oono and M. Kohmoto, Phys. Rev. Lett. 55 (1985) 2927.

[12] K. Kaneko, Symplectic cellular automata, Phys. Lett. A 129 (1988) 9.

[13] T. Toffoli, Cellular automata mechanics, Ph.D. Thesis, University of Michigan (1977).

[14] J. Crutchfield and K. Kaneko, Phenomenology of spatial-temporal chaos, preprint.

[15] G. Cosenza and F. Neri, Cellular automata and nonlinear dynamical models, Physica 27 D (1987) 357–372.

[16] S. Takesue, Phys. Rev. Lett. 59 (1987) 2499.

[17] A.W.M. Dress et al., Some proposal concerning the mathematical modeling of oscillating heterogeneous catalytic reactions on metal surfaces, in: Temporal Order, eds. L. Rensing and N. I. Jaeger (Springer, Berlin, 1984).

[18] B. Chopard and M. Droz, Cellular automata approach to non-equilibrium phase transitions in a surface reaction model: static and dynamic properties, J. Phys. A 21 (1988) 205–211.

[19] H. Hartman, Speculations on the origin and evolution of metabolism, J. Mol. Evolution 4 (1975) 359–370.

[20] H. Hartman, Speculations of the evolution of the genetic code, Origins of Life, part I: 6 (1975) 423–427, part II: 9 (1978) 133–136, part III: 14 (1984) 643–648.

[21] A.G. Cairns-Smith, The origin of life and the nature of the primitive gene, J. Theor. Biol. 10 (1966) 53–88.

[22] C. Langton, Studying artificial life with cellular automata, Physica D 22 (1986) 120–149.

[23] P. Tamayo and H. Hartman, Cellular automata, reaction–diffusion systems and the origin of life, in: Artificial Life, ed. C.G. Langton (Addison–Wesley, New York, 1988).

[24] P. Coullet, C. Elphick and D. Repaux, Nature of spatial chaos, Phys. Rev. Lett. 58 (1987) 431–434.

[25] J.P. Boon and A. Noullez, in: On Growth and Form, eds. E. Stanley and N. Ostrowsky (Nijhoff, The Hague, 1986).

[26] H. Meinhardt, Models of Biological Pattern Formation (Academic Press, New York, 1982).

[27] C.H. Bennett, Dissipation, information, computational complexity and the definition of organization, in: Emerging Synthesis in Science, ed. D. Pines (Addison–Wesley, New York, 1985).

[28] G.Y. Vichniac, Cellular automata models of disorder and organization, in: Disordered Systems and Biological Organization, eds. E. Bienenstock, F. Fogelman and G. Weisbuch (Springer, Berlin, 1985).

[29] N. Wiener and A. Rosenblueth, Arch. Inst. Cardiol. Mexico 16 (1946) 205–265.

[30] A.T. Winfree, Rotating solutions to reaction/diffusion equations in simply connected media, SIAM–AMS Proceedings 8 (AMS, Providence, RI, 1974); Organizing centers in a cellular excitable medium, Physica D 17 (1985) 109–115; and in: Oscillations and Traveling Waves, eds. R. Field and M. Burger (Wiley, New York, 1985).

[31] J.P. Allouche and C. Reder, Spatio–temporal oscillations of a cellular automaton in excitable media, Lecture Notes on Biomathematics, Vol. 49 (Springer, Berlin, 1983).

[32] R. Kapral and G. Oppo, Physica D 23 (1986) 455–463.

[33] N. Margolus, Physics and computation, Ph.D. Thesis, MIT (1987).

[34] Y. Pomeau, J. Phys. A 17 (1984) L415.

[35] K.G. Beauchamp, Applications of Walsh and Related Functions (Academic Press, New York, 1984).

Physica D 45 (1990) 307–327
North-Holland

SOLITON TURBULENCE IN ONE-DIMENSIONAL CELLULAR AUTOMATA

Y. AIZAWA [1], I. NISHIKAWA
Department of Physics, Kyoto University, Kyoto, Japan

and

K. KANEKO
Institute of Physics, College of Art and Sciences, University of Tokyo, Tokyo, Japan

Received 20 January 1990
Revised manuscript received 7 April 1990

Statistical properties of cellular automata which support simple solitary waves are numerically studied. Under these rules, spatio-temporal patterns are sensitively dependent on the collision processes of solitons, and other elementary excitations such as breathers, kinks, and nuclei. Patterns typically become randomized as collisions proceed. Turbulent states are realized after many collisions. The resulting global patterns can be classified by the types of elementary excitations they contain. In many turbulent states, the spectra reveal a long-range order with a $k^{-\nu}$ anomaly for $k \ll 1$. The irreversible process leading to turbulent equilibrium is characterized by the transient spectrum together with the Allan variance. The mean free motion of a soliton in the turbulent states is measured by the mutual information flow, and the information loss is shown to obey an inverse power law. The transient time to reach an attractor grows exponentially with system size, which suggests that in the thermodynamic limit soliton turbulence is not an attractor but rather a transient state.

1. Introduction

Recently, the dynamics of spatially extended objects and fields has been studied in many branches of physical sciences. It has become clear that nonlinear instability in spatial coupling often induces unpredictable phenomena [1]. Among the most striking of these phenomena is self-organization of turbulence or statistical order in space and time. Fundamental features of turbulence have been understood in terms of chaotic dynamics [2], e.g., strange attractors, non-periodic motions, etc. It seems, however, that the concept of turbulence must be recast in much wider framework than the dynamical theory of lumped systems allows. This will involve taking account of the essential role of spatial degrees of freedom.

The main purpose of this paper is to elucidate some of the effects induced by spatial interactions. We employ for this purpose simple one-dimensional cellular automata (CA) [3]. If local patterns are stable, cellular automata can be successfully used to simulate self-organization processes in a continuous field. This is in spite of the fact that CA operate on discrete variable in discrete space and time. Indeed, various regular structures are well represented in CA [4]. When a globally unstable process is supported in a CA, however, discreteness can be of crucial importance. In particular, turbulence in CA, as studied here, reveals novel features, not present in continuous systems [5]. In this context, the most pronounced differences between CA and continuous systems are as follows:

(a) CA have no dissipative regime in the high-wavenumber limit, so that short-wavelength excitations couple strongly with longer waves.

<hr>

[1] Present address: Department of Applied Physics, School of Science and Engineering, Waseda University, Tokyo, Japan.

(b) CA have no differentiable structure, so that the perturbational approach breaks down in general, even in nearly integrable cases.

(c) CA have no exponential instability, so that characterization in terms of chaotic dynamical systems theory is quite insufficient.

Due to these extraordinary restrictions inherent in the CA approach, we cannot make definitive statements concerning the nature of turbulence in real fluid or chemical systems. Nonetheless, we attempt to describe here some new generic features of turbulence induced by spatial interactions.

In recent studies [3], the large-time patterns generated by CA are classified into four types: (I) vanishing patterns, (II) non-propagating localized patterns, (III) spreading triangular patterns, and (IV) long-lived irregular patterns. The limit sets of type I and II are simple attractors, because the spatial interaction inhibits the propagation of local excitations. On the other hand, when propagation occurs in type III, it is known that patterns are generally chaotic or turbulent and that spatial correlations decay quickly with a short coherent length. Type IV patterns are more complicated, however. In this case, spatial interaction can lead to competitive excitatory and inhibitory effects. For instance, short-range interactions may be inhibitory while long-range interactions are excitatory or vice versa. As a result, type IV patterns become irregular and may either propagate slowly or eventually vanish. Theoretical analysis of types I–III has been partly successful, but type IV patterns remain poorly understood [6]. Many interesting turbulent processes have been found in type IV turbulence. Rules of this type may support a particular type of localized excitation called solitons. These solitons are often created by the competition between excitatory and inhibitory interactions. It is clear that both the slowly varying and soliton modes of behavior in type IV rules are due to the competition between excitatory and inhibitory interactions. However the direct interrelation between these modes is still not understood. In the present paper we classify turbulence in a special subgroup of type IV, in which a certain definite type of soliton always exists.

1.1. Soliton CA

In this section we introduce the class of soliton CA studied in the following. Define a binary variable at each lattice site i by S_i^t (= 0 or 1), where t stands for time. Then a CA is described by a mapping F representing the time development

$$S_i^{t+1} = F(\{S_j^{t'}\}, t' = t), \quad i, j = 1, 2, \ldots, N, \quad (1.1)$$

where N is the system size and F is independent of i. The simplest soliton in the 1D symmetric CA is $S1 = \cdots 011010 \cdots$ or $S2 = \cdots 010110 \cdots$, and the dynamics is realized at least by 5 neighboring sites, that is, $S_i^{t+1} = F(S_{i-2}^t, S_{i-1}^t, S_i^t, S_{i+1}^t, S_{i+2}^t)$. Each soliton can be transferred under the following rule:

$$F_{S1} = f_0 + f_1 + f_2 + f_3 + g_1 + g_2 + g_3,$$
$$F_{S2} = f_1 + f_2 + g_2 + g_3 \quad (\text{mod } 2), \quad (1.2)$$

where

$$f_0 = S_{i-2} + S_{i+2},$$
$$f_1 = S_{i-2}S_{i-1} + S_{i+1}S_{i+2},$$
$$f_2 = S_{i-1}S_i + S_iS_{i+1},$$
$$f_3 = S_{i-2}S_{i+1} + S_{i-1}S_{i+2},$$
$$g_1 = S_{i-2}S_{i-1}S_i + S_iS_{i+1}S_{i+2},$$
$$g_2 = S_{i-2}S_{i-1}S_{i+2} + S_{i-2}S_{i+1}S_{i+2},$$
$$g_3 = S_{i-2}S_{i-1}S_{i+1} + S_{i-1}S_{i+1}S_{i+2}. \quad (1.3)$$

In both rules, the solitons disappear after a collision, but by introducing a creative interaction, $F = F_{S1,2} + F_{int}$, a variety of patterns can be generated. For example, type A introduced in section 2 is represented by

$$F_{int} = f_4 + g_1 + g_4 + h_1 + k \quad (\text{mod } 2), \quad (1.4)$$

where

$$f_4 = S_{i-2}S_{i+2},$$
$$g_4 = S_{i-2}S_iS_{i+2},$$
$$h_1 = S_{i-2}S_{i-1}S_iS_{i+1} + S_{i-1}S_iS_{i+1}S_{i+2},$$
$$k = S_{i-2}S_{i-1}S_iS_{i+1}S_{i+2}. \quad (1.5)$$

Here one can see that the long-range cooperative interaction plays an essential role in reproducing second-generation solitons after a collision. Until now, the various collision patterns generated by these rules have not been explained in terms of the algebraic structure of the perturbation function F_{int}. In the following only the phenomenology of the turbulence in our systems will be studied.

This paper is organized as follows. In the next section (section 2) the global patterns of the turbulent states are classified into the following types: soliton turbulence, soliton–nucleus turbulence, soliton–breather turbulence, and soliton–kink turbulence.

Randomization of patterns, the origin of these turbulent states, is discussed in section 3. There it is shown, on the basis of a typical model, that phase sensitivity of the collisions of elementary excitations induces global randomization.

In section 4 the power-spectrum density is calculated. This characterizes the formation of the spatial order in the system. $k^{-\nu}$ spectral order in the small-wavenumber regime is observed in the turbulent equilibrium and even in transient stages. The detailed structure of the spectrum depends on the model chosen. The long-range order revealed by these spectra is due to a collision cascade of elementary excitations. In this irreversible process, the Allan variance $\sigma_{\text{A}}^2(n) \sim n^{\tau}$, together with the transient spectrum, is also obtained. The general relation $\nu = \tau + 1$ is demonstrated.

Next we study the mean free motion of a soliton in the turbulent states in section 5. Here the mutual information flow is calculated in a running coordinate system with constant velocity. Information loss is shown to obey an inverse power law, which is consistent with long-lived order which survives even in turbulent equilibrium.

Section 6 treats the turbulent attractors of the systems studied. The number of attractors is shown to depend exponentially on system size. We do not find any particular complex attractor which corresponds to turbulence in our model. Moreover, the exponential dependence of the turbulent transient time on system size strongly suggests that the turbulent state is transient.

Finally, section 7 contains some general remarks concerning these systems. We point out that the global chaos observed in these systems is sustained by a balance between creation and annihilation of localized turbulence. Turbulence in CA is compared with global chaos in partial differential equations.

2. Phenomenological classification

2.1. Description of soliton systems

Denote the local rule F by

$$F(V, W, X, Y, Z) = F(R), \qquad (2.1)$$

with $R = V \cdot 2^4 + W \cdot 2^3 + X \cdot 2^2 + Y \cdot 2 + Z$. The function $F(R)$ is defined on the 20 coordinates when the 12 symmetric conditions are satisfied; $F(1) = F(16), F(2) = F(8), F(3) = F(24)$, etc. Furthermore, the systems investigated in this paper are restricted to those where the simple soliton $S1$ or $S2$ can propagate, i.e. $(F(0), F(1), F(2), F(3), F(5), F(6), F(11), F(13)) = (0, 0, 0, 1, 0, 1, 1, 0)$ or $(0, 0, 1, 0, 0, 0, 1, 1)$, respectively. Therefore, the number of admissible rules in our study is 2^{13}, from which we have selected the following types for study.

Type A: The number of solitons increases after a two-body collision.

Type B: The number of solitons remains constant after a two body collision but is increased by multiple collisions.

Type C: There exists a stable nucleus which emits a number of solitons periodically in time.

Type D: There exists a non-propagating localized pattern. This pattern is called a breather since the pattern is usually oscillating in time.

Type E: There exists a giant soliton which emits a lot of breathers or solitons. This giant soliton is called a moving nucleus.

Type F: There exist kink patterns such as observed in the type III of Wolfram's classification.

In general, some of typical excitations mentioned above can coexist in a large system. As the classification A–F is phenomenological, it is plausible to add a last type G, which contains behaviors not included in the above.

Type G: There are no recognizable compact excitations.

We have simulated all 2^{13} rules. Many of these rules belong to types I, II, and III. However, only solitonic rules are studied here.

Table 1
Local rules for soliton turbulence. The propagation of a soliton is guaranteed by ∗-rules.

Coordinate R:						Mapping $F = S_i^{t+1}$									
	$(S_{i-2},$	$S_{i-1},$	$S_i,$	$S_{i+1},$	$S_{i+2})^t$	A(A′)	B(B′)	C_1	C_2	D_1	D_2	E	F	B_3	B_4
∗	(0,	0,	0,	0,	0)	0	0	0	0	0	0	0	0	0	0
∗	(1,	0,	0,	0,	0)	0	0	0	0	0	0	0	0	0	0
∗	(0,	1,	0,	0,	0)	0	1	0	0	0	0	0	0	0	0
	(0,	0,	1,	0,	0)	0	0	0	0	1	1	0	0	0	0
∗	(1,	1,	0,	0,	0)	1	0	1	1	1	1	1	1	1	1
∗	(0,	1,	1,	0,	0)	1	0	1	1	1	1	1	1	1	1
	(1,	0,	0,	0,	1)	1	1	1	1	1	1	1	0	0	0
∗	(1,	0,	1,	0,	0)	0	0	0	0	0	0	0	0	0	0
	(1,	0,	0,	1,	0)	0	1	0	1	0	1	0	1	1	1
	(0,	1,	0,	1,	0)	0	0	1	1	0	0	0	0	0	0
∗	(1,	1,	0,	1,	0)	1	1	1	1	1	1	1	1	1	1
	(1,	1,	0,	0,	1)	1	0(1)	1	0	1	0	0	0	0	1
	(0,	1,	1,	1,	0)	0(1)	1	0	1	0	1	1	0	0	0
∗	(1,	0,	1,	1,	0)	0	1	0	0	0	0	0	0	0	0
	(1,	0,	1,	0,	1)	0	0(1)	1	1	0	0	1	0	0	0
	(1,	1,	1,	0,	0)	1	1	1	1	0	1	1	1	1	0
	(1,	1,	0,	1,	1)	1	1	1	1	1	1	1	1	1	1
	(1,	1,	1,	0,	1)	1	1	0	0	1	0	1	0	0	0
	(1,	1,	1,	1,	0)	0	0	1	0	0	0	0	0	0	0
	(1,	1,	1,	1,	1)	1	0	0	0	0	0	1	1	0	1

Some rules typical of each type (A–F) are shown in table 1.

2.2. Evolution in time

The pattern dynamics generated by each rule in table 1 is illustrated in figs. 1.1–1.9. Each system has 400 sites with periodic boundary conditions. In these figures, black represents the site value 1, and white represents the site value 0. The initial configurations are random.

The time variation of the density (probability of 1) is shown in figs. 2.1–2.7. Here the system size N is 10^4 with periodic boundary conditions.

Fig. 1.1. Model A pattern. The number of solitons increases after the two-body collision of solitons. The vertical axis is site and the horizontal axis is time.

Fig. 1.2. Model B pattern for low-density case. The number of solitons is invariant under the two-body collision.

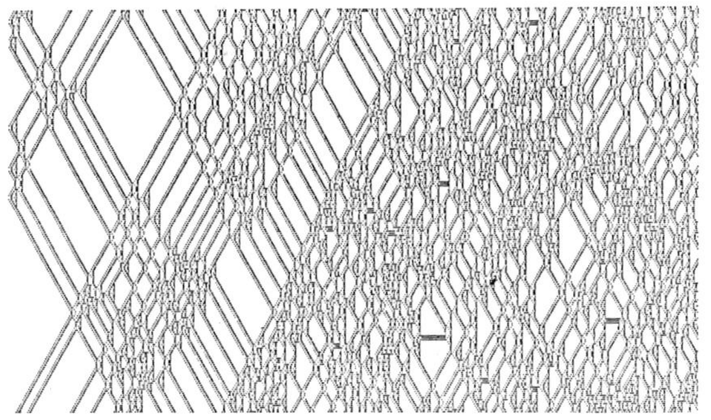

Fig. 1.3. Model B pattern for high-density case. The number of solitons increases under the n-body collision of solitons ($n \geq 3$).

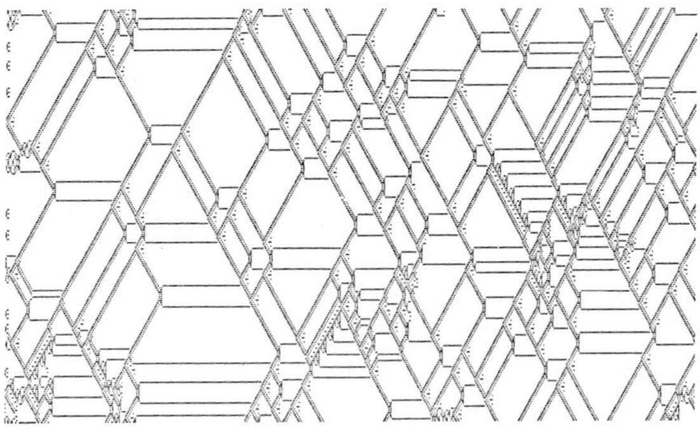

Fig. 1.6. Model D_1 pattern. Simple breathers (or localized patterns) are created after the collision of two solitons, and a moving nucleus is observed.

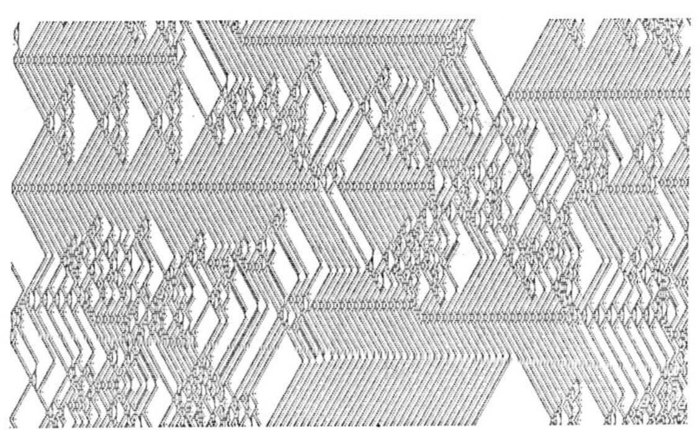

Fig. 1.4. Model C_1 pattern. Soliton chains are created by a nucleus.

Fig. 1.7. Model D_2 pattern. Incomplete kinks are generated after the collision of two solitons.

Fig. 1.5. Model C_2 pattern. Two kinds of solitons are observed. A nucleus emits solitons periodically.

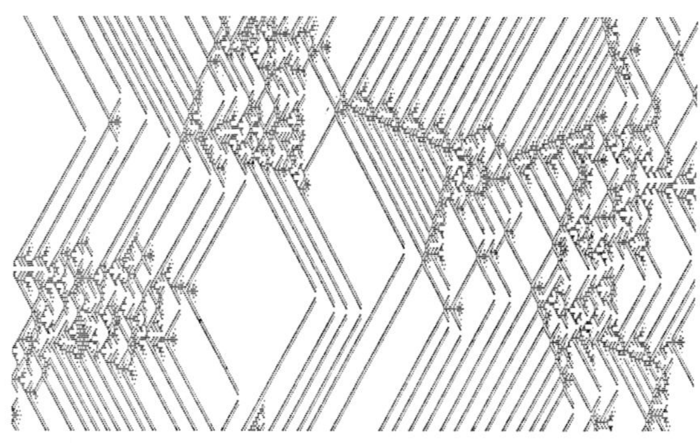

Fig. 1.8. Model E pattern. A giant moving nucleus is observed.

Fig. 1.9. Model F pattern. Kinks, antikinks, and solitons are mutually transformed after their collisions.

The time necessary to reach a stationary state is strongly dependent on the initial condition as well as on the system size.

Model A

The temporal evolution of the model is composed of three stages as shown in fig. 2.1. In the first stage, only a few solitons are created from the initial state and the number of solitons increases very slowly due to 2-body collisions. The total number of solitons is approximately given by (density) $\times N/3$. The second stage exhibits rapid multiplication of solitons due to the many-body collisions. The final stage is stationary turbulence, in which creation and annihilation of solitons is balanced. The equilibrium density is close to 0.5. Many simulations were conducted with different initial conditions. Each of these evolved to the same turbulent state. The behavior of the modified model A′ (see table 1) is almost the same as the behavior of model A.

Model B

In this model, solitons pass through each other in 2-body collisions, as would be the case in an integrable system. However, new solitons can be created by n-body collisions ($n \geq 3$). When the initial density is low enough to permit only 2-body collisions to occur, the density remains almost constant. If the initial density is higher than a certain threshold, however, the density increases due to the generation of solitons in multiple collision. The evolution of the density is then quite similar to model A (fig. 2.2). An important feature of type B is that the phase space of the system is composed of two stable subspaces. One contains a unique turbulent state, and the other contains the almost periodic states which correspond to a dilute gas of solitons. As the initial density increases, the system moves from the latter to the former subspace.

The coexistence of these two subspaces resembles a generic feature of Hamiltonian systems where a subspace of KAM tori coexists with stochastic subspace [2]. The equilibrium density is about 0.4 in the turbulent state. The behaviors of the modified models B′ are nearly the same as in model B itself.

Models C_1, C_2

In these models, one or two kinds of stable nu-

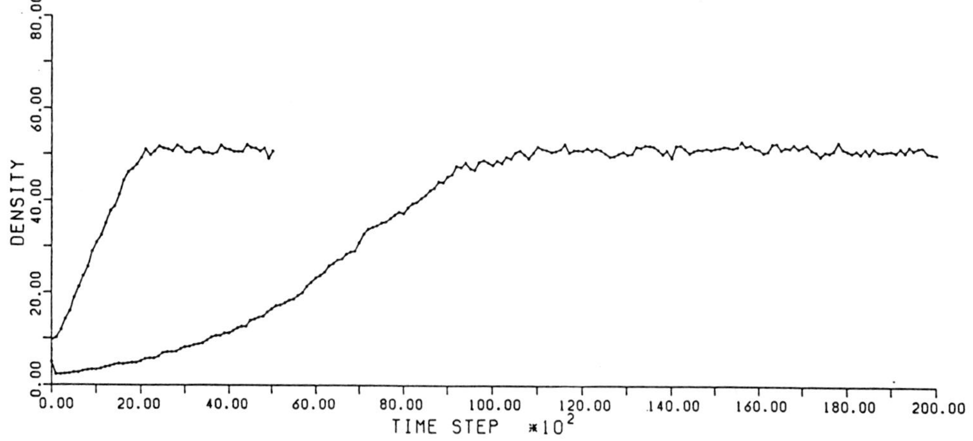

Fig. 2.1. Time courses of the density in model A.

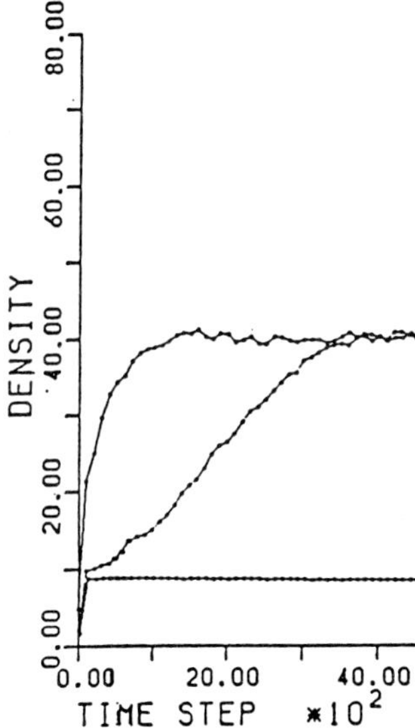

Fig. 2.2. Time courses of the density in model B.

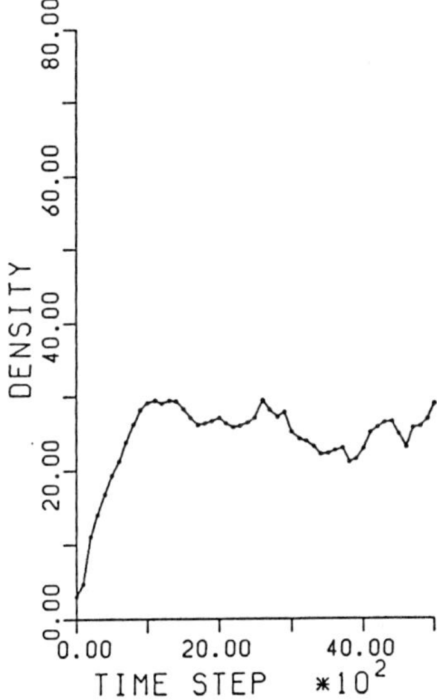

Fig. 2.4. Time courses of the density in model C_2.

cleus are generated by the collision of solitons, followed by the emission of a soliton chain from the nucleus. This system rapidly reaches a turbulent state (figs. 2.3, 2.4). Figs. 1.4, 1.5 illustrate the propagation of a giant soliton. The density undergoes large fluctuations as a result of the creation and the annihilation of nuclei.

Models D_1, D_2

Here, the multiplication of solitons is inhib-

ited by the creation of a breather. Solitons and breathers are statistically balanced in the turbulent state. A pair of breathers are generated after 2-body collisions of solitons (figs. 1.6 and 2.5). Furthermore, in this figure one can see a moving nucleus, or a giant soliton, which emits the breathers periodically. Fig. 1.7 corresponding to fig. 2.6 reveals the coexistence of breathers, solitons, and incomplete kinks, where a turbulent attractor seems to coexist with regular basins.

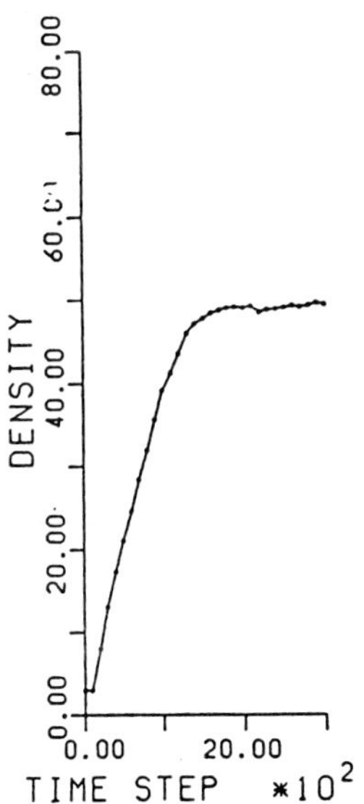

Fig. 2.3. Time courses of the density in model C_1.

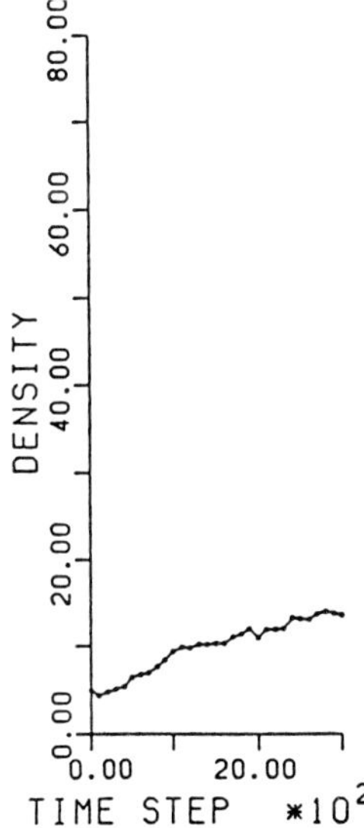

Fig. 2.5. Time courses of the density in model D_1.

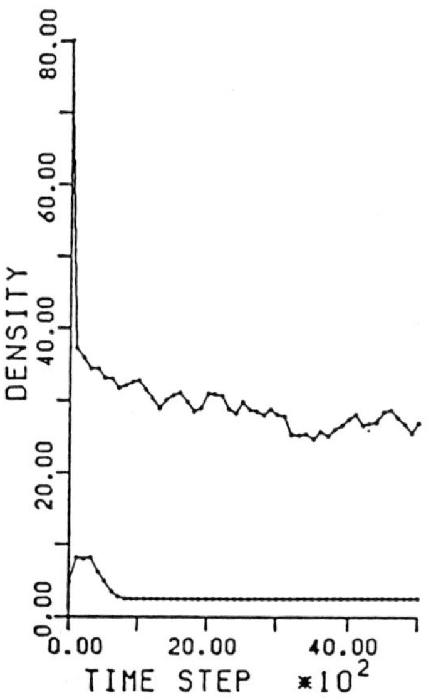

Fig. 2.6. Time courses of the density in model D_2.

Model E

In this model, a moving nucleus is generated after a three-soliton collision, which then emits simple soliton chains bi-laterally. The emission frequency is different in the two directions, a sort of Doppler effect (fig. 1.8). The density fluctuates very slowly and requires a long time to reach equilibrium.

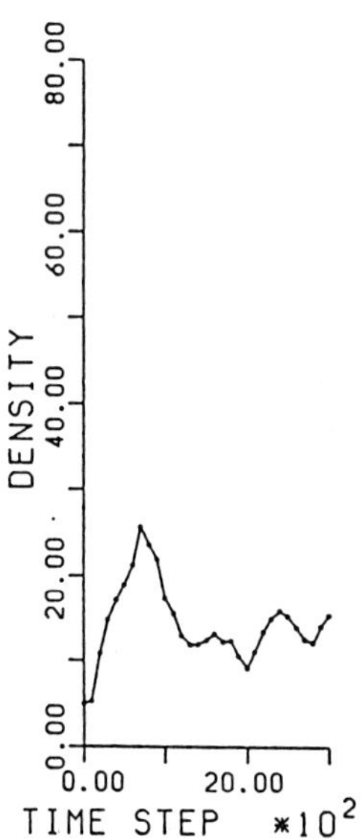

Fig. 2.7. Time courses of the density in model E.

Model F

In model F, solitons, kinks, and antikinks coexist stably. A kink has the structure $\cdots 1110 \cdots$, while an antikink has the structure $\cdots 010111 \cdots$. A soliton–kink collision generates a soliton–antikink pair, and a kink–kink collision produces an antikink pair. Inversely, the collision of two antikinks create a kink pair. The irreversibility of this model comes from the conversion of a pair of antikinks into a quasi-periodic pattern. This is shown in fig. 1.9.

Other types of kinks and antikinks are easily created by only slight changes in the CA rule. For example, $\cdots 0110101 \cdots$ or $\cdots 01101010 \cdots$ can propagate if $F(10)=1$ and $F(21)=0$. In addition, a soliton-train, or a soliton-chain, e.g. $\cdots 011010011010 \cdots$, is a kind of kink, which is always generated by a nucleus as shown in figs. 1.4, 1.5.

3. Phase sensitivity of collision patterns

As shown in the previous section, a variety of patterns are created by the interplay of fundamental excitations. Global randomization of patterns is induced by the collision cascade, which propagates local turbulence through the entire space. In this section we examine the early stage of pattern randomization in model A from the viewpoint of phase sensitivity of collisions.

There are two phases in two-soliton collisions, that is, an even phase ($\cdots 010110011010 \cdots$) and an odd phase ($\cdots 0101100011010 \cdots$). Therefore, the number of possible phases in the collision of two n-soliton chains is bounded below by 2^{2n-1}. Small perturbations in a soliton-chain are amplified by multiple collisions. Fig. 3.1 shows a few examples of collision patterns for $n = 5$. The initial distances between neighboring solitons in each chain are 4 or 5 (lattice units). Not only the phase, but the number of solitons depends sensitively on the initial conditions. Fig. 3.2 shows the distribution of the total number of solitons in the second generation, where all the initial distances of solitons are taken to be 4 or 5 (lattice units) with equal probability while the initial number of solitons is fixed at 10 ($= 2n$). Thus the variance of distribution of number of solitons in the out-going

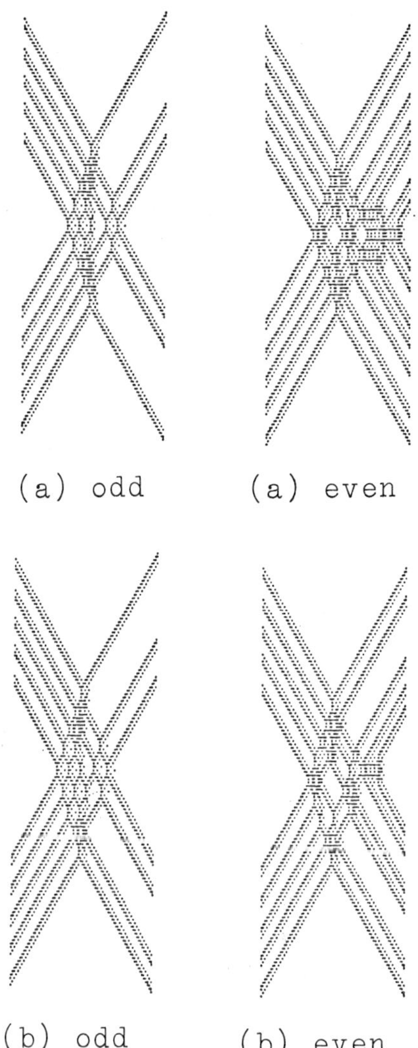

Fig. 3.1. Phase sensitivity of the collision processes in model A. (a) $S_1 \int^5 S_1 \int^4 S_1 \int^4 S_1 \int^5 S_1$ odd / even $S_1 \int^5 S_1 \int^4 S_1 \int^4 S_1 \int^5 S_1$; (b) $S_1 \int^5 S_1 \int^4 S_1 \int^4 S_1 \int^5 S_1$ odd / even $S_1 \int^5 S_1 \int^5 S_1 \int^4 S_1 \int^5 S_1$. $\int^n$ denotes the n blank sites.

pattern can be used as a measure of the phase sensitivity. That the average number of solitons in the next generation is larger than the initial number (approximately 12.5 in this case) implies that solitons are multiplied probabilistically by collisions.

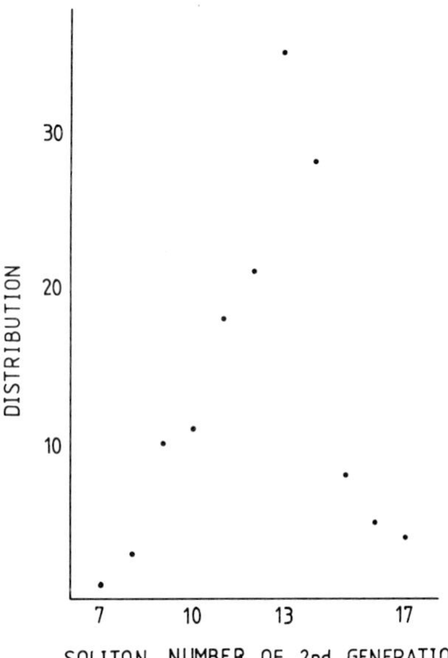

Fig. 3.2. Distribution function of the soliton number in the second generation after collisions. Ensemble is explained in the text.

4. Spectra of turbulence

The spatial correlation function $C(l)$ is defined by

$$C(l) = \frac{\langle S_{n+l}^t S_n^t \rangle - \langle S_n^t \rangle^2}{\langle S_n^{t\,2} \rangle - \langle S_n^t \rangle^2}, \tag{4.1}$$

where $\langle X_n \rangle$ is a spatial average:

$$\langle X_n \rangle = N^{-1} \sum_{n=1}^{N} X_n.$$

The spatial power spectral density (PSD) function is

$$\tilde{P}(k,t) = \left\langle \left| \sum_{m=1}^{L} e^{-i2\pi mk/L} S_l^t \right|^2 \right\rangle, \tag{4.2}$$

where $l = (n-1)L + m$ (n is a positive integer). Here L is the sample length and $\langle X_n \rangle$ is the sample average: $\langle X_n \rangle = M^{-1} \sum_{n=1}^{M} X_n$, and M is the largest integer satisfying $ML \leq N$. In the following, the system size is $N = 10^4$ and the sample length is $L = 2^{10}$. In order to measure mean tran-

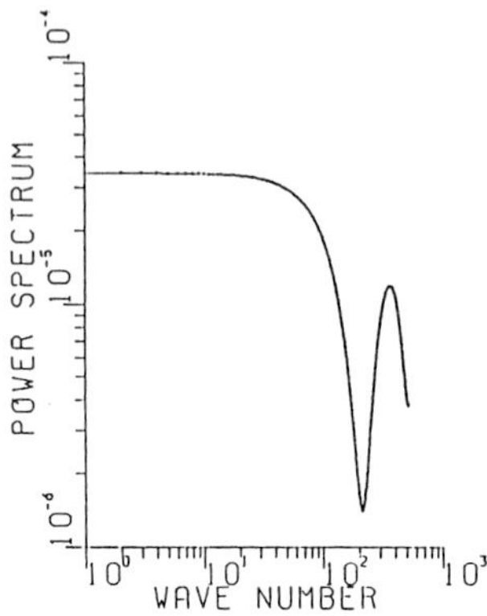

Fig. 4.1. The PSD function of one soliton.

sient behaviors, the PSD function is averaged over 500 time steps,

$$P(k,T) = \frac{1}{500} \sum_{t=1}^{500} \tilde{P}(k, T+t). \qquad (4.3)$$

The normalized correlation function and the PSD function for a single soliton $(\cdots 010110 \cdots)$ are approximately estimated as

$$C(l) \simeq \frac{1}{3}\left(2\delta_{l,0} + \sum_{j=0}^{3} \delta_{l,j} \right),$$

$$P(k) \simeq 1 + \frac{1}{3} \sum_{j=1}^{3} \cos\left(\frac{2\pi}{L} jk \right). \qquad (4.4)$$

Fig. 4.1 shows the above PSD function for a single soliton. The dominant peak around $k \simeq 0$ is due to the localized pattern itself. The second peak at $k = L/3$ $(\simeq 341)$ indicates a typical soliton-mode which characterizes the internal structure of the soliton [#1].

In general, propagation and collision of many solitons creates long-range long-time order in the system. The existence of such coherent structures is usually characterized by a power law decay of the correlation function: $C(l) \sim l^{-\beta}$, with a simultaneous power law decay of the PSD: $P(k) \sim k^{-\nu}$ (apart from logarithmic corrections). These singular behaviors are called infrared anomalies. The Wiener–Khinchin theorem implies that at station-

---

[#1] Although wavenumber is defined as k/L, henceforth k will be referred to as the wavenumber.

arity $(\nu < 1)$, $\beta = 1 - \nu$. However in the non-stationary case $(\nu \geq 1)$ $C(l)$ no longer decays as a power law, and more unusual relaxation occurs. In a statistical sense non-stationarity breaks translation-invariance symmetry.

The Allan variance supplies one of the most useful characterizations of the non-stationary regime. The Allan variance, $\sigma_A^2(n)$, is defined by

$$\sigma_A^2(n) = \frac{1}{2} \left\langle\!\left\langle \left(\frac{1}{n}\sum_{i=1}^{n} S_{m+i}^t - \frac{1}{n}\sum_{i=1}^{n} S_{m+n+i}^t \right)^2 \right\rangle\!\right\rangle, \qquad (4.5)$$

where $\langle\!\langle \ldots \rangle\!\rangle$ is the average over m and t [7]. It is well known that $\sigma_A^2(n) \sim n^\tau$ $(n \gg 1)$ with $\tau = \nu - 1$. The plateau for $\tau = 0$ is called the flicker floor corresponding to the typical k inverse, and the normal Gaussian noise reveals $\tau = -1$. The anomaly of the index $(\tau > -1)$, in general, represents the breakdown of the weak law of large numbers.

Model A

The change of the transient PSD is shown in fig. 4.2, where the initial density is 0.05. In the early stage $(T < 500)$, a Lorentzian spectrum appears in the large-scale structure. Then the soliton mode is quickly excited. In the intermediate stage $(T < 7500)$ the intensity of the soliton mode increases due to soliton multiplication, and a characteristic infrared anomaly rapidly appears: $P(k,T) \sim k^{-\nu(T)}$. This anomalous enhancement of large-scale structure is gradually suppressed as stationary turbulence is achieved in the final stage, $T > 20000$. It seems, however, that a weak anomaly remains even in the final equilibrium stage. Indeed, the anomaly index is $\nu \simeq 1/7$ at $T > 20000$.

Since a soliton mode is clearly observed even in the final stage, it is reasonable to call this stage soliton turbulence. The appearance of the transient infrared anomaly suggests that the collision cascade of solitons is self-similar as in the case of the Kolmogorov vortex cascade in the inertial regime in the fully developed fluid turbulence. In soliton turbulence, excitation energy flows from the soliton mode to low-wavenumber modes through the collision processes, the reverse of the energy flow in the Kolmogorov cascade in

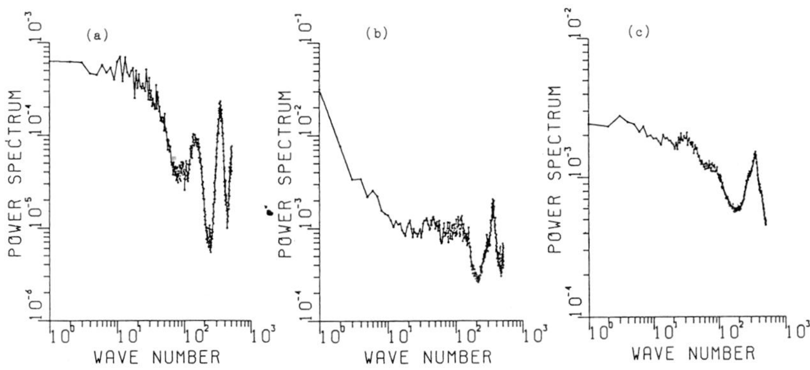

Fig. 4.2. $P(T, k)$ for model A. (a) $T = 500$, (b) $T = 7500$, (c) $T = 20000$.

fluid turbulence. Fig. 4.3 shows the Allan variance at time T. The general relation $\tau = \nu - 1$ seems to be satisfied even in the equilibrium state.

The modified model A$'$ reveals almost the same spectral behaviors as model A.

Model B

When the soliton density is low enough, the PSD is a simple superposition of pure soliton spectra (see fig. 4.1). In this regime, many-body collisions do not occur and solitons are randomly distributed in space. On the other hand, when soliton turbulence develops, a typical infrared anomaly is generated as discussed for model A (fig. 4.4). The anomaly index is $\nu \simeq 1/9$ in the turbulent equilibrium. The corresponding Allan variance is $\sigma_A^2(n) \simeq n^\tau$ with $\tau = \nu - 1$ (fig. 4.5). Qualita-

tively same results are obtained for the modified models B$'$.

Models C_1, C_2

In these models, long-range correlations are rapidly generated due to the appearance of nuclei and kinks (fig. 4.6). The transient infrared anomaly of model C$_1$ becomes $P(k, T) \simeq k^{-2}$ and $\sigma_A^2(n) \simeq n^1$ at $T = 500$. At a certain time the large-scale structure disappears and the spectrum becomes nearly white in the low-wavenumber regime, while sharp lines remain at the soliton mode wavenumber and its $1/2$ subharmonic. The line spectrum at $k \simeq 512$ reveals the internal structure of the nucleus. This chaotic state may be called the soliton–nucleus turbulence.

In model C$_2$, the wavelength of the soliton-chain

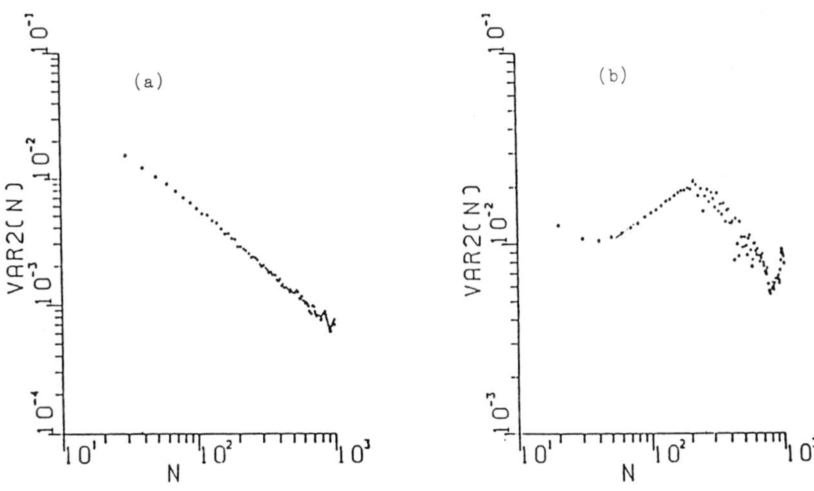

Fig. 4.3. $\sigma_A^2(N)$ for model A. (a) $T = 5500$, (b) $T = 11500$.

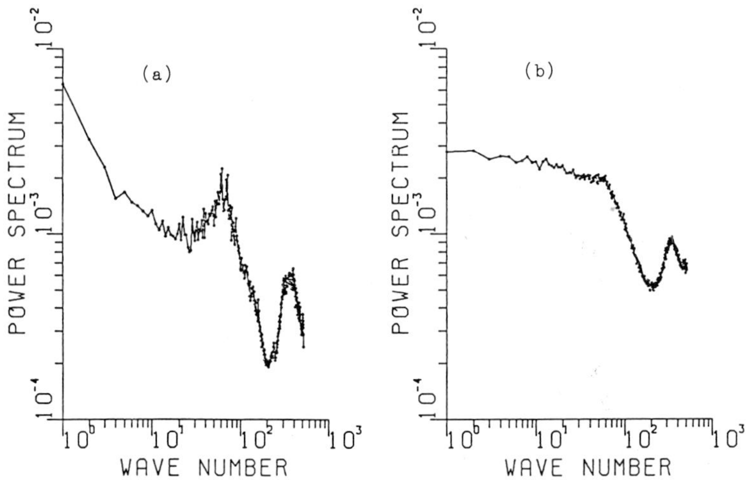

Fig. 4.4. $P(T, k)$ for model B. (a) $T = 1000$, (b) $T = 4000$.

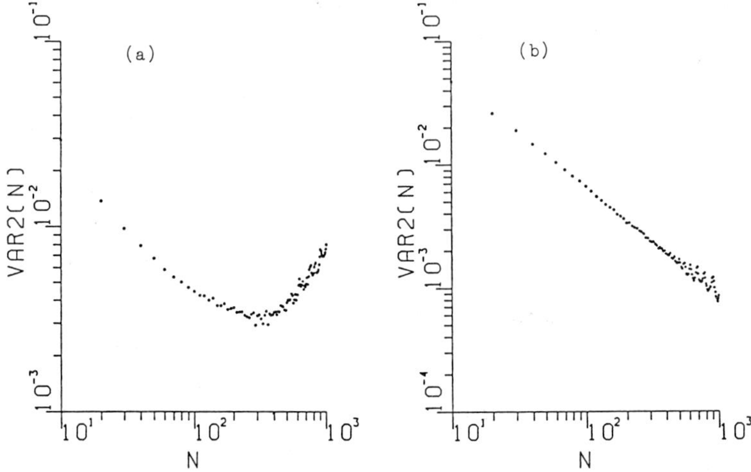

Fig. 4.5. $\sigma_A^2(N)$ for model B. (a) $T = 1000$, (b) $T = 4000$.

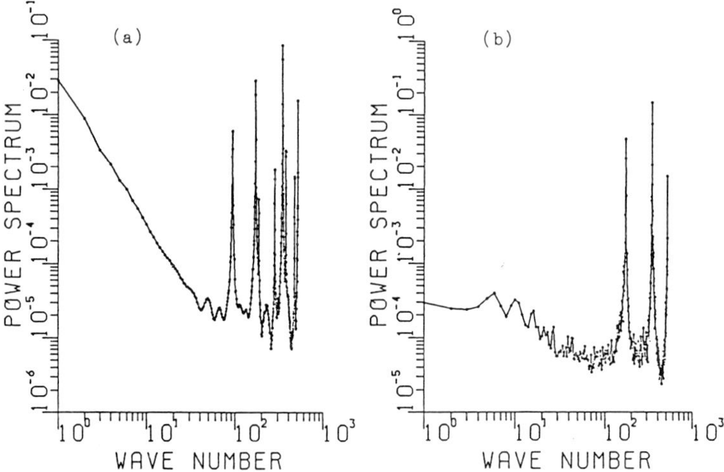

Fig. 4.6. $P(T, k)$ for model C$_1$. (a) $T = 500$, (b) $T = 7500$.

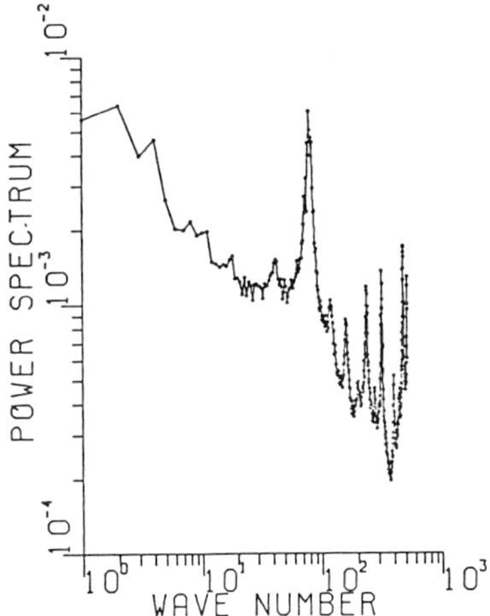

Fig. 4.7. $P(T, k)$ for model C_2 at $T = 4500$.

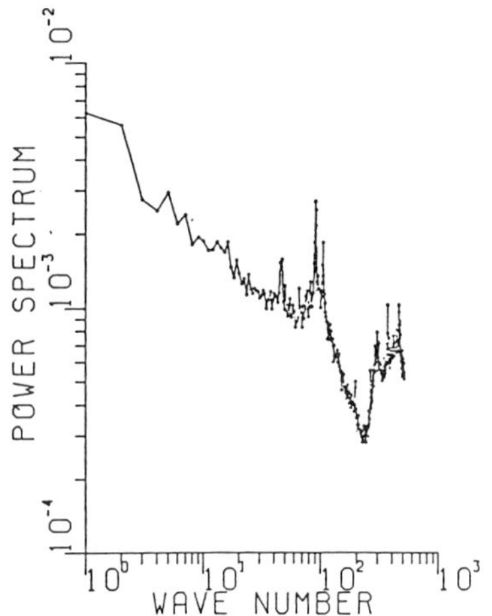

Fig. 4.9. $P(T, k)$ for model D_2 at $T = 4500$.

is twice that of model C_1. Therefore a sharp line appears at the 1/4 subharmonic of the soliton mode (fig. 4.7). Other sharp lines apparently correspond to the nucleus. While the transient has a strong anomaly, the equilibrium anomaly index is only approximately $\nu \simeq 2/3$.

Models D_1, D_2

The only turbulent state of model D_1 is soliton–breather turbulence (fig. 1.6). The equilibrium anomaly indices are $\nu \simeq 1/2$ and $\tau \simeq -1/2$ (fig. 4.8). Unlike models C_1 and C_2 above, transient anomaly enhancement is not observed. Three typical broad peaks are observed at $k \simeq 341$, 100, and 50. The latter two peaks describe the breather pair and/or the soliton-chain.

In model D_2, the breather is unstable and the turbulence is mainly sustained by soliton–soliton collisions (fig. 1.7). Here, the equilibrium anomaly is approximately $\nu \simeq 7/12$ if it exists at all (fig. 4.9).

Model E

It is quite difficult to see the internal structure of a moving nucleus from the spectra of model E (fig. 4.10). In this model, giant solitons with very small velocity are randomly distributed in space. Therefore, the spectrum is expected to be Lorentzian in the low-wavenumber regime. Indeed, the Allan variance reveals a flicker floor ($\tau \simeq 0$) in the intermediate scale ($n < 500$). However in the large scale ($500 < n < 1000$) it appears to decay

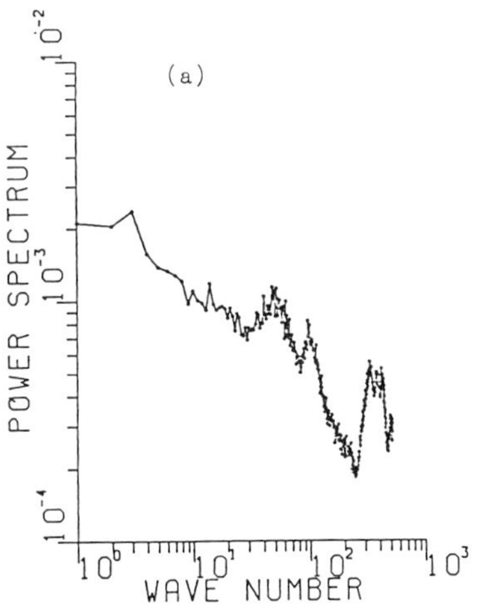
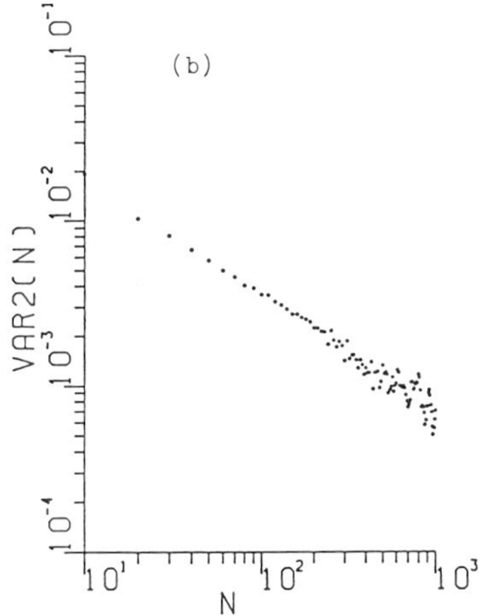

Fig. 4.8. $P(T, k)$ and $\sigma_A^2(N)$ for model D_1 at $T = 2500$; (a) $P(T, k)$, (b) $\sigma_A^2(N)$.

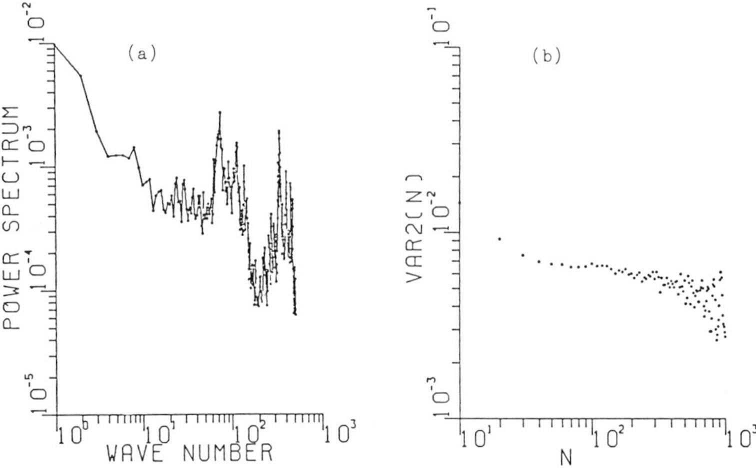

Fig. 4.10. $P(T,k)$ and $\sigma_{\mathrm{A}}^2(N)$ for model E at $T = 2500$; (a) $P(T,k)$, (b) $\sigma_{\mathrm{A}}^2(N)$.

as with an exponent $\tau \simeq -1$, which corresponds to Gaussian noise.

5. Free motion of a soliton and information flow

A soliton has the ability to act as an information carrier. If the soliton is stable under perturbations or collisions, it can transfer exact information without any loss, that is, without production of entropy. On the other hand, in soliton turbulence information can always be created by pattern randomization due to the phase sensitivity of collisions. In this section, we investigate the mechanism of solitonic information transfer.

The co-moving mutual information flow (COMIF) is defined as follows. First, we take an n-word at i-site: $\tau = \{S_i^k, S_i^{k+1}, \ldots, S_i^{k+n-1}\}$, and calculate the probability $P_i(\tau)$ of the word τ. Next, we introduce the conditional probability function $P_{i \to j}(\tau_1, \tau_2; v)$. For the two sequential words τ_1 and τ_2,

$$\tau_1 = \{S_i^k, S_i^{k+1}, \ldots, S_i^{k+n-1}\},$$
$$\tau_2 = \{S_j^l, S_j^{l+1}, \ldots, S_j^{l+n-1}\}, \qquad (5.1)$$

$P_{i \to j}(\tau_1, \tau_2; v)$ is defined as the joint probability that a site i takes the word τ_1 for time steps k to $k + n - 1$ *and* the site j takes the word τ_2 for delayed time steps l to $l+n-1$. The delay between

the two is given by $l = k + [|\,j - i\,|\,/v]$, with $[x]$ for the largest integer not greater than x. In other words v is the speed of the running coordinate. Finally, the COMIF $I(i \to j; v)$ is defined by

$$I(i \to j; v)$$
$$= \sum_{\tau} P_i(\tau) \log P_i(\tau) + \sum_{\tau} P_j(\tau) \log P_j(\tau)$$
$$- \sum_{\tau} \sum_{\sigma} P_{i \to j}(\tau, \sigma; v) \log P_{i \to j}(\tau, \sigma; v), \qquad (5.2)$$

where each summation is taken over all the possible n-words. COMIF is the information transferred from the i-site to the j-site with a constant velocity v [8].

Three models (A, C_2 and D_1) are studied here for the system with $N = 10^4$ and $n = 6$.

Model A

The COMIF has a peak at $1/v \simeq 0.6$. Hence the speed of the information carrier, which is the soliton in our system, is about 1.67 (lattice units) per time step. Since the soliton is accelerated by the collision, this speed is larger than that of a single soliton (see fig. 1.1). The COMIF decays very rapidly as the relative distance $|j - i|$ increases.

Model C_2

A sharp peak is observed at $v \simeq 1$ at large distance. This implies that a soliton, once emitted by a nucleus, can propagate at a constant speed for long time without collision. The peak height of

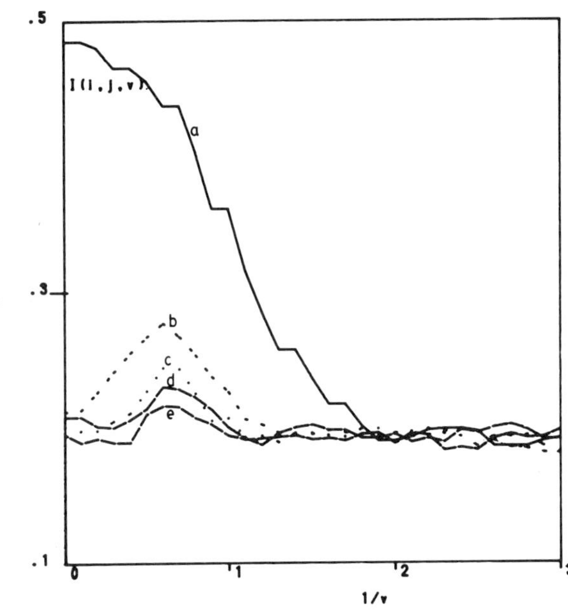

Fig. 5.1. COMIF for model A. (a) through (e) correspond to $|i - j| = 8, 13, 18, 23,$ and 28.

the COMIF decays as $I(i \to j, v) \simeq |i - j|^{-\alpha}$ with $0.7 \leq \alpha \leq 1$.

Model D_1

The effective speed is $v \simeq 1$ for long distance

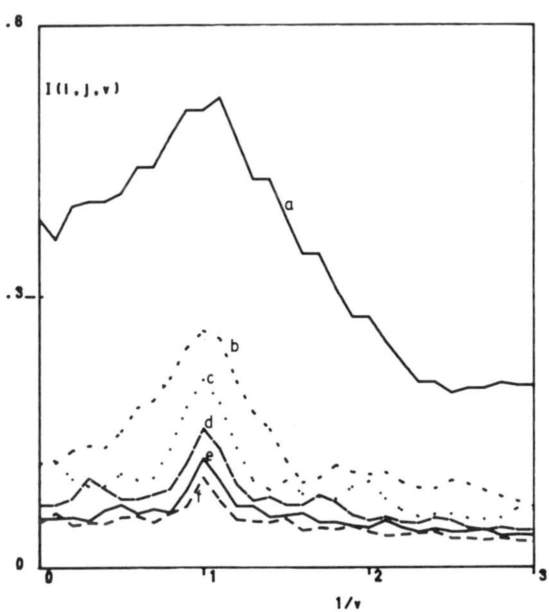

Fig. 5.2. COMIF for model C_2. (a) through (f) correspond to $|i - j| = 8, 18, 28, 38, 48,$ and 58.

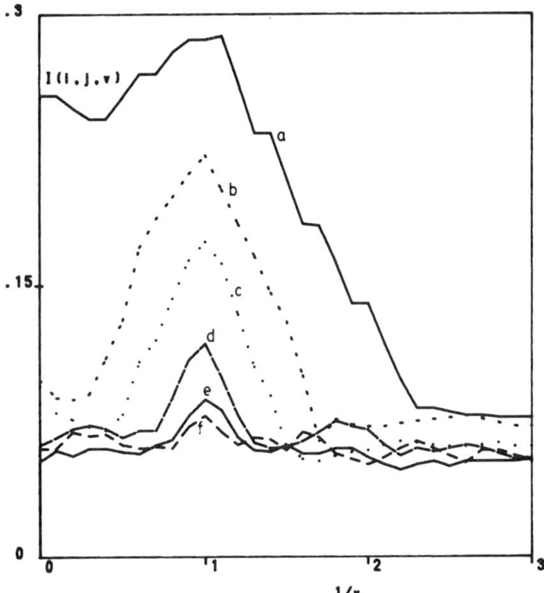

Fig. 5.3. COMIF for model D_1. (a) through (f) correspond to $|i - j| = 8, 13, 18, 23, 28, 38,$ and 48.

$|i - j| \leq 40$), and the COMIF shows the power decay also in this model with $0.8 \leq \alpha \leq 1.2$.

It is remarkable that the local information of only 6 bits can be clearly detected at the receptor point far from the emitter point in spite of the large information loss. In a sense, the COMIF indicates the intensity of the informational correlation. Therefore, the slow decay of COMIF in the model C_2 and D_1 is consistent with the observation of the infrared anomalies discussed in section 4. The free running time τ_f of a soliton is also believed to obey an inverse power law distribution.

6. Attractor of soliton turbulence

As long as the CA system size is finite, a periodic attractor will eventually be reached. An interesting question arises when the size is very large, but not infinite. That is, one may ask whether the CA turbulence observed above is a complex attractor or a complex transient motion approaching a simple attractor. In this section we study the dependence of attractors and their basins of attraction on system size. These studies focus on model B. Model B has many simple attractors in which only two-soliton collisions occur. Here,

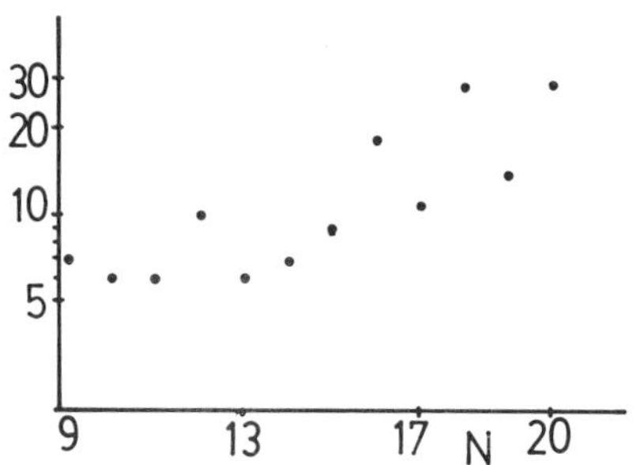

Fig. 6.1. Exponential growth of the number of attractors (model B).

such attractors are simply called regular attractors. In the same way, other attractors such as n-body collision ($n \geq 3$) etc. are called non-regular. The basins of regular attractors are called regular basins and so on.

In a regular attractor, the number of solitons is conserved. The motion of a soliton is quite similar to that of integrable Hamiltonian systems such as Toda lattice, although time reversibility is not satisfied in our CA system. Fig. 6.1 shows the exponential size dependence of the total number of attractors. Here attractors related by a translation

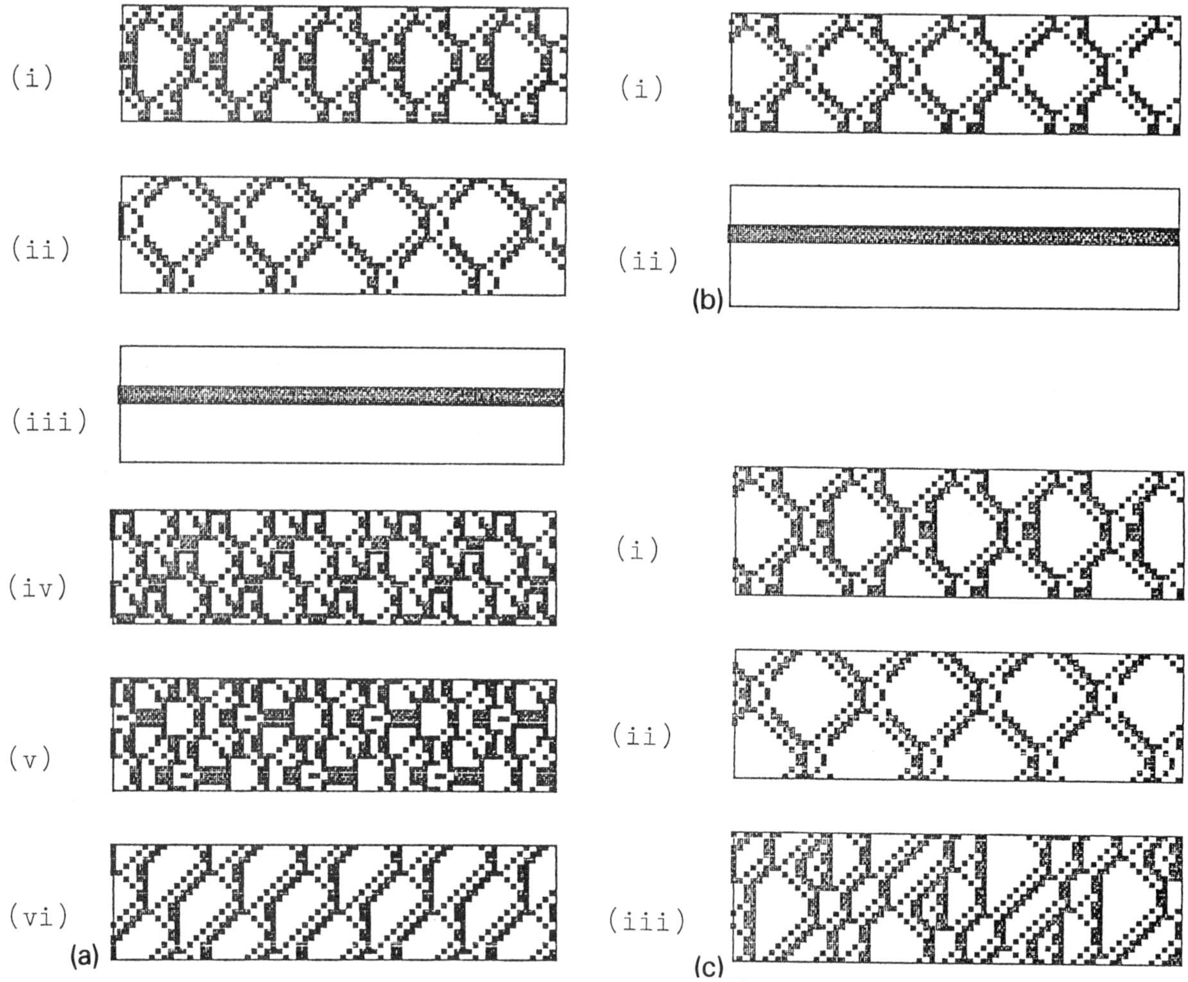

Fig. 6.2. (a) Dominant attractor patterns and their basin volume (%). System size $N = 18$. (i) 33%, (ii) 28.6%, (iii) 13%, (iv) 2.9%, (v) 1.9%, (vi) 1.8%. (b) Dominant attractor patterns and their basin volume (%). System size $N = 19$. (i) 78.7%, (ii) 10.8%. (c) Dominant attractor patterns and their basin volume (%). System size $N = 20$. (i) 57.4%, (ii) 16.2%, (iii) 1.2%.

are identified. The same exponential growth is observed in other CA systems studied in this paper. All of these attractors are very simple, hence we have not found complex attractors which would correspond to turbulence. Some of the attractors are illustrated in figs. 6.2. These have relatively large basin volume. The basin of each attractor seems to be quite intricately intertwinned. Here one can see that the volume fraction of the regular basins is very large, e.g. 0.75, 0.79, and 0.74 for $N = 18$, 19, and 20, respectively. If these features persist for even large systems, $N \gg 20$, where can we discover turbulent behavior? It must be in the transients to these attractors. Fig. 6.3 shows an example of such a chaotic transient for $N = 50$. After a long transient, the pattern settles down onto a regular attractor. This kind of long chaotic transients is also observed in the basins of other simple attractors [9]. Fig. 6.4 shows the exponential size dependence of the average transient time T_t, obtained from 50 randomly chosen initial conditions. The data are roughly fit by $T_t \propto 10^{0.02N}$. Therefore, the turbulent transient state can continue for an extremely long time, if the system size is large. For instance, $T_t \simeq 10^{200}$ for $N = 10^4$.

The above results strongly suggest that soliton turbulence is not an attractor but rather a transient state. We have to solve another question: By what mechanism, then, is the system kept almost stationary for a such a long time? Although we have no definitive answer to this yet, the following explanation is suggested. As discussed in section 3, instability is always generated in the local regions where sensitive collisions occur (see fig. 1.3). Local turbulence disappears quickly as soliton-chains are

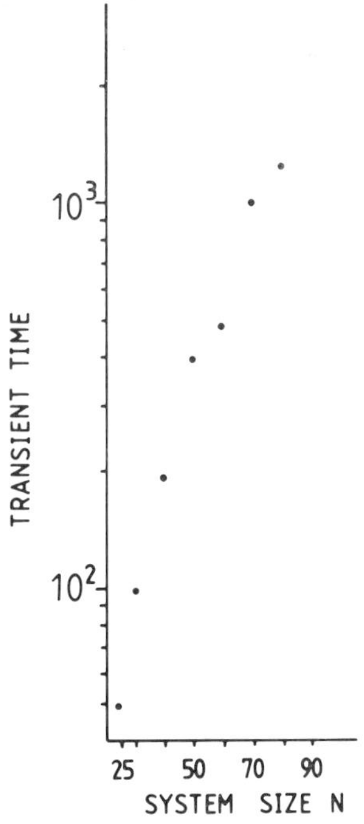

Fig. 6.4. System size dependence of transient chaotic time (model B).

emitted in the second generation. These solitons then propagate and create new local turbulence by collisions. Global turbulence is maintained by such alternation of generation. Moreover, global turbulence is the random superposition of such local turbulence. Therefore, even if the time necessary to realize such local equilibrium is relatively short in general, the length of time to reach equilibrium is very long. It takes a long time for every local region to simultaneously reach a phase sta-

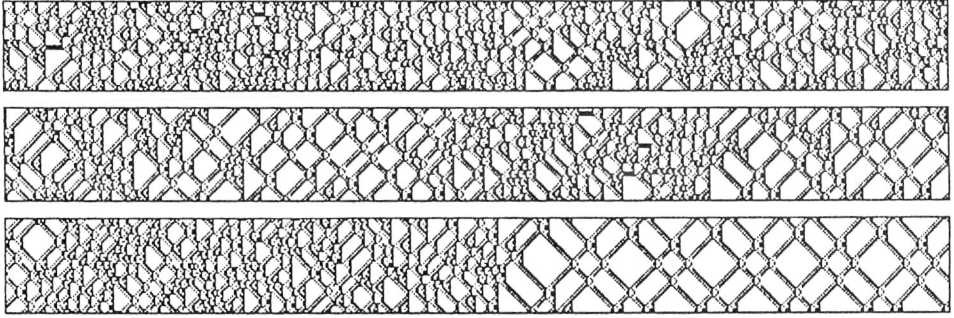

Fig. 6.3. An example of chaotic transient for $N = 50$ (model B).

ble against collisions. In other words, the soliton turbulence studied in this paper is not a true equilibrium state in the large, but is a superposition of local turbulence distributed in parallel.

The possibility that turbulence in general is not an attractor but rather a transient state has been discussed in ref. [10]. In that paper some classes of turbulent behavior in coupled map lattices [11] have been identified as type II supertransient turbulence. The type II supertransient is characterized by the exponential (or faster) divergence of transient length with system size and the existence of a quasi-stationary measure. We suggest that the CA soliton turbulence studied here belongs to this class of supertransient turbulence.

7. Discussions and summary

7.1. Origin of the long-range order in CA turbulence

In hydrodynamic turbulence the energy spectrum often decays as an inverse power law in the wave number, $k^{-\nu}$. The Kolmogorov spectrum has a power $\nu = 5/3$ for three-dimensional flows and $\nu = 3$ or 4 for two-dimensional flows [12]. A cascade process is the origin of these kinds of long-range order. In a cascade process a large-scale eddy is decomposed into smaller eddies. In two dimensions, the enstrophy, the strength of vortex, cascades inversely to the energy cascade. The self-similarity laws are in any case self-organized in each inertial regime, due to the nonlinearity of Navier–Stokes equation.

In CA models, we observe phenomena similar to an energy cascade. To see this, we review the typical stages of the development of soliton turbulence in CA.

(i) First stage: the growth of coherent structures. The number of solitons increases due to the collision of solitons as well as by the emission of soliton-chains from nuclei. This stage is dominant when the soliton gas is dilute. In this case the soliton-chain propagates exactly like a kink solution described by

$$S_i^t \simeq \Theta(n + ct) - \Theta(n - ct), \tag{7.1}$$

where $c \, (= 1)$ is the velocity of the kink and $\Theta(\cdot)$ is the Heaviside function. The power spectrum of such a traveling soliton-chain $S(k,t)$ is approximated by

$$
S(k,t)
$$
$$
\simeq \left(\sum_{-L/2}^{L/2} [\Theta(n - ct) - \Theta(n + ct)] \, e^{-i2\pi kn/L} \right)^2
$$
$$
\simeq o(k^{-2}), \tag{7.2}
$$

when k is small $(L/ct < k \ll L)$. This explains the occurrence of the large-scale similarity which is observed typically in figs. 4.2b, 4.4a, and 4.6a, though the spectral indices are not exactly 2. The essential feature of this stage is the formation of large-scale structure, which is always accompanied by the enhancement of the soliton mode and its subharmonics.

(ii) Second stage: the formation of local turbulent clusters. As the density of the soliton gas becomes high, the coherent structures created in the previous stage are destroyed by multiple collisions with other excitations, and complex clusters are generated. The spectral power in the low-wavenumber regime gradually decreases, though the intensity of soliton modes remains almost constant. This process is also observed in figs. 4.2c, 4.4b, and 4.6b. As a result, in this stage there coexist two different phases: one is the coherent phase which consists of regular solitons and the other is a phase of complex clusters created by many soliton collisions. The complex clusters are considered to be a kind of resonance state generated by soliton aggregates. These can be called local turbulent clusters. These local turbulent clusters break up easily due to their internal instability to create stable elementary excitations such as solitons, nuclei and breathers. The fine structure of local turbulence is characterized by line spectra in the high-wavenumber regime.

In the equilibrium state of global turbulence, both stages mentioned above are balanced in a statistical sense. Although the propagation of solitons is partially interrupted by the creation of local turbulence, a phase coherence of in-coming solitons can be maintained within the out-going solitons emitted by local turbulence. Such long-time memory in the propagation process of solitons is confirmed in section 5, where the existence of long-range correlation is discussed in terms of the propagation of mutual information. As discussed previously, the time scale necessary to establish global turbulence is rather short since global turbulence is simply a superposition of the local turbulence in these systems. On the other hand, global turbulence survives for an extremely long period before the system collapses onto a simple attractor by matching phases among neighboring turbulent clusters.

7.2. Comparison with solitons in Hamiltonian systems

Some of the CA solitons discussed in this paper can be compared with those in Hamiltonian systems, e.g. solitons in Toda lattice [13]. Systems of type B are especially good examples of integrable-like behaviors, that is, two solitons which pass through after collision. Such motions are observed in model B_3, as shown in figs. 7.1a and 7.1b, even if a nucleus exists in particular initial conditions. When the number density of solitons is low enough the system clearly exhibits regular behavior, while weak turbulence develops at high density (see fig. 7.2).

The power spectrum in the low-wavenumber region seems to obey a $k^{-\nu}$ decay with $\nu \simeq 1$ for the low-density case. A Lorentzian spectrum ($\nu \simeq 0$) is observed for the high-density turbulent state (figs. 7.3a and 7.3b). This stochastic transition

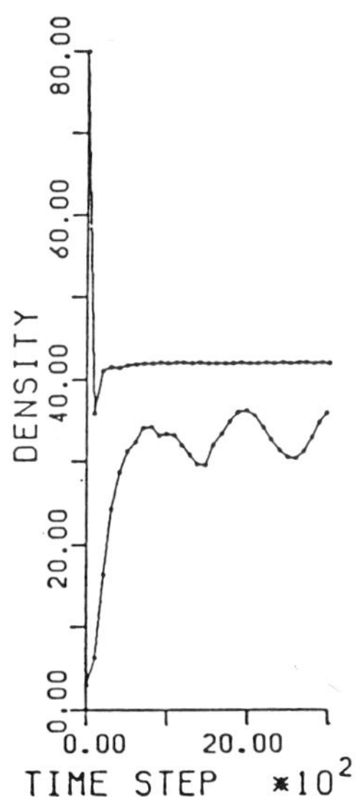

Fig. 7.2. Time courses of the density in model B_3.

from a regular to turbulent state strongly resembles Hamiltonian systems, which exhibit the coexistence in phase space of KAM tori with chaotic sea [2]. Our results give the first illustration of a stochastic transition in spatially extended dissipative dynamical systems.

Model B_3 exhibits strong integrability, where the final equilibrium states, or attractors, are always composed of integrable-like solitons. Turbulent clusters are destroyed in finite time (fig. 7.2). This behavior can be termed "integrability on an attractor". In model B_4, a modification of model B_3, a stable breather pattern confines and reflects

Fig. 7.1. (a) Model B_3 pattern for low-density case. The number of solitons is invariant under the two-body collision. (b) Model B_3 pattern for high-density case. The integrable-like behavior is slightly disturbed by the creation of nuclei, but the asymptotic aspect seems to tend the integrable-like state.

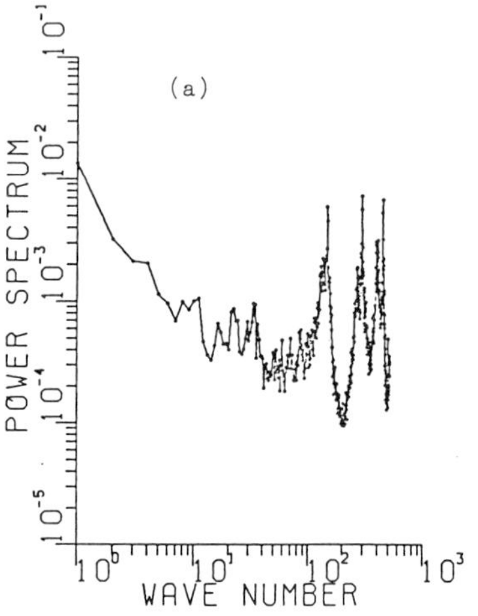

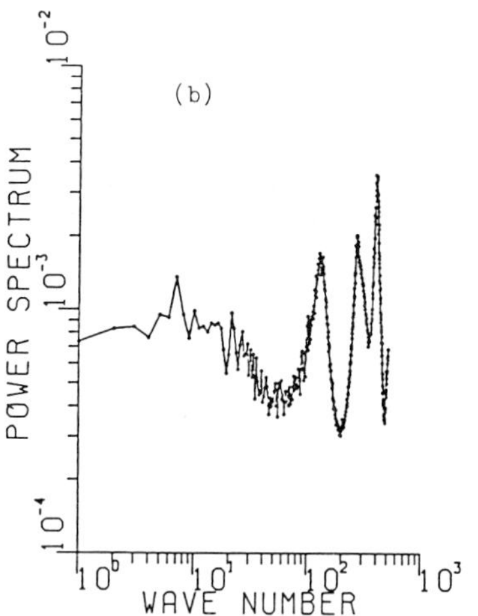

Fig. 7.3. $P(T, k)$ for model B_3. (a) $T = 500$, (b) $T = 2500$.

Fig. 7.4. Model B_4 pattern.

solitons in a box, where all the solitons exhibit integrability in the above sense (fig. 7.4).

This paper only treats the phenomenology of CA solitons. A theoretical description has not yet been developed [#2]. One of the most difficult problems here is that almost all perturbation approaches break down in the study of CA systems since an appropriate continuous metric does not exist. Soliton turbulence seems to reveal an essential difference between differentiable PDE systems and non-differentiable systems. Algebraic approaches should be developed in future to elucidate the diversity of CA systems.

Acknowledgements

The authors thank Dr. H.A. Gutowitz for his useful comments and critical reading of the manuscript.

[#2] Ref. [14] treats a special type of solitons.

References

[1] Y. Kuramoto, Chemical Oscillations, Waves, and Turbulence (Springer, Berlin, 1985).

[2] A.J. Lichtenberg and M.A. Lieberman, Regular and Stochastic Motion (Springer, Berlin, 1983).

[3] S. Wolfram, ed., Theory and Applications of Cellular Automata (World Scientific, Singapore, 1986).

[4] J. Greenberg and S. Hastings, SIAM J. Appl. Math. 34 (1978) 515.

[5] Y. Oono and M. Kohmoto, Phys. Rev. Lett. 55 (1985) 2927;
R.Z. Sagdeev, D.A. Usikov and G.M. Zaslavsky, Nonlinear Physics (Hardwood, New York, 1988) p. 470.

[6] Y. Aizawa and I. Nishikawa, in: Dynamical Systems and Nonlinear Oscillations, ed. G. Ikegami (World Scientific, Singapore, 1986) p. 210.

[7] Y. Aizawa, Non-stationary chaos revisited from large deviation theory, Prog. Theor. Phys. Suppl. 99 (1990), in press.

[8] K. Kaneko, Physica D 23 (1986) 436.

[9] K. Kaneko, in: Theory and Applications of Cellular Automata, ed. S. Wolfram (World Scientific, Singapore, 1986) p. 367.

[10] J.P. Crutchfield and K. Kaneko, Phys. Rev. Lett. 60 (1988) 2715;
K. Kaneko, Physica D 34 (1989) 1;
K. Kaneko, Prog. Theor. Phys. Suppl. 99 (1990), in press.

[11] K. Kaneko, Collapse of Tori and Genesis of Chaos in Dissipative Systems (World Scientific, Singapore, 1986);
J.P. Crutchfield and K. Kaneko, in: Direction in Chaos, ed. H. Bailin (World Scientific, Singapore, 1987) p. 272.

[12] J.O. Hinze, Turbulence (McGraw-Hill, New York, 1959);
J.R. Herring, S.A. Orszag, R.H. Kraichnan and D.G. Fox, J. Fluid Mech. 66 (1974) 417.

[13] M. Toda, Phys. Rep. 18 (1975) 1.

[14] J.K. Park, K. Steiglitz and W.P. Thurston, Physica D 19 (1986) 423.

Physica D 45 (1990) 328–344
North-Holland

KNOT INVARIANTS AND CELLULAR AUTOMATA [*]

Brosl HASSLACHER

*Department of Physics, University of California/San Diego, La Jolla, CA 92093, USA
and Theoretical Division and Center for Nonlinear Studies, Los Alamos National Laboratory,
Los Alamos, NM 87545, USA*

and

David A. MEYER

*Department of Physics and Institute for Pure and Applied Physical Sciences, University of California/San Diego,
La Jolla, CA 92093, USA*

Received 23 February 1990
Revised manuscript received 1 April 1990

There is a deep connection between the theory of invariants of knots in three dimensions and certain classes of solvable models in statistical mechanics. We study the consequences of this for a class of cellular automaton systems. Knot invariants furnish both a non-perturbative framework for the classification and analysis of such models and a primary tool with which to study the emergence of thermodynamic behavior in extended systems.

1. Introduction

Since the discovery of the Jones [10], HOMFLY [7] and Kauffman [14] knot invariants, a deep two-way connection – that seems to be more than formal – has emerged between knot theory and statistical mechanics [1,5,11,18,26,28,29]. At minimum, knot theory provides a fundamentally new and rich framework for treating non-perturbatively solvable statistical mechanics models and for detecting dynamical symmetries in the associated evolution equations.

In the simplest cases, the evaluation of certain invariant knot polynomials is equivalent to the evaluation of the partition function for exactly solvable two-dimensional lattice models at criticality, e.g., the 8-vertex SOS models (and other IRF models) in terms of modular functions [1,18]. By using the rich set of results one has for solvable lattice models in statistical mechanics, one can derive new results in the theory of knot invariants [5]. Moreover, the physical consequences of this connection can be explored.

It is yet an open and deep question why either a family of link invariants or solvable two-dimensional lattice models exists at all, but the existence of such a connection is already remarkable. Because the theory of three-dimensional knot invariants is being used to make statements about integrable two-dimensional statistical mechanics models, several authors have observed that integrability in two dimensions, as well as other statistical mechanical consequences of knot theory, may originate from various shadowings of three-dimensional field theories onto two dimensions [20,29], an intriguing and beautiful possibility.

Now, it is natural to use statistical mechanics to study cellular automaton models [30]. For probabilistic cellular automata one can, in fact, identify some models with particular statistical mechanics systems such as Ising spin systems [6]. Conversely, one can also construct cellular automata, either probabilistic

[*] Supported in part by INCOR and DOE grant DE FG03-89ER-13991.

or deterministic, which are dynamical models for such statistical mechanics systems [4]. Moreover, for reversible deterministic cellular automata (with which we shall be concerned in this paper) it is possible to find globally conserved quantities which can be interpreted as energy functionals [21,23] and used to define Boltzmann weights for an associated statistical mechanics model. One can naturally define partition functions for such associated models. If these partition functions be equivalent to invariant knot polynomials, one may, given the connection described above, use this information to analyze these models. We expect, therefore, that a knot theory viewpoint on cellular automaton systems may create for this field, where powerful analytical tools have been notably missing, a rich set of results.

This association between reversible cellular automata and statistical mechanics systems is (not accidentally) analogous to the connection between $(1 + 1)$-dimensional integrable quantum systems and solvable two-dimensional statistical mechanics models with its quantum inverse scattering transformations and tier of conserved integrals of the motion in involution as consequences of non-perturbative dynamical symmetry [1,33]. Such connections offer a possible line of attack on a central problem in nonlinear dynamics: understanding the behavior of extended systems, in particular the emergence of thermodynamic behavior.

Like lattice statistical mechanics models, cellular automaton systems are discrete and so there is no natural concept of a symplectic form on them. Even for reversible cellular automata, which conserve phase space, there is no exact Hamiltonian structure on the system. We will see, however, that there are knot theoretic arguments which can distinguish which of these cellular automaton systems can model integrable systems in the thermodynamic limit. These arguments do not require evolving the system or taking the thermodynamic limit directly, but rather can already be seen at the level of the specification of globally conserved quantities.

More generally, one would also like to study the emergence of thermodynamic properties in large (not necessarily integrable) systems; cellular automata have special advantages for this. Minimal cellular automaton models can be constructed for large systems which support thermodynamic behavior and the different degrees of onset of ergodicity can be studied as the system approaches the thermodynamic limit. In this sense, one can see how well statistical mechanics encodes the dynamics by checking the agreement between ensemble averages and time averages.

Our mode of investigation is quite different from typical methods which assume ergodicity and invoke the entire apparatus of kinetic theory, an essentially perturbative procedure, to find the behavior of collective modes. This work indicates how an approach informed by knot theoretic considerations may obtain non-perturbative information on how thermodynamic behavior is approached in large systems – an open question at the foundations of statistical mechanics itself.

2. Knot invariants and statistical mechanics

In this section we begin with a diagrammatic introduction to knot theory, focussing on the construction of polynomial knot invariants. This graphical calculus is largely due to Kauffman [12–14]. Calculations on particular cellular automata examples in the next section will force us to go from such a script calculus to the category of abstract tensors and use a mixed representation for knots and links. We introduce the abstract tensor formulation of knots because the conserved quantities we compute with are complicated knot theoretical objects when written as pure knot script and difficult to work with as such. Once we have the topological picture clear as well as its algebraic formulation, we describe the connection between knot theory and statistical mechanics.

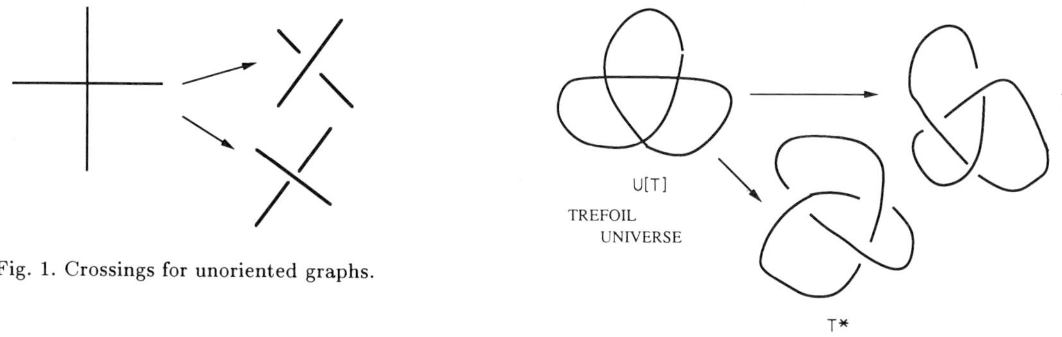

Fig. 1. Crossings for unoriented graphs.

Fig. 2. Trefoil universe and chiral states.

2.1. Basic knot theory

We begin with some definitions: A link in three-space is a particular embedding of a set of circles into three-dimensional Euclidean space. A circle and its topological deformations are called a component and a one-component link is called a knot if it cannot be deformed into a circle. The standard circle is called the unknot. In three-space, we can continuously deform objects into one another; restricted to the class of links, this equivalence relation is called ambient isotopy. Here deformations consist of parametric families of embeddings, depending on $t \in [0,1]$ and we will require that as a function of t, the deformations are continuous and tame.

One way (not the only way) to study ambient isotopy is to imagine a light shining onto a knot or link K' and look at the shadows it throws on the plane. This gives us a projection from the collection of knots and links in three-space to the collection of knot and link diagrams in two-space. Each diagram is a 4-valent transverse graph with the extra structure of crossing choices corresponding to passing over or under a strand. If we examine a vertex from a link graph K, ignoring the rest of K, we have the crossing possibilities illustrated in fig. 1.

For unoriented links and knots, the special shadowing that forgets about crossing information is called the universe of the knot or link $U[K]$. So for the classic example of the trefoil knot we have fig. 2. A universe $U[K]$ can be expanded to link diagrams K in 2^n ways where n is the number of vertices in $U[K]$. These are called states $S[K]$ of $U[K]$. We might imagine that information on K can be represented as a generalized sum over products of vertex weights corresponding to crossing choices and some global valuation of the state $S \in S[K]$ corresponding to the number of components in S. This actually happens and is in direct analogy to the partition function for a system in statistical mechanics.

Once we agree on how to represent projections $K' \to K$ onto the plane, it is important to know whether there is a sequence of deformations or "moves" we can make in the plane itself on K that would guarantee that the corresponding links in three-space were ambient isotopic. Reidemeister [22] showed that the following three moves in the plane, along with planar isotopy, are necessary and sufficient to do this. Two diagrams K_1 and K_2 in the plane represent ambient isotopic links K_1' and K_2' if and only if they can be transformed into one another by a finite sequence taken from the moves shown in fig. 3. Remarkably, these sliding and untwisting moves are enough.

Moves $R2$ and $R3$ already bear a resemblance to S matrix diagrams if we assign them an orientation and identify line labels with spin variables. For oriented graphs, the analogous Reidemeister moves are slightly more involved. First, we define the sign convention for oriented crossings as in fig. 4. Then the Reidemeister moves for oriented graphs are as shown in fig. 5. The obvious similarity of the $R3$ move to the Yang–Baxter relations in statistical mechanics is quite striking, although at this stage purely formal.

For the connection to statistical mechanics and the discovery of invariants on knots and links, it is

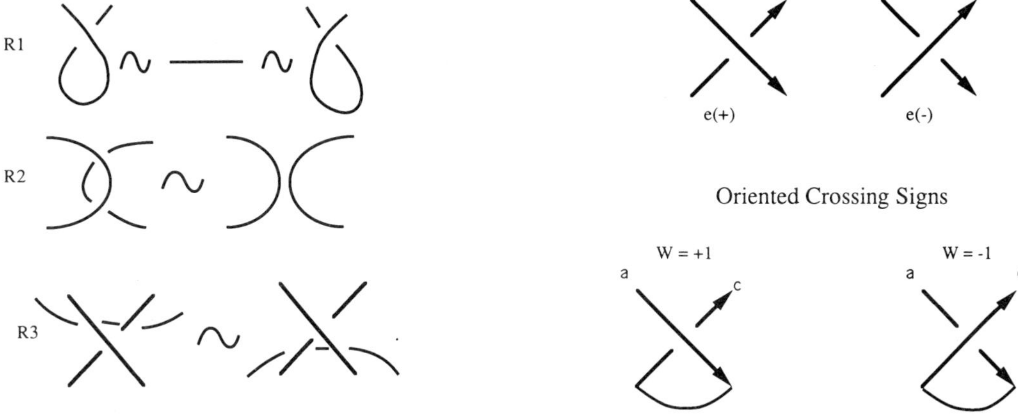

Fig. 3. Reidemeister moves.

Oriented Crossing Signs

Writhe Conventions

Fig. 4. (a) Oriented crossing signs. (b) Writhe conventions.

convenient to introduce another category of topological equivalence on knot diagrams, regular isotopy. Reidemeister's first move, $R1$, has a fundamentally distinct character from $R2$ and $R3$ and can often be factored out of invariance arguments [25]. Regular isotopy is defined to be invariance under $R2$ and $R3$ moves alone. To deal with the $R1$ move, we introduce the concept of the writhe of a knot diagram K. Let $\mathrm{Cr}[K]$ be the set of crossings of the diagram K. Then the twist number or writhe $w(K)$ is defined, for any oriented diagram K, by

$$w(K) := \sum_{p \in \mathrm{Cr}[K]} \epsilon(p).$$

The writhe $w(K)$ is an invariant of $R2$ and $R3$, but not of $R1$. This fact makes it possible to compensate for lack of $R1$ invariance by a normalizing writhe factor which allows one to recover ambient isotopy.

We illustrate this whole complex of ideas: ambient and regular isotopy, knot and link invariants, and writhe normalization, by introducing a construction due to Kauffman [12] called the bracket polynomial, $\langle \cdot \rangle$. These ideas will be used later when we discuss regular isotopy invariants defined by classes of cellular automata; in the bracket construction they appear in their simplest form.

The Kauffman bracket $\langle \cdot \rangle$ is defined for unoriented knot diagrams and produces a polynomial $\langle K \rangle(A, B, d) \in \mathbf{Z}[A, B, d]$ in three commuting variables (A, B, d). It satisfies the following recursion axiom:

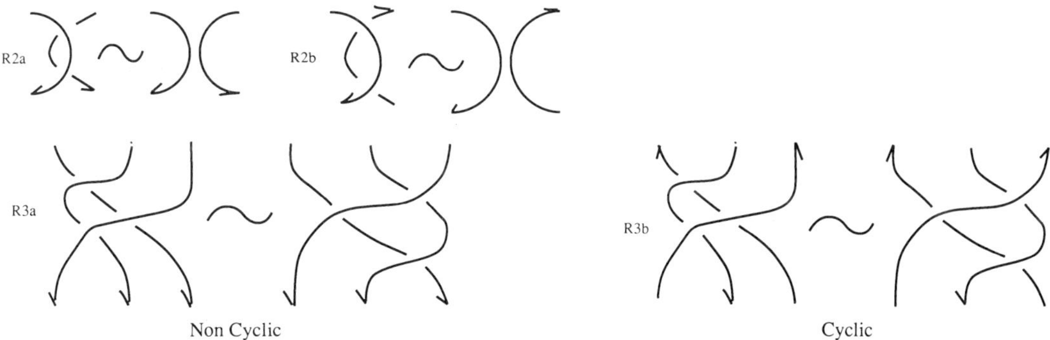

Non Cyclic　　　　　　　　　　　　　　　　　　　Cyclic

Fig. 5. Oriented Redemeister moves. (a) Non-cyclic. (b) Cyclic.

$$\left\langle \mathcal{X} \right\rangle = A\left\langle\, \asymp \,\right\rangle + B\left\langle\, \supset\, \subset \,\right\rangle, \tag{A1}$$

(where the $\mathcal{X}$ on the left-hand side of the equation denotes the crossing shown in fig. 6) as well as two axioms for the unknot:

$$\left\langle \bigcirc K \right\rangle = d\left\langle K \right\rangle, \tag{A2}$$

and

$$\left\langle \bigcirc \right\rangle = 1, \tag{A3}$$

The Kauffman bracket is a functional on the collection of link diagrams since the functional equation (A1) stands for insertions or crossings within larger diagrams that differ only by the (A1) brackets. The splicings of the crossing indicated in (A1) refer to the labelling of the two distinct regions defined by an unoriented crossing. If one turns the overcrossing line in a counter-clockwise direction to align it with the remaining line, the region swept out is called the A channel. The other region is the B channel. Fig. 6 shows this in script form.

These are already quite complex statements about what one is doing to K' since the operations which define $\langle K \rangle$ are operations on a projection or shadow of the knot and open to several splicing interpretations in three dimensions. For the present discussion, we simply take these definitions as axiomatic on knot script in two dimensions.

Now the idea is to construct an ambient isotopy invariant on knots using the bracket $\langle \cdot \rangle$, but restricting the possible relationships among (A, B, d) in order to satisfy the Reidemeister moves. We first look at $R2$ and see that the diagram on the left-hand side of fig. 3 expands as

$$\mathrm{LHS}(R2) = AB\left\langle\, \supset\, \subset \,\right\rangle + AB\left\langle\, \overset{\asymp}{\bigcirc} \,\right\rangle + (A^2 + B^2)\left\langle\, \asymp \,\right\rangle.$$

Note that

$$AB\left\langle\, \overset{\asymp}{\bigcirc} \,\right\rangle = dAB\left\langle\, \asymp \,\right\rangle,$$

so we get

$$\mathrm{LHS}(R2) = AB\left\langle\, \supset\, \subset \,\right\rangle + (dAB + A^2 + B^2)\left\langle\, \asymp \,\right\rangle.$$

To satisfy $R2$, we must have $AB = 1$ and $dAB + A^2 + B^2 = 0$. There is a one-parameter family of solutions to these equations: $B = A^{-1}$ and $d = -(A^2 + A^{-2})$.

It is easily checked that given this parameterization of (A, B, d), the bracket $\langle K \rangle$ is also invariant under $R3$. Thus we have found a regular isotopy invariant of knots and links. Under the $R1$ move, however, $\langle K \rangle$ is not invariant, a typical situation for such constructions; this is shown in fig. 7.

If we orient K, we can obtain an invariant of ambient isotopy by using the writhe $w(K)$ to normalize the coefficient of the $R1$ move to unity: Let $\alpha := -A^3$ and define $F[K] := \alpha^{-w(K)}\langle K \rangle$ where $\langle K \rangle$ is computed as above, ignoring orientations. $F[K]$ is clearly an invariant of ambient isotopy. Alternatively, after orienting K, we can insert a "writhe compensator" into the diagram to make the total writhe of the

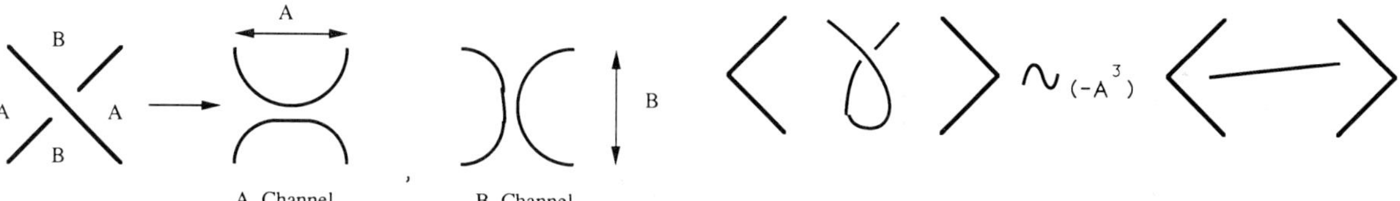

Fig. 6. Unoriented crossing splices.   Fig. 7. Bracket under $R1$.

new diagram K_0 vanish. Fig. 14 shows this for the trefoil T of fig. 2. It is an easy exercise to show that $\langle T \rangle = A^{-7} - A^{-3} - A^5$ and thus the ambient isotopy invariant $F[T] = -A^{-16} + A^{-12} + A^{-4}$. Also $\langle T_0 \rangle = F[T]$. Note that this calculation shows that the trefoil knot is chiral, for $F[T^*] = A^4 + A^{12} - A^{16} \neq F[T]$.

This simple construction illustrates all of the concepts we have introduced so far and is actually important in its own right, being directly related to the Jones polynomial invariant [13].

2.2. Abstract tensors and Reidemeister moves

The type of topological knot script calculation performed above is sufficient for simple examples illustrating concepts and for proving theorems in knot theory, but is too unwieldy to use on generic examples related to statistical mechanics. Thus we introduce the simple notion of abstract tensors and define the Reidemeister moves in these terms as a prelude to proceeding to the statistical mechanics case.

Diagrammatic abstract tensor formalism, first developed by Penrose [19], can be used as a way of identifying tensors having upper and lower indices with oriented link diagrams. For our purposes a tensor A^{ab}_{cd} is represented by the diagram in fig. 8a while formal sums over repeated indices are written $A^a_b B^b_c$ and drawn as in fig. 8b; fig. 8c shows the trace A^a_a. A line without boxes will always be taken to be a Kronecker delta.

Now we identify the box with the S matrix connecting the in and out states of quantum process and suppress the dependence on momentum. Formally we identify S^{ab}_{cd} with the box in fig. 9a and identify its inverse $\bar{S}^{ab}_{cd}$ with the crossed box in fig. 9b. Then unitarity of the S matrix is the relation

$$S^{ab}_{cd}\bar{S}^{cd}_{ef} = \delta^a_e \delta^e_f,$$

which is shown diagrammatically in fig. 10, on the left. Associate an S matrix to a crossing $\epsilon(+)$ and its inverse $\bar{S}$ to $\epsilon(-)$. Then the formal tensor diagram becomes the $R2a$ move for oriented links in knot theory: the diagram on the right of fig. 10. Replacing every crossing of a link or knot diagram by the appropriate S matrix gives a number, which we would like to be a scalar under the Reidemeister transformations.

Imposing the oriented $R2$ and $R3$ moves this way gives a set of relations which can be identified as the

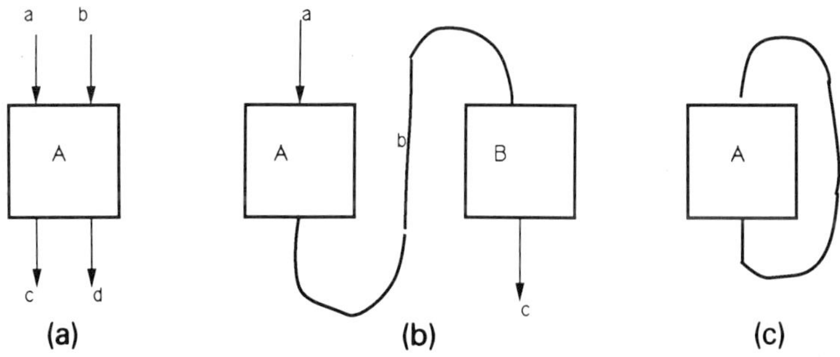

Fig. 8. (a) A diagram for the tensor A^{ab}_{cd}. (b) Contracting tensors: $A^a_b B^b_c$. (c) The trace A^a_a.

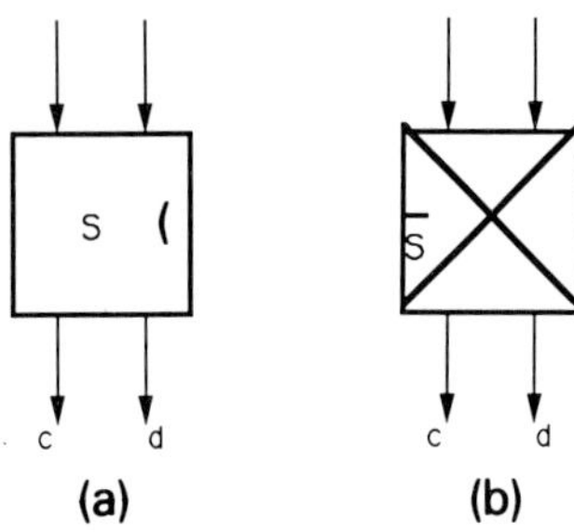

Fig. 9. (a) The scattering matrix S_{cd}^{ab}. (b) The inverse $\bar{S}_{cd}^{ab}$.

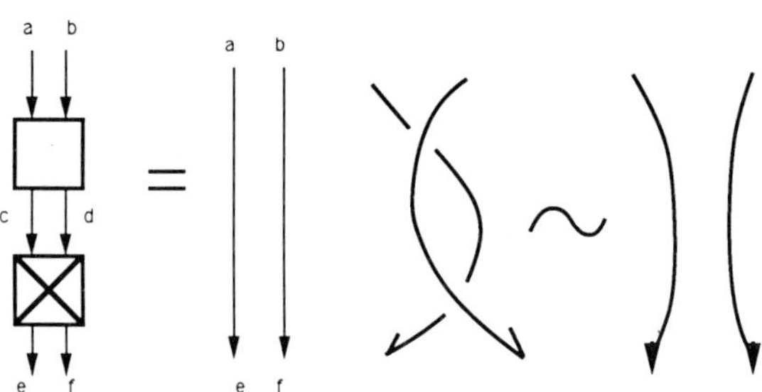

Fig. 10. Abstract tensors; Reidemeister move.

conditions of unitarity (fig. 11):

$$S_{ij}^{ab}\,\bar{S}_{cd}^{ij} = \delta_c^a \delta_d^b,$$

cross-channel unitarity (fig. 11):

$$S_{fd}^{ec}\,\bar{S}_{ac}^{bd} = \delta_a^e \delta_f^b,$$

and finally one version of $R3$ (fig. 12):

$$S_{rt}^{\alpha\beta}\,S_{ls}^{t\gamma}\,S_{\lambda\mu}^{rl} = S_{rt}^{\beta\gamma}\,S_{\lambda l}^{\alpha r}\,S_{\mu s}^{lt},$$

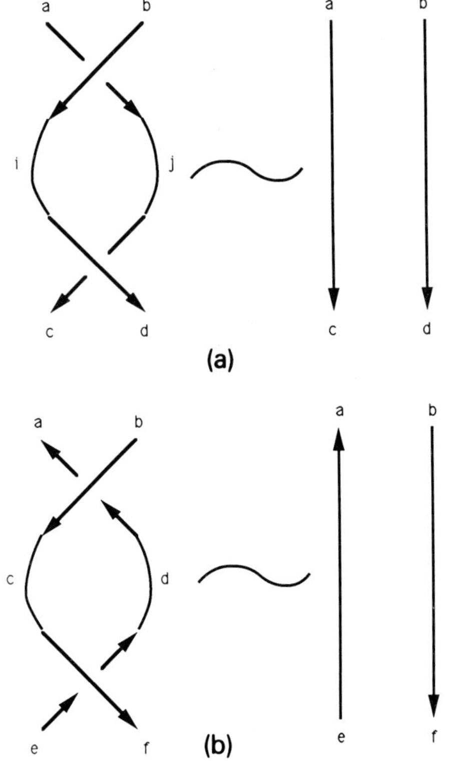

Fig. 11. (a) Unitary. (b) Cross-channel unitary.

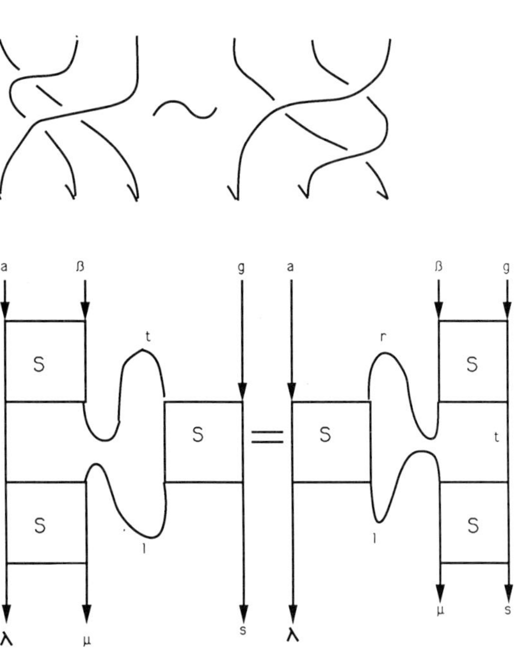

Fig. 12. $R3a$ Yang–Baxter relation.

which we recognize as the momentum forgetting version of the Yang–Baxter equation for vertex models in statistical mechanics.

This correspondence so far is formal. Nevertheless, if we have an S matrix which satisfies unitarity, cross-channel unitarity and the Yang–Baxter equations, we also have at least a regular isotopy invariant of knots by this correspondence.

2.3. Solvable models, critical behavior and knot invariants

Assuming some familiarity with modern statistical mechanics, we briefly review some concepts and results which at first sight seem unrelated to what has gone before. We will need them, however, in what follows. Integrable quantum field theories in $1 + 1$ dimensions and soluble lattice statistical mechanics models in 2 dimensions are intimately connected. In each case we have an infinite number of conserved quantities which cause the models to be either integrable or solvable. The most direct interpretation of the Yang–Baxter equations is in $(1 + 1)$-dimensional field theory where they are a factorization condition for a general S matrix [33]. If we denote by u the rapidity (in the literature it is also called a spectral parameter) we have the relation

$$S_{ij}^{ab}(u)\, S_{kf}^{jc}(u + v)\, S_{de}^{ik}(v) = S_{ij}^{bc}(v)\, S_{dk}^{ai}(u + v)\, S_{ef}^{kj}(u),$$

which is the quantum Yang–Baxter condition, where the indices correspond to spins. Fig. 13 shows this condition diagrammatically. Any solution to the factorized S matrix equation can also be thought of as the Boltzmann weight of a soluble vertex model. In this case the quantum Yang–Baxter equations correspond to the commutativity of the transfer matrices in the model. In another class of statistical mechanics models, the IRF (interaction round a face) models, the corresponding relation is called the star–triangle relation [2]. The IRF models are more general, since they contain a larger class of soluble models and include spin conserving vertex models via the Wu–Kadanoff–Wegner transformation (see ref. [1] or ref. [28]), but for this discussion we will not really need to go into their structure. Regarding the Yang–Baxter equations as a set of functional equations for Boltzmann factors, solutions to the Yang–Baxter system generate classes of soluble models. In two dimensions, there is an unbounded tower of such models [1,28].

To pass to the knot invariant/Reidemeister move aspect of the Yang–Baxter equation, we must suppress the rapidity (u, v) dependence of these functional equations. Even though one suspects that some information may be lost in doing so, it is a remarkable fact that a hierarchy of two-dimensional solvable statistical mechanics vertex models exists under these conditions [1]. If all three conditions – unitarity, cross-channel unitarity and the quantum Yang–Baxter equations – are satisfied, one is led to a class of models solved by modular functions. The rapidity dependence can be eliminated only if these models are critical; at criticality, modular functions become trigonometric or hyperbolic functions and one can proceed, by considerations on braid groups, Markov moves and Markov traces, to construct link and

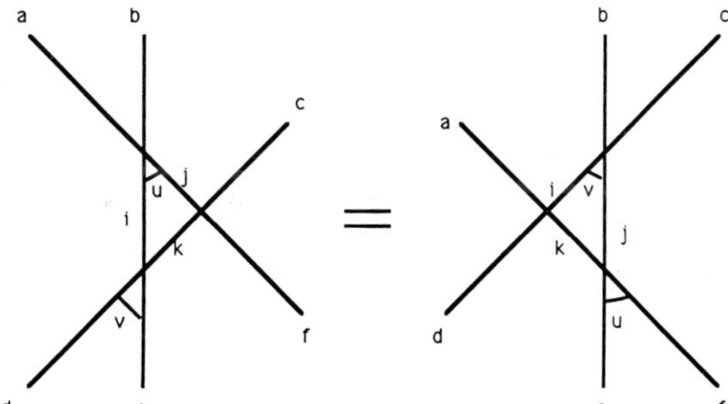

Fig. 13. Quantum Yang–Baxter relation.

knot polynomial invariants [1,5,10,18,26,28]. We do not want to go into the details of how one uses braid groups to do this since we will not need it for the present study; it is technically somewhat involved and, as Kaufmann has emphasized [13], it is not clear that braid groups must play as essential a role in the eventual emergence of knot invariance as they have heretofore.

The crucial point is that one must first move onto the critical hypersurface in the space of coupling constants for the original model before suppressing the rapidity dependence. This is the reason that Jones found the Temperley–Lieb algebra playing a central role in his braid group construction of the Jones invariant [10]. Why criticality is so important for knot invariants is still an open question, for there are deformations of the Reidemeister moves that still generate invariant knot polynomials. These models all fail crossing symmetry, but satisfy the Yang–Baxter relations. An example is the IRF $A^{(1)}_{m-1}$ affine Lie algebra model [18]. But these models are special in having additional structures on them. All of these exceptional models also turn out to sit at their critical points in knot invariant constructions.

3. Reversible cellular automata

As remarked in section 1, Liouville's theorem holds for reversible cellular automata: since the state space is discrete and each state has a unique predecessor, phase space volume is invariant under the evolution. Thus it is conceivable that some conserved quantity might function as an energy, namely govern the evolution as does a Hamiltonian in classical (or quantum) mechanics. ('t Hooft has, in fact, shown how to derive an approximate Hamiltonian for a particular class of reversible cellular automata in ref. [9].) At the very least, it might determine the form of the thermodynamic behavior of the model. For these reasons, the ideas described in the previous section apply directly to reversible cellular automata [16,17,21,23,24].

Let σ_i^t denote the state of site i at time t in some cellular automaton. Given any neighborhood U_i of the site i and any function $f(\{\sigma_j^t \mid j \in U_i\})$ taking values in the state space, Fredkin's construction [3,27] provides a reversible (second-order) rule:

$$\sigma_i^{t+1} := f\left(\{\sigma_j^t \mid j \in U_i\}\right) - \sigma_i^{t-1}.$$

Takesue [23] has recently begun a study of the thermodynamics of this type of cellular automata in the case when the σ_i^t are Boolean variables (taking the values 0 and 1) and the neighborhood has radius one: $U_i := \{i-1, i, i+1\}$. After partitioning the set of 2^8 such rules (each referred to by the number

$$\sum_{\sigma_1, \sigma_2, \sigma_3} 2^{4\sigma_1 + 2\sigma_2 + \sigma_3} f(\sigma_1, \sigma_2, \sigma_3)$$

followed by an R, for reversible) into equivalence classes under the operations of reflection and Boolean conjugation he proceeds to search for additive conserved quantities of the form:

$$\Phi(\sigma^t, \sigma^{t-1}) := \sum_i F(\sigma_i^t, \sigma_{i+1}^t, \sigma_i^{t-1}, \sigma_{i+1}^{t-1}),$$

where σ^t is the state of the whole cellular automaton at time t and where periodic boundary conditions have been imposed.

In light of the discussion in the previous section it is natural to define a vertex model associated to a cellular automaton with such a conserved quantity since the transfer matrix can be identified with

$$\exp\left[-\beta\Phi(\sigma^t, \sigma^{t-1})\right] = \prod_i \exp\left[-\beta F(\sigma_i^t, \sigma_{i+1}^t, \sigma_i^{t-1}, \sigma_{i+1}^{t-1})\right].$$

and the Boltzmann weights are thus

$$S_{cd}^{ab} := \exp[-\beta F(a,b,c,d)].$$

Testing ergodicity in the usual way one compares the one-dimensional statistical mechanics model with these weights with the dynamical evolution of the cellular automaton [23].

There are two possibilities, however, which guarantee immediately that ergodicity will fail. If some locally defined quantity is conserved as well as its sum over sites, it is said to satisfy a local conservation law. Typically such a law will cause the system to be partitioned by walls formed by configurations of states which are invariant or periodic in time. For example, in rule 73R, the three-site sequence

$$\begin{pmatrix} \sigma_{i-1}^{t-1} & \sigma_i^{t-1} & \sigma_{i+1}^{t-1} \\ \sigma_{i-1}^{t} & \sigma_i^{t} & \sigma_{i+1}^{t} \end{pmatrix} = \begin{pmatrix} 0 & 1 & 0 \\ 0 & 1 & 0 \end{pmatrix}$$

is invariant under the evolution [23]. No information or energy can be transmitted from one side of such a wall to the other so the system cannot have the mixing property and thus cannot be ergodic.

A local conservation law can also prevent ergodicity even without the existence of such invariant configurations, by inhibiting the propagation of energy directly. Takesue [23] demonstrates that when an additive conserved quantity of the form of Φ exists, there is a current J satisfying

$$F(\sigma_i^{t+1},\sigma_{i+1}^{t+1},\sigma_i^{t},\sigma_{i+1}^{t}) = F(\sigma_i^{t},\sigma_{i+1}^{t},\sigma_i^{t-1},\sigma_{i+1}^{t-1}) + J(\sigma_{i-1}^{t},\sigma_i^{t},\sigma_{i+1}^{t},\sigma_i^{t-1}) - J(\sigma_i^{t},\sigma_{i+1}^{t},\sigma_{i+2}^{t},\sigma_{i+1}^{t-1}).$$

If J vanishes (rule 46R), or is restricted (rule 24R), ergodic behavior is again impossible.

There are, however, some elementary reversible cellular automata with no local conservation laws, *i.e.*, which have a propagative energy. The additive invariants $F(a,b,c,d)$ which occur in these cellular automata are

1. $(a-d)^2 + (b-c)^2$.
2. $(b-c)^2 - (a-d)^2$.
3. $ad(1 - 2c - 2b) - bc(1 - 2a - 2d)$.
4. $1 + ac + bd - [1 - 2(1-a)(1-d)][1 - 2(1-b)(1-c)]$.

Using each of these invariants to define an energy produces a statistical mechanics model whose thermodynamics can be studied by simulation. Our approach is somewhat different: we ask whether these models are solvable models at criticality; if so, their thermodynamic behavior can be analytically computed rather than simulated.

To test for solvability, the criteria of the preceding section must be checked. That is, we will define each of these conserved quantities to be the energy in a Boltzmann weight to be associated with a knot crossing:

$$S_{cd}^{ab} := \exp[-F(a,b,c,d)/kT] =: x^{F(a,b,c,d)}$$

and test whether the (oriented) Reidemeister moves hold. We do this for each invariant separately.

Case 1. The Boltzmann weight is

$$S_{cd}^{ab} := x^{(a-d)^2 + (b-c)^2} = \begin{array}{c} \\ 00 \\ 01 \\ 10 \\ 11 \end{array} \begin{pmatrix} 1 & x & x & x^2 \\ x & x^2 & 1 & x \\ x & 1 & x^2 & x \\ x^2 & x & x & 1 \end{pmatrix},$$

where the indices ab and cd label the rows and columns, respectively. The unitarity condition defines $\bar{S}$:

$$\bar{S}^{ab}_{cd} = \frac{1}{(1-x^2)^2}\,
\begin{array}{c|cccc}
 & 00 & 01 & 10 & 11 \\\hline
00 & 1 & -x & -x & x^2 \\
01 & -x & x^2 & 1 & -x \\
10 & -x & 1 & x^2 & -x \\
11 & x^2 & -x & -x & 1
\end{array}\;.$$

For symmetry, we renormalize by dividing S^{ab}_{cd} by $1-x^2$ and multiplying $\bar{S}^{ab}_{cd}$ by the same factor. To check cross-channel unitarity we verify:

$$\delta^a_d \delta^b_c = S^{ai}_{cj} \bar{S}^{bj}_{di}$$

$$= \frac{1}{(1-x^2)^2}
\begin{array}{cccc}
1 & x & x & x^2 \\
x & x^2 & 1 & x \\
x & 1 & x^2 & x \\
x^2 & x & x & 1
\end{array}
\cdot
\begin{array}{cccc}
1 & -x & -x & x^2 \\
-x & 1 & x^2 & -x \\
-x & x^2 & 1 & -x \\
x^2 & -x & -x & 1
\end{array}
=
\begin{array}{cccc}
1 & 0 & 0 & 0 \\
0 & 0 & 1 & 0 \\
0 & 1 & 0 & 0 \\
0 & 0 & 0 & 1
\end{array}\;.$$

Finally, we check the Yang–Baxter equation:

$$S^{ab}_{ij} S^{jc}_{kf} S^{ik}_{de} = (1-x^2)^{-3} \sum_{i,j,k} x^{(a-j)^2+(b-i)^2+(j-f)^2+(c-k)^2+(i-e)^2+(k-d)^2}$$

$$= (1-x^2)^{-3} \sum_{i,j,k} x^{(b-i)^2+(c-k)^2+(a-j)^2+(k-d)^2+(j-f)^2+(i-e)^2}$$

$$= S^{bc}_{ki} S^{ak}_{dj} S^{ji}_{ef}\;.$$

Note that since $\bar{S}$ is obtained from S by replacing x with $-x$, the Yang–Baxter equation holds for $\bar{S}$ by the analogous computation.

We conclude that this model is the limit of an exactly solvable model at criticality. This means that there is some solvable model whose Boltzmann weights at $T = T_c$ limit to S^{ab}_{cd} when the rapidity is supressed. In terms of the parameters of this model, $x \neq \exp(-1/kT)$, but rather depends on what is known as the crossing point parameter. The explicit form of this critical solvable model should give further insight into the behavior of cellular automata in case 1, but for the moment it is enough to note that it is a model at its critical point. For further discussion see section 5.

Case 2. The Boltzmann weight is

$$S^{ab}_{cd} := x^{(b-c)^2-(a-d)^2} =
\begin{array}{c|cccc}
 & 00 & 01 & 10 & 11 \\\hline
00 & 1 & x^{-1} & x & 1 \\
01 & x & 1 & 1 & x^{-1} \\
10 & x^{-1} & 1 & 1 & x \\
11 & 1 & x & x^{-1} & 1
\end{array}\;.$$

Again, the unitarity condition defines $\bar{S}$:

$$\bar{S}^{ab}_{cd} = \frac{1}{2 - x^{-2} - x^2}\,
\begin{array}{c|cccc}
 & 00 & 01 & 10 & 11 \\\hline
00 & 1 & -x & -x^{-1} & 1 \\
01 & -x^{-1} & 1 & 1 & -x \\
10 & -x & 1 & 1 & -x^{-1} \\
11 & 1 & -x^{-1} & -x & 1
\end{array}\;,$$

and we make the situation more symmetric by dividing S by $x - x^{-1}$ and multiplying $\bar{S}$ by the same factor. As before we can verify that cross-channel unitarity holds as well as the Yang–Baxter equations for S and $\bar{S}$.

Thus this model is also the limit of some solvable model at criticality.

Case 3. The Boltzmann weight is now

$$S_{cd}^{ab} := x^{ad(1-2c-2b)-bc(1-2a-2d)}$$

and the computations become somewhat more complicated. Using the symbolic manipulation program *Mathematica* [31] this case can also be checked. Here we find that cross-channel unitarity fails, although the Yang–Baxter equation is satisfied. Hence we cannot conclude that this model is exactly solvable.

Case 4. The Boltzmann weight is

$$S_{cd}^{ab} := x^{1+ac+bd-[1-2(1-a)(1-d)][1-2(1-b)(1-c)]}$$

and again the computations are complicated enough to do the algebra by machine. The results are the same as in case 3; we cannot conclude that this model is solvable either.

It should be noted, however, that although the Boltzmann weights in cases 3 and 4 do not satisfy the condition of cross-channel unitarity, they do satisfy the Yang–Baxter relation. There are thus two reasons to believe that these cases may indeed correspond to exactly solvable models: First, it may be that adding "surface terms" to the conserved quantity may produce weights which do satisfy cross-channel unitarity. In cases 1 and 2, this can be done. Second, one should check whether the additional symmetries which allow models such as the IRF $A_{m-1}^{(1)}$ model mentioned at the end of section 2 to be solved even without cross-channel unitarity obtain in these two cases. These investigations are in progress.

4. Knot invariants from reversible cellular automata

We have just seen that in cases 1 and 2, the Boltzmann weights associated to knot crossings satisfy unitarity, cross-channel unitarity and the Yang–Baxter equations. Thus by the association explained in section 2, the tensor weights in these cases must generate regular isotopy invariants of links. In this section we investigate these invariants.

Proposition 1. *Let $P_1(\mathcal{L})$ be the polynomial obtained by evaluating the invariant of case 1 on an untwisted diagram of a link $\mathcal{L} = \mathcal{L}_1 \cup \cdots \cup \mathcal{L}_k$. Then*

$$P_1(\mathcal{L}) = \prod_{i=1}^{k} \frac{(1+x)^{|m_i|} + (1-x)^{|m_i|}}{(1-x^2)^{|m_i|}},$$

where $m_i := \mathrm{lk}(\mathcal{L}_i, \mathcal{L} \setminus \mathcal{L}_i) := 1/2 \sum \epsilon(p)$, the sum being over all crossings p between $\mathcal{L}_i$ and the rest of $\mathcal{L}$.

Proof. Note that S and $\bar{S}$ can be factored:

$$S_{cd}^{ab} = \frac{1}{1-x^2} A_d^a A_c^b, \qquad \text{where} \qquad A_d^a = \begin{array}{c} {}^{0} \\ {}^{0}_{1} \end{array}\!\begin{pmatrix} 1 & x \\ x & 1 \end{pmatrix}$$

$$\bar{S}_{cd}^{ab} = \frac{1}{1-x^2} \bar{A}_d^a \bar{A}_c^b, \qquad \text{where} \qquad \bar{A}_d^a = \begin{array}{c} {}^{0} \\ {}^{0}_{1} \end{array}\!\begin{pmatrix} 1 & -x \\ -x & 1 \end{pmatrix}.$$

Thus, to evaluate the invariant associated with a knot diagram we start anywhere on the diagram and travel in the direction of the orientation, multiplying by A or $\bar{A}$ as we pass through each crossing point, and then trace when we return to our starting point. But A and $\bar{A}$ commute:

$$A\bar{A} = \begin{pmatrix} 1 - x^2 & 0 \\ 0 & 1 - x^2 \end{pmatrix} = \bar{A}A.$$

This means we can commute the matrices in the result into the form

$$\mathrm{Tr}[(A\bar{A})^{2\bar{s}} A^{2(s-\bar{s})}] \qquad \text{or} \qquad \mathrm{Tr}[(A\bar{A})^{2s} \bar{A}^{2(\bar{s}-s)}],$$

depending on whether there are more S crossings s or more $\bar{S}$ crossings $\bar{s}$. Since $A\bar{A}$ is proportional to the identity, we can evaluate its power and factor the result out of the trace to get

$$(1 - x^2)^{2\bar{s}} \, \mathrm{Tr}\, A^{2w} \qquad \text{or} \qquad (1 - x^2)^{2s} \, \mathrm{Tr}\, \bar{A}^{-2w},$$

where w is the writhe of the diagram. But the eigenvalues of both A and $\bar{A}$ are $\{1 \pm x\}$. Thus the traces can be evaluated easily, giving

$$(1 - x^2)^{2\bar{s}}[(1 + x)^{2w} + (1 - x)^{2w}] \qquad \text{or} \qquad (1 - x^2)^{2s}[(1 + x)^{-2w} + (1 - x)^{-2w}].$$

Finally, recall the factors of $(1 - x^2)^{-1}$ in S and $\bar{S}$; including them gives

$$(1 - x^2)^{-w}[(1 + x)^{2w} + (1 - x)^{2w}] \qquad \text{or} \qquad (1 - x^2)^{w}[(1 + x)^{-2w} + (1 - x)^{-2w}].$$

Note that both these expressions are subsumed by the expression

$$(1 - x^2)^{-|w|}[(1 + x)^{2|w|} + (1 - x)^{2|w|}]$$

and when the writhe vanishes they coincide at the value 2.

Now consider an untwisted diagram of the link $\mathcal{L}$. Since it is an untwisted diagram, each component $\mathcal{L}_i$ of the link has writhe zero [32]. Nevertheless, the components may be linked with each other, in which case there will be an imbalance of A and $\bar{A}$ factors in the trace associated with each linked component. More precisely, if $m_i := \mathrm{lk}(\mathcal{L}_i, \mathcal{L} \setminus \mathcal{L}_i)$, then each component $\mathcal{L}_i$ will contribute a factor $(1 - x^2)^{-|m_i|}[(1 + x)^{2|m_i|} + (1 - x)^{2|m_i|}]$ to the link invariant. $\qquad\qquad\square$

Proposition 2. *Let $P_2(\mathcal{L})$ be the polynomial obtained by evaluating the invariant of case 2 on an untwisted diagram of a link $\mathcal{L} = \mathcal{L}_1 \cup \cdots \cup \mathcal{L}_k$. Then*

$$P_2(\mathcal{L}) = \prod_{i=1}^{k} \frac{(2 + x + x^{-1})^{|m_i|} + (2 - x - x^{-1})^{|m_i|}}{(x - x^{-1})^{|m_i|}},$$

where $m_i := \mathrm{lk}(\mathcal{L}_i, \mathcal{L} \setminus \mathcal{L}_i)$.

Proof. Again S and $\bar{S}$ can be factored:

$$S^{ab}_{cd} = \frac{1}{x - x^{-1}} B^a_d A^b_c, \qquad \text{where} \quad B^a_d = \begin{matrix} & \overset{\bullet}{0} & \overset{\bullet}{1} \\ \overset{0}{\underset{\bullet}{}} \\ \overset{1}{\underset{\bullet}{}} \end{matrix} \begin{pmatrix} 1 & x^{-1} \\ x^{-1} & 1 \end{pmatrix}$$

$$\bar{S}^{ab}_{cd} = \frac{1}{x^{-1} - x} \bar{B}^b_c \bar{A}^a_d, \qquad \text{where} \quad \bar{B}^b_c = \begin{matrix} & \overset{\bullet}{0} & \overset{\bullet}{1} \\ \overset{0}{\underset{\bullet}{}} \\ \overset{1}{\underset{\bullet}{}} \end{matrix} \begin{pmatrix} 1 & -x^{-1} \\ -x^{-1} & 1 \end{pmatrix}$$

and A, $\bar{A}$ are as in proposition 1. Again everything commutes because everything is symmetric. Evaluating the invariant on a knot as above, we get

$$\mathrm{Tr}[(A\bar{A})^{\bar{s}}(B\bar{B})^{\bar{s}}(AB)^{s-\bar{s}}] \quad \text{or} \quad \mathrm{Tr}[(A\bar{A})^{s}(B\bar{B})^{s}(AB)^{\bar{s}-s}],$$

using the same notation as in proposition 1. Pulling out the matrices proportional to the identity gives

$$(1-x^{-2})^{\bar{s}}(1-x^2)^{\bar{s}}\,\mathrm{Tr}[(AB)^w] \quad \text{or} \quad (1-x^{-2})^{s}(1-x^2)^{s}\,\mathrm{Tr}[(AB)^{-w}].$$

The eigenvalues of both AB and $\bar{A}\bar{B}$ are $\{2 \pm (x + x^{-1})\}$ so we can evaluate the traces to get

$$(1-x^{-2})^{\bar{s}}(1-x^2)^{\bar{s}}[(2 + x + x^{-1})^w + (2 - x - x^{-1})^w] \quad \text{or}$$

$$(1-x^{-2})^{s}(1-x^2)^{s}[(2 + x + x^{-1})^{-w} + (2 - x - x^{-1})^{-w}].$$

Thus the final results are

$$(x - x^{-1})^{-w}[(2 + x + x^{-1})^w + (2 - x - x^{-1})^w] \quad \text{or}$$

$$(x^{-1} - x)^{w}[(2 + x + x^{-1})^{-w} + (2 - x - x^{-1})^{-w}].$$

As before, both these expressions are subsumed by a single expression:

$$(x - x^{-1})^{-|w|}[(2 + x + x^{-1})^{|w|} + (2 - x - x^{-1})^{|w|}].$$

which is again 2 when the writhe vanishes.

Finally, it is clear that the same argument as above gives the desired result for links.　　□

In cases 3 and 4, although we know that the tensor weights do not satisfy cross-channel unitarity and hence cannot generate true regular isotopy invariants, it is still possible to evaluate the product trace associated with a given writhe normalized knot diagram. It is of interest to notice that both cases do distinguish chirality in the case of the trefoil knot diagrams. As an example we display these product trace polynomials up to overall identical factors for the case 4 invariant:

$$S^{ab}_{cd} := x^{1 + ac + bd - [1 - 2(1-a)(1-d)][1 - 2(1-b)(1-c)]}$$

evaluated on the writhe normalized trefoil diagrams of both chiralities shown in fig. 14. For T we get

$$4 + 17x + 40x^2 + 71x^3 + 112x^4 + 102x^5 + 132x^6 + 252x^7 + 268x^8 + 165x^9 + 100x^{10} + 17x^{11},$$

and for T^* the polynomial evaluates to:

$$-1 - 8x - 46x^2 - 113x^3 - 268x^4 - 514x^5 - 690x^6 - 686x^7 - 499x^8 - 210x^9 - 48x^{10} + 11x^{11}.$$

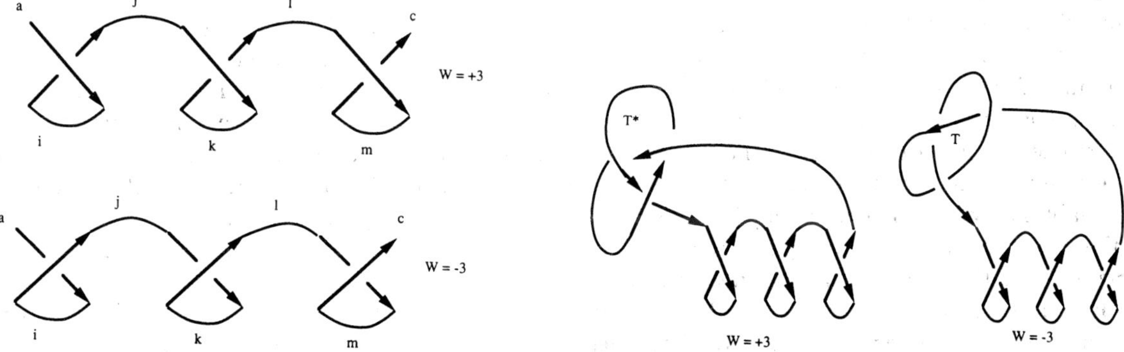

Fig. 14. Writhe normalized trefoil diagrams, both chiralities.

5. Conclusion

Cellular automata dynamics can be connected with statistical mechanical models in a variety of ways. Whether this has analytic advantages depends on how much we know about these statistical mechanical models and how that knowledge can be connected back to the evolution equations. In this paper our goal has been to show how the theory of knot and link invariants allows us to extract the non-perturbative result that many interesting classes of cellular automata systems in $1+1$ dimensions correspond to solvable two-dimensional statistical mechanics models at criticality.

The particular models we analyzed in detail are the special reversible models classified and studied by Takesue. These are some of the simplest models that could conceivably lead to thermodynamic behavior. It is in this setting that we can hope to see in detail how, and by what mechanisms, thermodynamic behavior evolves. For these models we were able to prove that the statistical mechanics systems generated by using global additive invariants to define Boltzmann weights, can, in some important cases, be shown to lead to solvable models at their critical points. This result follows immediately once it is established that the Boltzman weights attached to these models generate a link invariant, as a consequence of the powerful results concerning the modular function solutions to the Yang–Baxter functional equations. We have shown here that the cellular automata of cases 1 and 2 correspond to solvable critical statistical mechanics models. In fact, these models can be identified explicitly as asymmetric 6-vertex models [8]. We suspect that cases 3 and 4 may also correspond to critical statistical mechanics models since they already satisfy the Yang–Baxter equations in their current form (without deformation by surface terms) and are able to generate polynomials that distinguish chirality in writhe normalized trefoil knots.

Rule 26R is a particularly good example of these ideas. In a series of papers Takesue finds, by direct simulation, that with even a relatively small number (of order $N = 50$) points, rule 26R systems are in every sense thermodynamic. They go to correct ensemble distributions quickly, satisfy Fourier's law for heat conduction and their thermodynamic fluctuations obey the Green–Kubo prediction [23]. We call this type of behavior precociously thermodynamic. But rule 26R belongs to one of the classes of cellular automata that has the conserved quantity of case 1 and hence the link invariant generated thereby; thus it is associated to a solvable critical statistical mechanics system. We believe, but cannot make precise at this point, that the precocious thermodynamic behavior in one (spatial) dimension has its origins as a shadowing of the critical behavior in this two-dimensional statistical mechanics model and that this phenomenon is generic.

More generally, we would like to speculate that solvability/integrability and criticality of knot associated statistical mechanics models have direct consequences for the dynamical systems whose behavior we would like to understand. It seems likely that the exponential increase in the number of cycles (with the attendant relative increase in the influence of long cycles) as the size of the system increases [23] should be related to the infinite correlation lengths and scale invariance of an associated critical system, moderated by diminishing finite size effects.

At criticality, a statistical mechanics model has a functional universality property; it forgets the exact form of the action from which it came. This should have a corresponding reflection in the effect of the cycle structure of the associated dynamical systems. In simulations, an individual system is not truely ergodic – one typically sees an extraordinary degeneracy of cycle structures confined to a given energy surface. Nevertheless, one observes robust thermodynamics which may have its origin in the universality of the corresponding critical action. Rather than being properties of the particular dynamics, the thermodynamic observables may become normal in Kinchin's sense [15]. Then the energy surface would be indecomposable in the extended sense, a given physical state being represented by a point on each orbit on the energy surface. Thus the system could be effectively ergodic despite the cycle structure degeneracy.

The question of early and robust thermodynamic behavior is of interest far beyond hyper-discrete systems such as cellular automata. We have begun to explore other dynamical systems for the same properties: irreversible deterministic cellular automata, probabilistic cellular automata, lattice gas models and the intriguing thermodynamic formalism for iterated maps. In each case, we expect to be able to decide by non-perturbative knot invariant considerations whether the associated statistical mechanics model is solvable and lies on a critical hypersurface, and then explore the consequences of these facts for the original non-linear dynamical system. By doing this for systems where considerable analytic information about the dynamical system is already available, we hope to determine precisely the consequences of criticality for the associated dynamical system, testing the conjectures of the two preceding paragraphs.

From the discussion in this paper it seems immediately feasible to investigate these ideas for one-dimensional dynamical systems. It will be much more difficult to make progress in higher dimensions: here we have applied three-dimensional topology (perhaps best described by something like Witten's topological quantum field theory) to two-dimensional solvable statistical mechanics models associated with one-dimensional dynamical systems. This is a remarkable connection among diverse systems having completely different origins. A similar reduction ending in a higher-dimensional dynamical system would imply correspondingly higher dimensions for the statistical mechanics and topology. This seems to be quite difficult; very little work has been done on the tetrahedron (and higher-dimensional) functional equation generalizing the Yang–Baxter equation. Nevertheless, should the consequences of criticality and solvability prove as relevant as we expect for one-dimensional dynamical behavior, the same framework should help us understand higher dimensions as well.

References

[1] Y. Akutsu and M. Wadati, Knot invariants and the critical statistical systems, J. Phys. Soc. Japan 56 (1987) 839–842; Knots, braids and exactly solvable models in statistical mechanics, Commun. Math. Phys. 117 (1988) 243–259.

[2] R.J. Baxter, Exactly Solvable Models in Statistical Mechanics (Academic Press, New York, 1982).

[3] C. Bennet, Logical reversability of computation, IBM J. Res. Devel. 6 (1973).

[4] M. Creutz, Microcanonical Monte Carlo simulation, Phys. Rev. Lett. 50 (1983) 1411–1414; Deterministic Ising dynamics, Ann. Phys. 167 (1980) 62–72.

[5] T. Deguchi, M. Wadati and Y. Akutsu, Link polynomials constructed from solvable models in statistical mechanics, J. Phys. Soc. Japan 57 (1988) 2921–2935.

[6] E. Domany and W. Kinzel, Equivalence of cellular automata to Ising models and directed percolation, Phys. Rev. Lett. 53 (1984) 311–314.

[7] P. Freyd, D. Yetter, J. Hoste, W.B.R. Lickorish, K.C. Millet and A. Ocneanu, A new polynomial invariant of knots and links, Bull. Am. Math. Soc. 12 (1985) 239–246.

[8] B. Hasslacher and D.A. Meyer, Knots, criticality and thermodynamics in discrete systems, UCSD preprint (1990).

[9] G. 't Hooft, Equivalence relations between deterministic and quantum mechanical systems, J. Stat. Phys. 53 (1988) 323–344.

[10] V.F.R. Jones, A polynomial invariant for links via von Neumann algebras, Bull. Am. Math. Soc. 12 (1985) 103–111.

[11] V.F.R. Jones, On knot invariants related to some statistical mechanics models, Pacific J. Math. 137 (1989) 311–334.

[12] L.H. Kauffman, On Knots, Annals of Mathematics Studies, Vol. 115 (Princeton Univ. Press, Princeton, 1987).

[13] L.H. Kauffman, State models and the Jones polynomial, Topology 26 (1987) 395–407; Knots, abstract tensors and the Yang–Baxter equation, in: Knots, Topology and Quantum Field Theories, Proceedings of the Johns Hopkins Workshop on Current Problems in Particle Theory 13, ed. L. Lussana (World Scientific, Singapore, 1990) pp. 179–334.

[14] L.H. Kauffman, An invariant of regular isotopy, Trans. Am. Math. Soc. 318 (1990) 417–471.

[15] A.I. Khinchin, Mathematical Foundations of Statistical Mechanics, translated from the Russian by G. Gamow (Dover, New York, 1949).

[16] N. Margolis, Physics-like models of computation, Physica D 10 (1984) 81–95.

[17] J. von Neumann, Theory of Self-Reproducing Automata, edited and completed by A. Burks (University of Illinois Press, Urbana, 1966).

[18] M. Okado, M. Jimbo and M. Tetsuji, Solvable lattice models in two dimensions and modular functions, Sugaku Expositions 2 (1989) 29–54.

[19] R. Penrose, Applications of negative dimensional tensors, in: Combinatorial Mathematics and its Applications, ed. P.J.A. Welsh (Academic Press, New York, 1971) pp.221–244.

[20] A.M. Polyakov, Fermi–Bose transmutations induced by gauge fields, Mod. Phys. Lett. 3A (1988) 325–328.

[21] Y. Pomeau, Invariant in cellular automata, J. Phys. A 17 (1984) L415–418.

[22] K. Reidemeister, Knotentheorie (Chelsea, New York, 1948).

[23] S. Takesue, Reversible cellular automata and statistical mechanics, Phys. Rev. Lett. 59 (1987) 2499–2502; Ergodic properties and thermodynamic behaviour in elementary reversible cellular automata. I. Basic properties, J. Stat. Phys. 56 (1989) 371–402; Fourier's law and the Green–Kubo formula in a cellular automaton model, Phys. Rev. Lett. 64 (1990) 252–255.

[24] T. Toffoli, Cellular automata mechanics, University of Michigan Computer and Communication Sciences Department Technical Report 208 (1977).

[25] B. Trace, On the Reidemeister moves of a classical knot, Proc. Am. Math. Soc. 89 (1983) 722–724.

[26] V.G. Turaev, The Yang–Baxter equations and invariants of links, Inv. Math. 92 (1988) 527–553.

[27] G. Vichniac, Simulating physics with cellular automata, Physica D 10 (1984) 96–116.

[28] M. Wadati, T. Deguchi and Y. Akutsu, Exactly solvable models and knot theory, Phys. Rep. 180 (1989) 247–332.

[29] E. Witten, Quantum field theory and the Jones polynomial, Commun. Math. Phys. 121 (1989) 351–399; Gauge theories and integrable lattice models, Nucl. Phys. B 322 (1990) 629–697.

[30] S. Wolfram, Statistical mechanics of cellular automata, Rev. Mod. Phys. 55 (1983) 601–644.

[31] S. Wolfram, Mathematica: A system for doing mathematics by computer (Addison–Wesley, New York, 1988).

[32] S. Yamada, An operator on regular isotopy invariants of link diagrams, Topology 28 (1989) 369–377.

[33] A.B. Zamolodchikov and Al.B. Zamolodchikov, Factorized S-matrices in two dimensions as the exact solutions of certain relativistic quantum field theory models, Ann. Phys. 120 (1979) 253–291.

Physica D 45 (1990) 345–354
North-Holland

CRITICAL DYNAMICS OF ONE-DIMENSIONAL IRREVERSIBLE SYSTEMS

Olivier MARTIN

Department of Physics, City College of New York, New York, NY 10031, USA

Received 12 March 1990
Revised manuscript received 27 March 1990

Reversible and irreversible systems in the universality class of the kinetic Ising model with a non-conserved order parameter are studied. In the scaling regime, I show how to compute the connection between the bare and renormalized dynamics and thus show how to extract the correlation length and auto-correlation time.

1. Introduction

There has been much controversy recently over whether non-equilibrium irreversible models must have the same critical properties as equilibrium models [1,2,6]. In this paper, I present a calculational tool which can be used to study efficiently the critical behavior of non-equilibrium irreversible systems with the hope that this may help resolve the controversy. To make this paper self-contained, I first review the meaning of equilibrium, reversibility, irreversibility, and criticality for stochastic systems. The rest of the paper shows how to get a coarse-grained description of the dynamics (and thus statics) in some irreversible systems. In section 5, I show how to coarse-grain the kinetic Ising model. There I describe the use of the droplet picture and some random-walk techniques. Then, in section 6, I generalize the method to arbitrary irreversible models in the universality class of the kinetic Ising model with a non-conserved order parameter. Note that a brute force numerical approach is not well suited to the determination of the correlation length ξ because domain nucleations are rare, leading to large statistical fluctuations. However, when ξ diverges, the gas of domain walls becomes dilute and I provide a short-cut. Instead of having to deal with a many-body problem as in a simulation, I use the droplet picture which becomes asymptotically correct as $\xi \to \infty$. Determining ξ is then usually reduced to simulating a particle diffusing out of a well. As an extra bonus, this picture gives the critical properties of the dynamics, e.g., the auto-correlation time τ and its exponent z defined by $\tau = \xi^z$. Finally, illustrative examples are given in section 7.

2. Reversible and irreversible dynamics

It is well known that the fundamental equations of classical mechanics (Newton's equations) are time-reversal invariant. This corresponds to a symmetry of the equations of motion: if one reverses the direction of time, $t \to -t$, and thus also reverses the direction of the velocities, the equations of motion are unchanged. This symmetry can be understood at a pictorial level: a film of some mechanical motion shown backwards also describes a realization of Newton's equations. It is in this sense that time has no arrow, so the dynamics is called reversible. Any dynamics which is not reversible is called irreversible. For instance, motion where energy is dissipated by friction is irreversible. Thus systems which have irreversible dynamics are often called dissipative.

The systems which are of interest to us in this paper are probabilistic, i.e., contain noise terms. Thus, instead of "deterministic" equations like Newton's, consider a stochastic equation. The simplest example is the Langevin equation for a single particle moving in one dimension:

$$\dot{x}(t) = \eta(t); \quad \langle \eta(t)\,\eta(t') \rangle = 2kT\delta(t - t'),$$

where T is a temperature, η is a Gaussian white noise term of zero mean, $x(t)$ is the coordinate of the particle at time t, and $\dot{x}(t)$ its velocity. This equation describes a particle undergoing a random walk in the continuum, i.e. diffusion. Because of the noise term, any continuous trajectory $x(t)$ is possible, albeit with low probability. What then does one mean by reversible or irreversible dynamics? Instead of looking at individual trajectories as one does for deterministic equations, one has to consider an ensemble of trajectories. If one starts with a peaked distribution $P(x, t = 0)$ for x, then with time, the distribution spreads out since the above Langevin equation simply describes diffusion. Clearly the spreading does have a directionality with time, so one is tempted to say that the dynamics *is* irreversible. However, were x to be constrained to remain in a bounded interval, $P(x, t)$ would settle into a time-independent distribution $P_s(x)$ and the directionality with time would disappear. Thus for stochastic equations, one needs a criterion for reversibility which does not depend on initial conditions. (Note that with deterministic equations of motion, the behavior of the system can also look irreversible even with reversible equations of motion. This occurs when e.g. a cup breaks; the arrow of time is due to the non-equilibrium nature of the initial conditions.)

Given the above remarks, one can now define reversibility for probabilistic dynamics. For concreteness, take a system with discrete time and discrete variables, so that the dynamics is specified by a transition probability matrix T, and suppose that T allows a time-independent probability distribution P_s. Workers in the field of "interacting particle systems" [13] customarily define the dynamics to be reversible if and only if

$$P_s(C_1)\,T(C_1 \to C_2) = P_s(C_2)\,T(C_2 \to C_1),$$

where C_1 and C_2 are two arbitrary configurations. This condition is called the *detailed balance* or time-reversal condition and one says that P_s is an *equilibrium* distribution. Not surprisingly, the reversibility of T can be expressed without explicit reference to P_s (which need not be unique anyway) by requiring T to be of the form

$$T = S^{-1}DS,$$

where S is a symmetric matrix and D is a diagonal matrix. (Then in fact there exists at least one time-independent distribution P_s e.g., $D_{ii} = P_s(C_i)$.) The detailed balance equation can be interpreted by saying that when one has reached equilibrium, all currents or flows in configuration space are zero.

A consequence of detailed balance is *balance*, a much weaker condition:

$$P_s(C_1) \sum_{C_2} T(C_1 \to C_2)$$

$$= \sum_{C_3} P_s(C_3)\,T(C_3 \to C_1).$$

This condition simply means that P_s is a stationary (or steady-state) distribution, i.e. it is invariant under the dynamics. In mathematical language, one says that P_s is an invariant measure of T. If the dynamical variables are compact, then there exists at least one invariant measure. Furthermore, if the volume is finite (there are a finite number of degrees of freedom), and if T or some power thereof is fully probabilistic (all transitions occur with some non-zero probability), then necessarily the invariant measure is unique. This need not be true however when the volume is infinite: for certain critical parameter values, one may go from a single invariant measure to several, giving rise to a phase transition, a concept of great importance in statistical mechanics.

3. Statistical mechanics of probabilistic cellular automata

Statistical mechanics deals with static *equilibrium* properties. Typically an equilibrium system is specified by a finite range Hamiltonian H_{eq} which determines the relative probability distribution of the degrees of freedom. Thus if $\{\phi_j\}_{j \in I}$ is a finite subset of these degrees of freedom, one has

$$P_{eq}(\{\phi_j\}_{j \in I}) \sim \exp[-\beta H_{eq}(\{\phi_j\}_{j \in I})],$$

where $\beta = 1/kT$ and the proportionality constant is just a normalization factor (which depends on the (fixed) values of the variables not in the set I). If the system has a finite number of degrees of freedom, then the equilibrium distribution is unique and given by the Boltzmann factor:

$$P_{\mathrm{eq}}(C) = \exp[-\beta H_{\mathrm{eq}}(C)]/Z \,,$$

where C is a configuration with all the ϕ_j specified and Z is the partition function

$$Z(\beta) = \sum_C \exp[-\beta H(C)] \,.$$

If there are an infinite number of degrees of freedom, there may be several equilibrium distributions, all compatible with the above relative probability distribution. The main focus of statistical mechanics is to determine all the equilibrium distributions and their properties.

If one is dealing with a *non-equilibrium* system, instead of equilibrium distributions, one must find the invariant distributions, and in general these cannot be described by a Boltzmann factor with a local Hamiltonian. Thus non-equilibrium or irreversible systems are usually formulated in terms of their dynamics rather than their statics. Since most probabilistic cellular automata (PCA) are irreversible, one might not expect there to be much of a connection between PCA and statistical mechanics. However this is not true: even though the *spatial* degrees of freedom of the invariant measure are not described by a Hamiltonian, there exists a space–time Hamiltonian $H_{\mathrm{s-t}}$ which describes the probability distribution of the *space–time* patterns generated by PCA dynamics. The existence of such a Hamiltonian follows from the fact that the transition matrix T is a transfer matrix [11]. In principle, this Hamiltonian property gives one access to a wide variety of calculational tools such as low- or high-temperature expansions when the noise level is low or high. However, in practice, these methods are not useful for most PCA. *Low-temperature* expansions are generally not calculable, first, because the ground states of $H_{\mathrm{s-t}}$ are difficult to determine, and second, because the excitations from the ground states are very extended (if not unbounded!) in space–time. The use of *high-temperature* expansions requires that the PCA be fully probabilistic, i.e., all elementary transition probabilities must be non-zero. Many PCA of interest have some transition probabilities equal or close to zero; their space–time

Hamiltonians are unbounded, and this invalidates any high temperature expansion. Finally, there is the mean field approach. It is of limited reliability especially when the correlation length gets large. Also deriving systematic corrections to the mean field approximation is very tedious.

Nevertheless, statistical mechanical approaches can be useful for some very special cases of PCA [4,12,17] (of measure zero in the space of all PCA): affine PCA and reversible PCA. The first are exactly soluble, and the second have *equilibrium* distributions for the *spatial* degrees of freedom at fixed time as defined in the previous section. In particular, if the PCA has range r and is reversible, then the associated spatial Hamiltonian is of range r at most. From this Hamiltonian picture, objects such as spatial correlation functions can be calculated reliably within the framework of statistical mechanics. If a PCA is irreversible but nevertheless is very close to being reversible, it is possible to expand the dynamics about the reversible limit and to get perturbative expressions for the invariant distribution [4]. Unfortunately such expansions are very tedious. Also they are not applicable to most PCA as irreversibility is usually large. A method however which *is* applicable to most PCA is a kind of mean field theory in which the probability distribution of blocks of *spatial* variables are followed as a function of time [7]. Since this approach is relatively new, it is unclear at this time how far it can be pushed. It would be interesting to determine whether reliable answers can be obtained when the correlation length is large without using too large size blocks. (Note that the amount of work required grows exponentially with the block size.) With the possible exception of this last method, none of the approaches are reliable when the correlation length gets large, and this prevents one from studying the limit ($\xi \to \infty$). As mentioned in section 1, our long-term goal is to understand whether reversible and irreversible systems can have different critical behavior. In sections 5 and 6, I provide one calculational method which does not break down as $\xi \to \infty$.

4. Criticality and irreversibility

Central to the idea of criticality is the notion of length scales. If one considers a statistical mechanical lattice model, one length scale is given by the lattice spacing; other length scales may be introduced in the Hamiltonian for instance through its interaction range. Furthermore, any model *dynamically* generates a length scale called the *correlation length* which describes the distance over which variables are strongly correlated. For most temperatures and parameters in the Hamiltonian, this correlation length will be on the order of the interaction range, but for certain values, it may be much larger. When the correlation length is *infinite*, the system is said to be critical. Such a phenomenon occurs when a system undergoes a second-order phase transition. Near criticality, one has collective behavior and the various thermodynamic quantities exhibit "scaling", i.e., they become singular and typically diverge as a power of the correlation length [14]. In this limit, the diverging correlation length becomes the only relevant scale: the microscopic scale is irrelevant and one can introduce a continuum picture. More remarkable is the fact that at criticality, the large scale properties are insensitive to the details of the lattice Hamiltonian: all the terms (coupling constants) defined on the lattice scale can be summarized by their effect on just a few renormalized parameters describing the system on much larger scales. This is called universality: an arbitrary microscopic (lattice scale) Hamiltonian "flows" at large scales to a renormalized Hamiltonian parametrized by just a few renormalized coupling constants.

Discussion of universality has mainly been concerned with Hamiltonian systems. However a broader perspective is to consider the critical behavior of *invariant probability distributions* (not necessarily given by the usual Boltzmann weight) and to ask whether one has universality there too. It is now known that non-equilibrium systems can have very different behavior from equilibrium systems. A good example of this is the generic power-law form of correlation functions in driven diffusive systems [19]. Nevertheless it is tempting to suppose that the critical behavior of systems does not depend on whether they are Hamiltonian or

not. This is currently a hotly debated question [1,2,6], and one is still far from a resolution. There is numerical evidence both for universality [9,18] and against it [1,2]. On the theoretical side, it has been pointed out that within the framework of time-dependent Ginzburg–Landau theory, small amounts of irreversibility in the dynamics are irrelevant near four dimensions [6]. My hope is that one may resolve this universality issue (at least in certain cases) by first obtaining a description of the dynamics at large scales. Given this macroscopic dynamics, one may then ask whether it falls in a known Hamiltonian universality class. Clearly one must be able to accomplish this for systems in which the irreversibility is not necessarily small and for PCA which are not fully probabilistic. Perhaps one of the most challenging aspects of these systems is their combination of deterministic and stochastic dynamics.

5. The kinetic Ising model and the droplet picture

The (static) Ising model is a lattice statistical mechanical model of magnetism. On each site i of the lattice there is a spin S_i which can take on the values ± 1. In one dimension, the Hamiltonian for the nearest-neighbor Ising model is

$$H(\{S_i\}) = \sum_j S_j \cdot S_{j+1} .$$

The partition function

$$Z(kT) = \sum_{\{S_i\}} \exp[-H(\{S_i\})/kT]$$

determines the statics of the model and can be used to extract spatial correlation functions as a function of temperature. The standard correlation length ξ is defined by the asymptotic behavior of the spin–spin correlation function

$$\langle S_i \cdot S_{i+m} \rangle \approx \exp(-m/\xi),$$

where $\langle \cdots \rangle$ denotes the ensemble average. An exact calculation gives $\xi^{-1} = -\log(\tanh \beta)$ where $\beta = 1/kT$. Thus the correlation length diverges as $T \to 0$, and $T_c = 0$ is the only critical point of the model in one dimension. (Higher-dimensional models can have critical points at finite temperature, but a general result shows that Hamiltonian models in one dimension with finite interaction range can have critical points only at zero temperature.)

To move on to the *kinetic* Ising model, we need to introduce dynamics which specify the time evolution of the Ising spins. Note that a Hamiltonian determines the static properties of a model, but none of the dynamic properties. In general, we want to find some stochastic dynamics which will give rise to the correct invariant measure or probability distribution, as given by the statics. Obviously, a static model can have many dynamics. The most commonly chosen dynamics are single spin flips with transition probabilities depending on the neighboring spins in such a way as to satisfy detailed balance. There are also methods which use multi-spin flips such as the exchange of nearest-neighbor site values in Kawasaki [10] dynamics. Spin-flips can occur in continuous time as in Glauber [5] dynamics, or in discrete time as in most Metropolis Monte Carlo algorithms [15].

Almost all statistical mechanical models have "order parameters" which distinguish the various equilibrium distributions. Thus one commonly classifies the various possibilities for dynamics according to whether they do not preserve (class A) or do preserve (class B) the order parameter. For the Ising model, the order parameter is the total magnetization (the sum of all the spin values). Glauber and Metropolis dynamics do not conserve the order parameter, whereas Kawasaki dynamics does. These two categories of dynamics lead to very different dynamical critical behavior [8]. In what follows, I will only consider dynamics which do not conserve the order parameter.

Consider what happens in all kinetic Ising models in class A as one approaches the critical point. Since $T_c = 0$, the size of blocks of up or down spins diverges as $\xi \to \infty$. These blocks are called droplets. The droplets are separated by domain walls which are also called kinks, the density of which goes to 0 at zero temperature. In fact, the equilibrium properties of the *nearest-neighbor* Ising model can be described exactly by a grand canonical gas of non-interacting kinks with a chemical potential $\exp(-4\beta)$. (This kink picture can also be introduced for more compli-

cated Hamiltonians, although the kinks are no longer non-interacting.) The kinks undergo random walks with a diffusion constant $D/2$. If two kinks touch, they annihilate, thereby increasing the size of blocks of aligned spins. There is also a spontaneous creation of pairs of kinks in the bulk at a rate $\eta \ll 1$. This process corresponds to the nucleation of new droplets and happens most often in very large blocks, thereby decreasing the size of blocks. It is the competition between these two processes which determines the correlation length. By balancing the creation and annihilation rates of kinks, one can get a relation between η and ξ. However, to do so, one must assume in practice that the kinks are non-interacting and this is not the case for many of the more complicated models. Thus one needs a quantity analogous to η which can be defined on a length scale larger than the interaction length, but still much smaller than the correlation length. The best way to generalize η is to consider a continuum picture. The continuum limit can be reached by fixing ξ in physical units and taking the lattice spacing to zero. The correlation length in lattice units diverges, thus the lattice η must go to zero. The continuum nucleation rate has to be defined at a certain scale which does not change as the lattice spacing goes to zero. Essentially one has to renormalize the theory and make η scale dependent.

Modulo this technical difficulty, one expects the diffusion, annihilation, and nucleation processes to be the only relevant ones for determining the large scale properties of the dynamics. Thus the universality class of the kinetic Ising model class A should form a two-dimensional family of dynamics parametrized by D and some "nucleation" rate η, or equivalently in terms of directly measurable quantities such as D and ξ. Now notice that the two parameters D and η also determine the nucleation and growth in time of a *single* domain of down spins inside a sea of up spins. The method I propose here uses the inverse relation, i.e. the fact that the single droplet dynamics determine the many droplet dynamics in the critical region because domain walls are dilute. Thus two models (reversible or not) will have the same large scale behavior, correlation length ξ, and autocorrelation time τ if their single droplet dynamics are identical at scales much larger than the lattice spacing. If a numerical study of this single droplet dynamics is necessary, it can be done in a volume much smaller than the correlation length, a method far more efficient than a brute force numerical simulation of the original model. Further, for the kinetic Ising model and most other models of interest, the one droplet dynamics is so simple that it can be investigated at the master equation level rather than by direct simulation of the stochastic process. This eliminates any noise fluctuations and allows for an algorithm which does not deteriorate as $\eta \to 0$, i.e., one does not have to wait long times for rare events to occur.

Let me now illustrate how to solve the single domain dynamics. To be specific, I consider continuous-time Monte Carlo dynamics: attempt at each site to flip the spin with a probability proportional to the time interval dt. Accept the flipped spin with probability $\min\{1, \exp(-\beta\delta H)\}$ if the energy changes by a non-zero amount δH, and with probability $D/2$ if the energy is unchanged. To go to the single droplet dynamics, allow at most one domain of down spins in the system at any time. Spin flips are limited to either nucleation of this one domain or to the spins on the edges of the domain if it already exists. This allows one to obtain an equation of motion for the domain length L, which is represented schematically as

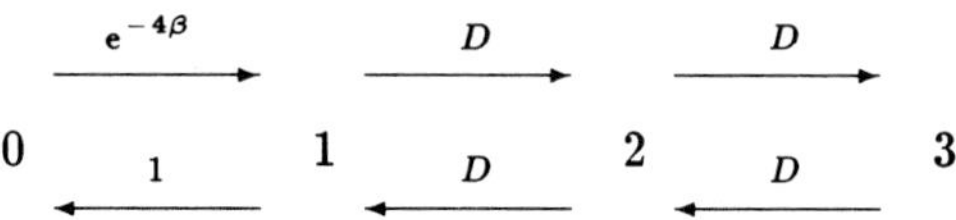

The values above the arrows are the $L \to L \pm 1$ transition rates, and D is the diffusion constant which simply sets the time scale. (The factor 2 in the hopping probability is due to the fact that both ends of the domain diffuse.) It is possible to formally solve for the time dependence of $P(L,t)$, the probability distribution of L at time t. A given realization of the stochastic process can be thought of as a random walk with the above rates. It is useful to break up the walk in time according to the first and last passage through the site

1, and to sum over all possible histories for each category. One then has

$$P(L,t) = \int dt_1 \int dt_2 \, g(L: 0 \to 1, 0 \to t_1)$$

$$\times G(L: 1 \to 1, t_1 \to t_2) \, Dg'(L: 2 \to L, t_2 \to t).$$

Here, g is the probability of first hopping to the site 1 at time t_1. G is the probability of returning to site 1 within time t_2 given that the particle is at the same site at time t_1. It can also be thought of as a Green's function. Finally, g' is the Green's function for going from site 2 to site L without passing through site 1. The above convolution is best evaluated using Laplace transforms. If $\tilde{P}(L,s)$ is the Laplace transform of $P(L,t)$, then

$$\tilde{P}(L,s) = g(0 \to 1) \, G(1 \to 1) \, Dg'(2 \to L),$$

where the expressions on the right-hand side are Laplace transforms of the above introduced quantities, but the s argument has been omitted. Using standard methods [16], one finds for arbitrary r_0

$$g'(r_0 + 1 \to L) = a(s)\,\lambda(s)^{L-r_0-1}\,(1 - \lambda^2),$$

where

$$a^{-1}(s) = 2D\sqrt{(1 + s/2D)^2 - 1}\,,$$

$$\lambda(s) = 1 + s/2D - \sqrt{(1 + s/2D)^2 - 1}\,.$$

Furthermore,

$$G(r_0 \to r_0) = \gamma_{r_0}/[1 - g(r_0 \to r_0)],$$

with

$$\gamma_1 = 1/(D + 1)s.$$

In the above expression, g has been generalized to be the Green's function for first passage through the second argument. $g(1 \to 1)$ can be further decomposed into those paths which go right rather than left: $g = g_{\rm L} + g_{\rm R}$. Both of these quantities can be explicitly computed in terms of D and the amplitude for hopping from the origin. Finally, to get $P(L,t)$, one has to do the integral corresponding to the inverse Laplace transform. Most of the contribution to the integral comes from small s arguments. By expanding the integrand near $s = 0$, one gets

$$P(L,t) = \int \mathrm{d}s \, \frac{\exp[Dts - (L-2)\sqrt{s}\,]}{\sqrt{s} + s/\eta},$$

where $\eta = \exp(-4\beta)$ is defined as the factor dividing s in the denominator given the form of the numerator. Not surprisingly, P corresponds to a two-parameter family of distributions; it specifies the large-scale behavior of the droplet model.

For all D and β, one knows the exact value of the correlation length and auto-correlation time in the full many-droplet system. This leads to the following relation between η and ξ for large correlation lengths:

$$\xi = 1/2\sqrt{\eta}\,.$$

Note that the parameter η is determined solely by the single droplet dynamics as demonstrated above, but nevertheless, it can be used to extract the correlation length. In addition, the value of the auto-correlation time follows from the relation

$$\tau = \xi^2/D.$$

These relations will be used in the non-soluble (irreversible) models: the main problem there will be to determine η.

6. General formulation

I begin with a stochastic process defined on a one-dimensional lattice, each site having an Ising variable. The process is taken to be local and homogeneous in space, though the method can be extended to processes which are periodic in space. Assume that there is a rule for flipping a spin which depends only on the current values of the spins in some finite neighborhood of size r_0. The transition rates are given and time-independent, but need not be non-zero, i.e., the formalism does not require a fully probabilistic process. For ease of presentation, I consider the spatial period to be one and time to be continuous, though this does not affect the conclusions. If the dynamics is to be in the same universality class as the kinetic Ising model with a non-conserved order parameter (class A), the spins must form large blocks, domain walls must be relatively sharp (much smaller than the correlation length) and they should execute unbiased random walks at large scales. In addition, two domain walls must annihilate when they cross, and the *dominant* mechanism for the formation of new domains must be nucleation in the *bulk*.

The first step towards obtaining a macroscopic picture is to identify the degrees of freedom corresponding to domain walls which can be identified with the kinks of the effective continuum model. If the walls are sharp, this identification is straightforward, but this is not always the case. The second step is to *reduce* the infinite-body stochastic process to a few-body process describing the dynamics of these new degrees of freedom when the correlation is large. One can usually do this by applying the stochastic process to an initial configuration of up spins and allowing at most one down domain to appear. This dilute gas, droplet, or one domain approximation is the source of the simplification. Since by hypothesis nucleation is very difficult ($\eta \ll 1$), this last step introduces only a small error. The third and last step is to use this *reduced* stochastic process to obtain $P(L,t)$, the probability that the down domain has size L at time t. We will see that P becomes universal as the nucleation time diverges, or equivalently, as $\xi \to \infty$: all systems in the universality class of the kinetic Ising model class A have P's belonging to a two-parameter family which can be labeled by the correlation length ξ and auto-correlation time τ.

Consider now the equation of motion for L, the size of the down droplet. Suppose that if two domain walls are further than r_0 apart they execute independent random walks with a diffusion constant $D/2$. Thus consider an evolution equation for the length L which is arbitrary for $r < r_0$, but which corresponds to nearest-neighbor hops with diffusion constant D for $r > r_0$. It is the coefficient D together with the dynamics in the region $r < r_0$ which determine the "nucleation" rate, and thus ξ and τ, in the full model. Since the dynamics in the region $r < r_0$ can have many parameters, one might expect $P(L,t)$ to have a complicated and non-universal structure. I will show why, as the nucleation rate becomes small, $P(L,t)$ collapses to a two-parameter distribution when $L \gg r_0$ and $t \gg 1$: the (irreversible or not) dynamics for $r < r_0$ is renormalized and can be modeled by just a single coefficient as in the kinetic Ising model.

Again, compute $P(L,t)$ by using Laplace transforms in t, and by writing P as a sum over all histories. A given history has a probability given by a product of weights:

– a staying probability $\gamma_n(t)$ each time the particle rests at site n for exactly time t;

– a hopping amplitude given by the stochastic process whenever the particle hops.

For $r > r_0$, only one-step hops are allowed, each carrying a weight D. An arbitrary path weight is factored according to the first and last passage through the site r_0, leading to

$$\tilde{P}(L,s) = g(0 \to r_0)\, G(r_0 \to r_0)\, Dg'(r_0 + 1 \to L).$$

On the right-hand side, the variable s is implicit and the functions are the same as defined in the previous section: g refers to the Green's function for first passage through r_0, g' for no passage through r_0, and G is the general Green's function. g' and g_R are the same as for the nearest-neighbor kinetic Ising model. Note that all the details relevant to nucleation occur only in the quantities $g(0 \to r_0)$ and g_L, which can be calculated by solving the following lattice diffusion equation (written in Laplace transform space)

$$s\tilde{P} - P(t = 0) = -M\tilde{P},$$

where M is the matrix of hopping weights inside $r < r_0$ and zero boundary conditions should be imposed on $\tilde{P}$ at r_0. This equation describes the process of a particle initially at site 0 diffusing and finally escaping from the region $r < r_0$. M generically has r_0 eigenvectors ψ_i and eigenvalues λ_i in terms of which one can express g_L and $g(0 \to r_0)$. In particular,

$$g_L = \gamma_{r_0} D^2 \sum_i \frac{\langle \psi_i | \delta_{r_0 - 1}\rangle}{s + \lambda_i}\psi_i(r_0 - 1).$$

$\langle \psi_i | \delta_{r_0 - 1}\rangle$ is the *parallel* projection of $|\delta_{r_0 - 1}\rangle$ onto $|\psi_i\rangle$ in the basis system formed by the eigenvectors, not a scalar product. The eigenvalues give the decay rates of the eigenvectors in the original time-dependent problem. As the nucleation rate goes to 0, there will be *one slowest decaying wavefunction* ψ_1 (associated with λ_1) which will dominate the long-time behavior. Thus, for small s, one can expand the above quantities leading to

$$\tilde{P}(L,s)$$

$$= \frac{\exp[-(L - r_0 - 1)\sqrt{s/D}][1 + \mathrm{O}(Ls/D)]}{\sqrt{s/D} + Ds[\langle \psi_1 | \delta_{r_0 - 1}\rangle / \lambda_1][1 + \mathrm{O}(s/\lambda_1)]}.$$

After a rescaling which makes the D dependence more explicit:

$$P(L,t) = \int ds\, \frac{\exp[Dts - (L - r_0 - 1)\sqrt{s}\,]}{\sqrt{s} + s/\eta}$$

$$\times [1 + \mathrm{O}(s^{3/2}/\eta^2 + Ls)],$$

where I have used the normalization convention introduced in the previous section. Then one has

$$\eta = \lambda_1 / \langle \psi_1 | \delta_{r_0 - 1}\rangle.$$

For large Dt, the contribution at small s dominates, so that asymptotically, one has

$$P(L,t) = \frac{\eta}{\pi}\exp\left(-\frac{L^2}{4Dt}\right)\int_{-\infty}^{+\infty} \frac{\exp(-\alpha^2 u^2)\,du}{1 + u^2},$$

with

$$\alpha = \eta\sqrt{Dt} + L/2\sqrt{Dt}.$$

This is the promised two-parameter (D and η) family from which one can extract ξ and τ. To determine the coefficient η, one could do a simulation of the master equation and thereby obtain P numerically, but it is simpler and more efficient to solve the diffusion equation and thereby get the eigenvalues and eigenvectors directly. The interpretation of η is obtained by performing the analysis for a soluble kinetic model. This is precisely what was done in the previous section, giving the asymptotic relation between η and the correlation length. This approach thus enables one to go from the droplet dynamics of an irreversible PCA to the critical behavior of the full PCA for infinite volume.

7. Results

This section illustrates the application of the method to two exactly soluble Hamiltonian models and to one irreversible model.

7.1. Next-nearest neighbor Ising model

The Hamiltonian for this model is

$$H = J_1 \sum S_i \cdot S_{i+1} + J_2 \sum S_i \cdot S_{i+2},$$

and the correlation length is

$$\xi^{-1} = -\log[\tanh 2\beta(J_1 + 2J_2)].$$

For the kinetics, use continuous-time Monte Carlo dynamics. The transition rates for a single domain are given by the following figure:

Since the model has only single hops, one can write a recursion relating $g_{2,L}$ to $g_{1,L}$, and I find $\eta = \exp[-4\beta(J_1 + 2J_2)]$. This is exactly what is expected from an expansion of the exact value:

$$\xi^{-1} = 2 \exp[-2\beta(J_1 + 2J_2)]$$
$$\times \{1 + O(\exp[-4\beta(J_1 + 2J_2)])\}.$$

7.2. The alternating Ising model

The second soluble model is a modified Ising Hamiltonian:

$$H = J_1 \sum_i S_{2i} \cdot S_{2i+1} + J_2 \sum_i S_{2i+1} \cdot S_{2i+2},$$

where without loss of generality one can take $J_1 > J_2$. The underlying lattice periodicity is two. A transfer matrix solution shows that $\xi^{-1} = -\log[\tanh(\beta J_1)\tanh(\beta J_2)]/2$. The model is of interest both because of this spatial periodicity and because for Glauber or Monte Carlo dynamics, the diffusion constant varies with temperature, $D = \exp[-2\beta(J_1 - J_2)]$. This variation [3] changes the dynamic critical exponent from $z = 2$ to $z = 1 + J_1/J_2$. The analysis of section 6 has to be modified slightly to accommodate the spatial periodicity. One obtains

$$\eta = [\exp(-2\beta J_1) + \exp(-2\beta J_2)]^2/4,$$

so that

$$\xi_\eta^{-1} = \exp(-2\beta J_1) + \exp(-2\beta J_2).$$

This is indeed the leading term in the expansion of the exact ξ at $T \to 0$.

In all cases one sees that the dilute gas approximation gives the correlation length with relative errors which go as ξ^{-2} as expected from the non-interacting gas picture.

7.3. One-sided nearest neighbor systems

Finally, what about irreversible stochastic processes such as probabilistic cellular automata or continuous-time lattice gases? For many such systems, it is unknown whether there exists a unique equilibrium phase even in one dimension. However, a number of results are known for special classes of such systems. In particular, it has been proved that additive one-sided nearest neighbor systems are ergodic [13]. I thus chose to study the following such process: a spin has a flipping probability which depends only on its current state and that of its right neighbor. If the spins are anti-parallel, the left spin flips with probability u per unit time. If the spins are parallel, the same spin flips with probability v per unit time. The system has a correlation length which goes to infinity as v goes to zero. An interesting feature of this model is that domain walls have a net average velocity in addition to diffusing. My formalism is readily adapted to this problem. In the dilute limit, the master equation for L is identical to an appropriately chosen kinetic Ising model, and this leads to

$$\xi_\eta = \tfrac{1}{2}(u/v)^{1/2}.$$

8. Conclusion

I have illustrated how to obtain the static and dynamic critical properties of irreversible systems in the universality class of the one-dimensional kinetic Ising model with a non-conserved order parameter. For large correlation lengths, the kinetics can be parametrized by a diffusion constant and a nucleation rate parameter η which can be determined by solving a simple few-body problem. This provides a method for calculating the connection between the bare and the coarse-grained dynamics, and directly gives the quantities ξ and τ. The method as it stands is applicable to a wide variety of PCA as long as they fall in this universality class. Clearly the work should be extended to other classes such as percolation where universality is very much debated. Finally, even for systems where universality may not hold, this approach provides a very efficient numerical tool since one does not need system sizes on the order of the correlation length.

Acknowledgements

Special thanks goes to Yu He, who participated in the beginning of this project. I am grateful to Hugues Chaté, Richard Friedberg, Nigel Goldenfeld, Howard Gutowitz, and Yoshi Oono for criticism and encouragement. This work was supported in part by the NSF, grant NSF-ECS-8909127.

References

[1] R. Bidaux, N. Boccara and H. Chaté, in: Cellular Automata and Modeling of Complex Physical Systems, eds. P. Manneville et al., Springer Proceedings in Physics, Vol. 46 (Springer, Berlin, 1989).

[2] H. Chaté and P. Manneville, Physica D 32 (1988) 409.

[3] M. Droz, J. Kamphorst Leal Da Silva and A. Malaspinas, Phys. Lett. A 115 (1986) 448.

[4] A. Georges and P. Le Doussal, J. Stat. Phys. 54 (1989) 1011.

[5] R.J. Glauber, J. Math. Phys. 4 (1963) 294.

[6] G. Grinstein, C. Jayaprakash and Y. He, Phys. Rev. Lett. 55 (1985) 2527.

[7] H.A. Gutowitz, J. Victor and B.W. Knight, Physica D 28 (1987) 18.

[8] B. Halperin and P. Hohenberg, Rev. Mod. Phys. 59 (1977) 435.

[9] Y. He, K. Chen and C. Jayaprakash, Phys. Rev. A 36 (1987) 2999.

[10] K. Kawasaki, in: Phase Transition and Critical Phenomena, Vol. 4, eds. C. Domb and M.S. Green (Academic Press, London, 1972).

[11] W. Kinzel, Z. Physik B 58 (1985) 229.

[12] J.L. Lebowitz, C. Maes and E.R. Speer, J. Stat. Phys., to appear.

[13] T.M. Liggett, Interacting Particle Systems (Springer, Berlin, 1985).

[14] S-K. Ma, Modern Theory of Critical Phenomena (Benjamin, New York, 1976)

[15] N. Metropolis, A.W. Rosenbluth, M.N. Rosenbluth, A.H. Teller and E. Teller, J. Chem. Phys. 21 (1953) 1087.

[16] E.W. Montroll and M.F. Shlesinger, in: Non Equilibrium Phenomena II: From Stochastics to Hydrodynamics, eds. J.L. Lebowitz and E.W. Montroll, (North-Holland, Amsterdam, 1984).

[17] P. Rujan, J. Stat. Phys. 49 (1987) 139.

[18] J-S. Wang and J.L. Lebowitz, J. Stat. Phys. 51 (1988) 893.

[19] M.Q. Zhang, J.S. Wang, J.L. Lebowitz and J.L. Valles, J. Stat. Phys. 52 (1988) 1461.

CHAPTER 5

COMPUTATION THEORY OF CELLULAR AUTOMATA

Physica D 45 (1990) 357–378
North-Holland

COMPUTATION THEORETIC ASPECTS OF CELLULAR AUTOMATA

K. CULIK II
Department of Computer Science, University of South Carolina, Columbia, SC 29208, USA

L.P. HURD
Laboratory for Plasma Research, University of Maryland, College Park, MD 20742, USA

and

S. YU [1]
Department of Computer Science, University of Western Ontario, London, Ontario, Canada N6A 5B7

Received 15 January 1990
Revised manuscript received 15 March 1990

Cellular automata may be viewed as computational models of complex systems. They also can be seen as continuous functions on a particular family of compact metric spaces. This paper surveys results largely motivated by an attempt to answer questions posed by the latter approach by means of techniques from the former.

1. Introduction

Cellular automata are models of complex systems in which an infinite lattice of finite state machines updates itself in parallel according to an isotropic local rule. Cellular automata were introduced by von Neumann [1] in an attempt to model self-reproducing systems such as biological systems; recently they have been used as models of complex systems from biology to highly parallel computers.

The dynamics of cellular automata are based on the following observations about physical systems:

(i) Information travels a finite distance in finite time.

(ii) The laws of physics are independent of the position of the observer.

To this list, von Neumann added the simplifying assumptions of discrete time, space, a local discrete state space, and a regular grid topology.

He used two dimensions, but the definition can be extended to any number of dimensions. Here we will mainly focus on the one-dimensional or linear case (see section 2).

Differences from more traditional models of computation (i.e. Turing machines) are immediately apparent. Cellular automata have no distinguished "head"; the computation proceeds in parallel across the entire lattice. Also, cellular automata need not have a distinguished terminal alphabet or halting state. This means that it is difficult to separate the dynamics of the cellular automaton from the computation.

What distinguishes the work in this survey from some other studies of cellular automata as computational models, is an attempt to take this degree of non-specialization one step further. Each of these sections relates to the global behavior of cellular automata considering the evolution of all possible initial configurations. This is sharply distinct from Turing machines, where very few possible tape-head configurations represent legal starts.

Each cellular automaton has at its base a lat-

[1] On leave from Kent State University, Kent, OH 44242, USA.

tice of dimension $d \geq 1$. However, the number of results which can be stated in this generality is somewhat small. Section 2 defines d-dimensional cellular automata and discusses some consequences of the restriction to the linear case which will be our primary focus.

One interesting aspect of cellular automata, and as we have seen the motivation for their first definition, is their ability to simulate computational processes. Section 3 defines what it means for a (one-dimensional) cellular automaton to be universal. It defines two normal forms: one-way cellular automata and totalistic cellular automata, and shows that each type is universal in the class of all one-dimensional cellular automata. It is shown that there exists a universal cellular automaton with fourteen states.

Section 4 reviews Wolfram's four classes of linear cellular automata [2], and their formalization by Culik and Yu (CY) [3]. Membership in each of these classes is shown to be undecidable and the degree of undecidability is discussed. Other classification schemes are mentioned.

Infinite words defined by the state of a one-dimensional lattice have a close relationship with the usual finite words of formal language theory [4]. Section 5 defines the language of finite words associated with sets of infinite configurations (subshifts) arising from cellular automata. Results are shown relating language-theoretic constructions such as GSMs [5] and cellular automata maps.

Another interpretation of cellular automata (in any dimension) is as continuous maps on a compact metric space commuting with certain shift operators. Section 6 casts the definition of cellular automata in the terminology of topological dynamical systems. In the case of linear cellular automata, standard constructions from topological dynamics, attractors, periodic sets, and non-wandering sets, are shown to give rise to sets of high language theoretic complexity as defined in the preceding section.

Section 7 discusses variants of the injectivity and surjectivity problem for cellular automata, the relationships among these properties and the possibility to test them algorithmically. Here a sharp distinction exists between dimension one and dimensions two and greater, and they are

dealt with separately.

2. Definitions

2.1. Cellular automata

Given a finite set S and a dimension d one can consider a d-dimensional lattice in which every point has a label from the set S. Formally, the lattice is the set $L = \mathbf{Z}^d$ and the *shift space*, the topological product S^L (i.e. the set of functions from $L \rightarrow S$), where S is given the discrete topology, and the product is given the product topology. This space is compact by Tychonoff's theorem [6]. What is more, each of the lattice directions determines a natural shift map σ_i which is a homeomorphism of the space.

Given the d-dimensional shift space, one can form a purely dynamical definition of cellular automata.

Definition 1 *A cellular automaton is a continuous map* $G: S^L \rightarrow S^L$ *which commutes with* σ_i *($1 \leq i \leq d$).*

This definition, however, is not useful for computations. Therefore we consider an alternate characterization. Given S a finite set, and d-dimensional shift space S^L, consider a finite set of translations, $N \subseteq L$. Given a function $f: S^N \rightarrow S$, called a local rule, the global cellular automaton map is given by

$$G_f(c)_v = f(c_{v+N})$$

where $v \in L$ and $v + N$ consists of the set of translates of v by elements in N.

Definition 2 *A cellular automaton is determined by a quadruple,* $A = (S, d, N, f)$ *where S is a finite set, d a positive integer, $N \subset \mathbf{Z}^d$ a finite set, and* $f: S^N \rightarrow S$ *an arbitrary (local) function.*

The global function $G_f: S^L \rightarrow S^L$ *is defined by* $G_f(c)_v = f(c_{v+N})$.

A result of Hedlund (see ref. [7]) shows that all cellular automata arise in this fashion, so that the two definitions are equivalent. The proof hinges on the fact that every continuous function on a compact space is uniformly continuous.

2.2. Linear cellular automata

If $d = 1$, the shift space on a finite set S is called the full shift on S, denoted $S^{\mathbf{Z}}$. In general, if $c \in S^{\mathbf{Z}}$ we will write $c(i)$ as c_i. The product topology is induced by the metric

$$d(x, y) = \sum_{i=-\infty}^{\infty} \delta(x_i, y_i)\, 2^{-|i|},$$

where $\delta(a, b) = 0$ if $a = b$ and 1 otherwise.

There is a single natural (left) shift map σ acting on $S^{\mathbf{Z}}$ defined by $\sigma(x)_i = x_{i+1}$. For each $c \in S^{\mathbf{Z}}$, the whole class $\{\sigma^i(c)\,|\,i \in S^{\mathbf{Z}}\}$ corresponds to one bi-infinite word. A closed, shift-invariant subset of $S^{\mathbf{Z}}$ is referred to as a *subshift*.

Different local functions can give rise to the same rule. Therefore, in some sections the notation $G: S^{\mathbf{Z}} \to S^{\mathbf{Z}}$ is used when the result is not dependent on the details of the specific local function. Also if the neighborhood of the cellular automaton consists of the points $\{-r, \ldots, r\}$, then the rule will be said to have *radius r*.

2.3. Enumerating cellular automata

An important aspect of cellular automata, and the reason they form such an interesting starting point for the merger of dynamical systems and computation theory, arises from the fact that all cellular automata can be effectively enumerated. To each cellular automaton, we can associate two numbers. The first, $k = |S|$, is the cardinality of the state space at each site. The second, $r = \mathrm{diam}(N)$, is the maximum distance (in the square metric) of an element of the neighborhood N from the origin. It is the farthest information can travel along any axis in one iteration of the map.

The important facts, are:

(i) For any given cellular automaton, k and r exist and are finite,

(ii) For any fixed k and r there are only finitely many possible cellular automaton rules.

This distinction will play a special role in section 7, where the surjectivity and injectivity of cellular automata are considered.

3. Universality and normal forms

3.1. Introduction

A cellular automaton (CA) is said to be *universal* if it can simulate every Turing machine or, even stronger, if it can simulate every CA of the same dimension. The first universal cellular automaton was given in the famous work of von Neumann [1] on the simulation of self-reproduction. He gave a two-dimensional universal and self-reproducing CA with 29 states. This was improved to 20 states in ref. [8].

Also well known is the work on small universal Turing machines, cf. refs. [9–11]. In this case the goal is to minimize simultaneously the number of states and the number of tape symbols. In ref. [10] it is shown that there is a universal Turing machine with either 4 symbols and 7 states or with 6 symbols and 6 states. Smith [12] has shown that any Turing machine with m symbols and n states can be simulated by a one-dimensional CA with $m + 2n$ states. Thus either one of the universal TM from ref. [10] yields a one-dimensional CA with $4 + 2 \times 7 = 6 + 2 \times 6 = 18$ states. In ref. [13] this result has been improved to 14 states and the stronger universality, that is the ability to simulate any other one-dimensional CA rather than just TM. The latter result is based on two normal forms for one-dimensional CA, the *totalistic* CA and the *one-way* CA, which are of interest on their own.

A cellular automaton with local rule (function) such that the next state of a cell is dependent on the state of the cell itself and its right neighbor only has been introduced in ref. [14] and called a one-way CA. The rather surprising, but not difficult result that, for bi-infinite configurations, every one-dimensional CA can be simulated, step by step, by a one-way CA has been shown in refs. [15–17].

Another special form of CA has been introduced by Wolfram [18]. A CA is called totalistic if its states are integers and the rule is of the form that the next state of a cell depends only on the sum of all the states in the neighborhood, including its own. Note that the totalistic form is more restrictive than the symmetric one, in which case a cell gets a set of inputs without information which one is coming from which neighbor. All the examples in section 4 are of totalistic CA. Wolfram asked whether totalistic automata are universal. Gordon [19] answered this positively by showing that every Turing machine can be simulated by a one-dimensional totalistic CA. This result has been strengthened in ref. [13], where it was shown that every one-dimensional CA can be simulated by a totalistic one, and further in ref. [20], where it was shown that every network of finite automata with underlying graph of bounded degree can be simulated by a semitotalistic one, and a network with uniform degree by a totalistic one. A semitotalistic CA (or network) in which states are non-negative integers and whose local rule (function) is such that the next state of each cell only depends on its own current state and the sum of the states of all its neighbors. An example of two-dimensional semitotalistic CA is the well known Game of Life. It has been shown that the Game of Life can simulate arbitrary logical networks (switching circuits), which is a form of universality [21].

In the remaining part of this section we will derive the two rather surprising but not difficult normal form results, and then show the existence of a universal CA with small number of states.

3.2. Universality of one-way CA

First we make more precise what it means for one cellular automaton to simulate another. We say that a cellular automaton A_1 with global function $G_1: S_1^Z \to S_1^Z$ is simulated by a cellular automaton A_2 with global function $G_2: S_2^Z \to S_2^Z$, if there exist $k \geq 1$ and a locally defined one-to-one mapping $g: \Omega_1 \to \Omega_2$ such that for each configuration $w \in \Omega_1$

$$g(G_1(w)) = G_2^k(g(w)).$$

We say that A_2 is k times slower than A_1.

A locally defined mapping g is a continuous map which satisfies $\sigma g = g \sigma^n$ for a fixed n. In formal language terms, g is a morphism on configurations in Ω_1 viewed as bi-infinite words. In subsections 3.2 and 3.3 this morphism is alphabetical, i.e. one site of Ω_1 is mapped to one site of Ω_2 (n may be chosen to be 1). In subsection 3.4 g maps one site of Ω_1 to several sites of Ω_2, however, their number

depends only on the cellular automaton A_1 and not on the site's state.

In this section we will consider one-dimensional CA with the neighborhood of each cell consisting of the cell itself and at most one neighbor on the left and on the right. Therefore the local rule is of the form $f: Q \times Q \times Q \to Q$ for a set of states Q. The CA is *one-way* if $f(x_1, y, z) = f(x_2, y, z)$ for all $x_1, x_2 \in Q$, i.e. the rule is independent of the left neighbor. Such a rule can be given in the simplified form $f': Q \times Q \to Q, f'(y, z) = f(x, y, z)$ for all $x, y, z \in Q$.

Theorem 1 *For every cellular automaton A with k states there exists a one-way cellular automaton A' which simulates A twice slower and A' needs at most $k^2 + k$ states.*

The construction of A' can be described by:

(i) merging of each state with the state of its right neighbor;

(ii) shifting to the left during the state transition.

Let $s_1, s_2, \ldots, s_k$ be states in a configuration, such that $s_1 \neq q$ and let $t_0, t_1, \ldots, t_{k+1}$ be the states after one transition as shown below

$$\cdots \quad s_0 \quad s_1 \quad s_2 \quad s_3 \quad \cdots \quad s_k \quad s_{k+1} \quad \cdots$$
$$\cdots \quad t_1 \quad t_2 \quad t_3 \quad \cdots \quad t_k \quad \cdots$$

Then it takes two time steps in our simulating automaton to produce the transitions

$$\cdots \quad s_0 \quad s_1 \quad s_2 \quad s_3 \quad \cdots \quad s_k \quad s_{k+1} \quad \cdots$$
$$\cdots \quad t_1 \quad t_2 \quad t_3 \quad \cdots \quad t_k \quad \cdots$$

For the given cellular automaton A let Q be the set of states and $f: Q \times Q \times Q \to Q$ the local rule. As defined above, in the simulating one-way cellular automaton A' the local rule f' will consider only the states of its own cell and its right neighbor cell.

Let $Q' = Q \cup Q \times Q$ and define $f': Q' \times Q' \to Q'$ as follows

$$f'(u, v) = uv,$$

$$f'(uv, vw) = f(u, v, w)$$

for all $u, v, w \in Q$.

Thus, the transition shown above is completed in two time steps by A' in the following way:

$$\ldots \quad s_0 \quad s_1 \quad \ldots \quad s_{k-1} \quad s_k \quad s_{k+1} \ldots$$
$$\ldots \quad s_0 s_1 \quad s_1 s_2 \quad \ldots \quad s_{k-1} s_k \quad s_k s_{k+1} \quad \ldots$$
$$\ldots \quad t_1 \quad t_2 \quad \ldots \quad t_k \quad \ldots$$

It is left to the reader to check the correctness of our construction. An element of Q' is either an element of Q or an ordered pair of elements from Q; so, if $|Q| = k$, $|Q'| = k^2 + k$.

3.3. Universality of totalistic CA

A cellular automaton A with set of states Q and local rule $f\colon Q \times Q \times Q \to Q$ is called totalistic, if $Q \subseteq \mathrm{N}$ (non-negative integers) and there exists a function $\phi\colon \mathrm{N} \to \mathrm{N}$ such that $f(x, y, z) = \phi(x + y + z)$ for all x, y, z in Q.

The transformation of an arbitrary cellular automaton whose states are identified with non-negative integers is now accomplished by a cyclic "coloring" of cells. We use four different powers of a basis such that in the number representation of the sum of three neighboring states left neighbor, right neighbor and own cell are still identifiable by the position of a missing entry.

Consider, for example, a cellular automaton A state set $Q = \{1, 2, \ldots, n\}$ and local rule $f\colon Q \times Q \times Q \to Q$. Let $B = n + 1$ be the basis for the coloring factors 10, 100 and 1000 (in B-ary notation). Then the configuration at the top of fig. 1 changes to the one at the bottom.

Thus our new set of states is

$$Q' = \{sm \mid s \in Q, m \in \{1, 10, 100, 1000\}\}.$$

So Q' contains $4n$ states and we can define now the (partial) totalistic local rule $f'\colon \mathrm{N} \to Q'$ as follows:

$$f'(xyz) = 10f(x, y, z),$$

$$f'(xyz0) = 100f(x, y, z),$$

$$f'(yzx) = 1000f(x, y, z),$$

$$f'(z0xy) = f(x, y, z)$$

for all x, y, z in Q.

Again $xyz, xyz0, yz0x, z0xy, 10, 100, 1000$ are to be interpreted as numbers in the B-ary system.

It is now straightforward that the cellular automaton A' with set of states Q' and totalistic rule f' correctly simulates A without loss of time.

Therefore we can state the following theorem that answers one of the questions posed by Wolfram in ref. [22]. Another question is considered in the next subsection.

Theorem 2 *For every cellular automaton A there exists a totalistic cellular automaton A' which simulates A without loss of time and has at most four times as many states as A.*

3.4. A universal cellular automaton

In the previous two subsections we have shown that the classes of one-way and totalistic one-dimensional CA are universal in the sense that any CA can be simulated by one in these classes. When we say that CA are capable of universal computation, we mean that for each TM M we can construct a one-dimensional CA A whose local rule depends on M and which simulates M in the following sense.

Each input w of M can be translated into an initial configuration w' of A so that the computation of A starting with w' encodes step by step the computation of M started with w on the tape. This is easy. We encode each cell of the tape of M by a cell of A. In addition, the cell over which the reading head of M is positioned also encodes this fact and the state of the final control of M. Now it is easy to translate the program of M into the local CA rule so that the computation of CA on the encodings of M step by step simulates the computation of M.

A universal TM is a TM M_u that simulates any other TM by getting as input the pair $\langle M, w \rangle$, the encoding of the program of the arbitrary TM M and an input string w. M_u is a model of a universal computer, M corresponds to a particular program and w to the input data. If the above construction of a simulating CA is done for the universal TM M_u, then we obtain a universal one-dimensional CA A_u. It can simulate any TM M on any input w by being started in the initial configuration that encodes the pair $\langle M, w \rangle$.

Similarly, a CA A_U is *strongly universal* in the following sense. Any CA A with local rule $f\colon Q^{2r+1} \to Q$ and any configuration w over Q can be encoded into an initial configuration of A_U so that each cell of the simulated CA is encoded as a

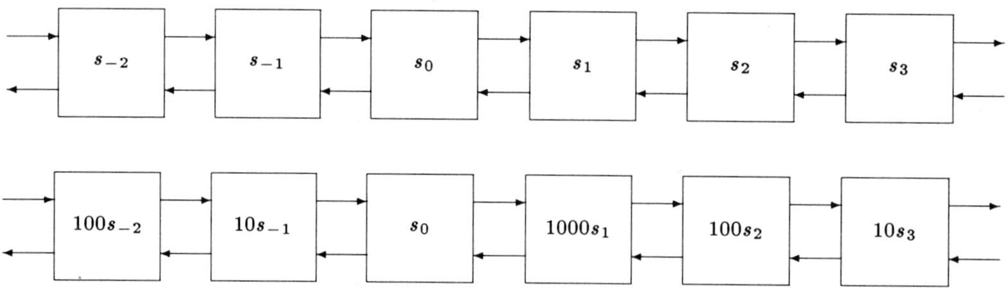

Fig. 1. Corresponding configurations of CA A and of the totalistic CA A' simulating A.

block of cells of A_U, where each of these blocks, except possibly those encoding the quiescent states, also encodes the local rule f. The computation of A_U starts with the initial configurations described above and then simulates step by step the computation of A started on configuration w.

In ref. [13] it is shown that there exists an A_U as described above with 14 states. This number almost certainly is not minimal; the minimal required number of states is not known.

Theorem 3 *There exists a universal CA A_U with 14 states which can simulate any given CA on any given initial configuration.*

In the proof of theorem 3 that is omitted here the normal forms from subsections 3.2 and 3.3 are used. In turn we can convert such an A_U into a one-way cellular automaton or one which is both one-way and totalistic. This normalization, however, would substantially increase the number of states.

4. Classification of linear cellular automata

Wolfram observed that the time–space diagrams of a simple class of two-state linear cellular automata seemed to fall into four easily distinguishable types. In ref. [2], he proposed a classification of cellular automata into four classes according to their observed behavior. Wolfram's first three classes heuristically correspond to the types of attractors of dynamical systems (fixed point, periodic, strange); the fourth class corresponds to well-organized systems like biological systems or computers.

Based on Wolfram's classification, Culik and Yu (CY) suggested a more specific definition of four classes of CA according to their behavior on finite configurations [3]. It was shown that the membership problem is undecidable in general for each of the four classes, even when only class 1 and 2 cellular automata are considered. Using techniques from the theory of recursive functions, Sutner was able to quantify this undecidability. He has shown more precisely that class I and II are Π_2^0-complete and class III is Σ_3^0-complete [23].

Other classification schemes have also been proposed, which include the classification according to their mean field approximations by Gutowitz et al. [24,25], two classification schemes for 3D semi-totalistic cellular automata by Bays [26], and a dynamical systems classification by Gilman [27,28].

In the following, we first introduce Wolfram's classification of four classes and corresponding definition by Culik and Yu. Then various undecidability problems concerning the four classes are discussed, and related results are introduced. At the end, other classification schemes are briefly introduced.

4.1. Wolfram's four classes

In ref. [2], Wolfram suggested a classification of cellular automata by the qualitative features of cellular automaton evolution. He observed that cellular automata appear to fall into the following four classes:

(1) Evolution leads to a homogeneous state.
(2) Evolution leads to a set of separated simple stable or periodic structures.
(3) Evolution leads to a chaotic pattern.

(4) Evolution leads to complex localized structures, sometimes long-lived.

In ref. [2], totalistic cellular automata are used as examples of these four classes. For a totalistic cellular automaton, k denotes the number of states, i.e. $S = \{0, 1, ..., k-1\}$, and r denotes the neighborhood radius, i.e. $N = \{-r, ..., 0, ..., r\}$.

When $k = 2$ and $r = 2$ (i.e. two states per site and a total neighborhood size of five), there are $2^6 = 64$ elementary totalistic rules. Making the assumption that the configuration which is all zeroes remains fixed, cuts this number in half. These rules are labeled by a function code which is a binary number $b_6...b_2 b_1 b_0$, for $b_i \in \{0, 1\}$. The interpretation is that

$$f(s_{-2}, ..., s_0, ...s_2) = b_i, \text{ if } s_{-2} + ... + s_0 + s_2 = i.$$

Usually, these binary function codes are converted into decimal numbers for convenience. The restriction to the case where the zero or quiescent state is fixed means that all the rules considered has even code numbers between 0 and 62.

In ref. [2], Wolfram considered all the cellular automata rules with $k = 2$ and $r = 2$. He classified them as follows:

Class 1: 0, 4, 16, 32, 36, 48, 54, 60, 62;

Class 2: 8, 24, 40, 56, 58;

Class 3: 2, 6, 10, 12, 14, 18, 22, 26, 28, 30, 34, 38, 42, 44, 46, 50;

Class 4: 20, 52.

Trying to follow Wolfram's informal specification closely, a formal definition of the four classes of cellular automata has been proposed by the first and third authors in ref. [3]. In CY's definition, CA are classified according to the evolutions of *finite* configurations even if the computations on all possible configurations are of interest. For convenience, the CY classes are defined as a hierarchy; each class contains all lower classes as subsets. However, it is easy to modify the classes to be mutually exclusive.

A formal description of the CY definition can be found in the next subsection. Informally, the CY definition of the four classes is the following: In class I, all the finite configurations evolve into a quiescent configuration. In class II, all the finite configurations have an ultimately periodic evolution. Class III consists of all CA for which it is decidable whether α ever evolves to β for two arbitrary finite configurations α and β. Class IV includes all CA. The first two CY classes appear to be larger than Wolfram's first two classes. However, it has been proved that, for the first three classes, CY's classification coincides with Wolfram's on all linear totalistic cellular automata with two states and radius two. It has not been formally proved that the cellular automata with code 20 and 52 are strictly in class IV according to CY's definition although it appears to be true.

4.2. CY definition of four classes

A state s of a cellular automaton A is called a *stable state* if the following condition holds:

$$f(s, s, ..., s) = s.$$

A configuration with all cells being in a stable state s is called a *homogeneous configuration* of s. A configuration with all but finitely many cells in the state s is called a *finite configuration* with respect to s.

One stable state q of A is distinguished and defined as the *quiescent state* of A. The homogeneous configuration of q is called the *quiescent configuration*. Finite configurations with respect to q are simply called *finite configurations*.

A *segment* is a finite, consecutive part of a configuration surrounded by quiescent cells containing no quiescent cells.

Let c_1 and c_2 be two configurations of A. If $G_f^i(c_1) = c_2$ for some $i \geq 0$, then we say that c_1 evolves to c_2. For a configuration c, the infinite sequence

$$c, G_f(c), G_f^2(c), ..., G_f^i(c), ...$$

is called the *evolution sequence* of c. If there exist positive integers $0 < m < n$ such that $G^m(c) = G^n(c)$, we say that c has an ultimately periodic evolution.

CY's definition of four classes is the following:

Definition 3 A cellular automata $A = (S, N, f)$ falls into one of the classes:

(i) There is a stable state $s \in S$ such that all the finite configurations of A with respect to s evolve to the homogeneous configuration of s.

(ii) There exists a stable state s such that every finite configuration of A with respect to s has an ultimately periodic evolution.

(iii) There exists a stable state s such that for any pair of finite configurations c_1 and c_2 with respect to s, it is decidable whether c_1 evolves to c_2.

(iv) All cellular automata.

Example 1 The totalistic cellular automaton with $k = 2, r = 2$ and function code 4 is in class I. It can be shown that every finite configuration of this CA evolves to a shorter one in at most 5 steps.

The cellular automata with $k = 2, r = 2$ and function code 0, 4, 16, 32, 36, 48, 54, 60, and 62, are all in class I. No other cellular automata with $k = 2$ and $r = 2$ are in class I.

Example 2 The cellular automaton with $k = 2, r = 2$ and function code 52 is not in class II since the finite configuration

110111100100001001111011

does not have a ultimately periodic evolution.

Note that it is not true in general for a class I cellular automaton, by either Wolfram's or this definition, that all configurations evolve to a homogeneous configuration. Consider at the following example.

Example 3 The following configuration of the simple class I cellular automaton with $k = 2, r = 2$ and function code 16 never evolves to a homogeneous configuration:

$$\ldots 0\,1\,0\,1\,0\,1\,0\,1\,0\,1 \ldots$$

Theorem 4 *For each cellular automaton, the following four statements are equivalent:*

(i) All configurations evolve to a homogeneous configuration.

(ii) All configurations evolve to a homogeneous configuration in a bounded (constant) time.

(iii) All finite configurations with respect to a stable state s evolve to the homogeneous configuration of s in a bounded (constant) time.

(iv) The limit set (see section 6.1) consists of the quiescent configuration only.

Proof: The equivalence of the first two statements can be easily proved by considering the stable state as the quiescent state and applying corollary 2 of ref. [29]. Statement 2 and 3 are clearly equivalent. Statement 2 implies statement 4. Statement 4 implies statement 1. □

Example 3 shows that if we defined class I to be the set of all cellular automata in which *all* configurations evolve to a homogeneous configuration then even very simple cellular automata would not be included. Although a cellular automaton satisfying the above conditions is trivially in class I, the converse is not true.

Theorem 4 says that in general there does not exist a constant bound for a class I cellular automaton such that every finite configuration evolves to a homogeneous configuration within the bound. But is there a linear bound or a polynomial bound in term of the lengths of (the non-s parts of) finite configurations with respect to s? We first look at an example.

Example 4 Define CA $A = (S, N, f)$ by $S = \{0, 1, -, q\}$, $N = (-1, 0, 1)$, and f:

$$f(q, 0, x) = q; \qquad f(q, -, x) = q;$$

$$f(x, 1, q) = 0; \qquad f(y, 0, q) = -;$$

$$f(y, -, q) = 0; \qquad f(y, -, z) = 1;$$

$$f(y, 0, -) = -; \qquad f(x, 1, -) = 0;$$

for $x = 0, 1, -,$ *or* q; $y, z = 0, 1,$ *or* $-$; *and*

$$f(a, b, c) = b, \quad \text{*for all other cases.*}$$

The reader can verify that a segment is simply treated as a binary number and it is decremented by one at each evolution step.

In the above example, a segment may evolve for expIntial number of steps in terms of its length before it becomes all quiescent. In fact, for any computable function $T(n)$, there exists a class I cellular automaton such that some of its finite configurations (with respect to s) need $T(n)$ steps to evolve to the quiescent configuration (homogeneous configuration of s).

4.3. Results related to the four classes

The set

$$D(c) = \{\, d \mid d = G_f^i(c) \text{ for some } i \geq 0 \,\}$$

is called the *descendant set* of c. For a configuration finite with respect to a stable state s we are only interested in the finite non-s part of it. Therefore, for each stable state s we define a mapping $\pi_s \colon S^{\mathbb{Z}} \to S^*$, which maps each configuration c into a word w, the minimum sequence of states in c that covers all the non-s state in c. If c is a finite configuration with respect to s then $\pi_s(D(c))$ is called the *descendant language* of c with respect to s.

It appears that the classification of four classes are not related to the limit sets or limit languages of CA (see section 6) but related to the descendant languages of CA. The following results can be found in ref. [3].

Theorem 5 *A cellular automaton $A = (S, N, f)$ is in class I or II if and only if there exists a stable state $s \in S$ such that the descendant language $D_s(c)$ is finite for all finite configurations c with respect to s. A is in class III if and only if there exists a stable state s such that the descendant language of every finite configuration with respect to s is recursive.*

Theorem 6 *There exists a one-dimensional class I cellular automaton that does not have a regular limit language.*

Proof: Let $A = (S, N, f)$ be a one-dimensional cellular automaton that has a nonregular limit language L. See ref. [4] for such a cellular automaton. We construct a cellular automaton $A' = (S', N, f')$, where $S' = S \cup \{q\}$, $q \notin S$ and q is the quiescent state of A'. The local function f' is the same as f on all the states in S, and maps any neighborhood with q to q. Obviously, all the finite configurations of A' evolve to the quiescent configuration. Therefore, A' is in class I. Let L' be the limit language of A'. Note that $L' \cap S^* = L$ and L is not regular. By the closure property of regular languages, L' is not regular. $\qquad\square$

Using essentially the same proof, one can show that the limit language of a class I cellular au-

tomaton can be as complicated as any cellular automaton limit language; in fact a class I cellular automaton can have a non-recursively enumerable limit language (see section 6).

The negative aspect of this classification scheme (and therefore informally also of Wolfram's) is that it is in general undecidable to which class a cellular automaton belongs, even when choosing only between class I and class II.

Lemma 1 *It is undecidable whether all the finite configurations of a given cellular automaton evolve to the quiescent configuration (in linear time).*

Corollary 1 *It is undecidable whether a given cellular automaton A is in class I, even when A is known to be in class II.*

Theorem 7 *It is undecidable whether a cellular automaton is in class II.*

Theorem 8 *It is undecidable whether a given cellular automaton is in class III.*

The proofs for the above undecidability results can be found in ref. [3].

Clearly, a cellular automaton that simulates the computation of a universal Turing machine is not in class III. Hence, universal cellular automata are not in class III. A cellular automaton with the property that every finite configuration either evolves to the quiescent configuration or evolves to a longer finite configuration in a uniformly bounded time is clearly in class III.

Recursive function theory provides a way of quantitizing undecidability. Sutner [23,30] has shown the position of the CA classes in the hierarchy of undecidability problems. In the notation of ref. [31] he has shown the following.

Theorem 9 *Class I and class II are Π_2^0-complete. Class III is Σ_3^0-complete.*

4.4. Products of cellular automata

Now, following ref. [3], we consider the products of cellular automata. It is interesting to know what class the product of two cellular automata is in, especially when the two cellular automata are in two different classes. But first we give a formal definition for the products of cellular automata.

Definition 4 Let $A_1 = (S_1, N_1, f_1)$ and $A_2 = (S_2, N_2, f_2)$ be two cellular automata of a same dimension with q_1 and q_2 being the quiescent state of A_1 and A_2, respectively. Then the product of A_1 and A_2 is the cellular automaton $A = (S, N, f)$, where

$$S = \{\langle s_1, s_2 \rangle | s_1 \in S_1, s_2 \in S_2\},$$

$$N = N_1 \cup N_2,$$

the quiescent state

$$q = \langle q_1, q_2 \rangle,$$

and A can be considered as having two tracks: f is the same as f_1 on the first track and f_2 on the second track.   □

Theorem 10 *Let A_1 and A_2 be two cellular automata in class I_1 and class I_2, respectively. Then the product of A_1 and A_2 is in class $max(I_1, I_2)$.*

In ref. [32], Wolfram discussed how a change of one site (cell) in the initial configuration propagates in the evolution of that configuration. He observed that the propagation of the difference is typical for each of his classes. It is easy to prove this as a corollary of theorem 7 (see ref. [3]).

4.5. Other classification schemes

A classification of cellular automata according to their mean field approximation has been introduced in refs. [24,25]. Here we give a brief description of this classification scheme. For convenience, we only consider one-dimensional CA with two states, 0 and 1, and neighborhood radius r.

Let P_β^t denote the probability of a block β appearing at the time t of a CA evolution. Let G be a block mapping defined by a CA. Then

$$P_\beta^{t+1} = \sum_{\alpha : G(\alpha) = \beta} P_\alpha^t.$$

We use $\#_0(\beta)$ and $\#_1(\beta)$ to denote the number of cells in state 0 and 1, respectively, in a block β. Under the assumption that the probability for a cell to be in a given state is uncorrelated with the states of other cells, we have

$$P_1^{t+1} = \sum_{\alpha : G(\alpha) = 1} (P_1^t)^{\#_1(\alpha)} (1 - P_1^t)^{\#_0(\alpha)}.$$

Note that P_1^t is the probability of a cell to be in state 1 at time t. This equation is called the mean field equation of the given CA.

Consider all the blocks of size $2r + 1$. Let a_i be the number of blocks that map to 1 and contain i cells in state 1 . Then the mean field equation can be written as

$$P^{t+1} = \sum_{i=0}^{2r+1} a_i (P^t)^i (1 - P^t)^{2r+1-i}$$

$$= f(a_0, \ldots, a_{2r+1}; P^t),$$

where P^x is P_1^x in the previous equation.

A mean field class consists of all cellular automata that have the same coefficient $\boldsymbol{a}$ of the mean field equation. For example, the mean field class of $r = 1$ and $(a_0, a_1, a_2, a_3) = (0, 1, 0, 0)$ consists of all CA of $r = 1$ with rules that map all neighborhood blocks to 0 except that one of 001, 010, and 100 to 1.

The fixed point solution p^* such that $p^* = f(\boldsymbol{a}; p^*)$ is an estimate of an invariant measure of the cellular automata in the class.

In ref. [24], Gutowitz has made comparisons between mean field classes and Wolfram's four classes. He has shown that rules in Wolfram's class 1 have a trivial invariant measures (only one state has a positive probability). Rules in class 2 have simple invariant measures which depend on the initial measure. Rules in class 3 have complex invariant measures which do not depend on the initial measure. Rules in class 4 may not have invariant measures. Even if a class 4 rule has an invariant measure, convergence may be arbitrarily slow.

Bays has given two observational classification schemes of three-dimensional semitotalistic CA in ref. [26]. The first scheme is according to behavior of isolated forms (clusters of nonquiescent cells). He has proposed three classes according to this scheme: (1) isolated forms never interact; (2) isolated forms interact; and (3) isolated forms grow without bounds. His second scheme, classification according to primordial evolution, is essentially a variation of Wolfram's classification.

Gilman has introduced a dynamical-system-based classification of CA in ref. [27], see also ref. [28].

5. Cellular automata and formal language theory

The study of cellular automata have been linked with language theory in two ways.

First, the sets of finite configurations of one-dimensional CA are themselves languages, and CA mappings on finite configurations are *GSM* mappings ([33]) on finite words. CA with restrictions on time, space, communication, or input and output form, e.g. real-time CA, linear-bounded CA, one-way CA, and iterative arrays, have been studied as language acceptors. See, for example, refs. [34–37] for details.

Second, CA are dynamical systems. The shift closure of a CA configuration is a bi-infinite word in language theory. Tools and results on bi-infinite words in language theory have been used to study various shift-invariant sets of CA [38]. In addition, a subshift is uniquely determined by the set of all finite substrings of its configurations. Thus, the study of subshifts has been directly linked to languages of finite words [39,4].

In this paper we consider CA as dynamical systems. Hence, only the second link above is of interest here. This link works in both directions: the techniques and results of language theory are useful in the study of CA and in turn language theory draws motivation and new techniques from the theory of dynamical systems.

5.1. Sofic systems and $\omega\omega$-regular sets

Recall that a closed, shift-invariant subset of $S^{\mathbf{Z}}$ is called a *subshift*. Two particularly important classes of subshifts are called subshifts of finite type, and sofic systems.

Definition 5 *A subshift of finite type consists of all configurations which do not contain blocks from a finite excluded list. A sofic system is a subshift of the form $G(K)$ where G is a cellular automaton map and K is a subshift of finite type (see ref. [40]).*

The set of finite substrings of configurations in $C \subseteq S^{\mathbf{Z}}$ is denoted $\mathcal{L}(C)$ The following theorems have been proved by the second author in ref. [4].

Theorem 11 *If K_1 and K_2 are subshifts, $\mathcal{L}(K_1) = \mathcal{L}(K_2) \Leftrightarrow K_1 = K_2$.*

Theorem 12 *For any subset $K \subseteq S^{\mathbf{Z}}$,*

$$\mathcal{L}(K) = \mathcal{L}(\overline{\bigcup_{i=-\infty}^{\infty} \sigma^i(K)})$$

Here σ^i denotes powers of the left shift map, and the bar represents topological closure.

A characterization theorem for sofic systems has been suggested in ref. [40] and proved in ref. [4]. It is stated as follows:

Theorem 13 *$C \subseteq S^{\mathbf{Z}}$ is a sofic system if and only if C is a subshift and $\mathcal{L}(C)$ is a regular language.*

Bi-infinite words and their recognition have been studied as a part of language theory. See, for example, refs. [41,29,42] for details. In ref. [42], Nivat and Perrin defined "a bi-infinite word as the equivalence class under the shift of a two-sided infinite sequence". They studied $\omega\omega$-regular sets and other properties of bi-infinite words based on results on one-way infinite words, for which many deep results have already been available.

A bi-infinite word v is a shift closure of a configuration $c \in S^{\mathbf{Z}}$, i.e. $v = \{\sigma^i(c) \mid -\infty < i < \infty\}$, where $\sigma \colon S^{\mathbf{Z}} \to S^{\mathbf{Z}}$ is a shift function such that $(\sigma(c))_{i+1} = c_i$ for $-\infty < i < \infty$.

Finite state machines that recognize sets of bi-infinite words have been defined and studied in refs. [42,41,29]. Formally, an $\omega\omega$-*finite automaton* ($\omega\omega$-FA) M is a quintuple $(Q, S, \delta, Q_{\mathrm{L}}, Q_{\mathrm{R}})$, where Q is the finite set of states; S is the input alphabet; δ is the transition function; $Q_{\mathrm{L}} \subseteq Q$ is the set of left (accepting) states; and $Q_{\mathrm{R}} \subseteq Q$ is the set of right (accepting) states.

A bi-infinite word v is said to be recognized by M if there is a mapping $\mathbf{Z} \to Q$, i.e. a bi-infinite sequence of states

$$\ldots q_{-2}, q_{-1}, q_0, q_1, q_2, \ldots$$

and a configuration c in v such that, for all $j \in \mathbf{Z}$,

(i) $\delta(q_j, c_j) = q_{j+1}$; and
(ii) there exist $m, n \in \mathbf{Z}$, $m \le j \le n$, such that $q_m \in Q_{\mathrm{L}}$ and $q_n \in Q_{\mathrm{R}}$.

In other words, v is said to be recognized by M if there is a bi-infinite computation of M on a

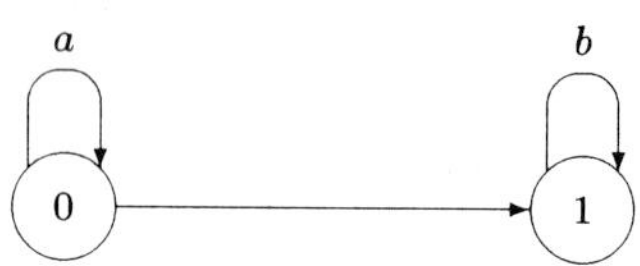

Fig. 2. An $\omega\omega$-FA M.

configuration c in v such that there is a left state appearing arbitrarily early, and there is a right state appearing arbitrarily late in the computation. Such a computation is called an *accepting computation*.

The set of bi-infinite words recognized by M is denoted $B(M)$. We call $B(M)$ an *$\omega\omega$-regular set*. Clearly, every $\omega\omega$-regular set is shift-invariant.

Example 5 Let $M = (Q, S, \delta, Q_L, Q_R)$ be an $\omega\omega$-FA, where $Q = \{0,1\}$, $S = \{a,b\}$, $Q_L = \{0\}$, $Q_R = \{1\}$, and δ is given in fig. 2. The set of bi-infinite words recognized by M is the set of all words which have infinitely many a's followed by infinitely many b's, i.e. $^\omega a\, b^\omega$.

Finite or one-way infinite words can be considered as special cases of bi-infinite words in the following sense: A special quiescent symbol is specified such that a one-way infinite word (ω-word) is a bi-infinite word with infinitely many quiescent symbols on the left end, and a finite word is a bi-infinite word with a finite consecutive nonquiescent subword.

Theorem 14 *Each sofic system is a $\omega\omega$-regular set.*

A finite state transducer [43] is the most general nondeterministic finite state device that maps finite words into finite words. The following is its generalization for bi-infinite words.

An *$\omega\omega$-finite transducer* T is a 6-tuple $(P, S, S', \rho, P_L, P_R)$, where P is the finite set of states; S is the input alphabet; S' is the output alphabet; $\rho: P \times (S \cup \{\lambda\}) \to P \times S'^*$ is the transition function; $P_L \subseteq P$ is the set of left (accepting) states; and $P_R \subseteq P$ is the set of right (accepting) states.

A bi-infinite word d is an output on input c under the $\omega\omega$-finite transducer T if there is a bi-infinite sequence of states

$$\dots, p_{-2}, p_{-1}, p_0, p_1, \dots$$

and configurations $x \in c$ and $y \in d$ such that for all $j \in Z$

   (i) $\rho(p_j, x_j) = (p_{j+1}, y_j)$; and
   (ii) there exist $m, n \in Z$, $m \le j \le n$, such that $p_m \in P_L$ and $p_n \in P_R$.

The relation defined by an $\omega\omega$-finite transducer is called an *$\omega\omega$-rational relation*. The next theorem has been proved in ref. [38].

Theorem 15 *The family of $\omega\omega$-regular sets is closed under $\omega\omega$-rational relations.*

The following results have been given in refs. [42,44]. A different proof for the closure property under complementation has also been given in ref. [38].

Theorem 16 *The family of $\omega\omega$-regular sets is closed under union, complementation and intersection.*

If $C \subseteq S^Z$ is $\omega\omega$-regular, then $\mathcal{L}(C)$ is a regular language. If $R \subseteq S^*$ is a regular set, then the set $\{c \in S^Z \mid L[c] \subseteq R\}$ is $\omega\omega$-regular and topologically closed. Note that the fact that $\mathcal{L}(C)$ is regular does not implies that C is $\omega\omega$-regular. For example, let X be the set of all the bi-infinite words over $\{a, b\}$ that have a prime number of a's. Then $\mathcal{L}(X) = \{a, b\}^*$ is regular, but X is not $\omega\omega$-regular. However, if a set of bi-infinite words X (which is shift-invariant) is topologically closed, the $\omega\omega$-regularity of X is implied by the regularity of $\mathcal{L}(X)$.

Each CA map is an $\omega\omega$-rational relation. Let G be an arbitrary CA map. Since S^Z is an $\omega\omega$-regular set, the following theorem is an immediate consequence of theorem 15.

Corollary 2 *For each $i \ge 0$ $G^i(S^Z)$ is an $\omega\omega$-regular set.*

The concept of algebraic limits, in terms of bi-infinite words, of languages have been defined in refs. [42,41,29]. Intuitively, the algebraic limit set of a language (of finite words) is the set of bi-infinite words that are obtained by an infinite two-way extension into words in the language.

More formally, a bi-infinite word w is an algebraic limit of language L if there is a sequence of finite words $x_1, x_2, \dots$ such that $x_i \in L$, x_i is a proper prefix of x_{i+1} and x_i is a prefix of w for all $i \geq 1$. The collection of all limits of L is called the algebraic limit set of L.

The next theorem has been proved in ref. [42].

Theorem 17 *The family of algebraic limit sets of regular languages is a proper subset of the family of $\omega\omega$-regular sets.*

This theorem implies that there exists a set of bi-infinite words which is $\omega\omega$-regular but not an algebraic limit of any regular language. An example is shown in ref. [38].

5.2. Adherences and centers of languages

Let $x \preceq z$ denote that x is a finite substring of z where z is either a finite or infinite word.

Definition 6 *The adherence of a language $L \subseteq S^*$ is a subset of S^Z defined by*

$$\mathrm{Adh}(L) = \{c \in S^Z \mid \text{for all } w \preceq c \text{ there exist } x, y \in S^* \text{ such that } xwy \in L \}.$$

Equivalently, the adherence of a language L is the algebraic limit set of the language that consists of all substrings of words in L. Therefore by theorem 17 we have

Corollary 3 *The adherence of any regular language is $\omega\omega$-regular.*

Theorem 18 *For any language L, $\mathrm{Adh}(L)$ is a topologically closed set.*

Note that the algebraic limit sets of a regular language is not necessarily topologically closed. For example, $\mathrm{limit}(a^+b^*c^+)$ is $\omega\omega$-regular but not topologically closed.

Corollary 4 *The family of adherence sets of regular languages is a proper subset of the family of algebraic limit sets of regular languages.*

Lemma 2 *For any $L \subseteq S^*$, $\mathrm{Adh}(L)$ is a subshift of S^Z.*

Using this lemma and the other results in the current and previous subsections the following result was obtained in ref. [38].

Theorem 19 *The following statements are equivalent:*

(i) $C \subseteq S^Z$ is a sofic system.

(ii) $C = \mathrm{Adh}(R)$ for some regular language R.

(iii) C is a topologically closed $\omega\omega$-regular set.

(iv) C is the topological closure of $\mathrm{limit}(R)$ for some regular language R.

A string w is said to be *bi-extensible* in a language L if for all $N > 0$ there exist strings $x, y \in S^*$ such that $xwy \in L$ and $|x|, |y| \geq N$.

Definition 7 *The center, $C(L)$, of a language $L \subseteq S^*$ is defined by*

$$C(L) = \{w \in S^* \mid w \text{ is bi-extensible in } L \}.$$

The concept of centers of languages was originally given by Nivat [45] in the context of one-way extension. Centers of $\omega\omega$-languages were introduced in ref. [46].

Definition 8 *A language $L \subseteq S^*$ is central if $L = C(L)$.*

Theorem 20 *For each subset $K \subseteq S^Z$ and language $L \subseteq S^*$:*

(i) $\mathcal{L}(\mathrm{Adh}(L)) = C(L)$.

(ii) $\mathrm{Adh}(\mathcal{L}(K)) = \overline{\bigcup_{i=-\infty}^{\infty} \sigma^i(K)}$.

It follows immediately that:

Corollary 5 *The operators $\mathcal{L}$ and Adh give a bijective correspondence between central languages over S^* and subshifts of S^Z.*

After introducing a new operator on languages, it is natural to ask about its closure properties with respect to the usual classes of formal languages. The first result along these lines is that regular languages are closed under the operator C.

Theorem 21 *If L is a regular language, $C(L)$ is regular.*

For context-free languages, the following result has been proved in ref. [38].

Theorem 22 *If L is context-free, $C(L)$ is context-free.*

Neither context-sensitive nor recursively enumerable languages are closed under C. In fact there exists a context sensitive language whose image under C fails to be recursively enumerable.

Theorem 23 *There exists a context-sensitive language, L_u, for which $C(L_u)$ is not recursively enumerable.*

5.3. Closure properties of CA mappings

By the correspondence set up in theorem 20, a family of languages gives rise to a unique family of subshifts which are its images under the operator Adh. It has been observed already that the subshifts corresponding to regular languages are exactly sofic systems. It is natural then to ask about subshifts generated by more complicated languages. Furthermore, the first question which arises in introducing a class of subshifts is whether they are closed under the images and inverse images of the natural class of maps, induced by cellular automata.

The answer for the forward image of a subshift is that under weak conditions on the family of languages the corresponding subshifts are closed under forward images. This comes from the fact that cellular automata behave like a well-known class of morphisms in language theory induced by ε-limited GSM mappings (see ref. [5]).

For more details and the discussion of the inverse images see ref. [47].

6. CA as dynamical systems

Cellular automata can be considered to be continuous maps on a compact metric space. In fact, the shift space upon which the map works is endowed with a natural measure, which allows the tools of ergodic theory to be brought into play [48]. In what follows, however, we will survey some properties related to the topological dynamics of cellular automata, considering them as metric spaces, and for the moment ignoring questions of measurability.

A good survey of the assumptions underlying this study is given in ref. [48], and several results are proved about specific cellular automata rules.

Other studies along these lines include refs. [49–51].

6.1. Limit sets

Cellular automata are not generally invertible. Most rules form a decreasing sequence of images

$$S^{\mathbf{Z}} \supseteq G(S^{\mathbf{Z}}) \supseteq \ldots$$

The *limit set*, $\Lambda(G)$, of a cellular automaton is the intersection of all forward images,

$$\Lambda(G) = \bigcap_{i=0}^{\infty} G^i(S^{\mathbf{Z}}).$$

The set $\Lambda(G)$ is the maximal attractor of the dynamical system specified by G in the sense of Conley [52].

The limit set is the set of all $c \in S^{\mathbf{Z}}$ which have an infinite sequence of pre-images. Often the easiest way to prove that a configuration is in $\Lambda(G)$ is to construct such a sequence.

A subset $Y \subseteq S^{\mathbf{Z}}$ is G-invariant if $G(Y) = Y$. The limit set, $\Lambda(G)$ is the maximal G-invariant subset of $S^{\mathbf{Z}}$. The G-invariance of $\Lambda(G)$ is straightforward, and the fact that it is maximal follows from the fact that if Y is G-invariant, then

$$Y = G(Y) = G^i(Y) = \bigcap_{i=0}^{\infty} G^i(Y) \subseteq \Lambda(G).$$

6.2. Other invariant subshifts

The statement that a configuration is in the limit set, means that it could be encountered at any iteration of the map. However, some points of the limit set may correspond to highly unlikely behavior. It is often useful to look at configurations with stronger recurrence properties. In particular, studies [53,54] have been done of the periodic set, $\Pi(G)$, and the non-wandering set, $\Omega(G)$.

Given a cellular automaton rule, G, a configuration, $c \in S^{\mathbf{Z}}$ is *periodic* if there exists $n > 0$ such that $G^n(c) = c$. In general, however, this set need not be closed. For example, if G is the left shift map, periodic points are dense in the whole space. The set $\Pi(G)$ is, therefore, defined to be the closure of the points periodic under G.

Periodicity is a strong form of recurrence, but often it is useful to examine weaker conditions. One type of condition that is especially important in topological dynamics is that of non-wandering. Consider an irrational rotation of the circle. There are no periodic points at all, but every orbit comes arbitrarily close to any given point infinitely often. Dealing with this weaker kind of recurrent behavior led to the introduction of the non-wandering set. A point is wandering if there is a neighborhood of it whose forward images never again intersect it. This condition clearly defines an open set, and the non-wandering set is its complement.

In general, if X is a compact metric space, and $G: X \to X$ is a continuous function, the non-wandering set [55] consists of all points $x \in X$ for which the inverse images of any neighborhood of x are not all disjoint.

Definition 9 *If $G: X \to X$ is a continuous map of a compact metric space, the non-wandering set of G, denoted $\Omega(G)$, consists of all points $x \in X$ such that for every neighborhood U of x, there exists an integer $n \geq 1$ such that $G^n(U) \cap U \neq \emptyset$.*

If $X = S^Z$, and $G: X \to X$ commutes with σ, it is clear from the definitions that $\Pi(G)$ and $\Omega(G)$ are closed and shift-invariant (i.e. subshifts).

Theorem 24 *For any cellular automaton G,*

$$\Pi(G) \subseteq \Omega(G) \subseteq \Lambda(G).$$

In refs. [51,50], it is shown that each of these inequalities can be strict.

6.3. Complexity of CA-invariant sets

In ref. [32] an algorithm is given to determine whether a block occurs in the time-n image of a cellular automaton map. This provides an upper bound on the complexity of a cellular automaton limit language; a semi-procedure which halts only if a string is not in this set. It is an easy consequence of compactness that if a block fails to occur in the infinite intersection of images, it must fail to appear in a particular image.

Theorem 25 *For every cellular automaton $G: S^Z \to S^Z$, the complement of the limit language, $S^* - \mathcal{L}(\Lambda(G))$, is recursively enumerable.*

Given this constraint there exist arbitrarily complicated cellular automaton languages. More exactly, given any language whose complement is recursively enumerable, one can construct a cellular automaton whose limit language yields the chosen language after intersection with a regular language, and ε-limited homomorphism.

Theorem 26 *If $L \subseteq A^*$ is a language whose complement is recursively enumerable, there exists a cellular automaton $G: S^Z \to S^Z$, a regular language $R \subseteq S^*$, and a homomorphism $\phi: S^* \to A^*$ such that*

$$\phi(\mathcal{L}(\Lambda(G)) \cap R) = L.$$

The procedure of intersection with a regular language, and homomorphisms both preserve the class of recursively enumerable languages. The proof of this theorem, coupled with the existence of languages which are not recursively enumerable although their complements are, yield an example of a cellular automaton whose limit language fails to be recursively enumerable.

Corollary 6 *There exists a cellular automaton $G_u: S^Z \to S^Z$ such that $\mathcal{L}(\Lambda(G_u)) \subseteq S^*$ is not a recursively enumerable language.*

Temporally periodic configurations (i.e. orbits satisfying $G^p(c) = c$) of a given period have a simple description (see ref. [32]). Using this procedure for each time step in turn gives rise to a procedure for determining if a given block occurs in a temporally periodic orbit of any period.

Theorem 27 *If G is any cellular automaton rule, $\mathcal{L}(\Pi(G))$ is recursively enumerable.*

Again, given this constraint, the behavior exhibited by the periodic set can be quite complex. Basically a cellular automaton is set up to simulate a Turing machine which upon halting starts the computation all over again.

Theorem 28 *If $L \subseteq A^*$ is any recursively enumerable language, there exists a cellular automaton rule $G: S^Z \to S^Z$, a regular language $R \subseteq S^*$, and a homomorphism $\phi: S^* \to A^*$ such that*

$$\phi(\mathcal{L}(\Pi(G)) \cap R) = L.$$

372   *K. Culik II et al. / Computation theoretic aspects of CA*

Table 1
Rule G_1.

.	.	W	.	.	$\mapsto$	W	
.	.	R	W	L	$\mapsto$	l	
R	W	L	.	.	$\mapsto$	r	
.	r	x	y	.	$\mapsto$	r	(x $\neq$ W x,y $\neq$ l,L)
.	R	x	y	.	$\mapsto$	R	(x $\neq$ W x,y $\neq$ l,L)
.	x	y	l	.	$\mapsto$	l	(y $\neq$ W x,y $\neq$ r,R)
.	x	y	L	.	$\mapsto$	L	(y $\neq$ W x,y $\neq$ r,R)
		otherwise			$\mapsto$	O	

(. = any symbol)
Highest applicable rule takes precedence.

```
R W r l  r LWRl  R       W       L  Wl  Wr W      lr W
  RW          W      R      W      L  W   W rW  l   rW
   W          W       R     W    L    W   W  W 1     W
   W          W         R    W    L   W   W  W 1     W
   W          W           R  W  L     W   W  Wl      W
   W          W             R W L     W   W   W      W
   W          W              RWL      W   W   W      W
   W          W              lWr      W   W   W      W
   W          W             l W r     W   W   W      W
   W          W            l  W   r   W   W   W      W
```

Fig. 3. Evolution of rule G_1 from a sample initial condition.

Theorems 26, 28 have the drawback that since each involves the complexity of a piece of the set in question, derived by intersection with a regular language, it is not very good in determining upper complexity bounds stronger than those of theorems 25, 27. To illustrate examples of say context-free or context-sensitive limit languages requires more work.

Examples are given in ref. [53] of rules with varying levels of complexity. By means of illustration, the first of these rules is defined in table 1 and its evolution from a typical initial state is shown in fig. 6.3.

In refs. [53,54] explicit examples are given of cellular automata for which:

(i) $\mathcal{L}(\Lambda(G)), \mathcal{L}(\Pi(G))$, and $\mathcal{L}(\Omega(G))$ are context-free.

(ii) $\mathcal{L}(\Lambda(G)), \mathcal{L}(\Pi(G))$, and $\mathcal{L}(\Omega(G))$ are context-sensitive.

(iii) $\mathcal{L}(\Lambda(G))$ is not recursively enumerable (although its complement is).

(iv) $\mathcal{L}(\Pi(G))$ is recursively enumerable but not recursive.

6.4. Limit sets and transition diagrams

A great deal of work on cellular automata limit sets stems from ref. [32] and a subsequent more formal treatment in refs. [4,29], but the first (independent) mention actually goes back to ref. [56] in a study of the graphs generated by the orbits of cellular automata.

Definition 10 *Given a cellular automaton* G: $S^Z \to S^Z$, *its* transition diagram Γ, *is the directed graph whose vertices are elements of* S^Z *and* $c \in S^Z$ *is connected by an edge to* $G(c)$.

This definition generalizes readily to cellular automata in higher dimensions, and it is the degree of generality in which it was first stated. The following theorem appeared in ref. [56].

Theorem 29 (Podkolzin) *If* G *is a cellular automaton, and* Γ *its associated transition diagram, then either:*

(i) Γ *has a single connected component.*

(ii) Γ *has an uncountably infinite number of connected components (in bijective correspondence with the continuum).*

The first case corresponds to the situation in which the finite image of the cellular automaton is a spatially homogeneous fixed point (everything is mapped to zero).

Definition 11 *If a cellular automaton* G *(dimension* $d \geq 1$*) has a transition diagram with a single connected component,* G *is said to be* nilpotent.

In ref. [56] it was shown that in dimensions two or greater, there was no finite procedure to determine whether the rule had one or uncountable many connected components.

Theorem 30 (Podkolzin) *For any fixed dimension* $d \geq 2$ *the question of cellular automaton nilpotency is undecidable.*

The theorem follows directly from the undecidability of the tiling or domino problem in dimensions two or greater [57,58]. For one dimension, however, the domino problem is trivially decidable, and more subtle techniques have to be brought to bear on the problem. The case $d = 1$

has only just been proved undecidable by Kari in ref. [59].

Theorem 31 (Kari) *For linear cellular automata ($d =1$) the nilpotency question is undecidable.*

Not every connected component of the transition diagram contains an element of the limit set. In fact the limit set of a cellular automaton can be countably infinite. Witness for example Wolfram's *elementary rule 128* [32] in which a site has value 1 at time t if and only if it and its nearest neighbors had value 1 at time $t - 1$. Thus zeroes grow linearly and segments of adjacent ones shrink from both sides at one site per time step. Its limit set [32,53] consists of all configurations with the property that no zero ever appears between two ones. This set is easily seen to be countably infinite.

However, if the limit set is finite, it must reduce to a single element, and once again the cellular automaton must be nilpotent. Since the limit set is shift-invariant, this is an immediate consequence of:

Lemma 3 *If $\Lambda(G)$ contains two elements, it contains a non-shift-invariant element.*

Proof: Let $x, y \in \Lambda(G)$, $x \neq y$ (this does not rule out the possibility that they are translates of each other under the shift). If either x or y is not spatially periodic (i.e. fixed by a finite power of the shift) the proof is done. Without loss of generality assume that x and y are spatially periodic, and in fact by taking the least common multiple of their periods, that they have the same period.

We can write $x = \ldots www|www\ldots$, $y = \ldots vvv|vvv\ldots$. Here the vertical bar indicates the position of site zero. Let N_w be the set of all configurations of the form $\ldots www|\overline{w}$ where $\overline{w}$ denotes any block except w, and the remaining symbols are arbitrary. This is a closed set, and no element of N_w is spatially periodic. It remains to show that its intersection with $\Lambda(G)$ is non-trivial.

By compactness, $N_w \cap \Lambda(G) \neq \emptyset$ if and only if $N_w \cap G^i(S^Z) \neq \emptyset$ for all i. Since x and y are both in $G^i(S^Z)$, one can glue together pre-images to form an element $z = \ldots wwws_1 \ldots s_n vvv \in G^i(S^Z)$ where $n = 2ir$. Some shift of z must be in N_w, hence the result follows. $\square$

Theorem 32 *If G is a cellular automaton, then either*

(i) $\Lambda(G)$ consists of a single element, and G is nilpotent, or

(ii) $\Lambda(G)$ contains an infinite number of elements.

7. Surjectivity and injectivity

Since Moore introduced the concept of Garden-of-Eden configuration [60,61], the surjectivity and injectivity of multi-dimensional CA mappings have been topics of great interest. A *Garden-of-Eden* is simply any configuration $c \in S^L$ which is not in the image of G. This means that once left, the configuration can never be regained.

Recall that a (set-theoretic) map $f: X \to Y$ is surjective if for every $y \in Y$ there is at least one $x \in X$ such that $f(x) = y$. It is injective if $f(x) = f(y)$ if and only if $x = y$ (i.e. the map is 1–1). A map satisfying both conditions is called bijective.

A number of interesting results have been obtained in the past three decades. For one-dimensional CA an algorithm for deciding both questions has been known for some time (see ref. [62]). Recently, the undecidability of the injectivity and surjectivity of two- (or higher-) dimensional CA maps has been shown by Kari [63]. In the following two subsections, we summarize the major results on the surjectivity and injectivity of CA maps. Definitions are introduced whenever necessary.

7.1. Relationship between surjectivity and injectivity

Already in 1961 Moore [60] proved that every CA that is not finitely injective (see below) has a Garden-of-Eden configuration, i.e. a configuration that cannot be reached from any other configuration. The converse has been shown by Richardson [65]. Clearly, a CA has a Garden-of-Eden configuration iff its global function is not surjective. After giving the necessary definitions we state these results together with additional ones in theorem 33 below.

Various restricted forms of surjectivity and injectivity have been introduced and studied [65,66]. In order to avoid confusion, we call the surjectivity and injectivity of a CA global map G defined on

$S^L \to S^L$ (with no restriction) the *global surjectivity and injectivity* of G. However, the word *global* will often be omitted when there is no confusion.

Next, we define a restricted form of surjectivity (injectivity) for CA maps, the *finite surjectivity (injectivity)*, which has been widely studied. Since sources fail to agree on the definition, we outline a definition below.

As in ref. [66], we define a relation R on $S^L \times S^L$ such that, for $c_1, c_2 \in S^L$, $c_1 R c_2$ holds if c_1 and c_2 are identical on all but a finite number of cells. The relation R is clearly an equivalence relation. Let $R[c] = \{x \in S^{Z^n} \mid xRc\}$ and for any two configurations c and c', let $G_{c,c'}$ be the global map G with domain restricted to $R[c]$ and range restricted to $R[c']$.

Definition 12 *A CA map G is* finitely injective (surjective) *if, for all $c, c' \in S^L$ such that $G(c) = c'$, the map $G_{c,c'}$ is injective (surjective) in the usual sense.*

Finite injectivity (surjectivity) is also called strong injectivity (surjectivity) in ref. [66].

The relationship between global and finite injectivity and surjectivity is summarized in the following theorem [65,66]:

Theorem 33 *Given a cellular automaton map G (of dimension $d \geq 1$) and the following properties of G:*

(GI)　　global injectivity,
(GS)　　global surjectivity,
(FI)　　finite injectivity,
(FS)　　finite surjectivity,

the following represents a complete set of valid implications among these properties:

$$(GI) \Rightarrow (FS) \Rightarrow (FI) \Leftrightarrow (GS).$$

Since global injectivity implies global surjectivity, global injectivity is a synonym for bijectivity. In fact it implies a stronger property.

The reversibility of CA maps has been frequently mentild in literature. We define the reversibility of a CA map as follows:

Definition 13 *A CA map G is* reversible *if there exists another CA map G' such that for any pair*

of configurations c and c' $G(c) = c'$ *if and only if* $G'(c') = c$.

On the surface the definition of reversibility seems to have two parts. First G must be a bijection as a set theoretic map, and thus have a formal inverse G'. The second statement is that this formal inverse G' is in fact another CA map. It turns out, however, that the two conditions are the same. In fact, the following stronger result has been proved by Richardson in ref. [65].

Theorem 34 *A cellular automaton map is globally injective if and only if it is reversible.*

Because of the above theorem there is no difference between studying reversibility, bijectivity, or injectivity of CA maps. The following results of Toffoli are in ref. [67].

Theorem 35 (Toffoli) *An arbitrary CA having k dimensions is embeddable in a reversible one having $k + 1$ dimensions.*

Theorem 36 (Toffoli) *There exist reversible universal cellular automata.*

We end this subsection with two open problems raised by Toffoli in ref. [67]:

(i) Does there exist a one-dimensional reversible universal cellular automaton?

(ii) Can an arbitrary cellular automaton be embedded in a reversible one having the same number of dimensions?

7.2. Deciding injectivity and surjectivity

The surjectivity and injectivity of linear cellular automaton maps was shown to be decidable by Amoroso and Patt [62] using a combinatorial approach. A formal language theoretic proof was later provided by the first author in ref. [68].

Theorem 37 (Amoroso, Patt) *For any* linear *(one-dimensional) CA map, G, there are finite decision procedures to determine (a) whether G is injective (hence reversible) and (b) surjective.*

The decidability of the one-dimensional injectivity question can be seen from the following theorem, which was proved by the first author in ref.

[68]. The theorem is stated for one-dimensional cellular automata, and indeed its generalization to higher dimensions is false. Recall that a configuration $c \in S^Z$ is spatially periodic if it is fixed by a power of the left shift ($\sigma^n(c) = c$ for some $n > 0$). A modified proof follows, see ref. [69].

Theorem 38 *For a one-dimensional CA, the global function $G\colon S^Z \to S^Z$ is injective if and only if it is injective on spatially periodic configurations.*

Proof: Since the *only if* part is trivially true, we now consider the *if* part.

Assume that G is not injective, that is, there exist two distinct configurations $\alpha, \beta \in S^Z$ such that $G(\alpha) = G(\beta) = \gamma$. It suffices to show that we can construct two distinct periodic configurations which are mapped to the same configuration. Since $\alpha \neq \beta$ and both α and β are infinite in both directions, we can find two substrings $\alpha(i) \dots \alpha(j)$ and $\beta(i) \dots \beta(j)$ of α and β, respectively, such that $\alpha(i) \dots \alpha(j) = sx_1 sx_2 tx_3 t$ and $\beta(i) \dots \beta(j) = uy_1 uy_2 vy_3 v$ where $|s| = |t| = |u| = |v| = 2r$ (note that we assume that r is the radius of a neighborhood, i.e. $2r + 1$ is the size of a neighborhood), $x_2 \neq y_2$, and $|x_k| = |y_k|$ for $k = 1, 2$, and 3. They are shown in fig. 4. Now, we consider the following two cases.

Case 1. $s \neq u$ or $t \neq v$. Without loss of generality we assume that $s \neq u$. Then we can form two periodic configurations α' and β' by repeating sx_1 and uy_1, respectively, bi-infinitely. Note that a period sx_1 in α' should start at the same position as a period uy_1 in β'. Clearly, these two periodic configurations are distinct but mapped onto the same configuration.

Case 2. Both $s = u$ and $t = v$. Then we can construct two distinct periodic configurations α' and β' by repeating the two patterns $p_1 = sx_2 t$ and $p_2 = uy_2 v$, respectively, bi-infinitely. Since $x_2 \neq y_2$ we have $\alpha' \neq \beta'$. In order to see that $G(\alpha') = G(\beta')$, we assume the contrary. Let $G(\alpha')$ and $G(\beta')$ differ at a position k, i.e. $G(\alpha')(k) \neq G(\beta')(k)$. Then the cell k can only be at a position such that its neighborhood (which determines the symbol at cell k under the mapping of G) must cross two periods of p_1 in α' and correspondingly two periods of p_2 in β' as well. But since $s = u$ and

$t = v$ and $|s| = |t| = 2r$, the two neighborhoods in α' and β', respectively, are exactly identical. They should not map to two different symbols. It is a contradiction. Therefore, $G(\alpha') = G(\beta')$.

We have considered all the possible cases. Therefore, G is not injective on periodic configurations. $\qquad\square$

A decision procedure then follows by setting up one semi-procedure to check for each cellular automaton whether it forms an inverse for G, in which case G is injective, or alternatively for each spatial period p check whether G send two periodic configurations to the same point, in which case G is not injective. In any case, one of these two procedures must halt.

Actually the problem is easier than the above argument would suggest, since the proof of theorem 38 contains an implicit limit on the size of periodic configurations which must be checked.

In two and higher dimensions the problems are much more difficult, and were only just recently solved. In refs. [63,64], Kari has shown that in general both the surjectivity and injectivity problem for cellular automata in dimension two (or greater) are undecidable.

Theorem 39 (Kari) *For any fixed dimension $d \geq 2$, it is undecidable whether an arbitrary CA map G is (a) injective (hence reversible) or (b) surjective.*

7.3. Other problems

Injectivity and surjectivity are examples of problems that are decidable for one-dimensional CA but undecidable for higher-dimensional CA. There are other problems which are decidable in dimension one and undecidable for dimensions two and greater. An example of such a problem is testing whether a given finite or spatially periodic configuration has a predecessor. It is easy to show an algorithm to test this for one-dimensional CA. However, for two-dimensional CA the question remains undecidable even if we restrict the problem to asking whether the everywhere quiescent configuration has a predecessor. It is not difficult to reduce the question to the tiling or domino problem [57,58].

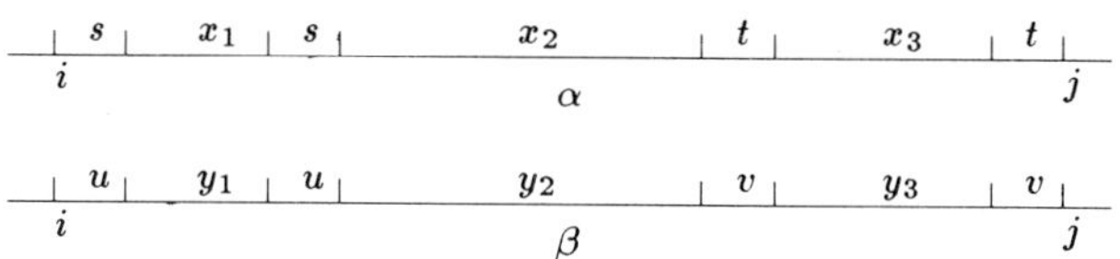

Fig. 4. Two configurations α and β.

Many problems that are undecidable for arbitrary (infinite) configurations or arbitrary long computations become decidable but intractable (NP-complete) for finite configurations (blocks of cells). For example, it has been shown in ref. [70] that the following problems are NP-complete for a fixed CA.

(i) Given a subconfiguration w of length n, is there a predecessor configuration that leads to w in n steps?

(ii) Given a block of cells of size n and its initial configuration will it be in the same configuration n-time steps later?

For more complexity results see ref. [71].

8. Conclusion

Cellular automata have the same computational power as Turing machines. However, they also have a natural definition in terms of dynamical systems. It is this interplay between dynamical properties and computational ones which makes cellular automata interesting.

This paper has outlined some of the properties of interest in examining cellular automata and outlined decidability and undecidability results about these properties. Much work, however, remains to be done in determining what one can decide about cellular automata. In refs. [63,72], Kari shows that, in analogy to Rice's theorem on properties of recursively enumerable sets specified by Turing machines, any non-trivial property of cellular automata limit sets is undecidable. This, however, does not affect the problem of finding particular examples of given complexity. Also left open are the computational power of sub-classes of cellular automata such as the surjective and reversible Is.

Acknowledgements

Reported research was supported by the Natural Sciences and Engineering Research Council of Canada grants OGPIN007 and the National Sciences Foundation Grant No. CCR-8702752, as well as DARPA under the Applied and Mathematics Computational Program. The second author would like to thank the Institute for Advanced Study for support during the time that some of this research was conducted.

References

[1] J. von Neumann, in: Theory of Self-Reproducing Automata, ed. A.W. Burks (Univ. of Illinois Press, Champaign, IL, 1966).

[2] S. Wolfram, Universality and complexity in cellular automata, Physica D 10 (1984) 1.

[3] K. Culik and S. Yu, Undecidability of CA classification schemes, Complex Systems 2 (1988) 177.

[4] L.P. Hurd, Formal language characterizations of cellular automaton limit sets, Complex Systems 1 (1987) 69.

[5] S. Ginsburg, Algebraic and Automata-Theoretic Properties of Formal Languages (North-Holland, Amsterdam, 1975).

[6] J.L. Kelley, General Topology (Van Nostrand, New York, 1955).

[7] G.A. Hedlund, Transformations commuting with the shift, in: Topological Dynamics, eds. J. Auslander and W.G. Gottschalk (Benjamin, New York,1968) p. 259; Endomorphisms and automorphisms of the shift dynamical system, Math. Syst. Theor. 3 (1969) 320.

[8] M.A. Arbib, Simple self-reproducing universal automata, Information Control 9 (1966) 177.

[9] H. Kleine Buening and Th. Ottmann, Kleine Universale Mehrdimensionale Turingmaschinen, Elektron. Informationsverarbeitung Kybern. 13 (1977) 179.

[10] M. Minsky, Computation: Finite and Infinite Machines (Prentice Hall, Englewood Cliffs, NJ, 1967).

[11] L. Priese, Towards a precise characterization of the complexity of universal and nonuniversal Turing Machines, SIAM J. Comput. 8 (1979) 508.

[12] A.R. Smith III, Simple computation-universal cellular spaces, J. ACM 18 (1971) 339.

[13] J. Albert and K. Culik II, A simple universal cellular automaton and its one-way and totalistic version, Complex Systems 1 (1987) 1.

[14] C.R. Dyer, One way bounded cellular automata, Information Control 44 (1980) 261.

[15] K. Culik II and I. Fris, Topological transformations as a tool in the design of systolic networks, J. Theoret. Computer Sci. 37 (1985) 183.

[16] K. Culik II and S. Yu, Translation of systolic algorithms between systems of different topology, Proc. 1985 IEEE and ACM International Conference of Parallel Processings (1985) p. 756.

[17] H. Umeo, K. Morita and K. Sugato, Deterministic one-way simulation of two-way real-time cellular automata and its related problems, Information Processing Lett. 14 (1982) 158.

[18] S. Wolfram, Statistical mechanics of cellular automata, Rev. Mod. Phys. 55 (1983) 601.

[19] D. Gordon, On the computational power of totalistic cellular automata, manuscript.

[20] K. Culik II and J. Karhumäki, On totalistic systolic networks, Information Processing Lett. 26 (1987/88) 231.

[21] E.R. Berlekamp, J.H. Conway and R.K. Guy, Winning Ways for Your Mathematical Plays, Vol. 2 (Academic Press, New York, 1982) ch. 25; M. Gardner, Wheels, Life and Other Mathematical Amusements (Freeman, San Francisco, 1983).

[22] S. Wolfram, Twenty problems in the theory of cellular automata, Physica Scripta T9 (1985) 170.

[23] K. Sutner, The computational complexity of cellular automata, Proceedings of Fundamentals of Computation Theory, Lecture Notes in Computer Science (Springer, Berlin, 1984) p. 451.

[24] H.A. Gutowitz, A hierarchical classification of cellular automata, Physica D 45 (1990) 136–156, these Proceedings.

[25] H.A. Gutowitz, J.D. Victor and B.W. Knight, Local structure theory for cellular automata, Physica D 28 (1987) 18.

[26] C. Bays, Classification of semitotalistic cellular automata in three dimensions, Complex Systems 2 (1988) 235.

[27] R. Gilman, Classes of linear automata, Ergodic Theor. Dynam. Systems 7 (1987) 105.

[28] R. Gilman, Periodic behavior of linear automata, in: Dynamical Systems, ed. J.C. Alexander, Springer Lecture Notes in Mathematics 1342 (Springer, Berlin, 1988) p. 216.

[29] K. Culik, J. Pachl and S. Yu, On the limit sets of cellular automata, SIAM J. Computing 18 (1989) 831.

[30] K. Sutner, A note on Culik–Yu classes, Complex Systems 3 (1989) 107.

[31] H. Rogers Jr., Theory of Recursive Functions and Effective Computability (MIT Press, Cambridge, MA, 1987).

[32] S. Wolfram, Computation theory of cellular automata, Commun. Math. Phys. 96 (1984) 15.

[33] M.A. Harrison, Introduction to Formal Language Theory (Addison–Wesley, Reading, MA, 1978).

[34] S.N. Cole, Real-time computation by n-dimensional iterative arrays of finite-state machines, IEEE Trans. Comp. C18 (4) (1969) 349.

[35] A.R. Smith III, Real-time language recognition by one-dimensional cellular automata, J. Computer System Sci. 6 (1972) 233.

[36] K. Culik II and S. Yu, Iterative tree automata, Theor. Computer Sci. 32 (1984) 227.

[37] S. Yu, On systolic automata, Ph.D. Dissertation, Department of Computer Science, University of Waterloo (1986).

[38] K. Culik II and S. Yu, Cellular automata, $\omega\omega$-regular sets, and sofic systems, Discrete Appl. Math., to appear.

[39] L.P. Hurd, The application of formal language theory to the dynamical behavior of cellular automata, Doctoral Thesis, Princeton University (1988).

[40] B. Weiss, Subshifts of finite type and sofic systems, Monatsh. Math. 77 (1973) 462.

[41] D. Beauquier, Ensembles reconnaissables de mots bi-infinis limite et déterminisme, in: Automata on Infinite Words, Lecture Notes in Computer Science, Vol. 192, eds. M. Nivat and D. Perrin (Springer, Berlin, 1984) p. 28.

[42] M. Nivat and D. Perrin, Ensembles reconnaissables de mots bi-infinis, in: Proceedings of the Fourteenth Annual ACM Symposium on Theory of Computing (1982) p. 47.

[43] J. Berstel, Transductions and Context-Free Languages (Teubner, Stuttgart, 1979).

[44] M. Nivat and D. Perrin, Ensembles reconnaisables de mots bi-infinis, Can. J. Math. 39 (1986) 513.

[45] M. Nivat, Sur les ensembles de mots infinis engendrés par une grammaire algébrique, RAIRO Informatique Théorique 12 (1978) 259.

[46] A. Restivo, finitely generated sofic systems, Research Report, Dipartimento di Matematica e Applicazioni, Università di Palermo (1986).

[47] K. Culik II, L.P. Hurd and S. Yu, Formal languages and global cellular automaton behavior, Physica D 45 (1990) 396–403, these Proceedings.

[48] D. Lind, Applications of ergodic theory and sofic systems to cellular automata, Physica D 10 (1984) 36.

[49] M. Hurley, Attractors in cellular automata, Ergodic Theor. Dynam. Systems, to appear.

[50] M. Hurley, Ergodic aspects of cellular automata, to appear.

[51] L.P. Hurd, The non-wandering set of a CA map, Complex Systems 2 (1988) 549.

[52] C. Conley, Isolated invariant sets and the Morse index, CBMS Regional Conference Series 38 (Am. Math. Soc., Providence, RI, 1978).

[53] L.P. Hurd, Recursive cellular automata invariant sets, Complex Systems, to appear.

[54] L.P. Hurd, Non-recursive cellular automata invariant sets, Complex Systems, to appear.

[55] P. Walters, An introduction to ergodic theory (Springer, Berlin, 1982).

[56] A.S. Podkolzin, On the behavior of homogeneous structures, Problemy Kibernetiky 31 (1976) 133.

[57] R. Berger, The undecidability of the domino problem, Mem. Am. Math. Soc. 66 (1966).

[58] R.M. Robinson, Undecidability and nonperiodicity of tilings of the plane, Inventiones Math. 12 (1967) 177.

[59] J. Kari, The nilpotency problem of one-dimensional cellular automata, to appear.

[60] E.F. Moore, Machine models of self-reproduction, Proc. Symp. Appl. Math. 14 (1961) 17.

[61] J. Myhill, The converse to Moore's Garden-of-Eden theorem, Proc. AMS 14 (1963) 685.

[62] S. Amoroso and Y.N. Patt, Decision Procedures for Surjectivity and Injectivity of Parallel Maps for Tessellation Structures.

[63] J. Kari, Decision problems concerning cellular automata, Doctoral Thesis, Department of Mathematics, University of Turku, Turku, Finland (1989).

[64] J. Kari, Reversibility and surjectivity problems of cellular automata, submitted to J. Computer System Sci.

[65] D. Richardson, Tessellation with local transformations, J. Computer System Sci. 6 (1972) 373.

[66] A. Maruoka and M. Kimura, Injectivity and surjectivity of parallel maps for cellular automata, J. Computer System Sci. 18 (1979) 47.

[67] T. Toffoli, Computation and construction universality of reversible automata, J. Computer System Sci. 15 (1977) 213.

[68] K. Culik II, On invertible cellular automata, Complex Systems 1 (1987) 1035.

[69] S. Dube, An investigation of cellular automata, Master's Thesis, Department of Computer Science, University of South Carolina (1988).

[70] F. Green, NP-complete problems in cellular automata, Complex Systems 1 (1987) 453.

[71] K. Sutner, The complexity of finite cellular automata, manuscript in preparation.

[72] J. Kari, Rice's theorem for limit sets of cellular automata, submitted to Theor. Computer Sci.

Physica D 45 (1990) 379–385
North-Holland

REVERSIBILITY OF 2D CELLULAR AUTOMATA IS UNDECIDABLE

Jarkko KARI

Mathematics Department, University of Turku, 20500 Turku, Finland

Received 8 January 1990
Revised manuscript received 20 February 1990

The problem whether a given two- or higher-dimensional cellular automaton is reversible is shown to be algorithmically undecidable. A cellular automaton is called reversible if it has an inverse automaton, that is a cellular automaton that makes the system retrace its steps backwards in time. The same problem is known to be decidable for one-dimensional automata.

1. Introduction

Cellular automata are discrete and deterministic dynamical systems. They are especially suitable for modeling natural systems that can be described as massive collections of simple objects interacting locally with each other. A d-dimensional cellular automaton consists of an infinite d-dimensional array of identical cells. Each cell is always in one state from a finite state set. The cells alter their states synchronously on discrete time steps according to a local rule. The rule gives the new state of each cell as a function of the old states of some nearby cells, its neighbors. The array is homogeneous so that all its cells operate under the same local rule. The states of all the cells in the array are described by a configuration. A configuration can be considered as the state of the whole array. The local rule of the automaton specifies a global function that tells how each configuration is changed in one time step.

Cellular automata provide simple models of complex natural systems encountered in physics, biology and other fields. Like natural systems they consist of large numbers of simple basic components that together produce the complex behavior of the system. A basic feature of microscopic mechanisms of physics seems to be reversibility [10]. It is possible for cellular automata to capture this important characteristic without sacrific-

ing other essential properties like computational universality. A cellular automaton rule is called reversible if there exists another rule, called the inverse rule, that makes the automaton retrace its computation steps backwards in time. The earlier configurations are uniquely determined by the present one and no information is lost during the computation. A cellular automaton defined by a reversible local rule is called reversible. It is known that a cellular automaton is reversible if its global function is one-to-one [8].

It is a natural question to ask what kind of local rules are reversible. No general characterizations are known. In fact, there does not exist an algorithm that would decide whether a given local rule is reversible or not. This fact is the main result reported in this article. If only one-dimensional rules are considered, then such an algorithm can be designed [1].

Another property a cellular automaton can have is surjectivity, that is the surjectivity of its global function. In a surjective cellular automaton every configuration can occur arbitrarily late during computations. It is known that an automaton is surjective if and only if the restriction of its global function to finite configurations is injective. This is the so-called Garden of Eden theorem proved by Moore [6] and Myhill [7]. A configuration is called finite if it has only finitely many cells in states different from one specific quiescent state. One can show, using a similar method as in connection with the reversibility, the algorithmic undecidability of the problem of testing whether a given cellular automaton is surjective. In the one-dimensional case the problem is, however, decidable [1].

The article is organized as follows. First the concepts used in the paper are formally defined. Then we outline the proof for the undecidability of the reversibility problem. The idea of the proof is very simple, but the complete proof contains a number of confusing details and it is unnecessary to go into these details in the present article. The full proof can be found in refs. [4,5]. In section 4 surjective cellular automata are considered. We briefly discuss the differences between the reversibility and surjectivity problems, and we present how the ideas of section 3 can be used to show the surjectivity problem undecidable. The final section contains some concluding remarks and consequences of the results.

2. Definitions

Let us first explain the notions used. Formally, a *cellular automaton* (CA) is a quadruple

$$\mathcal{A} = (d, S, N, f) \,,$$

where d is a positive integer indicating the *dimension* of $\mathcal{A}$, S is a finite *state set*, N is a *neighborhood vector*

$$N = (\bar{x}_1, \bar{x}_2, \ldots, \bar{x}_n)$$

of n different elements of Z^d and f is the *local rule* of the CA presented as a function from S^n into S. The cells are laid on an infinite d-dimensional array and their positions are indexed by Z^d, the set of d-tuples of integers. The *neighbors* of a cell situated in $\bar{x} \in Z^d$ are the cells in positions

$$\bar{x} + \bar{x}_i \,, \qquad \text{for } i = 1, 2, \ldots, n \,.$$

The local rule f gives the new state of a cell from the old states of its neighbors.

A *configuration* of a CA $\mathcal{A} = (d, S, N, f)$ is a function

$$c : Z^d \to S$$

that assigns states to all cells. Let $\mathcal{C}$ denote the set of all configurations. The local rule f determines the *global function*

$$G_f : \mathcal{C} \to \mathcal{C}$$

that describes the dynamics of the CA. At each time step a configuration c is transformed into a new configuration $G_f(c)$ where

$$G_f(c)(\bar{x}) = f(c(\bar{x} + \bar{x}_1), c(\bar{x} + \bar{x}_2), \ldots, c(\bar{x} + \bar{x}_n))$$

for all $\bar{x}$ in Z^d. For each state $s \in S$ let $conf(s)$ denote the homogeneous configuration where all the cells are in the same state s.

Sometimes a *quiescent state* q in S is distinguished. The quiescent state must have the property

$$f(q, q, \ldots, q) = q \,.$$

The configuration $conf(q)$ with all cells in the quiescent state is called the *quiescent configuration*. A configuration is called *finite* if it has only finitely many cells in non-quiescent states. Let $\mathcal{C}_F$ denote the set of finite configurations. It follows from the special property of the quiescent state that a finite configuration remains finite in the evolution of the CA. Let G_f^F denote the restriction of the global function G_f to the set of finite configurations.

In this work mainly two-dimensional cellular automata are considered. In this case the cells are laid on the plane. The following two-dimensional neighborhood vector will be frequently used :

$$N_{vN} = \big[(0,0), (1,0), (0,1), (-1,0), (0,-1)\big] \,.$$

This defines the so called *von Neumann neighborhood* – each cell has five neighbors: the cell itself and the four adjacent cells (see fig. 1). The four directions of the compass are used when we refer to the four adjacent neighbors.

A CA is called *injective* if its global function is one-to-one. Similarly, a CA is called *surjective* if its global function is surjective.

Example   Let

$$N = (\bar{x}_1, \bar{x}_2, \ldots, \bar{x}_n)$$

be any neighborhood vector of a d-dimensional CA,

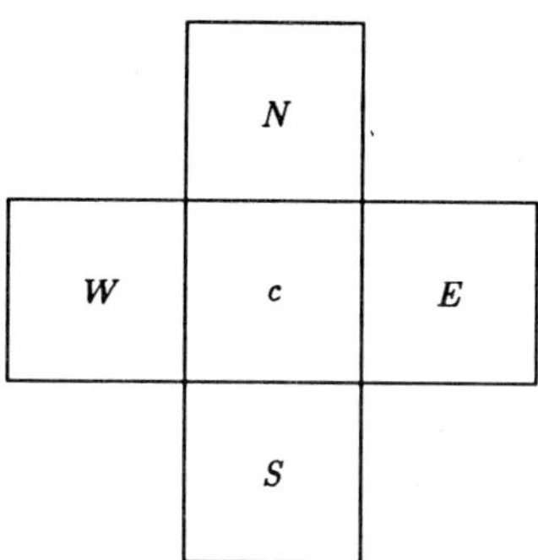

Fig. 1. The von Neumann neighborhood of cell c.

$$S = S_1 \times S_2 \times \ldots \times S_n$$

a Cartesian product of n finite sets and

$$\varphi\colon S \to S$$

a bijective function. Let

$$\pi_i\colon S_1 \times S_2 \times \ldots \times S_n \to S_i$$

denote the ith projection of S. Let f be a function from S^n into S defined by

$$f(s_1, s_2, \ldots, s_n) = \varphi(\pi_1(s_1), \pi_2(s_2), \ldots, \pi_n(s_n)).$$

It is easy to see that the d-dimensional cellular automaton $\mathcal{A} = (d, S, N, f)$ is injective. This follows from the bijectivity of φ – if you change its arguments then its value is changed as well. Define then another function g from S^n into S by

$$g(s_1, s_2, \ldots, s_n)$$

$$= (\pi_1(\varphi^{-1}(s_1)), \pi_2(\varphi^{-1}(s_2)), \ldots, \pi_n(\varphi^{-1}(s_n)))$$

and let N^{-1} be the neighborhood obtained from N by changing the signs of all of its coordinates. Then the CA $\mathcal{B} = (d, S, N^{-1}, g)$ is injective and, moreover, it is the inverse automaton of $\mathcal{A}$. This means that $G_g = G_f^{-1}$. $\square$

The injective CA in the example above were *reversible*, that is, they had an *inverse automaton*. Richardson [8] showed that this is the case with every injective CA.

3. The reversibility problem

The reversibility problem asks whether a given CA rule is reversible, or equivalently whether the

global function it defines is injective. In this section we present the basic ideas how one can prove the reversibility problem undecidable in the case of two-dimensional CA. The complete proof is given in refs. [4,5].

The proof is essentially a reduction from a well known undecidable problem, the so-called *tiling problem*. In the tiling problem we are given a finite set of unit squares with colored edges, the tiles, placed with their edges horizontal and vertical. We have infinitely many copies of all these tiles and we want to tile the entire plane using the copies, without rotating any of them. The tiles are placed to the unit square regions of the Euclidean plane bordered by the lines $y = n$ and $x = m$ for all integers n and m. In a valid tiling the abutting edges of the tiles must have the same color. The tiling problem consists of deciding whether the plane can be tiled with a given collection of tiles. The tiling problem was proved undecidable by Berger [2]. A simplified proof was given later by Robinson [9].

An easy application of König's infinity lemma shows that if one can tile arbitrarily large squares with a given tile set, then one can tile the whole plane as well. This fact will be used in our proof.

In the proof a specific set $\mathcal{D}$ of *directed tiles* is needed. The directed tiles are, like ordinary tiles, unit squares with colored edges. In addition, there is one direction from the set

$$\{N, E, S, W\}$$

attached to every tile. The direction refers to one of the four neighboring tiles lying to the north, east, south or west.

The directions define paths through the tiles on the plane in a natural way. The direction of each tile tells which way a path coming to the tile proceeds. The next tile on the path is the adjacent tile pointed by the direction. Every tile on the plane starts a path that follows the directions. The path is not necessarily infinite – it may come back to an earlier tile causing the path to fall into a loop.

Next we define a special property that the set $\mathcal{D}$ of directed tiles is supposed to satisfy. We call it the *plane filling property*. First, it is required that there exists a valid tiling of the plane with the tiles. Secondly, the essential requirement is that on every tiling of the plane with the tiles, valid or not, the directions can define only two different types of paths: Either there is a tile on the path where the tiling is incorrect, or otherwise the path visits all tiles of arbitrary large squares. So if the tiling property is not violated on the path then for each n there must be an $n \times n$ square each tile of which is on the path. Especially the directions cannot define any loop without a violation of the tiling property somewhere along the path.

The second requirement can be illustrated as follows. Consider a simple memoryless device that operates on the plane which is tiled with the directed tiles of $\mathcal{D}$. The device checks the tiling at the tile it is currently standing on. This means it looks at the four tiles that have an abutting edge with its own tile, and it checks whether the four pairs of abutting edges have the same color. If the tiling is proper, then the device goes to the neighboring tile that is pointed by the direction of its current tile. The operation is repeated on the new tile. If, on the other hand, the tiling property is violated at the tile, then the device halts, indicating that it has found an error.

If the tile set $\mathcal{D}$ has the plane filling property then, no matter how the plane is tiled with the tiles, and no matter which tile the device is started at, there are just two possible ways the device can operate. Either it halts because it finds a tiling error, or it visits all tiles of arbitrarily large squares.

The following proposition allows us to use the tile set $\mathcal{D}$ in our undecidability proof. The proof of the proposition is omitted. It can be found in refs. [4,5]. The proof consists of an explicit construction of the tile set with the plane-filling property. The plane-filling property is defined in a slightly different way in refs. [4,5], but the tile set can be modified rather easily to satisfy the formulation we have used here.

Proposition 1 *There exists a set $\mathcal{D}$ of directed tiles that satisfies the plane filling property.*

In fig. 2 there is a portion of the path defined by the directed tiles of $\mathcal{D}$. The path has the shape of the well known Peano curve.

Once we have the tile set $\mathcal{D}$ the reduction of the tiling problem to the injectivity problem of two-dimensional CA is very simple. All the complicated details of the reduction are contained in

Fig. 2. A part of the plane-filling curve defined by the directed tiles.

the construction of the tile set with the plane filling property.

Let $\mathcal{T}$ be any ordinary tile set, that is, a set of tiles without directions. One can construct a two-dimensional cellular automaton

$$\mathcal{A}_{\mathcal{T}} = (2, \mathcal{D} \times \mathcal{T} \times \{0,1\}, N_{\mathrm{vN}}, f_{\mathcal{T}})$$

such that $\mathcal{A}_{\mathcal{T}}$ is not injective if and only if the tile set $\mathcal{T}$ can be used to tile the plane.

The automaton $\mathcal{A}_{\mathcal{T}}$ uses the von Neumann neighborhood. Its states contain two tile components, one from $\mathcal{D}$ and one from $\mathcal{T}$. In addition there is a bit, 0 or 1, attached to each state. The local rule $f_{\mathcal{T}}$ may change only the bits. The tile components are never changed. At each cell the tilings with both the $\mathcal{D}$- and $\mathcal{T}$-components are checked. If there is a tiling error in either of the components then the state of the cell is not altered. If the tilings are correct then the bit component is changed by performing the exclusive or ($= \mathrm{XOR}$) operation with the bit that is attached to the cell pointed by the $\mathcal{D}$-component. The XOR operation is the same as addition of the bits modulo 2.

Suppose that there exists a tiling of the plane with the tile set $\mathcal{T}$. Construct two configurations c_0 and c_1 of $\mathcal{A}_{\mathcal{T}}$ as follows. The tile components of c_0 and c_1 constitute the same legal tilings of the plane with the tiles of $\mathcal{D}$ and $\mathcal{T}$. In c_0 all bits are

0 while in c_1 they equal 1. Because the tilings are correct everywhere each bit is changed using the XOR operation with the next bit on the path. This means that in both configurations c_0 and c_1 the bits are all changed to 0. So in $\mathcal{A}_T$ there are two different configurations c_0 and c_1 that are turned into the same configuration in one time step and $\mathcal{A}_T$ cannot be injective.

Conversely, suppose that $\mathcal{A}_T$ is not injective. Let c_0 and c_1 be two different configurations that $\mathcal{A}_T$ turns into the same configuration in one time step. The tile components of c_0 and c_1 must coincide. Consider a cell whose bit component is different in c_0 and c_1. The tilings must be correct at the cell, and the bits are different also in the cell pointed by the $\mathcal{D}$-component. The reasoning can be repeated for this next cell. By continuing through succeeding cells it is concluded that the tiling properties may not be violated at any of them. Since $\mathcal{D}$ has the plane filling property this path goes through arbitrary large squares. So arbitrary large squares can be tiled using the tiles of T. This means that the whole plane can be tiled.

We showed that there is a valid tiling of the plane using the tile set T if and only if the cellular automaton $\mathcal{A}_T$ is not reversible. If there were an algorithm for solving the reversibility problem, then this algorithm applied to $\mathcal{A}_T$ would solve the tiling problem. This is not possible because the tiling problem is undecidable.

Theorem 1 *It is undecidable whether a given two-dimensional cellular automaton with the von Neumann neighborhood is reversible.*

4. The surjectivity problem

A cellular automaton is called surjective if its global transition function is surjective. Surjectivity is a less restricted property of cellular automata than injectivity, since every injective rule is also surjective. This follows immediately from the well known Garden of Eden theorem of Moore [6] and Myhill [7], which states that the global function G_f is surjective if and only if its restriction G_f^{F} to finite configurations is injective. This means that instead of testing the surjectivity property one can test the injectivity of G_f^{F}.

The two injectivity problems – the problem of testing the injectivity of G_f on one hand, and the problem of testing the injectivity of G_f^{F} on the other hand – have one basic difference. The cellular automata with injective G_f can be *effectively enumerated* by listing all CA and checking for each pair if they are inverses of each other. If they are, they are added to the list of injective CA. Remember that a CA is injective if and only if it is reversible. On the other hand, we cannot enumerate the non-injective CA, because the injectivity problem is undecidable.

With the restriction G_f^{F} the situation is just the opposite: One can enumerate the CA having a non-injective G_f^{F} but, as theorem 2 below states, one cannot enumerate the CA with injective G_f^{F}. In the previous section the reduction from the tiling problem was done in such a way that the tiling problem had a solution iff the corresponding G_f was not injective. This reduction can be done since the tile sets that possess a solution cannot be effectively enumerated. Doing the same with G_f^{F} is not possible. A similar reduction would map the tile sets with valid tilings into the effectively enumerable set of non-injective G_f^{F}.

This problem is solved by introducing a new tiling problem, called *the finite tiling problem*. The tile sets that possess the finite tiling property can be effectively enumerated. This is why this problem seems better suited for our purpose.

In the finite tiling problem we are given a finite collection of tiles. The tiles have colored edges as before. One of the tiles is called *blank*. All of its edges have the same color. In a valid tiling of the plane the abutting edges of adjacent tiles must have the same color, just like before. There is of course at least one valid tiling – namely the one where all the tiles are blank. This tiling is called *trivial*. A tiling is called *finite* if only finitely many of the tiles on the plane are not blank. The problem is to decide whether there is a valid, finite and non-trivial tiling of the plane. This problem can easily be proved undecidable [4].

Proposition 2 *It is undecidable whether a given collection of tiles containing a blank tile can be used to form a finite tiling that is valid but not trivial.*

Reducing the finite tiling problem to the injectivity problem of G_f^{F} can be done using a similar method as in section 3. The set $\mathcal{D}$ of directed tiles must, however, be changed a little to allow finite tilings of the plane. Let $\mathcal{D}_0$ be the altered tile set (see ref. [4] for its construction). It contains one blank tile that does not have a direction, and instead of the plane-filling property the set satisfies the following property. For every non-trivial tiling of the plane and for every path defined by the directions of the tiles the following is true: If the tiling property is not violated on the path then either the path visits all tiles of arbitrarily large squares (as in section 3), or the path is a closed loop visiting all tiles of some square surrounded by blank tiles. In the latter case a specific tile d_0 is known to be at the center of the square, but not anywhere else inside the square. It is also known that for arbitrary large squares there are tilings with a closed path of the latter type through the square.

Let $\mathcal{T}$ be any tile set containing a blank tile. Let us construct a cellular automaton $\mathcal{B}_\mathcal{T}$ whose states have two tile components, one from $\mathcal{T}$ and one from $\mathcal{D}_0$. The components satisfy the following two restrictions:

(i) If the $\mathcal{D}_0$-component is blank then the $\mathcal{T}$-component is also blank, and

(ii) if the $\mathcal{D}_0$-component equals d_0 then the $\mathcal{T}$-component is *not* blank.

The state whose both components are blank is the quiescent state. Every non-quiescent state contains one bit.

The local rule of $\mathcal{B}_\mathcal{T}$ is defined in the same way as in section 3. At each cell the tilings with both $\mathcal{T}$- and $\mathcal{D}_0$-components are checked. If the tiling property is violated in either of the components, then the state of the cell is not changed. If, on the other hand, the tilings are valid at the cell, then the bit is changed by performing an exclusive or operation with the bit attached to the cell pointed by the direction of the $\mathcal{D}_0$-component. The quiescent state is never altered.

It is an easy matter to show that the restriction of the global function of $\mathcal{B}_0$ to finite configurations is injective if and only if the tile set $\mathcal{T}$ cannot be used to form a valid, non-trivial and finite tiling of the plane. Note how finite tilings correspond nicely to finite configurations. Since the finite tiling property is undecidable, and since a CA is surjective iff the restriction of its global function is injective, we have proved:

Theorem 2 *It is undecidable whether a given two-dimensional cellular automaton with the von Neumann neighborhood is surjective.*

5. Conclusion

Our main theorem, which states the undecidability of the reversibility problem, has interesting consequences. By definition, every reversible CA has an inverse automaton. But even though the original automaton uses the von Neumann neighborhood, its inverse may need a much wider neighborhood. In fact, no computable function of the size of the state set can be an upper bound for the inverse neighborhood. If there were a computable bound, then one could generate all candidates for the inverse CA and check one after the other whether some of them really is the inverse automaton. Note that it is an easy matter to check whether two given CA are the inverses of each other. This would yield an algorithm for the injectivity problem, which is not possible.

Theorem 1 implies that given a reversible two-dimensional CA rule finding its inverse is very difficult in general. There is no algorithm for finding the inverse rule with a time complexity bounded by a computable function.

This fact may have direct applications in public-key cryptography. Reversible CA rules can be used to encrypt messages in a natural way: The plain text is expressed as a configuration of the CA, and the encryption is done by applying the CA rule for a fixed number of time steps. The configuration obtained is the cryptotext. One may use periodic boundary conditions to make the configurations finite in size. The decryption is done simply by applying the inverse rule to the cryptotext for the same number of steps. The operations can be done very efficiently in parallel if proper hardware implementations are available.

If one is using an arbitrary two-dimensional reversible cellular automaton $\mathcal{A}$ to encrypt and its inverse automaton $\mathcal{A}^{-1}$ to decrypt messages, then one can make $\mathcal{A}$ public. Making $\mathcal{A}$ public does not reveal its inverse $\mathcal{A}^{-1}$.

One should be aware that theorem 1 does not deny the possibility that there exist simple characterizations of reversible CA rules. It would be very useful to have some general representation of reversible rules. However, such a representation must naturally be undecidable, so that we cannot test whether a given rule can be presented in the suggested form.

Acknowledgements

I am indebted to Arto Salomaa and Anatoli Bolotov for their comments and suggestions.

References

[1] S. Amoroso and Y. Patt, Decision procedures for surjectivity and injectivity of parallel maps for tessellation structures, J. Computer. System Sci. 6 (1972) 448–464.

[2] R. Berger, The undecidability of the domino problem, Mem. Am. Math. Soc. 66 (1966).

[3] K. Culik II, On invertible cellular automata, Complex Systems 1 (1987) 1035–1044.

[4] J. Kari, Decision problems concerning cellular automata, Ph.D. Thesis, University of Turku, Finland (1989).

[5] J. Kari, Reversibility and surjectivity problems of cellular automata, Computer System Sci., submitted for publication.

[6] E.F. Moore, Machine models of self-reproduction, Proc. Symp. Appl. Math. 14 (1962) 17–33.

[7] J. Myhill, The converse to Moore's Garden-of-Eden theorem, Proc. Am. Math. Soc. 14 (1963) 685–686.

[8] D. Richardson, Tessellations with local transformations, J. Computer System Sci. 6 (1972) 373–388.

[9] R.M. Robinson, Undecidability and nonperiodicity for tilings of the plane, Inventiones Mathematicae 12 (1971) 177–209.

[10] T. Toffoli and N. Margolus, Cellular Automata Machines (MIT Press, Cambridge, MA, 1987).

[11] S. Wolfram, Theory and Applications of Cellular Automata (World Scientific, Singapore, 1986).

Physica D 45 (1990) 386–395
North-Holland

CLASSIFYING CIRCULAR CELLULAR AUTOMATA

Klaus SUTNER

Stevens Institute of Technology, Hoboken, NJ 07030, USA

Received 16 January 1990
Revised manuscript received 15 March 1990

We introduce a hierarchy of linear cellular automata based on their limiting behavior on spatially periodic configurations. We show that it is undecidable to which class in the hierarchy a cellular automaton belongs. In particular, it is undecidable whether all spatially periodic configurations evolve to a fixed point. Furthermore, there is no computable bound on the period lengths of these configurations. Our arguments are based on a non-standard simulation of Turing machines on circular cellular automata.

1. Cellular automata and Turing machines

A *cellular automaton* (CA) is a discrete dynamical system that consists of a lattice of *sites*, a finite set of *states* and a *local map*. A function from the lattice of sites to the set of states is a *configuration* of the CA. Given a configuration the local map assigns a new state to each site c depending only on the states of finitely many sites in some fixed neighborhood of c. A *global map* from the space of all configurations to itself is obtained by applying the local map simultaneously to all the sites. Note that this map is continuous in the usual product topology and invariant under translations of the underlying lattice of cells.

In this paper we will study decidability questions relating to the classification of CA. Consider, for example, the question whether a CA is reversible. Reversibility is equivalent with the global map being injective. Amoroso and Patt showed that it is decidable whether the global map of a one-dimensional CA is injective, see ref. [2]. Alternative proofs were given by Culik [5] and Head [4] using automaton theoretic methods. Indeed, one can show that reversibility is decidable in quadratic time, i.e. there is an algorithm that determines whether a given one-dimensional CA is injective in a number of steps quadratic in the size of the automaton, see ref. [13]. The size of the CA here is the number of bits needed to specify the local map, a finitary object. In contradistinction, it was recently demonstrated by Kari [7] that reversibility is undecidable for two- and higher-dimensional CA.

Another simple property of some CA is surjectivity of the global map. The global map is surjective iff the limit set of the automaton consists of the space of all configurations. As with reversibility it turns out that surjectivity is decidable for dimension one but becomes undecidable for two- and higher-dimensional CA, see refs. [2,5,7].

In ref. [16] Wolfram suggested that many natural questions about the limiting behavior of CA are undecidable. Since Wolfram's heuristic hierarchy of CA in ref. [17] is concerned with limiting behavior, one might expect that it leads to undecidable problems.

Wolfram's classification was formalized by Culik and Yu in ref. [6]. It is shown there that it is indeed undecidable to which class a CA belongs. We briefly repeat Culik and Yu's definitions. They introduce three classes of CA (plus one default class comprising all CA) as follows.

– A CA is in class I iff every configuration ultimately evolves to a fixed point.
– A CA is in class II iff every configuration evolves to a periodic configuration.
– A CA is in class III iff it is decidable whether a configuration occurs in the orbit of another.

In other words, there is an algorithm that tests whether one configuration evolves to another.

We point out that only *finite* configurations are used in this classification. As usual, a configuration is called finite if it has finite support (i.e., only finitely many sites are in a state other than 0, the quiescent state). Infinite configurations do not in general possess finitary descriptions and can therefore not be handled in the framework of ordinary computability theory.

One should observe that there is a qualitative difference between class III and the other two classes. It is quite straightforward to construct CA in class II minus class I and in class III minus class II. On the other hand, to exhibit a CA not in class III one has to construct a local map ρ and a configuration X such that the orbit of X under ρ fails to be decidable. However, orbits are always semi-decidable: there exists a semi-algorithm to determine whether a configuration Y occurs in the orbit of X. A semi-algorithm is only required to correctly return the answer "Yes" after finitely many steps but will perform an infinite computation and not return any answer at all if the answer should indeed be "No". The semi-algorithm here consists simply of systematically generating all the configurations in the orbit of X. If Y is found, the semi-algorithm halts and returns "Yes", otherwise it continues forever. One can show that a set is decidable if and only if the set itself as well as its complement are semi-decidable. Thus the existence of a CA not in class III is contingent upon the existence of sets that are semi-decidable but not decidable. It requires a Cantor-style diagonalization argument and considerable coding machinery to show that such sets exist, see ref. [10]. The standard example is the sequence of internal states (state of processor, registers, memory and so forth) that form a computation of a general purpose computer. It is semi-decidable but not decidable whether a certain state ever occurs. This is essentially the famous Halting problem and the foundation of most undecidability proofs, see also the discussion of Turing machines below. We will show in section 2 how computations can be translated into orbits of suitable CA. Using an effective hierarchy of undecidable sets somewhat similar to the Borel hierarchy it is demonstrated

in ref. [15] that class III is indeed computationally harder than the other two classes in the Culik–Yu hierarchy.

In this paper we will study complexity questions for a similar classification.

However, we will consider spatially periodic configurations rather than finite ones. Note that a spatially periodic configuration possesses a finite description just like a finite configuration. To avoid permanent confusion between spatial and temporal periodicity we prefer to think of a spatially periodic configuration as a configuration over a suitable circular CA with periodic boundary conditions. The sites of the CA are arranged on a circle and the number of sites of the circular CA is the spatial period of the original configuration on the linear CA.

More formally, let $\mathcal{S}$ denote the *alphabet*, i.e., the collection of possible states of the automaton. For the sake of simplicity we will always assume that 0 is a quiescent state.

A function $X : \{0, 1, \ldots, n - 1\} \to \mathcal{S}$ from the set of all sites to the alphabet is a *configuration* of size n of the cellular automaton.

$\mathcal{C}_n$ denotes the space of all configurations of size n and $\mathcal{C} = \bigcup_n \mathcal{C}_n$ denotes the space of all configurations.

The *local map* has the form $\rho : \mathcal{S}^N \to \mathcal{S}$ where N is a finite set of integers, called the *basic neighborhood* of the CA. The map ρ is extended to a *global map* (which we again denote by ρ)

$$\rho : \mathcal{C} \to \mathcal{C}$$

as follows. Let X be an arbitrary configuration and n its size. Define for any site c, $0 \leq c < n$, the *local configuration* at c, $X_c : N \to \mathcal{S}$, by $X_c(z) := X(c + z)$. The arithmetic is supposed to be carried out modulo n. Then $\rho(X)(c) := \rho(X_c)$. The pair $\langle \mathcal{C}, \rho \rangle$ is a *circular cellular automaton*.

We will abuse notation and often refer to the local map ρ as a circular CA. The local map is a finitary object and thus suitable input for a decision algorithm. The reader should be aware, though, that a circular CA $\langle \mathcal{C}, \rho \rangle$ is an infinite object and questions pertaining to its classification may well be undecidable.

Now consider a one-dimensional CA ρ and a spatially periodic configuration X. We can write $X = {}^\omega Y^\omega$ where Y is a finite block of states of, say, length n (the superscripts ω denote infinitely many repetitions to the left and right). Let $\bar{\rho}$ be the circular CA with the same local map as ρ. Then clearly for all $t \geq 0$

$$\rho^t(X) = {}^\omega\left(\bar{\rho}^t(Y)\right)^\omega$$

and we may study the evolution of X under ρ by considering instead Y on $\bar{\rho}$.

Note that on any circular CA every configuration trivially evolves to a periodic configuration in finitely many steps. As we will see below, though, the analogue of the first Culik–Yu class remains interesting even for circular CA. To differentiate between types of circular CA we will consider the maximal temporal period length as a function of the size of the configuration. To this end define the *period* of a configuration X by

$$\operatorname{per}(X) := \min(p > 0 \mid \exists t\, \rho^t(X) = \rho^{t+p}(X)).$$

Also let

$$\operatorname{per}(\rho, n) := \max(\operatorname{per}(X) \mid X \in \mathcal{C}_n)$$

be the maximum period of any configuration of size n. A trivial upper bound for $\operatorname{per}(\rho, n)$ is $|\mathcal{S}|^n$, the cardinality of $\mathcal{C}_n$.

The dynamics of a circular CA are conveniently expressed in terms of the *transition graph* of the automaton. The transition graph is a directed graph whose vertices are configurations and whose edges are of the form $X \to \rho(X)$. Hence the transition graph of a circular CA is countably infinite but all its connected components are finite. Note that the out-degree of any vertex is exactly one. It follows that the components of the graph are unicyclic. Indeed, a component consists of a cycle with trees grafted onto it. Thus $\operatorname{per}(X)$ is the length of the limit cycle that X enters after a transient phase. Similarly $\operatorname{per}(\rho, n)$ is the size of the longest cycle in component $\mathcal{C}_n$ of the transient graph.

We mention in passing that a detailed analysis of the transition graph is rather complicated even for additive CA. Bounds on the lengths of transients and limit cycles typically depend on number theoretic properties of the size of the automaton. A number of results can be found in refs. [9,14].

1.1. A hierarchy of circular CA

We denote class $\mathbf{I}_p$ the class of all circular CA ρ such that every configuration evolves to a fixed point under ρ. Equivalently, one may think of class $\mathbf{I}_p$ as the collection of all linear CA ρ with the property that every spatially periodic configuration on ρ evolves to a fixed point.

Let f be any non-decreasing function from the natural numbers to the natural numbers. Define class $-\mathbf{O}(f)$ to be the collection of all CA with the property that every configuration of size n evolves to a periodic configuration with period length $\mathbf{O}(f(n))$:

$$\text{class}-\mathbf{O}(f) := \{\, \rho \mid \exists c\, \forall n\, \big(\operatorname{per}(\rho, n) \leq c \cdot f(n)\big)\,\}$$

In particular letting $f(n) = n^d$ we obtain all maps whose limit cycles grow no faster than the dth power of n, the size of the configurations. Thus class $-\mathbf{O}(1)$ is the class of all circular CA whose limit cycles have length less than some fixed bound. Similarly, class $-\mathbf{O}(n)$ is the class of all circular CA with linear-size limit cycles.

1.2. Examples

It is clear that the shift over alphabet $\{0, 1\}$ is in class $-\mathbf{O}(n)$.

An example for a CA ρ_2 with quadratic-size limit cycles can be obtained as follows. Think of the sites as being split into two tracks. In both tracks particles move in an inactive background of b's (blanks). In the upper track particle a_1 moves at a constant speed of one site per time step, say, to the right. In the lower track particle a_2 remains stationary unless it is hit by particle a_1. Particle a_2 then becomes energized and, at the next time step, also moves one site to the right and returns to its normal state. A suitable alphabet for the automaton is $\mathcal{S} = \{a_1, b\} \times \{a_2, \bar{a}_2, b\}$. Two adjacent particles of the same type vanish (i.e., they turn into blanks). It is easy to see that $\operatorname{per}(\rho_2, n) = n(n+1)$. An analogous construction with k tracks will produce a CA ρ_k such that

$$\operatorname{per}(\rho_k, n) = \sum_{i=1}^{k} n^i.$$

Hence ρ_k is in class $-\mathbf{O}(n^k)$ but not in class

$-\mathbf{O}(n^{k-1})$. Note, though, that there are configurations with period $\mathbf{O}(n^i)$ for any $i < k$.

To obtain a CA whose period length is not polynomially bounded one can use particles that are reflected by walls. A suitable alphabet is $\mathcal{S} = \{b, \$, L, R\}$ and the basic neighborhood is $\{-1, 0, 1\}$. Here L represents a left moving particle and R a right moving particle. At a wall $\$$ the particle is reflected. Symbol b represents a location unoccupied by a particle or a wall. If a collision between particles occurs they annihilate each other. Define X_m to be the configuration of size $m + 1$ and form $\$Rb\ldots bb\$$. Here as well as in what follows we represent configurations as strings of symbols. The first symbol is repeated at the end to emphasize the fact that cyclic boundary conditions are used. Clearly the period of X_m is $2m$. Similarly the period of $X = X_{m_1} X_{m_2} \ldots X_{m_r}$ is the least common multiple of $2, m_1, \ldots, m_r$. In particular if all the m_i's are different primes then the period is in fact the product $m = 2m_1 \ldots m_r$. Since r is unbounded m is not polynomially bounded by $\sum_{i=1}^{r} 1 + m_i$, the size of configuration X.

The computational complexity of problems associated with finite CA is often analogous to the corresponding infinite problems. Typically, undecidable problems for infinite CA translate into decidable but intractable problems for finite CA. Similarly decidable problems for infinite CA often have polynomial time algorithms for finite CA. Consider for example the classical problem of determining whether a specific configuration X has a predecessor under the global map. Configurations that fail to have a predecessor are frequently referred to as a Garden-of-Eden, see ref. [3]. For infinite one-dimensional CA it is decidable whether a given finite configuration is a Garden-of-Eden or not. However, it is in general undecidable whether a predecessor exists in two- or higher-dimensional automata, see ref. [18]. Similarly one can determine the existence of a predecessor in polynomial time for finite one-dimensional automata. By contrast it is **NP**-complete to test for predecessors in two-dimensional finite cellular automata, see ref. [12]. It is highly likely – though not yet proven – that **NP**-complete problems do not yield to polynomial time algorithms. For a detailed discussion of the computational complexity of various problems associated with the local structure

of the transition graph of a finite CA see ref. [12].

In this paper we will show that it is undecidable whether a circular CA belongs to class $\mathbf{I}_p$. Similarly it is undecidable whether a circular CA belongs to class $-\mathbf{O}(1)$ and indeed to any of the polynomial classes $-\mathbf{O}(n^k)$. As a consequence, it is undecidable whether all spatially periodic configurations on a one-dimensional linear CA evolve to a fixed point, whether they evolve to periodic configurations whose period is linear in the size of the spatial period of the original configuration and so forth.

1.3. Turing machines

In order to establish the undecidability of, say, class $\mathbf{I}_p$, we need to concoct a circular CA such that no algorithm can determine whether all configurations evolve to a fixed point. Our argument is based on a somewhat non-standard simulation of Turing machines on circular CA. For a detailed discussion of Turing machines see ref. [10]. Briefly, a Turing machine consists of a finite set of internal states, a tape alphabet, an initial and final state and a transition function. At any time during a computation a Turing machine is in a specific internal state and scans one cell on its infinite storage tape. A single step in a computation of a Turing machine consists of the following: the symbol in the currently scanned tape cell is read. Depending on the symbol and the internal state the machine makes a transition into a new state, overwrites the tape symbol and moves the read/write head one cell to the left or to the right. To use a Turing machine as a computational device one writes the input on the otherwise blank tape, sets the machine to the initial state and positions the tape-head on the first symbol of the input string. The machine then goes through a sequence of steps as described above. If the machine ever reaches the final state. The computation is called convergent in this case and divergent otherwise. The output is the content of the tape at the time the machine halts. For our arguments below we will actually not be concerned with output but only with the question whether the machine halts at all.

Turing machines have been shown to be computationally equivalent to many other models of computation such as μ-recursive functions, while-programs or the λ-calculus, see refs. [10,8] for a more comprehensive discussion.

As a consequence, a great many questions about the behavior of Turing machines themselves are undecidable. The perhaps best known example is the Halting problem: given a Turing machine and some input x, will the computation of that machine on input x ever terminate? Similarly it is undecidable – but still semi-decidable – whether a given Turing machine on input x will generate output y.

Turing machines are the computational model of choice in connection with CA: there is a rather straightforward way to simulate Turing machines by CA. To see this, note that the state of affairs at any time during the computation of a Turing machine can be expressed by a so-called *instantaneous description* (or ID for short). An ID is a string of the form

$$a_1 a_2 \ldots a_n \, p \, b_1 b_2 \ldots b_m.$$

Here p is the current state of the machine and the a_i's and b_i's are tape symbols. This is interpreted as follows: the machine is in state p, the tape inscription is $a_1 a_2 \ldots a_n \, b_1 b_2 \ldots b_m$ (ignoring the infinitely many blank cells on either side) and the head is positioned at symbol b_1.

The changes that can occur in an ID during a computation are strictly local in nature: in order to read or overwrite the symbol in a distant tape cell the head has to be moved to that cell first, crossing all the intermediate cells.

Hence the step from one ID to the next can be accomplished by a suitably chosen local map. The alphabet of the simulating CA consists essentially of the set of states of the Turing machine plus the set of tape symbols. For a more detailed description see the next section. Simulations of Turing machines on one-dimensional CA using finite configurations can be found, e.g. in refs. [6,1]. The last reference in particular shows how to simulate a universal Turing machine on a totalistic one-way CA.

A computation of a Turing machine thus corresponds naturally to the evolution of a configuration on a suitable linear CA. Since questions like the Halting problem for Turing machines are undecidable one may expect similar questions about

CA also to be undecidable. We will show that membership of circular CA in class I_p is very closely related to the Halting problem and therefore undecidable. Unfortunately, there are two major difficulties with undecidability proofs based on this general approach. We will comment on these difficulties and their solutions in the next section.

2. Simulation techniques

2.1. Convergent computations

The first problem comes from the fact that we are dealing with circular CA rather than the customary linear ones. To simplify the discussion let us assume that the Turing machine we are trying to simulate uses only tape cells $0, 1, 2, \ldots$. This does not impede the computational power of the machines. During the computation on input x the machine will in general use more tape cells than the ones originally occupied by the input string x. Indeed, Turing machines that use only the space originally occupied by the input are referred to as linear bounded automata and are significantly weaker than general Turing machines. For example, it is easily decidable whether a linear bounded automaton accepts a given input. For arbitrary Turing machines there is no computable bound on the number of tape cells used during a computation. Since a configuration of size n can represent at most tape inscriptions of size $n-1$ it may happen that the simulating CA runs out of space. We will define the simulating local map in such a way that the configuration evolves to $\mathbf{0}$ in this case.

Here $\mathbf{0}$ is the *null configuration* defined by $\mathbf{0}(c) = 0$ for all sites c. Since state 0 is assumed to be quiescent, $\mathbf{0}$ is a fixed point. Indeed 0 will be an *annihilator*: any local configuration containing a site in state 0 is mapped to 0. Hence any global configuration containing a site in state 0 will evolve to $\mathbf{0}$ in a linear number of steps.

Returning to our simulation, the circular CA will have a special state $\$$ that is used as an end-of-tape marker. Only sites between markers can be used during the simulation. Should the Turing machine try to move its head past $\$$, the site changes state to 0. Hence, after finitely many steps, the null configuration appears.

We now give a more detailed description of the simulating CA. Suppose we wish to simulate a Turing machine $\mathbf{M}$ whose tape alphabet is Γ and whose internal state set is Q. The behavior of $\mathbf{M}$ is described by its transition function

$$\delta\colon Q \times \Gamma \to Q \times \Gamma \times \{L, R\}.$$

Suppose $\mathbf{M}$ is in state p and scans symbol a. Then $\delta(p, a) = (q, \bar{a}, S)$ indicates that in one step $\mathbf{M}$ changes its state to q, overwrites a by $\bar{a}$ and moves the tape head one cell to the left if $S = L$ and to the right if $S = R$.

Choose two new states 0 and $\$$, used as annihilator and separator mark respectively. The simulating local map $\rho = \rho(\mathbf{M})$ uses the alphabet

$$\mathcal{S} = Q \cup \Gamma \times \{L, R\} \cup \{0, \$\}.$$

We write Γ_L for $\Gamma \times \{L\}$ and similarly Γ_R for $\Gamma \times \{R\}$. The symbols in Γ_L are used to represent tape symbols to the left of the read/write head. Similarly symbols in Γ_R represent tape symbols to the right of the head. The configurations relevant for the simulation thus are of the form

$$X = \$a_1 a_2 \ldots a_n\, p\, b_1 b_2 \ldots b_m \$,$$

where $a_i \in \Gamma_L$, $b_i \in \Gamma_R$ and $p \in Q$. Recall our convention to repeat the first symbol at the end to emphasize cyclic boundary conditions.

The local map ρ has basic neighborhood $N = \{-1, 0, 1, 2\}$. One can divide the local configurations $Z : N \to \mathcal{S}$ into three groups. Call Z *admissible* iff it can occur as subconfiguration of a configuration X as above. We will define ρ for each group separately.

– For any inadmissible local configuration Z let $\rho(Z) = 0$. Hence any global configuration that contains an inadmissible subconfiguration evolves to $\mathbf{0}$.

– For any admissible local configuration Z that does not contain a symbol in Q let ρ act like identity: $\rho(Z) = Z(0)$.

– Lastly, let Z be an admissible local configuration that contains a symbol in Q. Since Z is admissible only one such symbol can occur. For $Z = xyp\$$ let $\rho(Z) = 0$. Table 1 covers all remaining cases. For the sake of notational simplicity we do not distinguish between a tape symbol

Table 1

Z	$\delta(p, a) = (q, \bar{a}, L)$	$\delta(p, a) = (q, \bar{a}, R)$
$paxy$	$\bar{a}$	q
$xpay$	x if $x \neq \$$, else 0	$\bar{a}$
$xypa$	q	y
$xyzp$	y	y

a and the corresponding CA states $(a, L) \in \Gamma_L$ and $(a, R) \in \Gamma_R$. Thus in table 1 $x, y \in \Gamma \cup \{\$\}$ and $a \in \Gamma$. Observe that the annihilator state 0 is generated whenever the tape head tries to move across the separator symbol $\$$ from the right or approaches it from the left.

In the standard simulation of a Turing machine on a linear CA the initial configuration consists essentially of one cell in state q_0, the initial state of the Turing machine, and a block of cells representing the input x. All other cells are in state 0. However, for a simulation on a circular CA the initial configuration must include necessary scratch space used in the computation and is of the form

$$\$q_0 x bb \ldots b \$.$$

Fortunately, this still allows one to simulate all convergent computations: clearly any convergent computation uses only a finite number of tape cells. Thus the number of blank symbols b can be chosen to be, say, the number of steps in the computation of machine $\mathbf{M}$ on input x.

Note, though, that this simulation is not effective: the appropriate number of blank symbols is in general not computable. However, since our classification of circular CA is in terms of their behavior on all configurations we can still obtain the desired hardness results.

2.2. Self-verifying Turing machines

The second obstacle is somewhat harder to overcome. It arises from the fact that a Turing machine operates only on reasonable tape inscriptions. Most notably, one only deals with tape inscriptions that can occur during a computation on some input.

In the classification of cellular automata, however, one has to contend with arbitrary configurations. Some of these configurations will not correspond to any ID of the Turing machine one is trying to emulate. For example a configuration in which two sites are in a state that corresponds to a machine state cannot represent an ID. The simulating CA defined above eliminates such syntactically incorrect configurations: they evolve to the null configuration in a linear number of steps.

However, there are other undesired configurations that are somewhat more difficult to deal with. These are configurations that appear to represent IDs but that in fact cannot occur during any computation on any input. A configuration (or the corresponding ID) that occurs in some computation will be called *accessible* and *inaccessible* otherwise. Accessibility is semi-decidable but in general not decidable. Hence one cannot effectively eliminate inaccessible IDs the way one can eliminate syntactically incorrect configurations.

To resolve this problem we use a variation of the *self-verifying* Turing machines introduced in ref. [15]. Given an arbitrary Turing machine $\mathbf{M}$ one can construct a self-verifying Turing machine $\mathbf{M}'$ that simulates the old machine in a somewhat circuitous fashion. The redundancy in the computation of $\mathbf{M}'$ ensures that every inaccessible ID will be recognized as such after finitely many steps.

To this end $\mathbf{M}'$ uses IDs of $\mathbf{M}$ equipped with an additional tag that indicates how many steps it takes to reach the ID in question from a certain input. The tape alphabet of $\mathbf{M}'$ consists of the tape alphabet of $\mathbf{M}$, the state set of $\mathbf{M}$ and a separator symbol $\#$. A tagged ID has the form $\#s\#x\#I\#$ where x is the input, s the number of steps in the computation of $\mathbf{M}$ and I the ID. A tagged ID is *correct* iff indeed the computation of $\mathbf{M}$ on input x produces ID I after s steps. Unlike accessibility correctness of a tagged ID is an easily decidable property: one can simply execute the first s steps of the computation of $\mathbf{M}$ on input x. If the resulting ID is equal to I the tagged ID is correct and incorrect otherwise.

The point of this construction is that the simulation of machine $\mathbf{M}$ on a CA via machine $\mathbf{M}'$ does not have to contend with inaccessible configurations: after finitely many steps these configurations will all evolve to $\mathbf{0}$. Hence membership of the simulating circular CA in class I_p is determined exclusively by configurations that correspond to accessible IDs of machine $\mathbf{M}$. The undecidability of certain questions about the behavior of the Turing machine is thus inherited by the CA.

3. The hardness proofs

We will now give a formal proof that it is undecidable whether all configurations of a given circular CA evolve to a fixed point.

Theorem 1 *It is undecidable whether a circular one-dimensional cellular automaton belongs to class I_p.*

Proof. Consider some arbitrary Turing machine $\mathbf{M}$. It is well known that it is undecidable whether machine $\mathbf{M}$ will halt on the empty tape, see ref. [10]. We will construct a circular CA ρ that belongs to class I_p iff $\mathbf{M}$ fails to halt on the empty tape. Since ρ can be constructed from $\mathbf{M}$ effectively this shows that membership in class I_p is also undecidable. For the reasons discussed in the last section the simulation will not use machine $\mathbf{M}$ directly but rather a self-verifying version of it. We now give a detailed description of a self-verifying machine $\mathbf{M}'$ associated with $\mathbf{M}$. After an initialization procedure the machine cycles through three phases. Contingent on the results of tests performed during these phases $\mathbf{M}'$ will take certain actions as explained below.

(i) Initialization
Replace the tape inscription by

$$W_0 := \#0\#q_0\#.$$

Here q_0 is the initial state of the old machine $\mathbf{M}$. This is the tagged ID corresponding to 0 steps in the computation of $\mathbf{M}$ on the empty tape. $\mathbf{M}'$ continues with tape verification.

(ii) Tape verification
At this stage the machine first tests whether the tape inscription has the form $\#s\#I\#$. Here s is the alleged number of steps and I presumably an ID of $\mathbf{M}$. If the test fails $\mathbf{M}'$ goes back to its initialization phase. Otherwise $\mathbf{M}'$ continues with the accessibility test.

(iii) Accessibility test
$\mathbf{M}'$ now verifies the correctness of the tagged ID on its tape. To this end, $\mathbf{M}'$ simply simulates the computation of $\mathbf{M}$ on the empty tape for exactly s steps using the part of the tape to the right of the original inscription as scratch space. If the tagged ID turns out to be correct, the scratch space is

erased and $\mathbf{M}'$ continues with the next ID phase. Otherwise $\mathbf{M}'$ goes back to its initialization phase.

(iv) Next ID

We may now assume that the tape of $\mathbf{M}'$ contains a correct tagged ID $\#s\#I\#$. Machine $\mathbf{M}'$ now determines whether the ID of $\mathbf{M}$ is halting. If so, $\mathbf{M}'$ again returns to its initialization phase. Otherwise $\mathbf{M}'$ computes the next ID J of $\mathbf{M}$ and increments the counter s so that ultimately its tape inscription is

$$\#s+1\#J\#.$$

$\mathbf{M}'$ then returns to its verification phase.

This completes the description of the self-verifying Turing machine $\mathbf{M}'$.

The crucial observation concerning the behavior of machine $\mathbf{M}'$ is that it cannot perform more than a finite number of steps before entering a verification and accessibility testing phase. Hence, regardless of the initial ID, after a finite number of steps $\mathbf{M}'$ will simulate the computation of the original machine on the empty tape as input. If that computation is convergent $\mathbf{M}'$ will ultimately return to its initialization phase. From then on it will run through this cycle ad infinitum. If the computation of $\mathbf{M}$ fails to converge $\mathbf{M}'$ will also perform a divergent computation. Thus as a Turing machine $\mathbf{M}'$ is utterly uninteresting: it diverges regardless of the behavior of $\mathbf{M}$. Note, however, that in the second case $\mathbf{M}'$ will never reset to ID W_0. As a consequence the step counter s will increase beyond any fixed bound.

This is crucial for our next claim. Let $\rho = \rho(\mathbf{M}')$ be the cellular automaton that simulates $\mathbf{M}'$ as described in the last section.

Claim: Machine $\mathbf{M}$ fails to halt on the empty tape iff the circular CA ρ is in class I_p.

To see this first suppose that $\mathbf{M}$ halts. Consider the configuration

$$X = \$q_0'\#0'\#q_0\#b^r\$.$$

Here q_0' is the initial state of $\mathbf{M}'$ and r is sufficiently large so that machine $\mathbf{M}'$ uses at most r tape cells in the simulation of $\mathbf{M}$. b^r denotes a sequence of r b's. The $0'$ in configuration X represents the value zero of counter s (recall that 0 is

reserved for the annihilator state). By definition, $\mathbf{M}'$ then resets its tape to W_0 and starts all over. Hence X is a periodic configuration with period length depending on the number of steps that it takes $\mathbf{M}$ to halt. At any rate, X will not be a fixed point. But then ρ cannot be in class I.

For the opposite direction suppose $\mathbf{M}$ diverges on the empty tape. Assume X is a periodic configuration but not a fixed point. Since all syntactically wrong as well as inaccessible configurations evolve to the null configuration, X must correspond to an ID of $\mathbf{M}'$. There is in fact a slight complication: X might consist of a sequence of such IDs separated by $\$$'s. Since each ID evolves separately it suffices to consider only a single one of them. By the definition of ρ, configuration X must evolve to a configuration

$$Y = \$p\#s\#I\#b^t\$$$

for some values of s and t. But since $\mathbf{M}$ diverges, $\mathbf{M}'$ will keep incrementing the step counter s until a collision with the end-of-tape marker occurs. After that, X will evolve to $\mathbf{0}$ in a linear number of steps and we have a contradiction.

As mentioned before, the Halting problem for Turing machines is undecidable. Note that the simulating circular CA ρ can be constructed effectively from the Turing machine $\mathbf{M}$. It follows from the claim that membership in class $-\mathbf{O}(1)$ is also undecidable. $\square$

We now show how to modify the last argument to establish the undecidability of class $-\mathbf{O}(1)$.

Theorem 2 *It is undecidable whether there exists a bound for the periods of all configurations of a circular cellular automaton. Hence it is undecidable whether a circular cellular automaton belongs to class $-O(1)$.*

Proof. The proof strategy is very similar to 1. However, we now consider computations of Turing machines on some input $x \geq 0$ rather than just the empty tape. Again let $\mathbf{M}$ be some arbitrary Turing machine. Define $\mathbf{M}_1$ to be the Turing machine that, on input x, runs machine $\mathbf{M}$ on all inputs z, $0 \leq z \leq x$, in parallel. Machine $\mathbf{M}_1$ halts on input x iff machine $\mathbf{M}$ halts on all inputs z, $0 \leq z \leq x$.

Now let $\mathbf{M}'$ be the self-verifying Turing machine for $\mathbf{M}_1$ defined completely analogous to the machine in the last proof. Lastly, let $\rho = \rho(\mathbf{M}')$ be the local map that simulates machine $\mathbf{M}'$ as described in section 2.

Consider some $x \geq 0$ such that machine $\mathbf{M}_1$ – and therefore also machine $\mathbf{M}'$ – halts on input x. For sufficiently large r, say $r \geq r(x)$, the configuration $X(x, r) = \$q_0'\#0'\#x\#q_0\#b^r\$$ will evolve to a cyclic configuration with period at least x. This follows from the fact that it will take $\mathbf{M}_1$ at least x steps to halt on input x: $\mathbf{M}_1$ has to verify that $\mathbf{M}$ halts for all $z \leq x$.

Claim: Machine $\mathbf{M}$ fails to halt on some input x iff the circular CA ρ is in class $-\,\mathbf{O}(1)$.

First assume that the original machine $\mathbf{M}$ halts on all inputs. Then, by definition, machine $\mathbf{M}_1$ also halts on all inputs. But then the configurations $X(x, r)$, $r \geq r(x)$, show that $\mathrm{per}(\rho, n)$ cannot be bounded where n is the size of $X(x, r)$. Hence ρ is not in class $-\,\mathbf{O}(1)$.

On the other hand assume that machine $\mathbf{M}$ fails to halt on some input. Let x be the least such input. Then $\mathbf{M}_1$ halts on all inputs $z < x$ and fails to halt for all inputs $z \geq x$. Hence all configurations associated with computations on some input $z \geq x$ evolve to $\mathbf{0}$: the counter is incremented every time the self-verifying machine runs through its main loop and ultimately must collide with the fixed size of the configuration. The only configurations that may not evolve to $\mathbf{0}$ must be associated with computations on inputs $z < x$. Thus they all have a configuration of the form $X(z, r)$ in their orbit for some $z < x$ and $r \geq r(z)$. But since $\mathrm{per}(X(z, r)) = \mathrm{per}(X(z, r'))$ for all $r, r' \geq r(z)$, this shows that $\mathrm{per}(X)$ can assume only finitely many possible values. Thus there is a fixed bound on $\mathrm{per}(\rho, n)$ and ρ belongs to class $-\,\mathbf{O}(1)$.

It is well known to be undecidable whether a Turing machine diverges on some input, see refs. [10,15]. Since the simulating circular CA can be constructed effectively from the Turing machine it follows that membership in class $-\,\mathbf{O}(1)$ is also undecidable. $\square$

Consider a function f with the property that $f(\rho)$ is a bound for the length of all limit cycles for any circular CA ρ in class $-\,\mathbf{O}(1)$. For ρ not in class $-\,\mathbf{O}(1)$ $f(\rho)$ is undefined. As an easy consequence of the last result one can see that no such bounding function can be computable.

Corollary 1 *There is no computable bound on the lengths of limit cycles of a circular cellular automaton.*

Lastly we show how to combine the technique of the last proof with the k-particle CA from the examples in section 1 to show that all of the classes $-\,\mathbf{O}(n^k)$ are undecidable. Note that this contains theorem 2 as the special case $k = 0$.

Theorem 3 *For any $k \geq 0$ it is undecidable whether a circular cellular automaton belongs to class $-\,O\,(n^k)$.*

Proof. The idea is to use the CA that simulates the self-verifying Turing machine from the last theorem and combine it with a k-particle CA as in the examples in section 1. To this end one can divide the sites of the CA into $k + 1$ tracks. The k lower tracks are occupied by moving particles whereas the upper track is used to carry out the simulation of the Turing machine. The CA uses an alphabet of the form $\mathcal{S} = \mathcal{S}_0 \times \mathcal{S}_1 \times \ldots \times \mathcal{S}_k \cup \{0\}$ where 0 is the annihilator. Whenever the Turing machine simulation in track number zero finds a syntactically incorrect ID or runs out of space the corresponding site changes its state to 0. Thus, after a linear number of steps, configuration $\mathbf{0}$ is generated. Furthermore, the movement of the tape head in track zero is restricted: it can move only when all k particles are currently located in the same site as the read/write head. Consequently the simulating CA takes $\mathbf{O}(n^k)$ steps to simulate one single move of the Turing machine.

Now define configurations $X(x, r)$ for $r \geq r(x)$ sufficiently large as in the last argument. That is, track number 0 contains

$$\$q_0'\#0'\#x\#q_0\#b^r\$,$$

and the ith track, $i = 1, \ldots k$, contains $ba_i bb \ldots bb$. Configuration $X(x, r)$ corresponds to the initial configuration of a computation on input x with all particles aligned on the site that represents the initial machine state q_0'. $X(x, r)$ will evolve to a cyclic configuration with period at least xn^k for all x such that $\mathbf{M}$ converges on x. It follows that one can verify the following claim just as in the last theorem.

Claim: Machine $\mathbf{M}$ fails to halt on some input x iff the circular CA ρ is in class $-\mathbf{O}(n^k)$.

As in theorem 3.2 this finishes the argument. $\square$

4. Conclusion

We have shown that a classification of circular CA based on the lengths of limit cycles in the transition diagram leads to undecidable problems. It follows that it is undecidable whether all spatially periodic configurations on a given one-dimensional CA evolve to a fixed point. Similarly it is undecidable whether all spatially periodic configurations evolve to periodic configurations of some polynomially bounded period. Hence there is no uniform method that allows to determine fundamental properties of a one-dimensional CA with respect to periodic configuration. Instead, ad hoc arguments must be used in every individual case.

For those familiar with the arithmetical hierarchy as presented for example in ref. [11], our arguments show the following. For circular CA class I is Π_1^0-complete. By comparison, it is shown in ref. [15] that class I is Π_2^0-complete for ordinary one-dimensional CA.

Similarly class $-\mathbf{O}(n^d)$ is Σ_2^0-complete for circular CA, even for $d = 0$. Using Σ_2^0-uniformization one can obtain a Σ_2^0 bounding function $c(\rho)$ such that for all configurations $X \in \mathcal{C}_n$ we have $\mathrm{per}(X) \leq c(\rho)n^d$ for all local maps in class $-\mathbf{O}(n^d)$. No such function can be computable.

References

[1] J. Albert and K. Culik II, A simple universal cellular automaton and its one-way and totalistic version, Complex Systems 1 (1989) 1–16.

[2] S. Amoroso and Y.N. Patt, Decision procedures for surjectivity and injectivity of parallel maps for tesselation structures, J. Computers System Sci. 6 (1972) 448–464.

[3] A.W. Burks, Essays on Cellular Automata (University of Illinois Press, Urbana, IL, 1970).

[4] T. Head, Linear CA: injectivity from ambiguity, to appear.

[5] K. Culik II, On invertible cellular automata, Complex Systems 1 (1987) 1035–1044.

[6] K. Culik II and S. Yu, Undecidability of CA classification schemes, Complex Systems 2 (1988) 177–190.

[7] J. Kari, Reversibility of 2D cellular automata is undecidable, Physica D 45 (1990) 379–385, these Proceedings.

[8] A.J. Kfoury, R.N. Moll and M.A. Arbib, A Programming Approach to Computability (Springer, Berlin, 1982).

[9] O. Martin, A.M. Odlyzko and S. Wolfram, Algebraic properties of cellular automata, Commun. Math. Phys. 93 (1984) 219–258.

[10] H. Rogers, Theory of Recursive Functions and Effective Computability (McGraw Hill, New York, 1967).

[11] J.R. Shoenfield, Mathematical Logic (Addison-Wesley, Reading, MA, 1967).

[12] K. Sutner, The complexity of finite cellular automata, submitted for publication.

[13] K. Sutner, DeBruijn graphs and linear cellular automata, submitted for publication.

[14] K. Sutner, On σ-automata, Complex Systems 2 (1988) 1–28.

[15] K. Sutner, A note on Culik–Yu classes, Complex Systems 3 (1989) 107–115.

[16] S. Wolfram, Computation theory of cellular automata, Commun. Math. Phys. 96 (1984) 15–57.

[17] S. Wolfram, Universality and complexity in cellular automata, Physica D 10 (1984) 1–35.

[18] T. Yaku, The constructibility of a configuration in a cellular automaton, J. Computer System Sci. 7 (1973) 481–496.

Physica D 45 (1990) 396–403
North-Holland

FORMAL LANGUAGES AND GLOBAL CELLULAR AUTOMATON BEHAVIOR

K. CULIK II

Department of Computer Science, University of South Carolina, Columbia, SC 29208, USA

L. P. HURD

Laboratory for Plasma Research, University of Maryland, College Park, MD 20742, USA

and

S. YU[1]

Department of Computer Science, University of Western Ontario, London, Ontario, Canada, N6A 5B7

Received 15 January 1990
Revised manuscript received 4 April 1990

A connection is made between the description of one-dimensional cellular automata as dynamical systems acting on a compact metric space, and as computational systems acting on strings. Operators are defined which determine a bijective correspondence between subshifts of infinite configurations and languages of finite strings. Under this correspondence, families of languages give rise to families of subshifts. Regular languages correspond to sofic subshifts. Families of subshifts corresponding to context-free, context-sensitive, and recursively enumerable languages are introduced and their closure properties under forward and backward cellular automaton images studied.

1. Introduction

A linear cellular automaton consists of an infinite (one-dimensional) lattice of sites taking values from a finite alphabet, which updates itself in accordance with a local rule. The same rule is applied at each site isotropically, and the new value of a site depends only on sites a bounded distance away.

A connection between cellular automaton dynamics and formal language theory was suggested by Weiss [1], who observed that the image set of a cellular automaton (a sofic system) corresponded in a natural way to the regular languages of formal language theory. This connection was explored by Wolfram in ref. [2], where an algorithm was presented to yield a canonical description of the regular language arising at a given finite time step in the evolution of a given cellular automaton. Wolfram conjectured that infinite time properties of cellular automata might give rise to more complicated languages.

This conjecture was settled in ref. [3], where it was shown that the limit sets (the intersection of all forward images) of cellular automata could give rise to non-regular, and non-context-free languages. In ref. [3] a proof was sketched, completed in refs. [4–6], that cellular automata can give rise to limit sets corresponding to non-recursively enumerable languages.

This paper attempts to formalize the connection between languages and subshifts. This work has two motivations. The first relates to an attempt to classify and understand cellular automaton dynamics in terms of the complexity of its invariant subshifts [3–7]. A survey of this approach to cellular automaton dynamics appears in ref. [8]. The second motivation arose independently in an

---

[1]   On leave from Kent State University, Kent, OH 44242, USA.

attempt to understand infinite formal languages [9–13].

While the results of this paper are stated in formal language terms, many of the techniques used arise from topology and dynamical systems. Section 2 reviews the definition of cellular automata and some of the needed constructs from topology. Unless stated otherwise, the definitions and constructs from formal language theory are standard, and can be found, for example, in refs. [14–17].

Operators $\mathcal{L}$ from subshifts to languages and Adh from languages to subshifts are defined in section 3. Also an operator C on languages is defined. In theorem 1, it is shown that these operators determine a bijective correspondence between subshifts (closed shift-invariant sets of configurations) and central languages, which are exactly the fixed points of the operator C.

Section 4 establishes results about the closure properties of the operator $C(L)$ defined in section 3. Aside from being of interest in their own right, these results are used later in examining the closure properties of subshifts under inverse cellular automaton maps.

The bijective correspondence of theorem 1 gives a classification of subshifts based on their language theoretic complexity. In section 5 it is shown that many of these classes are preserved under cellular automaton images, in fact cellular automata have the same closure properties as a well-known class of mappings on formal languages, ε-limited GSM mappings (definition 6). However, the situation for inverse images is somewhat more complicated. Subshifts giving rise to context-sensitive languages can have inverse images which give rise to languages that are not recursively enumerable.

2. Definitions

The full shift on a finite set S, denoted S^Z, is the set of functions from the integers to S. Equivalently, elements of S^Z are doubly infinite words in the symbols from S with a fixed base symbol. If $c \in S^Z$ we will write $c(i)$ as c_i. This space is topologized by the metric

$$d(x, y) = \sum_{i=-\infty}^{\infty} \delta(x_i, y_i)\, 2^{-|i|},$$

where $\delta(a, b) = 0$ if $a = b$ and 1 otherwise.

There is a natural (left) shift map, σ, acting on S^Z defined by $\sigma(x)_i = x_{i+1}$. A closed, shift-invariant subset of S^Z is referred to as a *subshift*.

The topological closure of a set $K \subseteq S^Z$ is denoted $\overline{K}$. The smallest subshift containing a set $K \subseteq S^Z$, is

$$\overline{\bigcap_{i=-\infty}^{\infty} \sigma^i(K)}.$$

Given a finite string $s \in S^*$, the *cylinder set* $\mathrm{Cyl}(s)$ consists of all configurations with s as their middle substring. If the length of s is odd, this is unambiguous. If the length of s is even we arbitrarily allow it to extend one site further to the right. Formally this definition may be stated:

$$\mathrm{Cyl}(s_1 \ldots s_n) = \{c \in S^Z \mid c_i = s_i$$

$$\text{for } -[(n-1)/2] \le i \le [n/2]\},$$

where $[x]$ denotes the greatest integer in x.

The cylinder sets are open and closed, and form a basis for the topology of S^Z. They are not subshifts (they are not shift-invariant).

Definition 1 *A one-dimensional (linear) cellular automaton is defined by a triple $A = (S, r, f)$ where S is the finite set (alphabet), r is a nonnegative integer (radius of the neighborhood), and $f \colon S^{2r+1} \to S$ is a (local) function.*

A cellular automaton with local function f determines a global function $G_f \colon S^Z \to S^Z$ defined by $G_f(c)_i = f(c_{i-r}, \ldots, c_{i+r})$. The map G_f is continuous and commutes with σ. Also, all continuous shift-commuting maps arise in this fashion [18,19].

Two particularly important classes of subshifts are called subshifts of finite type, and sofic systems. A subshift of finite type consists of all configurations which do not contain blocks from a finite excluded list. A sofic system is a subshift of the form $G_f(K)$ where G_f is a cellular automaton map and K is a subshift of finite type [1].

3. Adherences and centers of languages

In this section, we study the links between sets of configurations in S^Z and languages of finite words in S^*. A bijective correspondence is set up between subshifts and a class of languages called *central* (in ref. [5], these languages are called *admissible*), by means of set-valued operators:

$$\mathcal{L}: 2^{S^Z} \to 2^{S^*},$$

taking sets of configurations to languages and

$$\text{Adh}: 2^{S^*} \to 2^{S^Z},$$

which takes languages to sets of configurations. Also an operator on languages

$$C: 2^{S^*} \to 2^{S^*}$$

is introduced whose fixed points form the image of $\mathcal{L}$ and define central languages.

In the following, the notation $s \preceq s'$ means that s is a finite substring of s' where s' is either a finite string in S^* or an infinite configuration in S^Z.

We start with a definition of a mapping from sets of configurations to languages. Although later we will focus our attention on subshifts, at the moment no assumptions are made about the set K.

Definition 2 *The language, $\mathcal{L}(K)$, associated with a set $K \subseteq S^Z$ is the set of all finite substrings of configurations in K.*

$$\mathcal{L}(K) = \{s \in S^* | s \preceq c \text{ for some } c \in K\}.$$

If K is a subshift, one loses no information by considering $\mathcal{L}(K)$. The following is proved in ref. [3].

Lemma 1 *If K_1 and K_2 are subshifts, $\mathcal{L}(K_1) = \mathcal{L}(K_2) \Rightarrow K_1 = K_2$.*

Also, given an arbitrary subset $K \subseteq S^Z$ one can consider the smallest subshift containing it, by considering its entire orbit with respect to the shift operator σ, and taking the topological closure of the resulting set. This procedure does not introduce any new finite blocks.

Lemma 2 *For any subset $K \subseteq S^Z$,*

$$\mathcal{L}(K) = \mathcal{L}\left(\overline{\bigcup_{i=-\infty}^{\infty} \sigma^i(K)}\right).$$

Proof: Clearly applying the shift operator to K cannot introduce any new finite blocks. Also if a set is shift-invariant, the statement $s \in \mathcal{L}(K)$ is equivalent to the statement that K has a nontrivial intersection with the cylinder set $\text{Cyl}(s)$. All cylinder sets are open, and an open set intersects a set if and only if it intersects its closure; so taking the closure also does not introduce any new finite blocks. $\qquad\square$

The adherence [9,10] of a language is the (possibly empty) maximal set of configurations whose finite substrings are also substrings of words in L.

Definition 3 *The adherence of a language $L \subseteq S^*$ is the subset of S^Z defined by*

$$\text{Adh}(L) = \{c \in S^Z \mid \text{for all } w \preceq c \text{ there exist} \\ x, y \in S^* \text{ such that } xwy \in L \}.$$

In ref. [11], adherences of languages are defined using limits of languages. It is not difficult to show that these two definitions are equivalent. See also ref. [11] for the proof of the following.

Lemma 3 *For any $L \subseteq S^*$, $\text{Adh}(L)$ is a subshift.*

A string w is said to be *bi-extensible* in a language L if for all $N > 0$ there exist strings $x, y \in S^*$ such that $xwy \in L$ and $|x|, |y| \geq N$.

Definition 4 *The center, $C(L)$, of a language $L \subseteq S^*$ is defined by*

$$C(L) = \{w \in S^* | w \text{ is bi-extensible in } L \}.$$

The concept of centers of languages was originally given by Nivat [12] in the context of one-way extensions. For two-way extensions, centers of languages were introduced in ref. [13].

Definition 5 *A language is* central *if $L = C(L)$.*

A related set $\mathcal{E}(L)$, called the extensible subset of a language, was introduced in ref. [5] and studied in refs. [11,5]. The two are related by $\mathcal{E}(L) = L \cap C(L)$. The central languages are precisely the *admissible* languages of ref. [5].

Two lemmas which will prove useful later are:

Lemma 4

(1) If $K_1 \subseteq K_2$, $C(K_1) \subseteq C(K_2)$.

(2) For all subsets $K \subseteq S^Z$, $C(\mathcal{L}(K)) = \mathcal{L}(K)$.

Proof: Apparent from the definition. $\square$

Lemma 5 *If $L \subseteq S^*$ and $w \in C(L)$, then there exists a sequence of strings $w_i \in C(L)$ such that $w_0 = w$ and $w_i = x_i w_{i-1} y_i$ for symbols $x_i, y_i \in S$.*

Proof: Let $n = |S|$ where $L \subseteq S^*$. Consider the n^2 strings in SwS. Since w has an infinite number of extensions, at least one of these possible extensions must itself have an infinite number of extensions. Let $w_1 = awb$ be one such extension, and proceed by induction. $\square$

Theorem 1 *For each subset $K \subseteq S^Z$ and language $L \subseteq S^*$:*

(1) $\mathcal{L}(\mathrm{Adh}(L)) = C(L)$.

(2) $\mathrm{Adh}(\mathcal{L}(K)) = \overline{\bigcup_{i=-\infty}^{\infty} \sigma^i(K)}$.

Proof:

(1) To show $\mathcal{L}(\mathrm{Adh}(L)) \subseteq C(L)$ consider $w \in \mathcal{L}(\mathrm{Adh}(L))$. By the definition of $\mathcal{L}$ this means that $w = c_1 \ldots c_n \preceq c$ for some $c \in \mathrm{Adh}(L)$. Choosing $N > 0$ then $c_{-N} \ldots c_{n+N}$ is also in $\mathcal{L}(\mathrm{Adh}(L))$. By the definition of Adh, $w \preceq c_{-N} \ldots c_{n+N} \preceq w'$ for some $w' \in L$, which is to say that w can be extended by strings of length greater than N to a string in L. Since N was chosen arbitrarily, $w \in C(L)$.

To show that $C(L) \subseteq \mathcal{L}(\mathrm{Adh}(L))$ consider $w \in C(L)$. By lemma 5, one can find a series of nested strings $w_i \in C(L)$ such that $w_0 = w$ and $w_i = x_i w_{i-1} y_i$, $x_i, y_i \in S$. Let $W = \bigcap_{i=0}^{\infty} \mathrm{Cyl}(w_i)$. It follows from the compactness of S^Z that W is a non-empty compact set. If $c \in W$ and $w \preceq c$ then $w \preceq w'$ for some $w' \in C(L)$. But by the definition of $C(L)$, $w' \preceq w''$ for $w'' \in L$. It follows that $c \in \mathrm{Adh}(L)$ and $W \subseteq \mathrm{Adh}(L)$. Thus $w \in \mathcal{L}(\mathrm{Adh}(L))$.

(2) Let $K' = \overline{\bigcup_{i=-\infty}^{\infty} \sigma^i(K)}$, $K'' = \mathrm{Adh}(\mathcal{L}(K))$. By lemma 2 $\mathcal{L}(K') = \mathcal{L}(K)$. By part (1), $\mathcal{L}(K'') = \mathcal{L}(\mathrm{Adh}(\mathcal{L}(K))) = C(\mathcal{L}(K))$ and by lemma 4, $\mathcal{L}(K'') = \mathcal{L}(K)$. By lemma 3 K'' is a subshift and by lemma 1, $K' = K''$. $\square$

Corollary 1

(1) The image under $\mathcal{L}$ of all subsets of S^Z are precisely the set of central languages.

(2) The image under Adh of languages are precisely the (possibly empty) subshifts.

(3) The operators $\mathcal{L}$ and Adh determine a bijective correspondence between central languages over S^* and subshifts of S^Z.

Proof: Immediate from lemmas 3, 4 and the preceding theorem. $\square$

4. Closure properties of C

After introducing a new operator on languages, it is natural to ask about its closure properties with respect to the usual classes of formal languages. Aside from being of interest in its own right, these results will be needed in section 5 when we examine closure of families of subshifts under inverse cellular automaton maps. The first result along these lines is that regular languages are closed under the operator C. To prove this requires a couple of lemmas.

Lemma 6 *If $L \subseteq S^*$ is a regular language, there exists a constant $N > 0$ such that $w \in C(L)$ if and only if there exist $x, y \in S^*$ such that $xwy \in L$ and $N + 1 \leq |x|, |y| \leq 2N + 1$.*

Proof: Since L is a regular language, it can be recognized by some finite automaton [15] which may be viewed as a labeled directed graph with vertices Q, a start state $q_s \in Q$ and accept states $Q_f \subseteq Q$. The language determined by the finite automaton is simply the set of labels corresponding to finite paths from q_s to an element of Q_f.

Let $N = |Q|$ be the number of states in the finite automaton recognizing L. The lemma follows from the fact that x and y must be realized by paths of length greater than N in a graph with N vertices, and therefore, each of them must contain a loop. Also if x and y are any strings representing paths through this graph, one can find another path whose length is less than $2N + 1$ by possibly eliminating loops.

The string x can be written in the form $x = abc$ and $y = def$ such that $ab^i cw de^j f \in L$ for all $i, j > 0$. Therefore $s \in C(L)$. $\square$

The operator $\mathrm{Cut}_{p,q} \colon S^* \to S^*$ takes strings to the string obtained by dropping the first p characters and the last q characters.

$$\mathrm{Cut}_{p,q}(a_1 \ldots a_n) = \varepsilon \text{ (the empty string)}$$
$$\text{if } n < p + q + 1$$
$$= a_{p+1} \ldots a_{n-q}$$
$$\text{otherwise}$$

Lemma 7 *If $L \subseteq S^*$ is a regular language, $\mathrm{Cut}_{p,q}(L)$ is regular.*

Proof: Regular languages are closed under reversal ($L \to L^{\mathrm{R}}$), so it suffices to prove this lemma for $\mathrm{Cut}_{p,0}$, since:

$$\mathrm{Cut}_{p,q}(L) = \mathrm{Cut}_{p,0}((\mathrm{Cut}_{q,0}(L^{\mathrm{R}}))^{\mathrm{R}}).$$

The concept of generalized sequential machines (GSMs) has been widely used in formal language theory, cf. refs. [14,15]. Informally a GSM consists of a directed graph in which each edge has an input label from the input alphabet and an output label which is a string over the output alphabet. The set of output strings produced from a given input string is given by concatenating the output strings corresponding to walks from the start state whose input labels match the input string.

Definition 6 *Formally, a GSM, M, is a sextuple $(Q, A_I, A_O, \delta, \lambda, q_0)$, where Q is the finite set of states and $q_0 \in Q$ is the start state, A_I and A_O are the input and output alphabets, respectively, $\delta\colon Q \times A_I \to Q$ is the transition function and $\lambda\colon Q \times A_I \to A_O^*$ is the output function.*

The function $\mathrm{Cut}_{p,0}$ is induced by a GSM mapping. Set $Q = \{q_0 \ldots q_p\}$ and $A_I = A_O = S$. Define the function δ and λ by

$$\delta(q_i, a) = q_{i+1} \quad \text{if } i < p,$$
$$= q_p \quad \text{otherwise;}$$
$$\lambda(q_i, a) = \varepsilon \quad \text{if } i < p,$$
$$= a \quad \text{if } i = p.$$

$\square$

Theorem 2 *If L is a regular language, $C(L)$ is regular.*

Proof: By lemma 6, there exists $n > 0$ such that $C(L)$ consists of exactly those strings which may

be extended p units to the left and q units to the right where $n+1 \leq p, q \leq 2n+1$. This is equivalent to saying that $C(L)$ is a union of those strings which can be extended p to the left and q to the right for every such choice of p and q.

$$C(L) = \bigcup_{N+1 \leq p,q \leq 2N+1} \mathrm{Cut}_{p,q}(L).$$

By lemma 7, each of the sets $\mathrm{Cut}_{p,q}(L)$ is regular, and finite unions of regular languages are regular. Therefore $C(L)$ is a regular language. $\square$

For context-free languages, the following result has been proved [11].

Theorem 3 *If L is context-free, $C(L)$ is context-free.*

Neither context-sensitive nor recursively enumerable languages are closed under C. In fact there exists a context-sensitive language whose image under C fails to be recursively enumerable.

Theorem 4 *There exists a context-sensitive language, L_u, for which $C(L_u)$ is not recursively enumerable.*

Proof: Let $L \subseteq S^*$ be a recursively enumerable language whose complement is not recursively enumerable, and let M be a Turing machine which recognizes L. Define a new language L_u with the following properties:

(i) Every string in L_u has the form $\#^i a_1 \ldots a_n \#^j$, where $a_i \in S$.

(ii) A string $\#^i a_1 \ldots a_n \#^j \in L_u$ if and only if when M is started on a tape containing the string $a_1 \ldots a_n$, it will take at least $i + j$ steps without halting.

A Turing machine can visit at most $i+j$ squares in time $i + j$, so this language can be recognized by a deterministic linear bounded automaton, and is thus context-sensitive.

The fact that $C(L_u)$ is not recursively enumerable follows from the fact that the intersection $C(L_u) \cap \{\#S_u^*\#\}$ consists of strings of the form $\#a_1 \ldots a_n \#$ where $a_1 \ldots a_n \notin L$. Since the complement of L was by assumption not recursively enumerable, neither is this intersection, and therefore neither is $C(L_u)$. $\square$

5. Closure properties of CA mappings

By the correspondence set up in theorem 1, a family of languages gives rise to a unique family of subshifts which are its images under the operator Adh. It has been observed that the subshifts corresponding to regular languages are exactly sofic systems. It is natural to ask about subshifts generated by more complicated languages. Furthermore, the first question which arises in introducing a class of subshifts is whether they are closed under the images and inverse images of the natural class of maps, induced by cellular automata.

The answer for the forward image of a subshift is that under weak conditions on the family of languages the corresponding subshifts are closed under forward images.

The question of the inverse image of a subshift is more subtle. If the cellular automaton is invertible then its inverse is given by another cellular automaton (see ref. [20] for a language-theoretic proof of this fact), and the same closure properties apply as for forward images. If the map is not invertible, the inverse image still makes sense as a set, but in this case the behavior can be somewhat more complicated. The difficulty arises from the fact that the inverse image of a configuration is not just determined by the inverse images of its finite substrings.

For each cellular automaton $A = (S, r, f)$, we define a block map $B_f : S^* \to S^*$ where for $x = x_1 \dots x_n$:

$$B_f(x) = f(x_1, \dots, x_{1+2r}) \dots f(x_{n-2r}, \dots, x_n),$$

$$\text{if } n > 2r;$$

$$= \varepsilon, \quad \text{otherwise.}$$

Lemma 8 *If K is a subshift, $\mathcal{L}(G_f(K)) = B_f(\mathcal{L}(K))$.*

Proof: This is a consequence of the fact that every string of length l in $\mathcal{L}(G_f(K))$ is determined by a string of length $l + 2r$ in $\mathcal{L}(K)$. $\square$

Let M be a GSM [16], δ and λ be the transition and output function, respectively, and q_0 be the start state of M. M is ε-limited if there exists a constant k such that for any input string w there is no substring x of w, $w = uxv$, such that $|x| > k$ and $\delta(\delta(q_0, u), x) = \varepsilon$.

Lemma 9 *For each cellular automaton $A = (S, r, f)$, B_f is a ε-limited GSM mapping.*

Proof: Define a GSM $M = (Q, S, S, \delta, \lambda, q_0)$ where

$$Q = \bigcup_{i=0}^{2r} S^i,$$

$$q_0 = \varepsilon,$$

$$\delta(a_1 a_2 \dots a_i, b) = a_1 a_2 \dots a_i b \quad \text{if } i < 2r,$$
$$= a_2 \dots a_i b \qquad \text{if } i = 2r,$$

$$\lambda(a_1 a_2 \dots a_i, b) = \varepsilon \quad \text{if } i < 2r,$$
$$= f(a_1, a_2, \dots, a_i, b)$$
$$\text{if } i = 2r.$$

Informally, this GSM works as follows: M has a shift register of length $2r + 1$. It reads $2r$ symbols into the register without writing anything. From the next letter on, it outputs the value of f on the contents of the register and forgets the least recently read symbol.

M has the property that no string of length greater than $2r$ is sent to ε. So, M is a ε-limited GSM. It is easy to see that B_f is exactly the mapping defined by M by the above construction. Therefore, M is a ε-limited GSM mapping. $\square$

Theorem 5 *If $\mathcal{S}$ is a class of languages closed under ε-limited GSM mappings and G_f is a cellular automaton map, then for each $K \subseteq S^{\mathbb{Z}}$, $\mathcal{L}(K) \in \mathcal{S} \Rightarrow \mathcal{L}(G_f(K)) \in \mathcal{S}$.*

Proof: The theorem is a direct consequence of lemma 8 and lemma 9. $\square$

Corollary 2 *If $\mathcal{S}$ is a Chomsky class (regular, context-free, context-sensitive, or recursively enumerable) and G_f is a CA map, then for each $K \subseteq S^{\mathbb{Z}}$, $\mathcal{L}(K) \in \mathcal{S}$ implies $\mathcal{L}(G_f(K)) \in \mathcal{S}$.*

Proof: Each of the Chomsky classes is closed under ε-limited GSM mappings. $\square$

The question of inverse images is somewhat subtler. At first glance it would seem that all the Chomsky classes should be closed under inverse maps, since each of the classes is closed under inverse GSM mappings. The difficulty arises from attempting to generalize lemma 8. While $\mathcal{L}(G_f^{-1}(K)) \subseteq lB_f^{-1}(\mathcal{L}(K))$, in general the inclusion can be strict. The correct generalization of lemma 8 is:

Lemma 10 *If K is a subshift, $\mathcal{L}(G_f^{-1}(K)) = C(B_f^{-1}(\mathcal{L}(K)))$.*

Proof: Let $s \in \mathcal{L}(G_f^{-1}(K))$. By definition $s \preceq c$ for some configuration $c \in G_f^{-1}(K)$. Then $B_f(s) \preceq G_f(c) \in K$. Hence $s \in B_f^{-1}(\mathcal{L}(K))$, and $\mathcal{L}(G_f^{-1}(K)) \subseteq B_f^{-1}(\mathcal{L}(K))$. By lemma 4, $\mathcal{L}(G_f^{-1}(K)) \subseteq C(B_f^{-1}(\mathcal{L}(K)))$.

To show the opposite inclusion, let $w \in C(B_f^{-1}(\mathcal{L}(K)))$. By lemma 5 there exist a nested sequence of strings $w_i = x_i w_{i-1} y_i$. By compactness the intersection $\bigcap_{i=0}^{\infty} \mathrm{Cyl}(w_i) \neq \emptyset$ and it consists of exactly one element. This element is in $G_f^{-1}(K)$, hence the inclusion follows. $\quad\square$

Theorem 6 *If S is a class of languages closed under inverse GSM mappings and the operator C, G_f is a cellular automaton map, then for each $K \subseteq S^{\mathbb{Z}}$, $\mathcal{L}(K) \in S \Rightarrow \mathcal{L}(G_f^{-1}(K)) \in S$.*

Proof: This follows immediately from lemmas 9 and 10. $\quad\square$

Corollary 3 *For any subshift K, if $\mathcal{L}(K)$ is regular (context-free) then $\mathcal{L}(G_f^{-1}(K))$ is regular (context-free).*

In particular if $K \subseteq S^{\mathbb{Z}}$ is a sofic system then so is $G_f^{-1}(K)$.

The classes of subshifts corresponding to context-sensitive and to recursively enumerable languages are not closed under inverse images of cellular automata.

Corollary 4 *There exists a cellular automaton $A = (S, r, f)$ and a set $K \subseteq S^{\mathbb{Z}}$ such that $\mathcal{L}(K)$ is context-sensitive and $\mathcal{L}(G_f^{-1}(K))$ is not recursively enumerable.*

Proof: Let L_u be the language described in theorem 4 and S_u its associated alphabet. Define a cellular automaton rule $A = (S_u \cup \{b, b'\}, 0, f)$ where f is defined by the rule $f(b) = b'$ and otherwise $f(x) = x$. The inverse image $G_\phi^{-1}(K)$ of any set K consists of exactly those configurations in K which do not contain the symbol b.

Let K_0 consist of all configurations of the form

$$\ldots bbba_1 \ldots a_n bbb \ldots, \qquad \text{where } a_1 \ldots a_n \in L_u.$$

The set K_0 is not necessarily closed. Since taking the closure of a set does not add any new finite strings, define $K = \overline{K_0}$.

Let $L_{\mathrm{R}} = S_u^*$ (a regular language) and consider $\mathcal{L}(G_f^{-1}(K)) \cap L_{\mathrm{R}}$. The only configurations in $\mathcal{L}(G_f^{-1}(K))$ are those which fail to contain the symbol b. Such configurations can only arise as limit points of elements of K_0. In fact, $\mathcal{L}(G_f^{-1}(K)) \cap L_{\mathrm{R}} = C(L_u)$. Since $C(L_u)$ is not recursively enumerable, and L_{R} is a regular language, it must be the case that $\mathcal{L}(G_f^{-1}(K))$ is not recursively enumerable proving the theorem. $\quad\square$

6. Discussion

The correspondence set up in theorem 1 lays the basis for a language-theoretic classification of subshifts. Theorem 5 shows that in most cases this classification is natural, since it is preserved by cellular automaton images. Also, by considering various invariant subshifts arising naturally from the evolution of cellular automata, one can use this theorem to give a classification of cellular automata themselves, and provide a qualitative measure of the complexity of their dynamics.

This classification has been carried out most extensively for the limit sets of cellular automata; as they formed a natural generalization of work on finite time images [2,22].

Aside from the calculations of regular limit languages in ref. [22] and non-regular limit languages in ref. [7], few explicit calculations of limit sets exist. It is still an open question whether one of Wolfram's 256 *elementary* rules [2] gives rise to a non-regular limit set. This problem may be somewhat difficult due to the recent result of Kari [23] that classifying cellular automata by the language-theoretic complexity of their limit sets (and indeed any non-trivial classification scheme based on limit sets) is undecidable.

The limit set is far from the only invariant subshift of interest. Other dynamically interesting invariant subshifts such as the periodic set and non-wandering set ([5–7,21]), have also been shown to have varying degrees of language-theoretic complexity.

Acknowledgements

Reported research was supported by the Natural Sciences and Engineering Research Council of Canada Grants OGPIN007 and the National Sciences Foundation Grant No. CCR-8702752, as well as DARPA under the Applied and Computational Mathematics Program. The second author would like to thank the Institute for Advanced Study for support during the time that some of this research was conducted.

References

[1] B. Weiss, Monatsh. Math. 77 (1973) 462.

[2] S. Wolfram, Commun. Math. Phys. 96 (1984) 15.

[3] L.P. Hurd, Complex Systems 1 (1987) 69.

[4] K. Culik II, J. Pachl and S. Yu, SIAM J. Comput. 18 (1989) 831.

[5] L.P. Hurd, The application of formal language theory to the dynamical behavior of cellular automata, Doctoral Thesis, Princeton University (1988).

[6] L.P. Hurd, Non-recursive cellular automata invariant sets, Complex Systems, to appear.

[7] L.P. Hurd, Recursive cellular automata invariant sets, Complex Systems, to appear.

[8] K. Culik II, L.P. Hurd and S. Yu, Computation theoretic aspects of cellular automata, Physica D 45 (1990) 357–378, these Proceedings.

[9] M. Nivat and D. Perrin, Proceedings of the Fourteenth Annual ACM Symposium on Theory of Computing (1982) p. 47.

[10] L. Boasson and M. Nivat, J. Computer System Sci. 20 (1980) 285.

[11] K. Culik II and S. Yu, Cellular automata, $\omega\omega$-regular sets, and sofic systems, Discrete Appl. Math., to be published.

[12] M. Nivat, RAIRO Informatique Théorique 12 (1978) 259.

[13] A. Restivo, Research Report, Dipartimento di Matematica e Applicazioni, Università di Palermo (1986).

[14] M.A. Harrison, Introduction to Formal Language Theory (Addison–Wesley, Reading, MA, 1978).

[15] J.E. Hopcroft and J.D. Ullman, Introduction to Automata Theory, Languages, and Computation (Addison–Wesley, Reading, MA, 1979).

[16] S. Ginsburg, Algebraic and Automata-Theoretic Properties of Formal Languages (North-Holland, Amsterdam, 1975).

[17] A. Salomaa, Formal Languages (Academic Press, New York, 1973).

[18] G.A. Hedlund, in: Topological Dynamics, eds. J. Auslander and W.G. Gottschalk (1968) p. 259.

[19] G.A. Hedlund, Math. Syst. Theory 3 (1969) 320.

[20] K. Culik II, Complex Systems 1 (1987) 1035.

[21] L.P. Hurd, Complex Systems 2 (1988) 549.

[22] L.P. Hurd, Theory and Applications of Cellular Automata (World Scientific, Singapore, 1986) p. 523.

[23] J. Kari, Rice's theorem for limit sets of cellular automata, Theoret. Comput. Sci., submitted for publication.

Physica D 45 (1990) 404–419
North-Holland

A CHARACTERIZATION OF
CONSTANT-TIME CELLULAR AUTOMATA COMPUTATION *

Sam KIM and Robert McCLOSKEY

Computer Science Department, Rensselaer Polytechnic Institute, Troy, NY 12180, USA

Received 17 January 1990
Revised manuscript received 20 March 1990

This paper investigates three different classes of one-dimensional CA languages that can be accepted in constant time. For each class of the languages, two characterizations are given, one in terms of finite sets and the other in terms of a subclass of deterministic finite automata. It is also shown that the constant-time CA languages have very restricted closure properties.

1. Introduction

A cellular automaton is an n-dimensional array of identical finite state machines, called cells, which operate in a sequence of discrete time steps. At each time step, every cell reads the states of its neighbors and changes its own state. The cellular automaton model has been extensively studied for the exploitation of its parallelism in language recognition and image processing on the one hand [1–3] and for identifying its dynamic properties and simulating physical systems on the other [4,5].

This paper investigates the languages that can be recognized in constant time by one-dimensional cellular automata and introduces two characterizations of these languages, one in terms of finite sets and the other in terms of a subclass of deterministic finite automata. The constant-time cellular automata languages defined in this paper are different from the finite-time sets on cellular automata defined in refs. [4,6]. There, the finite-time set of a cellular automaton is defined to be the set of configurations of a cellular automaton at some constant time t that can be evolved from all possible initial configurations. Every finite-time set is regular [4,6].

This paper has five sections including this one. Section 2 introduces basic definitions and notation. Sections 3 and 4 show characterizations of constant-time cellular automata languages in terms of finite sets and deterministic finite automata, respectively. Section 5 gives some further remarks.

2. Notation and terminology

A one-dimensional cellular automaton (CA) is a linear array of identical finite state machines, called cells, that operate synchronously at discrete time steps (see fig. 1) [1]. Initially, all cells are in the quiescent state and an input of length n is applied to n consecutive cells. At each time step, each cell changes its state depending on its own state, its external input and the current states of its two neighbors. The external input is applied during the first step and remains null thereafter. The number of cells of a CA is equal to the length of the input, except that it is one for the null input. We also assume that every cell has a distinguished accepting state and a distinguished rejecting state. All other states are said to be neutral. Once a cell is in the rejecting or the accepting state, it cannot change state.

Conventionally, a CA, as a language recognizer,

* Partial support for this research was provided by the Directorate of Computer and Information Science and Engineering of the National Science Foundation under Institutional Infrastructure No. CDA-8 805910.

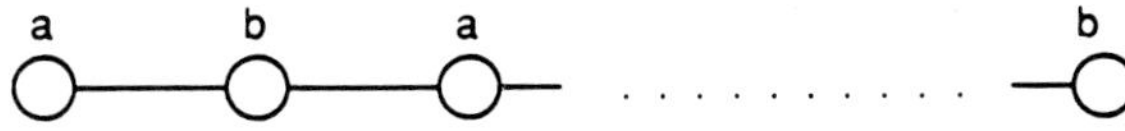

Fig. 1. A cellular automaton.

has a designated accepting cell (usually one of the boundary cells) [1]. An input is accepted iff the designated cell enters an accepting state. For a CA computation which allows linear- or higher-time complexity, it does not make any difference which cell is defined as the accepting cell because there is enough time to communicate across the entire array. However, for a sublinear- or constant-time CA computation, it does not make sense to have such a designated cell. Instead, the accepting condition must be defined differently. Various accepting conditions give rise to models with varying language recognition capabilities [7]. In this paper we study the following three CA models.

Definition 1 (constant-time CA). Let $k \geq 0$ be a fixed integer.

(i) ACA: The input is accepted iff all cells are simultaneously in the accepting state in k steps, with no cells having entered the rejecting state.

(ii) ECA: The input is accepted iff there is a cell in an accepting state in k steps.

(iii) MCA: The input is accepted iff there exists a cell in an accepting state with no cells in the rejecting state in k steps.

This definition is different from the one in ref. [8], where ACA, named 1-CSCA, was defined such that both accepting state and rejecting state can change to other states. An input is accepted iff all cells are simultaneously in an accepting state in k steps, for a constant k. Let XCA denote a class of CA defined above. By k-XCA we denote a constant-time XCA which accepts an input in some constant $k \geq 0$ steps. In this paper we only investigate deterministic CA.

Definition 2 (substring vector). Let Σ be a finite alphabet and $x \in \Sigma^*$. For an integer $k \geq 0$ and $x \in \Sigma^*$, define $f_k(x)$ to be the prefix of x of length k, $t_k(x)$ to be the suffix of x of length k, and $I_k(x)$ to be the set of substrings of x of length k, i.e. $\{v \mid v \in \Sigma^k$ and $x = uvw$, for some $u, w \in \Sigma^*\}$. If $|x| < k$, then $f_k(x) = t_k(x) = x$. For $k > 0$, the substring vector of order k of x is an ordered triple $SV_k(x) = (f_{k-1}(x), I_k(x), t_{k-1}(x))$.

Definition 3. For a finite alphabet Σ and integer $k > 0$, denote by $\Sigma^{<k}$ the set of strings in Σ^* of length less than k. Let $\alpha, \alpha', \gamma, \gamma' \subseteq \Sigma^{<k}$ and $\beta, \beta' \subseteq \Sigma^k$.

(i) $\hat{U}_k(\alpha, \beta, \gamma)$ denotes the language $L \subseteq \Sigma^*$ such that $x \in L$ iff $f_{k-1}(x) \in \alpha$, $I_k(x) \subseteq \beta$, and $t_{k-1}(x) \in \gamma$.

(ii) $\hat{E}_k(\alpha, \beta, \gamma)$ denotes the language $L \subseteq \Sigma^*$ such that $x \in L$ iff $f_{k-1}(x) \in \alpha$, $I_k(x) \cap \beta \neq \emptyset$, or $t_{k-1}(x) \in \gamma$.

(iii) $\hat{M}_k((\alpha, \beta, \gamma), (\alpha', \beta', \gamma'))$ denotes the language $L \subseteq \Sigma^*$ such that $x \in L$ iff

(a) $f_{k-1}(x) \in \alpha$, $I_k(x) \subseteq \beta$, and $t_{k-1}(x) \in \gamma$, and

(b) $f_{k-1}(x) \in \alpha'$, $I_k(x) \cap \beta' \neq \emptyset$, or $t_{k-1}(x) \in \gamma'$, i.e., $\hat{M}_k((\alpha, \beta, \gamma), (\alpha', \beta', \gamma')) = \hat{U}_k(\alpha, \beta, \gamma) \cap \hat{E}_k(\alpha', \beta', \gamma')$.

Definition 4 (LTSS) [9] . Let $L \subseteq \Sigma^*$. L is zero-testable in the strict sense (0-LTSS) iff it is Σ^* or $\emptyset$. For $k > 0$, L is k-LTSS iff there exist $\alpha \subseteq \Sigma^{<k}$, $\beta \subseteq \Sigma^k$, and $\gamma \subseteq \Sigma^{<k}$ such that $L = \hat{U}_k(\alpha, \beta, \gamma)$. A language is LTSS iff it is k-LTSS, for some $k \geq 0$.

Definition 5 (LTES). Let $L \subseteq \Sigma^*$. L is zero-testable in the existential sense (0-LTES) iff it is Σ^* or $\emptyset$. For $k > 0$, L is k-LTES iff there exist $\alpha \subseteq \Sigma^{<k}$, $\beta \subseteq \Sigma^k$, and $\gamma \subseteq \Sigma^{<k}$ such that $L = \hat{E}_k(\alpha, \beta, \gamma)$. A language is LTES iff it is k-LTES, for some $k \geq 0$.

Definition 6 (LTMS). Let $L \subseteq \Sigma^*$. L is zero-testable in the mixed sense (0-LTMS) iff it is Σ^* or $\emptyset$. For $k > 0$, L is k-LTMS iff there exist $\alpha, \alpha', \gamma, \gamma' \subseteq \Sigma^{<k}$ and $\beta, \beta' \subseteq \Sigma^k$ such that $L = \hat{M}_k((\alpha, \beta, \gamma), (\alpha', \beta', \gamma'))$. A language is LTMS iff it is k-LTMS, for some $k \geq 0$.

Definition 7 (locally testable languages) [9,10]. Let $L \subseteq \Sigma^*$.

(i) L is 0-testable iff it is Σ^* or $\emptyset$.

(ii) For an integer $k > 0$, L is k-testable iff, for all $x, y \in \Sigma^*$, $SV_k(x) = SV_k(y)$ implies that either both x and y are in L or neither is in L.

(iii) L is locally testable if it is k-testable for some $k \geq 0$.

Definition 8 (LTES, LTSS, LTMS and locally testable automata). A finite automaton is LTES (LTSS, LTMS, locally testable), if it accepts an LTES (respectively, LTSS, LTMS, locally testable) language.

3. Characterizations of constant-time CA languages

This section investigates the properties of the languages defined in section 2 and shows that they characterize constant-time CA languages.

Lemma 1. For any $k \geq 0$, L is k-LTSS iff $\overline{L}$ is k-LTES.

Proof. If L is 0-LTSS, then it is either Σ^* or $\emptyset$, making $\overline{L}$ either $\emptyset$ or Σ^*, respectively, and so $\overline{L}$ is 0-LTES by definition. If L is k-LTSS, for some $k > 0$, let α, β, and γ be such that $L = \hat{U}_k(\alpha, \beta, \gamma)$. Find $\alpha' = \Sigma^{<k} - \alpha$, $\beta' = \Sigma^k - \beta$ and $\gamma' = \Sigma^{<k} - \gamma$. Clearly, for any $x \in \Sigma^*$, $x \in \overline{L}$ iff $f_{k-1}(x) \in \alpha'$, $I_k(x) \cap \beta' \neq \emptyset$, or $t_{k-1}(x) \in \gamma'$. It follows that $\overline{L} = \hat{E}_k(\alpha', \beta', \gamma')$, and so $\overline{L}$ is k-LTES. The converse is proved similarly. $\square$

Lemma 2. If a language L is k-LTES (k-LTSS), then it is k'-LTES (k'-LTSS), for any $k' > k$.

Proof. Let L be k-LTES and let $k' > k$. If $k = 0$ then either $L = \hat{E}_{k'}(\emptyset, \emptyset, \emptyset) = \emptyset$ or $L = \hat{E}_{k'}(\Sigma^{<k'}, \Sigma^{k'}, \Sigma^{<k'}) = \Sigma^*$. Thus, L is also k'-LTES. If $k > 0$, let α, β, and γ be such that $L = \hat{E}_k(\alpha, \beta, \gamma)$. Construct the sets $\alpha' = \{xw | x \in \alpha, w \in \Sigma^*, |xw| < k'\}$, $\beta' = \{uxw | x \in \beta, u, w \in \Sigma^*, |uxw| = k'\}$, and $\gamma' = \{wx | x \in \gamma, w \in \Sigma^*, |wx| < k'\}$. We leave it for the reader to show that, for all $x \in \Sigma^*$, $f_{k-1}(x) \in \alpha$, $I_k(x) \cap \beta \neq \emptyset$, or $t_{k-1}(x) \in \gamma$ iff $f_{k'-1}(x) \in \alpha'$, $I_{k'}(x) \cap \beta' \neq \emptyset$, or $t_{k'-1}(x) \in \gamma'$. It follows that $L = \hat{E}_{k'}(\alpha', \beta', \gamma')$, i.e., L is k'-LTES. Now let L be k-LTSS. By lemma 1, $\overline{L}$ is k-LTES, and by the above, $\overline{L}$ is also k'-LTES. Then using lemma 1 again, we have that L is k'-LTSS. $\square$

Lemma 3. A language L is LTMS iff it is the intersection of an LTSS language and an LTES language.

Proof. Let L be k-LTMS. If $k = 0$, then either $L = \emptyset = \emptyset \cap \emptyset$ or $L = \Sigma^* = \Sigma^* \cap \Sigma^*$, from which it follows that L is the intersection of a 0-LTSS language and a 0-LTES language. If $k > 0$, let $\alpha, \beta, \gamma, \alpha', \beta'$, and γ' be such that $L = \hat{M}_k((\alpha, \beta, \gamma), (\alpha', \beta', \gamma'))$. Then L is the intersection of the k-LTSS language L_1 and the k-LTES language L_2, where $L_1 = \hat{U}_k(\alpha, \beta, \gamma)$ and $L_2 = \hat{E}_k(\alpha', \beta', \gamma')$. For the converse, let L_1 be k_1-LTSS and L_2 be k_2-LTES. By lemma 2, L_1 is k-LTSS and L_2 is k-LTES, where k is the maximum of k_1 and k_2. If $k = 0$, then $L_1 \cap L_2$ is either Σ^* or $\emptyset$, which is 0-LTMS. Otherwise, let α, β, γ, α', β', and γ' be such that $L_1 = \hat{U}_k(\alpha, \beta, \gamma)$ and $L_2 = \hat{E}_k(\alpha', \beta', \gamma')$. Then $L_1 \cap L_2 = \hat{M}_k((\alpha, \beta, \gamma), (\alpha', \beta', \gamma'))$, which is k-LTMS. $\square$

Lemma 4. For any $k \geq 0$, L is a k-ACA language iff $\overline{L}$ is k-ECA.

Proof. Let M be a k-ACA which accepts L. Construct a k-ECA M' which accepts $\overline{L}$ as follows. M' simulates M such that at the kth step each cell of M' enters the accepting state if its simulation is to enter a nonaccepting state. If the simulation is to enter the accepting state at the kth step, the cell stays in a neutral state. Conversely, let L be a language accepted by a k-ECA M'. Construct a k-ACA M which accepts $\overline{L}$. M simulates M' up to k steps and lets each cell enter the accepting state if its simulation is to enter a nonaccepting state.

$\square$

Theorem 1. A language L is LTSS iff it is accepted by an ACA.

Proof. Let L be k-LTSS. If $k = 0$, then L is either Σ^* or $\emptyset$. If L is Σ^*, then define the initial state of the cells of M to be the accepting state. If L is $\emptyset$, define the initial state of the cells to be the rejecting state. Clearly, M accepts L. If $k > 0$, let α, β, and γ be such that $L = \hat{U}_k(\alpha, \beta, \gamma)$. We construct a $(\lceil (k + 1)/2 \rceil)$-ACA M which accepts L as follows. The cells of M have three look-up tables for α, β, and γ, respectively. For an input $x = a_1 a_2 \ldots a_n$, let c_i be the cell which has a_i as its external input. Let $a_i = \epsilon$, for $i < 1$ or $i > n$. Construct M such that cell c_i finds the substring $w_i = a_{i - \lceil (k-1)/2 \rceil} \ldots a_i \ldots a_{i + \lfloor (k-1)/2 \rfloor}$ of x in $\lceil (k+1)/2 \rceil$ steps and enters the accepting state if any of the following conditions is satisfied.

(i) $|w_i| = k$ and $w_i \in \beta$.

(ii) w_i is a prefix of x, $|w_i| < k$ and $w_i y \in \alpha$, for some $y \in \Sigma^*$.

(iii) w_i is a suffix of x, $|w_i| < k$, and $yw_i \in \gamma$, for some $y \in \Sigma^*$.

Otherwise, c_i enters the rejecting state. Notice that by having the boundary cells send blank symbols to their neighbors, we can make it possible for cell c_i to determine whether w_i is a prefix or suffix of x. We leave it for the reader to show that $f_{k-1}(x) \in \alpha$, $I_k(x) \subseteq \beta$, and $t_{k-1}(x) \in \gamma$ iff, with input x, all cells c_i, $1 \le i \le n$ will be in the accepting state in $\lceil (k+1)/2 \rceil$ steps. Thus, M accepts L.

Now, let L be the language accepted by a k-ACA M. If $k = 0$, then L is either Σ^* or $\emptyset$. Thus, L is an LTSS language. Suppose $k \ge 1$. Find $L_0 = \{x| \; |x| \le 2k - 2$ and M accepts x in k steps$\}$ and construct the sets: $\alpha = L_0 \cup \{x| \;$ with input $x = a_1 \ldots a_{2k-2}$, cells $c_1, \ldots, c_{k-1}$ enter the accepting state in k steps$\}$, $\gamma = L_0 \cup \{x| \;$ with input $x = a_1 \ldots a_{2k-2}$, cells $c_k, \ldots, c_{2k-2}$ enter the accepting state in k steps$\}$ and $\beta = \{x| \;$ with input $x = a_1 \ldots a_{2k-1}$, cell c_k enters the accepting state in k steps$\}$.

We leave it for the reader to show that, for every word $w \in \Sigma^*$, w is accepted by M iff $f_{2k-2}(w) \in \alpha$, $I_{2k-1}(w) \subseteq \beta$ and $t_{2k-2}(w) \in \gamma$. Thus, L is the $(2k-1)$-LTSS language $\hat{U}_{2k-1}(\alpha, \beta, \gamma)$. $\quad\square$

Remark: Notice that the construction above does not necessarily give the smallest α, β and γ such that $\hat{U}_{2k-1}(\alpha, \beta, \gamma) = L$. For example, there may exist $w \in \beta$ such that, for all $x, y \in \Sigma^*$, $xwy \notin L$. Word w is useless for defining L. We can remove useless words from the sets by the following procedure.

(i) With α, β, and γ, construct a graph $G(V, E)$ as follows: $V = \{[w] | w$ is in either α or γ, or $wa \in \beta$ for some $a \in \Sigma\}$. For each pair $[w_1], [w_2]$ in V, assign a directed edge from $[w_1]$ to $[w_2]$ with label a, if $w_1 a \in \beta$ and $t_{2k-2}(w_1 a) = w_2$.

(ii) Search G and mark all nodes which are on a path from a node $[w_i]$, $w_i \in \alpha$, to a node $[w_j]$, $w_j \in \gamma$.

(iii) Delete w from α and γ if node $[w]$ has no mark in G.

(iv) If a node $[w_i]$ has no mark and has an outgoing edge labeled by $a \in \Sigma$, delete $w_i a$ from β.

Theorem 2. A language L is LTES iff it is accepted by an ECA.

Proof. The theorem follows from lemmas 1, 4 and theorem 1. $\quad\square$

Lemma 5. If a language L is LTMS, then it is accepted by an MCA.

Proof. If L is an LTMS language, then by lemma 3 there exist an LTSS language L_1 and an LTES language L_2 such that $L = L_1 \cap L_2$. Let M_1 and M_2 be a k-ACA and a k-ECA which accept L_1 and L_2, respectively. By theorems 1 and 2, such CA exist. Construct a k-MCA M which accepts L by simulating M_1 and M_2 simultaneously. At the kth step, each cell of M does the following.
– Enters the accepting state if both simulations are in the accepting state.
– Enters the rejecting state if the simulation for the cell of M_1 is in a nonaccepting state.
– Enters a neutral state, otherwise. $\quad\square$

Lemma 6. Let M be an MCA. Then $L(M)$ is an LTMS language.

Proof. Suppose M is a k-MCA, for some $k \ge 0$. If $k = 0$, then $L(M)$ is either Σ^* or $\emptyset$ and thus is an LTMS language. Suppose $k > 0$. Using the same idea as in the proof of theorem 1, we construct (α, β, γ) and $(\alpha', \beta', \gamma')$ such that $L(M) = \hat{M}_k((\alpha, \beta, \gamma), (\alpha', \beta', \gamma'))$. Find $L_0 = \{x| \; |x| \le 2k - 2$ and with input x, no cell of M enters the rejecting state in k steps$\}$ and construct the sets: $\alpha = L_0 \cup \{x| \;$ with input $x = a_1 \ldots a_{2k-2}$, $a_i \in \Sigma$, no cells among $c_1, \ldots, c_{k-1}$ enter the rejecting state at the kth step$\}$, $\gamma = L_0 \cup \{x| \;$ with input $x = a_1 \ldots a_{2k-2}$, $a_i \in \Sigma$, no cells among $c_k, \ldots, c_{2k-2}$ enter the rejecting state at the kth step$\}$ and $\beta = \{x| \;$ with input $x = a_1 \ldots a_{2k-1}$, cell c_k does not enter the rejecting state at the kth step$\}$.

Let $L_0' = \{x| \; |x| \le 2k - 2$ and M accepts x in k steps$\}$. We construct α', β' and γ' as follows: $\alpha' = L_0' \cup \{x| \;$ with input $x = a_1 \ldots a_{2k-2}$, $a_i \in \Sigma$, at least one cell among $c_1, \ldots, c_{k-1}$ enters the accepting state at the kth step$\}$, $\gamma' = L_0 \cup \{x| \;$ with input $x = a_1 \ldots a_{2k-2}$, $a_i \in \Sigma$, at least one cell among $c_k, \ldots, c_{2k-2}$ enters the accepting state at the kth step$\}$ and $\beta' = \{x| \;$ with input $x = a_1 \ldots a_{2k-1}$, $a_i \in \Sigma$, cell c_k enters the accepting state at the kth step$\}$.

We leave it for the reader to show that $L(M) = \hat{M}_{2k-1}((\alpha, \beta, \gamma), (\alpha', \beta', \gamma'))$. $\quad\square$

Theorem 3. A language L is an LTMS language

iff it is accepted by an MCA.

Proof. The theorem follows from lemmas 5 and 6. $\qquad\square$

Since any ACA or ECA can be simulated by an MCA, we have the following corollary.

Corollary 1. The class of LTMS languages includes the classes of LTES languages and LTSS languages.

As the following theorem shows, the subclasses of CA languages that we have introduced have very limited closure properties under Boolean operations.

Theorem 4.

(a) The class of LTMS languages is not closed under union, intersection or complementation.

(b) The class of LTSS languages is closed under intersection, but is not closed under union or complementation.

(c) The class of LTES languages is closed under union, but is not closed under intersection or complementation.

Proof.

(a) Let $\Sigma = \{a, b, c\}$. Consider $L_1 = b\{a, b, c\}^*$ and $L_2 = \{a, b\}^*$. Clearly, L_1 is LTES and L_2 is LTSS. By corollary 1 both L_1 and L_2 are LTMS languages. Suppose $L_3 = L_1 \cup L_2$ is an LTMS language which is accepted by a k-MCA M, for some $k \geq 0$. Let $x = bb^{2k+1}cb^{2k+1}$, $y = ab^{2k+1}bb^{2k+1}$, and $z = ab^{2k+1}cb^{2k+1}$. Clearly, x and y are in L_3, and and z is not. We show that if M accepts x and y, then it must also accept z. Let $C_w(xwy)$ denote the sequence of consecutive cells on a CA which get the substring w on an input xwy. We consider the following two cases:

Case (1). When M accepts x by having at least one cell in $C_{b^k cb^{2k+1}}(bb^{2k+1}cb^{2k+1})$ entering the accepting state in k steps, then, with input z, at least one cell in $C_{b^k cb^k}(ab^{2k+1}cb^{2k+1})$ will enter the accepting state in k steps with no cell in the rejecting state. Since y is accepted, no cells in $C_{ab^{k+1}}(ab^{2k+1}bb^{2k+1})$ will be in the rejecting state in k steps. It follows that M must also accept z, a contradiction.

Case (2). When M accepts x by having at least one cell in $C_{bb^k}(bb^{2k+1}cb^{2k+1})$ entering the accepting

state in k steps, with no cells in $C_{b^k cb^{2k+1}}(bb^{2k+1} \times cb^{2k+1})$ entering either the accepting or rejecting state. Since no cell in $C_{b^k cb^{2k+1}}(bb^{2k+1}cb^{2k+1})$ enters the accepting state in k steps, with input y no cells in $C_{b^{k+1}bb^{2k+1}}(ab^{2k+1}bb^{2k+1})$ will enter the accepting state in k steps. Hence, M accepts y by having at least one cell in $C_{ab^k}(ab^{2k+1}bb^{2k+1})$ entering the accepting state in k steps. Obviously, M must also accept z since at least one cell in $C_{ab^k}(ab^{2k+1}cb^{2k+1})$ will enter the accepting state in k steps with no cells in $C_{b^k cb^{2k+1}}(ab^{2k+1}cb^{2k+1})$ in the rejecting state. Again we are in a contradiction. LTMS languages are not closed under union.

Now, consider $\overline{L}_1$ and $\overline{L}_2$ which are LTSS and LTES languages, respectively, by lemma 1. By lemma 3 $L_4 = \overline{L}_1 \cap \overline{L}_2$ is LTMS. We just proved that $\overline{L}_4 = L_1 \cup L_2$ is not an LTMS language. It follows that LTMS languages are not closed under complementation.

Finally, let $L_5 = \Sigma^* a \Sigma^*$ and $L_6 = \Sigma^* b \Sigma^*$. Both L_5 and L_6 are LTMS languages. Suppose $L_7 = L_5 \cap L_6$ is an LTMS language which is accepted by a k-MCA M, for some $k \geq 0$. Consider $w = c^n ac^n bc^n$ and $w' = c^n bc^n ac^n$, for some $n > k$. Both w and w' will be accepted by M. It follows that either $c^n ac^n ac^n$ or $c^n bc^n bc^n$, neither of which is in L_7, must be accepted by M, a contradiction. LTMS languages are not closed under intersection.

(b) Given two ACA M_1 and M_2, it is easy to construct an ACA M such that each cell simulates the cells of M_1 and M_2 in parallel and accepts the input iff both simulations accept the input. ACA languages are closed under intersection. Using the technique from part (a), we can prove that ACA languages are not closed under union or complementation.

Part (c) can be proved similarly. We leave the proofs for the reader. $\qquad\square$

By using the same idea as in ref. [8], it can be shown that a language is regular iff it is accepted by a nondeterministic ACA. This implies that nondeterministic constant-time CA have more computational power than their deterministic version. This is in contrast to the finite automata model.

4. Sequential machine characterization of constant-time CA

We know that there are two subclasses of computational models, parallel and sequential. An interesting question is: For a given parallel model, is there a sequential model which is computationally equivalent? By computational equivalence we mean that both models solve the same problems without regard to their time complexity. If two computing models are computationally equivalent, we say that each characterizes the other. If a parallel model, like unrestricted CA, can simulate a Turing machine, then by Church's thesis it is computationally equivalent to the Turing machine. We are interested in some subclasses of the models which are less powerful than the Turing machine. In ref. [11], it is shown that some subclasses of systolic models can be characterized in terms of sequential models. Section 3 showed that ACA, ECA, and MCA accept LTSS, LTES, and LTMS languages, respectively. All these languages are regular. This section identifies three subclasses of deterministic finite state automata which accept these languages, actually showing sequential model characterizations for the constant-time CA defined in section 2.

We shall introduce some definitions concerning deterministic finite automata before we describe the main result. Let $M = (Q, \Sigma, \delta, q_0, F)$ be a deterministic finite automaton. (Our notation follows ref. [12].) $L(M)$ denotes the language accepted by M, i.e., $\{x \mid \delta(q_0, x) \in F\}$. We sometimes use $L(M, p)$ to denote $\{x \mid \delta(p, x) \in F\}$. Thus, $L(M) = L(M, q_0)$. We assume that the automaton is given in terms of its state transition graph. By q_d and q_e we denote the rejecting sink state and the accepting sink state, respectively, i.e., $\delta(q_d, a) = q_d$ and $\delta(q_e, a) = q_e$, for all $a \in \Sigma$, q_e is in F and q_d is not in F. We shall use the terminology "component" to refer to any subgraph of a state transition graph whose underlying undirected subgraph is connected. When we show a state transition graph, we will omit q_d and all the edges leading to it. By SCC we mean a maximal strongly connected component of a state transition graph excluding q_d.

Definition 9 (reachable component). Let m_i

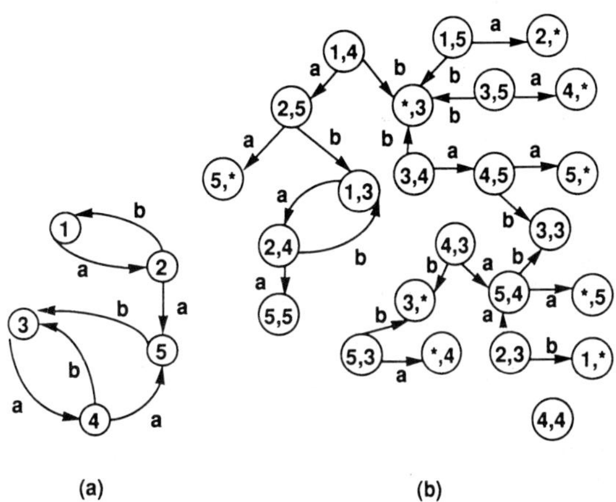

Fig. 2. (a) A state transition graph. (b) The pair graph on $Q_1 \times Q_2 = \{1, 2, 3, 4, 5\} \times \{3, 4, 5\}$.

be a component of a state transition graph. The *reachable component* from m_i, denoted by m_{i*}, is the subgraph which comprises m_i and all directed paths reachable from m_i. Likewise m_{p*} denote the reachable component from state p. Notice that if p is a state in an SCC m_i, then $m_{i*} = m_{p*}$.

Definition 10 (pair graph). Let $M = (Q, \Sigma, \delta, q_0, F)$ be an automaton. Let $Q_1, Q_2 \subseteq Q$, which are not necessarily disjoint, and $*$ be a symbol not in Q. The pair graph on $Q_1 \times Q_2$ is the edge-labeled graph $G(V, E)$, where $V = (Q_1 \cup *) \times (Q_2 \cup \{*\}) - \{(*, *)\}$ and E is defined as follows (see fig. 2, for an example):

Define $\delta_i : Q_i \times \Sigma \to Q_i \cup \{*\}$, $i = 1, 2$ such that for all $p \in Q_i$ and $a \in \Sigma$,

$$\delta_i(p, a) = q \quad \text{if } \delta(p, a) = q \in Q_i,$$
$$= * \quad \text{if } \delta(p, a) \notin Q_i.$$

Then $E = \{((p, q), a, (r, s)) \mid p \in Q_1, q \in Q_2, p \neq q, r \in Q_1 \cup \{*\}, s \in Q_2 \cup \{*\}, \delta_1(p, a) = r, \delta_2(q, a) = s, \text{ and } r \neq * \text{ or } s \neq *\}$. Notice that from a node $(p, *), (*, q),$ or (p, p) of the pair graph , no outgoing edge exists. A pair graph is not necessarily connected.

Definition 11 (input span). Let $M = (Q, \Sigma, \delta, q_0, F)$ be a deterministic finite automaton, and let m be a component of the state transition graph of the automaton. A word x is an input span for state p in m iff $\delta(q, x) = p$, for some q in m, such that, for every prefix w of x, $\delta(q, w)$ is in m. Note that p and q may be the same state.

Definition 12 (finite input memory) [13]. Let $M = (Q, \Sigma, \delta, q_0, F)$ be a deterministic finite automaton. Let m_i be a component of the state transition graph of M, and let Q_i be the set of states in m_i. Component m_i has input memory of order k if k is the minimum nonnegative integer such that no two states in $Q_i - \{q_d\}$ have a common input span in m_i of length k. Component m_i has *finite input memory* iff it has input memory of order k for some $k \geq 0$. Let m_1 and m_2 be two components of the state transition graph. Then m_1 and m_2 have *pairwise finite input memory* iff there exists an integer k such that no states p and q, respectively, in m_1 and m_2 have a common input span of length greater than k on m_1 and m_2, respectively.

Notice that finiteness of input memory of a state transition graph is independent of the assignment of accepting states. It is not difficult to show that if a component m of a state transition graph has infinite input memory, then m has two states p and q which are not q_d such that $\delta(p, x) = p$ and $\delta(q, x) = q$, for some $x \in \Sigma^+$, and, for every prefix x' of x, both $\delta(p, x')$ and $\delta(q, x')$ are in m. Let S be the set of states (other than q_d) of a component m of a state transition graph. We can determine if component m has finite input memory by constructing the pair graph on $S \times S$ and testing if the graph is acyclic. If it is, then m has finite input memory, otherwise m does not. Specifically, if $|S| = 1$, then m has finite input memory of order zero. Otherwise, if the pair graph is acyclic, m has input memory of order $r + 1$, where r is the length of the longest path in the pair graph .

In ref. [14], we introduced the following characterization theorem for the class of locally testable deterministic finite automata and showed that, given a deterministic finite automaton, it is possible to determine if it is locally testable in $O(n^2)$ time, where n is the number of states of the automaton.

Theorem 5. Let $M = (Q, \Sigma, \delta, q_0, F)$ be a reduced deterministic finite automaton. The automaton accepts a locally testable language iff the transition graph satisfies the following properties.

(i) Every SCC has finite input memory.

(ii) For every pair of SCCs m_i and m_j such that m_i is an ancestor of m_j, either (a) they have pairwise finite input memory, or, otherwise (b) for every pair of states p in m_i and q in m_j, and $x \in \Sigma^+$ such that $\delta(p, x) = p$ and $\delta(q, x) = q$, $\delta(p, w) = r$ is an ancestor of q iff $\delta(q, w)$ is in m_j, for all $w \in \Sigma^*$.

Theorem 6. Let M be a deterministic finite automaton which is reduced. M has finite input memory iff $L(M)$ is LTSS.

Proof. Let $M = (Q, \Sigma, \delta, q_0, F)$ and let $Q_0 = Q - \{q_d\}$. Suppose that the state transition graph has finite input memory of order k. If $k = 0$, then $|Q_0| \leq 1$. If $|Q_0| = 0$, we have $L(M) = \emptyset$, which is 0-LTSS. If $|Q_0| = 1$, then $L(M) = A^*$ for some $A \subseteq \Sigma$. Thus $L(M) = \hat{U}_1(\{\epsilon\}, A, \{\epsilon\})$, which is LTSS. Suppose $|Q_0| \geq 2$. Assign each state in Q_0 by its input spans of length k, i.e., every state $p \in Q_0$ is assigned by words $x \in \Sigma^*$ satisfying either $|x| < k$ and $\delta(q_0, x) = p$, or $|x| = k$ and $\delta(q, x) = p$ for some $q \in Q_0$. Now, construct the sets $\alpha = \{x | x \in \Sigma^*, |x| \leq k$ and $\delta(q_0, x) \neq q_d\}$, $\beta = \{xa | x \in \Sigma^k, a \in \Sigma, x$ is assigned to some $q \in Q$ and $\delta(q, a) \neq q_d\}$, and $\gamma = \{x | x$ is assigned to some $q \in F\}$. We leave it to the reader to show that $L(M)$ is the $(k+1)$-LTSS language $\hat{U}_{k+1}(\alpha, \beta, \gamma)$.

Now, let L be the k-LTSS language $\hat{U}_k(\alpha, \beta, \gamma)$. We construct a finite state automaton $M' = (Q', \Sigma, \delta', q_0', F')$, where $Q' = \{[w] | wx \in \alpha,$ for some $x \in \Sigma^*\} \cup \{[w] | w \in \gamma\} \cup \{[w] | wa \in \beta,$ for some $a \in \Sigma\} \cup \{q_d\}$, $q_0' = [\epsilon]$, $F' = \{[w] \mid w \in \gamma\}$, and δ' is defined as follows. For every pair of states $[w_1], [w_2] \in Q'$, $\delta'([w_1], a) = [w_2]$ if $t_k(w_1 a) = w_2$, for some $a \in \Sigma$. Finally, make all transitions as yet undefined go to the dead sink state q_d. Notice that a state $[w] \in Q'$ has input span w and no two states have a common input span of length k. M' has finite input memory. We leave it to the reader to show that $x \in L$ iff M' accepts x. $\qquad \square$

Since the complement of an LTSS language is LTES, we have the following corollary.

Corollary 2. Let $M = (Q, \Sigma, \delta, q_0, F)$ be a deterministic finite automaton. $L(M)$ is LTES iff the complement $\overline{M} = (Q, \Sigma, \delta, q_0, Q - F)$ of M has finite input memory.

Now, we investigate deterministic finite automata which accept LTMS languages. Before we show the main theorem, we need the following lemmas.

Lemma 7. Let $M = (Q, \Sigma, \delta, q_0, F)$ be a deterministic finite automaton such that every SCC in the state transition graph of M has finite input memory. Let m be a component of the state transition graph and let S be the set of states in m. If there exists $p \in S$ such that $\delta(p, x^n)$ is in m, for some $x \in \Sigma^+$ and $n \geq |S|$, and, for every prefix w of x^n, $\delta(p, w)$ is in m, then $\delta(q, x) = \mathring{q}$, for some state q in m.

Proof. If $\delta(p, x^n)$ is in m and, for every prefix w of x^n, $\delta(p, w)$ is in m, it must be that $\delta(p, x^i) = \delta(p, x^j)$, for some $0 \leq i < j \leq n$. Let $\delta(p, x^i) = q$. We claim that $\delta(q, x) = q$. Suppose $\delta(q, x) = r \neq q$. Then $\delta(r, x^{j-i}) = r$ and $\delta(q, x^{j-i}) = q$. Clearly, q and r are in the same SCC, and, hence, the SCC has infinite input memory, a contradiction. $\square$

Lemma 8. Let $M = (Q, \Sigma, \delta, q_0, F)$ be a reduced deterministic automaton. Let $p, q \in Q - \{q_d\}$ be a pair of distinct states such that $L(M, p) \subset L(M, q)$. Then, for all $w \in \Sigma^*$, if $\delta(p, w) = r \neq \delta(q, w) = s$, then $L(M, r) \subset L(M, s)$.

Proof. Since M is a reduced automaton, $L(M, r) \neq L(M, s)$. Suppose $L(M, r) \supset L(M, s)$, and let $z \in L(M, r) - L(M, s)$. Then wz is in $L(M, p) - L(M, q)$, a contradiction. $\square$

In the following lemmas we shall use extensively the notation $C_w(xwy)$, first used in the proof of theorem 4, to denote the sequence of cells of a CA which get w on an input xwy.

Lemma 9. Let $M = (Q, \Sigma, \delta, q_0, F)$ be a reduced deterministic finite automaton which accepts an LTMS language. Let M' be an MCA which accepts $L(M)$ in k steps, for some $k \geq 0$. Let $p, q \in Q - \{q_d\}$ be such that $\delta(p, x) = p$ and $\delta(q, x) = q$, for some $x \in \Sigma^+$, and $L(M, p) \subset L(M, q)$. Then for any $w_1, w_2, v \in \Sigma^*$ such that $\delta(q_0, w_1) = p$ and $\delta(q_0, w_2) = q$, M' will have

(1) at least one cell in $C_{w_2 x^k}(w_2 x^{2k+1} v)$ entering the accepting state with no cells rejecting at the kth step, and

(2) all cells in $C_{w_1 x^k}(w_1 x^{2k+1} v)$ entering a neutral state at the kth step.

Proof. Since $L(M, p) \subset L(M, q)$, there exists $u \in \Sigma^*$ such that $\delta(p, u)$ is not in an accepting state, while $\delta(q, u)$ is. M' accepts $w_2 x^{2k+1} u$ and does not accept $w_1 x^{2k+1} u$. This implies that at the kth step M' will have at least one cell in $C_{w_2 x^k}(w_2 x^{2k+1} u)$ entering the accepting state and no cells rejecting. Thus, part (1) of the lemma is true. M' has no cells in $C_{w_1 x^k}(w_1 x^{2k+1} u)$ entering the accepting state at the kth step. Since $p \neq q_d$, there exists $u' \in \Sigma^*$ such that $\delta(p, u')$ is in an accepting state. M' accepts $w_1 x^{2k+1} u'$ and $w_2 x^{2k+1} u$, and does not accept $w_1 x^{2k+1} u$. For all $v \in \Sigma^*$, M' must have all cells in $C_{w_1 x^k}(wx^{2k+1} v)$ entering a neutral state at the kth step. Suppose not. If M' has a cell in $C_{w_1 x^k}(wx^{2k+1} v)$ entering the rejecting state at the kth step, then $w_1 x^{2k+1} u'$ cannot be accepted. If M' has a cell in $C_{w_1 x^k}(wx^{2k+1} v)$ entering the accepting state at the kth step with no rejecting state, then $w_1 x^{2k+1} u$ will be accepted. Part (2) of the lemma is true. $\square$

Lemma 10. Let $M = (Q, \Sigma, \delta, q_0, F)$ be a reduced deterministic automaton which has two distinct states $p, q \in Q - \{q_d\}$ such that $\delta(p, x) = p$ and $\delta(q, x) = q$, for some $x \in \Sigma^+$. If $L(M)$ is LTMS, then either $L(M, p) \supset L(M, q)$ or $L(M, p) \subset L(M, q)$. In particular, if p is an ancestor of q, then $L(M, p) \subset L(M, q)$.

Proof. Suppose that neither p nor q is an ancestor of the other and neither $L(M, p) \supset L(M, q)$ nor $L(M, p) \subset L(M, q)$. Let $w_1, w_2 \in \Sigma^*$ be such that $\delta(q_0, w_1) = p$, $\delta(q_0, w_2) = q$. Find $y, z \in \Sigma^*$ such that $\delta(p, y)$ is an accepting state, while $\delta(q, y)$ is not, and $\delta(p, z)$ is not an accepting state, while $\delta(q, z)$ is. Since M is reduced and neither $L(M, p) \subset L(M, q)$ nor $L(M, p) \supset L(M, q)$, there exist such words y and z. Suppose $L(M)$ is accepted by a k-MCA M', for some $k \geq 0$. Since $w_1 x^{2k+1} y$ is accepted by M', no cells in $C_{x^k y}(w_1 x^{2k+1} y)$ will enter the rejecting state in k steps. It follows that M' rejects $w_2 x^{2k+1} y$ by having either all cells entering a neutral state or at least one cell in $C_{w_2 x^k}(w_2 x^{2k+1} y)$ entering the rejecting state at the kth step. We consider the following two cases.

Case (1). When M' rejects $w_2 x^{2k+1} y$ by having at least one cell in $C_{w_2 x^k}(w_2 x^{2k+1} y)$ entering the rejecting state. Since $\delta(q, z)$ is in an accepting state, $w_2 x^{2k+1} z$ is in $L(M)$. However, M'

will reject $w_2 x^{2k+1} z$ because at least one cell in $C_{w_2 x^k}(w_2 x^{2k+1} z)$ will enter the rejecting state in k steps.

Case (2). When M' rejects $w_2 x^{2k+1} y$ by having all cells entering a neutral state at the kth step. M' must accept $w_2 x^{2k+1} z$ by having at least one cell in $C_{x^k z}(w_2 x^{2k+1} z)$ entering in the accepting state in k steps, with no cell in the rejecting state. Since M' accepts $w_1 x^{2k+1} y$, no cell in $C_{w_1 x^k}(w_1 x^{2k+1} y)$ will enter the rejecting state. It follows that M' accepts $w_1 x^{2k+1} z$ by having at least one cell in $C_{x^k z}(w_1 x^{2k+1} z)$ entering the accepting state in k steps. In either case we have $L(M) \neq L(M')$, a contradiction.

Now, suppose that p is an ancestor of q and it does not satisfy $L(M, p) \subset L(M, q)$. Find $w, y, z \in \Sigma^+$ such that $\delta(q_0, w) = p$, $\delta(p, y) = q$, $\delta(p, z)$ is an accepting state and $\delta(q, z)$ is not. Suppose that $L(M)$ is a k-LTMS language, for some $k \geq 0$. Since M is reduced and q is not q_d, there must be $u \in \Sigma^*$ such that $\delta(q, u)$ is an accepting state. M' accepts $wx^{2k+1} yx^{2k+1} u$, while $wx^{2k+1} yx^{2k+1} z$ is rejected, which implies that, for all $v \in \Sigma^*$, no cells in $C_{wx^{2k+1} yx^k}(wx^{2k+1} yx^{2k+1} v)$ enter the rejecting state in k steps. We consider two cases depending on whether or not any cells in $C_{wx^{2k+1} yx^k}(wx^{2k+1} yx^{2k+1} v)$ enter the accepting state in k steps. Suppose that no cells in $C_{wx^{2k+1} yx^k}(wx^{2k+1} yx^{2k+1} v)$ are in the accepting state after the kth step. Then, since $wx^{2k+1} z$ is accepted, at least one cell in $C_{x^k z}(wx^{2k+1} z)$ must enter the accepting state in k steps with no other cells entering the rejecting state, which implies that $wx^{2k+1} yx^{2k+1} z$ must be accepted, a contradiction. Now, suppose that at least one cell in $C_{wx^{2k+1} yx^k}(wx^{2k+1} yx^{2k+1} v)$ is in the accepting state after the kth step with no other cells rejecting. Since $wx^{2k+1} z$ is accepted by M', no cells in $C_{x^k z}(wx^{2k+1} z)$ enter a rejecting state in k steps, which implies that $wx^{2k+1} yx^{2k+1} z$ must be accepted. Again we are in a contradiction. $\square$

Lemma 11. Let $M = (Q, \Sigma, \delta, q_0, F)$ be a reduced deterministic finite automaton which accepts an LTMS language, and let $p, q \in Q - \{q_d\}$ be distinct states such that $\delta(p, x) = p$ and $\delta(q, x) = q$, for some $x \in \Sigma^+$, and $L(M, p) \subset L(M, q)$. If there exist $r, s \in Q - \{q_d\}$ such that

    (1) r is in the reachable component m_{q*} and s

is not, and

    (2) $\delta(r, y) = r$ and $\delta(s, y) = s$, for some $y \in \Sigma^+$, then $L(M, r) \supset L(M, s)$.

Proof. Let M' be an MCA which accepts $L(M)$ in k steps, for some $k \geq 0$. Find $w' \in \Sigma^*$ such that $\delta(q_0, w') = q$. Since $\delta(p, x) = p$ and $\delta(q, x) = q$ and $L(M, p) \subset L(M, q)$, by lemma 9 M' will have at least one cell in $C_{w' x^k}(w' x^{2k+1})$ entering an accepting state, with no cells rejecting, at the kth step. Let $z \in \Sigma^*$ such that $\delta(q, z) = r$. Then M' has at least one cell in $C_{w' x^{2k+1} zy^k}(w' x^{2k+1} zy^{2k+1})$ entering the accepting state at the kth step.

By lemma 10 either $L(M, r) \subset L(M, s)$ or $L(M, r) \supset L(M, s)$. Suppose $L(M, r) \subset L(M, s)$. By lemma 9, for any $w \in \Sigma^*$ such that $\delta(q_0, w) = r$, M' will have all cells in $C_{wy^k}(wy^{2k+1})$ entering a neutral state at the kth step. Let $w = w' x^{2k+1} z$. We showed above that M' has at least one cell in $C_{w' x^{2k+1} zy^k}(w' x^{2k+1} zy^{2k+1})$ entering the accepting state at the kth step. We are in a contradiction. $\square$

Lemma 12. Let $M = (Q, \Sigma, \delta, q_0, F)$ be a reduced deterministic automaton which has two distinct states $p, q \in Q - \{q_d\}$ such that $\delta(p, x) = p$, $\delta(q, x) = q$, for some $x \in \Sigma^+$, and $L(M, p) \subset L(M, q)$. If $L(M)$ is LTMS, then the reachable component m_{q*} has no pair of distinct states $r, s \in Q - \{q_d\}$ such that $\delta(r, y) = r$ and $\delta(s, y) = s$, for some $y \in \Sigma^+$.

Proof. Suppose that m_{q*} has such states r and s. Find $w_1, w_2, w_3, w_4 \in \Sigma^*$ such that $\delta(q_0, w_1) = p$, $\delta(q_0, w_2) = q$, $\delta(q, w_3) = r$ and $\delta(q, w_4) = s$. Now, suppose that $L(M)$ is an LTMS language accepted by an MCA M' in k steps, for some $k \geq 0$. By Lemma 9 M' will have at least one cell in $C_{w_2 x^k}(w_2 x^{2k+1})$ entering the accepting state at the kth step. Since $r \neq s$, there must be $u \in \Sigma^*$ such that either $\delta(r, u)$ is an accepting state and $\delta(s, u)$ is not, or vice versa. Without loss of generality suppose $\delta(r, u)$ is an accepting state and $\delta(s, u)$ is not. Consider words $w_2 x^{2k+1} w_3 y^{2k+1} u$ in $L(M)$ and $w_2 x^{2k+1} w_4 y^{2k+1} u$ not in $L(M)$. Clearly, M' must reject $w_2 x^{2k+1} w_4 y^{2k+1} u$ by having at least one cell in $C_{x^k w_4 y^k}(w_2 x^{2k+1} w_4 y^{2k+1} u)$ entering the rejecting state at the kth step, which implies that M' rejects $w_2 x^{2k+1} w_4 y^{2k+1} z$, for all $z \in \Sigma^*$. However, since s is not the rejecting sink state, there must be

$u' \in \Sigma^*$ such that $\delta(s, u')$ is an accepting state. Hence, $w_2 x^{2k+1} w_4 y^{2k+1} u'$ is in $L(M)$. It follows that $L(M) \neq L(M')$, a contradiction. $\square$

Lemma 13. Let $M = (Q, \Sigma, \delta, q_0, F)$ be a reduced deterministic finite automaton which accepts an LTMS language. Let M' be an MCA which accepts $L(M)$ in k steps, for some $k \geq 0$. Let $p, q \in Q - \{q_d\}$ be distinct states such that $\delta(p, x) = p$ and $\delta(q, x) = q$, for some $x \in \Sigma^+$. If there exist $r, s \in Q - \{q_d\}$ such that

(1) r is in the reachable component m_{q*} and s is not, and

(2) $\delta(r, y) = r$ and $\delta(s, y) = s$, for some $y \in \Sigma^+$, then, for every $z \in \Sigma^*$ such that $\delta(q, z) = r$, either $\delta(p, zy^{2k+1}) = s$ or $\delta(p, zy^{2k+1}) = r$.

Proof. Let M' be an MCA which accepts $L(M)$ in k steps. Find $w_1, w_2, w_3 \in \Sigma^*$ such that $\delta(q_0, w_1) = p$, $\delta(q_0, w_2) = q$, and $\delta(q_0, w_3) = s$. For any $z \in \Sigma^*$ such that $\delta(q, z) = r$, let $s' = \delta(p, zy^{2k+1})$ and suppose neither $s' = s$ nor $s' = r$. By lemma 10 either $L(M, p) \supset L(M, q)$ or $L(M, p) \subset L(M, q)$.

Case (1). When $L(M, p) \subset L(M, q)$. By lemma 11 $L(M, r) \supset L(M, s)$. By lemma 9 M' will have all cell in $C_{w_3 y^k}(w_3 y^{2k+1})$ entering a neutral state at the kth step. Hence, for all $v \in \Sigma^*$, M' will either accept or reject $w_3 y^{2k+1} v$ depending on the states of the cells in $C_{y^k v}(w_3 y^{2k+1} v)$ at the kth step. We consider the following two cases.

Case (1.1). When there exists $u \in \Sigma^*$ such that $\delta(q_0, w_1 x^{2k+1} z y^{2k+1} u)$ does not enter an accepting state and $\delta(q_0, w_2 x^{2k+1} z y^{2k+1} u)$ does. Since $\delta(p, x) = p$, $\delta(q, x) = q$ and $L(M, p) \subset L(M, q)$, by lemma 9 M' will have all cells in $C_{w_1 x^k}(w_1 x^{2k+1} v)$ entering a neutral state at the kth step, for all $v \in \Sigma^*$. Also all cells in $C_{x^k z y^k}(w_1 x^{2k+1} z y^{2k+1} u)$ enter a neutral state at the kth step. It follows that, for all $v \in \Sigma^*$, M' accepts $w_1 x^{2k+1} z y^{2k+1} v$ iff there exists at least one cell in $C_{y^k v}(w_1 x^{2k+1} z y^{2k+1} v)$ entering the accepting state at the kth step with no cells rejecting. Recall that, for all $v \in \Sigma^*$, M' accepts $w_3 y^{2k+1} v$ iff there exists at least one cell in $C_{y^k v}(w_3 y^{2k+1} v)$ entering the accepting state at the kth step with no cells rejecting. It follows that, for all $v \in \Sigma^*$ and $z \in \Sigma^*$ such that $\delta(q, z) = r$, $w_1 x^{2k+1} z y^{2k+1} v$ is in $L(M)$ iff $w_3 y^{2k+1} v$ is, which implies that s' and s are equivalent states. Since M is a reduced automaton, we have $s = s'$.

Case (1.2). When there is no $u \in \Sigma^*$ such that $\delta(q_0, w_1 x^{2k+1} z y^{2k+1} u)$ does not enter an accepting state and $\delta(q_0, w_1 x^{2k+1} z y^{2k+1} u)$ does. Since $L(M, p) \subset L(M, q)$, this case implies that for all $u \in \Sigma^*$, $w_1 x^{2k+1} z y^{2k+1} u$ is in $L(M)$ iff $w_2 x^{2k+1} z y^{2k+1} u$ is. It follows that s' and r are equivalent. Since M is a reduced automaton, we have $s' = r$.

Case (2). When $L(M, p) \supset L(M, q)$. By lemma 9 M' will have at least one cell in $C_{w_1 x^k}(w_1 x^{2k+1})$ entering the accepting state at the kth step with no cells rejecting. Suppose $s' \neq r$. By lemma 8 we have $L(M, s') \supset L(M, r)$. Find $u \in \Sigma^*$ such that $\delta(s', u)$ enters an accepting state and $\delta(r, u)$ does not. Then $w_1 x^{2k+1} z y^{2k+1} u$ is in $L(M)$ and $w_2 x^{2k+1} z y^{2k+1} u$ is not. It follows that M' will have all cells in $C_{x^k z y^k}(w_1 x^{2k+1} z y^{2k+1} v)$ entering a neutral state at the kth step, for all $v \in \Sigma^*$. It follows that, for all $v \in \Sigma^*$, $w_1 x^{2k+1} z y^{2k+1} v$ is in $L(M)$ iff M' has no cells in $C_{y^k v}(w_1 x^{2k+1} z y^{2k+1} v)$ entering the rejecting state at the kth step. Since $r \neq q_d$, it must be $s' \neq q_d$. By lemma 10 either $L(M, s) \supset L(M, r)$ or $L(M, s) \subset L(M, r)$.

Case (2.1). When $L(M, s) \supset L(M, r)$. By lemma 9 M' will have at least one cell in $C_{w_3 y^k}(w_3 y^{2k+1})$ entering the accepting state with no cells rejecting. It follows that, for all $v \in \Sigma^*$, $w_3 y^{2k+1} v$ is in $L(M)$ iff M' has no cells in $C_{y^k v}(w_3 y^{2k+1} v)$ entering the rejecting state at the kth step. We know that $w_1 x^{2k+1} z y^{2k+1} v$ is in $L(M)$ iff M' has no cells in $C_{y^k v}(w_1 x^{2k+1} z y^{2k+1} v)$ entering the rejecting state at the kth step. Hence, for all $v \in \Sigma^*$, $w_1 x^{2k+1} z y^{2k+1} v$ is in $L(M)$ iff $w_3 y^{2k+1} v$ is in $L(M)$. It follows that s and s' are equivalent. Since M is a reduced automaton, we have $s = s'$.

Case (2.2). When $L(M, s) \subset L(M, r)$. By the same argument for case (2.1), we can show that, for all $v \in \Sigma^*$, no cells in $C_{y^k v}(w_2 x^{2k+1} z y^{2k+1} v)$ enter the rejecting state at the kth step. It follows that, for all $v \in \Sigma^*$, $w_2 x^{2k+1} z y^{2k+1} v$ is in $L(M)$ iff $w_1 x^{2k+1} z y^{2k+1} v$ is in $L(M)$, which implies $s' = r$. $\square$

Now, we present the main theorem.

Theorem 7. Let $M = (Q, \Sigma, \delta, q_0, F)$ be a reduced finite deterministic automaton. $L(M)$ is LTMS iff M satisfies the following conditions.

(1) M is locally testable.

(2) Let $p, q \in Q - \{q_d\}$ any distinct states such that $\delta(p, x) = p$ and $\delta(q, x) = q$, for some $x \in \Sigma^+$.

(a) Either $L(M, p) \supset L(M, q)$ or $L(M, p) \subset L(M, q)$ if neither p nor q is an ancestor of the other. $L(M, p) \subset L(M, q)$ if p is an ancestor of q.

(b) Suppose $L(M, p) \subset L(M, q)$. If there exist $r, s \in Q - \{q_d\}$ such that r is in the reachable component m_{q*} and s is not, and $\delta(r, y) = r$ and $\delta(s, y) = s$, for some $y \in \Sigma^+$, then $L(M, r) \supset L(M, s)$.

(c) If $L(M, p) \subset L(M, q)$, then the reachable component m_{q*} has no pair of distinct states $r, s \in Q - \{q_d\}$ such that $\delta(r, y) = r$ and $\delta(s, y) = s$, for some $y \in \Sigma^+$.

(d) If there exist a pair of distinct states $r, s \in Q - \{q_d\}$ such that r is in the reachable component m_{q*} and s is not, and $\delta(r, y) = r$ and $\delta(s, y) = s$, for some $y \in \Sigma^+$, then, for every $z \in \Sigma^*$ such that $\delta(q, z) = r$, either $\delta(p, zy^n) = s$ or $\delta(p, zy^n) = r$, for some constant $n \geq 0$.

Proof (of necessity). By ref. [9], the class of locally testable languages is the Boolean closure of the class of LTSS languages. Thus by lemmas 1 and 3, every LTMS language is locally testable. Hence, part (1) holds. Parts (2a), (2b) and (2c) follow from lemmas 10, 11 and 12, respectively. Let M' be an MCA which accepts $L(M)$ in k steps, for some $k \geq 0$, and let $n = 2k + 1$. Then by lemma 13, part (2d) is true.

For the proof of sufficiency of the theorem, we will show that if M satisfies all the conditions of the theorem, then we can effectively construct an LTES automaton M_1 and an LTSS automaton M_2 such that $L(M) = L(M_1) \cap L(M_2)$. By lemma 3, $L(M)$ is LTMS. Before we show an algorithm which constructs M_1 and M_2 and prove that $L(M) = L(M_1) \cap L(M_2)$, we introduce the following lemmas.

Lemma 14. Let $M = (Q, \Sigma, \delta, q_0, F)$ be a deterministic automaton which is reduced. If M satisfies all the conditions of theorem 7, then it does not have distinct states $p, q, r \in Q - \{q_d\}$ such that $\delta(p, x) = p$, $\delta(q, x) = q$ and $\delta(r, x) = r$, for some $x \in \Sigma^+$.

Proof. Suppose M has such states p, q, and r. No two states of p, q and r can be in the same SCC. Otherwise, it violates condition (1) of the theorem. By condition (2c) of the theorem, p, q

and r cannot appear on a single path of the state transition graph. Let p, q, q and r, respectively, be the states p, q, r and s of condition (2d) of the theorem. Then either q or r must be a descendant of p by condition (2d). Suppose r is a descendant of p. Then by condition (2a) of the theorem we have $L(M, p) \subset L(M, r)$. By condition (2a) of the theorem, either $L(M, p) \supset L(M, q)$ or $L(M, p) \subset L(M, q)$. If $L(M, p) \supset L(M, q)$, then the reachable component m_{p*} has two states p and r such that $\delta(p, x) = p$ and $\delta(r, x) = r$, which violates condition (2c) of the theorem. Suppose $L(M, p) \subset L(M, q)$. Let q, r, r and p be, respectively, the states p, q, r and s of lemma 13. Then either p or r must be a descendant of q. If p is a descendant of q, then we cannot have $L(M, p) \subset L(M, q)$ by condition (2a) of the theorem. If r is a descendant of q, then it violates the condition (2c) of the theorem. $\square$

Definition 13 (product graph). Let G be a state transition graph of a deterministic automaton $M = (Q, \Sigma, \delta, q_0, F)$, and let $p, q \in Q$ be a pair of distinct states. The product graph with respect to the ordered pair (p, q), is a state transition graph $G_{p \times q} = (Q', \Sigma, \delta', F')$, where $Q' = \{[r, s] \mid r, s \in Q, r \neq s, \delta(p, w) = r \text{ and } \delta(q, w) = s, \text{ for some } w \in \Sigma^*\} \cup \{q_e\}$, $F' = \{[r, s] \mid [r, s] \in Q', r \in F\} \cup \{q_e\}$, and δ' is de fined as follows: For $[r, s] \in Q'$ and $a \in \Sigma$, let $\delta(r, a) = t$, $\delta(s, a) = t'$. Then

(a) $\delta'([r, s], a) = [t, t']$, if $t \neq t'$,

(b) $\delta'([r, s], a) = q_e$, if $t = t'$.

Notice that the product graph has no start state, which is the only difference from the state transition graph of an automaton.

Lemma 15. Let $M = (Q, \Sigma, \delta, q_0, F)$ be a reduced deterministic automaton which satisfies the conditions of theorem 7. Let $p, q \in Q - \{q_d\}$ be any pair of distinct states such that $\delta(p, x) = p$ and $\delta(q, x) = q$, for some $x \in \Sigma^+$. Then the product graph $G_{p \times q} = (Q', \Sigma, \delta', F')$ has no pair of states $[p_1, q_1], [p_2, q_2] \in Q'$ such that $\delta'([p_1, q_1], x) = [p_1, q_1]$ and $\delta'([p_2, q_2], x) = [p_2, q_2]$, for any $x \in \Sigma^+$.

Proof. By condition (2a) of theorem 7, either $L(M, p) \subset L(M, q)$ or $L(M, p) \supset L(M, q)$. Without loss of generality suppose that $L(M, p) \subset L(M, q)$. Since $[p_1, q_1] \neq [p_2, q_2]$, either $p_1 \neq p_2$ or

$q_1 \neq q_2$. Notice that by the definition of the product graph, p_1 and p_2 are descendants of p, and q_1 and q_2 are descendant of q. Since $L(M,p) \subset L(M,q)$, neither q_1 nor q_2 is q_d. Otherwise, either $p_1 = q_d$ or $p_2 = q_d$, implying that either $[p_1,q_1] = [q_d,q_d]$ or $[p_2,q_2] = [q_d,q_d]$. Either $[p_1,q_1]$ or $[p_2,q_2]$ should have been defined as q_e by the definition of the product graph . If $q_1 \neq q_2$, then q has two distinct descendant states q_1 and q_2 such that $\delta(q_1,x) = q_1$ and $\delta(q_2,x) = q_2$, which violates condition (2c) of theorem 7.

Suppose $q_1 = q_2$ and $p_1 \neq p_2$. If both p_1 and p_2 are not q_d, then we have three states p_1, q_1 and p_2 such that $\delta(p_1,x) = p_1$, $\delta(q_1,x) = q_1$ and $\delta(p_2,x) = p_2$, which violates lemma 14. Either p_1 or p_2 is q_d and the other is not q_d. Without loss of generality suppose $p_1 = q_d$ and $p_2 \neq q_d$. State p_2 is not a descendant of q. Otherwise, q and p_2 belong to the reaching component m_{q*}, which violates condition (2c) of theorem 7. Find $w_1, w_2 \in \Sigma^*$ such that $\delta'([p,q],w_1) = [p_1,q_1]$ and $\delta'([p,q],w_2) = [p_2,q_2]$. Let p, q, q_1 and p_2 be, respectively, the states p, q, r and s of condition (2c) of theorem 7. We have $\delta(q,w_1x^n) = \delta(q,w_2x^n) = q_1 = q_2$, for all $n \geq 0$, while $\delta(p,w_1x^n) = p_1 = q_d$ and $\delta(p,w_2x^n) = p_2$. It violates the condition. $\square$

Now, for an automaton M which satisfies all the conditions of the theorem, let M_1 and M_2 be the automata constructed by algorithms 1 and 2 in figs. 3 and 4, respectively. Lemmas 16, 17 and 18 below show that M_1 is LTES, M_2 is LTSS and $L(M) = L(M_1) \cap L(M_2)$.

Lemma 16. Let $M = (Q, \Sigma, \delta, q_0, F)$ be a reduced deterministic automaton which satisfies the conditions of theorem 7. The automaton M_1 which is constructed from M by algorithm 1 of fig. 3 is LTES.

Proof. Let $M_1 = (Q_1, \Sigma, \delta_1, q_{01}, F_1)$. We assume that the states in Q_1 have the same labels assigned by the algorithm, i.e., $Q_1 = Q_{11} \cup Q_{12} \cup \{q_e\}$, where $Q_{11} \subseteq Q$ and $Q_{12} \subseteq Q \times Q$. Notice that the states in Q_{12} are from product graphs constructed by the algorithm. For the convenience of identification, we will use indexed state labels, like $[p_i,q_i]$, for a state in Q_{12}.

By corollary 2 M_1 is LTES if its complement is LTSS. Notice that by the construction M_1 has no q_d. Hence, to prove M_1 is LTES, it is enough

to show that M_1 has no two nonsink states which have identical loop transitions. We will show that there exists no $x \in \Sigma^+$ and a pair of distinct states from among $r, s \in Q_{11}$ and $[p_1,q_1],[p_2,q_2] \in Q_{12}$ satisfying any of the following conditions.

(a) $\delta_1(r,x) = r$ and $\delta_1(s,x) = s$.

(b) $\delta_1(r,x) = r$ and $\delta_1([p_1,q_1],x) = [p_1,q_1]$.

(c) $\delta_1([p_1,q_1],x) = [p_1,q_1]$ and $\delta_1([p_2,q_2],x) = [p_2,q_2]$.

Suppose M_1 has two distinct states r and s which satisfy (a). They belong to an SCC. By condition (1) of theorem 7, they cannot be in the same SCC. Let m_i and m_j be the SCCs which has r and s, respectively. Then step (4) of algorithm 1 must have deleted r and s when it deletes the reachable components m_{i*} and m_{j*}. Hence, M_1 cannot have such states r and s. Suppose M_1 has r and $[p_1,q_1]$ which satisfy (b) above. Since $[p_1,q_1]$ is from a product graph, it must be $p_1 \neq q_1$, $r \neq p_1$ and $r \neq q_1$. We have $\delta(r,x) = r$, $\delta(p_1,x) = p_1$ and $\delta(q_1,x) = q_1$, which contradicts lemma 14. Finally, suppose M_1 has $[p_1,q_1],[p_2,q_2] \in Q_{12}$ which satisfy (c) above. Since $[p_1,q_1] \neq [p_2,q_2]$, either $p_1 \neq p_2$ or $q_1 \neq q_2$. Since both $[p_1,q_1]$ and $[p_2,q_2]$ are from a product graph, it must be $p_1 \neq q_1$ and $p_2 \neq q_2$. We claim that either $p_1 \neq q_2$ or $q_1 \neq p_2$. Suppose not, i.e., $p_1 = q_2$ and $q_1 = p_2$. Then we have $[p_1,q_1] = [p_1,p_2]$ and $[p_2,q_2] = [p_2,p_1]$. By Lemma 8 $L(M,p_1) \subset L(M,p_2)$ because $[p_1,p_2]$ is a state from a product graph, say $G_{p \times q}$, such that $L(M,p) \subset L(M,q)$ and $p_1 = \delta(p,w)$ and $p_2 = \delta(q,w)$, for some $w \in \Sigma^*$. By the same argument, we have $L(M,p_2) \subset L(M,p_1)$. This is a contradiction.

We have that either $p_1 \neq p_2$ or $q_1 \neq q_2$, and that either $p_1 \neq q_2$ or $q_1 \neq p_2$. Combining these, there are four possible cases to be considered. In each case, there are at least three distinct states from among p_1, p_2, q_1, and q_2. We know that $\delta(p_1,x) = p_1$, $\delta(p_2,x) = p_2$, $\delta(q_1,x) = q_1$, and $\delta(q_2,x) = q_2$. This violates lemma 14. $\square$

Lemma 17. Let $M = (Q, \Sigma, \delta, q_0, F)$ be a reduced deterministic automaton which satisfies the conditions of theorem 7. The automaton M_2 which is constructed from M by algorithm 2 of fig. 3 is LTSS.

Proof. Let $M_2 = (Q_2, \Sigma, \delta_2, q_{02}, F_2)$. Notice that going top-down algorithm 1 finds an SCC m_i

Procedure *Construct-LTES;*
$//G$ is the state transition graph of M.$//$
$G' := G;$
Repeat
    Going top down on G, for each SCC m_i which has no mark and has pairwise
    infinite input memory with respect to an SCC m_j **do**
        if $L(M,p) \subset L(M,q)$, for a pair of states p and q, resp ectively, in m_i and m_j,
        such that $\delta(p,x) = p$ and $\delta(q,x) = q$, for some $x \in \Sigma^+$, **then**
        begin
            (1) Identify the reachable components m_{i*} and m_{j*};
            Let Q_{i*} and Q_{j*} be the sets of states, respectively, in m_{i*} and m_{j*};
            (2) Construct the sets of edges:
$$S_i = \{(r,a,s)|a \in \Sigma, r \in Q - (Q_{i*} \cup Q_{j*}), s \in Q_{i*} - Q_{j*}, \delta(r,a) = s\}.$$
$$S_j = \{(r,a,s)|a \in \Sigma, r \in Q - (Q_{i*} \cup Q_{j*}), s \in Q_{j*}, \delta(r,a) = s\}.$$
            (3) Construct the product graph $G_{p \times q}$;
            (4) Mark m_{i*} and m_{j*} in G and delete m_{i*} and m_{j*} from G';
            (5) Attach-and-merge the product graph $G_{p \times q}$ to G' as follows:
                $(//S_{ij}$ will be used in algorithm 2.$//)$
                $S_{ij} := \emptyset$;
                For each edge (r,a,s) in S_i **do**
                begin
                    Find a node $[s,t]$ in $G_{p \times q}$, for some $t \in Q_{j*}$;
                    Put edge $(r,a,[s,t])$ on G';
                    $S_{ij} := S_{ij} \cup (r,a,[s,t])$;
                end;
                Merge all states in G' which have the same label into a single state;
            (6) **For** each edge $(r,a,s) \in S_j$, put edge (r,a,q_e) on G';
        end
until (there is no SCC in G which has no mark and has infinite input memory
        with respect to some other SCC)
Change q_d in G', if any, into q_e;
Let q_0 or $[q_0,p]$, for some $p \in Q$, whichever exists in G', be the start state;
Let M_1 be the automaton having G' as the state transition graph;

Fig. 3. Algorithm 1 (constructing LTES M_1).

Procedure *Construct-LTSS;*
$//G$ is the state transition graph of M.$//$
$G' := G;$
for each set S_{ij} constructed by step (5) of algorithm 1 **do**
begin
    Delete $m_{i*} - m_{j*}$ from G';
    for each edge $(r,a,[s,t])$ in S_{ij}, put edge (r,a,t) on G';
end;
If $[q_0,p]$ is the start state of M_1, for some $p \in Q$, **then**
        let p be the start state of G' **else** let q_0 be the start state of G';
Let M_2 be the automaton having G' as the state transition graph;

Fig. 4. Algorithm 2 (constructing LTSS M_2).

which has pairwise infinite input memory with respect to an SCC m_j such that $L(M, p) \subset L(M, q)$, for a pair of states p and q in m_i and m_j, respectively. Constructing M_2, algorithm 2 deletes $m_{i*} - m_{j*}$ from G' and diverts all the edges coming into $m_{i*} - m_{j*}$ (i.e. S_{ij}) onto m_{j*}. Let $Q_2 = Q_{21} \cup Q_{22}$, where Q_{21} is the set of states which do not belong to any such reachable component m_{j*} and $Q_{22} = Q_2 - Q_{21}$.

Now, suppose that M_2 is not LTSS by having two distinct states $r, s \in Q_2 - \{q_d\}$ such that $\delta_2(r, y) = r$ and $\delta_2(s, y) = s$, for some $y \in \Sigma^+$. Both r and s cannot be in Q_{21}. Otherwise, by condition (2a) of theorem 7 either $L(M, r) \supset L(M, s)$ or $L(M, r) \subset L(M, s)$, and algorithm 2 must have deleted either s or r when it deleted $m_{i*} - m_{j*}$. Suppose r is in Q_{21} and s is in Q_{22}. By condition (2c) of theorem 7 $L(M, r) \subset L(M, s)$. The algorithm must have deleted r constructing M_2. Finally, suppose r and s are in Q_{22}. By condition (2c) of theorem 7, r and s cannot belong to the same reachable component m_{j*} described above. Suppose r is in m_{j*} and s is in the reachable component m_{k*} from an SCC m_k which has pairwise infinite input memory with respect to an SCC m_l such that $L(M, t) \subset L(M, t')$, for a pair of states t and t' in m_l and m_k, respectively. By condition (2c), we have $L(M, r) \supset L(M, s)$ and $L(M, r) \subset L(M, s)$, which implies $r = s$, a contradiction. $\square$

Lemma 18. Let $M = (Q, \Sigma, \delta, q_0, F)$ be a deterministic finite state automaton which satisfies the conditions of theorem 7. Let M_1 and M_2 be the automata which are constructed by algorithms 1 and 2, respectively. Then $L(M) = L(M_1) \cap L(M_2)$.

Proof. Let $M_1 = (Q_1, \Sigma, \delta_1, q_1, F_1)$ and $M_2 = (Q_2, \Sigma, \delta_2, q_2, F_2)$. For the proof we need the following observations:

(a) $Q_2 \subseteq Q$ and $Q_1 = Q_{11} \cup Q_{12} \cup \{q_e\}$, where $Q_{11} \subseteq Q$ and $Q_{12} \subseteq Q \times Q$. Notice that M_1 does not have q_d.

(b) If $[q_0, p]$ is the start state of M_1, then p is the start state of M_2. Otherwise, q_0 is the start state of both M_1 and M_2.

(c) For all $p, q \in Q_{11}$ and $a \in \Sigma$, if $\delta_1(p, a) = q$ in M_1, then $\delta_2(p, a) = q$ in M_2.

(d) If M_1 has $\delta_1(p, a) = q_e$, for some $p \in Q_{11}$ and $a \in \Sigma$, then M and M_2 have $\delta(p, a) = q$ and $\delta_2(p, a) = q$, respectively, for some $q \in Q$.

(e) For any $r, s, t \in Q$ and $a \in \Sigma$, if M_1 has $\delta_1(r, a) = [s, t]$, then M_2 has $\delta_2(r, a) = t$.

(f) If M_1 has $\delta_1([q, r], a) = [s, t]$, then M_2 has $\delta_2(r, a) = t$.

(g) If M_1 has $\delta_1([r, s], a) = q_e$, then M has $\delta(r, a) = \delta(s, a) = t$, for some $t \in Q$, and M_2 has $\delta_2(s, a) = t$.

Observation (a) is obvious because M_2 has no product graph attached, and M_1 may have components from both M and a product graph attached by step (5) of algorithm 1. If the start state q_0 of M is in an SCC m_i which has pairwise infinite input memory with respect to an SCC m_j, then algorithms 1 and 2 define $[q_0, p]$ and p to be the start states of M_1 and M_2, respectively, for some $p \in Q$. Otherwise, q_0 will be the start state of the automata. Hence, part (b) of the observation is true. Since the algorithms do nothing with the states which are not involved with the product graphs constructed by step (5) of algorithms 1, such states and the transitions between them will appear in both M_1 and M_2. Hence, observation (c) holds. Algorithm 1, by step (5), attaches the product graph $G_{p \times q}$ by adding an edge $(r, a, [s, t])$, for each edge (r, a, s) in S_i. Notice that S_i is the set of edges which link a state in $Q - (Q_{i*} \cup Q_{j*})$ to a state $Q_{i*} - Q_{j*}$, and S_j is the set of edges which link a state $Q - (Q_{i*} \cup Q_{j*})$ to a state in Q_{j*}. Step (6) of algorithm 1, constructing M_1, puts edge (r, a, q_e), for every edge $(r, a, s) \in S_j$, and q_d, if any, is changed to q_e. Clearly, constructing M_2 the algorithms do nothing with the edges in S_j. All the transitions corresponding to the edges in S_j still exist in M_2. Hence, part (d) of the observations follows. Observation (e) follows from steps (5) of algorithm 1 and step (2) of algorithms 2. Obviously, if M_1 has $\delta_1([q, r], a) = [s, t]$, then the transition is from a product graph constructed by step (5) of algorithm 1, which implies that $\delta_2(r, a) = t$ in M_2. Hence, observation (f) follows. Finally, for part (g) of the observations, notice that $[r, s]$ is a state from a product graph. By definition 13, $[r, s]$ takes transition to q_e on an input $a \in \Sigma$ if $\delta(r, a) = \delta(s, a)$.

Now, we show that $L(M) = L(M_1) \cap L(M_2)$. Construct $M' = (Q_1 \times Q_2, \Sigma, \delta', [q_1, q_2], F')$ such that, for $[p, q] \in Q_1 \times Q_2$ and $a \in \Sigma$, $\delta'([p, q], a) =$

$[\delta_1(p, a), \delta_2(q, a)]$, and $F' = \{[r, s] | r \in F_1$ and $s \in F_2\}$. We prove by induction that, for all $x \in \Sigma^*$, $x \in L(M')$ iff $x \in L(M)$. For convenience we assume that M_1 and M_2 have the state labels that were assigned when they were constructed from M by algorithms 1 and 2, respectively. Recall that by definition 13 a node $[r, s]$ of a product graph is an accepting state iff r is an accepting state in M. Notice that every state $[r, s] \in Q_{12}$ is derived from a product graph $G_{p \times q}$ with some $p, q \in Q - \{q_d\}$ such that $\delta(p, x) = p$, $\delta(q, x) = q$, for some $x \in \Sigma^+$, and $L(M, p) \subset L(M, q)$. Hence, $\delta(p, w) = r$ and $\delta(q, w) = s$, for some $w \in \Sigma^*$, and by lemma 10 $L(M, r) \subset L(M, s)$. If r is an accepting state, then so is s. Hence the start state of M', which is either $[q_0, q_0]$ or $[[q_0, p], p]$, is an accepting state iff q_0 is. It follows that M' accepts the null string ϵ iff M accepts. Suppose that, for all words w of length less than n, M accepts w iff M' does. Let $\delta(q_0, w) = p$ in M. Then, with input w, M' will enter one of the following states: $[p, p]$, $[q_e, p]$, $[[p, r], r]$, for some $r \in Q$.

Now, suppose that, for $a \in \Sigma$, M takes transition $\delta(p, a) = q$. M' will take one of the following transitions, for some $r, s \in Q$.

   (a) $\delta'([p, p], a) = [q, q]$.
   (b) $\delta'([p, p], a) = [q_e, q]$.
   (c) $\delta'([p, p], a) = [[q, r], r]$.
   (d) $\delta'([p, r], r], a) = [[q, s], s]$.
   (e) $\delta'([q_e, p], a) = [q_e, q]$.
   (f) $\delta'([[p, r], r], a) = [q_e, q]$.

Transitions (a), (b), (c), (d) and (f) follow from the observations (3), (4), (5), (6) and (7), respectively. Transition (e) is a consequence of transition (b) or (f). Clearly, q is an accepting state iff $[q, q]$, $[q_e, q]$, $[[q, r], r]$ and $[[q, s], s]$ are.   □

Now, we are ready to prove the sufficiency of theorem 7.

Proof (of sufficiency). Lemmas 16, 17 and 18 show that if M satisfies all the conditions of the theorem, then we can construct an LTES automaton M_1 and an LTSS automaton M_2 such that $L(M) = L(M_1) \cap L(M_2)$.   □

5. Conclusion

We have identified three subclasses of finite state automata that characterize the constant-time CA languages introduced in this paper. If we change the definition of a CA so that its cells can have a transition from the rejecting state to a non-rejecting state as in ref. [8], then, under this new definition, the ACA and MCA languages would be closed under union. But still the ACA languages would not be closed under complementation, and the MCA languages would not be closed under intersection or complementation. The class of locally testable languages is the Boolean closure of each of the LTSS, LTES, and LTMS classes of languages [9]. Interesting questions concerning the constant-time CA languages are:

(1) Given a deterministic finite automaton M, is it possible to decide whether $L(M)$ is LTMS in polynomial time?

(2) Is there any natural CA model that accepts locally testable languages in constant time?

We believe that (1) can be answered affirmatively by using an algorithm similar to that found in ref. [14]. For (2), it seems to be impossible unless we are allowed to take a complementary interpretation on an outcome of a CA computation.

Acknowledgements

The authors are grateful to Professor Robert McNaughton for his many valuable comments and suggestions.

References

[1] A.R. Smith, Real-time language recognition by one-dimensional cellular automata, J. Computer System Sci. 6 (1972) 23–253.

[2] A. Rosenfeld, Picture Languages (Academic Press, New York, 1979).

[3] M.J.B. Duff and T.J. Fountain, Cellular Logic Image Processing (Academic Press, New York, 1986).

[4] S. Wolfram, Theory and Applications of Cellular Automata (World Scientific, Singapore, 1986).

[5] J. Demongeot, E. Goles and M. Tchuente, Dynamical Systems and Cellular Automata (Academic Press, New York, 1985).

[6] M.G. Nordahl, Formal languages and finite cellular automata, Complex Systems 3 (1989) 63–78.

[7] O.H. Ibarra and S.M. Kim, Fast parallel language recognition by cellular automata, Theor. Computer Sci. 41 (1985) 231–246.

[8] R. Sommerhalder and S.C. Westrhenen, Parallel language recognition in constant time by cellular automata, Acta Informatica 19 (1983) 397–407.

[9] R. McNaughton and S. Papert, Counter-free Automata (MIT Press, Cambridge, MA, 1971).

[10] J. Brzozowski and I. Simon, Characterizations of locally testable events, Discrete Math. 4 (1973) 243–271.

[11] O.H. Ibarra, S.M. Kim and M. Palis, Some results concerning linear iterative (systolic) arrays, J. Parallel Distributed Computing 2 (1985) 182–218.

[12] J.E. Hopcroft and J.D. Ullman, Introduction to Automata Theory, Languages, and Computation (Addison–Wesley, Reading, MA, 1979).

[13] R. Martin, Studies in Feedback-Shift-Register Synthesis of Sequential Machines (MIT Press, Reading, MA, 1968).

[14] S.M. Kim, R. McNaughton and R. McCloskey, A polynomial time algorithm for the local testability problem of deterministic finite automata, Lect. Notes Computer Sci. 382 (1989) 420–436.

Physica D 45 (1990) 420–427
North-Holland

CONSTRUCTIVE CHAOS BY CELLULAR AUTOMATA AND POSSIBLE SOURCES OF AN ARROW OF TIME

Karl SVOZIL

Institute for Theoretical Physics, Technical University Vienna, Karlsplatz 13, A-1040 Vienna, Austria

Received 13 September 1989
Revised manuscript received 20 March 1990

Present notions of chaos and their characterization by complexity measures are critically reviewed. The possible causes for the induction of an arrow of time are discussed: First, the *intrinsic undecidability* for the determination of complexity and the associated entropy measures may result in a time asymmetry even for finite reversible evolution laws. Second, the employment of transfinite resources (such as the infinite time limit) may result in an uncomputable final state from reversible evolution laws and computable initial values. Third, the existence of recursively enumerable reversible cellular automata rules whose inverse are not recursively enumerable is conjectured.

1. Introduction

Since CA are universal computational devices as well as excellent representations for specific physical systems [1], they are the models of choice to study the *computational aspect of physical system performance*[#1]. In particular, the characterization of physical processes by the associated complexity as well as entropy measures and a study of their interrelation, as well as an investigation of undecidability constraints, may yield new insights into longstanding issues, such as the possible "induction" of an arrow of time by reversible physical evolution laws.

If the world is a machine, then which type of machine is it? – By inventing mathematical models for physical applications one has to make implicit or explicit assumptions about the underlying set-theoretical as well as functional and computational models. For instance, by allowing continua in classical physics as well as by postulating randomness in quantum theory, those models require *transfinite resources*. In this view, both classical

continuum physics and quantum mechanics correspond to *oracle* machine models. Whether transfinite resources can be legitimately applied is a longstanding concern in the foundations of mathematics [5]. For physics one may hold the pragmatically oriented opinion that the question of whether to employ transfinite methods is relevant only insofar as they can be falsified. However, tests of transfinite means can hardly be operationalized. Therefore, the motivation to search for appropriate mathematical models for physics which, such as finite-size CA, do *not* require transfinite resources, seems quite justified.

The statistical mechanics properties of CA have been discussed earlier [6] in terms of computational complexity measures, i.e., the amount of time and storage space necessary to perform a computation. Here a more comprehensive approach to characterize CA evolution is pursued, which also deals with algorithmic complexity measures, i.e., with the minimal program length necessary to perform a computation. (In the context of *symbolic dynamics* [7], in particular of CA, the terms *evolution* and *computation* are equivalent.) First a short review of complexity measures and the associated definitions of randomness will be given. Then these techniques will be applied to a characterization of the evolution of determinis-

---

[#1] In a way the study of computational aspects of physical systems is complementary to the perception of a *computation* as a physical processes, such as *reversible computation* [2] and the study of the minimal dissipation [3,4] associated with particular tasks.

tic CA. Finally the induction of an arrow of time *via* "heuristic" complexity measures or in certain processes and/or limits is considered.

2. Complexity measures and randomness

In the following, *chaos* will be understood as the random (in a weaker form: undecidable [8]) evolution of a CA. Evidently it is of great importance which formal definition for the term "randomness" is envisaged. Here this term shall be based upon complexity measures [9].

Generally speaking, complexity is a measure of the resources necessary to perform a computation. These resources may be grouped into the following two categories. (i) Dynamic or computational complexity measures characterize the minimal amount of time (and storage capacity), whereas (ii) static or algorithmic complexity measures specify the minimal program size (and loop depth) necessary to perform a computational task. The associated definitions of randomness are based upon intuitive notions of incompressibility – either in time resources, such that no computational "shortcut" exists [6], or in program size, such that no shorter description interpretable as generating law [9] may reproduce a random timeflow. The formal definitions will be given next. They rest upon the representation of CA states in a symbolic string $x = x_1 x_2 x_3 \ldots$ [7]. For simplicity it is assumed that x is a binary sequence (i.e., $x_i \in [0, 1]$).

The *static complexity* $H(x)$ of a string x is defined to be the length of the shortest program p which runs on a computer C and generates the output x, i.e.,

$$H(x) = \min_{C(p)=x} \text{length}(p).$$

If no program makes a computer output x, then $H(x) = \infty$.

A sequence x is *absolutely Chaitin random* (ACR) [9] if the static complexity of the initial segment $x(n) = x_1 \ldots x_n$ of length n does not drop arbitrarily far below n, i.e.,

$$\liminf_{n \to \infty} [H(x(n)) - n] > -\infty.$$

(Heuristically speaking, there is no "generating law" capable of "compressing" any such sequence. A description of an ACR sequence amounts to mere enumeration.)

It has been proven [9] that an ACR sequence *passes all statistical tests* of randomness, such as frequency tests and the like. Therefore, ACR is equivalent to previous notions of randomness proposed by Martin-Löf, Solovay and others, based upon statistical criteria.

For physical applications, *normalized* ACR, henceforth called CR randomness, is very important [7]. An infinite sequence x is CR random, if

$$K(x) = \liminf_{n \to \infty} [H(x(n))/n] > 0.$$

The *normalized dynamical (computational) complexity* $K_{\mathrm{D}}(x(n))$ of a sequence $x(n) = x_1 \ldots x_n$ is the number of computing steps it takes for the fastest program p running on a computer C to calculate $x(n)$ divided by n, i.e.,

$$K_{\mathrm{D}}(x(n)) = \min_{C(p)=x(n)} [\text{computing steps}(p)/n].$$

An infinite sequence x is *T-random* (TR) if for the fastest program running on C the number of computing steps necessary to calculate an arbitrary position x_n of x is of the order of or greater than n; that is

$$K_{\mathrm{D}}(x) = \liminf_{n \to \infty} K_{\mathrm{D}}(x(n))$$

$$= \liminf_{n \to \infty} \min_{C(p)=x(n)} [\text{computing steps}(p)/n] > 0.$$

(Heuristically speaking, if $x(n)$ is TR there is no computational "shortcut" to calculate $x(n)$ in a cycle time less than a value of the order of n.)

Table 1 schematically shows the various forms of complexity and the associated types of randomness [#2]. So far, no computer model C has been specified. Clearly, the complexity measures as well as randomness are machine-dependent concepts.

[#2] Several attempts have been made in the literature to propose complexity measures which grasp the intuitive notion of "organization". These measures shall not be critically discussed here, but their enumeration seems in order. The notion of "logical depth" was introduced by Bennett [20]. A notion of "self-generating" complexity was proposed by Grassberger [21]. A criterion called "thermodynamic depth" has been introduced by Lloyd and Pagels [22] and is critically reviewed in ref. [12].

Table 1
Forms of complexity and their associated types of randomness.

complexity	static	algorithmic/ program size	Chaitin/Martin-Löf/Solovay randomness
		loop depth	–
	dynamic	computational/ time	T–randomness
		storage size	–

One may imagine C as a Turing machine (embeddable in a CA), bearing in mind that there exist calculations which can be speeded up by local parallel processing, in particular on CA.

Both K and K_D are intractable, i.e., there exists no "systematic" way to derive them. This is ultimately a consequence of Gödel's celebrated incompleteness theorem [10,11] [#3]. Although no proof will be given here, the following *gedankenexperiment* illustrates the intractability of the concept of randomness, which cannot be operationalized. Consider a physical system Σ producing numbers on a display. Assume an observer A, for whom Σ for all practical purposes is a "black box", i.e., despite the display A has no knowledge of Σ. Assume a second observer B, who by intuition or other insight knows that Σ calculates the digits of π, displays them, and in doing so has arrived at a specific nth digit. In this case one may ask the following questions. (i) How does A without communicating with B learn about the "meaning" of Σ, i.e., how could A find out that Σ outputs the digits of π ? (ii) To what extend is the predictive power of B restricted by finite computational resources ? – What sense makes any "knowledge" claimed by B that Σ has arrived at the $n = 10^{200}$th decimal place of π ? For even if one uses a whole galaxy as computer, and even if one is willing to wait for the result of the computation for a time comparable to the age of the universe, at least with present-day mathematical means, it is impossible to confirm this statement and to predict the $(10^{200} + 1)$th decimal place of π [13,14] [#4]. Although ideally π can be calculated deterministically to an arbitrary precision, one is forced to a probabilistic description by restrictions in computational resources and intuition – this was, after all, the perception yielding Laplace's "old" probability theory.

Statistical tests sometimes are very weak hints on the stochastic nature of the underlying evolution. Take for instance the simplest nontrivial sequence built from natural numbers 1, 2, 3, ..., enumerated in binary notation: 11011.... It can be shown that it is a Bernoulli sequence [19], i.e. any arbitrary partial sequence occurs with the expected limiting frequency. The same has been demonstrated numerically [15] for the decimal expansion of π up to 26 million places and for partial sequences of length 6. What can be learned from these examples is that *sequences looking rather chaotic may originate from a deterministic evolution which itself has extremely low complexity.*

3. Classification of chaos

In what follows, a classification of chaotic motion with respect to the computability of the initial values and the evolution functions together with the type of randomness will be given. Table 2 lists the various aspects of the four chaos classes.

(i) Chaos I is generated by a computable evolution of CA with uncomputable (CR) initial values. The randomness of the initial value "unfolds" in the deterministic time evolution. As an example, assume a random real $x = x_1 x_2 x_3 \ldots$ in binary

---

[#3] Gödel's incompleteness [10] theorem defines a universal recursively enumerable (and by Church's hypothesis) computable set S and states that for every finitely axiomatized formal theory T (assumed to be sound) there is a number n for which the fact $n \notin S$ is *true but not provable* in T.

[#4] More emphatically, although A may have no access to a Cray 2 supercomputer, he might be willing to believe Bailey's claim [15] that the next ten digits following the 29 359 000th digit in the decimal expansion of π are 3, 4, 1, 9, 2, 8, 4, 1, 7, 8, but he will not accept a claim such as *"with a probability greater than 1/10, the 10^{200}th digit in a decimal expansion of π is 7 "* on a rational (i.e., non-intuitive) basis.

Table 2
Schematic representation of the features of chaos classes.

	Deterministic CA rule	Indeterministic CA rule
computable initial values	**chaos IV** *T-random* *(Chaitin random)* recursive resources	**chaos II** *Chaitin random* nonrecursive resources
uncomputable initial values	**chaos I** *Chaitin random* nonrecursive resources	**chaos III** *Chaitin random* nonrecursive resources

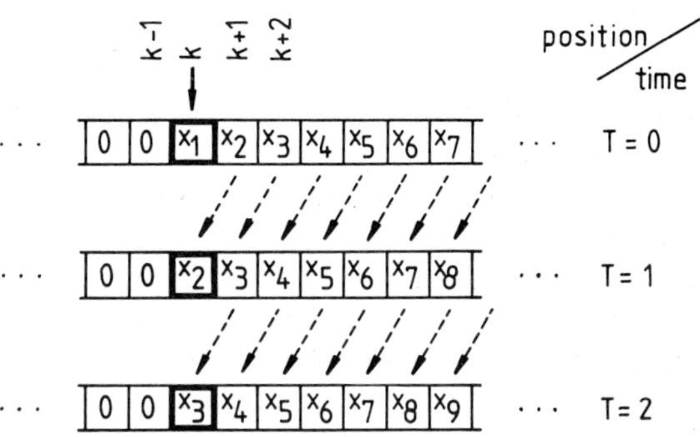

Fig. 1. Evolution of one-dimensional 2-state CA with random initial configuration.

notation which is encoded in a one-dimensional 2-state CA starting from an arbitrary position, say the cell labeled by k. Define subsequent measurement results (this, of course, is a very naive perception of a measurement process, since it does not include its self-referential character) as the value of this cell (number k) at subsequent time cycles. Assume further as evolution rule a left shift of all cell values followed by a single left truncation at cell number $k - 1$. One then effectively obtains a CR measurement sequence [5] $x_1, x_2, x_3, \ldots$ (see fig. 1).

This is precisely the signature for chaos in classical, deterministic continuum mechanics – the evolution of suitable (positive Lyapunov exponents) deterministic (= computable) functions with initial values from a continuous spectrum, which serves as a kind of "pool of random reals": The randomness of a classical deterministic chaos resides in its initial configuration.

In contrast to classical continuum physics, quantum mechanics is a theory operating with a *discrete phase space* which comes close to a tesse-

[5] This can be seen by considering a specific form of the logistic equation $X_{n+1} = \alpha X_n (1 - X_n)$; for $\alpha = 4$ and after the variable transformation $X_n = \sin^2(\pi x_n)$ one obtains a map f: $x_n \rightarrow x_{n+1} = 2x_n \pmod 1$, where (mod 1) means that one has to drop the integer part of $2x_n$. By assuming a starting value x, the formal solutions to n iterations is $f^{(n)}(x) = x_n = 2^n x \pmod 1$. Notice that f is computable and linear. If x is in binary representation, this is just n times a left shift of the digits of x, followed by a left truncation before the decimal point. Now assume $x \in (0,1)$ is CR. Then the computable function $f^{(n)}(x)$ yields a random map onto itself. The point is that although the evolution function f itself is recursive and linear, x is random, and f "unfolds" the "complexity" contained in x.

lation of state space, as required by CA modeling. Thus a CA modeling of quantum processes might be based upon discrete phase space rather than on configuration space, as has been suggested before [16,17]. In this context we only refer to important results concerning the conservation of computability for bounded systems [18]. (Quantum modeling, modeling of the measuring process as well as recursive analysis are not subject of this paper and will be treated elsewhere.)

The following two chaos classes II and III are, strictly speaking, not allowed for CA, since by definition CA are deterministic computers, i.e., they represent computable algorithms. However, since stochastic CA modeling has been used for quantized systems [23], they are mentioned here. These stochastic CA would then correspond to a network of "oracles", one oracle per CA cell, outputting random numbers on recursively enumerable local (i.e., nearest-neighbor) input.

(ii) Chaos II is generated by the uncomputable evolution of CA with computable initial values. It operates with computable initial values and uncomputable evolution laws.

(iii) Chaos III is generated by the uncomputable evolution of CA with uncomputable initial values.

Whereas chaos of class I, II and III supports CR, infinite complexity has to be assumed, either in the initial values, or in the time evolution, or in both. The following chaos class can, for finite times, only support TR. It has the advantage of requiring merely computability, and in some cases only finite resources.

(iv) Chaos IV is generated by the computable

evolution of CA with computable initial values. The (incompressible) dynamic complexity of TR sequences is generated by the unfolding of a computable, but TR initial value (the associated evolution law must have positive Lyapunov exponent(s)), or by a dynamically incompressible algorithm and arbitrary initial values. As for chaos I, chaos IV may originate in a TR initial value which is "unfolded in time" by a suitable evolution function. Moreover, the evolution function itself may give raise to a TR output on non-TR input. Examples for the latter situation is a CA computation of some number whose dynamical complexity increases at least as strong as the number of digits computed.

Note that *computable algorithms may have uncomputable limits* [11]. One example for this fact is Chaitin's random real Ω [9]. The seemingly paradoxical conclusion that "a Chaitin random sequence can be obtained *via* a constructive computable algorithm" is resolved by the inability to give a recursively enumerable radius of convergence for Ω. What is important here is the fact that with "suitable" evolution functions and in the limit of infinite time, chaos IV is capable of becoming CR random [#6].

4. Possible induction of an arrow of time

One central result of symbolic dynamics [7] can be formulated as follows. For chaos I and with probability one (i.e., for random initial values), the normalized static complexity $K(x)$ of single trajectories (representable as symbolic string $x = x_1 x_2 x_3 \cdots$ of CA states in a finite size observation interval, as indicated in fig.1) is equal to

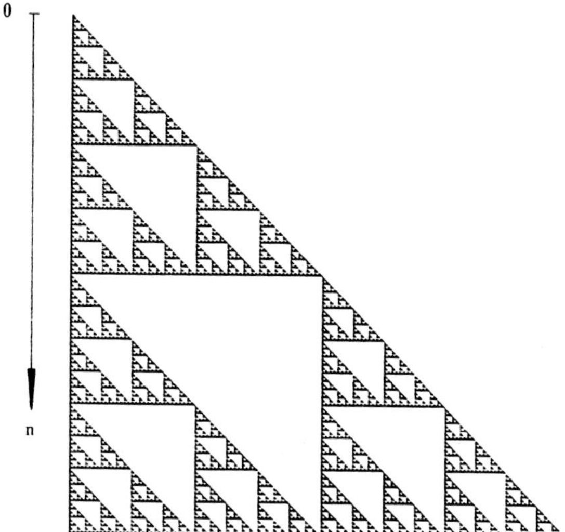

Fig. 2. Pascal's triangle (mod 2) is equivalent to the evolution of a one dimensional state-2 CA. Dots and blanks indicate the digits 0 and 1, respectively.

the overall metric entropy measure $h_\mu(f)$ of a dynamic system with evolution function f [#7], and to the sum of all positive Lyapunov exponents,

$$K(x) = h_\mu(f) = \sum_{\lambda^+ > 0} \lambda^+.$$

This connection between entropy measure and the algorithmic complexity measure of single trajectories provides a powerful link of algorithmic information theory and the theory of computability on the one hand and statistical physics and thermodynamics on the other hand [#8].

4.1. Arrow of time by uncomputable limit

In what follows, I shall concentrate on the above connections for the constructive chaos IV. The capability of computable functions to "produce" uncomputable output on computable initial values

---

[#6] By the introduction of two time scales and the method of *Zenon squeezing* it is possible to arrive at uncomputable output, i.e., irreversibility, even for finite proper times. Consider the following two time scales. (i) The *proper time* τ measures the physical system time by clocks in a way similar to the usual operationalizations; whereas (ii) a discrete *cycle time* t characterizes a sort of "intrinsic" time scale for a process to run on a machine model, say a CA. For some (here unspecified) reason one could "squeeze" τ with respect to t in the following way: $\tau_0 = 1$ and $\tau_n = \sum_{t=0}^{n} k^t$, where $k < 1$. In the limit of infinite cycle time $t \to \infty$, the proper time $\tau_\infty = 1/(1 - k) < \infty$ remains finite.

[#7] f has to be a C^2-diffeomorphism preserving an ergodic Borel measure μ.

[#8] As was shown by Atmanspacher and Scheingraber [24], if ρ stands for a distribution function, $\hat{L}$ is the Liouville operator acting on ρ, i.e., $i\,\partial\rho/\partial t = \hat{L}\rho$, and $\hat{M}$ is an information operator whose eigenvalue is the information contained in the state ρ, then $[\hat{L}, \hat{M}] = \hat{L}\hat{M} - \hat{M}\hat{L} = -i\,\mathbf{1}K$, i.e., information is "gained" during the evolution if the complexity $K > 0$. However, as has been pointed out earlier, an interpretation of this information gain, i.e. its decoding, is in general intractable.

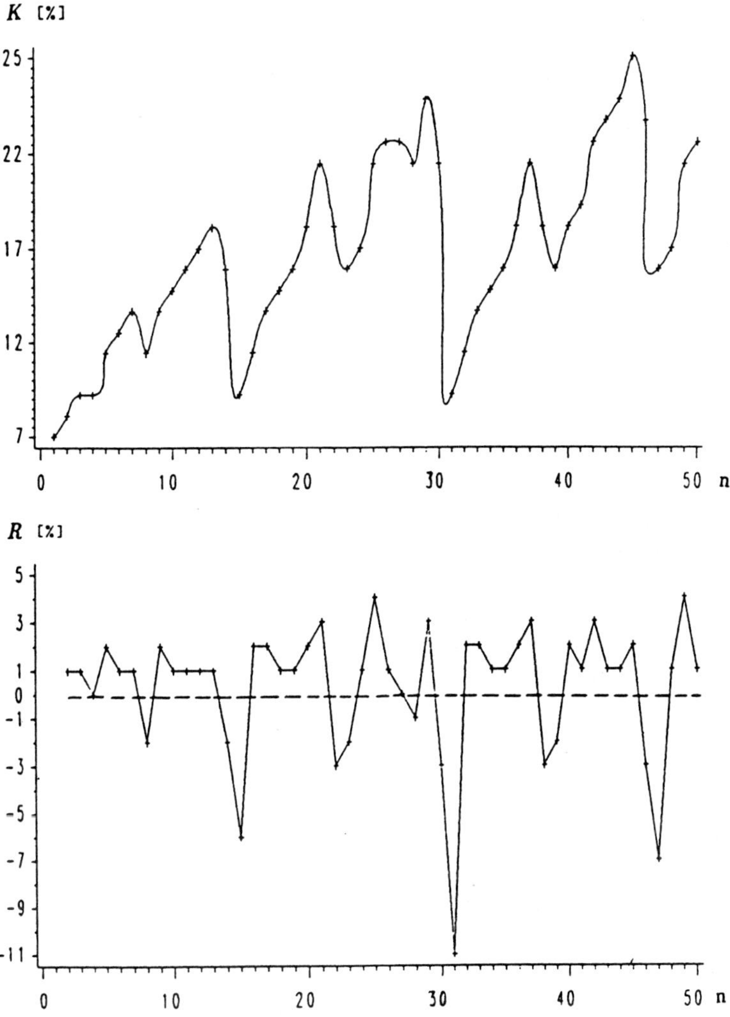

Fig. 3. Heuristic toy model study of K and $R = \delta K/\delta K_{\mathrm{D}}$ on Pascal's triangle (mod 2), drawn in fig. 2.

may have some far-reaching consequences in the physical perception of reversibility. As has been pointed out already, algorithmic complexity may be "created" by an investment of dynamic complexity, for instance by an infinite amount of time. One may therefore define the ratio $R = \delta K/\delta K_{\mathrm{D}}$ and call a system for which, in the infinite time limit, $R > 0$, "creative", expressing the fact that algorithmic complexity is created. For such processes, the above equivalence between K of single trajectories and the overall entropy measure h_μ in the infinite time limit also holds for a suitable, i.e. "creative", chaos IV. Creativity induces a unique time direction even for deterministic

CA with computable initial configurations. In this sense the creation of algorithmic complexity may be perceived as a formal aspect of irreversibility.

4.2. Arrow of time via undecidability

One can motivate the induction of an arrow of time even for *finite* processes as follows. Since both K and K_{D} are uncomputable, R is uncomputable. One may, nevertheless, employ heuristic measures obtained from standard-type compression algorithms for a bound from above on K and K_{D}, and hence approximate R. Such a heuristic approach will most likely utilize *suboptimal* compression algorithms and will thus yield too high complexity measures. More precisely, finding the optimal compression algorithm (corresponding to physical laws) by deductive methods is generally provable impossible. An arrow of time is then induced by *ignorance*, i.e. by the incomputability of the "true" shortest law.

As an example, fig. 2 shows Pascal's triangle, mod 2, representing even and odd binomial coefficients, which may be locally generated by an asymmetric CA with the following nontrivial rules (all others zero) $1/0/0 \to 1$, $0/1/0 \to 1$, $1/0/1 \to 1$, $0/1/1 \to 1$. Fig. 3 shows a heuristic study of K_H and R_H (the index "H" stands for "heuristic") on this structure. A standard (Huffman-type) compression algorithm using lookup tables has been introduced to compress the first successive horizontal lines of constant length of fig. 2. Since there is no guarantee that this compression algorithm is optimal, only bounds from above on K are derived. R_H has been calculated under the assumption that the dynamic complexity K_D increases with the number of lines, i.e., with the CA cycle time. In this case, R_H is obtained by taking the difference between two consecutive K_H's.

On the average, there is an increase of K_H, corresponding to a positive R_H. Hence, the heuristic compression algorithm for the determination of K and R induces a time arrow through undecidability of the "true" quantities. Of course this is due to the fact that the Huffman-type compression algorithm used in the example is suboptimal. A better compression algorithm would use the above listed CA rules generating Pascal's triangle (or even shorter codes, if they exist).

A third, technically rather subtle and speculative contribution to an arrow of time would be a reversible CA whose time evolution is recursively enumerable but whose inverse is uncomputable. This would correspond to a computable bijection $f(x) = y$ whose inverse $f^{-1}(y)$ is uncomputable. To the author's knowledge the existence of such a function can only be conjectured. (That similar concepts are important in the theory of recursive functions can be seen by the constructive existence of recursively enumerable subsets of the natural numbers whose complements cannot be recursively enumerated.) [#9]

For deterministic reversible systems with computable initial values it can be expected that further investigations along these lines will yield new insight into the second law of thermodynamics.

Acknowledgements

This work was supported in part by the Erwin-Schrödinger-Gesellschaft.

[#9] Another nonconstructive possibility leading to an arrow of time would be a CA composed out of *oracles* (such as mentioned for chaos II and III), outputting random numbers with every cycle, thereby increasing the complexity measure K.

References

[1] T. Toffoli and N. Margolus, Cellular Automata Machines (MIT Press, Cambridge, MA, 1987).

[2] C.H. Bennett, IBM J. Res. Dev. 17 (1973) 525.

[3] R. Landauer, IBM J. Res. Dev. 5 (1961) 183

[4] C.H. Bennett and R. Landauer, Sci. Am. 253 (1985) 48;
D. Deutsch, Proc. R. Soc. (London) Ser. A 400 (1985) 97;
A. Peres, Phys. Rev. A 32 (1985) 3266.

[5] A. Fraenkel, Y. Bar Hillel and A. Levy, Foundations of Set Theory, 2nd Ed. (North-Holland, Amsterdam, 1973).

[6] S. Wolfram, Phys. Rev. Lett. 54 (1985) 735; Physica D 10 (1984) 1; Rev. Mod. Phys. 55 (1983) 601.

[7] V.M. Alekseev and M.V. Yakobson, Phys. Rep. 75 (1981) 287.

[8] K. Svozil, Measurement uncertainty due to intrinsic undecidability, TU Vienna preprint (January 1990).

[9] G.J. Chaitin, Algorithmic Information Theory (Cambridge Univ. Press, Cambridge, 1987);
A.K. Zvonkin and L.A. Levin, Russian Math. Surv. 25 (1970) 83.

[10] K. Gödel, Monatsh. Math. Phys. 38 (1931) 173.

[11] H. Rogers, Theory of Recursive Functions and Effective Computability (McGraw-Hill, New York, 1967);
C. Calude, Theories of Computational Complexity (North-Holland, Amsterdam, 1988).

[12] K. Svozil, Randomness and Undecidability in Physics (World Scientific, Singapore), in preparation; Are chaotic systems dynamically random?, Phys. Lett. A 140 (1990) 5.

[13] R.O. Gandy, Limitations to mathematical knowledge, in: Logic Colloquium '82, eds. D. van Dalen, D. Lascar and J. Smiley (North-Holland, Amsterdam, 1982).

[14] R.O. Gandy, Church's thesis and principles for mechanics, in: The Kleene Symposium, eds. J. Barwise, H.J. Kreisler and K. Kunen (North-Holland, Amsterdam, 1980).

[15] D.H. Bailey, Math. Comput. 60 (1988) 283.

[16] M. Minsky, Int. J. Theor. Phys. 21 (1982) 537.

[17] K. Svozil, Phys. Lett. A 119 (1986) 153.

[18] M.B. Pour-El and J.I. Richards, Computability in Analysis and Physics (Springer, Berlin, 1989).

[19] D.G. Champernowne, J. London Math. Soc. 8 (1933) 254.

[20] Ch. H. Bennett, in: Emerging Synthesis in Science, ed. D. Pines (Academic Press, New York, 1985).

[21] P. Grassberger, Int. J. Theor. Phys. 25 (1986) 907.

[22] S. Lloyd and H. Pagels, Ann. Phys. (NY) 188 (1988) 186.

[23] G. Grössing and A. Zeilinger, Physica B 151 (1988) 366; Physica D 31 (188) 70; G. Grössing, Phys. Lett. A 131 (1988) 1.

[24] H. Atmanspacher and H. Scheingraber, Found. Phys. 17 (1987) 939.

CHAPTER 6

GENERALIZATIONS OF CELLULAR AUTOMATA

Physica D 45 (1990) 431–440
North-Holland

CELLULAR AUTOMATA AND DISCRETE NEURAL NETWORKS

Max GARZON

Department of Mathematical Sciences and Institute for Intelligent Systems, Memphis State University,
Memphis, TN 38152, USA

Received 16 January 1990
Revised manuscript received 5 March 1990

Recent results about generalizations of the cellular automaton model to arbitrary digraphs and quasi-linear transition functions are surveyed. Computations by both types of devices can be abstractly regarded as dynamical systems on the Cantor set. This formulation is used as a basis for a systematic comparison of their computational power and their classification. Classifying cellular automata and neural networks on finite graphs by isomorphism of their dynamics digraphs gives rise to difficult **NP**-hard and *co***NP**-complete problems. On infinite locally finite graphs one obtains a proper generalization of Turing computability. Neural and automata networks on graphs of finite bandwidth are just as powerful as cellular automata. Computable self-maps of the Cantor set play here the role that partial recursive functions play in classical computability. A characterization of functions computable by neural networks (or equivalently, cellular automata) on a Cantor set in one time step is given. Finally, some open problems are briefly discussed.

1. Introduction

The purpose of this paper is to survey some recent results on *discrete synchronous neural networks* (or simply, neural networks), a generalization of cellular automata. Neural networks have been successfully applied to create computational models of a cognitive universe [1] and their phenomena (associative memory [2], learning [3] and adaptation [4], pattern recognition [5], speech [6], artificial life [7], etc.). Their similarity to vertebrate nervous systems appeals to cognitive intuition. Neural networks appear, in many respects, just as suitable for cognitive modeling as cellular automata are for physical modeling. Moreover, combinatorial problems that are hard in the classical sense can be given fast, approximate solutions using neural networks [8,9].

One can regard a cellular automaton as a network of automata built on a graph whose sites are located at integer lattice points of Euclidean space and whose links are the edges in the corresponding grid. The neural networks considered below differ from cellular automata in two important aspects. First, neural network cells are located at vertices of an arbitrary (possibly infinite) graph with arbitrary communication links (see ref. [10] for graph-theoretic notions.) Second, the local transition function of a neural network is *not* homogeneous as it varies from site to site. On the other hand, the states, then called *activation levels*, of the network must have particular algebraic operations of addition and multiplication satisfying the ordinary properties of the integers. Moreover, the local transition functions are of a very restricted type (a slight perturbation of an additive cellular automaton rule).

A second generalization, *automata networks*, obtained by just relaxing the restriction of homogeneity on the underlying lattice, will also be considered. Although the proofs of some of the results below carry over to neural networks with arbitrary activations (including real numbers) and to automata networks, attention will be focused on discrete neural networks. *Sequential* computability over the real numbers has been studied, e.g. by Blum, Schub and Smale [11].

Two chief questions will be explored in the following sections. First, how does the computational capability of a neural network compare to that of a cellular automaton and classical computational devices such as Turing machines? Second, what is the computational difficulty of solving basic questions about neural networks, both finite and infinite, such as classification according to their long-term behavior? Given the rather disjoint coverage of phenomena and success of the two models, one may hope to exploit any relationship between cellular automata and neural networks to exchange tools (e.g. for analytical study of long-term behavior) and applications (e.g. physical modeling vs cognitive modeling) between the two fields.

Section 2 provides basic definitions of the two models. Section 3 presents some discussion on classification of neural networks, and some results about classification of finite networks by isomorphism as discrete dynamical systems. Section 4 is concerned with the problem of comparing their computational power. Section 5 explores the problem of characterizing those transformations which are computable by either of the two. Finally, section 6 points out some related issues and problems of interest for further research.

2. Generalizations of cellular automata

It is natural to generalize the concept of a cellular automaton by putting the sites at the vertices of a digraph (in lieu of the ordinary Euclidean lattice of integer points) so that both communication and local transitions may depend *nonuniformly* on neighbor sites. The specific type of generalization defined below possesses properties which are a compromise between the highly discrete and finitary Turing machine model, on the one hand, and highly infinite and physically unrealizable models, on the other. A brief justification of each of the basic features of the model considered here is given next.

First, it is desirable to have a realistic model of computation. In order for the generalization to be *physically realizable*, each vertex of the underlying digraph must have only finitely many incoming and outgoing arcs, i.e. the digraph must be *locally finite*. Second, there is the issue of our ability to *control* and *fully utilize* a physical machine constructed according to a model of this type. (The ensuing important issue of *measurement* is ignored

here but see section 6.) In order to make the model as general as possible we will simply assume that the activation levels of cells at the sites of the network can have either discrete or continuous values, although our discussion will emphasize discrete activations from fixed sets A (each containing the *quiescent* state 0), the set of the possible activation levels (or states) of each cell at site i. Further, the evolution of the network is assumed to take place in discrete time steps. Thirdly, neural networks on digraphs with a bounded finite number of vertices and discrete activation levels are easily seen to be equivalent to finite state machines, a very restricted model of computation. Neural networks capable of simulating general purpose devices such as Turing machines require a finite number of cells that dynamically increases without bound. (Simulations of this type appear in refs. [12,13].) Thus, in order to have *full computational ability* we will simply assume a *countably* infinite number of sites.

Under these assumptions, a neural network operates as follows. Each arc in a digraph D from vertex j to vertex i is labeled with a weight w_{ij}. Cells sum their weighted inputs and apply an *activation function* f_i: A $\rightarrow$ A. Each f_i satisfies $f_i(0) = 0$ (in order to avoid spontaneous generation of activation). At any time $t + 1$ (t a nonnegative integer) the net input of each cell is the weighted sum of the inputs from its neighbors given by the function

$$\text{net}_i(t + 1) = \sum_j w_{ij} a_j(t), \tag{1}$$

where

$$a_j(t) = f_j(\text{net}_j(t)) \tag{2}$$

is the activation of the cell j at time t, and the sum is taken over all cells j supporting links into i. The composition of net_i followed by the activation function f_i plays a role analogous to that of the local rule δ of a cellular automaton.

Definition 1 *A (discrete) neural network is a triple* $\mathcal{N} = \langle \text{D}, \text{A}, \{f_i\} \rangle$ *consisting of a locally finite, arc-weighted, countable (finite or infinite) digraph* D, *a family of weights* w_{ij} *for each vertex* i *and adjacent vertices* j, *a subset of site activation (state) values of a set* A *allowing multiplication by weights and addition of products, and a family* *of activation functions $f_i \colon \mathrm{A} \to \mathrm{A}$, one for each vertex i in D. The local dynamics of $\mathcal{N}$ is defined by eqs. (1) and (2) above.*

The local transitions of a neural network applied in parallel and synchronously to all cells lead to a global transformation of the activation vector describing the total state of the network (hereafter referred to as a *configuration*) similar to the global map of a cellular automaton. Regarding a neural network as a dynamical system, at time t the network has at each cell i its net input $\mathrm{net}_i(t)$ and its activation $a_i(t)$. The vector of activations x has in its ith component an activation value $x_i(t)$ corresponding to the state of the ith cell of D. At each tick of a global clock, the change of activation vector according to the local transition defines a *global dynamics* $T \colon \mathcal{C} \to \mathcal{C}$ on the set of all configurations $\mathcal{C}$ given by

$$T(x)_i = f_i(\mathrm{net}_i(x)). \tag{3}$$

A further justification for consideration of this model is its advantage in handling problems that are provably unsolvable in the framework of classical computation. The point is illustrated with the halting problem for Turing machines (hereforth referred to as the halting problem). Recall that the halting problem asks for a general algorithm to determine whether or not any given Turing machine will eventually halt when given some specified, but arbitrary, input. Turing showed that no such algorithm exists [14]. Nonetheless, the halting problem, when suitably encoded on a neural network, can be solved by a potentially infinite neural network.

Theorem 1 [15, theorem 3.2] *The membership problem for any countable set of finite strings is neurally solvable by an infinite neural network of bandwidth 2. In particular, the halting problem for Turing machines is neurally solvable.*

This result may be a bit surprising and it will be discussed briefly. A similar argument can be used to show that the membership problem for any countable set of finite strings of digits from a fixed finite set is solvable by a neural network.

The halting problem is formulated as a decision problem for neural networks as follows. An instance $\langle M, x \rangle$ consisting of a Turing machine M and its input x is encoded as a configuration in a cellular space consisting of two disconnected one-way infinite paths with activations from the set of integers modulo 6. The proof requires the construction of a network $\mathcal{N}$ and two configurations x_{yes} and x_{no} such that $\mathcal{N}$ stabilizes on every input at the correct answer of the given instance, i.e. at either x_{yes} or x_{no}, according as whether or not M halts on x. Details of a specific network to accomplish precisely this are discussed in ref. [15].

Despite its appearance, theorem 1 is *not* a counterexample to the Church–Turing thesis since the thesis concerns only algorithmic problems solvable by *finitary* means, whereas the network involved in the proof of the theorem requires the use of an essentially infinite object, namely the underlying digraph. The point here is that a simple infinite object of *finitary type* solves the problem.

There is yet another generalization of cellular automata that will be referred to below. An *automata network* is defined on an arbitrary locally finite digraph with state sets Q_i by a local rule

$$\delta_i \colon \prod_{j \to i} Q_j \to Q_i \, .$$

This model is somewhat closer to a cellular automaton than a neural network since the local transition δ_i, although not necessarily of the special neural network type, is the nonhomogeneous version of a cellular automaton rule.

Although other types of networks (asynchronous, continuous, adaptive, and/or probabilistic) lead to important models (see e.g. ref. [16]), we have restricted our attention to cellular automata and networks that are discrete, synchronous, and deterministic.

The following result shows that neural and automata networks are indeed generalizations of cellular automata. Its proof, which appears in ref. [15], is reproduced below.

Theorem 2 *Every cellular automaton is a neural network. Each is an automata network.*

Proof: Let Q be the state set of a cellular automaton $\mathcal{M}$ and $\delta \colon Q \times Q^d \to Q$ be its local dynamics, where $d := |N|$ is the size of its neighborhood N. Without loss of generality assume that $Q = \{0, \ldots, m - 1\}$, where m is the number of states in Q and 0 is the quiescent state. Construct a neural network $\mathcal{N}$ with activations from the set $A := Z_{m^d}$ of integers modulo m^d whose underlying digraph is the lattice of $\mathcal{M}$ as follows. Assign a fixed translation-invariant numbering to the neighbors of each vertex i on $\mathcal{M}$'s digraph. Let the weight w_{ij} simply be $w_{ij} := m^j$ (independently of i) and define the activation function f_i by $f_i(r) := \delta(q_{d-1}, \ldots, q_0)$, where the q_j's are the digits of the m-ary expansion of r so that $r = \sum_j q_j m^j$. Thus $\mathcal{N}$ induces a global dynamics according to eqs. (1) and (2), i.e. if $a_j(t) := q_j$, the next activation level of cell i is given by

$$a_i(t+1) = f_i\left(\sum_j m^j q_j\right) = \delta(q_{n-1}, \ldots, q_0)$$

$$= f_i(\mathrm{net}_i(t)) = f_i\left(\sum_j w_{ij} a_j(t)\right),$$

and hence the new activation level of a cell corresponds precisely to the new state of the site in the cellular automaton $\mathcal{M}$.

The second assertion is obvious since a neural network cell is a finite state machine with local transition $\delta_i(x_i, x_{i_1}, \ldots, x_{i_d}) := f_i(\sum_j w_{ii_j} x_{i_j})$, and a cellular automaton has a homogeneous local transition rule $\delta_i := \delta$. $\qquad\square$

Some of the results presented below, particularly the comparisons of computational power of neural networks and cellular automata, are proven with the additional hypothesis of finite bandwidth. This graphical property is satisfied by the communications digraphs of virtually all discrete neural networks of practical interest. (This assumption may not be necessary. See section 6.)

Definition 2 *A labeling of a digraph* D *is a one-one mapping* $\ell: \mathrm{V}(\mathrm{D}) \to \mathbf{N}$ *that assigns a positive integer to each vertex of* D. *The bandwidth of a labeling* ℓ *is the maximum of the differences* $|\ell(i) - \ell(j)|$ *for every pair of adjacent vertices* i, j, *or* ∞ *if there is no maximum. The bandwidth of a (di)graph is the minimum bandwidth of any of its labelings. The class* $\mathbf{NN_0}$ *(*$\mathbf{CA_0}, \mathbf{AN_0}$, *respectively) is the subclass of* $\mathbf{NN}$ *(*$\mathbf{CA}, \mathbf{AN}$*) consisting of global dynamics defined on digraphs of finite bandwidth induced by neural networks (cellular automata, automata networks, respectively).*

Note that finite bandwidth in a digraph implies bounded-degree. That is, there is a uniform bound on the number of links entering or leaving every vertex.

3. Classifying neural networks

Cellular automata have been classified into Wolfram classes according to the type of dynamical systems to which they give rise. No such classification exists for neural networks. Current classifications are made in terms of their global dynamics or the nature of the activation values or functions. Wolfram's classification extends via theorem 6 below to networks of finite bandwidth, but the characterization would be complete for infinite networks only if that theorem can be extended to arbitrary cellular spaces (see section 6).

A satisfactory classification of neural networks requires a natural and precise definition of equivalence. Among the several possible such notions at different levels of abstraction, equivalence in terms of identical or very similar transitions appears somewhat more natural for finite networks, whereas equivalence as dynamical systems appears more suited for infinite networks. A more semantic classification is provided by so-called *functional isomorphism* in which the number of cells or the exact transitions are not as important as the satisfaction of an overall function that the corresponding dynamical systems are supposed to accomplish. For instance, a primitive organism's brain may be regarded as a neural network of some sort, and two organisms of the same species are considered functionally identical because they exhibit very similar behavior. Although important from a biological viewpoint, this criterion seems to lead too far afield and will not be discussed here.

In this section we consider the classification problem for *finite* neural networks. The following sections will consider infinite neural networks. In order to make the notion of equivalence precise, it is necessary to introduce some terminology. A finite threshold neural network can be represented by a pair $\langle \mathrm{W}, \boldsymbol{b} \rangle$ consisting of an $n \times n$ real weight matrix $\mathrm{W} := [w_{ij}]$ of "synaptic strengths" and a real threshold vector $\boldsymbol{b} \in \mathrm{R}^n$ defining its activation functions. A neural network is *symmetric* if and only if the weight matrix W is symmetric. A configuration of an n-cell network reduces to a vector $x \in \boldsymbol{B}^n$, where $\boldsymbol{B} := \{0, 1\}$. In this section $\mathcal{C}_n$ denotes the corresponding set of n-bit configurations.

The global dynamics of an n-cell neural network $\mathcal{N}$ with real weight matrix $\mathsf{W} = [w_{ij}]$ and activation vector $\boldsymbol{b}$ can be naturally represented by a *dynamics digraph* with vertex set $\mathcal{C}_n$ and arcs connecting each x to $T(x)$. Each cell updates its activation values for the next time step according to

$$x_i(t+1) := \mathbf{1}_i\left(\mathsf{W}x(t) - b\right)$$
$$= \mathbf{1}\left(\sum_{j=1}^{n} w_{ij}x_j(t) - b_i\right), \qquad (4)$$

where the ith component of the threshold function $\mathbf{1}$ is given for a vector $\boldsymbol{x} \in \mathrm{R}^n$ by

$$\mathbf{1}_i(\boldsymbol{x}) := 0 \quad \text{if } \boldsymbol{x}_i < 0, \qquad (5)$$
$$:= 1 \quad \text{else.} \qquad (6)$$

The criterion for classification discussed below is based on isomorphism of global dynamics. Neural networks with different weights and thresholds may give rise to identical dynamics digraphs. These networks are said to be *isomorphic*. It would be ideal to have a classification of isomorphism of finite networks in terms of well-known invariants of the weight matrices and activation vectors. Unfortunately, it is easy to see that isomorphic networks may differ by more than relabeling of their cells. In particular, the weight matrices of isomorphic networks need not have the same eigenvalues. Further evidence given below suggests that a simple condition characterizing isomorphism of finite neural networks is highly impractical or even nonexistent.

The best object next to a simple formula is an algorithm to test finite neural network isomorphism. Such an algorithm exists, for it is in principle possible to make an exhaustive systematic comparison of the transitions in the dynamics digraphs of two given networks. Thus the fundamental question concerning a classification by isomorphism is whether it is computationally feasible. When confronted with a complexity question of this kind, one has only two possibilities. Either find an *efficient* algorithm to solve the problem or prove that such an algorithm does not exist. Computational complexity is the area of computation theory concerned with questions of this kind. An algorithm is considered efficient when the number of instructions executed to solve an instance is polynomial in the size of the given instance. The class of problems admitting a polynomial time solution is denoted $\mathbf{P}$. It is an important open problem whether this class coincides with the apparently larger class $\mathbf{NP}$ of problems whose solutions can only be *verified* in polynomial time, or with the apparently larger class of problems $co\mathbf{NP}$ whose purported solutions can be *refuted* in polynomial time. (See ref. [17] for background definitions and results on computational complexity.)

Finite threshold networks possess structural features that might give rise to fast isomorphism testing. For instance, dynamics digraphs are "functional" digraphs of out-degree 1, and more generally, each weakly connected component is a unicyclic digraph (see ref. [10] for graph-theoretic notions) with a unique directed shortest path between every pair of vertices. Also, symmetric network dynamics (with a symmetric weight matrix) admit an energy (or Lyapunov) function that decreases with time [18], which implies that they always converge to fixed points or 2-cycles (although the networks may take up to $2^{n/3}$ iterations to stabilize [19]). Further, it is well known that every neural network on n cells is isomorphic to one with integer weights and thresholds [20], [21, theorem 4.6.1] which are expressible in $\mathrm{O}(n \log n)$ bits. For the purpose of network isomorphism testing, instances are given as an integer weight matrix and an integer threshold vector of size polynomial in the number of cells (which can then be taken as the input size). Thus, despite appearances, the dynamics isomorphism problem is a purely combinatorial one.

Testing isomorphism of dynamics digraphs can be interpreted in two different ways, as labeled (preserving vertex labels from $\mathcal{C}_n$) or as unlabeled digraphs. The corresponding decision problems are called SIT (*strong-isomorphism* testing) and WIT (*weak-isomorphism* testing) in ref. [22]. These algorithmic problems can be precisely cast as follows.

STRONG-ISOMORPHISM TESTING – SIT
(WEAK-ISOMORPHISM TESTING – WIT)

INSTANCE: two networks $\langle W_k, \boldsymbol{b}_k \rangle$ $(k = 1, 2)$
QUESTION: are they (weakly) isomorphic?

As a byproduct of the study of the complexity of neural network isomorphism, one obtains a result on the following related problem of practical interest dealing with finite cellular automata and neural networks.

PREDECESSOR COUNTING

INSTANCE: a neural network $\langle W, \boldsymbol{b} \rangle$ on n cells and a configuration $x \in C_n$
QUESTION: how many predecessors does x have in the dynamics of $\langle W, \boldsymbol{b} \rangle$?

The classification problem of *finite* neural networks has been investigated from a computational viewpoint in ref. [22]. As it turns out, the problem is somewhat different from the ordinary graph isomorphism problem in at least two aspects. First, in the case of SIT one does not have to contend with a factorial number of permutations of the vertex labels, but rather with an *exponential* number of vertices in the size of the input (which is essentially the square of the number of cells in the network). On the other hand, WIT is the usual type of combinatorial isomorphism problem compounded with the basic difficulty of SIT. Moreover, in both cases equivalent dynamics can be succinctly represented by augmented matrices which may differ by more than a permutation of rows and columns or, more generally, may not even have the same eigenvalues. For these reasons one would expect these problems to be of higher complexity.

In fact, however, isomorphism testing of neural networks belong to the family of hard combinatorial problems, of which the *traveling salesman* is a well-known example. The significance of this property lies in the fact that a fast, efficient solution to any one of them would provide a similar solution to hundreds of problems of great importance in optimization and other areas of computation. The precise results are as follows.

Theorem 3 [22, theorem 3.2] *SIT is coNP-complete, even if restricted to symmetric networks.*

Recall that if $\mathbf{C}$ is a complexity class (such as $\mathbf{NP}$), $co\mathbf{C}$ stands for the class of complements of problems in $\mathbf{C}$. Therefore the question of whether SIT itself is in $\mathbf{P}$ (respectively, $\mathbf{NP}$) is equivalent to the well-known open problem $\mathbf{P} \overset{?}{=} \mathbf{NP}$ (respectively, $\mathbf{NP} \overset{?}{=} co\mathbf{NP}$). It is believed that $\mathbf{P} = \mathbf{NP}$ is a highly unlikely event.

In contrast, the situation for WIT may be entirely different from that in theorem 3 given the current belief that even $co\mathbf{NP} \neq \mathbf{NP}$.

Theorem 4 [22, theorem 3.4] WIT *is* $\mathbf{NP}$-*hard, even if restricted to symmetric networks.*

A slight generalization of combinatorially hard problems deals with counting the number of positive answers to given instances (e.g. *how many tours visit all cities in the traveling salesman problem?*). Some problems present the same phenomenon of completeness in the corresponding class #$\mathbf{P}$ observed in $\mathbf{NP}$ and $co\mathbf{NP}$. The following result thus provides a measure of difficulty of the predecessor counting problem for neural networks dynamics.

Theorem 5 [22, theorem 3.5] PREDECESSOR COUNTING *is* #$\mathbf{P}$-*complete, even if restricted to symmetric networks.*

In view of theorem 6 below, the results of this section also hold for finite cellular automata.

We now turn to the chief questions in this paper for infinite neural networks.

4. Computational power

McCulloch and Pitts, in their seminal paper on neural networks [23], asserted that neural networks are computationally universal, suggesting that a neural implementation of the finite control of a Turing machine be somehow attached to a read–write head and tape. A concrete and detailed neural network implementation of a Turing machine was provided in ref. [13]. The *Turing* unsolvability of the stability problem for a relatively simple class of neural networks (synchronous, binary, linear-threshold) is a consequence of this effective implementation. The primary consequence is, of course, that any algorithmic problem that is Turing solvable can be encoded as a problem solvable by neural networks: *neural networks are at least as powerful as Turing machines.*

The converse has been widely presumed true [12], since computability has become synonymous with Turing computability. Theorem 1, however, shows that *infinite neural networks are strictly more powerful computing devices than Turing machines*, even if we are willing to allow *only* potentially infinite digraphs of finitary type.

Since Turing machines can be simulated by one-dimensional cellular automata, the class of Turing computable (partial recursive) functions **TM** is included in **CA**. This inclusion is proper for the class **CA** considered here since the prototypical recursively unsolvable algorithmic problem, the halting problem, is solvable by a simple neural network. In a different vein, it is not too hard to prove that a one-dimensional cellular automaton can take as input a real number (as an *infinite* binary expansion) and stabilize (not necessarily at the correct answer) if and only if it is an integer. (A Turing machine cannot solve this problem since the integer 1 could be given as $0.9999\ldots$ and hence the machine cannot even finish reading its input.) This problem is thus solvable by a neural network *in a weak sense*, in contrast to the strong solution in the proof of theorem 1 where the network stabilizes at the correct answer of the given instance. In conjunction with theorem 2 it follows that $\mathbf{TM} \subset \mathbf{CA} \subseteq \mathbf{NN} \subseteq \mathbf{AN}$.

A systematic comparison of the computational power of neural networks to cellular automata requires a precise definition of equivalence. Toward this end one is led to the common notion of a cellular space. A *cellular space* $\langle \mathrm{D}, \{\mathbf{Q}_i\} \rangle$ consists of a countable (finite or infinite), locally finite, directed graph D (which may be a symmetric digraph, i.e. just a graph, in the case of a cellular automaton) and a family of finite sets $\mathbf{Q}_i$ (each containing the *quiescent* state 0), the states of the cell at site i.

Associated to every cellular space there is a *configuration space* $\prod_i \mathbf{Q}_i$ (or simply $\mathcal{C}$) consisting of all *configurations* (or *total states*) of the space, i.e. choice functions $x\colon \mathrm{V} \to \bigcup_i \mathbf{Q}_i$ that associate a state (level of activation) $x_i \in \mathbf{Q}_i$ to each i in the vertex set V of the digraph D. The set $\mathcal{C}$ of all configurations is also refered to as activation space. A configuration is written as a vector of activations of all the individual cells at a time, ordered in a (fixed) ordering. The all-zero configuration is denoted O. The configuration space of cellular automata and neural networks with countably many cells is essentially the Cantor middle-third set of real numbers. Recall that the Cantor set is the set of points left in the unit interval $[0, 1]$ of the real line after first deleting its middle third $[\frac{1}{3}, \frac{2}{3}]$, and then continuing to delete the middle thirds of all remaining intervals *ad infinitum*. It is sometimes refered to as the Cantor dust. Its elements are characterized by ternary expansions not containing the digit 1. The Cantor set is characterized topologically as a perfect, totally disconnected, compact metric space [24, p. 97].

Thus a neural network and a cellular automaton can each be regarded as being characterized by the self-map of the Cantor set defined by their global dynamics. Iteration of this map gives rise in a natural way to a dynamical system (see theorems 8,9). Their computational power can be compared by examining the corresponding self-maps of the Cantor set or dynamical system, entities endowed with a topological structure where it is possible to deal precisely with dynamical system notions such as convergence and continuity, etc.

Theorem 1 raises the possibility that these classes of abstract dynamics $\mathbf{CA} \subseteq \mathbf{NN} \subseteq \mathbf{AN}$ are one and the same, e.g. that every neural network can be simulated on a cellular automaton. A systematic comparison of the computing power of neural networks has been carried out in ref. [15]. One of the main results in that paper is that the converse of theorem 2 is essentially true for infinite networks of finite bandwidth.

Theorem 6 [15, theorem 3.5] *Any neural network on a space of finite bandwidth can be simulated by a three-dimensional Euclidean cellular automaton.*

In fact, ref. [15] gives a somewhat stronger result.

Theorem 7 [15, theorem 3.6] *All three models, neural networks, cellular automata and automata networks, are computationally equivalent when restricted to digraphs of finite bandwidth, i.e.*

$$\mathbf{CA}_0 = \mathbf{NN}_0 = \mathbf{AN}_0.$$

The key fact in the proofs of these results is the existence of a *universal neural network* capable of simulating any neural network of finite bandwidth (in particular, one-dimensional cellular automata. See section 6 about higher-dimensional cellular automata.) Full proofs of the results in this section appear in ref. [15]. A full-fledged description of the universal network is contained in ref. [25].

5. Neurally computable functions

The results of the previous section raise the problem of identifying the nature of those self-maps of the Cantor set determined by cellular automata and its generalizations. An early theorem of Richardson [26] (predated by a one-dimensional version by Hedlund [27]) gives a topological characterization of global dynamics of cellular automata.

Theorem 8 [26, corollary 2] *A map $T: C \to C$ is realizable by a Euclidean cellular automaton if and only if*

(1) $T(O) = O$;
(2) T is continuous;
(3) T commutes with shifts.

Analogous theorems characterizing the computability of self-maps of the Cantor set by global dynamics of neural networks appear in refs. [28,29]. In the following result, a global activation function $F : C \to C$ is said to be *strictly local* if $F(x)_i = F(x_i e^i)_i$ for all x and i. A pixel e^k is a configuration with value 1 at cell k and 0 elsewhere.

Theorem 9 [29, theorem 3.2] *A self-map $T: C \to C$ is realizable as a global dynamics of neural network activations if and only if*

(1) $T(O) = O$;
(2) T is continuous;
(3) $T(e^k)$ has finite support for each pixel configuration e^k; and
(4) $T = F \circ L$, where L is a linear self-map of C and F is strictly local.

In theorem 9, condition (1) disallows spontaneous generation within the network. Condition (2) al-

lows the recovery of the underlying network structure. Condition (3) reflects the local finiteness of the network. Condition (4) mirrors the type of local dynamics of the network. Thus, as one would expect, the first two conditions are two of those from Richardson's theorem, while the other two result from the less regular type of architecture of the network, and from the characteristic form of its local dynamics. Note that condition (4) holds with linear activation map F if and only if T is linear.

Theorem 9 makes it easy to find self-maps of the Cantor set that are not neurally computable *in a single time step*. The mapping defined by $T(x) = 1 - x$ reflects the Cantor set about $x = \frac{1}{2}$. It fails to satisfy condition (1). The mapping defined by $T(x) = 3x$ mod 1 usually shifts the digits in the ternary representation of x one place to the left, then dropping the left-most. Since T is not continuous at $x = \frac{1}{3}$, it fails to satisfy condition (2). Finally the self-map that converts from ternary to binary representation of points of the Cantor set fails to satisfy condition (3) since $\frac{1}{3}$ is a ternary pixel $1000\ldots$ with an infinite binary representation $0.010101\ldots$.

These characterizations are satisfactory for computation of global dynamics in a single time step. The more general problem of characterizing the dynamical systems arising from iteration of these global dynamics is, of course, one of the fundamental concerns of current research in cellular automata and neural networks. In particular, it is not clear that there exist continuous self-maps of the Cantor set which are not computable by a cellular automaton dynamical system. Apparently the first example of such a problem, the *stability problem* for neural networks, appears in ref. [15].

As mentioned in section 4, the stability problem of a certain relatively simple class of neural networks is Turing undecidable [13]. In view of the neural solvability of the halting problem, a more germaine question is *whether it is neurally unsolvable* as well. This problem is formally analogous to the classical halting problem for Turing machines.

Every pair consisting of a finite bandwidth network $\mathcal{N}$ and its input x has been encoded in ref. [15] as a configuration $\langle \mathcal{N}, x \rangle$ of the universal neural network $\mathcal{U}$. This encoding has the following properties:

(i) The universal network $\mathcal{U}$ can simulate $\mathcal{N}$ on x when given $\langle \mathcal{N}, x \rangle$ as initial configuration.

(ii) It preserves finiteness of $\mathcal{N}$, i.e. it has finite support if $\mathcal{N}$ has only finitely many cells.

(iii) $\mathcal{N}$ stabilizes on input x if and only if $\mathcal{U}$ stabilizes on input $\langle \mathcal{N}, x \rangle$.

Now consider the problem of membership in the set of configurations

$$\mathrm{L}'_s := \{\langle \mathcal{N}, x \rangle : \mathcal{N} \text{ has finite bandwidth and}$$
$$\mathcal{N} \text{ does } not \text{ stabilize on } x\}.$$

This problem is called the *instability problem* for neural networks of finite bandwidth. The complement L_s of this set is called *the stability problem.* Recall that a neural network weakly solves this problem in the sense that it stabilizes on $\langle \mathcal{N}, x \rangle$ if and only if $\langle \mathcal{N}, x \rangle \in \mathrm{L}_s$, although the stable configuration is not necessarily a (fixed) encoding of the correct answer.

Theorem 10 *The problem of instability for neural networks of finite bandwidth is not weakly solvable (i.e., no neural network of finite bandwidth with a given set of activations takes as input some encoding of an arbitrary network with activations from the same set and stabilizes if and only if its input does not).*

Since neural solvability of a problem clearly implies weak solvability of the problem and its complement it follows that

Corollary 1 *The problem of stability for neural networks of finite bandwidth is neurally unsolvable (i.e. no neural network with a given set of activations takes as input some encoding of an arbitrary network with activations from the same set and decides whether it eventually stabilizes).*

Unlike the situation for recursively enumerable sets, problems which are not weakly solvable can be made so by a *finite* extension of the set of activation values (here playing the role of the character set of a Turing machine's tape).

Theorem 11 *The problem of stability for neural networks of finite bandwidth with activation set A is weakly solvable by a neural network with activations from some extended set A'.*

Theorem 10 suggests that this solution cannot be of finite bandwidth. The proof also yields

Corollary 2 *The problem of stability for neural networks of finite bandwidth is weakly solvable by a three-dimensional Euclidean cellular automaton.*

6. Conclusions and open problems

First, one word of caution is necessary concerning the scope of the results in sections 4 and 5 Viewing the dynamics of any of these three types of networks as a self-map of the Cantor set makes the comparison of their computational power possible. Cellular automata and neural and automata networks have been shown equivalent as dynamical systems on digraphs of finite bandwidth. Cellular spaces of finite bandwidth are based on (di)graphs with essentially one degree of freedom. It appears plausible that the proof technique used to establish the results in refs. [15] and [25] can be modified to generalize them to arbitrary digraphs. However, the question of their relative power on arbitrary cellular spaces remains open.

Second, the results in section 5 can be regarded as characterizing self-maps of the Cantor set computable in a *single* time step. The problem of characterizing and classifying problems solvable by *dynamical* systems defined by iteration of local rules of cellular automata and/or neural networks is a difficult and important problem of current interest. The notion of isomorphism in section 3 is a first approach to the problem. The weak isomorphism problem for infinite spaces is a discrete analog of the important problem of isomorphism of symbolic dynamics.

Third, the notions of problems solvable and weakly solvable by (infinite) neural networks can be thought of as natural analogs of the classical notions of recursive and recursively enumerable sets. Pursuing the analogy with recursive and recursively enumerable sets one might conjecture that a decision problem is solvable if it and its complement are weakly solvable.

Finally, the less theoretical question of *physical* implementation of the models considered here has been ignored in this paper. That this should be a proper concern in theoretical considerations is not currently accepted. However, this is in fact a very important issue, as argued in ref. [30].

Acknowledgements

This work was performed with partial support from the National Science Foundation grant NSF-8602319. The author would like to thank Stan Franklin and Ming Zhang for stimulating conversations on the research reported here. Thanks also go to Michael Angerman and an anonymous referee for comments that enhanced the quality of exposition of a preliminary version of the paper.

References

[1] D.E. Rumelhart and J.L. McClelland, eds., Parallel Distributed Processing, Vols. 1, 2 (MIT Press, Cambridge, MA, 1986).

[2] B. Kosko, Constructing an associative memory, BYTE (September 1987) 137–144.

[3] D.E. Rumelhart, G. Hinton and J.R.J. Williams, Learning internal representations by error propagation, in: Parallel Distributed Processing, eds. D.E. Rumelhart and J.L. McClelland, Vol. 1 (MIT Press, Cambridge, MA, 1986).

[4] S. Grossberg, Competitive learning: from interactive activation to adaptive resonance, in: Connectionist Models and their Applications, eds. D. Waltz and J.A. Feldman (Ablex, Norwood, 1988).

[5] T. Kohonen, An introduction to neural computing, Neural Networks 1 (1988) 3–16.

[6] T.J. Sejnowski and C.R. Rosenberg, Parallel networks that learn to pronounce English text, Complex Systems 1 (1987) 145–168.

[7] C.G. Langton, Artificial Life (MIT Press, Cambridge, MA, 1989).

[8] J.J. Hopfield, Neural networks and physical systems with emergent collective computational abilities, Proc. Natl. Acad. Sci. US 79 (1982) 2554–2558.

[9] J.J. Hopfield, Computing with neural circuits: a model, Science 233 (1986) 625–633.

[10] G. Chartrand and L. Lesniak, Graphs and Digraphs, 2nd Ed. (Wadsworth and Brooks/Cole, Monterey, CA, 1986)

[11] L. Blum, M. Shub and S. Smale, On a theory of computation over the real numbers; **NP**-completeness, recursive functions and universal machines, Bull. Am. Math. Soc. 21 (1989) 1

[12] R. Hartley and H. Szu, A comparison of the computational power of neural network models, in: Proc. IEEE First Int. Conf. on Neural Networks III (1987) pp. 17–22.

[13] S.P. Franklin and M. Garzon, Neural computability, in: Progress in Neural Networks, Vol. 1, ed. O. Omidvar (Ablex, Norwood, NJ, 1990).

[14] A. Turing, On computable numbers, with an application to the Entscheidungunsproblem, Proc. London Math. Soc. 2 (1936) 230–265; 43 (1936) 544–546 (erratum).

[15] M. Garzon and S.P. Franklin, Neural computability II, submitted, Extended Abstract in Proc. 3rd Int. Joint Conf. on Neural Networks, Vol. I, Washington DC (1989) pp. 631–637.

[16] P. Rujan, Cellular automata and statistical mechanical models, J. Stat. Phys. 49 (1987) 139–232.

[17] M.R. Garey and D.S. Johnson, Computers and Intractabillity: A Guide to the Theory of NP-Completeness (Freeman, San Francisco, 1978).

[18] E. Goles, Lyapunov functions for parallel neural networks, Proc. AIP Conference 151, Neural Networks for Computing, ed. J.S. Denker, American Institute of Physics, Salt Lake City, Utah (1986) pp. 165–181.

[19] E. Goles and J. Olivos, Long transients in neural networks, Inf. Control 6 (1981) 96–103.

[20] S. Muroga, I. Toda and S. Takasu, Theory of majority decision elements, J. Franklin Inst. 271 (1961) 376–418.

[21] I. Parberry and G. Schnitger, Parallel computation with threshold functions, J. Comput. Syst. Sci. 36 (1988) 278–302.

[22] M. Garzon and M. Zhang, Classifying neural networks, in: Proc. IEEE Southeast Conference '90, New Orleans, ed. R.E. Trahan Jr.

[23] W.S. McCulloch and W. Pitts, A logical calculus of the ideas immanent in nervous activity, Bull. Math. Biophys. 5 (1943) 115–33.

[24] J.G. Hocking and G.S. Young, Topology (Addison-Wesley, Boston, 1969).

[25] M. Garzon and S.P. Franklin, AMNIAC: A Universal Neural Network, forthcoming.

[26] D. Richardson, Tessellation with local transformations, J. Comput. Syst. Sci. 6 (1972) 373–388.

[27] G.A. Hedlund, Endomorphism and automorphism of the shift dynamical system, Math. Syst. Theory 3 (1969) 320.

[28] S.P. Franklin and M. Garzon, Global dynamics in neural networks, Complex Systems 3 (1989) 29–36.

[29] M. Garzon and S.P. Franklin, Global dynamics in neural networks II, in: Proc. 1st IEEE Annual Symp. Par. Distr. Comput., Dallas, Texas, 1989, ed. B. Shirazi, pp. 292–299.

[30] C. Fields, Consequences of nonclassical measurement for the algorithmic description of continuous dynamical systems, J. Expt. Theor. Artif. Intell. 1 (1989) 171–178.

Physica D 45 (1990) 441–451
North-Holland

ATTRACTOR DOMINANCE PATTERNS
IN SPARSELY CONNECTED BOOLEAN NETS

C.C. WALKER

Department of Information Management, University of Connecticut, 368 Fairfield Rd., Storrs, CT 06269, USA

Received 18 December 1989
Revised manuscript received 19 April 1990

An attractor "dominates" those states in phase space that lie in its basin of attraction. This paper examines attractor dominance as a function of distance from the attractor. When trajectories of sparsely connected Boolean nets are aggregated suitably, attractor dominance is a reasonably smooth function of distance. The shape of the dominance function varies widely, depending on the automaton rule in place. In this paper I order the dominance functions of the rules of a family of sparsely connected Boolean nets. These nets are structurally generalized cellular automata. Two classes contain most of the variability in attractor dominance patterns. Within these classes, two dimensions locate dominance functions and reveal changes in phase space dominance with lattice size. Increasing lattice size appears to move most, but, importantly, not all dominance patterns toward the extremes shown by the three trivial rules. Implications of these findings are discussed.

1. Introduction

This paper examines a family of automata that are straightforward generalizations of cellular automata (CA). CA are regular lattices of sites which interact according to identical, simple rules. During the past few years CA have received wide attention because they nicely demonstrate the important thesis that complex behavior can be produced by a very simple substrate, and they serve as instructive models of a variety of dynamic systems [1].

In the past, CA have been studied primarily as objects apart, not as components of larger things. However, as their modeling scope widens with further work, it seems reasonable to expect CA to play an increasing role in models of composites such as evolutionary and adaptive systems. Such systems differ from autonomous systems in at least two respects. Their dynamics are affected by sources of influence located outside the system, and the potential for interaction between environment and system typically exists for extended periods of time. Some work with this thrust in research on cellular and related automata has al-

ready been reported [2–4].

Among the questions that become important when CA are used to model long-term processes is the general nature of their persistent dynamics, namely, the dynamics produced by the attractors in CA phase spaces.

When modeling systems with input over the long term, restricting our attention here to systems in which relaxation times are short compared to the length of time between pertubations which affect the systems, the question of *where* attractors are located in phase space becomes especially important. In particular, it can be useful to know to what extent system states measurably close to an attractor are also dynamically related to that attractor.

The metric used to asses distance from an attractor is naturally a better choice when it reflects the modeled system's environmental circumstances. For example, one might grade noise, pertubations, or other environmental events impinging on a system as to strength, and use an appropriately defined strength as the distance measure. If the effects of some kind of environmental influence, say, noise, are to be predicted, the effects noise has on the model then need to be given in terms of how relevant characteristics of the phase space are related to noise level.

Some work on metricized phase spaces and attractors in CA and in CA-like systems exists. Kauffman [5] studied attractor-to-attractor transitions in functionally heterogeneous Boolean nets under low noise conditions. Wolfram, Hurd, Li, and Kaneko [6–9] studied attractors in one-dimensional CA. Kaneko, in particular, examined attractor-to-attractor transitions in a low noise regime for a limited set of CA rules.

There seems to be less known about more fully metricized automaton phase spaces. I examined [10] the dominance of attractors over states in the phase space in a family of sparsely connected Boolean nets over the full range of distances from the attractor. The dominance functions in this family of automata showed wide variability in shape, but also enough similarities to suggest relatedness or progression of some kind. The present paper takes up for analysis those possibilities suggested by the earlier data.

An additional issue arises in studying CA phase spaces: How can the multiplicity of CA rules be dealt with effectively?

To reduce the dimensionality of CA rule spaces, several rule classification and grouping schemes have been proposed. These schemes include the behavioral equivalence classes discussed in refs. [11–13] as well as measures seeking to directly characterize rules properties, such as internal homogeneity [11], forcibility [5], λ [14], local structure classification [15], and relational measures making explicit use of rule space structure [16]. Such categorization can be of interest in itself, but as property characterizing schemes in particular often map rules into a number line, investigators can in that case go farther and assess the relation between those numbers and CA behavioral characteristics. This approach is perhaps most appropriate where the behaviors to be explained are themselves unidimensional, e.g., length of transients, degree of trajectory convergence, and so forth.

Where the dynamics of interest are intrinsically qualitative or multidimensional, work can efficiently proceed in the opposite direction, with behaviors being grouped first. Then, if further explanation is sought, it is available through the systematizing of similarities and differences among rules grouped by the behavioral categories. Wolfram followed this approach in developing his four category grouping of one-dimensional CA [6].

Since I am concerned in the present paper with relatively complex behavioral characteristics, namely those exhibited in attractor dominance functions, I also follow the second approach. Using data from ref. [10] I develop groups of CA rules that have similar dominance characteristics, and then make use of a method that orders dominance patterns in two dimensions so as to allow the effects of lattice size to be seen clearly.

2. The automata examined

The generalized CA discussed in the present paper are nets or *lattices*, the customary CA term, of N sites which are affected by themselves and two other, not necessarily different, sites. The state of site i at time $t + 1$ is given by

$$a_i^{t+1} = T\{a_{L_i}^t, a_i^t, a_{R_i}^t\},$$

where T is a binary function of its three arguments. L_i and R_i are respectively the sites that affect the ith site's "left" and "right" inputs. T can be given in the format of table 1. In table 1, entries $a, b, \ldots, h$ give a_i^{t+1}. In a particular realization of T, a zero or a one replaces each entry. A particular T is referred to in what follows as $T(I, J)$ where I is the decimal equivalent of the binary integer $abcd$, the left column of T in table 1, and J is similarly derived from the right column. Thus, $T(0001, 0110)$ is $T(1, 6)$, and is equivalent to Wolfram rule number 104 for binary, one-dimensional, nearest-neighbor CA [17]. The Wolfram rule num-

Table 1
General form of the CA rule T.

Inputs		a_i^t	
$a_{L_i}^t$	$a_{R_i}^t$	0	1
0	0	a	e
0	1	b	f
1	0	c	g
1	1	d	h

ber of T is the decimal value of the binary integer $hgdcfeba$.

Time is discrete, indexed by t, and all update operations occur simultaneously. In a given automaton all sites compute T, and connections among sites are fixed. But for the provision that site i is affected by arbitrary sites, and not adjacent sites $i-1$, and $i+1$, the automata examined here would be the one-dimensional, binary, nearest neighbor CA of ref. [17].

The automata of the present study can be described accurately as synchronous, sparsely connected, random nets of identical Boolean functions. Since CA are synchronous automata whose dynamics are determined by strictly local phenomena on regular lattices, the present automata can be more succintly said to be random structure CA, or complex structure CA, or "partially non-local" CA. (The latter term is due to Li [18].) These automata have also been called "Ashby nets" [19]. For brevity in what follows, I refer to the automata of interest here simply as "CA," qualifying the term when clarity requires it.

While their behavior must reflect their typically irregular structure, the CA examined here have dynamics whose elementary characteristics (described below) are identical to those of conventional, regular-structure CA.

The elementary dynamics of CA are as follows. The binary N-tuple which gives the state of all sites in the lattice at time t is the state of the CA, or just *state*, at t. After being started at an arbitrary state, a particular CA proceeds stepwise in a fully determined fashion from one state to another, finally producing two sequences of states. The first sequence, the transient or run-in, is followed by a second, the cycle, in repetitions of which the system is then permanently trapped, barring environmental disturbance.

The number of distinct states in a sequence is the length of that sequence. The length of the run-in may be zero or more. Cycle length may be one or more.

A disclosure length, the amount of automaton time required for a CA to disclose the cycle reachable from a given initial state, is the length of the run-in from that state plus the length of the following cycle.

In CA, a cycle, of any length, is an attractor, and an attractor's basin of attraction is the cycle and all run-ins to that cycle.

3. Attractor dominance

In the present paper we examine the dominance of attractors over phase space at various distances from the attractor. Attractor dominance, D_Δ, is the relative number of states at a given distance Δ from an attractor that are in the attractor's basin of attraction.

The phase space characteristic measured by D_Δ can be interpreted in several ways, each appropriate in a different modeling context. For example, D_Δ can be seen as measuring the stability of phase space cycles with respect to displacements in phase space. More concretely, D_Δ can reflect the tendency of computations or other activities to resist the effects of disturbance or noise. Still more concretely, D_Δ can represent the probability that a cell will produce its customary tissue type following a random biological shock of given magnitude [5], or the likelihood of exact reproduction of a class examplar following presentation of a corrupted stimulus probe, or the average fraction of potential energy actually available at a given energy contour, and so forth. I use "attractor dominance" to name the concept here because that term seems to suggest a relevant phase space property without making a commitment to a narrow, albeit possibly significant, interpretative domain.

D_Δ is reported here in percentage units. Since we are using sampled data, D_Δ is an estimate of a corresponding population percentage.

Distance Δ is in units of percent Hamming distance. Δ can be interpreted as the strength of an environmenal pertubation (such as a noise, an error, a stimulus presentation, etc.) that is the given percent of the largest pertubation possible, applied to a system that has reached an attractor. In different terms, distance from the attractor refers to an event that changes the values of exactly Δ percent of the sites in the lattice of a cycling CA.

For a fuller understanding of the data dealt with here, I review the procedure used in ref. [10] to produce dominance functions ("stability graphs" in that paper).

For a lattice size N, and rule T, a lattice connection table was chosen at random as follows. Each site's "left" input was assigned to an affecting site using equiprobable, independent, with-replacement sampling from the N sites. The "right" connection table was chosen similarly.

Following the above definition of a particular automaton, a state was chosen equiprobably from the 2^N states. The CA was started at the chosen state, and allowed to proceed to a cycle. One of the states in the cycle was chosen equiprobably from among the cyclic states. Then a state at Δ percent Hamming distance from the chosen cyclic state was picked equiprobably from among the states on the same locus, and the CA was restarted at the "displaced" state.

If the trajectory from the displaced state re-encountered the cycle from which the displacement was made, the displaced state was counted as dominated by the attractor. If the trajectory ran to a different attractor, it was not counted. The procedure described was repeated $m \geq 1000$ times for each Δ, N, and T combination, and D_Δ is the percent of the m displaced states at Δ that were dominated by the attractor.

Given that the data are aggregated over initial state and lattice structure table ensembles, it is appropriate to see a D_Δ versus Δ plot, the dominance function, as describing characteristics of a virtual phase space for the given rule at N.

All 88 independent rules in the family of CA were examined, thus estimating dominance functions for all rules at N. All rules were examined at $N = 10$, and at one other lattice size $N' > N$. For most rules $N' = 20$. Where disclosure lengths allowed, N' was increased beyond 20. A few rules have disclosure lengths sufficiently long that N' could only be raised to 12 or 13.

4. Dominance pattern classes

The dominance functions of the present family of CA are quite varied in shape [10]. Only a very few are nearly duplicates, but some broad similarities exist. We first consider the similarities and define two well populated classes of dominance functions, each class characterizing a pattern of dominance localization in phase space.

The first class, exhibiting *coherent* localization,

Table 2
Number of rules in dominance
localization classes.

Class	$N = 10$	$N' > 10$
bipolar	63	56
coherent	22	25
diffuse	3	7

is such that the attractor primarily dominates local space. Attractor dominance is approximately a decreasing function of distance from the attractor. (See fig. 1, top functions.)

The second class, showing *bipolar* localization, consists of U- or (reverse) J-shaped functions. Bipolar functions are such that the attractor dominates local space and increasingly dominates very distant phase space as well. (See fig. 1, bottom functions.)

A relatively small residual class of dominance patterns which do not fit coherent or bipolar classes well is also used. I call this class *diffuse*. Diffuse dominance functions show no localization of dominance at all or one locally dominated neighborhood and no narrowly dominated alternative regions. (See fig. 1, middle functions.)

Fig. 1 gives typical examples from each of the three pattern classes, and the three trivial T's. In fig. 1, dominance D_Δ, in percent, is plotted vertically, against distance from the attractor Δ, in percent Hamming distance. $W(\ldots)$ is the Wolfram rule number equivalent of T. The larger of the N values in a graph is N'. The trivial rules are $T(0,0)$, $T(0,15)$, and $T(15,0)$. Lattice sizes of the trivial rules can be ignored. m is the number of replications at each graphed point: $m = 1000$ unless noted otherwise.

In fig. 1, for both lattice sizes, dominance functions of $T(3,3)$ and $T(0,15)$ are classed as coherent, those of $T(2,8)$ and $T(0,0)$ as diffuse, and those of $T(1,8)$ and $T(15,0)$ as bipolar.

Fig. 2 gives examples of less common dominance functions. Except for $T(0,14)$ at $N = 10$, which is (just barely) bipolar, the top four T's show coherent functions. The bottom two show borderline cases classed as bipolar, instead of diffuse, because of the weakness of their changes in curvature.

In what follows, I adopt the convention of speaking of a T as if it exhibits the characteristics of its

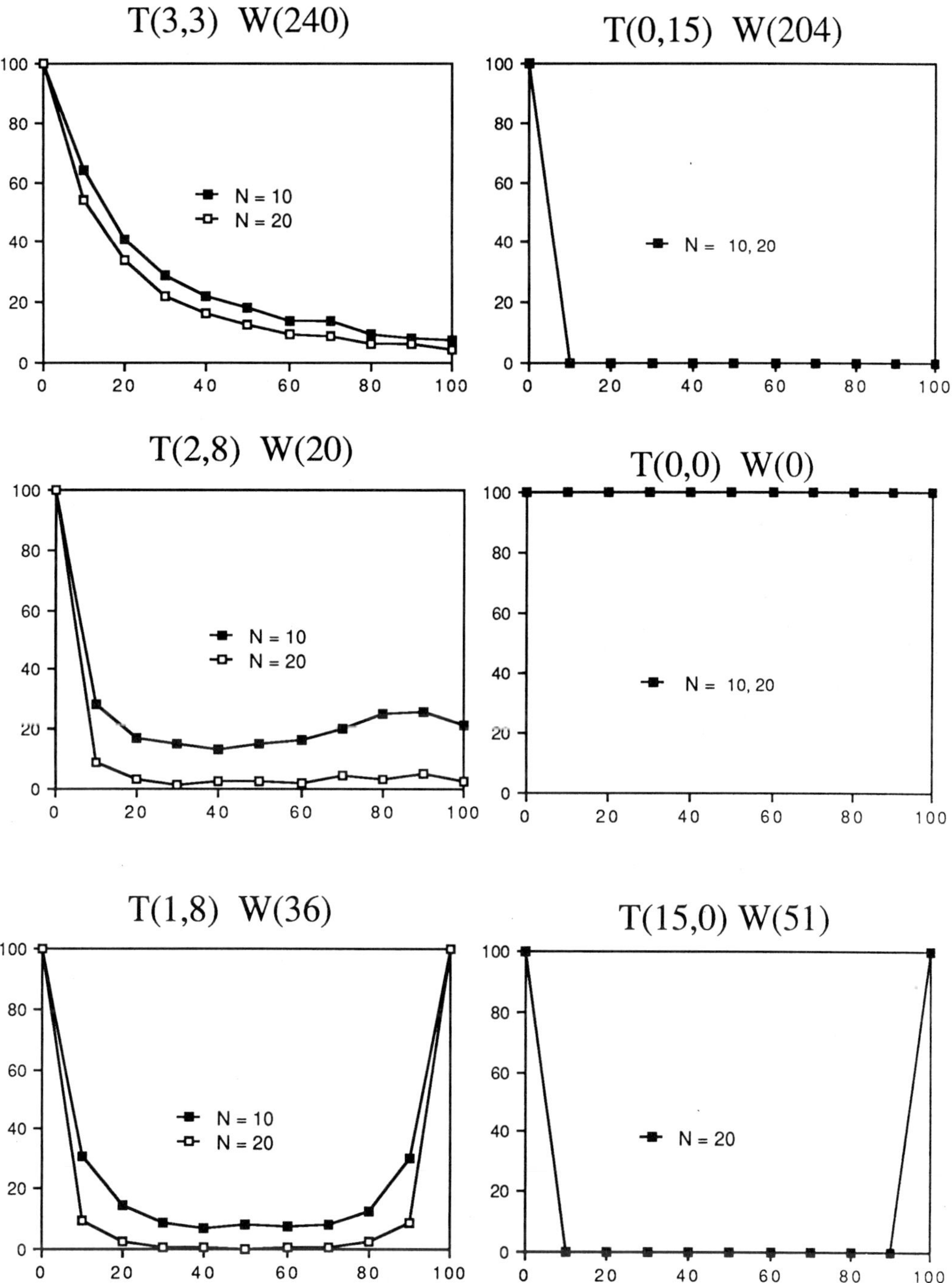

Fig. 1. Examples of typical dominance functions from each pattern class, and the three trivial T's. The top two T functions are classed as coherent, those of the middle two as diffuse, and those of the bottom two as bipolar, at N and at N'. In each case, the ordinate is attractor dominance in percent; the abscissa is distance from the attractor in percent Hamming distance.

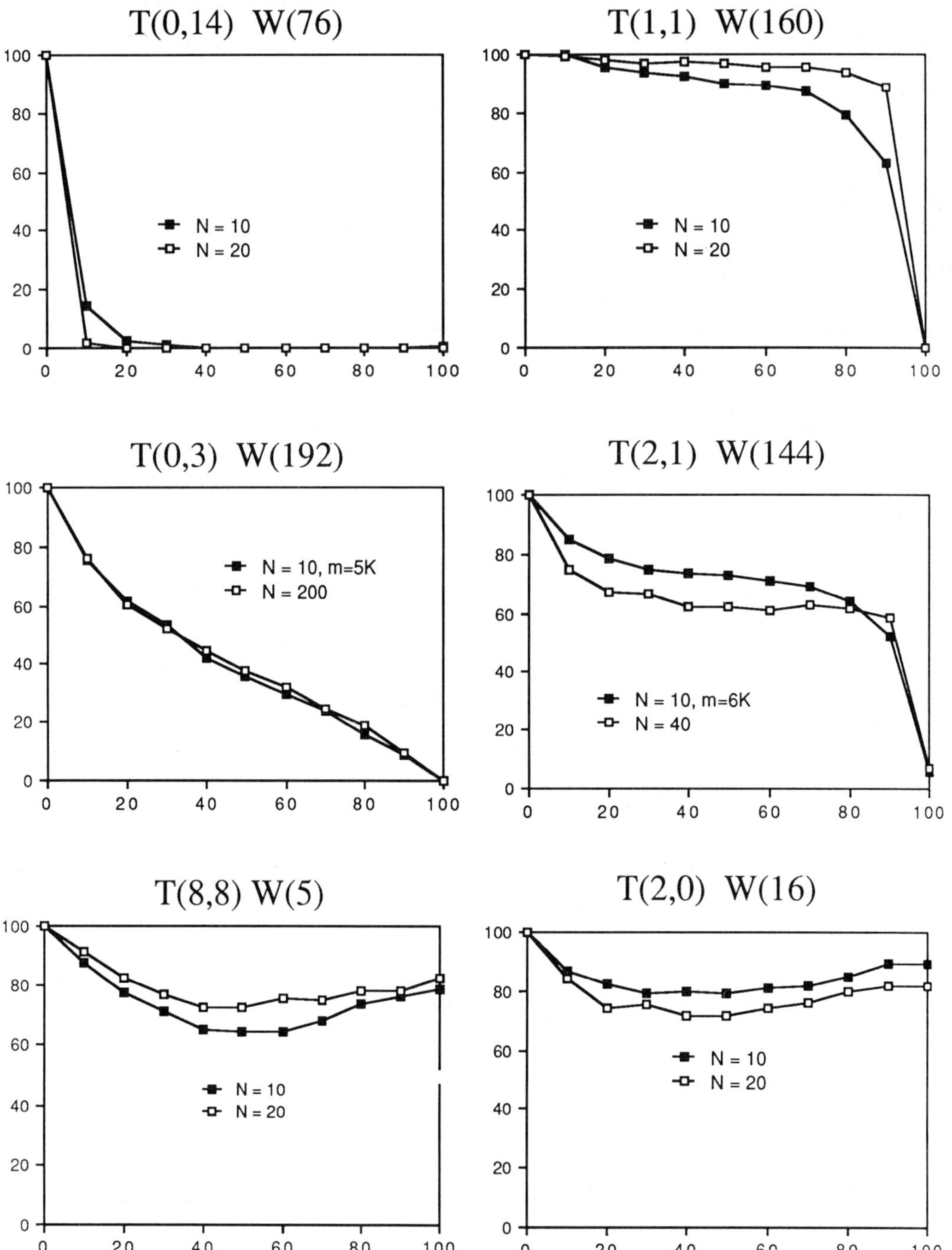

Fig. 2. Other dominance function shapes. The top four T functions are classed as coherent, except for that of $T(0,14)$ at $N = 10$, which is (just barely) bipolar. The bottom two T's show (borderline) bipolar functions. The ordinate is attractor dominance in percent; the abscissa is distance from the attractor in percent Hamming distance.

dominance function at a given lattice size or sizes. For example, "$T(1,1)$ is coherent at $N = 10$" means that the dominance function of $T(1,1)$, $N = 10$, is coherent.

Table 2 shows the class membership counts for the three dominance localization classes at N and N'.

Recalling Li and Packard's characterization of h in table 1 as one of the "hot bits" in circular CA (a is the other) [16], it may not be surprising that the main feature distinguishing T's in coherent and bipolar classes is whether they are odd or even, that is, whether h is 1 or 0.

In more detail (also see table 3), at N, with one exception – $T(8,14)$ – all coherent T's are odd; with two exceptions – $T(1,9)$ and $T(7,1)$ – all bipolar T's are even. Diffuse T's are $T(0,0)$, $T(2,8)$ and $T(6,1)$.

Moving from N to N', the coherent class loses one odd T to diffuse and picks up four additional even T's: $T(0,14)$, $T(1,14)$, $T(2,14)$, and $T(8,10)$, to become 20, out of 25, odd. The bipolar class loses four even T's to coherent and four to diffuse, while picking up one odd T from diffuse, to become 53, out of 56, even.

5. How lattice size affects attractor dominance

Dominance functions change with lattice size, as can be seen in figs. 1 and 2. As lattice size increases $T(0,14)$ tends toward $T(0,15)$, and $T(1,1)$ tends toward an extreme pattern not shown by the three trivial rules in the CA class examined. $T(0,3)$ and $T(2,1)$ are examples of rules that may not tend toward the shapes of trivial rules as lattice size increases. $T(8,8)$ and $T(2,0)$ show, respectively, increased and decreased overall levels of dominance in larger lattices.

To better display the detailed constellation of dominance function shapes, and to make the change in dominance with lattice size more apparent, two dimensions – mean dominance and polar salience – can be defined to locate the functions within each pattern class.

Mean dominance for a given T at a given N is

$$\overline{D} = (D_{10} + D_{20} + \cdots + D_{100})/10.$$

In the definition above I neglect the entirely predictable $D_0 = 100$. Mean dominance is a straightforward measure of overall attractor dominance.

Polar salience quantifies in a rough but direct way the intensity of the attractor's distinctive dominance. For coherent functions, polar salience is the salience of the attractor "pole" (the attractor itself). Given the generally well behaved nature of coherent functions, a very simple measure of the (negative) gradient at D_0 appears adequate to quantify polar salience. Accordingly, for coherent functions polar salience is measured by the salience of the attractor, defined as

$$S_\mathrm{A} = D_0 - D_{10} = 100 - D_{10}.$$

Bipolar functions appear to be such that they can be graded by measuring only the salience of the contrary pole, those states at farthest remove from the attractor. Salience of the contrary pole is defined as

$$S_\mathrm{C} = D_{100} - \min(D_{10}, D_{20}, \ldots, D_{90}).$$

Plotting the 88 pairs of points resulting from the definitions yields fig. 3. In fig. 3 mean dominance, $\overline{D}$, is plotted vertically against polar salience. The three dominance localization pattern classes are plotted separately in the figure. While the classes share a common ordinate, their horizontal dimensions are different. S_A is the abscissa for coherent patterns, and S_C for bipolar patterns. No attempt was made to define polar salience for diffuse patterns. Despite these dimensional differences, it appears to make sense to align the plots for the three classes as shown.

In fig. 3, a solid rectangle locates a dominance pattern for a particular T at N. An open rectangle locates a pattern for a T at N'. An arrow joining two rectangles indicates a change in phase space dominance patterns for a given T produced by the N to N' change in lattice size. An arrow re-entering a rectangle indicates that there was little or no change in patterns. The locations of the three trivial rules' patterns are marked.

The vectors in fig. 3 give some idea of how quickly T's dominance functions are changing with lattice size. However, it needs to be kept in mind that the $N' - N$ difference is not the same for all T's, and that vectors which indicate movement

Table 3

Data. Dominance localization class, mean dominance, and polar salience for each rule. T's are ordered by S at N.

T	at $N=10$			N'	at N'			T	at $N=10$			N'	at N'		
	Class	$\overline{D}$	S		Class	$\overline{D}$	S		Class	$\overline{D}$	S		Class	$\overline{D}$	S
1, 1	c	79	2	20	c	86	1	6, 8	b	40	16	20	d	42	–
0, 1	c	80	3	20	c	86	1	3, 10	b	43	16	20	b	36	3
2, 1	c	65	15	40	c	58	25	10, 12	b	56	16	20	b	47	12
3, 1	c	59	22	20	d	54	–	9, 10	b	60	16	20	b	61	4
1, 3	c	35	23	200	c	36	28	0, 8	b	18	17	20	b	4	10
0, 3	c	35	25	200	c	36	24	10, 4	b	54	17	20	b	39	9
3, 5	c	40	27	40	c	32	46	6, 0	b	53	19	20	b	45	8
2, 3	c	32	31	20	c	30	33	3, 14	b	22	20	20	b	11	7
3, 3	c	23	36	20	c	17	46	0, 6	b	44	20	100	b	40	22
2, 5	c	32	42	20	c	28	46	6, 2	b	59	21	20	b	59	5
1, 7	c	15	43	20	c	9	58	1, 6	b	52	22	40	b	49	22
3, 9	c	36	48	20	c	27	64	7, 6	b	57	22	20	b	57	10
8, 14	c	17	54	20	c	4	83	3, 6	b	49	23	20	b	42	18
0, 7	c	11	60	20	c	9	67	7, 2	b	73	24	20	b	69	14
1, 11	c	9	63	20	c	4	80	2, 6	b	44	29	20	b	38	26
2, 11	c	8	67	20	c	2	87	9, 6	b	18	30	13	b	15	29
2, 9	c	12	70	20	c	2	93	7, 10	b	44	32	20	b	35	18
0, 9	c	7	73	20	c	1	92	9, 8	b	63	33	20	b	60	21
0, 11	c	3	78	20	c	1	95	10, 0	b	57	35	20	b	46	33
6, 9	c	9	79	13	c	7	85	3, 4	b	63	36	20	b	59	36
2, 13	c	3	81	20	c	0	98	3, 2	b	64	37	20	b	55	41
0, 15	c	0	100	–	c	0	100	11, 4	b	30	39	20	b	12	11
0, 0	d	100	–	–	d	100	–	1, 10	b	10	40	20	b	2	12
2, 8	d	20	–	20	d	4	–	9, 2	b	40	40	20	b	34	24
6, 1	d	47	–	20	b	42	4	10, 10	b	44	40	20	b	33	35
0, 14	b	2	1	20	c	0	98	10, 2	b	65	40	20	b	59	43
1, 0	b	100	1	100	d	100	–	3, 0	b	70	41	20	b	60	55
1, 14	b	4	4	20	c	0	99	10, 8	b	70	41	40	b	66	50
2, 14	b	10	4	20	c	2	85	7, 8	b	34	43	20	b	14	13
1, 9	b	15	4	20	b	3	0	11, 8	b	65	46	20	b	58	52
2, 10	b	20	4	20	b	8	1	2, 4	b	64	48	20	b	59	52
0, 10	b	5	6	20	b	1	0	9, 0	b	29	55	20	b	13	58
8, 10	b	40	8	20	c	27	51	7, 14	b	48	62	20	b	49	60
6, 10	b	47	8	20	d	48	–	3, 8	b	34	65	20	b	21	44
2, 0	b	83	10	20	b	77	10	6, 6	b	41	71	12	b	40	74
1, 2	b	79	11	200	b	77	14	7, 1	b	33	75	20	b	16	89
0, 2	b	72	12	200	b	69	14	7, 0	b	40	76	20	b	18	95
6, 14	b	44	14	20	b	41	3	14, 8	b	38	85	20	b	28	93
8, 6	b	46	14	20	d	37	–	14, 0	b	38	87	20	b	28	94
8, 8	b	73	14	20	b	79	10	11, 2	b	25	90	20	b	14	99
8, 2	b	65	14	20	b	60	13	11, 0	b	25	92	20	b	14	100
2, 2	b	68	14	20	b	62	14	1, 8	b	23	93	20	b	13	100
8, 0	b	81	14	20	b	86	11	3, 12	b	17	98	20	b	10	100
2, 12	b	9	15	20	b	1	1	15, 0	b	10	100	–	b	10	100

between localization classes are likewise not comparable.

The general impression given by fig. 3 is that, with some exceptions, primarily near the center of the figure, there is implied movement of patterns toward the lower center of the figure, and toward the lower right. Most coherent T's appear to be moving toward the trivial $T(0,15)$. Bipolar T's split, some moving toward $T(0,15)$, others toward $T(15,0)$, and a very few toward $T(0,0)$.

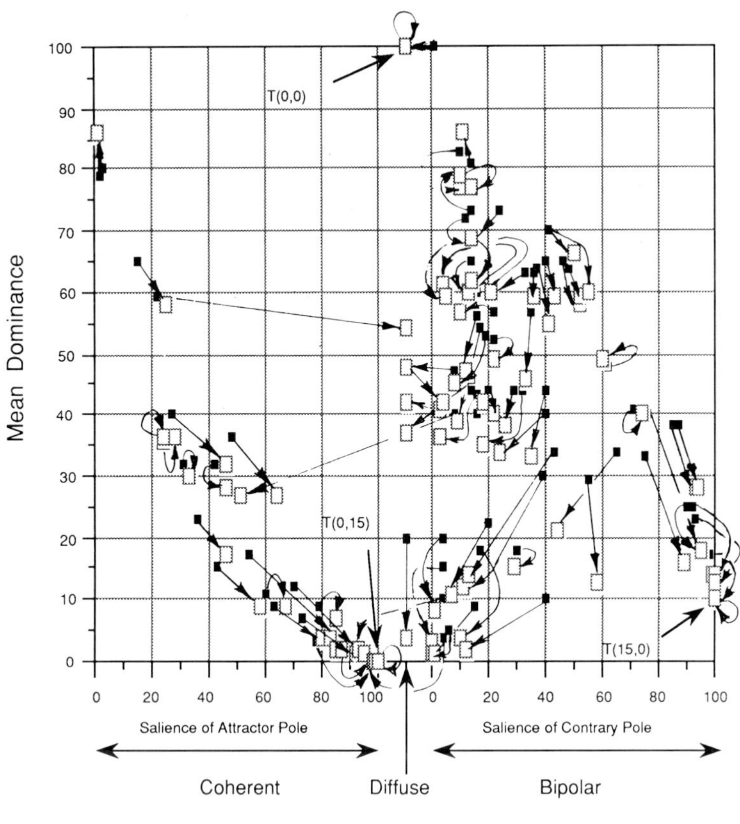

Fig. 3. Attractor dominance patterns. A solid rectangle locates a dominance pattern at N; an open rectangle, at N'.

Notice that some T's do not move much at all.

To list the main conclusions drawn from fig. 3:

– For relatively small lattice sizes, attractor dominance patterns fall into two main classes.

– For most T's, overall attractor dominance *decreases* with increasing lattice size. Only ten T's show increased $\overline{D}$: 0,1; 0,3; 1,1; 1,3; 6,8; 6,10; 7,14; 8,0; 8,8; 9,10.

– There appears to be a widespread trend toward trivialization of attractor dominance patterns as lattices become larger. That is, the trend is toward attractor dominance patterns shown by trivial rules.

– An implication of the attractor dominance trivialization described above is that when measured by percent Hamming distance, T's that trivialize to $T(0,15)$ or to $T(15,0)$ have attractors whose phase space dominance is increasingly localized.

– Those T's that do become more like the trivial

rules appear do so along common routes that lie within the dominance localization classes. If this observation holds in more generality, it may provide a way of predicting dominance pattern characteristics.

– An extreme dominance pattern unlike those of the three trivial rules in this CA family is approached by $T(0,1)$ and $T(1,1)$.

– While triviality of their attractor dominance patterns may be the fate of most T's in large lattices, there are some T's which are affected differently by increased lattice size, namely, those which show either very slow change with N, or where distance from the attractor strongly interacts, in the statistical sense, with lattice size. $T(0,3)$, and $T(2,1)$, shown in fig. 2, are, respectively, examples of slow change and interaction. Rules that resemble these should at least resist trivialization at moderate N values. Some of them *may* resist in the limit, in the sense of tending toward stable, non-extreme patterns as lattices become very large.

– In any event, the T's of the item just above provide a set of rules which may have interesting large-lattice phase spaces. The 21 rules so identified are: 0,2; 0,3; 0,6; 1,0; 1,2; 1,3; 1,6; 2,1; 2,3; 3,1; 3,5; 6,1; 6,2; 6,8; 6,10; 6,14; 7,6; 7,14; 9,8; 9,10; and 10,8.

6. Conclusion

We have seen that some phase space characteristics of the attractors of a random-structure CA family can be conceptualized in a space that has few dimensions.

An implication of this work is that it may be possible to characterize families of random-structure CA rules as to attractor dominance by reference to a pattern space and to a set of patterns that persists in that space as lattice size becomes arbitrarily large. For the CA family of interest here the reference set appears to consist of the dominance patterns of the three trivial rules, one new extreme, trivial-like pattern, and a subset of not more than 21 intermediate patterns.

Naturally, experimentation with larger lattices and more complex automata is needed to test the generality of the method proposed here.

The pattern space dimensions proposed in this paper were developed intuitively. I trust that theoretical analysis will provide dimensions that are more profoundly descriptive.

Since the dominance functions examined here are constructed from aggregated data, a natural and important question to ask is: How much variability would be attributable to lattice structure if we were to de-aggregate the functions? Or, put differently, how well do aggregated dominance functions represent patterns of attractor dominance in individual phase spaces? Similarly, although regular-structure CA such as the circular CA dealt with in much of the literature (e.g., refs. [6,17]) are a subset of random-structure CA, we can reasonably ask how much the attractor dominance functions of circular CA differ from random-structure functions.

The use of a metric scaled relative to lattice size (percent Hamming distance) leaves open the question of how attractor dominance changes with distance measured in absolute units. This question is related to the work cited above on attractor-to-attractor transitions in low noise, and is a matter of considerable interest.

Acknowledgements

I thank the anonomous referees for close readings of the paper and for very helpful commentary.

Some of the work on this paper was accomplished while the author was at the School of Business, University of Southern Indiana, Evansville, Indiana, U.S.A.

References

[1] S. Wolfram, ed., Theory and Applications of Cellular Automata (World Scientific, Singapore, 1986).

[2] S. Kauffman and R. Smith, Adaptive automata based on Darwinian selection, Physica D 22 (1986) 68–82.

[3] Y. Lee, Adaptive non-deterministic CA, presented at Workshop: Cellular automata, theory and experiment, Los Alamos, NM, Sept. 9–12, 1989.

[4] H. Sieburg, Adaptive processes in anticipatory systems, presented at Workshop: Cellular automata, theory and experiment, Los Alamos, NM, Sept. 9–12, 1989.

[5] S. Kauffman, Metabolic stability and epigenesis in randomly constructed genetic nets, J. Theoretical Biology 22 (1969) 436–467.

[6] S. Wolfram, Universality and complexity in cellular automata, Physica D 10 (1984) 1–84.

[7] L. Hurd, Formal language characterizations of cellular automata limit sets, Complex Systems 1 (1987) 69–80.

[8] W. Li, Power spectra of regular languages and cellular automata, Complex Systems 1 (1987) 107–130.

[9] K. Kaneko, Attractors, basin structure and information processing in cellular automata, in: Theory and Applications of Cellular Automata, ed. S. Wolfram (World Scientific, Singapore, 1986) pp. 367–399.

[10] C. Walker, Stability of equilibrial states and limit cycles in sparsely connected, structurally complex Boolean nets, Complex Systems 1 (1987) 1063–1087.

[11] C. Walker and W. Ashby, On temporal characteristics of behavior in a class of complex systems, Kybernetik 3 (1966) 100–108.

[12] C. Walker, Behavior of a class of complex systems: The effect of system size on properties of terminal cycles, J. Cybern. 1 (1971) 55–67.

[13] L. Hurd, Table 1 in Appendix: Properties of the $k = 2$, $r = 1$ cellular automata, in: Theory and Applications of Cellular Automata, ed. S. Wolfram (World Scientific, Singapore, 1986) p. 485–492.

[14] C. Langton, Studying artificial life with cellular automata, Physica D 22 (1986) 120–149.

[15] H.A. Gutowitz, J. Victor and B. Knight, Local structure theory for cellular automata, Physica D 28 (1987) 18–48.

[16] W. Li and N. Packard, The structure of the elementary cellular automata rule space, Technical Report CCSR-89-8, Beckman Institute, University of Illinois, Urbana (1989)

[17] S. Wolfram, Statistical mechanics of cellular automata, Rev. Mod. Phys. 55 (1983) 601–644.

[18] W. Li, personal communication, 1989.

[19] A. Gelfand and C.C. Walker, Ensemble Modeling (Dekker, New York, 1984).

Physica D 45 (1990) 452–460
North-Holland

PERIODIC ORBITS AND LONG TRANSIENTS IN COUPLED MAP LATTICES

R. LIVI[a, b], G. MARTÍNEZ-MEKLER[b, c] and S. RUFFO[b, d]

[a] *Dipartimento di Fisica, Università di Firenze, Largo E. Fermi 2, 50125 Florence, Italy*
[b] *Istituto Nazionale di Fisica Nucleare, Sezione di Firenze Largo E. Fermi 2, 50125 Florence, Italy*
[c] *Instituto de Fisica, UNAM, Apdo. Postal 20-364, 01000 Mexico D.F., Mexico*
[d] *Facoltà di Scienze M.F.N., Università della Basilicata, Potenza, Italy*

Received 17 January 1990
Revised manuscript received 27 February 1990

A coupled map lattice model is studied which presents transient and asymptotic chaotic states depending on the value of the control parameters. A first-order approximation detects the presence of asymptotic periodic attractors. The statistics of transient times as well as their dependence on lattice size are investigated.

1. Introduction

The study of complex dynamical behavior is currently of interest in many research fields ranging from physics to chemistry and biology. This has induced a demand for sufficiently general and simple models, which reproduce the main features of such dynamical regimes. Coupled map lattices (CML) have been proposed by various authors [1] as a promising approach in this direction. They are usually defined as discrete (both in space and time) dynamical systems, where replicas of a map are coupled via a nearest-neighbor interaction, reproducing a discrete version of a diffusive coupling. Temporal and spatial discreteness allow natural computer implementation of such models. On the other hand, due to the diffusive coupling, CML are expected to be a simplified version of systems, where complex evolution may arise as a product of a collective behavior originated by coherent interactions among degrees of freedom. For instance, this is a typical situation for the transition to turbulence via spatiotemporal intermittency in fluid systems.

Moreover, recent results (third paper of ref. [1]) have shed new light on the possibility of a statistical description of CML. It is expected that the crossover between different global dynamical regimes can be described in terms of a phase transition mechanism. In a recent paper [2] this conjecture has been supported by numerical analysis for a purely expanding linear piecewise map and for the logistic map. Both cases are characterized by the appearance of new ground states as the strength of the coupling is varied.

A similar crossover between "turbulent" and "soliton" phases was studied in a one-dimensional cell dynamical model for chemical turbulence [3]. The aim was to describe a typical reaction–diffusion process by the introduction of a three-state cellular automaton (CA) cell dynamics. This appeared to be a necessary ingredient, because complex behaviors in reaction–diffusion processes are expected to appear from the interaction of a great number of orbitally stable limit cycles.

In this paper we study a CML model which describes reaction–diffusion processes. Its definition and main features are presented in section 2. In the following sections the reader will realize that the proposed model is characterized by a very interesting and rich variety of dynamical behaviors. It is worth stressing that knowledge of the spatially homogeneous dynamics discussed in section 3 represents a sort of "first-order" dynamical description of our model. More precisely, the bifurcation diagram characteristic of the superstable version of our map shows that only periodic orbits are present. This provides a guide for the description of the asymptotic dynamics in the inhomogeneous case, as discussed in section 4. In fact, the main feature of this model is the presence of periodic attractors (typical of the homogeneous dynamics) towards which the CML may relax after some transient time. We have observed numerically that the statistics of the transient times is typically Poisson-like. This appears to be a generic feature of the model. This indicates the presence of characteristic escape times, depending on the values of the parameters. As discussed in section 4, Lyapunov characteristic exponent (LCE) analysis has provided a sound description for the case of inhomogeneous initial conditions. In particular the dynamical analysis of the LCE spectrum clearly characterizes:

– the various dynamical mechanisms which lead to periodic attractors (when they are present);

– the "chaoticity" of the transients towards these attractors.

Here the word "chaoticity" is used in order to stress the fact that, for any finite array of maps, the CML dynamics relaxes to a LCE spectrum with a positive component, before reaching over longer times its asymptotic value. We want to stress that this is a well-defined concept. In analogy with low-dimensional repellers, the positive part of the LCE spectrum can be interpreted as a measure of the chaotic component of the invariant set characterizing the transient dynamics. We also report some results on the exponential growth of the average transient time for increasing lattice dimension, thus showing the relevance of transient dynamics to the properties of the system in the thermodynamic limit. Finally, we report some simple analytic calculations of the LCE spectrum when the dynamics is attracted towards a homogeneous fixed point.

2. The model

CML models are defined on a regular lattice where the state x of a variable at site (or cell) i evolves according to a map F applied to its neighborhood ν_i:

$$x_i^{t+1} = F\left(\sum_{j \in \nu_i} \alpha_j x_j^t\right),\tag{1}$$

where t represents the discrete time step and $i = 1, 2, .., N$ with periodic boundary conditions.

In one dimension the α_j's are usually chosen as follows:

$$\alpha_i = 1 - \varepsilon, \quad \alpha_{i\pm1} = \varepsilon/2, \quad \varepsilon \in [0, 1].\tag{2}$$

This rule mimics a space–time discrete version of a nonlinear diffusive process as observed, for instance, in reaction–diffusion systems. In such cases for $\varepsilon = 0$ (isolated cells) the dynamics (1) should generate a globally stable limit cycle. The simplest choice that complies with this constraint is the reduction of dynamics of an isolated cell to a cyclic visitation of discrete levels [3]. This actually reduces CML to CA.

One of the motivations behind this work is to study a CML model in which the periodic behavior is a consequence of the coupling and not intrinsic to the single map dynamics. Keeping this in mind we have chosen a map of the interval with a fixed-point dynamics, in which an expanding and a contracting component can be identified. We shall call these components "turbulent" and "quiescent" phases, respectively. The mapping is given by

$$F_{a,b}(x) = x/a, \qquad\qquad x < a,$$
$$\qquad\quad = a + b(x - a), \quad x \geq a,\tag{3}$$

with $0 < a, b < 1$. Any initial condition (apart from $x = 0$) will eventually be mapped into the stable fixed point $x = a$.

For the coupled problem our choice of parameters α_j is

$$\alpha_i = 1, \quad \alpha_{i\pm1} = \varepsilon/2, \quad \varepsilon \in [-1, \infty).\tag{4}$$

Notice that this parameter selection requires the introduction of a modulo 1 operation in the argument of F. This corresponds to a shift operation, which, as we shall see in the following sections, gives rise to different global periodic behaviors. Our coupled map dynamics may now be summarized by the following formula:

$$x_i^{t+1} = G_{a,b}(\tfrac{1}{2}\varepsilon D x_i^t + \eta x_i^t),\tag{5}$$

where $G_{a,b}(x) = F_{a,b}([x] \bmod 1)$, D is the discrete Laplace operator and η is the reaction constant. The argument of G makes explicit the relation with a reaction–diffusion system; in our case, both the reaction and the diffusion constants are related to the coupling parameter ε, the former being $\varepsilon/2$ and the latter $1 + \varepsilon$. We will see that the model so defined exhibits extremely rich dynamical behavior, with periodic and chaotic regimes arising as collective phenomena. In this respect the relevance of the model goes beyond the modeling of a reaction–diffusion process.

3. Homogeneous dynamics

Homogeneous initial conditions, i.e. $x_i^0 = \bar{x}^0 \ \forall i$, $\bar{x}^0 \in [0, 1]$, reduce eq. (5) to

$$\bar{x}^{t+1} = G_{a,b}[(1 + \varepsilon)\bar{x}^t], \tag{6}$$

where the explicit dependence on the space index i is lost but the ε dependence is maintained. This persistence of the ε dependence is a feature of our model, at variance with the usual CML treatments reported so far in the literature [1] [#1]. It allows us, as we shall show further on, to construct a "first-order" approximation of the collective behavior of eq. (5) with generic initial conditions, in terms of the analysis of the homogeneous dynamics (6). We shall devote this section to the study of the homogeneous dynamics (6), for which some exact results can be established, with the aim of gaining some insight of the behavior for random initial conditions.

3.1. The case b = 0

In section 2 we mentioned that the modulo-1 operation is the origin of multiperiodic dynamics. The study of eq. (6) for $b = 0$, which corresponds to the map with a superstable fixed point $\bar{x} = a$ (see fig. 1), provides a first understanding of this statement. It is straightforward to observe that when $\varepsilon \in [-1, a - 1)$, for any initial condition,

---

[#1] Let us observe that such a dependence is maintained also for diffusive coupling (see ref. [2]), when spatially periodic solutions of period larger than 1 are considered.

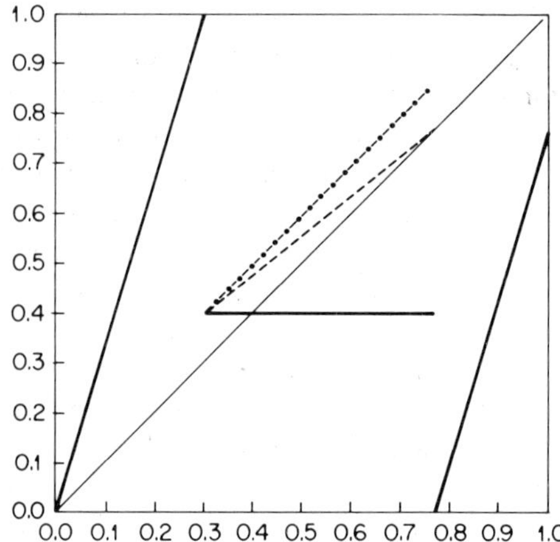

Fig. 1. The thick solid line segments are the graph of eq. (6) for $(a, \varepsilon, b) = (0.4, 0.3, 0.0)$; when $b = b_c = 0.61538$ the horizontal line should be substituted by the dashed line; when $b = b_r = 0.76923$ it should be substituted by the dash–dotted one.

$\bar{x}^0$, the system will relax to the stable fixed point $\bar{x} = 0$, while at $\varepsilon = a - 1$, the system remains frozen in $\bar{x}^0$. For $\varepsilon > a - 1$, there is always a stable periodic orbit present. This can be proved by the following:

Lemma 1 *For any $\bar{x}^0$, $0 < a < 1$ and $\varepsilon \in (a - 1, \infty)$ the preimages in eq. (6) of $\bar{x} = a$ have Lebesgue measure $\mu_L = 1$, i.e. they coincide with the whole interval $[0, 1]$ apart from a set of zero Lebesgue measure.*

Proof: When $\varepsilon \in (a - 1, \infty)$ the first preimage of the point $\bar{x} = a$ is the set $A = (a/(1 + \varepsilon), 1/(1 + \varepsilon))$; the following preimages are then some intervals I_k such that $\mu_L(I_k) \neq 0$ and $I_k \bigcap A = \emptyset$, i.e. the I_k belong to the set where $G_{a,0}$ is expanding. Let us suppose now that there exists a set J, $\mu_L(J) \neq 0$ such that $J^n \bigcap A = \emptyset \ \ \forall n = 1, 2, \ldots$ (where $J^n \equiv G_{a,0}^n(J)$ stands for the nth image of J). By definition, $J^n \bigcap(\bigcup_k I_k) = \emptyset \ \ \forall n$, but this is impossible. According to the hypothesis, J and its images must belong to the expanding component of $G_{a,0}$, and there will exist an $N < \infty$ such that $J^N \bigcap(\bigcup_k I_k) \neq \emptyset$. $\square$

Theorem 1 *For any $0 < a < 1$ and $\varepsilon \in (a - 1, \infty)$ the map (6) has only periodic orbits for μ_L-almost all $\bar{x}^0$.*

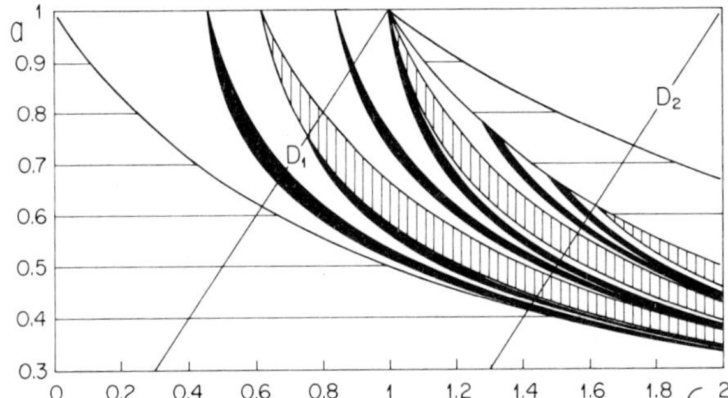

Fig. 2. Partial (a, ε) bifurcation diagram for eq. (6), showing some of the regions of stable periodic dynamics (horizontal stripes indicate period 1, vertical period 2 and black period 3). To the left of line D_1 (1-reinjection zone) all period-1,2,3 tongues are indicated. In between D_1 and D_2 all period-1 tongues and all period-2 and -3 tongues bounded by the first two period-1 tongues are shown. To the right of D_2, all period-2 and some of the period-3 delimited by the first two period-1 tongues are depicted.

Proof: As a consequence of lemma 1 for μ_{L}-almost all $\bar{x}^0$ the dynamics will eventually reach the point $\bar{x} = a$, whose image μ_{L}-almost surely belongs to the set of preimages of a. This proves the theorem and moreover implies that $\bar{x} = a$ belongs to every stable periodic orbit, the period of which depends on the values of a and ε. $\qquad\square$

As we vary ε from $a - 1$ to 0 we find stable solutions of period $n + 1$ for $a^{n/(n+1)} - 1 \le \varepsilon \le a^{(n-1)/n} - 1$, $n = 1, 2, \ldots$. When $\varepsilon = 0$ we have the decoupled case with the fixed point solution $\bar{x} = a$. Once ε is positive then the coupling is diffusive, which is physically more interesting. In this case the (a, ε) parameter space is partitioned into tongues of different periodicity. As a consequence of theorem 1, these tongues fill the (a, ε) space and do not overlap. This last property is due to the non-intersection of the preimages of a for mapping (6), i.e., multistability is ruled out for $b = 0$.

In fig. 2, the plot of just some of the low-period tongues already gives an indication of the complexity of the full bifurcation diagram. A useful classification of this diagram can be made in terms of the number of reinjections, i.e. the number of plateaus in the graph of mapping (6), for a given value of (a, ε).

For $a > \varepsilon$, we are in the 1-reinjection zone of the (a, ε) diagram (region on the left of D_1 in fig.

2). Within this region, the tongues are disposed hierarchically: each tongue of period $n > 1$ has on both sides one tongue of period $n + 1$, with the separatrices of each tongue corresponding to infinitely periodic solutions.

As we enter the 2-reinjection zone bounded by the straight lines $a = \varepsilon$ and $a = 1 + \varepsilon$ (D_1 and D_2 of fig. 2, respectively), the above structure is strongly enriched by the generation of new tongues. Moreover, new features appear: for example, a period-3 tongue shares a boundary with the second period-1 tongue. In general, this trend toward increasing complexification is observed as the number of reinjections grows.

It is not our aim in this work to give a detailed description of the bifurcation diagram but rather focus our attention on the dynamical behavior of the CML in some regions of the 1- and 2-reinjection zones between the first and second period-1 tongues.

3.2. The case $b \neq 0$

When $b \neq 0$, the overall periodic behaviour is modified by the appearance of chaotic dynamics. Some features of the bifurcation diagram of the now three-dimensional parameter space (a, ε, b) can be studied by looking at the evolution of the $b = 0$ plane for increasing values of b. As an example of this procedure we shall consider the first period-1 tongue.

In fig. 1, we plot the mapping (6) for the (a, ε, b) parameter values: $(0.4, 0.3, 0)$, $(0.4, 0.3, b_{\mathrm{c}})$ and $(0.4, 0.3, b_{\mathrm{r}})$, where $b_{\mathrm{c}}(a, \varepsilon) = [1/(1+\varepsilon) - a]/(1-a)$ is the value of b at which the mapping looses its fixed point and $b_{\mathrm{r}}(a, \varepsilon) = 1/(1+\varepsilon)$ is the minimum value of b for which the mapping becomes fully unstable.

Since we have chosen (a, ε) within the $b = 0$ first period-1 tongue, as long as $0 < b < b_{\mathrm{c}}$ we will always have a stable fixed point $\bar{x}_{\mathrm{fp}} = (b - 1)a/[b(1 + \varepsilon) - 1]$ within the interval A of lemma 1. Moreover, the fixed point is stable and its domain of attraction is the whole interval, since the preimages of A are the same as in the $b = 0$ case. However, in this case the inverse image of a is no longer the interval A. This shows that the system evolves asymptotically to a stable fixed point

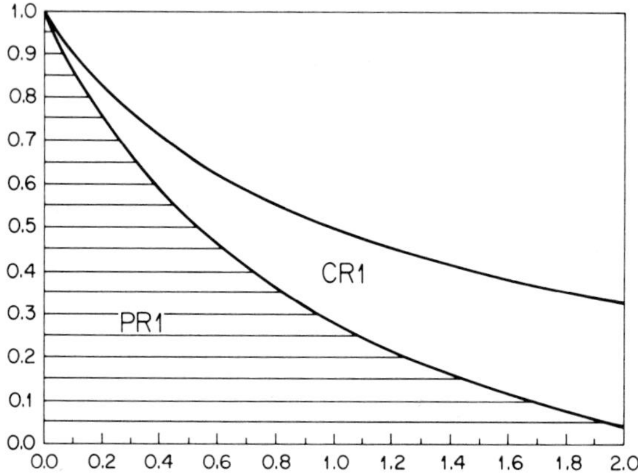

Fig. 3. Diagram of the first period-1 tongue in the (a, ε) parameter plane for different b values: when $b = 0$ the tongue is the sum of regions PR1 and CR1; for $b = 0.2$ it has been compressed to region PR1. In this last case region CR1 corresponds to a chaotic behavior.

instead of reaching in a finite number of steps a period-1 orbit.

When $b > b_r$ the mapping is unstable since all the slopes are larger than one. For the remaining case of $b_r > b > b_c$ we conjecture, on the basis of numerical evidence, that stable periodic orbits are not present. So far our calculation of the Lyapunov exponent within this region suggests a chaotic dynamics. The above information is summarized in fig. 3. For $b \neq 0$, the $b = 0$ first period-1 tongue delimited above by $a(\varepsilon, 0) = 1/(1 + \varepsilon)$ is compressed to the striped region PR1 bounded above by $a(\varepsilon, b) = [(1/(1 + \varepsilon)) - b]/(1 - b)$. Within this region the stable period-1 dynamics is preserved. Region CR1 on the other hand corresponds to a chaotic behavior.

The mechanism we have so far described for the evolution of the first period-1 tongue takes place for any tongue of the $b = 0$ bifurcation diagram. The tongue reduces in size as b grows, leaving behind a region with no stable periodic orbits. Namely, if we take any set of parameters $(a, \varepsilon, b = 0)$, it must belong to a tongue, say of period n; we may then define the values b_c^n and b_r^n for the graph of the nth iteration of the mapping in analogy to b_c and b_r of fig. 1. Note that $b_r(\varepsilon) \geq b_r^n(a, \varepsilon)$, which is in agreement with the fact that for any $\varepsilon \geq (1 - b)/b$ the dynamics is globally unstable.

4. Inhomogeneous initial conditions

The dynamical scenarios which characterize the homogeneous map (6) are ruled by different kinds of attractors. For any fixed value of b the (a, ε) space is partitioned into zones dominated by periodic or chaotic dynamics. Here we want to study how these dynamical regimes change when the coupled map dynamics (5) is considered in full generality. Even when relatively simple local dynamics is considered, the coupling of a large number of spatial degrees of freedom may produce complex collective behaviors. We will stress in this paper the presence of chaotic transients which grow with system size (this feature has already been noted [4] for other models). The knowledge of the partition of the $(a, \varepsilon, b = 0)$ space into periodic attractors for the homogeneous state provides a good approximation for the orbits of lower period towards which an inhomogeneous or random initial state eventually converges. A more complicated situation arises for $b \neq 0$, when chaotic attractors and chaotic transients may coexist, which require a more refined description of the invariant and transient measures. For fixed N (≤ 20) and fixed parameter values a, ε by varying b we have observed a value $\bar{b}(N, a, \varepsilon)$, below which the initially inhomogeneous state relaxes to a periodic attractor, showing a transient chaotic behavior. Above $\bar{b}$ numerical experiments indicate the presence of a chaotic attractor. We will not analyze this case in this paper, but rather concentrate on the study of chaotic transients leading to periodic attractors. From the analysis of the previous section we know that the necessary condition for this to happen is $b < b_c$.

Most of the results that we are going to discuss have been obtained by numerical simulations. The analysis of a chaotic transient requires two ingredients. First, one must compute a transient mean time over different initial conditions, when the dynamics relaxes on an attractor. We shall see that the probability distribution of transient times decreases exponentially at large times. Hence, a typical escape time can be defined, as in the theory of low-dimensional repellers [5]. Second, one must characterize the chaoticity within the transient. For this purpose we have chosen to study the finite time spectrum of Lyapunov characteristic exponents (LCE), defined as follows:

$$\lambda_{i+1}(T) = \eta_{i+1}(T) - \eta_i(T),$$

where

$$\eta_i(T) = \frac{1}{T} \sum_{t=1}^{T} \ln \frac{\|\boldsymbol{\xi}_1^{t+1} \wedge \boldsymbol{\xi}_2^{t+1} \wedge \ldots \wedge \boldsymbol{\xi}_i^{t+1}\|}{\|\boldsymbol{\xi}_1^{t} \wedge \boldsymbol{\xi}_2^{t} \wedge \ldots \wedge \boldsymbol{\xi}_i^{t}\|},$$

$\wedge$ being the external product and $\boldsymbol{\xi}_j^t, j = 1, \ldots, N$ a set of N-dimensional linearly independent vectors chosen to be initially an orthonormal basis in the tangent space of the dynamical system (5). The linear dynamics of the $\boldsymbol{\xi}_j^t$ is therefore

$$\boldsymbol{\xi}_j^{t+1} = \mathbf{A}^t \boldsymbol{\xi}_j^t,$$

with

$$(\mathbf{A})_{ij}^t = \frac{\partial F}{\partial y_i^t} \frac{\partial y_i^t}{\partial x_j^t}, \quad y_i^t = \sum_{j \in \nu_i} \alpha_j x_j^t. \tag{7}$$

The existence of a positive component of the LCE spectrum as $T \to \infty$ guarantees the presence of a chaotic attractor [5]. In the presence of a chaotic transient we observe the relaxation within a finite time to a LCE spectrum having a positive component, before the latter asymptotically converges to the full negative LCE spectrum typical of a periodic attractor.

4.1. The case b = 0

In the superstable case $b = 0$, for finite N, the LCE spectrum collapses to $-\infty$. This is easily revealed numerically by the convergence to the null vector of the set of tangent vectors $\boldsymbol{\xi}_i$. We have observed two main mechanisms for this convergence, depending on the chosen initial condition. If the LCE spectrum uniformly collapses to $-\infty$ following the largest LCE, then, the dynamics converges to the corresponding homogeneous periodic orbit in the (a, ε) parameter space. New periodic solutions, unexpected from the analysis based on the homogeneous state, arise when a subset of the LCE spectrum first collapses to $-\infty$, followed later by the largest LCEs. In this case the asymptotic state of the lattice is inhomogeneous in space and periodic in time. This multistability is not present in the first period-1 tongue. In fact $0 < \varepsilon < (1 - a)/a$ is a sufficient condition for the

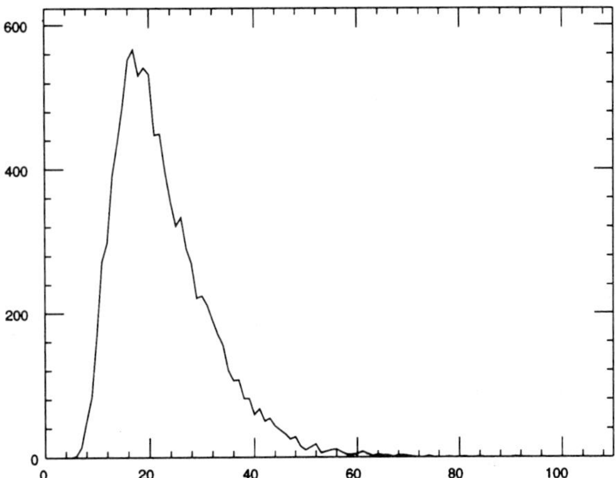

Fig. 4. Unnormalized frequency distribution of transient times for $N = 50$ and $(a, \varepsilon, b) = (0.75, 0.3332, 0)$ inside the first period-1 tongue of fig. 2. The statistical sample has been obtained by considering 10^4 different random initial conditions.

dynamics to relax towards the superstable fixed point for any finite N. This is a straightforward consequence of the fact that the argument of $G_{a,0}$ in eq. (5) is at most 1, rendering the modulo-1 operation ineffective. This implies that, no matter what the values of $x_{i\pm1}^t$ are, some $x_i^{t+\tau}$ will certainly reach the fixed point for a finite τ. From then on, whatever $x_{i\pm1}^{t+\tau}$ is, the argument of $G_{a,0}$ always belongs to the attracting superstable fixed point, thus producing a dynamically frozen situation. We have numerical evidence that as $N \to \infty$, $\tau \to \infty$ in a complicated way (for small N a logarithmic tail is observed in the first period-1 tongue). Therefore, in this limit the transient becomes the asymptotic state and the lattice will never satisfy the conditions discussed above which guarantee the convergence to the fixed point.

Returning to the finite N case, fig. 4 shows the frequency of transient times over a sample of 10^4 different random initial conditions for $(a, \varepsilon, b) = (0.75, 0.3332, 0)$ inside the first period-1 tongue. Similar frequency distributions have been observed for increasing N (up to $N = 400$). During this process the most probable transient time value increases and the variance is approximately constant. The frequency distributions appear to be Poisson-like, with an exponential tail for large transients.

Repeating the same analysis just outside this

tongue, $(a, \varepsilon, b) = (0.75, 0.3335, 0)$, we observe again Poisson-like distributions of transient times and a very slow increase of the average transient with N for small N. However, at variance with the previous case, multistability is present, i.e. the lattice evolves towards different periodic attractors varying the initial condition. In any case, the existence of a typical escape time characteristic of the statistics is a common feature of our model for any $(a, \varepsilon, 0)$ values, although the mechanisms of convergence to asymptotic attractors may be different. This is reminiscent of chaotic repellers, e.g. Cantor unstable invariant sets found in some low-dimensional maps. However, while such sets are in general difficult to reveal numerically, in our case most of the dynamics is ruled by this kind of transient chaos if N is sufficiently large.

In the language introduced in ref. [4] the observed transients look "quasi-stationary", although in our model we observe only an exponential increase of transient times with system size, at variance with the superexponentials found in ref. [4].

We have already observed that the modulo-1 shift operation should be at the origin of the complex behaviors of this model. We expect, for instance, that increase in the number of reinjections should increase transient times. In order to verify this hypothesis we have repeated the same numer-

ical simulations in some points inside the second period-1 tongue in the 2-reinjection zone. In fig. 5 we show the increase of the mean transient time with N, which is typically well fit by an exponential at large N.

Another effect of increasing number of reinjections is the convergence towards multiple periodic attractors. For instance in a period-3 tongue of the 2-reinjection zone for $(a, \varepsilon, b) = (0.7692307, 1.3, 0)$ we have observed convergence to three distinct attractors of periods 3 (homogeneous), 45, 243 (periodic inhomogeneous states), over a number of 30 trials for $N = 9$. Before converging to these attractors the dynamics develops on a chaotic transient. In fig. 6 we show the transient LCE spectrum for $N = 50$ (crosses) and $N = 100$ (full triangles) for the same parameter values, in both cases we plot the largest $N/2$ exponents λ_i as a function of i/N. Two features are relevant: the convergence to a limit distribution of the LCE spectrum as N increases (typical of chaotic dynamical systems [6,7]) and the presence of a positive component. Another interesting effect is that the limiting distribution of LCE extends to $-\infty$. This can be proved rigorously since the determinant of the matrix $\mathbf{A}^t$ in eq. (7) is zero due to the presence of a zero derivative attractive interval.

We have seen that the greater the number of reinjections, the longer the transients. This effect

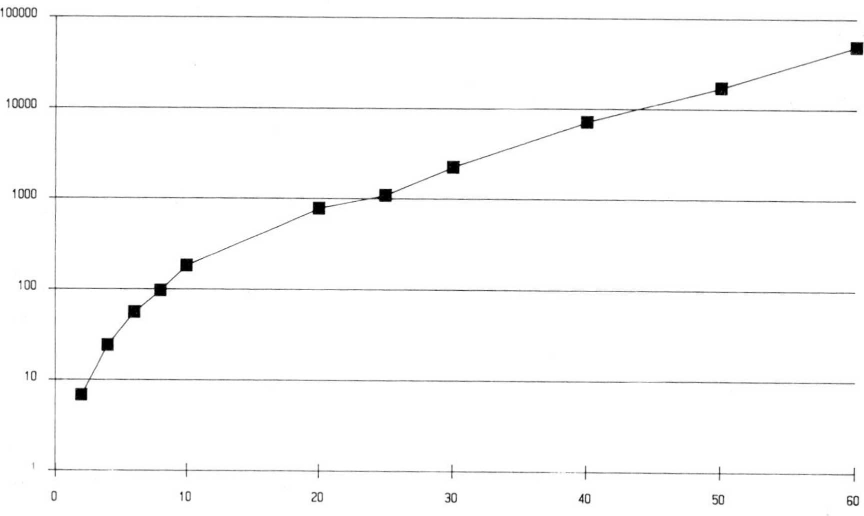

Fig. 5. Logarithm of the average transient time as a function of the lattice size N for $(a, \varepsilon, b) = (0.81, 1.3, 0)$. Each point is obtained by taking the mean over 100 different random initial conditions.

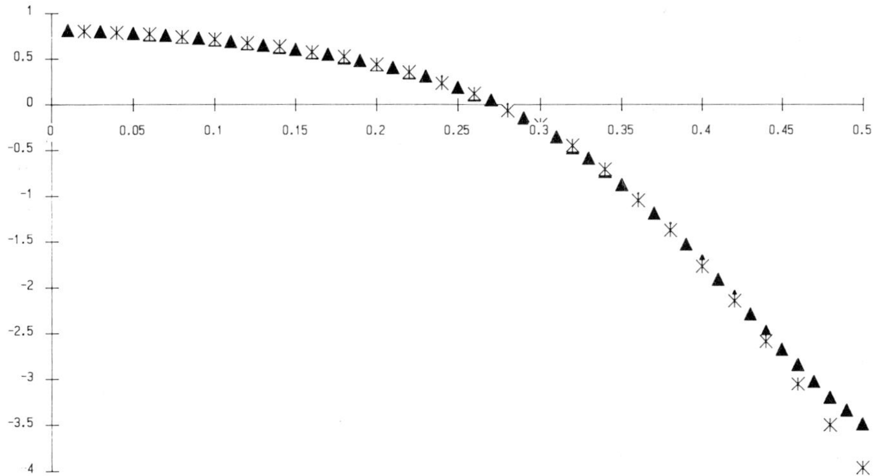

Fig. 6. Transient spectrum of Lyapunov characteristic exponents for $(a, \varepsilon, b) = (0.7692307, 1.3, 0)$ for $N = 50$ (crosses) and $N = 100$ (full triangles), in both cases we plot the largest $N/2$ exponents as a function of i/N. Observe the convergence to a limiting transient LCE distribution as $N \to \infty$. The spectrum is independent of the initial condition.

remains in the $\varepsilon \to \infty$ limit since the Lebesgue measure of the plateaus in the graph of the map corresponding to the homogeneous solution is finite and tends to $1 - a$.

A simple qualitative characterization of space–time patterns is obtained by partitioning the interval into two regions, corresponding to the expanding and contracting components, respectively. Chaotic transients show symbolic patterns typical of spatiotemporal intermittency [8] (see fig. 3 in ref. [9] for a typical pattern), both for cases leading to homogeneous and inhomogeneous periodic states.

4.2. A simple analytical result for $b \neq 0$

We have not yet analyzed numerically in detail the evolution of the lattice from an inhomogeneous initial state for $b \neq 0$. However a simple result is obtained when the dynamics converges to a homogeneous fixed point. In this state the eigenvalues of matrix $\mathbf{A}^t$ in (7) are given by

$$\mu_k = b \left[1 + \varepsilon \cos \left(\frac{2\pi(k-1)}{N} \right) \right], \qquad k = 1, \ldots, N.$$

One can therefore obtain rigorously the asymptotic LCE spectrum, by taking the logarithms of the moduli of the μ_k's. The previous formula was

obtained also in ref. [10] for constant expansion rates. This is for instance the case in the first periodic region PR1 of fig. 3 when $b < b_c$. If this happens one can extend the definition of chaotic transient used in section (4.1) to the $b \neq 0$ case, when the asymptotic solution is a homogeneous fixed point. In fact one can define transient time as the time needed to reach the known asymptotic LCE spectrum.

Little is known about the $b > b_c$ case. As we have anticipated in section 4 the value $\bar{b} > b_c$, above which a chaotic attractor exists, depends on N in a nontrivial way; a preliminary analysis shows a slow decrease of $\bar{b}$ with N, presumably reaching b_c as $N \to \infty$. While the study of transient chaos by the analysis of the LCE spectrum can be reproduced for $b < \bar{b}$, the characterization of chaoticity for $b > \bar{b}$ represents a much more difficult task. In the latter case the LCE spectra do not allow a clear qualitative discrimination between transient chaos and chaotic attractors, since the shape of LCE spectra does not change significantly with respect to that obtained in a regime of chaotic transient.

Interesting effects should be present at the border of the region PR1 with the region CR1 of fig. 3, and in general at the transition from a region where a chaotic transient is present to one where the dynamics evolves towards a chaotic attractor.

Spatiotemporal symbolic patterns reproduce those of fully developed turbulence [11] only far above $\bar{b}$, while near $\bar{b}$ a large degree of spatial correlation is still present (see fig. 4a in ref. [9] for an example).

5. Conclusions

In this paper we have investigated the dynamical properties of a CML model for reaction–diffusion processes. The presence of chaotic behavior characterizing the transient dynamics towards periodic states has been analyzed in detail. In particular a study of the homogeneous dynamics has provided, through the identification of some properties of the asymptotic periodic state, a "first order" description of the dynamical system. In the second paper of ref. [1] the stability of an initial spatially periodic state with coupling was numerically investigated; our approach is complementary, in the sense that we study analytically the periodic time evolution of a homogeneous initial state.

The main results that we have obtained concern:

– the statistics of transient times, which has been observed to be Poisson-like; this suggests the existence of characteristic escape times from the chaotic transient;

– the presence over typical transient times of a relaxed spectrum of LCE with a positive component detecting chaotic states;

– an exponential growth with lattice size of the average transient times. This demonstrates the relevance of chaotic transients to the properties of our model in the thermodynamic limit.

Acknowledgements

GM thanks the DGAPA of the UNAM, CNR/CONACYT of Mexico City and the INFN in Florence for financial support. Moreover RL and SR thank the CNR/CONACYT support to their visits in Mexico City. We all thank F. Bagnoli, S. Isola and A. Politi for useful discussions.

References

[1] I. Waller and R. Kapral, Phys. Rev. A 30 (1984) 2047;
     K. Kaneko, Progr. Theor. Phys. 72 (1984) 480;
     L.A. Bunimovich and Ya.G. Sinai, Nonlinearity 1 (1989) 491.

[2] L.A. Bunimovich, A. Lambert and R. Lima, J. Stat. Phys., in press.

[3] Y. Oono and C. Yeung, J. Stat. Phys. 48 (1987) 593.

[4] J.P. Crutchfield and K. Kaneko, Phys. Rev. Lett. 60 (1988) 2715.

[5] L.P. Kadanoff, Applications of scaling ideas to dynamics, in: Regular and Chaotic Motions in Dynamic Systems, eds. G. Velo and A.S. Wightman, NATO ASI Series, Vol. 118 (Plenum Press, New York, 1985).

[6] J.P. Eckmann and D. Ruelle, Rev. Mod. Phys. 57 (1985) 617.

[7] R. Livi, A. Politi and S. Ruffo, J. Phys. A 19 (1986) 2033.

[8] K. Kaneko, Prog. Theor. Phys. 74 (1985) 1033.

[9] F. Bagnoli, S. Isola, R. Livi, G. Martinez-Mekler and S. Ruffo, Periodic orbits in a coupled map lattice model, in: Cellular Automata and Modeling of Complex Systems, eds. P. Manneville, N. Boccara, G.Y. Vichniac and R. Bideaux, Springer Proceedings in Physics, Vol. 46 (Springer, Berlin, 1990) p. 282.

[10] K. Kaneko, Physica D 23 (1986) 436.

[11] K. Kaneko, Physica D 34 (1989) 1.

APPENDICES

Physica D 45 (1990) 463–476
North-Holland

APPENDIX I
A BRIEF REVIEW OF CELLULAR AUTOMATA PACKAGES

David HIEBELER
Center for Nonlinear Studies, MS B258, Los Alamos National Laboratory, Los Alamos, NM 87545, USA

1. Introduction

A very frequently asked question is "what CA simulation packages are available?" Of course, the answer depends largely on what features you are looking for, and how much you are willing to pay. I would like to briefly mention here a few of the simulation packages that I have encountered. This is by no means a complete list, and I apologize to those authors whose packages I failed to mention. Please feel free to contact me about omissions, so that future versions of this article will be more complete.

2. Available packages

2.1. CA Lab

CA Lab is a software package which runs on IBM PC/XT/AT, PS/2, or compatible, with CGA, EGA, or VGA graphics adaptor. It is targeted primarily toward hobbyists and scientists, and costs approximately $60. The package comes with two main programs: RC, a low-resolution interactive tool, and CA, a high-resolution (320×200) programmable tool. The CA program supports 8-bit cells, and a 16-bit lookup table.

Many CA rules and images come with CA Lab; you can also write your own rules in C, Basic, Pascal, or assembly language. Contact:
Autodesk, Inc.
2320 Marinship Way
Sausalito, CA 9496, USA

2.2. LCAU

LCAU is one of a collection of programs, which are aimed mainly toward simulation and analysis of 1D CA. These programs run on IBM-PC compatible computers. Among the features of LCAU that separate it from other packages are provisions for mean field theory and local structure theory analysis, and the generation of de Bruijn diagrams (see for example ref. [8]). These programs can be obtained by sending the appropriate number of IBM-PC diskettes and return postage to author, H.V. McIntosh. Before sending diskettes, contact him for details. Contact:
Harold McIntosh
Departamento de Aplicacion de Microcomputadoras
Instituto de Ciencias, Universidad Autonoma de Puebla
Apartado postal 461, 72000 Puebla
Puebla, Mexico.

2.3. CAM-PC

CAM-PC is a hardware/software combination. It is a plug-in board which goes into an IBM PC/XT/AT or compatible, and gives you Cray-1 speeds for doing CA updates (60 frames/s on a 256×256 array). CAM-PC is compatible with CAM-6, but has many new features such as extended neighborhoods and more data-analysis capabilities (see ref. [9], and Mark Smith's article elsewhere in these Proceedings [10] for several experiments performed with CAM-6). CAM-PC uses a 16-bit lookup table, and supports complex programmable run-cycles, enabling you to work with multi-step CA rules and to perform real-time data analysis and logging. Cost is approximately $1900. CAM-PC is first and foremost a physical modeling package; many modeling applications come with the software, as well as an excellent book by Toffoli and Margolus introducing techniques for physical modeling with CA [9]. CAM-PC, and a purely software counterpart for those unable to afford the hardware, will be commercially available in fall 1990. Contact:
Automatrix, Inc.
P.O. Box 196
Rexford, NY 12148-0196, USA

2.4. Cellsim

Cellsim is a public-domain, window-based CA simulator which runs on Sun workstations (using Sun-View, currently), and which can also run on a Connection Machine (CM2). An X-windows version is in the works and should be available by the fall of 1990.

The primary applications of Cellsim are in CA theory and physical modeling. The Connection Machine interface is sufficiently flexible that Cellsim can be extended to applications beyond traditional CA. Cellsim allows researchers to experiment with the general class of lattice-based dynamical systems, computing arbitrary real-valued functions at each lattice site and allowing the use of global as well as local information in the update function. The dynamics may be viewed in real-time on either the Sun screen or on the CM2 frame-buffers.

Cellsim currently supports one- and two-dimensional arrays. Version 3.0 (due out September 1990) will support 3D arrays of up to 128 sites on a side.

Cellsim has been picked up by many institutions, and there is a growing community of researchers who are trading rules and images among themselves.

For further information or to obtain Cellsim, contact:
Chris Langton
Complex Systems Group, MSB213
Theoretical Division, T-13
Los Alamos National Laboratory
Los Alamos, NM 87545, USA
email: cgl@LANL.GOV

2.5. Mathematica

Mathematica is a symbolic mathematics package, available on a wide variety of platforms. While the current version only has provisions for running 1D CA, you can do very flexible symbolic manipulation and analysis of the CA, which other packages typically do not offer. The cost of the package varies with the system you purchase it for. Contact:
Wolfram Research, Inc.
P.O. Box 6059
Champaign, IL 61826-6059, USA

Here is the complete CA package in Mathematica. It is used to generate the figures shown below. (Mathematica figures by Matthew Cook, courtesy of Wolfram Research Incorporated).

```
(** Cellular Automata **)

UpdateCA::usage =
        "UpdateCA[list, rule] applies one time step of a CA rule to list.
        The CA rule is given as a list of replacements for neighborhoods,
        such as {{1,1,1}->1, {1,1,_}->0, {1,0,1}->1, {0,a_,b_}:>Mod[a+b,2]}."

EvolveCA::usage =
        "EvolveCA[init, rule, t] evolves a CA for t time steps, giving a
        list of the configurations produced."

ShowCA::usage =
        "ShowCA[list] displays a cellular automaton evolution history
        graphically."

NumberedRule::usage =
        "NumberedRule[n] gives a rule table for a k=2, r=1 rule specified
        by the integer n."

CenterSpot::usage =
        "CenterSpot[list, n] generates a list of n zeroes, with the
        elements of list in the middle."

UpdateCA[list_List, rule_] :=
        Block[ { t },
                t = Join[ {Last[list]}, list, {First[list]} ] ;
                t = Partition[t, 3, 1] ;
                Map[Replace[#, rule]&, t]]

Pad[list_List, n_Integer] :=
        If[Length[list] < n, Join[Table[0, {n-Length[list]}], list], list]

Neighborhoods[k_Integer:2, rr_Integer:3] :=
        Block[ { t },
                t = Range[k^rr - 1, 0, -1] ;
                Map[ Pad[ Digits[#, k], rr ] &, t ]]

EvolveCA[init_List, rule_, t_Integer] :=
        Block[{ drule = Dispatch[rule] },
                NestList[UpdateCA[#, drule]&, init, t]]

ShowCA[list_List] :=
        Show[Graphics[CellArray[
                Reverse[ 1 - list ]
        ]],AspectRatio -> 1]
```

```
NumberedRule[n_Integer, k_Integer:2, rr_Integer:3] :=
        Thread[ Neighborhoods[k, rr] -> Pad[Digits[n, k], k^rr] ]

CenterSpot[list_List, n_Integer] :=
        Block[ { i1 },
                i1 = Floor[ (n - Length[list])/ 2 ] ;
                Join[ Table[0, {i1}],
                        list,
                        Table[0, {n-i1-Length[list]}]]]]
```

3. Comments

A CA electronic mailing list currently exists, where novices and veterans in the CA field participate
in discussions on a wide range of topics. Some "home-brewed" CA simulators are sometimes posted or
announced on the list as well. Send e-mail to `cellular-automata-request@think.com` for information
on joining the list. The mailing list is connected bidirectionally with the `comp.theory.cell-automata`
Usenet newsgroup as well, so either method may be used to read and submit messages to the list.

The figures on the following two pages were generated by CA Lab, captions by Rudy Rucker.

Fig. 1. The Ranch Rule. This a symbiotic cross between three cellular automata: Conway's Life, Silverman's Brain, and Vichniac's Vote. In each rule, the cells look only at the lowest bit of their neighbors' states, forming a "firing eight-sum." Vote and Life have two states: 0 and 1, while Brain has an additional state: 2. In the Vote rule, a cell's new state is 0 or 1 according to whether the firing eight-sum is 0, 1, 2, 3, or 5 on the one hand, or 4, 6, 7, 8, or 9 on the other. In the Life rule a state 0 cell goes to state 1 iff its firing eight-sum is 3, and a state 1 cell is allowed to stay in state 1 iff its firing eight-sum is 2 or 3. In the Brain rule, a state 0 cell goes to state 1 iff its firing eight-sum is 2. A Brain cell in state 1 always goes to state 2, and a Brain cell in state 2 always goes to state 0. In the Ranch rule, Vote is run on some extra background states in order to divide the background up into yellow "land" and black "sea." The land cells obey the Life rule, and the sea cells obey the Brain rule. Activity flows back and forth between land to sea and can last a very long time. See ref. [2] for details.

Fig. 2. Faders. This and the next three figures show "nluky rules", which represent a genetic cross between Life and Brain. If N is positive integer and L, U, K, and Y are positive integers between 0 and 8, the "nluky rule" determined by N, L, U, K, and Y is specified as follows:

(i) The cells have $N + 2$ states: $0, 1, 2, 4, \ldots, 2N$.

(ii) At each update each cell forms the firing eight-sum, which counts how many of its neighbors are in state 1.

(iii) If a cell is in state 0 then its new state is 1 if its firing eight-sum is between L and U (inclusive). Otherwise the cell's new state is 0.

(iv) If a cell is in state 1, then its new state is 1 if its firing eight-sum is between K and Y (inclusive). Otherwise the cell's new state is 2.

(v) If a cell has an even state S between 2 and $2N - 2$ (inclusive), its new state is $S + 2$.

(vi) If a cell has state $2N$ its new state is 0.

Faders is the nluky rule with the N, L, U, K, Y values 127, 2, 2, 2, 2. To be more concise, we say Faders is nluky(127, 2, 2, 2, 2). Patterns like this can be started with a single "Fader egg" which is a small L made of three cells in state 1. It sustains interesting activity for a very long time.

Fig. 3. Rainzha, or nluky(7, 2, 3, 2, 2). This is one of the simplest rules which produces a Zhabotinsky-style reaction, characterized by the spontaneous organization of single and double scrolls.

Fig. 4. Nluky(6, 2, 6, 4, 7). Here the choice of 4 and 7 for K and Y allows certain cells to stay in state 1 if they organize themselves into a grid pattern. These grids grow to take over the space, although in some variations of this rule they do not.

Fig. 5. The Hodge rule. This rule, invented by Gerhardt and Schuster, produces very strong Zhabotinsky-style patterns. The states here range from 0 to 31. A state 0 cell gets put into state 2 or into state 3 if the full sum of its eight neighbors' states is greater than 4 or greater than 99, respectively. A nonzero cell with state less than 31 gets set equal to the average of its neighbors plus 5, but if this new value is greater than 31 it is set equal to 31. A cell in state 31 goes to state 0. The specific numerical values used are not terribly crucial. See refs. [3,4] .

Fig. 6. 128-state Eat rule. Griffeath and Fish have investigated "cyclic competition rules" which are N-state rules in which each cell changes to the next higher state mod N if one of its neighbors is already in that higher state. A randomizer is used to decide which of its neighbors a cell looks at. In this implementation I used 128 states, and used the one-dimensional CA rule with Wolfram code 30 for my randomizer. For a start screen, I randomize all cells to values between 0 and 127, and then set a large central ellipse of cells all to state 0. See ref. [5].

Fig. 7. Three-species Wator. Here there are three "species" of cells: the shrimps, fish and sharks. In the update cycle each cell uses rule 30 to select a random neighbor, and then BOTH cells are updated according to the outcome of their interaction. Thus if a fish selects a shrimp, the fish's hunger counter is reduced, and the shrimp is changed to a background cell. In order for motion to take place, the new cell positions are swapped, so the ex-shrimp becomes a less hungry fish and the ex-fish becomes a background cell.

Fig. 8. Vant lace. Here there are three "vants" or "tur-mites." (See ref. [6] or [7].) Each vant has three bits of direction, and two bits to tell what kind of vant it is. In the update, each vant looks at the cell it is about to move to, moves to the cell, paints it a new color, and changes direction. A typical vant might, on entering a red cell, paint it gray and turn right, and on entering a black or gray cell paint it red and turn left. In this CA Lab simulation, three different vants are started on the boundaries of three rectangles.

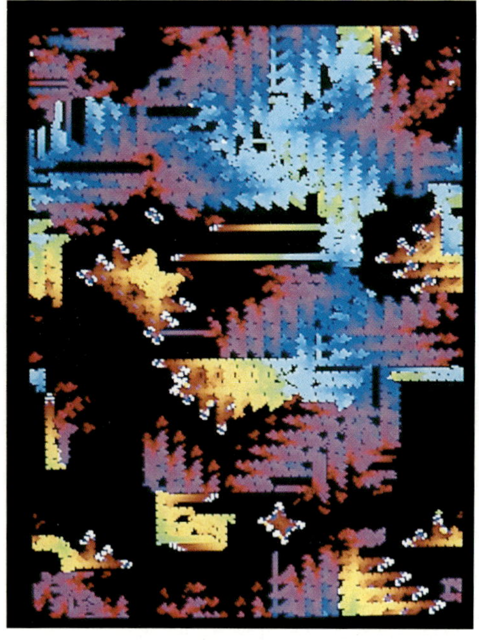

Fig. 2.

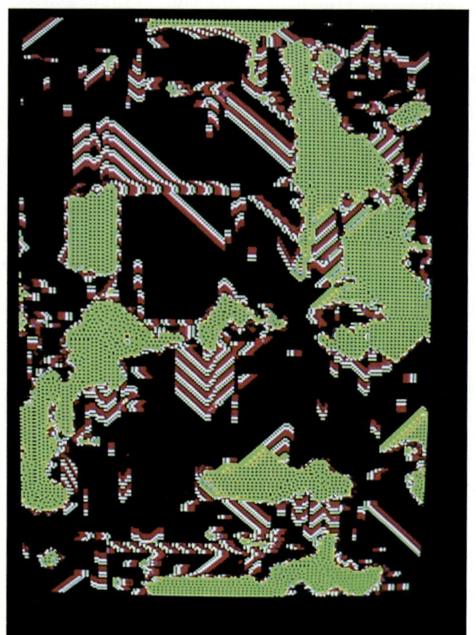

Fig. 4.

Fig. 1.

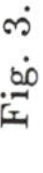

Fig. 3.

Fig. 6.

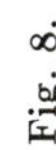

Fig. 8.

Fig. 5.

Fig. 7.

The three figures on this page were generated on a CAM.

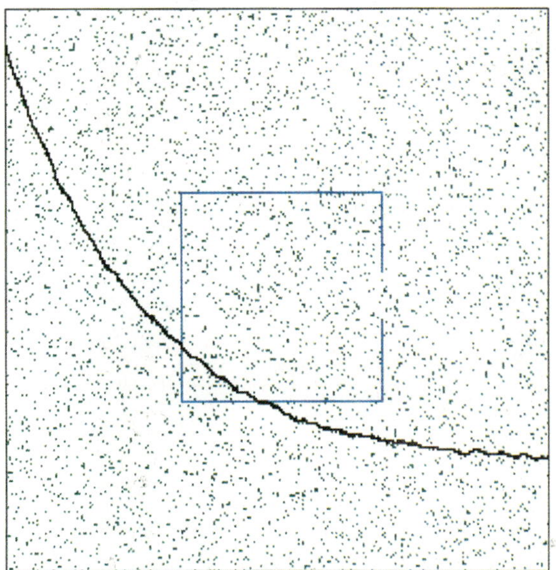

Fig. 9. Self-diffusion of a simple (HPP) lattice gas. There were red and green particles in the simulation (only green particles are shown here, for clarity). At the beginning of the experiment, all of the green particles were inside the box, and all red particles were outside. During every time step, the number of green particles in the box is counted, and plotted over the display. Particles enter and leave the box through the opening in its side. Basic kinetic theory (see e.g. ref. [11]) predicts an exponential decay of the number of green particles in the box, which is the behavior observed here. This experiment, which simulated approximately 32 000 particles, demonstrates CAM's ability to do real-time simulations and analysis (60 steps/s).

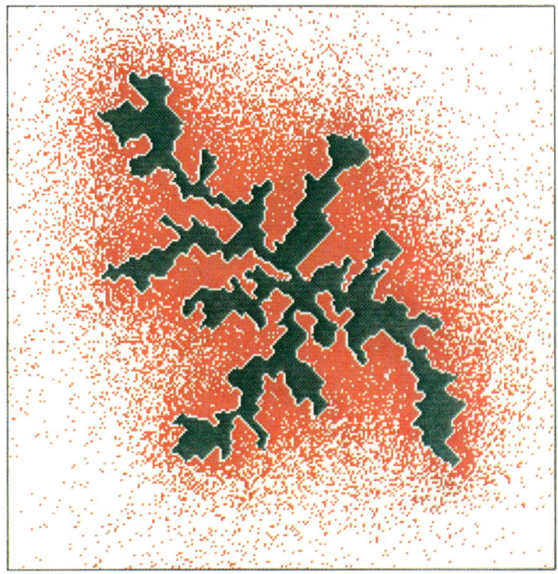

Fig. 10. A simple heat-of-formation solidification model. Green represents solid, red is heat. An empty site next to a solid will itself solidify, unless there is too much heat in the vicinity. When a site solidifies, it releases some heat. See ref. [12] for a similar but more sophisticated model.

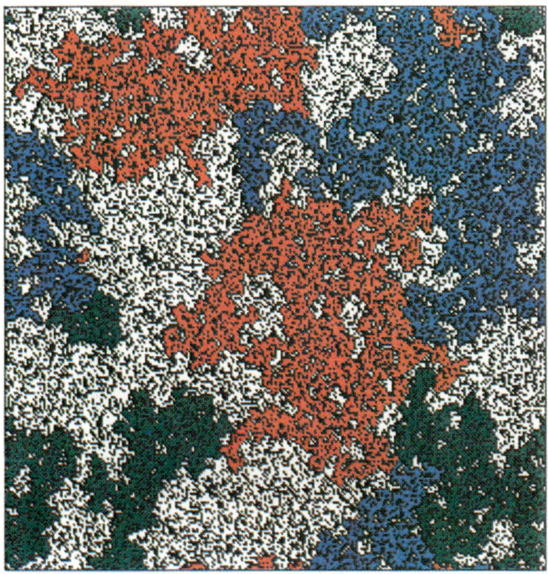

Fig. 11. Percolation clusters (see e.g. ref. [13]). Cells are either empty or occupied; a random configuration is generated to begin with. Then, several "seeds" are scattered in the array. Any occupied cell next to a colored seed will become a seed itself. Different colors represent clusters that began from different seeds.

The two figures on this page were generated by Cellsim.

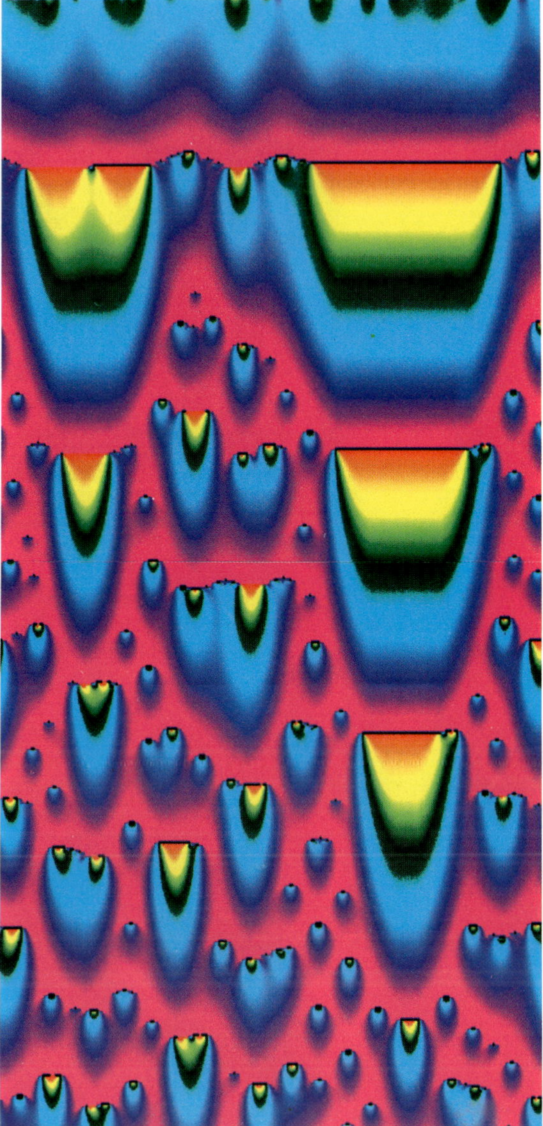

Fig. 12. This figure shows 512 time steps in the evolution of a solution to the classic "firing-squad synchronization" problem in a one-dimensional array of 256 sites. The solution to the problem illustrated here is due to Balzer [14], and requires only 8 states per "soldier'.' The problem is to get an arbitrarily long line of "soldiers" to all fire at the same moment sometime after a command to fire is given by a general standing at one end of the line. As the soldiers can only communicate with their immediate neighbors, it takes some time for the line to organize itself by passing messages back and forth. When each soldier in a "blue" state observes that both of his neighbors are also in the "blue" state, then he knows that he can fire on the next time step. Notice that all of the cells go to the "black" state simultaneously at the bottom of the evolution. The recursive nature of this solution is illustrated beautifuly here.

Fig. 13. This figure shows 512 time steps in the evolution of a simple model of driven heat flow exhibiting chaotic "boiling". In this 1D simulation, temperature is represented by a spectrum of 256 colors, with red being cold and violet being hot. Each cell determines its temperature at the next time step by averaging over a neighborhood template of radius 2, adding a constant which is the driving heat (2 in this case), and taking the result modulo 256. When taking the modulus causes a site (or group of sites) to wrap around from a high temperature to a low temperature, it is as if the system loses heat locally to the outside world in the form of a burst of steam, lowering the local temperature. In this simulation, a line of 256 sites is initialized to random temperatures on the top line . The system rapidly heats up and almost equilibriates as the randomness averages out, in which case the system would simply cycle periodicaly. However, several sites "boil" before others, setting the stage for the chaotic behavior which follows.

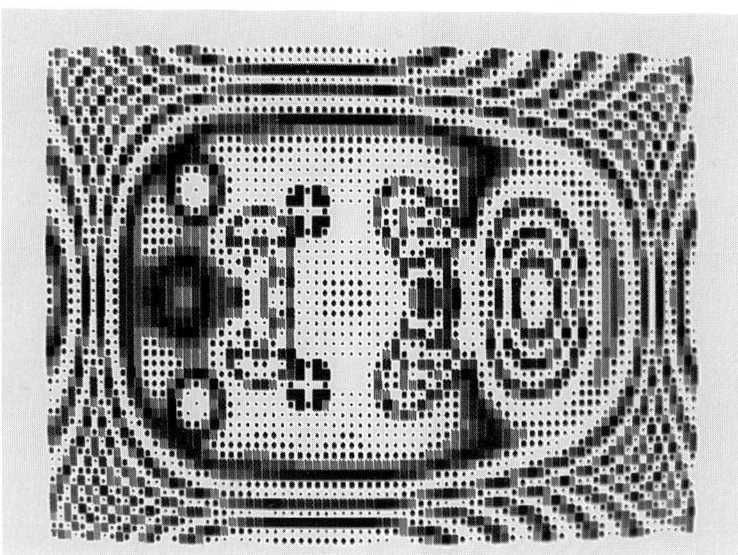

Fig. 14. A long-lived oscillation in the Rug rule. Here cells have 256 possible states, and at each update each cell's new state becomes the average of its eight nearest neighbors plus one, modulo 256. This pattern is made by starting with a blank screen whose edges are frozen at zero, and by then breaking one of the symmetries by shifting the pattern a few cells to the side [1]. (Caption by R. Rucker.)

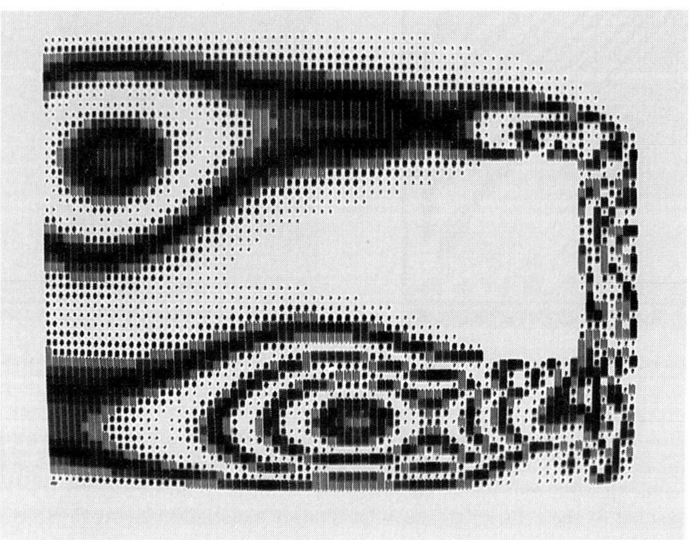

Fig. 15. A pattern resembling a wind tunnel. Here cells have 64 K possible states, and at each update each cell's new state becomes the average of its eight nearest neighbors minus 16, modulo 64 K. In addition, the whole pattern is shifted one cell sideways with each update, the black block is continually masked to state 0, and random cell values are fed in from one side. Like the preceding figure, this is at resolution 80 by 50 cells. (Caption by R. Rucker.)

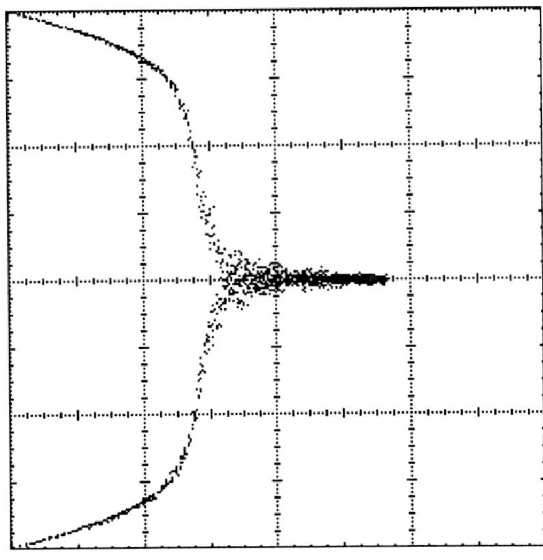

Fig. 16. A plot of magnetization versus energy in an Ising model. A better figure of this, as well as much discussion of several CAM-based Ising models, appears in ref. [9]. (Figure generated on a CAM.)

Here is a sample Mathematica session which uses the CA package to produce pictures such as those shown on the next page (fig. 17). These patterns are all generated from the same rule, using different initial patterns.

```
Mathematica 1.2 (August 30,  1989) [With pre-loaded data]
by S. Wolfram,  D. Grayson,  R. Maeder,  H. Cejtin,
   S. Omohundro,  D. Ballman and J. Keiper
with I. Rivin and D. Withoff
Copyright 1988, 1989 Wolfram Research Inc.

In[1]:= <<CellularAutomata.m

In[2]:= rule = {{a_,  b_,  c_} :> Mod[a + b + c,  2]}

Out[2]= {{a_,  b_,  c_} :> Mod[a + b + c,  2]}

In[3]:= ProduceSquare[pattern_,  size_] :=
          ShowCA[EvolveCA[CenterSpot[pattern,  size],  rule,  size]]

In[4]:= ProduceSquare[ {1} ,  2250 ]

Out[5]= -Graphics-
```

Fig. 17.

This picture was generated by Mathematica using the rule

```
{{a_, b_, c_} :> Mod[a + b + a c, 2]}.
```

Acknowledgements

I am very grateful to those who kindly supplied me with their cellular automata packages. Many thanks to Sam Kim, Henry Hollinger, and Bob Tatar, for helping me pursue work in the field of cellular automata, and to Chris Langton and Gary Doolen for bringing me to Los Alamos.

References

[1] R. Rucker and J. Walker, CA Lab: Rudy Rucker's Cellular Automata Laboratory, Autodesk, Inc., Sausalito, CA (1989).

[2] R. Rucker, Symbiotic programming: crossbreeding cellular automaton rules on the CAM-6, Complex Systems 3 (1989) 79–90.

[3] A.K. Dewdney, The Armchair Universe (Freeman, San Francisco, 1988).

[4] A.K. Dewdney, Computer recreations: the hodgepodge machine makes waves, Sci. Am. (August 1988) 104–107.

[5] D. Griffeath, Cyclic random competition, Not. Am. Math. Soc. 35 (1988) 1472–1480.

[6] C.G. Langton, Studying artificial life with cellular automata, Physica D 22 (1986) 120–149.

[7] A.K.Dewdney, Two-dimensional Turing machines and tur-mites make tracks on a plane, Sci. Am. (September 1989) 180–183.

[8] H. Gutowitz, J. Victor and B. Knight, Local structure theory for cellular automata, Physica 28 D (1987) 49–79.

[9] T. Toffoli and N. Margolus, Cellular Automata Machines: A New Environment for Modeling (MIT Press, Cambridge, MA, 1987).

[10] M.A. Smith, Representations of geometrical and topological quantities in cellular automata, Physica D 45 (1990) 271–277, these Proceedings.

[11] F. Sears and G. Salinger, Thermodynamics, Kinetic Theory, and Statistical Thermodynamics (Addison–Wesley, Reading, MA, 1975).

[12] N.H. Packard, Lattice models for solidification and aggregation, reprinted in: Theory and Applications of Cellular Automata, ed. S. Wolfram (World Scientific, Singapore, 1986).

[13] D. Stauffer, Introduction to Percolation Theory (Taylor and Francis, London, 1985).

[14] R. Balzer, An 8-state minimal time solution to the firing squad synchronization problem, Information and Control 10 (1967) 22–42.

Physica D 45 (1990) 477–479
North-Holland

APPENDIX II

MAPS OF RECENT CELLULAR AUTOMATA AND LATTICE GAS AUTOMATA LITERATURE

Howard A. GUTOWITZ

*Center for Nonlinear Studies and Complex Systems Group, Los Alamos National Laboratory,
MS-B258, Los Alamos, NM 87545, USA*

The purpose of this appendix is to give those new to cellular automata a quick overview of some of the interconnections between subdisciplines in the field.

The tool used for this is a set of maps which show the relationship between the literature in subdisciplines of the fields of cellular automata and lattice gas automata in the scientific literature during the years 1986–1988 (fig. 1). These maps were prepared by David Pendelbury and staff at the Institute for Scientific Information (ISI) in Philadelphia. Each node in a map represents a book or research article which is often cited. The length of the lines which connect these "core documents" give their distance in a co-citation metric. Two documents are close in this metric if they are often cited together. The size of the circle surrounding each core document indicates the rela-

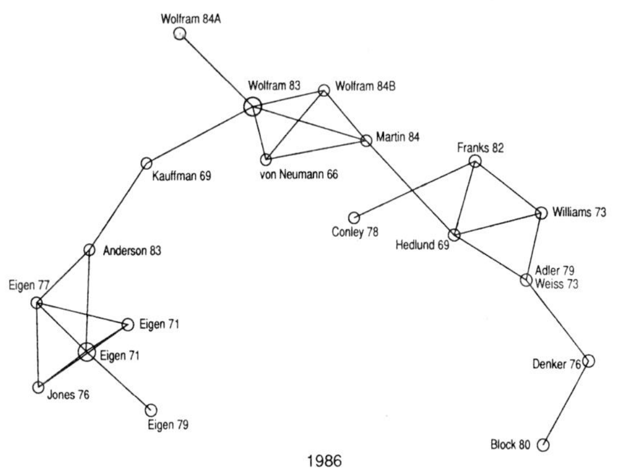

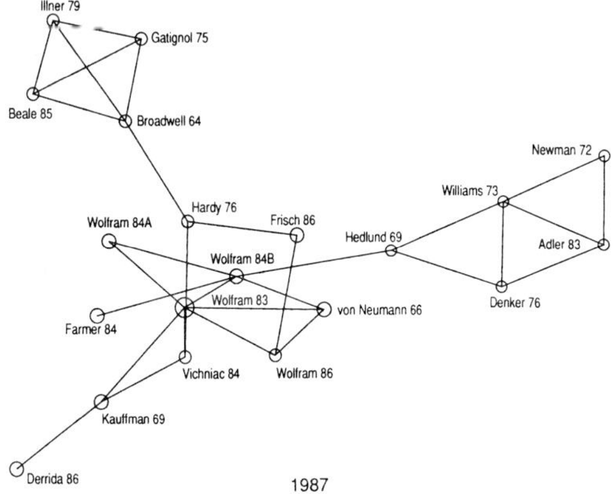

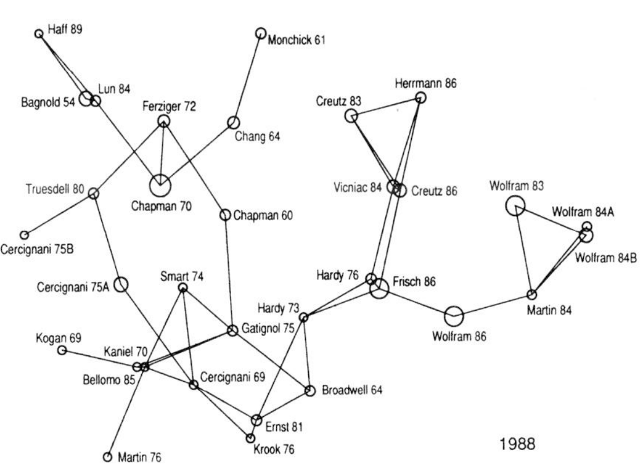

Fig. 1. Citation maps of CA and LGA literature in the years 1986–1988. Courtesy of the Institute for Scientific Information, 3501 Market Street, Philadelphia, Pennsylvania 19104.

tive number of citations that article received in the given year. The full citations of the articles which appear as nodes in these maps are listed in the bibliography. Some additional, more recent, reviews and books are listed at the end of the bibliography.

It is evident that these maps do not directly indicate *conceptual* connections between documents. They only reflect the citation behavior of scientists working in the field. It is equally evident, however, that articles which are often cited together may be expected to bear on similar subjects. This expectation is supported by research conducted at the ISI [44]. In any event, these core documents are useful keys for bibliographic database searches.

Some of the main features of these maps to observe are:

In 1986, the field of lattice gas automata (represented e.g. by the papers by Frisch et al. and Hardy et al.) had just emerged. It was not connected to cellular automata and hence not shown in 1986. The field of cellular automata (represented e.g. by papers by Wolfram and von Neumann) was connected in 1986 on the right to work in pure mathematics on "shift dynamical systems" (cellular automata), via the paper by Martin et al. (algebraic properties of cellular automata). Cellular automata were connected on the left through the field of random Boolean networks (Kauffman) to studies on the origin of life (Eigen).

In 1987, research in lattice gas automata and cellular automata were strongly interconnected, and closely connected as well with random Boolean networks. At a further remove, they were connected to shift dynamical systems on the right, and kinetic theory (Broadwell, Gatignol) above.

In 1988, lattice gas automata and cellular automata had diverged somewhat, though they remained interconnected. Random Boolean nets (not shown) had diverged to form a separate subfield. Lattice gas automata were connected to spin systems (Creutz) above, and kinetic theory to the left.

References

[1] R.L. Adler and B. Marcus, Topological entropy and equivalence of dynamical systems, Mem. Am. Math. Soc. 219 (1979) .

[2] R.L. Adler, D. Coppersmith and M. Hassner, Algorithms for sliding block codes: An application of symbolic dynamics to information theory, IEEE Trans. Information Theory 29 (1983) 5.

[3] P.W. Anderson, Suggested model for prebiotic evolution: The use of chaos, Proc. Natl. Acad. Sci. US 80 (1983) 3386.

[4] R.A. Bagnold, Experiments on a gravity-free dispersion of large solid spheres in a Newtonian fluid under sheer, Proc. R. Soc. London Ser. A 225 (1954) 49.

[5] J.T. Beale, Large-time behavior of the Broadwell model of a discrete velocity gas, Commun. Math. Phys. 102 (1985) 217.

[6] N. Bellomo and G. Toscani, On the Cauchy problem for the nonlinear Boltzmann equation: global existence, uniqueness and asymptotic stability, J. Math. Phys. 26 (1985) 334.

[7] L. Block, Periodic points and topological entropy of one-dimensional maps, in: Global theory of dynamical systems, Lecture Notes in Mathematics No. 819 (1980) p. 18.

[8] J. Broadwell, Shock structure in a simple discrete velocity gas, Phys. Fluids 7 (1964) 1243–1247.

[9] C. Cercignanni, Mathematical Methods in Kinetic Theory (Plenum Press, New York, 1969).

[10] C. Cercignanni, Theory and Applications of the Boltzmann equation (Scottish Academic Press, Edinburgh, 1975).

[11] C.S.W. Chang, G.E. Uhlenbeck and J. de Boer, The head conductivity and viscosity of polyatomic gases, in: Studies in Statistical Mechanics, Vol. II, eds. J. de Boer and G.E. Uhlenbeck (North-Holland, Amsterdam, 1964).

[12] S. Chapman and T.G. Cowling, Mathematical Theory of Nonuniform Gases (Cambridge Univ. Press, Cambridge, 1960, 1970).

[13] C. Conley, Isolated invariant sets and the Morse index, Am. Math. Soc., Regional Conference Series 38 (1978).

[14] M. Creutz, Microcanonical Monte Carlo simulation, Phys. Rev. Lett. 50 (1983) 1411.

[15] M. Creutz, Deterministic Ising dynamics, Ann. Phys. 167 (1986) 62.

[16] M. Denker, C. Grillenberger and K. Sigmund, Ergodic Theory on Compact Spaces, Lecture Notes in Mathematics No. 527 (Springer, Berlin, 1976).

[17] B. Derrida, J. Phys. (Paris) 47 (1986) 1297.

[18] M. Eigen, Selforganization of matter and the evolution of biological macromolecules, Naturwissenschaften 58 (1971) 465.

[19] M. Eigen, The hypercycle, part A: Emergence of the hypercycle, Naturwissenschaften 64 (1977) 541.

[20] M. Eigen, The hypercycle, part C: The realistic hypercycle, Naturwissenschaften 65 (1978) 341.

[21] M. Eigen and P. Schuster, The Hypercycle: A principle of Natural Self-Organization (Springer, Berlin, 1979).

[22] M.H. Ernst, Nonlinear model – Boltzmann equations and exact solutions, Phys. Rep. 78 (1981) 1.

[23] J.D. Farmer, T. Toffoli and S. Wolfram, Cellular Automata, Proceedings of an Interdisciplinary Workshop, Physica D 10 (1984).

[24] J.H. Ferzinger, Mathematical Theory of Transport Processes in Gases (North-Holland, Amsterdam, 1972).

[25] J.M. Franks, Homology and dynamical systems, Am. Math. Soc. Conf. Board 49 (1982).

[26] U. Frish, B. Hasslacher and Y. Pomeau, Lattice gas automata for the Navier–Stokes equation, Phys. Rev. Lett. 56 (1986) 1505.

[27] R. Gatignol, Théorie Cinétique des Gaz à Repartition Discrete de Vitesses, Lecture Notes in Physics No. 36 (Springer, Berlin, 1975).

[28] P.K. Haff, Grain flow as a fluid-mechanical phenomenon, J. Fluid Mech. 134 (1983) 401.

[29] J. Hardy, Y. Pomeau and O. de Pazzis, Time evolution of a two-dimensional model system. I. Invariant states and time correlation functions, J. Math. Phys. 14 (1973) 1746.

[30] J. Hardy, O. de Pazzis and Y. Pomeau, Molecular dynamics of a classical lattice gas: transport properties and time correlation functions, Phys. Rev. A 13 (1976) 1949.

[31] G.A. Hedlund, Endomorphisms and automorphisms of the shift dynamical system, Math. Systems Theor. 3 (1969) 320.

[32] H.J. Herrmann, Fast algorithm for the simulation of Ising models, J. Stat. Phys. 45 (1986) 145.

[33] R. Illner, Global existence for two-velocity models of the Boltzmann equation, Math. Meth. Appl. Sci. 1 (1979) 187.

[34] B.L. Jones, R.H. Enns and Rangnekar, On the theory of selection of coupled macromolecular systems, Bull. Math. Biol. 38 (1976) 15.

[35] S. Kaniel and M. Shinbrot, The Boltzmann equation, 1. Uniqueness and local existence, Commun. Math Phys. 58 (1978) 65.

[36] S.A. Kauffman, Metabolic stability and epigenesis in randomly constructed genetic nets, J. Theor. Biol. 22 (1969) 437.

[37] M.N. Kogan, Rarefied gas dynamics (Plenum Press, New York, 1969).

[38] M. Krook and T.T. Wu, Formation of Maxwellian tails, Phys. Rev. Lett. 36 (1976) 1107.

[39] C.K.K. Lun, S.B. Savage, D.J. Jeffrey and N. Chepurny, Kinetic theories for granular flow: inelastic particles in Couette flow and slightly inelastic particles in a general flowfield, J. Fluid Mech. 140 (1984) 223.

[40] O. Martin, A. Odlyzko and S. Wolfram, Algebraic properties of cellular automata, Commun. Math Phys. 93 (1984) 219.

[41] R.H. Martin, Nonlinear Operators and Differential Equations in Banach Spaces (Wiley, New York, 1976) (Reissue: Krieger, New York, 1987).

[42] L. J. Monchick and E.A. Mason, Transport properties of polar gases, Chem. Phys. 35 (1984) 1676.

[43] M. Newman, Integral Matrices (Academic Press, New York, 1972).

[44] H. Small and E. Garfield, The geography of science: disciplinary and national mappings, J. Info. Sci. 11 (1985) 147–159.

[45] D.R. Smart, Fixed Point Theorems (Cambridge Univ. Press, Cambridge, 1974).

[46] C. Truesdell and R.G. Muncaster, Fundamentals of Maxwell's Kinetic Theory of a Simple Monatomic Gas; Treated as a Branch of Rational Mechanics (Academic Press, New York, 1980).

[47] G.Y. Vichniac, Simulating physics with cellular automata, Physica D 10 (1984) 96.

[48] J. von Neumann, Theory of Self-reproducing Automata, ed. A.W. Burks (Univ. of Illinois Press, Champaign, IL, 1966).

[49] B. Weiss, Subshifts of finite type and sofic systems, Monatsh. Math. 77 (1973) 462.

[50] R.F. Williams, Classification of shifts of finite type, Ann. Math. 98 (1973) 120.

[51] S. Wolfram, Statistical mechanics of cellular automata, Rev. Mod. Phys. 55 (1983) 601.

[52] S. Wolfram, Cellular automata as models of complexity, Nature 311 (1984) 419 [Wolfram 84a in 1986 and 1987].

[53] S. Wolfram, Computation theory of cellular automata, Commun. Math. Phys. 96 (1984) 15 [Wolfram 84a in 1988].

[54] S. Wolfram, Universality and complexity in cellular automata, Physica D 10 (1984) 1 [Wolfram 84b in 1986 and 1987, Wolfram 84a in 1988].

[55] S. Wolfram, Theory and Applications of Cellular Automata (World Scientific, Singapore, 1986).

Some additional books and reviews

[56] H.J. Herrmann, Cellular automata, in: Nonlinear Phenomena in Complex Systems, ed. A.N Proto (North-Holland, Amsterdam, 1989).

[57] C.G. Langton, Artificial life, Santa Fe Institute Studies in the Sciences of Complexity (Addison–Wesley, New York, 1988).

[58] R. Livi, S. Ruffo, S. Ciliberto and M. Buiatti, Chaos and Complexity (World Scientific, Singapore, 1988).

[59] P. Manneville, N. Boccara, G.Y. Vichniac and R. Bidaux, Cellular automata and modeling of complex physical systems, Springer Proceedings in Physics No. 46 (Springer, Berlin, 1989).

[60] P. Rujan, Cellular automata and statistical mechanical models, J. Stat. Phys. 49 (1987) 139.

[61] T. Toffoli and N. Margolus, Cellular Automata Machines – A New Environment for Modeling (MIT Press, Cambridge, MA, 1987).

Physica D 45 (1990) 481
North-Holland

LIST OF CONTRIBUTORS

Physica D 45 (1990) 482–483
North-Holland

ANALYTIC SUBJECT INDEX